Davydov's Soliton Revisited

Self-Trapping of Vibrational Energy in Protein

NATO ASI Series

Advanced Science Institutes Series

A series presenting the results of activities sponsored by the NATO Science Committee, which aims at the dissemination of advanced scientific and technological knowledge, with a view to strengthening links between scientific communities.

The series is published by an international board of publishers in conjunction with the NATO Scientific Affairs Division

A	Life Sciences	Plenum Publishing Corporation
B	Physics	New York and London
C	Mathematical and Physical Sciences	Kluwer Academic Publishers
		Dordrecht, Boston, and London
D	Behavioral and Social Sciences	
E	Applied Sciences	
F	Computer and Systems Sciences	Springer-Verlag
G	Ecological Sciences	Berlin, Heidelberg, New York, London,
H	Cell Biology	Paris, and Tokyo

Recent Volumes in this Series

Volume 238—Physics, Geometry, and Topology
edited by H. C. Lee

Volume 239—Kinetics of Ordering and Growth at Surfaces
edited by Max G. Lagally

Volume 240—Global Climate and Ecosystem Change
edited by Gordon J. MacDonald and Luigi Sertorio

Volume 241—Applied Laser Spectroscopy
edited by Wolfgang Demtröder and Massimo Inguscio

Volume 242—Light, Lasers, and Synchrotron Radiation: A Health Risk Assessment
edited by M. Grandolfo, A. Rindi, and D. H. Sliney

Volume 243—Davydov's Soliton Revisited: Self-Trapping of
Vibrational Energy in Protein
edited by Peter L. Christiansen and Alwyn C. Scott

Volume 244—Nonlinear Wave Processes in Excitable Media
edited by Arun V. Holden, Mario Markus, and Hans G. Othmer

Volume 245—Differential Geometric Methods in Theoretical Physics:
Physics and Geometry
edited by Ling-Lie Chau and Werner Nahm

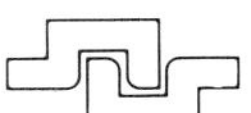

Series B: Physics

Davydov's Soliton Revisited

Self-Trapping of Vibrational Energy in Protein

Edited by

Peter L. Christiansen

The Technical University of Denmark
Lyngby, Denmark

and

Alwyn C. Scott

The Technical University of Denmark
Lyngby, Denmark and
The University of Arizona
Tucson, Arizona

Plenum Press
New York and London
Published in cooperation with NATO Scientific Affairs Division

Proceedings of a NATO Advanced Research Workshop
on Self-Trapping of Vibrational Energy in Protein,
held July 30–August 5, 1989,
in Thisted, Denmark

Library of Congress Cataloging-in-Publication Data

Davydov's soliton revisited : self-trapping of vibrational energy in
   protein / edited by Peter L. Christiansen and Alwyn C. Scott.
        p.   cm. -- (NATO ASI series. Series B, Physics ; vcl. 243)
     "Published in cooperation with NATO Scientific Affairs Division."
     "Proceedings of a NATO Advanced Research Workshop on Self-Trapping
   of Vibrational Energy in Protein, held July 30-August 5, 1989, in
   Thisted, Denmark"--Copr. p.
     Includes bibliographical references and index.
     ISBN 0-306-43734-1
     1. Proteins--Congresses. 2. Solitons--Congresses. 3. Energy
   transfer--Congresses.   I. Christiansen, Peter L., 1937-   .
   II. Scott, Alwyn, 1931-   . III. North Atlantic Treaty
   Organization. Scientific Affairs Division. IV. NATO Advanced
   Research Workshop on Self-Trapping of Vibrational Energy in Protein
   (1989 : Thisted, Denmark) V. Series: NATO ASI series. Series B,
   Physics ; v. 243.
   QP551.D36 1991
   574.19'121--dc20   90-20930
                                                             CIP

SPECIAL PROGRAM ON CHAOS, ORDER, AND PATTERNS

This book contains the proceedings of a NATO Advanced Research Workshop
held within the program of activities of the NATO Special Program on
Chaos, Order, and Patterns.

In Memory of
Stephanos Pnevmatikos
1957—1990

Preface

It was just over ten years ago, at Aspenäsgården near Gothenburg, Sweden, that Professor Alexandr Sergeevich Davydov presented his soliton theory for the storage and transport of biological energy in protein to scientists from Europe, North America and Japan. Since then, his ideas have been vigorously studied and investigated throughout the world. Many feel that Davydov's theory is an important contribution to biomolecular dynamics, but others caution that neglected dispersive effects may destroy the energy localization that arises in his theory. It was to discuss these differences of opinion that we organized a NATO Advanced Research Workshop on "Self-trapping of Vibrational Energy in Protein" from July 30 to August 5, 1989 at Hanstholm, Denmark.

In addition to substantial financial support from the Special Programme on "Chaos, Order and Patterns" of the NATO Scientific Affairs Division, we received a generous grant from the Danish Natural Science Research Council. We also acknowledge invaluable assistance provided by the interdepartmental center of nonlinear studies ("MIDIT" is the Danish acronym) as well as the Laboratory of Applied Mathematical Physics, both at the Technical University of Denmark. It is a particular pleasure to thank Lise Gudmandsen and Dorthe Thøgersen for many forms of assistance before, during, and after the workshop. All the participants express their thanks to the management and staff of the Hotel Hanstholm for the excellent service and facilities provided throughout the meeting, and the editors are grateful to Barra Press International and The Plenum

Publishing Corporation for their careful preparation and efficient publication of these Proceedings.

There were forty-three participants from fourteen countries (Canada, the People's Republic of China, Denmark, France, Greece, Italy, Japan, Portugal, the Soviet Union, Turkey, the United Kingdom, the United States, West Germany, and Yugoslavia) all of whom are actively engaged in research on the Davydov soliton. We attribute the nearly perfect attendance throughout the week, not only to the inclement weather typical of the northwest coast of Denmark, but also to the intense interest of the participants. We believe that the vigorous and uninhibited discussions among the participants have contributed substantially to our collective understanding of this subject.

These Proceedings present the variety of opinions that developed before and during the workshop; in fact several of the chapters were written after the meeting. It is our hope that this volume will become a source book for future research in nonlinear biomolecular dynamics and provide inspiration to young scientists entering the field.

As noted above, the main theme at the workshop was soliton stability in the presence of various dispersive effects. This question can be resolved into three distinct issues: i) Fundamental theory for the localization of vibrational and electronic energy in biomolecules, ii) Strengths of the nonlinear parameters that lead to energy localization, and iii) Dispersive effects at physiological temperature. It is evident from the chapters of Section I that realistic parameter values are necessary for the reliable evaluation of theoretical predictions. The chapters of Section II demonstrate the enormous difficulties of calculating relevant nonlinear parameters from first principles, but fortunately, some experimental values are available. The question of thermal dispersion of Davydov's soliton, discussed in Section III is—in our opinion—not yet resolved.

In order to arrive at the truth of the matter, experimental evidence is required. Relevant data from typical studies of "model proteins" (crystalline, hydrogen bonded amides) are presented in Section IV, and more direct experiments on realistic protein structures (synthetic alpha-helix) are planned for the near future. The volume also contains related studies, gathered in Section V, and a simple theoretical description for self-trapping of energy, discussed in Section VI. Each section is preceded by some introductory remarks which we hope will be helpful.

Finally we are pleased that a Scandinavian country, once again, could contribute to the exchange of ideas among the world scientific community. Such a contribution is particularly important at the present time when opportunities for international scientific collaborations seem to be growing.

Peter Leth Christiansen
Alwyn Scott

Lyngby, March 1990

Contents

Section V: Related Topics   413

Section VI: The Discrete Self-Trapping Equation   469

Section I

Low Temperature Theory

A thing may look specious in theory and yet be ruinous in practice;
a thing may look evil in theory, and yet be in practice excellent.

Edmund Burke

In this first section of these proceedings, we include papers that are primarily concerned with the theory of Davydov's soliton at zero degrees Kelvin. The basic ideas behind self-trapping of vibrational energy on alpha-helix in protein (see Figure 1) can be readily explained in terms of the following parameters.

The exciton-phonon coupling constant (χ). This parameter indicates how strongly localized vibrational energy will induce distortion of the alpha-helix, and also how strongly alpha-helix distortion will trap localized vibrational energy. Attempts to determine the magnitude of χ are discussed below in Section II, but, briefly, it arises from modulation of amide-I (CO stretching) energy by stretching of the adjacent hydrogen bond.

The longitudinal spring constant(w). This is the proportionality constant between the longitudinal displacement and the corresponding restoring force.

The dipole-dipole interaction energies (J_{mn}). These arise from electromagnetic interactions between pairs (labeled m and n) of amide-I oscillators. Often it is convenient, if somewhat inaccurate, to assume only a single, nearest neighbor coupling, J.

In terms of these parameters and the alpha-helix model of Figure 1, Davydov's mechanism for self-trapping of amide-I (CO stretching) vibrational energy can be explained as follows:

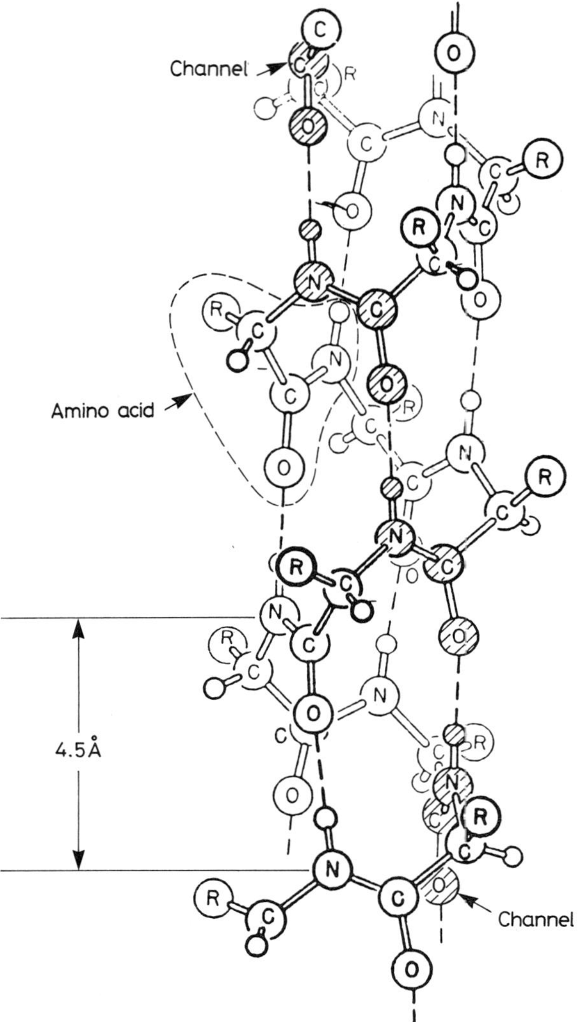

Figure 1. An atomic model of the alpha-helix. One "channel" is cross-hatched.

i) Localized vibrational energy acts, through χ, to cause a local distortion of the alpha-helix structure.

ii) The magnitude of this distortion is inversely proportional to w.

iii) The localized alpha-helix distortion **reacts**, again through χ, to trap the localized vibrational energy preventing its dispersion.

Thus it should not be surprising that the tendency for energy to remain localized is proportional to χ^2/w. The tendency for energy to disperse, on the other hand, is proportional to the nearest neighbor dipole-dipole coupling energy, J (in the simplest theoretical model). Thus the

$$soliton\ size\ \propto\ Jw/\chi^2. \tag{1}$$

A characterizing feature of the alpha-helix shown in Figure 1 is the three longitudinal channels (one of which is indicated by cross-hatching) with the atomic sequence

$$\cdots \mathbf{H} - \mathbf{N} - \mathbf{C} \overset{\chi}{=} \mathbf{O} \cdots \mathbf{H} - \mathbf{N} - \underset{J}{\mathbf{C} \overset{\chi}{=} \mathbf{O}} \cdots$$

Theoretical studies of Davydov's soliton often focus attention on a single channel and the single (nearest neighbor) dipole-dipole interaction energy, J. The motivation for this procrustean approximation is the convenience of representing the alpha-helix by the Fröhlich Hamiltonian [1], but, as has been emphasized [2,3], numerically accurate results require consideration of the full alpha-helix with several dipole-dipole interaction terms.

The single channel model of alpha-helix leads to a set of Fröhlich Hamiltonian parameters which are often described as "standard" or "widely accepted". The source of this confusion seems to be Reference [4] in which a single channel was treated for analytical convenience. In this reference, however, it was explicitly stated: "For a detailed analysis it is necessary to consider the interaction of all three channels, but one is sufficient to lay out the basic ideas" [5].

To gain some numerical perspective, let us assume that self-trapping on the alpha-helix results only from interactions with longitudinal sound waves. Then an appropriate mechanical representation of the helix is as in Figure 2.

Table 1. A comparison of the widely used parameter values derived from a single channel model of the alpha-helix with more realistic values derived from a three channel model.

Parameter	Single channel model	Three channel model	Units
w	13	39–58.5	newtons/meter
M	1.9×10^{-25}	5.7×10^{-25}	kg
χ	62	35–62	piconewtons
J	1.55×10^{-22}	1.55×10^{-22}	joules
Jw/χ^2	0.52	1.6–7.4	—

In Table 1, we compare the parameters of the single channel model which have been widely used for numerical evaluations of Davydov's soliton with more realistic (or less unrealistic!) parameter values derived from the three channel model.

Several comments concerning this table are appropriate.

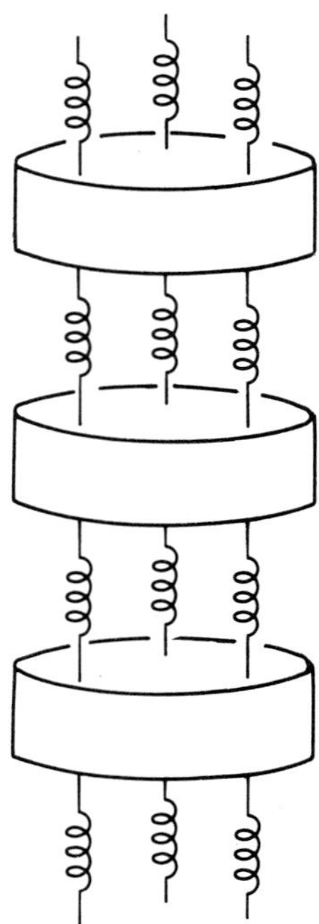

Figure 2. A mechanical model of the alpha-helix that represents three channel propagation with self-trapping by longitudinal sound.

1. The parameters for the three channel model are derived from Figure 2, but with more realistic estimates of uncertainties.

2. The hydrogen bond spring constant $w = 13$ newtons/meter was introduced in Reference [6] based upon experimental measurements by Itoh and Shimanouchi [7]. Three times this value appears in the three channel model.

3. The hydrogen bond spring constant $w = 19.5$ newtons/meter was introduced in Reference [2] based upon *ab initio* calculations by Kuprievich and Kudritskaja [8]. Three times this value is also entered in the three channel model to give an estimate of the uncertainty involved.

4. The mass M in the single channel model corresponds to 114 m_p which is the average mass of amino acids in the alpha-helix of myosin [2]. In the three channel model, it is three times as large.

5. The exciton-phonon coupling constant χ in the three channel model is chosen to include the currently available experimental values. (See the following section for details.) This provides an estimate of the range of uncertainty.

6. In both the single channel and the three channel models, dipole-dipole coupling is restricted to nearest neighbor interactions. This is not because we think it is the best approximation. The numerical results presented in Reference [2], for example, show that several such interactions should be included. This approximation is tolerated in the three channel model to facilitate comparison with the single channel model. It is discussed further below.

One of the most striking differences between the two models of Table 1 is in the soliton size parameter: Jw/χ^2. Reference to the last row of the table indicates that the single channel model may underestimate the soliton size by a factor somewhere between 3 and 14. This underestimation would appear even larger if more than the nearest neighbor dipole-dipole coupling energies were considered. Thus statements concerning soliton size which are based on the single channel parameter values must be treated with caution.

The theorist should also bear in mind that the real alpha-helix (see Figure 1) has many more phonon degrees of freedom than are captured in the simple picture (Figure 2) upon which the three channel model is based [9]. On the one hand, these additional degrees of freedom may reduce soliton stability by providing additional channels for soliton scattering. On the other hand, they may increase stability by providing some subtle combination of stretching, twisting and bending for which the corresponding values of χ, w, and J particularly favor soliton formation.

With these caveats in mind, let us briefly review the theoretical studies in this section. In the first paper, Davydov provides a summary account of his original soliton theory and indicates how it has been extended to a theory for the transport of electronic charge pairs along alpha-helix via a "bisoliton". This latter idea leads naturally to a theory of "high-T_c superconductivity" which has been of great interest over the past few years.

In the second paper Kerr and Lomdahl present the details of a fully quantum mechanical derivation of Davydov's starting equations. Although this derivation is only for the simple model of a single channel with nearest neighbor dipole-dipole interactions, the authors show how several quanta of amide-I energy can be included in the analysis.

At this point, it seems appropriate to emphasize a feature of the interaction Hamiltonian [see Equation (1) of the Kerr-Lomdahl (**KL**) chapter] that is sometimes overlooked. In the above definition of χ, it was said to arise from modulation of amide-I energy by stretching of the **adjacent** hydrogen bond. This restriction seems reasonable in the context of the atomic model of alpha-helix, and it has been confirmed by *ab initio* calculations (see Reference [8] and the following section). Thus Equation (1) of KL should be amended to read

$$\hat{H}_{int} = \chi \sum_n (\hat{u}_{n+1} - \hat{u}_n)\hat{B}_n^\dagger \hat{B}_n. \tag{2}$$

We note that this correction does not appear in the works of Davydov [10] nor was it included in the numerical calculations of Reference [2], but it was introduced into the calculations of Reference [3]. With this change, **Davydov's equations** [Equations (41) and (23) of KL] become

$$i\hbar\dot{a}_n = E_0 a_n - J(a_{n+1} + a_{n-1}) + \chi(\beta_{n+1} - \beta_n)a_n \tag{3}$$

$$M\ddot{\beta}_n = w(\beta_{n+1} - 2\beta_n + \beta_{n-1}) + \chi(|a_n|^2 - |a_{n-1}|^2) \tag{4}$$

where $|a_n|^2$ is the probability of finding a quantum of amide-I energy at site n, β_n is the longitudinal displacement of the nth amino acid, and $E_0/\hbar$ is the site frequency of an amide-I oscillator. Equations (3) and (4) are a simplified version of the KL formalism that describe the dynamics of only a single amide-I quantum.

If we assume $\ddot{\beta}_n = 0$ (the so-called adiabatic approximation), Equations (3) and (4) reduce to

$$(i\hbar\frac{d}{dt} - E_0)a_n + J(a_{n+1} + a_{n-1}) + \frac{\chi^2}{w}|a_n|^2 a_n = 0 \tag{5}$$

$$\beta_{n+1} - \beta_n = -\frac{\chi}{w}|a_n|^2. \tag{6}$$

Equation (5) is a form of the **discrete self-trapping (DST)** system which is discussed in some detail in Section VI of these proceedings. The Hamiltonian for Equation (5) is

$$\tilde{H} = \sum_n \left[E_0 |a_n|^2 - J(a_{n+1}^* a_n + a_n a_{n-1}^*) - \frac{\chi^2}{2w} |a_n|^4 \right] \qquad (7)$$

and stationary solutions [i.e. solutions for which $a_n(t) = \phi_n\, exp(-i\omega t)$] are readily obtained using a "shooting" method [11]. Two limiting cases are of particular interest

1. $Jw/\chi^2 \ll 1$. In this case all of the probability for finding a quantum of amide-I vibrational energy can be localized on a single site. The energy is

$$\tilde{H} = E_0 - \frac{\chi^2}{2w}. \qquad (8)$$

2. $Jw/\chi^2 \gg 1$. This is the case treated by Davydov in the first chapter. Equation (5) reduces to the nonlinear Schrödinger equation [12] with the familiar hyperbolic secant (or "soliton") shape and energy

$$\tilde{H} = E_0 - 2J - \frac{\chi^4}{48w^2 J}. \qquad (9)$$

But we digress.

The Hamiltonian used by Davydov conserves the exciton **number** (i.e. the number of amide-I bosons in the vibrational soliton or electrons in the electrosoliton). Takeno's chapter, however, argues that it is unrealistic to assume amide-I boson conservation. Thus he constructs a theory that is conceptually similar to Davydov's, but which does not include this restriction. Takeno calls these "vibron solitons" and pays particular attention to analytically describing the effects of lattice discreteness. Lindenberg et al. pick up this theme and analyzes the vibron soliton in some detail, pointing out several differences from and similarities to Davydov's soliton.

The next two chapters, by Brown et al. and by Wang et al. should be considered together. In the first, an alternative formalism is developed to follow the progression from the case $Jw/\chi^2 \ll 1$ (called the "self-trapped" state) through intermediate values to the case $Jw/\chi^2 \gg 1$ (called the "soliton" state). In the second, a quantum Monte Carlo technique is used to find thermal equilibrium properties of eigenstates in Davydov's model. Since numerical estimates in these chapters are based on the single channel model (see Table 1), they tend to underestimate the size parameter, Jw/χ^2. It is our hope that the following suggestions are considered for future work.

1. Develop the theory of Brown et al. and the quantum Monte Carlo calculations of Wang et al. for the interaction energy operator indicated in Equation (2). Numerical estimates should be based on the parameters of the three channel model in Table 1.

2. Compare results of the theory of Brown et al. and the quantum Monte Carlo calculations of Wang et al. with a direct analysis of Equation (5) [11]. Carried through in detail, this would provide an estimate of the numerical reliability of Davydov's theory.

Bolterauer presents detailed calculations of the "quantum lifetime" of Davydov's soliton that are based on the requirement that nondegenerate eigenstates of the Fröhlich Hamiltonian must also be eigenstates of the translation operator. He comes to two conclusions: i) Davydov's ansatz is exact for a classical lattice, and ii) The quantum lifetime of Davydov's soliton for the single "single channel" parameter values (see Table 1) is 3.2×10^{-13} seconds.

The second conclusion is surprising for the following reason. If we assume that the exciton-phonon coupling constant (χ) is zero, then the exciton dynamics is governed by the Schrödinger equation

$$(i\hbar \frac{d}{dt} - E_0)a_n + J(a_{n+1} + a_{n-1}) = 0 \tag{10}$$

where the parameters E_0 and J have the same meaning as in Equation (3). Being linear, however, Equation (10) no longer depends upon Davydov's product ansatz, and a_n is an exact quantum mechanical probability amplitude. Furthermore it can be exactly solved [16]. Assuming an infinite lattice indexed from the center as $n = 0, \pm 1, \pm 2, \cdots$ with initial conditions $a_0(0) = 1$, $a_n(0) = 0$ for $n \neq 0$, the quantum probabilities evolve as

$$|a_n(t)|^2 = J_n^2(2\varepsilon t) \tag{11}$$

where $\varepsilon = J/\hbar$ and J_n is the Bessel function of the first kind of order n. Noting that half of the initial probability is propagated in each direction and that $J_2(3.05) \doteq 0.486$, the half-life (T) arising from quantum dispersion is

$$T = \frac{1.5}{\varepsilon}. \tag{12}$$

With a nearest neighbor dipole-dipole interaction $J = 1.55 \times 10^{-22}$ joules (the value used by Bolterauer), Equation (12) implies a half-life $T = 10^{-12}$ seconds. The second conclusion of Bolterauer is therefore equivalent to stating that turning on the exciton phonon coupling decreases the quantum lifetime of a wave packet by a factor of three.

The chapter by Škrinjar et al. complements those of Kerr and Lomdahl and of Brown et al. They are concerned with the accuracy of Davydov's original trial wave function and conclude: i) It should be reliable for computing the average values of the energy and number operators, but   ii) It may be unreliable for computing transfer rates. The following chapter by Kapor et al. provides a theoretical perspective for the experimental observations of self-trapped in crystalline acetanilide (ACN). It should be studied in relation to the experimental papers by Careri and by Bigio et al. in Section IV.

One of the striking features of soliton equations is that exact solutions to the initial value problem can be constructed using a linear technique known as the inverse scattering method (**ISM**). For the nonlinear Schrödinger equation [which is a continuum approximation to Equation (5) in the limit $Jw/\chi^2 \gg 1$], the appropriate ISM was developed in an important paper by Zakharov and Shabat [12]. Brizhik uses this ISM in the next chapter to determine requirements on initial conditions for soliton generation. (Some complementary calculations of this sort have been reported in Reference [6].) For particular initial conditions one finds a threshold requirement on the anharmonic parameter (χ^2/Jw) below which solitons will not appear. It is important to note, however, that the value of this threshold depends strongly upon the shape of the initial conditions. Brizhik shows that **the threshold value of the anharmonic parameter (χ^2/Jw) approaches zero as the initial condition approaches the hyperbolic secant shape of a soliton.** This fact should be considered when one studies whether solitons will be launched on a real alpha-helix [13].

Whether a stationary self-trapped state should be considered an exact eigenfunction of the Fröhlich (or Takeno) Hamiltonian remains an open question. Such states may be "quasi-modes" in the sense introduced by Arnol'd [14,15]. These quasimodes appear to be exact eigenfunctions under a semiclassical analysis, but they are not necessarily eigenfunctions of symmetry operators that commute with the Hamiltonian. One might suppose that an initial condition for Brizhik's ISM calculation results from a quantum transition into a quasimode. The corresponding threshold value of anharmonicity would then be zero.

The next chapter, by Savin and Zolotaryuk, assumes an anharmonic Einstein oscillator as the source of self-trapping. Davydov's theory is generalized to include more than one quantum of amide-I energy, again complementing the work of Kerr and Lomdahl. Their numerical results are important for understanding the "bisoliton" concept introduced by Davydov in the first chapter.

Mašković presents, for the first time in these proceedings, some analytical results that take account of the three dimensional geometry of alpha-helix (see Figure 1). She shows how the energy levels of various soliton states depend upon geometrical assumptions and how the longitudinal dimension of a soliton is influenced by the number of chains considered.

Enol'skii considers in detail a self-trapping mechanism that acts through dispersive optical phonons. This analysis, therefore, differs from that of Kapor et al. which assumed self-trapping by Einstein oscillators. For optical mode dispersion of the form $\omega^2 = \omega_0^2 + k^2 V_0^2$, the soliton speed must be less than V_0. Finally Zolotaryuk et al. discuss the effects of modifying the standard Davydov model by adding an on site harmonic potential for each molecule, thus complementing the model studied by Enol'skii. Numerical studies suggest that the resulting solitons should be more stable against thermal fluctuations.

In summary, the chapters in this section provide a variety of evidence for mobile, self-trapped vibrational states of the Fröhlich Hamiltonian (and its more realistic cousins) at low temperature and over some reasonable range of the parameter values. Whether this range of parameters corresponds to an accurate description of real alpha-helix may continue to be a subject of discussion, but the "single channel values" (see Table 1) lead to a substantial underestimation of the size of a self-trapped state. These values should not be considered "standard" or "widely accepted."

References

[1] H. Fröhlich, Adv.Phys. **3**, 325 (1954).

[2] A.C. Scott, Phys. Rev. A **26**. 578 (1982); [Errata: ibid **27**, 2767 (1983)].

[3] L. MacNeil and A.C. Scott, Physica Scripta **29**, 284 (1984).

[4] A.C. Scott, Phil. Trans. R. Soc. Lond. A **315**, 423 (1985).

[5] Reference [4], page 424, lines 25 and 26.

[6] A. C. Scott, Physica Scripta **29**, 279 (1984).

[7] K. Itoh and T. Shimanouchi, J. Molec. Spec. **42**, 86 (1972).

[8] V. A. Kuprievich and Z.G. Kudritskaya, Preprint, Inst. Theoretical Phys. , Kiev, ITP-82-64E (1982).

[9] O.H. Olsen, M.R. Samuelsen, S.B. Petersen and L. Nørskov, Phys. Rev. A **38**, 5856 (1988).

[10] A. S. Davydov, **Solitons in Molecular Systems**, D.Reidel Publishing Co., Dordrecht, Boston, Lancaster (1983); page 4.

[11] A.C. Scott and L. MacNeil, Phys. Lett. A **98**, 87 (1983).

[12] V.E. Zakhorov and A.B. Shabat, Soviet Phys. JETP **34**, 62 (1972) [Zh. Eksp. Theor. Fiz. **61**, 118 (1971)].

[13] B. Mechtly and P.B. Shaw, Phys. Rev. B **38**, 3075 (1988).

[14] V.I. Arnol'd, Funct. anal. applic. **6**, 94 (1972).

[15] M.V. Berry, J. Phys. A **10**, L193 (1977).

[16] V.M. Kenkre and S.M. Phatak, Phys. Lett. A **100**, 101 (1984).

SOLITONS IN BIOLOGY AND POSSIBLE ROLE OF BISOLITONS IN HIGH-T_C SUPERCONDUCTIVITY

A.S. Davydov

Institute for Theoretical Physics

Academy of Sciences of the Ukrainian SSR, Kiev

1 Solitons and Bisolitons in Biology

1.1 Introduction

Many processes in living organisms are associated with a space propagation of energy or electrons along protein molecules. Investigations carried out at the Institute for Theoretical Physics in Kiev starting from 1973 have shown that the effective transport of energy or electrons is due to certain properties of excited states of alpha-helical protein molecules in which peptide groups are arranged as three parallel quasiperiodic chains [1]. Peptide groups have a constant electric dipole moment of 3.5 Debye; therefore they form a potential well capable of retaining an extra electron. The overlap of wave functions of the extra electron and the resonance dipole-dipole interaction of amide-I vibrations of the neighboring peptide groups, J, explains the collective nature of the corresponding excitations. Quantized, these excitations have effective mass $m = \hbar^2/2a^2 J$ and are called quasi-particles.

1.2 One Soliton in a Chain

The creation of a quasi-particle in the chain induces a local deformation of the chain. Excited states of a protein molecule are, therefore, of a complicated character due to the interaction between quasi-particles and local deformation of the chain. The excitation propagates along the chain with a constant velocity V. This state is described by a nonlinear equation for the function

Davydov's Soliton Revisited, Edited by P.L. Christiansen and A.C. Scott
Plenum Press, New York, 1990

$$\psi(x,t) = \Phi(x,t)\exp\{i(kx - Et/\hbar)\}, \quad k = \frac{mV}{\hbar}. \tag{1}$$

When we introduce the dimensionless variable $\xi = (x - Vt)/a$, the envelope function $\Phi(\xi)$ is defined by the equations

$$\left[\frac{d^2}{d\xi^2} + \varepsilon + 2g\Phi^2(\xi)\right]\Phi(\xi) = 0, \quad \int \Phi^2(\xi)d\xi = 1. \tag{2}$$

The field of the local deformation $\rho(\xi)$ is

$$\rho(\xi) = \frac{\chi}{w(1 - s^2)}\Phi^2(\xi), \tag{3}$$

where $s^2 = V^2/V_0^2 \ll 1$, V_0 is the velocity of sound in the chain, w is its elasticity coefficient, χ is the parameter of deformational exciton-phonon interaction, and

$$g = \frac{\chi^2}{2wJ(1 - s^2)} \approx \frac{\chi^2}{2wJ} = g_0. \tag{4}$$

The solution of (2) is the soliton function

$$\Phi(\xi) = \tfrac{1}{2}\sqrt{g}\,\operatorname{sech}\left(\tfrac{1}{2}g\xi\right) \tag{5}$$

$$\varepsilon = -\tfrac{1}{4}g^2. \tag{6}$$

The total energy of the soliton including the energy of local deformation, at $V^2 < V_0^2$ below the bottom of the energy band of a free quasi-particle, is given by the expression

$$E_{sol}(V) = \tfrac{1}{2}M_{sol}V^2 - \tfrac{1}{12}g_0^2 J. \tag{7}$$

The quantity

$$M_{sol} \approx m + g_0^2 J/3V_0^2 \tag{8}$$

may be called the effective soliton mass. There are four basic reasons for the high stability of isolated solitons.

1. They do not interact with acoustic phonons because this interaction is completely taken into account when a quasi-particle is bound to a local deformation. Moving always with a velocity less than that of longitudinal sound, they do not emit phonons, i.e. they do not transform their kinetic energy into thermal energy.

2. The envelope of the soliton wave function (5) keeps its form and does not smear, as distinct from a wave packet.

3. The energy (7) must be consumed to decompose the soliton into its components.

4. The solitons have topological stability. After the soliton has passed, all unit cells (molecules) behind the soliton remain displaced from their initial equilibrium positions. To destroy the soliton, they must return to the initial positions.

To illustrate the considerable change in the lifetime of a pair of particles when they unite to form a more complex one, we can consider a deuteron consisting of a neutron and a proton. Even if the proton is comparatively weakly bound to the neutron (its lifetime, when in free state, is $\approx 10^3$ s), it lives inside the deuteron for an infinitely long time.

1.3   Bisolitons in a Chain

Every quasi-particle in a soliton state, taken in a moving coordinate system ξ, is in the potential well (3), generated by the local displacement in equilibrium positions of the unit cells [2,3,4]. When two solitons with oppositely directed spins come together to a distance $d \approx 4\pi/g$, they unite (pairing), forming a single Bose particle—a *bisoliton* with zero spin and a double electric charge. This particle moves in the common potential well

$$U_{\alpha\beta}(\xi) = \frac{\chi^2}{w(1 - s^2)} \int \left[\Phi_\alpha^2(\xi) + \Phi_\beta^2(\xi) \right] d\xi \tag{9}$$

which deepens with decreasing distance d.

When the Coulomb repulsion of quasi-particles is disregarded, the deepest well corresponds to $d = 0$. In this case the orbital functions $\Phi_\alpha(\xi)$ and $\Phi_\beta(\xi)$ coincide (the subscripts α and β are therefore omitted below) and satisfy the equation

$$\left[\frac{d^2}{d\xi^2} + \varepsilon_0 + 4g\,\Phi_0^2(\xi) \right] \Phi_0(\xi) = 0. \tag{10}$$

This nonlinear equation has a solution

$$\Phi_0(\xi) = \left(\frac{g}{2} \right)^{1/2} \operatorname{sech}(g\xi) \tag{11}$$

normalized by the condition

$$\int_{-\infty}^{\infty} \Phi_0^2(\xi) d\xi = 1 \tag{12}$$

which corresponds to describing the state of one bisoliton in an infinite chain.

Both quasi-particles in a bisoliton thus move in one potential well

$$U_{\uparrow\downarrow}(\xi) = -2g^2 J \operatorname{sech}^2(g\xi) \tag{13}$$

that is deeper and narrower than the potential wells of individual solitons. The distribution of each of two quasi-particles in a bisoliton is defined by the function

$$\Phi_0^2(\xi) = \tfrac{1}{2} g \operatorname{sech}^2(g\xi). \tag{14}$$

Consequently, the deformation region and the quasi-particle distribution region are characterized by the length

$$\Delta\xi_{bs} \approx \frac{2\pi}{g}. \tag{15}$$

The energy of such a bisoliton moving with constant velocity V (measured from the conduction band bottom of free quasi-particles) is defined by

$$E_{bs}(V) = 2 \left[\frac{m^* V^2}{2} - \frac{g^2 J(1 - 5s^2)}{3(1 - s^2)} \right]. \tag{16}$$

Thus the creation of a bisoliton by two solitons moving in parallel with velocity V is accompanied by the emission of the energy

$$E_\infty(V) - E_{bs}(V) = \frac{g_0^2 J(1 - 5s^2)}{2(1 - s^2)^3}. \tag{17}$$

This energy is positive if the bisoliton velocity does not exceed $V_0/\sqrt{5}$.

At small velocities, the energy (16) can be approximated as

$$E_{bs}(V) \approx E_{bs}(0) + \tfrac{1}{2} M_{bs} V^2; \qquad V^2 \ll V_0^2, \tag{18}$$

where

$$E_{bs}(0) \approx -\frac{2g_0^2 J}{3}, \tag{19}$$

$$M_{bs} \approx 2\left(M_{sol} + \frac{g_0^2 J}{V_0^2} \right). \tag{20}$$

The effective bisoliton mass M_{bs} is greater than the sum of the masses of two individual solitons, because the motion of a bisoliton is connected with the movement of a deeper local deformation of a chain.

So the energy of a soliton at rest is below the conduction band bottom of two quasi-particles at the level (19) which is four times deeper than the position of the energy level of one soliton. Consequently, the formation of a resting bisoliton incorporating two solitons is accompanied by the emission of energy

$$\Delta E_1 \approx \tfrac{1}{2} g_0^2 J, \tag{21}$$

and the energy ΔE_1 must be spent for the bisoliton to decay into two solitons.

The great stability of isolated bisolitons seems to explain the very high efficiency in the utilization by living organisms, in ATP molecule synthesis, of electron transfer processes through mitochondrial and chloroplast membranes. It is established experimentally [5] that in these processes the electrons are transferred in pairs, but not individually. These processes proceed at physiological temperatures ≈ 310 K. The chemical energy stored in ATP molecules through food product oxidation is the basic source of energy used by living organism in all their functions.

2 Nonlinear Bisoliton Model of High-Temperature Superconductivity of Ceramic Compounds

2.1 Introduction

In 1986, J.G. Bednorz and K.A. Müller in Switzerland, and later researchers in other countries, produced ceramic oxide materials containing copper, oxygen, lanthanum, and barium or strontium ($L_{2-\delta}\, Ba_\delta\, Cu\, O_4$) that go into a superconducting state when cooled below 40 K. In February 1987, C.W.P. Chu, M.K. Wu, J.B. Ashburn et al., of the University of Houston showed that the critical temperature can be raised to 93 K by replacing lanthanum with yttrium ($Y\, Ba_2\, Cu_3\, O_{7-\delta}$). These results have been confirmed by studies carried out in many laboratories the world over.

The basic properties of the new high-temperature compounds have not been explained by the generally accepted metal superconductivity theory formulated by J. Bardeen, L.N. Cooper, and J.R. Schriffer (**BCS** theory) thirty years ago. The new high T_c superconductors have very significant features.

1. They possess a very great anisotropy. Figure 1 reproduces the scheme of ion arrangement in a $Y\, Ba_2\, Cu_3\, O_{7-\delta}$ crystal. It follows from the figure that the crystal contains (in the b-direction) quasi-one-dimensional chains of alternating ions of copper and oxygen. In a certain approximation, such quasi-one-dimensional chains may be considered

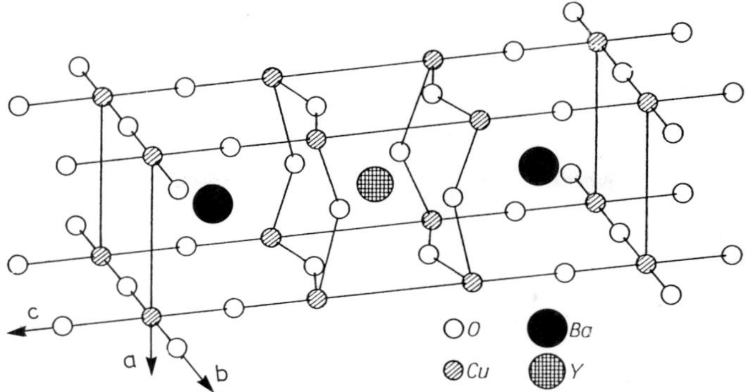

Figure 1. Crystal structure of the unit cell of a $Y^{3+} Ba_2^{2+} Cu_3^{2+} O_{7-\delta}^{2-}$ crystal.

independently, making it possible to study the conductivity caused by the movement of quasi-particles along one of the chains.

2. Superconductivity provides a low density of carriers of double electron charges.

3. Cooper pairs in such materials have a small radius (about 15Å–30Å). The radius of the pair of electrons in superconducting metals is much bigger $\approx 10^{-4}$cm.

4. In ceramic superconductors a very small isotope effect is observed.

In the next section, we will discuss the main properties of a set of bisolitons at a temperature close to zero.

2.2   Bose-Condensation of Bisolitons

In a quasi-one-dimensional chain that consists of a great number of quasi-particle pairs, the pairing is realized at low temperature only from the states of free quasi-particles with the wave numbers lying near the wave number k_F of the Fermi surface.

Owing to the quasi-momentum conservation law, the paired particle moving with velocity $V = \hbar k/m$ can be generated from two free quasi-particles with wave numbers $k_1 = 2k - k_F$ and $k_2 = k_F$, or $k_1 = k - k_F$ and $k_2 + k_F$. In a singlet spin state, the symmetric coordinate wave function $\Psi(\xi_1, \xi_2)$ of a quasi-particle pair has the form

$$\Psi(\xi_1, \xi_2) = \sqrt{2}\Phi(\xi_1)\Phi(\xi_2)\cos\left[(k - k_F)(\xi_1 - \xi_2)\right] \cdot \exp\left\{ik(\xi_1 + \xi_2) - i\mathcal{E}t/\hbar\right\} \tag{22}$$

as has been shown by V.N. Ermakov and this author [3]. Here $\mathcal{E}$ is the energy of two paired quasi-particles in a potential field of local deformation of the chain caused by the interaction between these particles and displacements from equilibrium positions of unit cells of the mass M; $\xi_i = (x_i - Vt)/a$, and V is the velocity of motion connected with the wave number k by the relation $k = mV/\hbar$. The amplitude functions $\Phi(\xi_i)$ satisfy the nonlinear differential equations:

$$\left[\frac{\partial^2}{\partial \xi_i^2} + 4g\Phi^2(\xi_i) - \varepsilon\right]\Phi(\xi_i) = 0, \qquad i = 1, 2 \tag{23}$$

Here the dimensionless parameter g characterizes the binding between a quasi-particle and the field of chain deformation $\rho(\xi)$.

Equation (22) can be considered as a function describing two coupled wave fields characterized by the coordinates ξ_1 and ξ_2. With N quasi-particle pairs available in the infinite chain, this function is normalized by the condition

$$\int_{-\infty}^{\infty} |\Psi(\xi_1,\xi_2)|^2 d\xi_1 d\xi_2 = N. \tag{24}$$

In this case, (23) admits periodic solutions corresponding to a uniform distribution of quasi-particle pairs along the chain. The real functions $\Phi(\xi_i)$ of these solutions should satisfy the periodicity condition with the period L

$$\Phi(\xi_i) = \Phi(\xi_i + L) \tag{25}$$

and normalization

$$\int_0^L \Phi^2(\xi)d\xi = 1 \tag{26}$$

in order that each quasi-particle pair be represented in every period. Then in each period the quasi-particles pair energy $\mathcal{E}(V)$ in the field of deformation $\rho(\xi)$ induced by quasi-particles is determined by

$$\mathcal{E}(V) = \mathcal{E}_F + \varepsilon J + mV^2 - \hbar V k_F \tag{27}$$

where $\mathcal{E}_F = \hbar^2 k_F^2/m$ is the energy of two quasi-particles at the Fermi level, and ε is the eigenvalue of a dimensionless equation (23).

The local deformation field $\rho(\xi)$ and the chain deformation energy W in each period are expressed through the functions $\Phi(\xi)$ by the equations

$$W = 2G\left(1 - s^4\right)\int_0^L \Phi^4(\xi)d\xi, \qquad s = V/V_0 \tag{28}$$

$$\rho(\xi) = -2\frac{G}{\chi}\Phi^2(\xi) \tag{29}$$

in which

$$G = \frac{\chi^2}{w\left(1 - s^2\right)}, \quad \text{and} \quad g = \frac{G}{2J} \tag{30}$$

$\alpha\chi$ is the energy of the interaction between a quasi-particle and the chain, w is its elasticity coefficient, V_0 is the velocity of a longitudinal sound, and α is the lattice constant.

Periodic solutions of (23) are expressed through Jacobi elliptic functions $dn(u,q)$ by

$$\Phi_q(\xi_i) = \left(\frac{g}{2}\right)^{1/2} E^{-1}(q)\, dn(u_i,q), \quad u_i = \frac{g\xi_i}{E(q)}. \tag{31}$$

The explicit form of the Jacobi functions $dn(u,q)$ depends on the modulus q that takes continuous values from zero to unity. These values are expressed by a product of a period L (determined by the linear quasi-particle pairs density) and a dimensionless electron-phonon interaction parameter g by means of the expression

$$gL = 2E(q)\, K(q) > \pi/2 \tag{32}$$

in which the functions $K(q)$ and $E(q)$ are the complete elliptic integrals of the first and second kind.

The periodic solutions (23) correspond, according to (29), to the periodic deformation field

$$U_q(\xi) \equiv \chi\rho_q(\xi) = -2G\Phi_q{}^2(\xi). \tag{33}$$

This field realizes an indirect interaction between bisolitons binding them in a single unit Bose-condensate. All bisolitons in this condensate travel with the same velocity V. In each

16

period L the symmetric coordinate wave function of bonded quasi-particle pairs (bisolitons) is determined by the expression

$$\Psi_q(\xi_1,\xi_2,t) \;=\; \sqrt{2}E^{-2}(q)\cos\{(k-k_F)(\xi_1-\xi_2)\}\times$$
$$\times\, dn(g\xi_1/E(q),q)\,dn(g\xi_2/E(q),q)\times$$
$$\times\,\exp\{i\,[\varphi(k)-\mathcal{E}t/\hbar]\}\,. \tag{34}$$

Then the phase $\varphi(k)$, coherent for the whole chain, is determined by

$$\varphi(k) = 2k\,[R-Vt]\,. \tag{35}$$

It varies smoothly with the bisoliton coordinate $R = (x_1+x_2)/2$. The velocity of condensate motion is expressed through a derivative of the function (34) by means of the equality

$$V = \frac{\hbar}{2m}\frac{d\varphi}{dR} = \frac{\hbar k}{m}\,. \tag{36}$$

Due to coherence, bisolitons are in a state of a homogeneous motion in the condensate. An abrupt change in the velocity of a bisoliton should cause changes in the character of motion of all bisolitons that compose the condensate. To each function (31) at a fixed modulus q there corresponds the energy eigenvalue of (23) which equals

$$\varepsilon_q = \frac{-q^2 g^2}{E(q)}\,. \tag{37}$$

According to (27) and (28), the total bisoliton energy including that of deformation W takes the form

$$\Theta_q(V) = \mathcal{E}_F + mV - \hbar k_F V - \frac{q^2 g^2 J}{E(q)} + W_q\,. \tag{38}$$

At low velocities $(V \ll V_0)$ the total energy (38), transferred by every bisoliton in the condensate, can be written in the form

$$\Theta_q(V) = \mathcal{E}_F - D_0(q) + \frac{M_{bs}V^2}{2} - \hbar V k_F \tag{39}$$

where the function $-D_0(q)$ determines the lowering in energy of two quasi-particles when they are paired accounting for the positive energy of the local chain deformation

$$-\,D_0(q) = W_q - \frac{g^2 q^2}{E(q)}\,. \tag{40}$$

Calculating from (28) we get

$$D_0(q) = g^2 J F(q) \tag{41}$$

where

$$F(q) \equiv \frac{2}{E^3(q)}\left[2 - \frac{q^2\Xi(q)}{E(q)}\right] \tag{42}$$

$$\Xi(q) = \tfrac{1}{3}\left[2(2-q^2)E(q) - (1-q^2 K(q)\right]\,. \tag{43}$$

A bisoliton mass M_{bs} exceeds that of two quasi-particles $2m$,

$$M_{bs} = 2m + \frac{4Jg^2\Xi}{V_0^2 E^3(q)}\,. \tag{44}$$

The value $D_0(q)$ determines the energy of pairing at the absolute zero, i.e. the energy released when a bisoliton is formed of two free quasi-particles. Then the value $\Delta \equiv D_0(q)/2$ characterizes the energy gap in a quasi-particle spectrum that appears in a quasi-particles spectrum while pairing.

Equation (41) is obtained without regard to the screened Coulomb repulsion between quasi-particles. Such an approximation is justified only under the condition of bell-shaped elevations with the width $\Delta\xi = 2\pi/g$ which are separated by the distance L. All of them characterize the position of bisolitons moving with the same velocity.

With an increase in quasi-particles density ($gL \to \pi^2/2$), the modulus q tends to zero. Then $\Phi(\xi)$ is expressed through trigonometric functions with a modulation amplitude $\approx q^2$. In this case, the function (42) takes the value $16/\pi^3$, and the pairing energy $D_0(q)$ is determined by

$$D_0(q) \approx \frac{16g^2 J}{\pi^3}, \qquad q^2 \ll 1. \tag{45}$$

At the critical value $gL = \pi^2/2$ the modulus q^2 turns to zero. Consequently modulations of the function $\Phi(\xi)$ vanish, i.e. the periodicity of the solutions that provides stability of condensate-bisolitons vanishes.

Thus, according to a bisoliton model, the conditions favoring superconductivity correspond to an optimal concentration of quasi-particles when $F(q)$ attains its maximal value.

The dependence of superconductivity (obtained by us on the basis of a bisoliton model) on the concentration of charge carriers is confirmed by experimental studies. In a review by Khurana [10], it is noted that no experiments have shown superconductivity of these compounds at concentrations δ less than 0.05 and higher than 0.4. The maximal critical temperature is attained at concentrations of carriers near 0.15.

2.3   Stability Conditions of a Bisoliton Condensate

The availability of a charged Bose-particles condensate is a necessary but not sufficient condition for the appearance of superconductivity. At small velocities, bisoliton condensate stability is ensured by:

1. Other one-particle excitations lying lower than the bisoliton energy band are absent, and

2. The Landau superfluidity condition $[d\Theta_q(V)/dV]_{V \to 0} > 0$ is satisfied.

For $k_F \neq 0$, this condition follows directly from (39). The first condition in our model, as well as in that of BCS, is satisfied since the quasi-particle pairing is guaranteed by the electron-phonon mechanism.

To destroy a bisoliton at absolute zero with creation of two quasi-particles at the Fermi level, the pairing energy D_0 must be expended (41). With increasing temperature, the pairing energy $D(T)$ decreases. The critical temperature T_c of superconductivity break-down is determined as the temperature at which $D(T)$ vanishes.

Instead of the pairing energy denoted by D_0, in the BCS theory one usually employs the representation of an energy gap Δ in a quasi-particle spectrum where $D_0 = 2\Delta$. In the BCS theory, the critical temperature T_c is directly determined from the energy gap by $k_B T_c = 3.5\Delta$. The proportionality between the pairing energy D_0 and critical temperature will probably be preserved in our bisoliton superconductivity model; therefore we assume the relation

$$k_B T \sim D_0 \equiv g^2 J F(q) \tag{46}$$

to be satisfied. The temperature T_c is independent of the ion mass M. This mass is involved only in (44) that determines, in (40), the kinetic energy of bisolitons. Thus, the bisoliton model explains naturally why there is almost no isotopic effect [6,7] in the new superconductors.

With increasing bisoliton concentration in the chain, the elliptic function modulus q varies from unity to zero. When quasi-particles density $(gL \ll 1)$ is relatively small, the value $q \approx 1$, and the function $F(q)$ involved in (41) takes the form

$$F(q) = \tfrac{2}{3}\left[1 + 4gL \, \exp(-gL)\right]. \tag{47}$$

In this case, the pairing energy (41) increases and bisoliton effective mass (44) decreases with increasing quasi-particles density.

As the bisolitons condensate velocity rises, there are restrictions on upper velocity values. In as much as bisolitons are formed when two quasi-particles unite with a local chain deformation, their combined motion, as a single entity, is possible if its velocity does not exceed the lesser of two velocities:

1. The velocity of longitudinal sound $V_0 \approx 10^5$ cm/sec, or

2. The maximal group velocity $V_g = 2aJ/\hbar$ of free quasi-particles in their conductivity band.

When $\alpha \approx 7.8 \times 10^{-8}$ cm, $J \approx 0.01$ eV, the value $V_g \approx 2 \times 10^6$ cm/sec is obtained.

The fact that the energy (39) involves a linear term with respect to the product of the velocity and k_F stipulates the second restriction on the condensate velocity. The availability of such a term at velocities exceeding the critical value V_{cr} results in a bisoliton decay into the two free quasi-particles with the total quasi-momentum being conserved. This process can be considered a fast transition in which the local deformation W is unable to change and quasi-particles in a bisoliton moving with velocity $V = \hbar k/m$ go to free states with quasi-momenta $\hbar(2k - k_F)$ and $\hbar k_F$. As has been shown by V.N. Ermakov and this author [4], such a decay does not take place if the condensate velocity is less than the critical value V_{cr} defined by

$$V_{cr} = \frac{D(q)}{\hbar k_F}. \tag{48}$$

Consequently, the superconductivity of ceramic superconductors is preserved (at $T = 0$) when current densities are less than the critical value

$$J_{cr} = \frac{2eD_0(q)}{\hbar k_F}. \tag{49}$$

In ceramic superconductors, the soliton density and the Fermi momentum k_F are small, but the critical current density may be large.

2.4   Direct Energy Gap Measurements

There are two ways of measuring the energy gap Δ in superconductors:

1. From metal to superconductor electron tunneling, and

2. From reflection and absorption of infrared radiation by superconductors.

Infrared measurements [8] in superconducting $L_{2-\delta}\, Sr_\delta\, Cu\, O_4$ have shown that the dimensionless ratios $2\Delta/k_B T_c$ lie in the region of values 1.3 to 2.7 which is less than the value of 3.53 that follows from the BCS theory. However tunneling measurements with the same specimens have exhibited values within 3.5 to 6, i.e. twice as great as the values obtained in infrared studies.

Studying superconducting $Y\,Ba_2\,Cu_3\,O_{7-\delta}$ $(T_c = 87$ K), Y.R. Kirtley et al. [8] have confirmed that the energy gap measured in tunneling experiments is larger than that when infrared spectroscopy was used. Thus the infrared spectroscopy has given, for the dimensionless relation $2\Delta/k_B T_c$, values that lie within the interval 1.6 to 3.4 ($\Delta = 6$ to 13×10^{-3} eV). At the same time the tunneling studies of the same crystals have given the values from 3.7 to 5.6 ($\Delta_{tunn} = 13.7$ to 20.7×10^{-3} eV).

Such a substantial difference in the gap widths determined by different methods was not explained theoretically. However, this can be interpreted on the basis of the high-temperature superconductivity bisoliton model suggested here.

As has already been noted, bisolitons of a superconducting condensate are three component entities; each pair of quasi-particles moves in the local deformation field created by them. Absorbing far infrared radiation ($\approx 100 cm^{-1}$), a quasi-particles pair breaks down and simultaneously the local deformation is annihilated (slow process). That is why the required energy is determined by (41).

The same energy is expended in thermal destruction of a bisoliton. This is why the critical temperature of transition to a superconduction state is determined by the gap $\Delta(q)$ by means of (46) taking into account that $D_q = 2\Delta(q)$.

In the tunneling process, a quasi-particle pair breaks down preserving the local deformation well (rapid process) therefore the larger energy

$$2\Delta_{tunn}(q) = 2\Delta(q) + W = \frac{q^2 g^2}{E(q)} \tag{50}$$

must be expended.

2.5   The Superconducting Bisoliton Condensate in a Magnetic Field

In Reference [9] it was shown that the magnetic field influence on ceramic high-temperature superconductors differs qualitatively from the effect on superconductors satisfying the BCS theory according to a bisoliton model. The energy gap and critical temperature change weakly with increasing field and jump to zero when the fields exceed the critical values.

2.6   Conclusion

To study the high-temperature superconductivity of ceramic materials, a model of quasi-one-dimensional chains of alternating ions Cu^{2+} and O^{2-} composing these crystals was used. Proceeding from a representation based on moderately strong electron-phonon binding between a quasi-particle (arising when copper ions valency is being changed) and displacements of unit cells containing these ions and the oxygen ions surrounding them, the conditions for generating a bisoliton Bose-condensate that induces superconductivity are obtained. The development of this theory of high-temperature superconductivity mechanism may account for qualitative singularities.

For illustration we present estimates of some parameters of the theory. If we take the value of the lattice constant α equal to 7.76Å and the bisoliton diameter equal to 30Å, the dimensionless binding parameter value $g \approx 1.62$ follows from (32). Equation (46) then gives that the critical temperature $T_c \approx 100$ K is possible when the exchange interaction energy value J equals ≈ 0.05eV. Knowing g and J we can evaluate, using (30), the ratio χ^2/w determining the energy of binding between quasi-particles and displacements of unit cells equilibrium positions. It equals 0.15eV. If $J = 0.05$eV, the quasi-particle effective mass in the conductivity band is about $\approx 1.2m$.

So the nonlinear bisoliton model, based on strong electron-phonon coupling, allows one to explain the following peculiarities of ceramic high-temperature superconductors:

1. The small correlation radius of paired quasi-particles.

2. The very small isotopic effect.

3. Nonmonotone dependence of temperature of the transition to a superconductivity that vanishes at very large and very small concentrations.

4. The different energy gap width values Δ obtained in tunneling studies and in infrared spectroscopy methods. The tunneling measurements exhibit the larger width values Δ_{tunn} than the infrared ones Δ.

Acknowledgment

The author is thankful to his colleagues L.S. Brizhik and V.N. Ermakov who participated in developing the concept of a bisoliton model of high-temperature superconductivity of ceramic compounds.

References

[1] A.S. Davydov, **Solitons in molecular systems**, D. Reidel Pub. Co., Dordrecht, Boston, Lancaster (1987).

[2] A.S. Davydov, Bisoliton Mechanism of High-Temperature Superconductivity. **phys. stat. sol. (b) 146**, 619 (1988).

[3] A.S. Davydov and V.N. Ermakov, Stability of Superconducting Condensate of Bisolitons. **phys. stat. sol. (b) 148**, 305 (1988).

[4] A.S. Davydov and V.N. Ermakov. Soliton Generation at the Boundary of Molecular Chain. **Physica D 32**, 318 (1988).

[5] P.G. Hinkle and R.E. McCarty, How cells makes ATP. **Sci. Amer. 288**, no. 3, 104 (1978).

[6] B. Batlog, Isotope Effect in the High-T_c Superconductors. **Phys. Rev. Lett. 58**, 2333 (1987).

[7] H. Katayama. Isotope Effect. **Physica C. 156**, 481 (1988).

[8] J.K. Kirtley, Tunneling and Infrared Measurements of the Energy Gap in the High-Critical Temperature Superconductors. **Phys. Rev. B35**, no. 16, 8846 (1987).

[9] L.S. Brizhik, A.S. Davydov and V.N. Ermakov, The Magnetic Field Influence on a Superconduction Bisoliton Condensate: Preprint ITP-88-8E Inst. Theor. Phys. Ukr. SSR, Kiev, 1988.

[10] A. Khurana. Superconductivity seen above the boiling point of nitrogen. **Physics Today 40**, no. 4, 17 (1987).

Question by Kenkre: Solitons are, as is generally known, one-dimensional objects. High-T_c superconductors, on the other hand, have 2-dimensional and 3-dimensional structure. Why does this not present problems for your theory? Is it perhaps that by bisolitons you merely mean bipolarons?

Reply: Solitons are quasi-one-dimensional objects. They describe the movement of excitation in one direction having one phase in perpendicular cross-section. As is known, a plane wave can move in a three dimensional medium. (Laser beams reach the Moon from Earth.) Creation of solitons in new superconductors is facilitated by the strong asymmetry of crystals—presence of one-dimensional chains of alternating ions Cu and O. Solitons and polarons are very different: **Polarons** are created in the medium (ionic crystals) through long-range interactions with **optical phonons**. They can move only with a velocity that is slower than the group velocity of the optical phonons. This velocity equals zero if the optical phonons are nondispersive. Solitons are created in the medium through short-range interactions with acoustic phonons. They can move with a velocity that is slower than the acoustic phonon velocity ($\sim 10^5$ cm/sec).

Question by Lomdahl: The pairing mechanism in superconductivity is usually studied within the framework of models that allow for strong electron-electron interactions, e.g. the Hubbard model. In your approach, the electron-electron correlations enter only indirectly via the electron-phonon coupling. Could you elaborate on the justification for neglecting the direct electron-electron interaction in your Hamiltonian?

Reply: I think a model that assumes only electron-electron interactions does not explain the pairing of electrons. In the Hubbard model, the energy of a pair on one site is higher than the energy of free particles. So one finds not stationary but virtual states, which decay. The electron-electron interaction in our model is screened because the distance between the particles in a pair is bigger than α. That electron-electron repulsion decreases effectively the attraction via the electron-phonon coupling.

Question by Pnevmatikos: In the vibrational part of your Hamiltonian one expects to have a damping term. This term will seriously affect the perfectness that we need in order to have superconductivity. Would you discuss the fact that you neglect damping in you model?

Reply: In our model as in the BCS model, all bisolitons in a superconducting state make up a single condensate. They are described by a unique wave function and all are moving with a single velocity. If we change the velocity of one of them, we must change the velocity of all others. This is the main reason for the phenomenon of superconductivity.

2

QUANTUM-MECHANICAL DERIVATION OF THE DAVYDOV EQUATIONS
FOR MULTI-QUANTA STATES

W. C. Kerr

Olin Physical Laboratory
Wake Forest University
Winston-Salem, NC 27109-7507

P. S. Lomdahl

Theoretical Division
Los Alamos National Laboratory
Los Alamos, NM 87545

Davydov and Kislukha[1] suggested in the 1970's that nonlinear self-trapping could serve as a method of energy transport along quasi-one-dimensional chains of molecules. The problem was to explain how the energy released by hydrolysis of adenosine triphosphate and transferred to proteins in biological systems remains localized and moves along the protein chains at a reasonable rate to perform useful biological functions. The α-helix protein structure was considered, which consists of three chains of hydrogen-bonded peptide groups (HNCO) with associated side groups which contribute to the molecular mass but are assumed dynamically inert. The coupled fields which they suggested are relevant in this problem are a high frequency intramolecular vibration of the peptide groups (the Amide-I or C=O stretch mode, at about 1665 cm^{-1}), and the low frequency vibrations of the entire peptide groups (and associated side groups). These fields are coupled through the dependence of the Amide-I energy on the length of the hydrogen bond coupling neighboring peptide groups.[2] The Hamiltonian Davydov used to describe this situation is the same as that used for the polaron problem (the Froehlich Hamiltonian for electron-phonon interactions) with some changes in the meaning of the symbols. Davydov's method of analysis[1,3] of this Hamiltonian led to connections with ideas of soliton propagation in other physical systems.[4]

Following Davydov's original suggestion, the model has been elaborated by Scott and collaborators to describe more accurately the three-chain structure of α-helix, and numerical calculations have been carried out which verify the existence of self-trapped states in this model.[5]

Our purpose here is to present a derivation of the Davydov equations which employs only quantum-mechanical techniques. The derivation here is more general than our previous treatment of this problem[6] because we use an *Ansatz* which has present several quanta of the high frequency oscillator system rather than just one quantum.[7]

Davydov's Soliton Revisited, Edited by P.L. Christiansen and A.C. Scott
Plenum Press, New York, 1990

Since some steps of the calculation that follows are the same as those in our paper[6] which treats the single quantum case, reference will be made to that paper for some of the those details.

Davydov's Hamiltonian is[1,3]

$$H = \sum_n \left[E_0 B_n^\dagger B_n - J(B_{n+1}^\dagger B_n + B_n^\dagger B_{n+1}) \right] \tag{1}$$

$$+ \sum_n \left[\frac{p_n^2}{2m} + \frac{1}{2} w (u_{n+1} - u_n)^2 \right] + \chi \sum_n (u_{n+1} - u_{n-1}) B_n^\dagger B_n$$

$$= H_v + H_p + H_{int}.$$

Here, $B_n^\dagger$ and B_n are boson creation and annihilation operators for quanta of intramolecular vibrations with energy $E_0 = 1665$ cm^{-1} at site n (the C=O stretch mode), u_n and p_n are the molecular displacement and momentum operators for the molecule at site n, m and w are the molecular mass and intermolecular force constant, and J is the intersite transfer energy produced by dipole-dipole interactions. The non-linear coupling constant χ arises from modulation of the on-site energy by the molecular displacements. The vibrational part H_v, the phonon part H_p, and the interaction part H_{int} are defined to be the individual terms in (1).

The phonon part of the Hamiltonian can be cast into familiar form in terms of phonon creation and annihilation operators by the use of the standard transformation

$$u_n = \sum_q \left(\frac{\hbar}{2Nm\omega_q} \right)^{1/2} e^{iqnl} (a_{-q}^\dagger + a_q), \tag{2a}$$

$$p_n = \sum_q \left(\frac{m\hbar\omega_q}{2N} \right)^{1/2} e^{iqnl} i (a_{-q}^\dagger - a_q). \tag{2b}$$

In these formulas l is the lattice spacing (the distance between peptide groups), and

$$\omega_q = 2(w/m)^{1/2} |\sin(ql/2)| \tag{3}$$

is the dispersion relation for H_p.

To understand the dynamics arising from the Hamiltonian (1), we make the *Ansatz* for the state vector

$$|\psi(t)\rangle = \frac{1}{\sqrt{Q!}} \left[\sum_n a_n(t) B_n^\dagger \right]^Q \exp\left\{ -\frac{i}{\hbar} \sum_j \left[\beta_j(t) p_j - \pi_j(t) u_j \right] \right\} |0\rangle, \tag{4}$$

where $|0\rangle$ is the ground state vector (i.e. it is annihilated both by B_n and by the phonon operators a_q). Davydov's original *Ansatz*[1,3] was the $Q = 1$ case of this formula. Assuming that the time evolution of this state vector is approximately the same as that of the (unknown) exact state vector, one can then understand the system behavior by finding the time evolution of the three sets of unknown functions $a_n(t)$, $\beta_n(t)$, and $\pi_n(t)$.

First, we establish the necessary conditions for this state vector to be normalized. The normalization is

$$\langle\psi(t)|\psi(t)\rangle = \frac{1}{Q!} \sum_{\substack{m_1 \cdots m_Q \\ n_1 \cdots n_Q}} a_{m_1}^* \cdots a_{m_Q}^* a_{n_1} \cdots a_{n_Q} \langle 0| B_{m_1} \cdots B_{m_Q} B_{n_1}^\dagger \cdots B_{n_Q}^\dagger |0\rangle. \tag{5}$$

By using mathematical induction one can show that the ground state expectation value appearing in (5) is equal to the $Q \times Q$ *permanent*

$$<0|B_{m_1}\cdots B_{m_Q}B_{n_1}^{\dagger}\cdots B_{n_Q}^{\dagger}|0> = per\,(\delta_{m_i n_j}); \quad i,j = 1,...,Q. \tag{6}$$

Here $\delta_{i,j}$ is the Kronecker delta. A permanent is evaluated similarly to a determinant except that all of the terms are taken with positive signs. When we substitute this permanent (6) of Kronecker deltas into (5), we find that (5) reduces to $Q!$ identical terms each of which has Q identical factors, so that

$$<\psi(t)|\psi(t)> = \frac{1}{Q!}\sum_{m_1\cdots m_Q}|a_{m_1}|^2\cdots|a_{m_Q}|^2 Q! = (\sum_m|a_m|^2)^Q. \tag{7}$$

Therefore a necessary and sufficient condition for the state vector to be normalized,

$$<\psi(t)|\psi(t)> = 1 , \tag{8}$$

is that the amplitudes satisfy

$$\sum_m|a_m|^2 = 1. \tag{9}$$

In order to derive the Davydov equations implied by the above *Ansatz*, we need to know the average number of quanta at any given site,

$$<N_p> = <\psi(t)|B_p^{\dagger}B_p|\psi(t)>. \tag{10}$$

From this we can subsequently determine the average total number of quanta present. We substitute (4) into (10) and then use the boson commutation relations to get

$$<N_p> = \frac{1}{Q!}\sum_{\substack{m_1\cdots m_Q \\ n_1\cdots n_Q}} a_{m_1}^*\cdots a_{m_Q}^* a_{n_1}\cdots a_{n_Q} \tag{11}$$

$$\times\left\{<0|B_{m_1}\cdots B_{m_Q}B_p B_{n_1}^{\dagger}\cdots B_{n_Q}^{\dagger}B_p^{\dagger}|0> - <0|B_{m_1}\cdots B_{m_Q}B_{n_1}^{\dagger}\cdots B_{n_Q}^{\dagger}|0>\right\}.$$

The two ground state expectation values in (11) are a $(Q+1)\times(Q+1)$ permanent and a $Q\times Q$ permanent, respectively. The evaluation of this quantity is explained in the appendix. The result is

$$<N_p> = Q\,|a_p|^2, \tag{12}$$

which implies

$$<\sum_p N_p> = Q \tag{13}$$

by (9).

The interpretations of $\beta_n(t)$ and $\pi_n(t)$ are obtained as follows. Davydov[3] points out that the part of $|\psi(t)>$ depending on the displacement and momentum operators is a coherent state of the normal mode creation and annihilation operators. A coherent state for the mode with wavevector q is[8]

$$|\alpha_q> = \exp(\alpha_q a_q^{\dagger} - \alpha_q^* a_q)|0>. \tag{14}$$

To see that (4) is a coherent state of all the normal modes, we use (2) to show that

$$-\frac{i}{\hbar}\sum_n(\beta_n p_n - \pi_n u_n) = \sum_q(\alpha_q a_q^{\dagger} - \alpha_q^* a_q), \tag{15}$$

where

$$\alpha_q = \left[\frac{m\omega_q}{2\hbar}\right]^{1/2}\beta_q + i\left[\frac{1}{2m\hbar\omega_q}\right]^{1/2}\pi_q. \tag{16}$$

[Here β_q is the spatial Fourier transform of β_n,

$$\beta_q = \frac{1}{\sqrt{N}}\sum_n e^{-iqnl}\beta_n, \tag{17}$$

and similarly for π_q.] We substitute (15) into (4) and get a factor of the form (14) for every normal mode. With the property

$$\langle\alpha_q|a_q|\alpha_q\rangle = \alpha_q, \tag{18}$$

and also using (2), (16), and (17), we straightforwardly obtain

$$\langle\psi(t)|u_n|\psi(t)\rangle = \beta_n(t), \tag{19a}$$

$$\langle\psi(t)|p_n|\psi(t)\rangle = \pi_n(t). \tag{19b}$$

The basic assumption in deriving the equations of motion is that $|\psi(t)\rangle$ is a solution of the time-dependent Schroedinger equation

$$i\hbar\frac{\partial}{\partial t}|\psi(t)\rangle = H|\psi(t)\rangle. \tag{20}$$

Since (19) identifies $\beta_n(t)$ and $\pi_n(t)$ as expectation values, standard quantum-mechanical procedure gives

$$\dot{\beta}_n(t) = \frac{1}{i\hbar}\langle\psi(t)|[u_n, H]|\psi(t)\rangle, \tag{21a}$$

$$\dot{\pi}_n(t) = \frac{1}{i\hbar}\langle\psi(t)|[p_n, H]|\psi(t)\rangle. \tag{21b}$$

The commutators are

$$[u_n, H] = i\hbar p_n/m, \tag{22a}$$

$$[p_n, H] = i\hbar w(u_{n+1} - 2u_n + u_{n-1}) + i\hbar\chi(B_{n+1}^\dagger B_{n+1} - B_{n-1}^\dagger B_{n-1}). \tag{22b}$$

Using (10), (12) and (19), we get one of Davydov's equations

$$m\ddot{\beta}_n = w(\beta_{n+1} - 2\beta_n + \beta_{n-1}) + Q\chi(|a_{n+1}|^2 - |a_{n-1}|^2). \tag{23}$$

The presence of Q quanta for the vibron oscillators increases the driving force on the phonon field by that factor, compared with the one-quantum case.

Next we derive the equation for $a_n(t)$. First we introduce a notation for the two parts of the state vector in (4):

$$|\psi(t)\rangle = |Q,a\rangle|\beta,\pi\rangle; \tag{24}$$

[for economy of notation the site and time-dependence of $a_n(t)$, $\beta_n(t)$ and $\pi_n(t)$ are left implicit]. The left-hand-side of the Schroedinger equation (20) is

$$i\hbar\frac{\partial}{\partial t}|\psi(t)\rangle = \left\{i\hbar\frac{\partial}{\partial t}|Q,a\rangle\right\}|\beta,\pi\rangle + |Q,a\rangle\left\{i\hbar\frac{\partial}{\partial t}|\beta,\pi\rangle\right\}. \tag{25}$$

Since all the operators defining $|Q,a\rangle$ commute, it is straightforward to show that

$$i\hbar\frac{\partial}{\partial t}|Q,a\rangle = \sqrt{Q}\left[\sum_n i\hbar\dot{a}_n B_n^\dagger\right]|Q-1,a\rangle. \tag{26}$$

The other time derivative appearing in (25) is evaluated in Ref. 6 (the coherent state lattice part of the wave function is the same in these two calculations).

$$i\hbar\frac{\partial}{\partial t}|\beta,\pi\rangle = \sum_n\left[\dot{\beta}_n p_n - \dot{\pi}_n u_n + \frac{1}{2}(\beta_n\dot{\pi}_n - \dot{\beta}_n\pi_n)\right]|\beta,\pi\rangle. \tag{27}$$

Taking the inner product of (26) and (27) with $\langle\beta,\pi|$ and using (19) gives the reduction to vibron operators of the left-hand-side of the Schroedinger equation.

$$<\beta,\pi \mid i\hbar\frac{\partial}{\partial t} \mid \psi(t)> \tag{28}$$

$$= \sqrt{Q}\left[\sum_n i\hbar\dot{a}_n B_n^\dagger\right] \mid Q-1,a> + \mid Q,a>\frac{1}{2}\sum_m \left[\dot{\beta}_m \pi_n - \dot{\pi}_n \beta_n\right].$$

Similarly, for the right-hand-side of the Schroedinger equation

$$<\beta,\pi \mid H_v + H_p + H_{int} \mid \psi(t)> = H_v \mid Q,a> + \mid Q,a>W(t) \tag{29}$$

$$+ \chi\sum_n (\beta_{n+1} - \beta_{n-1})B_n^\dagger B_n \mid Q,a>.$$

The quantity $W(t)$ is the phonon energy; the evaluation of this quantity given in Ref. 6 also applies here.

$$W(t) = <\beta,\pi \mid H_p \mid \beta,\pi> = \sum_n \left[\frac{1}{2m}\pi_n^2 + \frac{1}{2}w(\beta_{n+1} - \beta_n)^2\right] + \sum_q \frac{1}{2}\hbar\omega_q . \tag{30}$$

We combine (28) and (29) to get the reduction of the time-dependent Schroedinger equation to vibron operators.

$$\sqrt{Q}\,[i\hbar\sum_n \dot{a}_n B_n^\dagger] \mid Q-1,a> = H_v \mid Q,a> \tag{31}$$

$$+ \chi\sum_n (\beta_{n+1} - \beta_{n-1})B_n^\dagger B_n \mid Q,a> + \left\{W(t) - \frac{1}{2}\sum_n (\dot{\beta}_n \pi_n - \dot{\pi}_n \beta_n)\right\}\mid Q,a>.$$

Using (4) to write the state $\mid Q-1,a>$ and renaming the summation variable n on the left-hand-side of (31) as n_Q, we get

$$\sqrt{Q}\left[i\hbar\sum_n \dot{a}_n B_n^\dagger\right] \mid Q-1,a> = \frac{Q}{\sqrt{Q!}}\sum_{n_1\cdots n_Q} a_{n_1}\cdots a_{n_{Q-1}}(i\hbar\dot{a}_{n_Q})B_{n_1}^\dagger \cdots B_{n_Q}^\dagger \mid 0>. \tag{32}$$

The evaluation of the right-hand-side of (31) requires operating with H_v and H_{int} on the vibron state $\mid Q,a>$. Both of these terms require application of two boson operators, for possibly different sites, to $\mid Q,a>$. This evaluation is effected by using the following identity, which can be proved by mathematical induction.

$$(B_{l_1}^\dagger B_{l_2})B_{n_1}^\dagger \cdots B_{n_Q}^\dagger \mid 0> \tag{33}$$

$$= (\delta_{l_2 n_1}B_{l_1}^\dagger B_{n_2}^\dagger \cdots B_{n_Q}^\dagger + \delta_{l_2 n_2}B_{n_1}^\dagger B_{l_1}^\dagger B_{n_3}^\dagger \cdots B_{n_Q}^\dagger + \cdots + \delta_{l_2 n_Q}B_{n_1}^\dagger \cdots B_{n_{Q-1}}^\dagger B_{l_1}^\dagger)\mid 0>.$$

There are Q terms here, each with Q factors. With this formula, the terms involved in the application of H_v can be shown to be

$$\sum_l B_l^\dagger B_l \mid Q,a> = \frac{Q}{\sqrt{Q!}}\sum_{n_1\cdots n_Q} a_{n_1}\cdots a_{n_Q}B_{n_1}^\dagger \cdots B_{n_Q}^\dagger \mid 0>, \tag{34}$$

and

$$\sum_l \left[B_l^\dagger B_{l+1} + B_{l+1}^\dagger B_l\right] \mid Q,a> \tag{35}$$

$$= \frac{Q}{\sqrt{Q!}}\sum_{n_1\cdots n_Q} \left[a_{n_1}\cdots a_{n_{Q-1}}a_{n_Q+1} + a_{n_1}\cdots a_{n_{Q-1}}a_{n_Q-1}\right]B_{n_1}^\dagger \cdots B_{n_Q}^\dagger \mid 0>.$$

The interaction term, also obtained by using (33), is

$$\sum_l \left[\beta_{l+1} - \beta_{l-1}\right]B_l^\dagger B_l \mid Q,a> \tag{36}$$

$$= \frac{Q}{\sqrt{Q!}}\sum_{n_1\cdots n_Q} \left[\beta_{n_Q+1} - \beta_{n_Q-1}\right]a_{n_1}\cdots a_{n_Q}B_{n_1}^\dagger \cdots B_{n_Q}^\dagger \mid 0>,$$

We now insert (32), (34) and (36) into (31), equate coefficients of $B_{n_1}^\dagger \cdots B_{n_Q}^\dagger |0\rangle$ in every term, cancel common factors of a_{n_i}, and arrive at an equation for a_n:

$$i\hbar\dot{a}_n = \left\{ E_0 + \frac{1}{Q}\left[W(t) - \frac{1}{2}\sum_m \left(\dot{\beta}_m \pi_m - \dot{\pi}_m \beta_m \right) \right] \right\} a_n \tag{37}$$

$$- J(a_{n+1} + a_{n-1}) + \chi(\beta_{n+1} - \beta_{n-1})a_n .$$

By making use of the equations of motion (23) for π_m and β_m and (30) for the phonon energy, we can rewrite the quantity in square brackets in (37) as

$$\frac{1}{Q}\left[W(t) - \frac{1}{2}\sum_m \left(\dot{\beta}_m \pi_m - \dot{\pi}_m \beta_m \right) \right] \tag{38}$$

$$= \frac{1}{Q}\sum_q \frac{1}{2}\hbar\omega_q + \frac{1}{2}\chi\sum_m \beta_m \left(|a_{m+1}|^2 - |a_{m-1}|^2 \right) .$$

The factor of Q^{-1} multiplying the zero-point energy is the only place that the equation for $\dot{a}_n$ for the multi-quanta Davydov state differs from the corresponding equation for the single quantum case. It has been pointed out previously[6,9] that some physically measurable quantities, e.g. optical spectra, are sensitive to this phase of $a_n(t)$. Therefore it is conceivable that such measurements might distinguish different values of Q.

We now perform a phase change on the amplitude a_n

$$a_n(t) \rightarrow a_n(t)\exp\left[-\frac{i}{\hbar}\int\gamma(t')dt' \right] , \tag{39}$$

where $\gamma(t)$ is the site-independent terms in (37),

$$\gamma(t) = E_0 + \frac{1}{Q}\sum_q \frac{1}{2}\hbar\omega_q + \frac{1}{2}\chi\sum_m \beta_m \left(|a_{m+1}|^2 - |a_{m-1}|^2 \right) . \tag{40}$$

The equation of motion for the redefined a's is

$$i\hbar\dot{a}_n = -J(a_{n+1} + a_{n-1}) + \chi(\beta_{n+1} - \beta_{n-1})a_n , \tag{41}$$

which is the other Davydov equation.

To summarize, the Davydov equations for the multi-quantum state (4) are equations (23) and (41). The multi-quantum property of the state results in a stronger driving force on the phonon modes (23) but *no* modification of the equation for the probability amplitudes.

ACKNOWLEDGMENTS

The authors wish to acknowledge helpful conversations with A. Clogston. One of the authors (WCK) wishes to thank the Aspen Center for Physics, where part of this work was done, for their hospitality. The work at Los Alamos was done under the auspices of the US DOE.

REFERENCES

1. A. S. Davydov and N. I. Kislukha, Zh. Eksp. Teor. Fiz. **71**, 1090 (1976) [Sov. Phys. JETP 44, 571 (1976)].

2. For a diagram of this system, see A. C. Scott, Phil. Trans. R. Soc. A **315**, 423 (1985).

3. A. S. Davydov, Usp. Fiz. Nauk **138**, 603 (1982) [Sov. Phys. Usp. 25, 898 (1982)].

4. A. C. Scott, F. Y. Chu and D. W. McLaughlin, Proc. IEEE **61**, 1443 (1973).

5. J. M. Hyman, D. W. McLaughlin, and A. C. Scott, Physica D **3**, 23 (1981); A. C. Scott, Phys. Rev. A **26**, 578 (1982); **27**, 2767 (1983); Phys. Scr. **25**, 651 (1982); L. MacNeil and A. C. Scott, ibid. **29**, 284 (1984).

6. W. C. Kerr and P. S. Lomdahl, Phys. Rev. B **35**, 3629 (1987).

7. This method for introducing multiple quanta of the vibron field was suggested to us by A. Clogston.

8. R. J. Glauber, Phys. Rev. **131**, 2766 (1963).

9. D. W. Brown, B. J. West, and K. Lindenberg, Phys. Rev. **33**, 4110 (1986).

Comment by Kapor: If excitons and phonons are not coupled ($\chi = 0$), Equation (37) should not depend on phonon parameters, yet it does— it includes phonon ground state energy. There are two ways to avoid this paradox. One can subtract phonon ground state energy from the Hamiltonian from the beginning of the calculation, or include $i\hbar \frac{\partial |0\rangle_{ph}}{\partial t} = \sum_q \frac{1}{2}\hbar\omega_q |0\rangle_{ph}$ while performing time derivatives. Then these terms cancel on both sides of the equation.

APPENDIX

The evaluation of the permanents appearing in (11) is explained here. The $(Q+1)\times(Q+1)$ permanent appearing first in that equation is

$$
per \begin{bmatrix}
\delta_{m_1 n_1} & \cdots & \delta_{m_1 n_Q} & \delta_{m_1 p} \\
\cdot & \cdot & \cdot & \cdot \\
\cdot & \cdot & \cdot & \cdot \\
\cdot & \cdot & \cdot & \cdot \\
\delta_{m_Q n_1} & \cdots & \delta_{m_Q n_Q} & \delta_{m_Q p} \\
\delta_{p n_1} & \cdots & \delta_{p n_Q} & 1
\end{bmatrix} . \tag{A1}
$$

The element 1 in the lower right corner is δ_{pp}, which appears because two of the operators in (11) have the subscript p. The $Q \times Q$ permanent appearing second in (11) is the first Q rows and columns of this one.

When (A1) is expanded by the minors of the bottom row (with all positive signs because it is a permanent), the $Q \times Q$ minor of the element 1 is cancelled by the second permanent in (11). Thus the difference of the two permanents in (11) is the following sum of Q terms:

$$
\delta_{p n_1} \, per \begin{bmatrix}
\delta_{m_1 n_2} & \cdots & \delta_{m_1 p} \\
\cdot & \cdot & \cdot \\
\cdot & \cdot & \cdot \\
\cdot & \cdot & \cdot \\
\delta_{m_Q n_2} & \cdots & \delta_{m_Q p}
\end{bmatrix}
+ \delta_{p n_2} \, per \begin{bmatrix}
\delta_{m_1 n_1} & \delta_{m_1 n_3} & \cdots & \delta_{m_1 p} \\
\cdot & \cdot & \cdot & \cdot \\
\cdot & \cdot & \cdot & \cdot \\
\cdot & \cdot & \cdot & \cdot \\
\delta_{m_Q n_1} & \delta_{m_Q n_3} & \cdots & \delta_{m_Q p}
\end{bmatrix}
$$

$$
+ \cdots + \delta_{pn_Q} \; per
\begin{bmatrix}
\delta_{m_1 n_1} & \cdots & \delta_{m_1 n_{Q-1}} & \delta_{m_1 p} \\
& \cdot & \cdot & \cdot \\
\cdot & \cdot & \cdot & \cdot \\
\cdot & \cdot & \cdot & \cdot \\
\delta_{m_Q n_1} & \cdots & \delta_{m_Q n_{Q-1}} & \delta_{m_Q p}
\end{bmatrix} .
\tag{A2}
$$

When we substitute this formula for the difference of the two permanents in (11) and use the Kronecker delta prefactors to replace the p subscripts in each permanent, all of these permanents become the same, and it is the same one as in (6), viz.

$$
per
\begin{bmatrix}
\delta_{m_1 n_1} & \cdots & \delta_{m_1 n_Q} \\
\cdot & \cdot & \cdot \\
\cdot & \cdot & \cdot \\
\cdot & \cdot & \cdot \\
\delta_{m_Q n_1} & \cdots & \delta_{m_Q n_Q}
\end{bmatrix}
\equiv per(\delta_{m_i n_j}).
\tag{A3}
$$

Thus

$$
\langle N_p \rangle = \frac{1}{Q!} \sum_{\substack{m_1 \cdots m_Q \\ n_1 \cdots n_Q}} a_{m_1}^* \cdots a_{m_Q}^* a_{n_1} \cdots a_{n_Q} \left[\delta_{pn_1} + \delta_{pn_2} + \cdots + \delta_{pn_Q} \right] per(\delta_{m_i n_j}).
\tag{A4}
$$

The permanent expands into $Q!$ terms in which every m_i is set equal to a different one of the n_j's. This generates Q factors of $|a_{m_i}|^2$ in each of the $Q!$ terms. Then the Q Kronecker deltas in the square brackets give Q terms in each of which one of the n's is set equal to p, and the other $Q-1$ are summed on, each giving the same result. The result is

$$
\langle N_p \rangle = \frac{1}{Q!} |a_p|^2 Q Q! \left[\sum_n |a_n|^2 \right]^{Q-1} .
\tag{A5}
$$

We use the normalization condition in (9) and obtain the result (12).

3

A CLASSICAL AND QUANTUM THEORY OF DYNAMICAL SELF-TRAPPING IN NONLINEAR SYSTEMS AND ITS IMPLICATION TO ENERGY TRANSFER IN BIOLOGICAL SYSTEMS

Shozo Takeno

Physics Laboratory, Faculty of Engineering and Design
Kyoto Institute of Technology
Kyoto 606, Japan

1. INTRODUCTION

Dynamical self-trapping of waves or particles, originally put forward by Landau long time ago, by their nonlinear self-interactions in classical and quantum systems has received much attention in recent years. The principal reason for the upsurge of the renewed interest is due to: (i) The development of the soliton theory in mathematical physics. (ii) Ever-lasting interest in Davydov solitons and their possible implication to biological energy transfer.[1] (iii) Generalization of the concept of Davydov solitons in various directions,[2] in which vibron solitons are one of such examples.[3] (iv) Several experiments which can be accounted for by the concept of the dynamical self-trapping, such as the shift of the infrared absorption spectra of molecular crystals acetanilide,[4] local modes in benzene,[5] the anomalous temperature dependence of the Raman spectra in ℓ-alanine,[6] and so on. (v) Numerical experiments which show the existence of dynamical self-trapped states under certain conditions in various model dynamical nonlinear systems.[7-9] (vi) Elucidation of the existence of intrinsic anharmonic localized or resonant modes in lattice dynamics of pure crystal lattices.[10,11]

In the context of the soliton theory, a spatially localized entity caused by the so-called dynamical self-trapping is categorized as an envelope soliton. Physically, this is caused by trapping of travelling waves by the intrinsic negative anharmonic potential of a dynamic system. Conceptually, the dynamical self-trapping can be classified into two types. The conventional one is the self-trapping of one wave by another. In plasma physics, this is well known as Zakharov solitons due to interactions of electron plasmons with ion-acoustic phonons.[12] Davydov solitons and vibron solitons are another examples, where Frenkel excitons and vibrons are modulated into envelope solitons by their interactions with acoustic phonons. Another type is a spatially or spatio-temporally localized entity caused by the intrinsic nonlinearity or anharmonicity in a single dynamical system. Anharmonic or resonant modes which exist in a pure anharmonic lattice is a typical example of such a truely intrinsic mode.[10,11] Due to the history of the problem of the dynamical self-trapping, which has its origin in the Landau polaron theory,[13] the first type has much longer history than the second.

In spite of much current interest and development, much has remained to be done concerning the problem of the dynamical self-trapping related to the

Davydov solitons, the vibron solitons, and other sorts of intrinsically
nonlinear particle-like entities of similar nature. These are: (a) The
quantum or classical structure of elementary excitations or collective modes
undergoing dynamical self-trapping due to the effect of self-trapping, (b)
the effect of environmental fields or the temperature effect on dynamical
self-trapping, (c) the lattice-discreteness effect, (d) the spatial dimen-
sionality, and so on.

It is the purpose of this paper to develop a theory of a classical and
quantum theory of dynamical self-trapping in nonlinear systems with attention paid to the above-mentioned four points. Of these four points an
emphasis is placed on the latter two points for which a number of novel
features are presented. For the first point, it is shown that Frenkel excitons with exciton transfer given by dipole-dipole interactions should be
treated as Paulions or spin 1/2 particles rather than as Boson, where simultaneous excitations or de-excitation of two molecules or atoms must also be
taken into account. For the second point, we show explicitly that the
effect of phonons on Frenkel excitons or vibrons can be taken into account
in a rigorous way by achieving an elimination of phonon-field variables and
thereby writing equations for these elementary excitations in the form of
the generalized Langevin equation.

2. ANHARMONIC LOCALIZED MODES OR SOLITONS IN DISCRETE, d-DIMENSIONAL NON-
   LINEAR SCHRÖDINGER EQUATION

As the simpliest illustration of the dynamical self-trapping in discrete lattice space which is also intimately related to the Davydov-soliton
problem, we consider a discrete version of the d-dimensional nonlinear
Schrödinger (NLS) equation

$$i \, du(n)/dt = W_0 u(n) - J\sum_j [u(n+e_j) + u(n-e_j)] - \lambda |u(n)|^2 u(n), \qquad (2.1)$$

where ω_0, J, and λ are an on-site constant frequency, a nearest-neighbour
coupling constant, and a self-interaction constant, respectively. In the
above equation, u(n) is a complex field variable associated with an nth
lattice site with lattice vector R_n in a d-dimensional-space version of a
simple cubic lattice. The quantities $n=(n_1,n_2,...,n_d)$ and e_j are d-dimensional vectors with n_j's (j=1,2,...,d) being integers and the unite vector
in the direction of the jth axis, respectively. We seek solutions to (2.1)
in the form

$$u(n) = b_k(n)\exp[-i(\omega t - k\cdot R_n)] \qquad (2.2)$$

to obtain

$$\omega_0 b_k(n) - J\sum_j \cos(k_j a)[b_k(n+e_j)+b_k(n-e_j)] - \lambda b_k(n)^3 = \omega b_k(n), \qquad (2.3)$$

$$db_k(n)/dt + \sum_j J\sin(k_j a)[b_k(n+e_j) - b_k(n-e_j)] = 0. \qquad (2.4)$$

We are concerned with the frequency eigenvalue ω which appears below the
bottom of the frequency band

$$\omega_k(q) = \omega_0 - 2J\sum_j \cos(k_j a)\cos(q_j a). \qquad (2.5)$$

The quantities ω and k are then the eigenfrequency of anharmonic localized
mode and its wave number, respectively, in the d-dimensional simple cubic
lattice with lattice constant a. The space-time evolution of the localized

mode described by solutions to Eq.(2.4) is given by

$$b_k(n,t) = \sum_m \prod_j J_{n_j-m_j}[2Jt\sin(k_ja)]\,b_k(m,0) \equiv \sum_m F_k(n,m;t)b_k(m,0), \quad (2.6)$$

where the J_n's are the Bessel function of the nth order, and $b_k(m,0)$ is solutions to Eq.(2.3). In terms of lattice Green's functions

$$G_k(n,m)\equiv G(n,m;\omega)=(1/N^d)\sum_q[\omega_k(q)-\omega]^{-1}\exp[iq\cdot(R_n-R_m)], \quad (2.7)$$

where the sum extends over the first Brillouin zone, Eq.(2.3) is rewritten as

$$b_k(n) = \lambda\sum_m G_k(n,m)b_k(m)^3. \quad (2.8)$$

Equations (2.8) constitute a set of simultaneous nonlinear equations of infinite dimension. However, the calculation of the nonlinear eigenvalue problem is facilitated by the fact that for ω lying outside the frequency band $\omega_k(q)$, $G_k(n,m)$ is a rapidly decreasing function of $|R_n-R_m|$. Thus, we actually need only consider a small number of the simultaneous equations. Inserting solutions to Eq.(2.8) back into Eq.(2.6) gives localized-mode solutions to Eq.(2.1) or Eqs.(2.3) and (2.4). It is seen from Eq.(2.6) that the moving-localized-mode profile undergoes dispersing described by the Bessel functions except for the case of a stationary or nonmoving localized mode satisfying k=0. It is only in the continuum limit for Eq.(2.4) that a dispersionless, soliton-like propagation of the localized mode is realizable.

We illustrate the above procedure by studying a one-localized-mode solution to Eq.(2.8). We separate $b_k(n)$ into the amplitude A and the shape function $S_k(n)$, assuming the central position of the mode to be n=0,

$$b_k(n) = AS_k(n), \qquad S_k(n) = \lambda A^2\sum_m G_k(n,m)S_k(m)^3, \qquad S_k(0)=1. \quad (2.9)$$

Equation (2.9) can be solved by using a successive approximation procedure. A detailed discussion on this method is omitted here. Using an exact identity relation satisfied by the $G_k(n,m)$'s, an <u>exact</u> expression for the dispersion relation of a localized mode can be obtained in terms of the shape functions. The result of such a calculation is written as

$$\omega = \omega_0 - J\sum_j\cos(k_ja)\bigl[S_{kj}^{(+)} + S_{kj}^{(-)}\bigr] - \lambda A^2, \quad (2.10)$$

where $S_{kj}^{(+)}$ and $S_{kj}^{(-)}$ are the values of $S_k(n)$ at e_j and $-e_j$, respectively. It is seen that eigenfrequency of the moving localized mode constitutes a frequency band.

The interrelationship between the anharmonic localized mode studied here and the conventional soliton can be established in the one-dimensional case in the continuum limit. In such a limit, the quantity $F_k(n,m;t)$ defined by Eq.(2.6) and $G_k(n,m)$ is given by $F_k(n,m;t)=\delta(x-vt)$ with $v=2Ja^2k$ and $G_k(n,m)\propto\exp(\gamma_k|x-x'|)$ with $\gamma_k=(\lambda A^2/2Ja^2)^{1/2}$, respectively. Inserting these expressions into Eqs.(2.6) and (2.9), we arrive at a result which is entirely identical with the conventional one-soliton solution of the one-dimensional NSL equation.[14]

The fundamental difference of the anharmonic localized mode of the discrete, d-dimensional NLS equation for $d\geqslant 3$ from that for d=1,2 is that in order for the localized mode in the former case to appear, the self interaction constant λA^2 must exceeds a critical value determined by the value of the $G_k(n,m)$ for $\omega = \omega_{kb}$, where ω_{kb} is the bottom of the frequency band $\omega_k(q)$. Mathematically, this is related to the numerical value of the following integral

$$G_k(n,m; \omega_{kb}) = (1/2J)\int_0^\infty \exp(-dz)\prod_j I_{n_j-m_j}\left[\cos(k_j a)z\right]\,dz, \tag{2.11}$$

Thus, we have shown that exact anharmonic-localized-mode solutions to the discrete, d-dimensional NLS equation are obtainable in terms of the lattice Green's functions, and that a stationary or nonmoving localized mode is stable while a moving one disperses, the rate of dispersing being smaller for a smaller moving velocity.[14]

3. AN EXACTLY SOLUBLE SOLITON MODEL OF DYNAMICAL SELF-TRAPPING IN DISCRETE LATTICE SPACE

As shown by Eq.(2.6), an anharmonic localized mode which exists for the d-dimensional discrete NLS equation cannot be qualified as a soliton in a strict mathematical sense, because it undergoes dispersing when it moves. It is shown here that a model nonlinear field equation defined in a one-dimensional (1d) lattice space,

$$i\,du_n/dt = \omega_0 u_n - J(u_{n+1}+u_{n-1}) - \lambda|u_n|^2(u_{n+1}+u_{n-1}), \tag{3.1}$$

which is a slightly modified version of Eq.(2.1) for d=1, provides us with an exactly soluble soliton model of dynamical self-trapping in discrete lattice space. Here u_n is a complex field variable associated with an nth lattice site, and the meaning of ω_0, J, and λ are the same as in the case of Eq.(2.1). Putting

$$u_n = b_n \exp\left[-i(\omega t - kna)\right], \tag{3.2}$$

where, as before, ω and k are the eigenfrequency of a soliton and its wave number, respectively, and a is a lattice constant, and assuming b_n to be rela, we obtain

$$(\omega_0 - \omega)b_n = \cos(ka)(b_{n+1} + b_{n-1})(J + \lambda b_n^2), \tag{3.3}$$

$$(db_n/dt) + \sin(ka)(b_{n+1} - b_{n-1})(J + \lambda b_n^2) = 0. \tag{3.3'}$$

It is shown that Eq.(3.3') admits an exact one-soliton solution

$$b_n = (J/\lambda)^{1/2}\sinh(Ka)\,\mathrm{sech}\left[Kna - 2J\sin(ka)\sinh(Ka)t\right] \equiv AS_n. \tag{3.4}$$

The corresponding solution to Eq.(3.3) is given by

$$b_n = (J/\lambda)^{1/2}\sinh(ka)\,\mathrm{sech}(Kna) \equiv A\,\mathrm{sech}(Ka), \tag{3.5}$$

provided the quantity satisfies the relation

$$\omega = \omega_0 - 2J\cosh(Ka)\cos(ka). \tag{3.6}$$

Equation (3.6) is the dispersion relation for the soliton. Equations (3.4), (3.5), and (3.6) are exact one-soliton solutions to Eq.(3.1), where A and S_n are identified as the amplitude of the localized mode and its shape function, respectively. Equation (3.1) is a slightly modified version of the Ablowitz-Ladik equation.[15] In order to make an appropriate generalization of a discretized version of the eigenvalue problem in the inverse scattering theory, these worker considered a nonlinear evolution equation

$$i\, du_n/dt = u_{n+1} + u_{n-1} - 2u_n \pm |u_n|^2 (u_{n+1} + u_{n-1}) \qquad (3.7)$$

to show that it yields exact solitons.[15] Our primary concern here is to look at Eq.(3.1) from the viewpoint of the dynamical self-trapping. A more physical insight into the nature of solitons here can be gained by rewriting the dispersion relation in terms of λ and A as

$$\omega = \omega_0 - 2J\cos(ka)\left[1 + (\lambda/J)A^2\right]^{1/2}. \qquad (3.8)$$

This is to be compared with the dispersion relation

$$\omega = \omega_0 - 2J + Ja^2k^2 - \lambda A^2 \qquad (3.9)$$

for a soliton governed by the continuum version of Eq.(3.1),

$$i\, du/dx = (\omega_0 - 2J)u + Ja^2(\partial^2 u/\partial x^2) - 2\lambda |u|^2 u. \qquad (3.10)$$

Though the soliton profile between these two cases looks similar to each other, the soliton dispersion relation (3.6) or (3.8) associated with Eq. (3.1) is different from Eq.(3.9) associated with the continuum NLS equation in the following important respects: (i) The soliton eigenfrequency of the former constitutes a soliton frequency band, while that of the latter extends to infinity. (ii) Comparison of Eq.(3.8) with Eq.(3.9) leads to conclusion that in Eq.(3.1) the effect of self-interaction constant λ and the finite soliton amplitude A have been taken care of to all orders of their strength. (iii) The binding energy of the soliton in the former case depends on wave number k in constrast to the case of the continuum model. (iv) The binding frequency of the soliton in the former decreases with increasing k, while that in the latter is entirely independent of k.

If we employ the reasoning used in Section 2, solitons here can also be regarded as anharmonic localized modes. We can make one-to-one correspondence between these two cases. Namely, Eq.(2.6) corresponds to Eq.(3.4), where no dispersive effect exists for the profile of the anharmonic local mode for the latter. Using the same argument as that employed in proceeding from Eq.(2.7) to Eq.(2.10), we obtain the exact formal expression for the dispersion relation of the anharmonic localized mode in terms of the shape functions of the localized mode:

$$\omega = \omega_0 - \cos(ka)\left[S_k^{(+)} + S_k^{(-)}\right](J + \lambda A^2). \qquad (3.11)$$

where $S_k^{(+)}$ and $S_k^{(-)}$ are the shape functions at n=1 and -1, respectively, in which the central position of the anharmonic localized mode has been assumed to be n=0. Insertion of

$$S_k^{(+)} = S_k^{(-)} = \mathrm{sech}(Ka), \quad A = (J/\lambda)^{1/2}\sinh(Ka) \qquad (3.12)$$

into Eq.(3.11) is shown to yield Eq.(3.6). Thus, the equivalnece of the anharmonic localized mode to the soliton can be established in the case of Eq.(3.1) without resorting to the continuum approximation. Observation of envelope solitons in the context of anharmonic localized modes considerably enhances our understanding of soiltons in discrete lattice space.

Here we have shown that by slightly modifying the Ablowitz-Ladik equation,[15] to which only mathematicians paied attention so far, we can devise an exactly soluble soliton model of dynamical self-trapping in lattice.

4. ANHARMONIC LOCALIZED MODES OR SOLITONS IN DISCRETE, d-DIMENSIONAL NONLINEAR KLEIN-GORDON EQUATION

When one encounters the nonlinear Klein-Gordon equations, continuous or discrete in form, the equations are often reduced to the corresponding NLS equation by assuming that envelope functions appearing therein are slowly-varying. Here we study anharmonic localized modes in the discrete, d-dimensional nonlinear Klein-Gordon equation in close analogy with those in the corresponding NLS equation to make an attempt to clarify the similarity and the difference of the properties of the localized mode between these two cases. The nonlinear Klein-Gordon (NKG) equation to be considered here is of the form

$$d^2 u(n)/dt^2 + W_0^2 u(n) - J\sum_j [u(n+e_j) + u(n-e_j)] - \lambda |u(n)|^2 u(n), \qquad (4.1)$$

where the meaning of $u(n)$, ω_0, J, λ, etc. are the same as those in Section 2. We insert Eq.(2.2) into Eq.(4.1) to obtain

$$d^2 b_k(n)/dt^2 + \omega_0^2 b_k(n) - J\sum_j \cos(k_j a)[b_k(n+e_j) + b_k(n-e_j)] - \lambda\, b_k(n)^3 = \omega^2 b_k(n),$$
$$(4.2)$$

$$db_k(n)/dt + \sum_j [J\sin(k_j a)/2\omega][b_k(n+e_j) - b_k(n-e_j)] = 0. \qquad (4.3)$$

Equation (4.3) is essentially of the same form as Eq.(2.4), so its formal solution is easily obtained from Eq.(2.6). In terms of the propagator function $F_k(n,m;t)$ defined there, the solution is written as

$$F_k(n,m;t) = \prod_j J_{n_j - m_j}\{[J\sin(k_j a)/\omega]\, t\} \equiv \prod_j J_{n_j - m_j}(U_{kj} t/a), \qquad (4.4)$$

where U_{kj} is the jth component of the velocity U_k of a moving anharmonic localized mode with wavenumber k. Because of the appearance of the Bessel functions, the spatio-temporal evolution of the anharmonic localized modes in the NKG equation is the same as that in the coreesponding NLS equation. The main difference of Eq.(4.2) from its counterpart, Eq.(2.3), is the existence of the term $d^2 b_k(n)/dt^2$. To make Eq.(4.2) similar to Eq.(2.3), we take time derivative of Eq.(4.3) and use it to express $d^2 b_k(n)/dt^2$ solely in terms of the $b_k(n)$'s. Inserting this into the first term on the left-hand side of Eq.(4.2), we obtain

$$\omega_0^2 b_k(n) - J\sum_j \cos(k_j a)[b_k(n+e_j) + b_k(n-e_j)] - \sum_{jm}(\nu_{kj}\nu_{km}/4)[b_k(n+e_j+e_m)$$

$$+ b_k(n-e_j-e_m) - b_k(n+e_j-e_m) - b_k(n-e_j+e_m)] - \lambda\, b_k(n)^3 = \omega^2 b_k(n), \qquad (4.5)$$

$$\nu_{kj} = J\sin(k_j a)/\omega \equiv U_{kj}/a. \qquad (4.6)$$

Solving the nonlinear eigenvalue problem posed by Eq.(4.5) and inserting the solutions obtained into Eq.(2.6) with $F_k(n,m;t)$ given by Eq.(4.4) give solutions to Eq.(4.1).

The mathematical procedure for solving Eq.(4.5) is essentially the same that in the case of Eq.(2.3). Namely, in terms of the lattice Green's function $G_k(n,m) \equiv G_k(n,m;\omega)$ defined by

$$G_k(n,m) = G_k(n,m;\omega) = (1/N^d)\sum_q [\omega_k(q)^2 - \omega^2]^{-1} \exp[iq\cdot(R_n - R_m)], \qquad (4.7)$$

where, as before, N is the total number of lattice points in each direction of the d-dimensional simple cubic lattice, Eq.(4.5) is rewritten as Eq.(2.8) with $G_k(n,m)$ given by eq.(4.7). For a one-localized mode problem, we obtain Eq.(2.9). This can also be solved iteratively. An exact formal expression for the dispersion relation of a moving localized mode corresponding to Eq.(2.10) can also be obtained. Let $S_{kj}^{(\pm)}$, $S_{kjm}^{(\pm,\pm)}$, $S_{kjm}^{(\pm,\mp)}$, and $S_{kj}^{(2\pm)}$ be the shape functions of the localized mode at its neighbouring lattice sites $\pm e_j$, $\pm e_j \pm e_m$, $\pm e_j \mp e_m$, and $\pm 2e_j$, $(j,m=1,2,\ldots,d)$, respectively, where the central position of the localized mode is taken to be at the origin of the coordinate. Then, using an identity relation satisfied by the $G_k(n,m)$'s, we obtain

$$\omega^2 = \omega_0^2 - J \sum_j \cos(k_j a)\left[S_{kj}^{(+)} + S_{kj}^{(-)}\right] - \lambda A^2 - \sum_j (\lambda_{kj}^2/2)$$

$$- {\sum_{jm}}' (\nu_{kj}\nu_{km}/4)\left[S_{kjm}^{(+,+)} + S_{kjm}^{(-,-)} - S_{kjm}^{(-,+)} - S_{kjm}^{(-,+)}\right]$$

$$+ \sum_j (\nu_{kj}^2/4)\left[S_{kj}^{(2+)} + S_{kj}^{(2-)}\right], \tag{4.8}$$

where the prime on the summation symbol excludes the case j=m. This is an exact formal expression for the dispersion relation of a moving localized mode of the NKG equation. Equation (4.8) is much more involved in form as compared with its counterpart, Eq.(2.10), in the NLS equation. The complexity comes from the terms containing the ν_{kj}'s or the U_{kj}'s. Taking all the k_j's to be zero, we obtain an exact expression for the eigenfrequency of the stationary localized mode

$$\omega^2 = \omega_0^2 - J \sum_j \left[S_{0j}^{(+)} + S_{0j}^{(-)}\right] - \lambda A^2. \tag{4.9}$$

In view of the relation $S_k(0)=1$ (cf. Eq.(2.9)), it is convenient to put

$$S_k(n) = 1 - C_k(n) \tag{4.10}$$

to rewrite Eq.(4.8) in terms of the $C_k(n)$'s as

$$\omega^2 = (1/2)\left\{\Omega_1(k) + \left[\Omega_1(k)^2 - \Omega_2(k)\right]^{1/2}\right\}, \tag{4.11}$$

with

$$\Omega_1(k) = \omega(k)^2 - \lambda A^2 + J \sum_j \cos(k_j a)\left[C_{kj}^{(+)} + C_{kj}^{(-)}\right], \tag{4.12}$$

$$\Omega_2(k) = J^2 {\sum_{jm}}' \sin(k_j a)\sin(k_m a)\left[C_{kjm}^{(+,+)} + C_{kjm}^{(-,-)} - C_{kjm}^{(+,-)} - C_{kjm}^{(-,+)}\right]$$

$$+ J^2 \sum_j \sin^2(k_j a)\left[C_{kj}^{(2+)} + C_{kj}^{(2-)}\right], \tag{4.13}$$

where

$$\omega^2(k) = \omega_0^2 - 2J \sum_j \cos(k_j a) \tag{4.14}$$

Here $C_{kj}^{(\pm)}$, $C_{kjm}^{(\pm,\pm)}$, etc. correspond to $S_{kj}^{(\pm)}$, $S_{kjm}^{(\pm,\pm)}$, etc., respectively. We close this section by giving a brief remark on the difference and the similarity existing for the dispersion relation of anharmonic localized modes in the NKG equation and the NLS equation. We note that Eq. (4.9) is similar to Eq.(2.10). This shows that in the stationary regime the anharmonic localized mode of the NKG equation is similar to that of the NLS equation. For moving anharmonic localized mode the dispersion relation for the NKG equation has a form different from that for the NLS equation. The difference increases as the wavenumber k of the moving anharmonic localized mode increases.

Since our discussion on anharmonic localized modes is based on the method that a harmonic part of a model nonlinear field equation can be regarded as a reference system, we can easily consider more general situations. We illustrate this by considering a generalized version of the discrete, d-dimensional version of the NKG equation:

$$d^2u(n)/dt^2 + \omega_0^2 u(n) - J\sum_j [u(n+e_j)+u(n-e_j)] - \sum_p \lambda_p |u(n)|^{2p+1} u(n) = 0. \quad (4.15)$$

Here Eq.(4.3) or (4.4) still holds, while the nonlinear eigenvalue problem determining the eigenfrequency of anharmonic localized modes is to solve the equations

$$b_k(n) = \sum_m G_k(n,m) \sum_p \lambda_p b_k(m)^{2p+1}. \quad (\lambda_1 > 0: \text{ const}, \quad p: \text{ integer}) \quad (4.16)$$

For example, for one-localized-mode problem, the nonlinear eigenvalue equations are given by Eq.(4.16) with $b_k(n)$ and λ_p replaced by $S_k(n)$ and $A^{2p}\lambda_p$, respectively. Then, Eqs.(4.8)~(4.14) remain unchanged except for the fact that A^2 should be replaced by $\sum_p \lambda_p A^{2p}$. A more detailed discussion on the result presented here is given in Ref. 16.

5. A THEORY OF VIBRON SOLITONS - A GENERALIZED AND IMPROVED VERSION OF THE DYNAMICAL-SELF-TRAPPING PROBLEM ORIGINALLY DUE TO DAVYDOV

Davydov has shown that in α helical proteins solitons can be formed by coupling of propagation of amide-I vibrations with longitudinal phonons along spines that such dynamical entities are responsible for mechanism of biological energy transfer in biological systems.- Theoretically, the Davydov soliton theory has been formulated as the dynamical self-trapping of Frenkel-excitons by acoustic phonons.[1] Since that time a large number of studies have been made to explore ramification of this germinal idea. The present author proposed the concept of vibron solitons as a possibly better candidate than the Davydov solitons for soliton-like entities in molecular, hydrogen-bonded and biological systems.[3] In this Section, we shall present a brief account of the vibron-soliton theory with due attention to the recent development along this line of approach.[17-19)] Generally speaking, the vibron soliton theory can be formulated from two sides, classical and quantum mechanical. In what follows, we discuss these two approaches in succession.

(A) The vibron-soliton theory, classical

To simulate dynamical self-trapping of amide-I vibrons by acoustic phonons along spines, we consider a 1d oscillator system in which each oscillator is linearly coupled with neighbouring ones and nonlinearly coucoupled with acoustic lattice phonons. The Hamiltonians for the oscillator system and the acoustic phonon systems, H_{osc} and H_{ph}, are taken to be

$$H_{osc} = \sum_n [(\mu \dot{q}_n^2/2) + v(q_n)] - (1/2) \sum_{nm} L(n,m) q_n q_m, \quad (5.1)$$

$$H_{ph} = \sum_n [(M\dot{u}_n^2/2) + (K/2)(u_{n+1}-u_n)^2], \quad (5.2)$$

with $v(q) = (\mu \omega_0^2/2)q^2 + (b/4)q^4 + (c/6)q^6, \quad c > 0. \quad (5.3)$

Here q_n and u_n are the displacement of an nth oscillator with effective
mass μ and eigenfrequency ω_0 in the oscillator system and that of an nth
atom or molecule with atomic or molecular mass M in the phonon system, respectively. The quantities $v(q_n)$, $L(n,m)$, and K are an on-site potential
for the nth oscillator, the dipole-dipole interaction energy between the
nth and mth oscillators, and the nearest-neighbour force constant of the
phonon system, respectively. The quantities b and c are constants. The
Hamiltonian H' describing vibron-phonon interactions is taken to be

$$H' = \sum_n V(q_n)(u_{n+1} - u_{n-1}). \tag{5.4}$$

This is caused by fluctuations of the on-site potential due to acoustic
phonons, where $V(q)$ is an even-function of q. The contribution of fluctuations of the $L(n,m)$'s are omitted for simplicity. Equations of motion for
the q-field and the u-field are written as

$$\mu \ddot{q}_n + v'(q_n) - \sum_m L(n,m)q_m + V'(q_n)(u_{n+1} - u_{n-1}) = 0, \tag{5.5}$$

$$M\ddot{u}_n - K(u_{n+1} + u_{n-1} - 2u_n) - V(q_{n+1} - q_{n+1}) = 0, \tag{5.6}$$

where primes on $v(q_n)$ and $V(q_n)$ denote the derivative with respect to q_n.
The eigenfrequency $\omega_v(q)$ of phonon-free vibrons and that of oscillator-free
phonons, $\omega_{ph}(q)$, are given by

$$\omega_v(q)^2 = \omega_0^2 - \sum_m L(n,m)\exp[iq(n-m)a], \quad \omega_0^2 \gg \sum_m L(n,m), \tag{5.7}$$

$$\omega_{ph}(q)^2 = (2K/M)[1-\cos(qa)] \equiv (\omega_1^2/2)[1 - \cos(qa)], \tag{5.8}$$

where a is the lattice constant. As shown in Ref. 19, we can solve Eq.(5.6)
to obtain an exact relationship between the q-field and the u-field:

$$u_{n+1} - u_{n-1} = -(1/K)[2V(q_n)+V(q_{n+1})+V(q_{n-1})] + (1/K)\sum_m M_{n-m}(t)V[q_m(0)]$$

$$+ (k/K)\sum_m \int_0^t d\tau M_{n-m}(t-\tau)\dot{V}[q_m(\tau)] + f_{n+1} - f_{n-1}, \tag{5.9}$$

where

$$M_n(t) = 2J_{2n}(\omega_1 t) + J_{2(n+1)}(\omega_1 t) + J_{2(n-1)}(\omega_1 t), \tag{5.10}$$

$$f_n(t) = \sum_m \left[J_{2(n-m)}(\omega_1 t)u_m(0) + \int dt\, J_{2(n-m)}(\omega_1 t)\dot{u}_m(0) \right]. \tag{5.11}$$

In the above equations the J's are the Bessel functions. Inserting Eq.
(5.9) back into Eq.(5.5), we obtain the equation for the q-field in the
form of a discrete, 1d KG equation having the form of the generalized Langevin equation, where the friction term and the random-force term are char-
actorized by the memory kernel $M_{n-m}(t-\tau)$ and the initial values of the u
field and the Bessel functions, respectively. Of various nonlinear effects
arising from nonlinear vibron-phonon interactions, we are particularly
interested in dynamical self-trapping of vibrons by phonons or vibron-polar-
ons. We separate q_n into a negative frequency part Q_n and a positive frequency one Q_n^*, where Q_n^* is the complex conjugate of Q_n, and seek solutions in the form

$$Q_n = b_{kn}\exp[i(kna - \omega t)] \equiv b_n\exp[i(kna - \omega t)], \tag{5.12}$$

The physical meaning of b_{kn}, k, and ω are the same as that in Eqs.(2.2) and (3.2). Then, we employ a rotating-wave approximation by taking $V(q)=(C/2)q^2$ to put

$$V(q_n) = (C/2)(Q_n + Q_n^*)^2 \rightarrow C|Q_n|^2 = C\, b_n^2, \qquad C>0. \tag{5.13}$$

Equation (5.9) is then rewritten as

$$u_{n+1} - u_{n-1} = -(C/K)(2b_n^2 + b_{n+1}^2 + b_{n-1}^2) + (C/K)\sum_m M_{n-m}(t)b_m(0)^2$$

$$+ (C/K)\sum_m \int d\tau\, M_{n-m}(t-\tau)(d/d\tau)b_n(\tau)^2 + f_{n+1} - f_{n-1}. \tag{5.14}$$

Inserting Eq.(5.14) into Eq.(5.5), using Eq.(5.12), and the rotating-wave approximation similar to (5.13) for the nonlinearity terms in $v'(q_n)$, we obtain

$$\ddot{b}_n + (\omega_0^2 - \omega^2)b_n - (1/\mu)\sum_{m \neq 0} L(m)\cos(kma)(b_{n+m} + b_{n-m}) - F(b_n)$$

$$+ (C^2/K\mu)\sum_m M_{n-m}(t)b_m(0)^2 b_n(t) + (C^2/K\mu)\sum_m \int_0^t d\tau M_{n-m}(t-\tau)(d/d\tau)b_m(\tau)^2 b_n(t)$$

$$= (C/\mu)(f_{n+1} - f_{n-1})b_n, \tag{5.15}$$

$$\dot{b}_n + (1/2\,\omega\mu)\sum_{m \neq 0} L(m)\sin(kma)(b_{n+m} - b_{n-m}) = 0, \tag{5.16}$$

where

$$F(b_n) = [(2C^2/K\mu) - (3b/\mu)]b_n^3 + (C^2/K\mu)(b_{n+1}^2 + b_{n-1}^2)b_n - (5c/\mu)b_n^5, \tag{5.17}$$

with $2C^2/K > 3b$. Within the framework of the rotating-wave approximation,[20] Eqs.(5.15) and (5.16) are exact equations describing dynamical self-trapping of vibrons by acoustic phonons in the context of classical nonlinear dynamics. Inspite of being of 1d nature, these equations have the properties of much extended version of Eqs.(2.3) and (2.4) in the NLS equation, Eqs.(3.3) and (3.3') in the case of the exactly solvable soliton model of dynamical self-trapping, and Eqs.(4.2) and (4.3) in the NKG equation. A discussion on approximate solutions to Eqs.(5.15) and (5.16) will be given later on.

(B) The vibron-soliton theory, quantum-mechanical

Let us call attention, first of all, to the fact that Frenkel excitons with exciton transfer by dipole-dipole interactions can be described by the Hamiltonian H_{ex} having the same form as H_{osc}. On the other hand, the model exciton Hamiltonian H_{exD} originally due to Davydov is written as

$$H_{exD} = \sum_n \mathcal{E}\, a_n^+ a_n - (1/2)\sum_{nm} J(n,m)a_n^+ a_m, \tag{5.18}$$

where $\mathcal{E}$ is the on-site exciton energy, and $J(n,m)$ is the exciton transfer energy. The quantities a_n^+ and a_n are the exciton creation and annihilation operators, respecteively, at an site n, which are assumed to satisfy Bose-type commutation relation $[a_n^+, a_m] = \Delta(n,m)$, etc. A number of researchers in the field of the Davydov soliton problem employed this model Hamiltonian from the outset.[1,2] For a molecular crystal described by the Heitler-London scheme, an <u>exact</u> formal expression for H_{ex} is written as

$$H_{ex} = \sum_{n\lambda} \mathcal{E}_\lambda^{(0)} \sigma_{n\lambda\lambda}^+ + (1/2)\sum_{nm}\sum_{\lambda\lambda'\mu\mu'} \langle \lambda_n \mu_m | V(n,m) | \lambda_n' \mu_m' \rangle \sigma_{n\lambda\lambda'} \sigma_{m\mu\mu'}, \tag{5.19}$$

with

$$H_0(n)|\lambda_n\rangle = \mathcal{E}_\lambda^{(0)}|\lambda_n\rangle. \tag{5.20}$$

Here $\mathcal{E}_\lambda^{(0)}$ and $|\lambda_n\rangle$ are the eigenvalue and the eigenfunction of the Hamiltonian $H_0(n)$ of an nth free molecule with quantum number λ_n. The symbol $\sigma_{n\lambda\lambda'}$ is a molecular operator associated with the nth molecule and defined by

$$\sigma_{n\lambda\lambda'} = |\lambda_n\rangle\langle\lambda_n'| \quad \text{with} \quad \sum_\lambda \sigma_{n\lambda\lambda}=1, \quad \sigma_{n\lambda\lambda'}\sigma_{n\mu\mu'}=\sigma_{n\lambda\mu'}\Delta(\lambda',\mu). \tag{5.21}$$

The second term in Eq.(5.19) is the matrix element of the di- and multi-pole interaction energy between and n and m molecules. In order to make Eq.(5.19) analytically tractable, we employ a <u>two-level approximation</u> to pick up only two states, the ground state $\lambda=0$ and a relevant excited state $\lambda=f$. Then, physically, σ_{nf0} and σ_{n0f} are exciton creation and annihilation operators at the site n, which correspond to a_n^+ and a_n, respectively. But $\sigma_{nf0}\equiv\sigma_n^+$ and $\sigma_{n0f}\equiv\sigma_n^-$ are neither Boson nor Fermion operators, but they can be identified as components of Pauli or spin-1/2 operators. In fact, we have

$$\sigma_n^x = (\sigma_{nf0}+\sigma_{n0f})/2, \quad \sigma_n^y=(\sigma_{nf0}-\sigma_{n0f})/2i, \quad \sigma_n^z=(\sigma_{nff}-\sigma_{n00})/2, \tag{5.22}$$

In terms of the σ_n^α's ($\alpha=x,y,z$), Eq.(5.19) is rewritten as[21]

$$H_{ex}=\sum_n \mathcal{E}\sigma_n^z - (1/2)\sum_{nm} J(n,m)\sigma_n^x\sigma_m^x + (1/2)\sum_{nm} I(n,m)\sigma_n^z\sigma_m^z, \tag{5.23}$$

where $\mathcal{E}$, $J(n,m)$ and $I(n,m)$ are the renormaloized on-site energy, the dipole-dipole interacton energy and the exciton-exciton interaction energy, respectively. Explicit expressions for $\mathcal{E}$, $J(n,m)$, and $I(n,m)$ in terms of $\mathcal{E}_\lambda^{(0)}$ and the matrix elements of $V(n,m)$ in Eq.(5.19) are omitted. In what follows, we neglect the last term in Eq.(5.22) on the assumption that the density of excitons is small. Equation (5.23) and its simplified version is the working model Hamiltonian for Frenkel excitons with dipole-dipole interactions. If we make the correspondence: $\sigma_{nf0}\to a_n^+$ and $\sigma_{n0f}\to a_n$, the the essential difference of Eq.(5.23) with the last term omitted from Eq.(5.18) is that the former contains the number-nonconversing terms $a_n^+a_n^+ +a_na_n$. A result similar to Eq.(5.23) was also obtained by Anderson long time ago.[22]

The Hamiltonian H_{ph} of the phonon system is also rewritten as

$$H_{ph} =\sum_q \hbar\omega_{ph}(q)\, b_q^+b_q, \quad \text{with} \quad u_n=(\hbar/2M)^{1/2}\sum_q e^{iqna}\omega_{ph}(q)^{-1/2}(b_q+b_{-q}^+), \tag{5.24}$$

where $\omega_{ph}(q)$ is given by Eq.(5.8), and b_q^+ and b_q are creation and annihilation operators of phonon with wave number q. The second of Eqs.(5.24) gives the relationship between the q-number displacement and the creation and annihilation operators. The exciton-phonon interaction arising from the fluctuation of on-site energy due to lattice vibrations is given by

$$H' =\sum_n g_n\sigma_n^z(u_{n+1}-u_{n-1}) = \sum_{qn} (\chi_{nq}b_q+\chi_{nq}^* b_q^+)\sigma_n^z, \tag{5.25}$$

where the g_n's and the χ_{nq}'s are interaction constants.

In studying quantum dynamics of the exciton-phonon system governed by the Hamiltonian $H_{ex} + H_{ph} + H'$, we are particularly concerned with coherent states associated with the exciton system and the phonon system. For this purpose, we introduce spin-coherent states $|\theta_n\varphi_n\rangle$ for the Pauli spin operator $\sigma_n=(\sigma_n^x,\sigma_n^y,\sigma_n^z)$ and phonon coherent states $|\beta_q\rangle$ by the equations[23-25]

$$|\theta_n \varphi_n\rangle = \exp(\zeta_n \sigma_n^- - \zeta_n^* \sigma_n^+)|\sigma_n\rangle \quad \text{with} \quad \zeta_n = (\theta_n/2)\exp(i\varphi_n), \quad (5.26)$$

$$b_q |\beta_q\rangle = \beta_q |\beta_q\rangle, \qquad \langle\beta_q| b_q^+ = \langle\beta_q|\beta_q^*. \qquad (5.27)$$

Here $|\sigma_n\rangle$ is the normalized "spin up" state, $\sigma_n^z|\sigma_n\rangle = (1/2)|\sigma_n\rangle$.[23,24] $|\beta_q\rangle$ and $\langle\beta_q|$ are the right and left eigenstates of b_q and b_q^+, respectively. Let us assume that the exciton-phonon system under consideration is in spin- and Boson-coherent states $|\alpha\rangle \equiv |\{\theta_n\}\{\varphi_r\}\{\beta_q\}\rangle$. The fundamental assumption in this paper is that the exciton-phonon system remains coherent as time evolves. Let $\langle H\rangle \equiv \langle\alpha|H|\alpha\rangle$ be the diagonal coherent-state representation of H. Then equations of motion obeyed by θ_n, φ_n, and β_q are written as [26]

$$\dot{\theta}_n = (2/\sin\theta_n)\,\partial\langle H\rangle/\partial\varphi_n, \qquad \dot{\varphi}_n = -(2/\sin\theta_n)\,\partial\langle H\rangle/\partial\theta_n, \quad (5.28)$$

$$i\dot{\beta}_q = \partial\langle H\rangle/\partial\beta_q^*, \qquad \text{and c.c.} \qquad (5.29)$$

These equations are essentially of classical nature except for the stationary cases for which the quations hold: $\dot{\theta}_n = \dot{\varphi}_n = \dot{\beta}_q = 0$. The stationary states obtainable by such a procedure amounts to using a variational method to seek approximate local energy minima of the exciton-phonon system. Equations (5.28) and (5.29) are equivalent to [27]

$$\dot{\bar{\sigma}}_n^x = -\left[\bar{\sigma}_n^y(\partial\langle H\rangle/\partial\bar{\sigma}_n^z) - \bar{\sigma}_n^z(\partial\langle H\rangle/\partial\bar{\sigma}_n^y)\right] \quad \text{and cyclically,} \qquad (5.30)$$

$$\dot{\bar{u}}_n = \partial\langle H\rangle/\partial \bar{P}_n, \qquad \dot{\bar{P}}_n = -\partial\langle H\rangle/\partial\bar{u}_n, \qquad \bar{P}_n = M\dot{\bar{u}}_n, \qquad (5.31)$$

with
$$\bar{\sigma}_n^\alpha = \langle\alpha|\sigma_n^\alpha|\alpha\rangle, \qquad a = x,y,z, \qquad \bar{u}_n = \langle\alpha|u_n|\alpha\rangle. \qquad (5.32)$$

Here $\bar{P}_n$ is a variable conjugate to u_n formed from the b_q's by the conventional procedure. By using such a quantum-classical correspondence, explicit expressions for $\langle H_{ex}\rangle \equiv \langle\alpha|H_{ex}|\alpha\rangle$, $\langle H_{ph}\rangle \equiv \langle\alpha|H_{ph}|\alpha\rangle$, and $\langle H'\rangle \equiv \langle\alpha|H'|\alpha\rangle$ are given by

$$\langle H_{ex}\rangle = \sum_n (\hbar^2/2\epsilon)\dot{q}_n^2 - \sum_n (\epsilon/2)(1-2q_n^2)^{1/2} - (1/4)\sum_{nm} J(n,m)q_n q_m, \qquad (5.33)$$

$$\langle H_{ph}\rangle = \sum_n (M\dot{\bar{u}}_n^2/2) + (K/2)(\bar{u}_{n+1} - \bar{u}_n)^2, \qquad (5.34)$$

$$\langle H'\rangle = -\sum_n (g_n \epsilon/2)(1-2q_n^2)^{1/2}(\bar{u}_{n+1} - \bar{u}_{n-1}), \qquad (5.35)$$

with
$$q_n = 2^{1/2}\bar{\sigma}_n^x. \qquad (5.36)$$

In obtaining Eqs. (5.33), (5.34), and (5.35), we have used [28]

$$\bar{\sigma}_n^z = -\left[(1/4) - (\bar{\sigma}_n^x)^2 - (\bar{\sigma}_n^y)^2\right]^{1/2} \cong -\left[(1/4) - (\bar{\sigma}_n^x)^2\right]^{1/2} + (\bar{\sigma}_n^y)^2$$

$$\cong -\left[(1/4) - (\bar{\sigma}_n^x)^2\right]^{1/2} + (\hbar^2/\epsilon^2)(\dot{\bar{\sigma}}_n^x)^2. \qquad (5.37)$$

In arriving at the second line of Eq. (5.37), we used $\hbar\dot{\bar{\sigma}}_n^x = -\epsilon\bar{\sigma}_n^y$, and in Eq. (5.35) we have neglected the contribution from the $\bar{\sigma}_n^y$-term. The last procedure is justified for large ϵ satisfying $(\hbar\dot{\bar{\sigma}}_n^x/\epsilon) \ll \bar{\sigma}_n^x$. It is immediately seen that Eq. (5.34) is entirely equivalent to Eq. (5.2), provided u_n is identified as the classical displacement field. Equation (5.33) reduces to Eq. (5.1) for v(q) having the form

$$v(q) = -(\epsilon/2)(1-2q^2)^{1/2} \cong \text{const} + (\epsilon/2)q^2 + (\epsilon/4)q^4 + (\epsilon/4)q^6 + \dots \quad (5.38)$$

while Eq.(5.35) is equal to Eq.(5.4) provided $V(q_n) = -(g_n \mathcal{E}/2)(1-2q_n)^{1/2}$. Thus, under the assumption of the Frenkel-exciton-phonon system being in coherent states, the model Hamiltonian for the system is equivalent under certain conditions to the classical vibron-phonon system governed by Eqs. (5.1)-(5.4). In contrast to the case of Davydov excitons, the number-non-conserving properties of the present exciton model make the existence of such coherent states much more feasible because it allows high quantum states as well.

6. VIBRON SOLITONS AS ANHARMONIC OR SELF—INDUCED LOCALIZED VIBRON MODES

As shown in the previous Section, Eqs.(5.15)-(5.17) can be used as basic equations for dynamical self-trapping of vibrons or Frenekl excitons by acoustic phonons. A wealth of physical contents has been contained in these equations in the form of the generalized Langevin equation resulting from elimination of the phonon coordinates. A key-factor in these equations is the $F(b_n)$-term representing a negative instantaneous force induced by phonons, leading to self-induced localized modes. The fundamental difference of the present vibron soliton theory from the Davydov soliton theory is that the former is described by the modified version of the NKG equation rather than the NLS equation. In order to reduce Eqs.(5.15(-(5.17) to a mathematically tractable form, we employ the nearest-neighbour-interaction approximation for the $L(n,m)$'s and discard all the transient-effect term, the friction term with the memory kernel, and the random-force term given by the last two terms on the left-hand side and the right-hand side in Eq. (5.15). Taking the nearest-neighbour-interaction constant as L, then we can reduce Eqs.(5.15) and (5.16) to

$$\ddot{b}_n + (\omega_0^2 - \omega^2)b_n - (L/\mu)\cos(ka)(b_{n+1}+b_{n-1}) - F(b_n) = 0, \tag{6.1}$$

$$\dot{b}_n + (L/2\omega\mu)\sin(ka)(b_{n+1}-b_{n-1}) = 0. \tag{6.2}$$

Equations (6.1) and (6.2) are obviously a 1d version of Eqs.(4.2) and (4.3), so the result obtained there can be used in the present case. Namely, the spatio-temporal evolution of self-induced localized modes is described by the equation

$$b_n(t) = \sum_m{}' J_{n-m}\left[\{L\sin(ka)/\mu\omega\}t\right]b_m(0), \tag{6.3}$$

where $b_m(0)$ is the solution to the following nonlinear eigenvalue equation:

$$\omega_0^2 b_n - (L/\mu)\cos(ka)(b_{n+1}+b_{n-1}) + (L'/2\mu\omega)^2(b_{n+2}+b_{n-2}-2b_n) - F(b_n) = \omega^2 b_n.$$
$$L' = L\sin(ka). \tag{6.4}$$

Using the same procedure as that employed in proceeding from Eq.(4.5) to (4.8), we obtain an exact formal expression for the dispersion relation satisfied by the eigenfrequency of a moving localized mode in terms of the shape functions $S_{\pm 1}$ and $S_{\pm 2}$ (we have omitted the subscript k):

$$\omega^2 = \omega_0^2 - (L/\mu)\cos(ka) - (C^2/K\mu)(S_1^2+S_{-1}^2)A^2 + (L'/2\mu\omega)^2(S_2+S_{-2}) - (L^2/2\mu^2\omega^2)$$
$$- \left[(2C^2/K\mu)-(3b/\mu)\right]A^2 + (5C/\mu)A^4. \tag{6.5}$$

Explicit expressions for $S_{\pm 1}$ and $S_{\pm 2}$ are obtained by solving iteratively the eigenvalue equations satisfied by the S_n's:

$$S_n = \sum_m{}' G_k(n,m)\left\{\left[(2C^2/K\mu)-(3b/\mu)\right]A^2 S_n^3 + (C^2/K\mu)A^2(S_{n+1}^2+S_{n-1}^2)S_n - (5C/\mu)A^4 S_n^5\right\}. \tag{6.6}$$

The result obtained here is very different from the conventional ones obtained by the continuum approximation. Namely, the spatio-time evolution of the self-induced localized mode is described by the Bessel functions, and the dispersion relation for the eigenfrequency of the localized mode constitutes a frequency band, where its binding energy generally depends on the wave number of the localized mode as well. It is of some interest to compare Eqs.(6.1) and (6.2) with Eqs.(3.3) and (3.3'). Although these two types of the equations are analytically solvable, the solutions to the latter for the soliton-bearing medel system are mathematically much neater than the latter.

7. SELF-SUSTAINED KINKS – NONLINEAR EXCITATIONS IN THE LARGE-AMPLITUDE REGIME

Anharmonic or self-induced localized modes associated with the dynamical self-trapping problem are relatively small-amplitude localized modes undergoing excursion not far from the absolute minimum of the potential function, though the excitations themselves are highly nonlinear. It is shown here that if we consider the high-amplitude regime, kink-type excitations emerge. To show this briefly, we insert Eq.(5.9) with the second and third terms on its right-hand side neglected into Eq.(5.5) to obtain

$$\ddot{q}_n - L(q_{n+1}+q_{n-1}) + v'_{eff}(q_n) = 0, \qquad \mu\omega_0^2 \gg 2L \qquad (7.1)$$

where

$$v_{eff}(q) = (\mu\omega_0^2/2)q^2 - [(C^2/2K)-(b/4)]q^4 + (c/6)q^6, \qquad (7.2)$$

with $(C^2/2K) > (b/4)$. The effective potential $v_{eff}(q)$ has a pair of degenerate relative minima at $q=\pm q_2$ with $v_{eff}(\pm q_2)>0$ and a pair of degenerate relative maxima at $q=\pm q_1$ with $0<v_{eff}(\pm q_2)<v_{eff}(\pm q_1)$, $0<q_1<q_2$ and $3q_1^2 q_2$, in addition to the absolute minimum at $q=0$. Self-or anharmonic-localized modes are the nonlinear excitation with amplitude q satisfying $0<|q|<q_1$. What we are concerned here are nonlinear excitations with amplitude satisfying $0<|q|<q_2$. Instead of Eq.(7.1), we consider a slightly modified version of Eq.(7.1):

$$\ddot{q}_n + 2\gamma\dot{q}_n - L(q_{n+1}+q_{n-1}) + v'_{eff}(q_n)=0, \qquad \mu\omega_0^2 \gg 2L, \qquad (7.3)$$

where $2\gamma\dot{q}_n$ is a friction term, which is either positive or negative for $q_n>0$ depending on the sign of a constant factor γ. It is shown by numerical analysis that Ex.(7.3) yields approximate self-sustained kink solutions: [17]

$$q_n = q_2\{1+ \exp[\pm(na-vt)/\ell_0]\}^{-1/2}, \qquad (7.4)$$

with

$$\gamma v = \mp\{2[(3/c)(c_0^2 - v^2)]^{1/2}/q_2^4\}v_{eff}(+q_2), \qquad (7.5)$$

$$\ell_0 = [3(c_0^2 - v^2)/c]^{1/2}/2q_2^2 \qquad (7.6)$$

Here the quantities ℓ_0, c_0, v, and q_2 are kink width, phonon velocity, kink velocity, and the kink amplitude, respectively. The type of the kink, kink or anti-kink, and its velocity are uniquely related to the friction constant by Eq.(7.5). The result of our numerical experiment, which is omitted here due to the limitation of the space, shows that the kink mode of the form (7.4) is a very good analytical approximation for the kink width satisfying $\ell_0 \gtrsim a$.

8. CONCLUDING REMARKS

We have presented a classical and quantum theory of dynamical self-trapping in nonlinear systems, paying particular attention to the lattice discreteness effect. A new concept emerging here is anharmonic local modes.

REFERENCES

1. A.S. Davydov and N.I. Kislukha, Sov. Phys.-JETP $\underline{44}$, 571 (1976).
    See, also A.S. Davydov, "Solitons in Molecular Systems", D. Reidel
    Publishing Company, Dordrect, Boston, and Lancaster, (1985), and also
    references cited therein.
2. D.W. Brown, K. Lindenberg, and B.J. West, Phys. Rev. $\underline{A33}$, 4104 (1986);
    D.W. Brown, B.J. West, and K. Lindenberg, Phys. Rev. $\underline{A33}$, 4110 (1988);
    Xidi Wang, D.W. Brown, K. Lindenberg, and B.J. West, Phys. Rev. $\underline{A37}$,
    3557 (1988).
3. S. Takeno, Prog. Theor. Phys. $\underline{69}$, 1798 (1983); $\underline{71}$, 395 (1984), $\underline{73}$, 853,
    (1985); $\underline{75}$, 1 (1986).
4. G. Careri, U. Buontempo, F. Carta, E. Gratton, and A.C. Scott, Phys.
    Rev. Lett. $\underline{51}$, 304(1983); G. Careri, U. Buontempo, F. Galluzzi, A.C.
    Scott, E. Gratton, and E. Shyamsunder, Phys. Rev. $\underline{B30}$, 4689 (1984).
5. R.H. Page, Y.R. Shen, and Y.T. Lee, Phys. Rev. Lett., $\underline{59}$, 1293 (1987).
6. A. Migliori, P.M. Maxton, A.M. Clogston, E. Zirngiebl, and M. Lowe,
    Phys. Rev. $\underline{B38}$, 13464 (1988).
7. A.C. Scott, Phys. Scr. $\underline{29}$, 279 (1984); L. MacNeil and A.C. Scott,
    Phys, Scr. $\underline{29}$, 284 (1984).
8. P.S. Lomdahl and W.C. Kerr, Phys. Rev. Lett., $\underline{55}$, 1235 (1985); in
    Physics of Many Particle Systems, edited by A.S. Davydov, Naukova Dumka,
    Kiev (1988).
9. Xidi Wang, D.W. Brown, and K. Lindenberg, Phys. Rev. Lett., $\underline{62}$, 1796
    (1989).
10. A.J. Sievers and S. Takeno, Phys. Rev. Lett., $\underline{61}$, 970 (1988); S.Takeno,
    Prog. Thoer. Phys. Suppl. $\underline{No.94}$, 242 (1988).
11. S. Takeno and A.J. Siever, Solid State Commun., $\underline{67}$, 1023 (1989).
12. V.E. Zakharov, Sov. Phys.-JETP $\underline{35}$, 908 (1972).
13. L.D. Landau, Phys. Zeit. Sowjetunion, $\underline{3}$, 664 (1933).
14. S. Takeno, J. Phys. Soc. Jpn., $\underline{58}$, 759 (1989).
15. M.J. Ablowitz and J.F. Ladik, J. Math. Phys., $\underline{17}$, 1011 (1976).
16. S. Takeno, Submitted to J. Phys. Soc. Jpn.
17. S. Takeno, J. Phys. Soc. Jpn., $\underline{57}$, 675 (1988).
18. Xidi Wang, D.W. Brown, and K. Lindenberg, Phys. Rev. $\underline{B39}$, 5366 (1989).
19. S. Takeno, J. Phys. Soc. Jpn., $\underline{58}$, 1639 (1989).
20. See, for example, W.H. Louisell, "Quantum Statistical Properties of
    Radiation", John Wiley & Sons, New York, London, Sydney and Toronto,
    (1973), p. 324.
21. S. Takeno and M. Mabuchi, Prog. Theor. Phys., $\underline{50}$, 1848 (1973).
22. P.W. Anderson, "Concepts in Solids", W.A. Benjamin, Inc., New York and
    Amsterdam, (1963), p. 132.
23. J.M. Radcliffe, J. Phys. $\underline{A4}$, 313 (1971).
24. F.T. Arecchi, E. Courtens, R. Gilmore, and H. Thomas, Phys. Rev. $\underline{A6}$,
    2211 (1972).
25. R.J. Glauber, Phys. Rev. $\underline{131}$, 2766 (1963).
26. S. Takeno, J. Phys. Soc. Jpn., $\underline{48}$, 1075 (1980),
27. For Boson systems, see for example, J.S. Langer, Phys. Rev. $\underline{167}$, 183
    (1968).
28. See also, S. Takeno and S. Homma, J. Phys. Soc. Jpn., $\underline{49}$, 1671 (1980).

Question by Eilbeck: Your model for ACN suggests a very localized soliton state, but other models give this result also. Does your model make any new predictions which would enable experimentalists to distinguish between these different models?

Reply: As far as the eigenfrequency of the localized mode is concerned, my model gives the same result as the conventional ones based on the Davydov theory, because the $C=O$ stretching vibration frequency ($\simeq 1665\text{cm}^{-1}$) is much larger than the dipole-dipole interaction energy between a pair of the $C=O$ units. The essential difference of my model from other models is the inclusion of the double excitation and de-excitation processes associated with dipole transitions. The difference may be observed by high-intensity infrared absorption measurements. To understand the experimental result, we need a theory of high-intensity infrared absorption in ACN. I have not yet investigated my model to take into account the effect of the radiation field, except the integrated absorption intensity.

Question by Kenkre: You mentioned for the Frenkel excitons interaction via dipole-dipole interactions are indeed treated in traditional theory and neglected on physical grounds: the double creation term on the grounds of low excitation density and double destruction term on the grounds of the disparity of time scales: probe time and lifetime of the excitation. Do you believe that those assumptions are invalid in the system you treat?

Reply: I believe that the neglect of the double creation term and the double destruction term is invalid, provided we are treating excitons with transfer by dipole-dipole interactions. These two terms are often neglected on the physical grounds you mentioned. I do not think that reasoning along these physical grounds is correct. In considering exciton-exciton interactions, two types of interactions exist, one is kinematical and the other dynamical. In the conventional treatment of the Davydov theory, there appears to exist confusion between these two points. To understand the physics of Frenkel excitons with dipole-dipole interactions, the book by P.W. Anderson **Concept in Solids** (1963) is very useful. The conventional exciton model in the Davydov theory can be applied only for exciton transfer by exchange interactions.

4

VIBRON SOLITONS: A SEMICLASSICAL APPROACH

Katja Lindenberg[a,b], Xidi Wang[c], and David W. Brown[a]

[a] Institute for Nonlinear Science, R-002
[b] Department of Chemistry, B-040, and
[c] Department of Physics, B-019
University of California at San Diego, La Jolla, CA 92093 U.S.A.

INTRODUCTION

The central topic of interest in this conference revolves around the soliton mechanism first proposed by Davydov and Kislukha [1] as a means of transporting energy in molecular (mainly biological) aggregates. The last few years have seen a large amount of activity in this area, and this workshop has gathered many of the principals in the discussion. The study of the Fröhlich Hamiltonian [2] which is the starting point of the Davydov theory has become quite sophisticated and has overcome many of the restrictive assumptions that were necessarily introduced when the subject began. Thus, this conference has served to clarify many questions involving the effects of quantum fluctuations (since most theories are semiclassical), the meaning of localization when the tendency of a translationally invariant system is to delocalize wave functions, the effect of temperature on the stability of the Davydov soliton, and questions surrounding the soliton lifetime. On a more subtle level, much insight has been gained into the precise meaning of the approximations made at various historical stages of the subject, a case in point being the various variational principles that have been used to derive equations of motion for the Davydov system. The considerable theoretical advances do not yet unequivocally answer the question of the possible importance of soliton mechanisms in the transport of biological energy since the starting model in most theories (one-dimensional Fröhlich Hamiltonian) may be well removed from what goes on in real proteins. However, *if* the model is indeed appropriate, then the weight of the evidence suggests that at least in the α-helix soliton transport at room temperature (or even at low temperatures) should not play an important role (cf. Brown *et al.*, this volume). Most of these issues and references to the work leading to our current understanding can be found throughout these proceedings.

The Fröhlich Hamiltonian is but one of the models that one can consider in this context. The Hamiltonian describes two linear fields, say an electronic excitation or an intramolecular vibration, and a collection of phonons, interacting nonlinearly with one another. There are several features of the model that one may wish to question if one is to relate it to a real physical or biological system. For example, in the Fröhlich interaction one field modulates the local site-energy of the other but not the transfer of excitation from one site to the other. The Fröhlich Hamiltonian as a whole conserves the number of excitations in one of the two fields. Number conservation may be appropriate when the excitation in question is electronic, but its appropriateness is less clear when the excitation is an intramolecular vibration, as it is in the case of the α-helix. In this paper we address this latter restriction and some of the consequences of relaxing it.

Takeno has introduced a Hamiltonian that describes a vibron field coupled to a phonon field [3-5]. In the Takeno Hamiltonian, the energy of the vibrons at each lattice site is modulated by the phonons, and, as in the Fröhlich Hamiltonian, the transfer of vibronic excitation from one site to another is not modulated. This Hamiltonian will serve as the model in our analysis [6].

Davydov's Soliton Revisited, Edited by P.L. Christiansen and A.C. Scott
Plenum Press, New York, 1990

In the next sections we spell out the detailed model to be considered and its relation to the Davydov system. Before doing that, however, some general remarks are warranted that will help to place later results in a clear context.

A simple qualitative picture of spontaneous localization of energy arises in the following scenario. Let us suppose that in the absence of an interaction between the two linear fields their dispersion relations are as sketched by the solid lines in Figure 1. The upper branch represents the electronic or vibronic excitation, while the lower branch corresponds to acoustic phonons. Suppose that a small region of the system is excited, now in the presence of the interaction between the two fields. Since it is a common finding that the bond potentials giving rise to vibron bands soften with increasing amplitude, it may happen that the finite amplitude of an oscillation causes a depression in the frequency of the excitation as shown by the dashed curve in the figure. If there are no other decay channels at the depressed frequency, then the energy can not escape from the excited region. Thus, the energy has become localized by virtue of the nonlinear interactions.

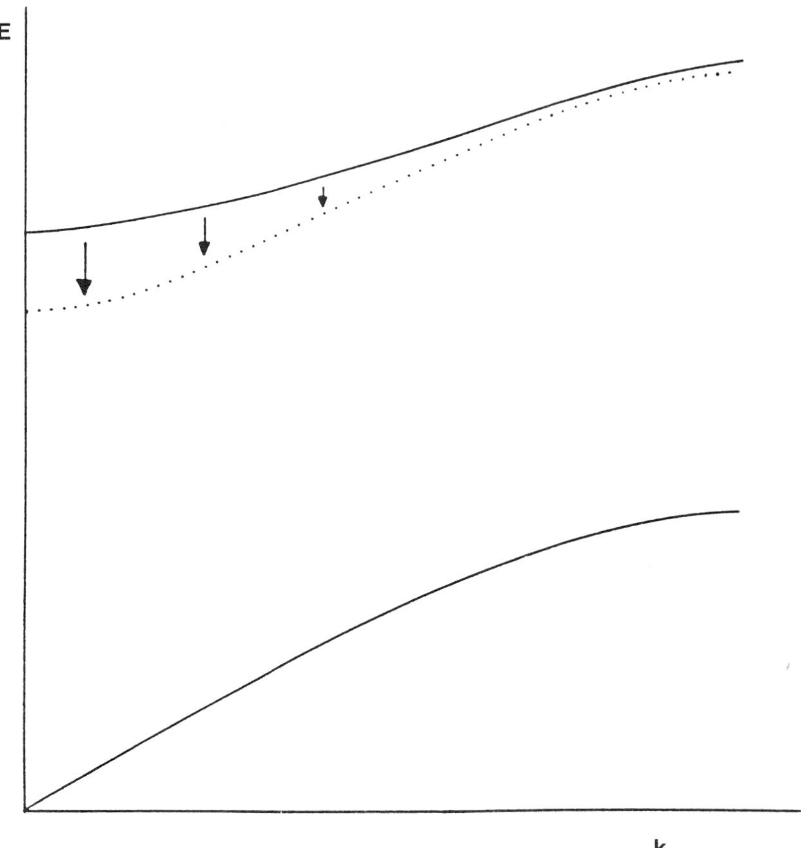

Fig. 1 Solid lines: dispersion relations for vibronic or electronic excitation (upper curve) and for acoustic phonons (lower curve) in the absence of an interaction between the two linear fields. Dashed curve: lowering of the excitation frequency caused by an interaction between the two fields.

A recent experiment by Migliori *et al.* [7] (reported in more detail elsewhere in this volume) has been interpreted in terms similar to those of this description. They study an *l*-alanine crystal (a low-symmetry crystal where anisotropies may allow essentially one-dimensional behavior) in which the lowest optical phonons (vibrons, upper branch in Figure 1) are coupled to acoustic phonons (lower branch). The vibrons arise from librational motions of the rigid l-alanine molecule and are sufficiently low in energy to be thermally populated at relatively low temperatures. These modes are Raman active, so that selection rules permit these excitations to be long-lived. There is evidence that these vibrons and the longitudinal acoustic phonons along the c-axis are in fact nonlinearly coupled.

The experimental observations made by Migliori $et\ al.$ include an anomalous temperature dependence of the Raman spectra of the two lowest frequency optical modes, which are centered at 42 cm^{-1} and 49 cm^{-1}. This anomaly arises as follows. The sum of the intensities of the two lines as a function of temperature follows a Bose distribution while each line separately does not. This behavior leads Migliori $et\ al.$ to conclude that both lines must arise from a single population of quanta of a single type of nonlinear oscillator. They then carefully construct a potential, qualitatively sketched in Figure 2, consisting of a superposition of two harmonic potentials, one characterized by 49 cm^{-1} energy-level spacings and the other by spacings of 42 cm^{-1}. The composite potential can now be analyzed and its energy levels can be found: one can calculate the distribution of energy as a function of temperature in the composite potential. In particular, one can now predict how much intensity is expected in the 49 cm^{-1} and 42 cm^{-1} lines at any given temperature under the assumption that the thermal energy is uniformly distributed throughout the solid. The $anomaly$ arises in the surplus of intensity (relative to this calculation) that is observed in the 42 cm^{-1} and the deficit of intensity in the 49 cm^{-1} line observed at very low temperatures where intensity should be observed mostly at 49 cm^{-1} and very little at 42 cm^{-1}. The resolution of this apparent anomaly involves the assumption of spontaneous localization of vibrational energy so that in these local regions the intensity of excitation is sufficiently high to cross the 42 cm^{-1} threshold at a much lower temperature than predicted.

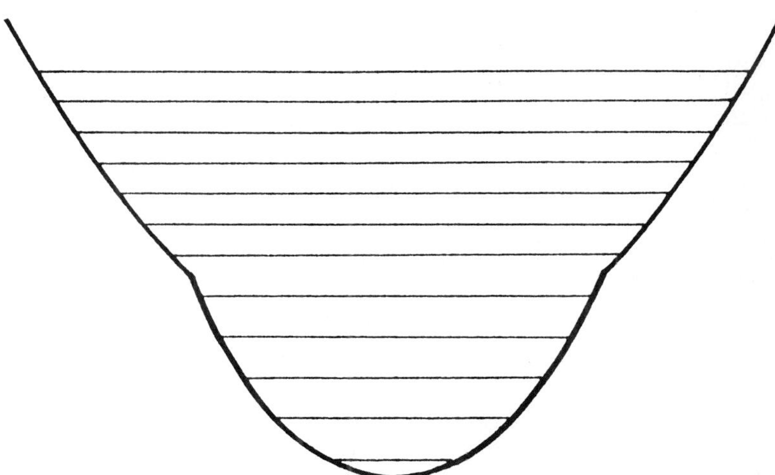

Fig. 2   Qualitative sketch of the potential constructed by Migliori $et\ al.$

COMPARISON OF TAKENO HAMILTONIAN AND DAVYDOV HAMILTONIAN

The Hamiltonian that we consider [6] was first proposed by Takeno [3-5] and has the form

$$H_T = H_{vib} + H_{ph} + H_{vib-ph} \ ,$$

(1)

where

$$H_{vib} = \sum_n \left[\frac{\hat{p}_n^2}{2m} + v(\hat{q}_n) \right] - \sum_{m,n} L_{mn} \hat{q}_m \hat{q}_n \quad , \tag{2}$$

$$H_{ph} = \sum_n \left[\frac{\hat{P}_n^2}{2M} + \frac{w}{2}(\hat{Q}_n - \hat{Q}_{n-1})^2 \right] \quad , \tag{3}$$

$$H_{vib-ph} = \sum_n (\hat{Q}_{n+1} - \hat{Q}_{n-1}) g(\hat{q}_n) \quad . \tag{4}$$

H_{vib} describes a set of oscillators of reduced mass m and whose position and momentum operators in the nth molecular unit are respectively $\hat{q}_n$ and $\hat{p}_n$. The local vibrational potential for these oscillators is $v(\hat{q}_n)$, and $L_{mn} = L_{nm}$ is the transfer coefficient due to the coupling of the local oscillators in the nth and m^{th} molecular units. The position and momentum operators of the nth molecular unit of mass M are respectively $\hat{Q}_n$ and $\hat{P}_n$. These units vibrate longitudinally with stiffness coefficient w, giving rise to acoustic phonons as described by H_{ph}. The interaction between the local oscillators and the acoustic phonons is contained in H_{vib-ph} and is assumed to be linear in the phonon coordinates but in general nonlinear via the force function $g(\hat{q}_n)$ in the oscillator coordinates. A useful special case of the Takeno model is the *quadratic Takeno Hamiltonian* in which both the local oscillator potentials and the coupling forces are quadratic functions of the oscillator coordinates:

$$v(\hat{q}_n) = \frac{1}{2}m\omega_v^2 \hat{q}_n^2 \quad , \qquad g(\hat{q}_n) = g\hat{q}_n^2 \quad . \tag{5}$$

The Davydov model Hamiltonian

$$H_D = H_{ex} + H_{ph} + H_{ex-ph} \quad , \tag{6}$$

with H_{ph} exactly as in (3), and

$$H_{ex} = \sum_n E a_n^\dagger a_n - \sum_{m,n} J_{mn} a_m^\dagger a_n \quad , \tag{7}$$

$$H_{ex-ph} = \sum_n \chi(\hat{Q}_{n+1} - \hat{Q}_{n-1}) a_n^\dagger a_n \quad , \tag{8}$$

can be viewed as a number-conserving truncation of the quadratic Takeno Hamiltonian. If one identifies the a-boson operators as those associated with the decoupled local oscillators, i.e.

$$\hat{q}_n = (\hbar/2m\omega_v)^{1/2}(a_n^\dagger + a_n) \quad , \tag{9}$$

$$\hat{p}_n = i(m\hbar\omega_v/2)^{1/2}(a_n^\dagger - a_n) \quad . \tag{10}$$

then the expectation values of the quadratic Takeno Hamiltonian and of the Davydov Hamiltonian in any state containing a definite number of a-bosons are the same. This equality requires the parameter identifications

$$J_{mn} = \frac{\hbar L_{mn}}{m\omega_v} \quad , \qquad E = \hbar\omega_v \quad , \qquad \chi = \frac{\hbar g}{m\omega_v} \quad . \tag{11}$$

The difference between the two Hamiltonians emerges in a number of obvious ways. First, the number-conserving character of the free excitation portion H_{ex} in the Davydov Hamiltonian leads to a Schrödinger equation description of the corresponding linear waves; for the quadratic Takeno Hamiltonian the corresponding linear waves are described by the Klein-Gordon equation. Second, the excitation-phonon interaction in the Davydov case is number-conserving while it is not in the Takeno case. These differences imply that whereas the Davydov Hamiltonian has eigenstates of a definite number of excitations, the Takeno Hamiltonian does not. A third obvious difference arises from the symmetries of the excitation-phonon coupling terms. In the Davydov case the coupling is number-conserving and hence only affects the phase of the excitation (cf. spins coupled to a phonon bath); in the Takeno model the coupling can affect both the phase and the amplitude of excitation (cf. a mechanical oscillator coupled to a phonon bath).

It is more difficult to speculate *a priori* what the effect of these differences might be on the evolution of the system and on the stability of possible soliton solutions. Number nonconservation on the one hand provides an additional channel for the decay of an excitation, perhaps enhancing soliton decay into some other forms of energy (and in this sense might serve to destabilize a Takeno soliton relative to a Davydov soliton). On the other hand, this same extra channel may allow the system to reach lower energies at which soliton solutions might be more stable. We argue that the latter occurs. The form of the excitation-phonon coupling affects the thermal stability of any coherent solution. We have argued on qualitative grounds that the coupling in the Davydov model produces fluctuations that tend to delocalize the excitation more than those that arise in the Takeno model.

LEVEL OF SOPHISTICATION

A great deal of discussion at this conference has dealt with the methods by which the evolution of the Davydov system is deduced from the Hamiltonian and the nature of the approximations introduced by the various methods. Many analytic methods rely on the construction of an Ansatz state followed by some procedure to obtain evolution equations for the time-dependent parameters appearing in the assumed Ansatz.

There is a great deal of discussion about the different Ansatz states and the approximations that are introduced by making specific assumptions as to the form of these states. The early work of Davydov (and even much contemporary work) begins from a state which is assumed to be a direct product of a single exciton state and a phonon coherent state [8-12]. The neglect of any phase mixing of these two states clearly eliminates certain quantum effects and makes this a "semiclassical" approximation. Davydov later introduced a more general Ansatz state that allows for some phase mixing [8,9,11]. More recently, products of partially dressed states (in the sense of polaron theory) have been considered [13] (cf. Brown *et al.*, this volume).

The second part of the discussion deals with the way in which the time dependent parameters in the Ansatz states are determined. Early approaches took these parameters to represent classical coordinates, and deduced their evolution on the assumption that they satisfy Hamilton's variational principle [8,9]. Equivalent equations are obtained by simply assuming that the operators in the *exact* Heisenberg equations of evolution that arise from the Davydov Hamiltonian are c-numbers.

More recent analyses recognize that the quantum mechanical variational principle is more appropriate and several authors have applied it to the various Ansatz states [11,14-16]. This improved procedure is generally accepted to lead to equations that contain more of the quantum aspects of the problem.

To avoid any approximations altogether, i.e. to escape the restrictions of approximate Ansatz states and approximate variational or perturbation methods, it has been necessary to solve the problem numerically using quantum Monte Carlo techniques [17] (cf. Wang *et al.*, this volume).

The result of this history has been a fairly clear understanding of the Davydov model and of the effect of different approximations on its solution. In particular, it is clear that the simplest (direct product form) Ansatz state neglects important quantum effects and becomes more and more accurate only as one approaches a classical limit. The actual system might perhaps approach this limit at high levels of excitation (correspondence principle).

In the case of the Takeno model, analysis relative to this historical context is still in its infancy. Specifically, we find ourselves at the "direct product Ansatz state" stage [6]. One therefore wonders how far the results of our analysis are from the actual evolution of a system described by the Takeno Hamiltonian. Our expectation is that the results may be somewhat closer to the "real" ones than in the Davydov model because of the number nonconservation character of the Takeno problem: since the number of excitations is not conserved, the actual state of the system will include admixtures of highly excited states, and these highly excited states predispose the system evolution towards classical behavior. Therefore, we expect the quality of the results to be gained through our present analysis to be comparable to or better than that of well-known results already established for the Davydov model. An analysis of more complex Ansatz states (e.g. partially dressed ones) and quantum Monte Carlo calculations on the Takeno system would be highly desirable but are not yet available.

EQUATIONS OF MOTION, CONTINUUM LIMIT, INITIAL CONDITIONS

As mentioned above, the simplest Ansatz state for the Davydov Hamiltonian is a direct product of a coherent phonon state and a single exciton state. A single exciton state is sufficient in the Davydov analysis because of the exciton number conservation property of the Hamiltonian: if only one exciton is present initially, then exactly one exciton is present at all times. The nonconservative Takeno Hamiltonian, on the other hand, must have eigenstates with an indefinite number of vibrons as well as an indefinite number of phonons. The natural generalization of the Ansatz for the Takeno problem is the direct product of two coherent states, one to describe the phonons ($| \beta(t) >$) and the other to describe the vibronic excitations ($| \alpha(t) >$):

$$| \psi(t) > \; \equiv \; | \alpha(t) > \times \, | \beta(t) > \tag{12}$$

where

$$| \alpha(t) > \; \equiv \; \exp \left\{ \frac{i}{\hbar} \sum_n [q_n(t) \hat{p}_n - p_n(t) \hat{q}_n)] \right\} | 0 > \; , \tag{13}$$

$$| \beta(t) > \; \equiv \; \exp \left\{ \frac{i}{\hbar} \sum_n [Q_n(t) \hat{P}_n - P_n(t) \hat{Q}_n)] \right\} | 0 > \; . \tag{14}$$

The time-dependent quantities $q_n(t), p_n(t), Q_n(t)$ and $P_n(t)$ must be determined, and some recipe or variational principle must be used to determine them.

Again in the spirit of Davydov's work, one way to obtain evolution equations for the time-dependent parameters is to assume that these quantities can be used to define classical coordinates, and then to construct Hamilton's equations for them using the expectation value of H_T of Eqs. (1)-(4) in the state (12) as the Hamiltonian function. Essentially equivalent equations are obtained by constructing the Heisenberg equations of motion for the position and momentum operators and then simply taking the operators to be c-numbers. Either procedure leads to $4n$ equations of motion, one for each of the $4n$ coefficients. The equations for the phonon position and momentum functions $Q_n(t)$ and $P_n(t)$ are formally linear and can be integrated explicitly. After substituting the result into the $q_n(t)$ and $p_n(t)$ equations, the lattice coordinates now appear only through their initial values. The resulting equations for the vibron coordinates are

$$\dot{q}_n(t) = p_n(t)/m \quad , \tag{15}$$

$$\dot{p}_n(t) = - \frac{\partial v [q_n(t)]}{\partial q_n(t)} + 2 \sum_m L_{nm} q_m - g^{-1} f_n(t) \frac{\partial g [q_n(t))]}{\partial q_n(t)}$$

$$- g^{-2} \sum_m \int_0^t d\tau \, \dot{K}_{nm}(t - \tau) g [q_m(\tau)] \frac{\partial g [q_n(t)]}{\partial q_n(t)} \; . \tag{16}$$

For the quadratic Takeno Hamiltonian the latter equation becomes

$$\dot{p}_n(t) = - m \omega_v^2 q_n(t) + 2 \sum_m L_{nm} q_m(t) - 2 f_n(t) q_n(t) - \sum_m \int_0^t d\tau \, \dot{K}_{nm}(t - \tau) q_m^2(\tau) q_n(t) \tag{17}$$

where

$$K_{nm}(t) \equiv 2 \sum_q g_n^q g_m^{-q} \hbar \omega^q \cos(\omega_q t) \tag{18}$$

with

$$g_n^q = \frac{g\,[-2i\sin(ql)]}{(2NM\hbar\omega_q^3)^{\frac{1}{2}}}\,e^{-iqR_n} \quad . \tag{19}$$

The coefficient $f_n(t)$ contains the initial values $Q_n(0)$ and $P_n(0)$ and is displayed below in the continuum limit. The uncertainty of these initial values, which must in general be drawn at random from an ensemble of such values, leads to the identification of $f_n(t)$ as a "fluctuating coefficient".

Since we are interested in soliton solutions that extend over a number of lattice sites (i.e. long-wavelength solitary-wave solutions), it is convenient to consider the continuum limit of the set of (15) and (17). The continuum equations are written in terms of bulk parameters constructed so as to remain finite as the lattice constant $l \to 0$:

$$\eta = M/l \quad , \quad \zeta = wl \quad , \quad \upsilon_a = \sqrt{\zeta/\eta} \quad , \quad \upsilon_f = (2Ll^2/m)^{\frac{1}{2}} \quad , \tag{20a}$$

$$\omega(k) = \sqrt{\omega(0)^2 + k^2\upsilon_f^2} \quad , \quad \omega(0) = \sqrt{\omega_v^2 - 4L/m} \quad , \tag{20b}$$

$$L_{nm} = L(\delta_{n,m+1} + \delta_{n,m-1}) \to 2L + Ll^2\frac{\partial^2}{\partial x^2} \quad . \tag{20c}$$

Note that υ_a is the speed of sound and $\omega(k)$ is the free vibron band frequency, the lowest vibron frequency in the band being $\omega(0)$. With $q_n(t) \to q(x,t)$ the continuum limit is

$$\left[\upsilon_f^2\frac{\partial^2}{\partial x^2} - \frac{\partial^2}{\partial t^2}\right]q(x,t) = \omega(0)^2 q(x,t) + \frac{2}{\mu}f(x,t)q(x,t) + \int_0^t d\tau\,\Phi(q;x,t,\tau)q(x,t) \quad . \tag{21}$$

The fluctuating coefficient in the continuum limit is given by

$$f_n(t) \to f(x,t) = g\left[\frac{\partial Q(y,0)}{\partial y} + \frac{P(y,0)}{\eta\upsilon_a}\right]_{y=x+\upsilon_a t}$$

$$+ g\left[\frac{\partial Q(y,0)}{\partial y} - \frac{P(y,0)}{\eta\upsilon_a}\right]_{y=x-\upsilon_a t} \tag{22}$$

and the kernel Φ by

$$\Phi(q;x,t,\tau) = \frac{4g^2}{\mu\zeta}\left\{\left[\frac{\partial}{\partial t}q^2(y,\tau)\right]_{y=x+\upsilon_a t} + \left[\frac{\partial}{\partial t}q^2(y,\tau)\right]_{y=x-\upsilon_a t}\right\} \quad . \tag{23}$$

In addition to the free-vibron Klein-Gordon wave equation contribution, Eq. (21) contains a contribution depending on the lattice initial conditions through which thermal fluctuations enter the dynamics, and it also contains a nonlinear intergal contribution through which damping and nonlinearity enter the dynamics.

Of all the possible solutions of Eq. (21) we are particularly interested in the existence of solitons. Therefore at this stage we only search for solutions that are of the modulated carrier wave form

$$q(x,t) = \phi[\kappa(x - \upsilon t)]\cos(kx - \tilde{\omega}t) \quad . \tag{24}$$

Here υ is the group velocity of the traveling wave, $\phi[\kappa(x - \upsilon t)]$ is an envelope function of width κ^{-1}, and $\tilde{\omega}$ is the carrier wave frequency. The relations among these parameters and their relation to the free vibron parameters such as the free-vibron dispersion relation $\omega(k)$ must still be determined.

Substitution of Eq. (24) into (21) immediately reveals that the latter can not be an exact solution of the former in most cases. In particular, the nonlinear form of the integral term causes the admixture of higher multiples of the carrier wave frequency. However, if there are many carrier wave oscillations modulating the envelope function then these higher frequency contributions may be ignored. This may occur when many carrier wave oscillations elapse during the time that it takes a sound wave to propagate across the envelope (temporal averaging), or when the envelope spans many wavelengths of the carrier (spatial averaging). In either case, the nonlinear contribution in the integral term can be simplified via the rotating wave approximation (RWA)

$$q^2(y,\tau) = \phi^2[\kappa(y - \upsilon\tau)]\cos^2(ky - \tilde\omega\tau) \approx \frac{1}{2}\phi^2[\kappa(y - \upsilon\tau)] \quad . \tag{25}$$

This substitution (into the integral only) leads to a simpler form for (21):

$$\left[\upsilon_f^2 \frac{\partial^2}{\partial x^2} - \frac{\partial^2}{\partial t^2}\right]q(x,t) = \omega(0)^2 q(x,t) - \frac{G(\upsilon)}{\mu}\phi^2[\kappa(x - \upsilon t)]q(x,t) + \frac{2}{\mu}f(x,t)p(x,t)$$

$$+ \frac{G(0)}{2\mu}\left\{\frac{\upsilon_a}{\upsilon_a - \upsilon}\phi^2[\kappa(x - \upsilon_a t)] + \frac{\upsilon_a}{\upsilon_a + \upsilon}\phi^2[\kappa(x + \upsilon_a t)]\right\}q(x,t) \tag{26}$$

where

$$G(\upsilon) = \frac{4g^2}{\zeta(1 - \upsilon^2/\upsilon_a^2)} \quad . \tag{27}$$

The form (22) can still not be a solution of Eq. (24) in the presence of the traveling potential terms that propagate with the speed of sound υ_a instead of the velocity υ [second line of (24)] and in the presence of the fluctuating coefficient whose time-dependence is also not of the D'Alembert form $(x - \upsilon t)$. It is possible that these "distorting" contributions cancel one another, leaving only the desired solution, as follows. First, it must be noted that the coefficient $f(x,t)$ determined by the lattice initial conditions contains not only thermal noise but also any initial systematic distortion of the medium. We call the former contribution f^{th} and the latter f^{sys}:

$$f(x,t) = f^{sys}(x,t) + f^{th}(x,t) \quad . \tag{28}$$

Next we restrict this analysis to zero temperature whence f^{th} vanishes. Finally, we note that the initial lattice configuration can be chosen *exactly* so as to cancel the traveling potential terms, i.e. f^{sys} can be chosen to exactly cancel the second line in (26). The remaining equation then is

$$\left[\upsilon_f^2 \frac{\partial^2}{\partial x^2} - \frac{\partial^2}{\partial t^2}\right]q(x,t) = \omega(0)^2 q(x,t) - \frac{G(\upsilon)}{\mu}\phi^2[\kappa(x - \upsilon t)]q(x,t) \quad . \tag{29}$$

This equation indeed has a solution of the assumed form (22):

$$q(x,t) = \phi_o \mathrm{sech}[\kappa(x - \upsilon t)]\cos(kx - \tilde\omega t) \quad , \tag{30}$$

with the accompanying lattice deformation

$$Q(x,t) = \frac{-1}{(1 - \upsilon^2/\upsilon_a^2)}\frac{g}{\kappa\zeta}\phi_o^2 \tanh[\kappa(x - \upsilon t)] \quad . \tag{31}$$

To complete the solution of the problem we need to assign values to the parameters occurring in (30) and (31). The solutions contain five parameters: ϕ_o, κ, υ, k and $\tilde\omega$. However, only two of these can be chosen freely since the set of five parameters is constrained by the three relations

$$\upsilon\left[\frac{\tilde{\omega}(k)}{k}\right] = \upsilon_f^2 \tag{32}$$

which relates the group and phase velocities,

$$\kappa = \left[\frac{\omega^2(k) - \tilde{\omega}^2(k)}{\upsilon_f^2 - \upsilon^2}\right]^{\frac{1}{2}} \tag{33}$$

which constrains the width of the envelope, and the boundary condition that insures that the solitary-wave form is chosen over all the other possible wave trains that are solutions of Eq. (29):

$$\phi_o = \left\{\frac{2\mu[\omega^2(k) - \tilde{\omega}^2(k)]}{G(\upsilon)}\right\}^{\frac{1}{2}} . \tag{34}$$

By way of contrast, we point out that in the Davydov problem the solutions are characterized by a single parameter rather than by two, and that single parameter is often chosen to be the soliton velocity. The appearance of a second parameter in the Takeno model is due to the relaxation of the number-conservation constraint that is inherent in the Fröhlich Hamiltonian and that here introduces an extra "degree of freedom" in the problem.

Note that the three parameters that one chooses to constrain by these relations depend on the remaining two. In particular, for example, the new dispersion relation $\tilde{\omega}(k)$ depends not only on k (as indicated explicitly) but also on ϕ_o, i.e. the new dispersion relation encompasses a band for each k. This is a reflection of the nonlinear nature of the problem that causes the energy of the excitation to depend on its amplitude ϕ_o.

ANALYSIS OF SOLUTION

To complete the analysis of the problem one should now compare the solution (30)-(31) with the behavior of the free excitation. To compare the behavior of the solitary-wave solution with that of the free excitation, the pair of parameters υ, k is not particularly useful because the free excitation is not characterized by any parameter comparable to υ. The pair ϕ_o, k is also not particularly useful because the amplitude ϕ_o changes as the vibron-phonon coupling changes. A more useful pair of parameters is one that is independent of this coupling strength so that a direct comparison can be made as the coupling is turned on. We have chosen to consider the pair n, k where n is the mean number of quanta, whose relation to the other parameters is yet to be established. Our reasoning is that if one turns on the coupling of the excitations and the phonons *slowly*, then the energy levels of the system change but the distribution of quanta over the energy levels (and, in particular, the mean number of quanta) does not. This is the quantum analog of the adiabatic invariance of the action in classical mechanics. One thus begins with a vibron state having a certain distribution of excitations over the different eigenstates that contribute to the coherent state and a corresponding mean number n of quanta, and assumes that these remain invariant as the coupling is turned on.

In the decoupled (linear) problem, one knows the relation between the amplitude ϕ_o of a vibron state and the mean number of quanta $n \equiv <\sum_k a_k^\dagger a_k>$ in the state:

$$\phi_o = \left[\frac{n\hbar}{2\mu\omega(k)L}\right]^{\frac{1}{2}} , \tag{35}$$

where L is the size of the system. For the fully-coupled nonlinear problem we do not know *a priori* the relation between the mean number of quanta and the other parameters of the solution. We assume a form that smoothly converges to (35) in the weak coupling limit:

$$\phi_o = \left[\frac{n\hbar\kappa}{\mu\tilde{\omega}(k)}\right]^{\frac{1}{2}} . \tag{36}$$

As coupling becomes weaker, the soliton width $\kappa^{-1} \to L$ and $\bar{\omega}(k) \to \omega(k)$. Note that $\bar{\omega}(k)$ (and υ and κ) also depends implicitly on n, a reflection of the dependence of the energy per quantum of excitation on the number of quanta present, a consequence of the nonlinearity of the problem.

First let us consider the k- and n-dependence of the carrier wave frequency $\bar{\omega}(k)$. It is convenient to express the n-dependence via the dimensionless parameter

$$\sigma \equiv \frac{n\hbar G(0)}{\mu^2 \upsilon_f \, \omega^2(0)} \quad . \tag{37}$$

Eqs. (34) and (36) can be converted into one relation to determine the k- and σ-dependence of $\bar{\omega}(k)$:

$$[\omega^2(k) - \bar{\omega}^2(k)]\,[\bar{\omega}^2(k) - k^2\upsilon_f^2] = \frac{1}{4}\,\frac{1}{(1 - \upsilon^2/\upsilon_a^2)^2}\,\sigma^2\omega^4(0) \quad . \tag{38}$$

Equation (38) has two solutions for $\bar{\omega}^2(k)$, one of which we have argued elsewhere [6] to probably be an artifact of our procedure (at least within the confines of our approximations) and hence unphysical. Below we discuss only the solution that we believe to be the physically relevant one (cf. Figure 3, where we show the behavior of $\bar{\omega}(k)/\omega(0)$ for various values of σ in the limit of an infinite acoustic speed, $\upsilon_a \to \infty$).

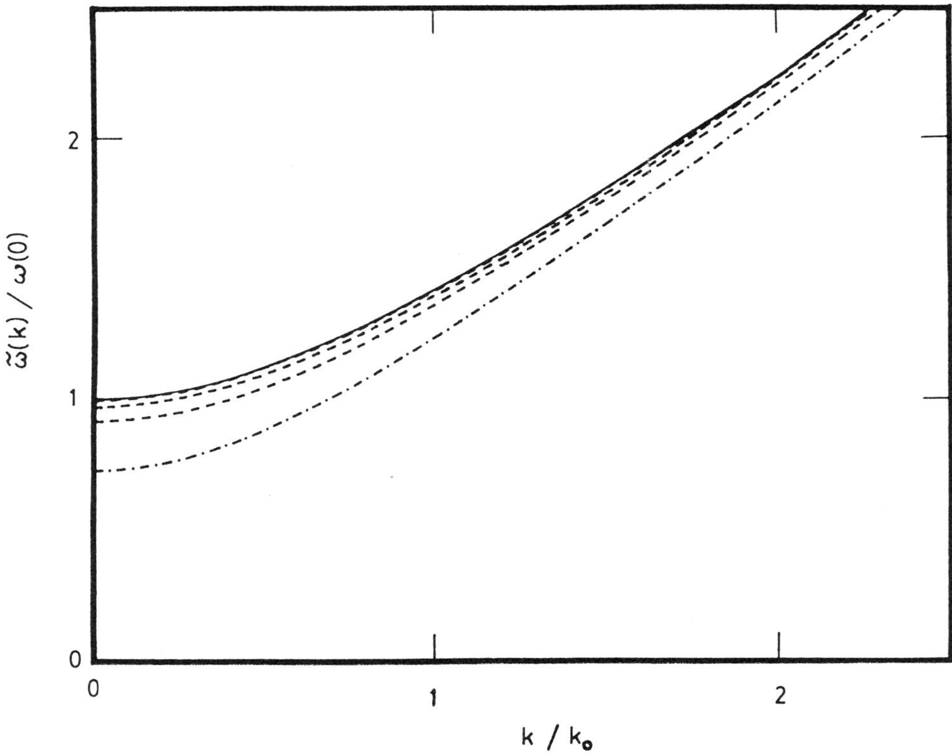

Fig. 3   Carrier wave frequency $\bar{\omega}(k)$ in the limit of infinite acoustic sound speed. Solid line: linear limit ($\sigma = 0$); dashed lines: $\sigma = \frac{3}{4}, \frac{1}{2}, \frac{1}{4}$; chain-dotted line: $\sigma = 1$. Note that for $\sigma = \frac{1}{4}$, $\bar{\omega}(k)$ lies too close to the linear limit curve $\omega(k)$ to be resolved in this graph.

First consider Eq. (38) in the limit of a large velocity of sound, $\upsilon_a \gg \upsilon$ (which is safe to assume if $\upsilon_f \ll \upsilon_a$, i.e. for "narrow-band" systems). In this case we find the solution

$$\bar{\omega}^2(k) = \bar{\omega}^2(0) + k^2 \upsilon_f^2 \ , \tag{39}$$

reminiscent of the free vibron dispersion relation in (20b), but now with

$$\bar{\omega}(0) = \omega(0)\cos(\tfrac{1}{2}\sin^{-1}\sigma) \ . \tag{40}$$

The soliton group velocity and width are determined directly from Eqs. (32) and (33).

Several features of the solution (39) are noteworthy. First, the carrier frequency is increasingly red-shifted with increasing σ because of the shift in $\bar{\omega}(0)$. Second, the frequency becomes complex if $\sigma > 1$ (indicating a real physical instability discussed subsequently) if $\sigma > 1$. Hence this analysis must be restricted to coupling strengths and/or occupation numbers such that $\sigma \le 1$. Moreover, since we must not forget the approximations that have led to these results, we must be particularly cognizant of the restrictions imposed by the RWA (25) which in fact impose a stronger restriction on σ. These restrictions are not uniform throughout the entire range of values of k, i.e. for a given value of σ our approximations may hold only a limited range of k. In particular, for large wave vectors ($\bar{\omega}(k) < k\upsilon_a$) the RWA is only valid if the wavelength k^{-1} of the carrier wave is much smaller than the width κ^{-1} of the soliton envelope. This can be translated into the condition

$$\frac{\upsilon}{\upsilon_f} \gg \tan(\tfrac{1}{2}\sin^{-1}\sigma) \tag{41}$$

(recall that $\upsilon < \upsilon_f$). For small wave vectors, on the other hand ($\bar{\omega}(k) > k\upsilon_a$) where the width of the soliton envelope is greater than the carrier wave length, validity of the RWA relies on the inequality

$$\frac{\upsilon_f}{\upsilon_a} \gg \tan(\tfrac{1}{2}\sin^{-1}\sigma) \tag{42}$$

(recall that we are considering for the moment $\upsilon_f < \upsilon_a$). Note that in Figure 3 only the restriction (41) is relevant since $\bar{\omega}(k)$ is smaller than $k\upsilon_a$ almost everywhere.

It is simple to reintroduce a finite sound speed υ_a at this point: its only effect is to replace every σ by $\sigma' \equiv \sigma/(1 - \upsilon^2/\upsilon_a^2)$, i.e. the coupling must now be reduced. The frequency is complex beyond the "critical value" $\sigma' = 1$ (again, the inherent instability reflected by this restriction is discussed subsequently). The restrictions (41) and (42) continue to hold but again with σ replaced by σ'. Further analysis of the solution shows the appearance of apparently spurious cutoffs that are avoided if we impose in place of $\sigma < 1$ the even stronger restriction $\sigma < 1 - \upsilon_f^2/\upsilon_a^2$.

If we consider broad-band systems ($\upsilon_a < \upsilon_f$) then only condition (42) with $\sigma \to \sigma'$ is relevant.

Consider now some of the other quantities that characterize the behavior of the excitation as a function of k and n. Of particular interest is the total energy

$$\bar{E} = \ <\psi(t)\,|\,H_T\,|\,\psi(t)> \ , \tag{43}$$

i.e. the expectation value of the Takeno Hamiltonian (1) in the Ansatz state (12). Quantities of interest related to this energy are the soliton binding energy E_{bind} and the soliton effective mass m_{sol}^T defined below. For a narrow-band systems ($\upsilon_a > \upsilon_f$) we find

$$\bar{E} = n\hbar\bar{\omega}(k)\{1 + \tfrac{1}{3}[\tan(\tfrac{1}{2}\sin^{-1}\sigma)]^2\} \ . \tag{44}$$

The energy as a function of k for various values of σ is shown in Figure 4 for an infinite sound speed ($\upsilon_a \to \infty$). Note the fact that for each σ there is a value of k at which $\bar{E}(\sigma)$ crosses $\bar{E}(\sigma = 0)$, i.e. the energy of the coupled nonlinear system is lower than that of the free vibron linear system up to some value of k but higher than that of the linear system beyond. The nature and possible implications of these crossings are discussed below.

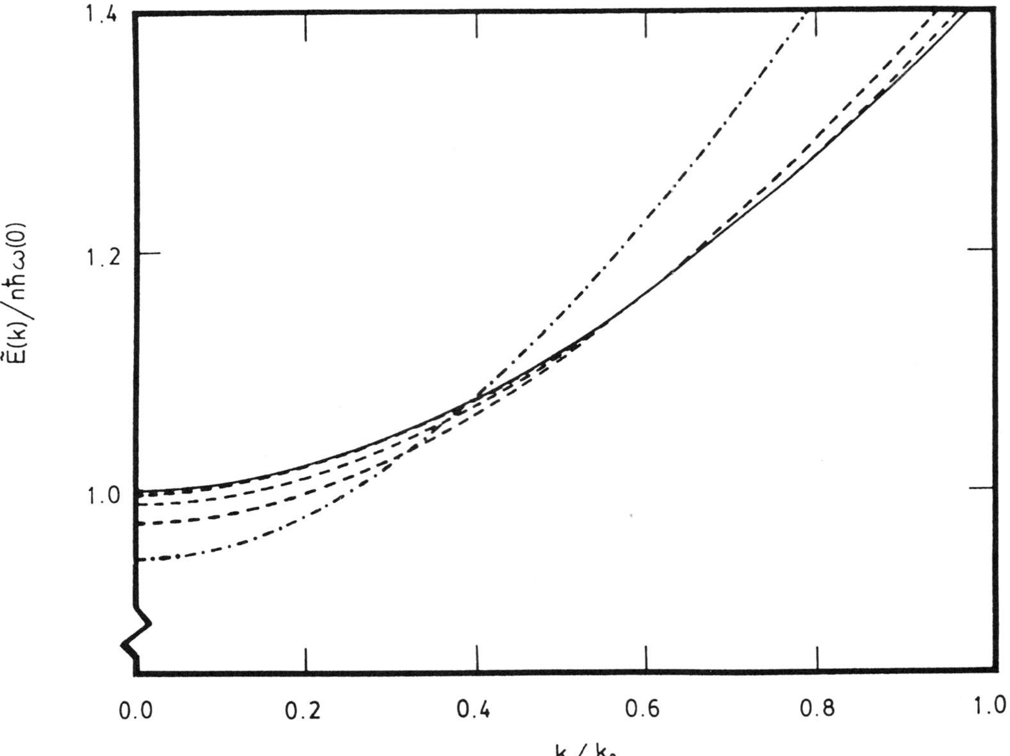

Fig. 4 Total system energy in the solitary-wave state $\bar{E}(k)$ in the limit of an infinite acoustic sound speed. Solid line: linear limit ($\sigma = 0$); dashed lines: $\sigma = \frac{1}{4}, \frac{1}{2}, \frac{3}{4}$; chain-dotted curve: $\sigma = 1$.

For finite sound speed the expression for the energy is considerably more complicated but exhibits the same qualitative behavior shown in Figure 4. It is instructive to consider the energy in the limit of small soliton speeds υ, where $\bar{E}$ can be written as

$$\bar{E} = n\hbar\omega(0) - E_{bind} + \frac{1}{2}m_{sol}^{T}\upsilon^2 \quad . \tag{45}$$

We find the expressions for the soliton binding energy and the soliton effective mass

$$E_{bind} = \frac{1}{3}\{3 - 2\cos(\tfrac{1}{2}\sin^{-1}\sigma) - [\cos(\tfrac{1}{2}\sin^{-1}\sigma)]^{-1}\}n\hbar\omega(0) \tag{46}$$

and

$$m_{sol}^{T} = \frac{n\hbar\omega(0)}{\upsilon_f^2}\left[[\cos(\tfrac{1}{2}\sin^{-1}\sigma)]^{-1} - \frac{2}{3}\left[1 + \frac{\upsilon_f^2}{\upsilon_a^2}\right]\tan^2(\tfrac{1}{2}\sin^{-1}\sigma)\right] \quad . \tag{47}$$

In the weak coupling limit, $\sigma \to 0$, the binding energy agrees with that obtained from the Davydov model,

$$E_{bind} \approx \frac{\sigma^2}{24}\, n\hbar\omega(0) \quad , \tag{48}$$

58

and the effective mass reduces to

$$m_{sol}^{T} \approx \frac{n\hbar\omega(0)}{\upsilon_f^2}\left[1 - \frac{\sigma^2}{24}\left[1 - 4\frac{\upsilon_f^2}{\upsilon_a^2}\right]\right] \; . \tag{49}$$

Equation (49) reduces to the corresponding Davydov result only for broad-band systems, i.e. when $\upsilon_f^2 \gg \upsilon_a^2$, whence

$$m_{sol}^{T} \approx \frac{n\hbar\omega(0)}{\upsilon_f^2}\left[1 + \frac{\sigma^2}{6}\frac{\upsilon_f^2}{\upsilon_a^2}\right] \; . \tag{50}$$

In this limit increasing the coupling strength leads to an increase in the effective mass of the soliton. On the other hand, for narrow-band systems with $\upsilon_f \ll \tfrac{1}{2}\upsilon_a$ Eq. (49) leads to a *decrease* of the effective mass with increasing coupling.

INSTABILITIES AND MINIMUM ENERGY

In the preceding analysis we observed the crossing of the nonlinear $[\bar{E}(\sigma)]$ and linear $[\bar{E}(\sigma = 0)]$ energy curves at some value of k. The precise crossing point k^* depends on the value of σ, but always lies in the range $k_1^* \leq k^* \leq k_0^*$, with

$$k_0^* = \frac{1}{\sqrt{2}}\frac{\omega(0)}{\upsilon_f} \; , \qquad k_1^* = \frac{1}{\sqrt{5}}\frac{\omega(0)}{\upsilon_f} \; . \tag{51}$$

Thus, all linear waves with wave vectors $k > k_0^*$ have energies lower than the solitary-wave solutions, while solitary-wave energies for wave vectors $k < k_1^*$ are always lower than their linear counterparts. Which is lower in energy in the range (k_1^*, k_0^*) depends on the value of σ. These energy relationships suggest that short-wavelength linear waves and broad, long-wavelength solitary waves should make up the dominant contributions to the spectral decomposition of a general state. We have interpreted the appearance of these crossings as a manifestation of a Benjamin-Feir instability [18-20]. This generic property of nonlinear wave equations involves the instability of plane waves in weakly nonlinear dispersive media to long-wavelength perturbations. In the present context, each complex-amplitude wave train is perturbed by the complex conjugate wave train with which it necessarily coexists. This leads to the instability of long-wavelength $(k < k^*)$ plane waves, while those of short wavelength remain stable.

The discussion up to this point has focused on the dependences of the vibron frequency and energy on the wave vector k and, through σ, on the mean occupation number n. At this point it is interesting to speculate on the implications that these observations might have for dynamical processes that lead to equilibration if the system is initially not in equilibrium. It is also interesting to compare these speculations with corresponding inferences about the Davydov model.

In the Davydov model the number of excitations is conserved and is usually taken to be unity $(n = 1)$. The energy function depends only on the wave vector (or equivalently, on the velocity), and changes in energy (e.g. energy relaxation of an initially epithermal state) can only take place via phonon scattering processes that on the average decrease the wave vector and broaden the state. In the Takeno model, on the other hand, there is an additional channel for relaxation introduced by the nonconservation of excitation number. In particular, we observe that for each value of k one can find the number of quanta n that corresponds to the state of lowest energy at that wave vector. As k is varied, so does the number of quanta that leads to the lowest energy of the system. To identify the minimal energy curve (along which both k and n vary) we must minimize the energy per quantum, $\bar{E}/n$, with respect to the number of quanta at each fixed value of k. The resulting value of σ then identifies the optimal value of n for that k. In the narrow-band limit we find explicitly

$$\left[\frac{1}{2} - \frac{2k^2}{(k_0^*)^2} + \left[\frac{1}{4} + \frac{2k^2}{(k_0^*)^2}\right]^{1/2}\right] \qquad 0 < k < k_0^*$$

The minimum energy envelope given by substituting (52) back into (44) for $k < k_0^*$ and using the linear wave dispersion relation for $k > k_0^*$ is shown in Figure 5. Note that the small wave vector, low energy states contain more quanta, i.e. σ *increases* as we move down the minimal energy curve towards $k = 0$. Note also that along the minimal energy curve the states are *narrowest* near $k = 0$, increase in width as k increases, and achieve infinite width at k_0^* and beyond.

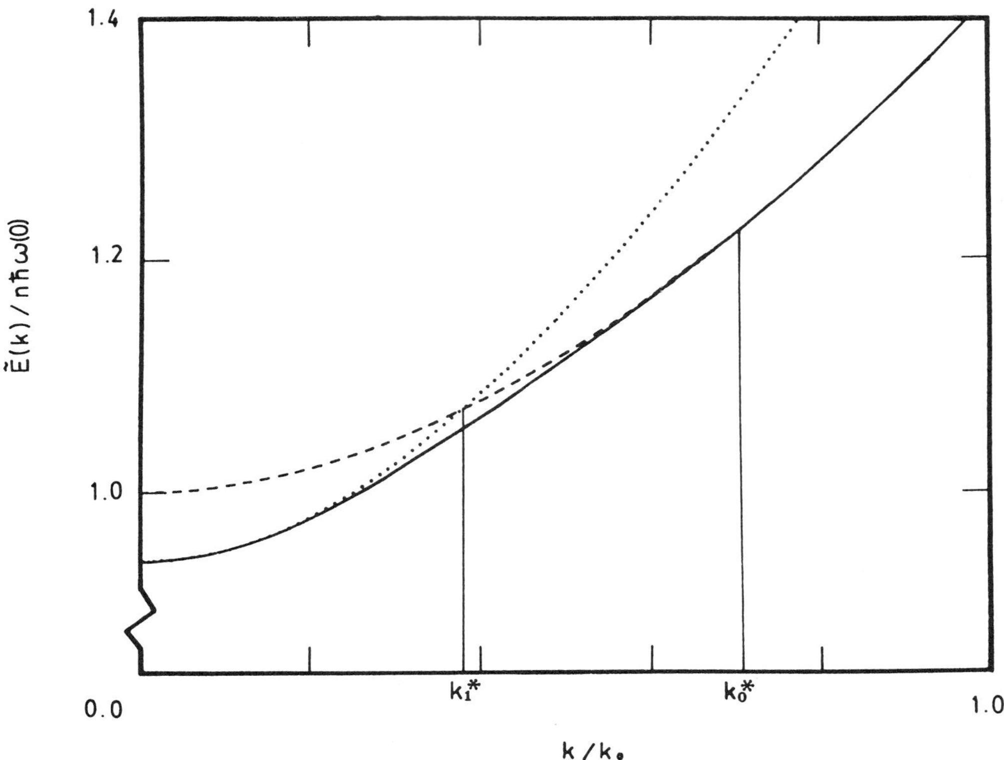

Fig. 5 Minimum energy envelope and related quantities. Solid line: minimum energy envelope; dashed line: linear energy ($\sigma = 0$); dotted line: energy for $\sigma = 1$. Note that the energy scale is broken; $[\hbar\omega(0) - \tilde{E}(\sigma = 1)]/\hbar\omega(0) \approx 5\%$.

The following picture of the dynamics of equilibration then emerges: an initially epithermal state might first relax quickly to a distribution along the minimum-energy envelope, resolving the itself into short-wavelength plane waves and long-wavelengths solitary waves. The plane waves then relax along the linear dispersion curve towards k_0^* through number-conserving scattering events that change the wave vector. Below k_0^*, further relaxation occurs via number non-conserving scattering events that tend to *increase* the number of vibron quanta since these processes now become energetically favored. Note also the interesting consequence that according to this view, energy relaxation favors energy localization.

INSTABILITY AT STRONG COUPLING

The analysis in the earlier sections points to an instability that occurs beyond the critical value $\sigma = 1$, i.e. if the vibron-phonon coupling is too strong or the mean number of vibrons is too high. Our solutions (because of the assumptions made in the approximations) are actually restricted to smaller

60

values of σ, and one might question whether $\sigma = 1$ indicates a real instability in the quadratic Takeno system or is merely an artifact of our approach. We argue that it is the former, and that it is a reflection of the effective potential in which the vibrons evolve. Thus, consider the nonlinear Klein-Gordon equation closely related to our Eq. (29):

$$\left[\upsilon_f^2 \frac{\partial^2}{\partial x^2} - \frac{\partial^2}{\partial t^2} \right] q(x,t) = \omega(0)^2 q(x,t) - \frac{G(\upsilon)}{\mu} q^3(x,t) \ . \tag{53}$$

Solutions of this equation are stable provided the amplitude $q(x,t)$ is smaller than the critical value $q_c = [\mu\omega^2(0)/G(\upsilon)]^{\frac{1}{2}}$. The effective potential qualitatively sketched in Figure 6 (solid curve) indicates the reason for the instability: if the amplitude is too large, then the restoring force becomes a repulsive force, causing the oscillation amplitude to grow without bound. It is interesting to speculate whether in some real systems such a run-away phenomenon might correspond to the breaking of chemical bonds. This instability disappears if instead the effective potential were that shown by dashed lines in Figure 6 (compare with the Migliori *et al.* potential of Figure 2). This stabilizing effective potential would indeed describe the system if one went beyond the *quadratic* Takeno Hamiltonian to include nonlinear potentials in the original Hamiltonian.

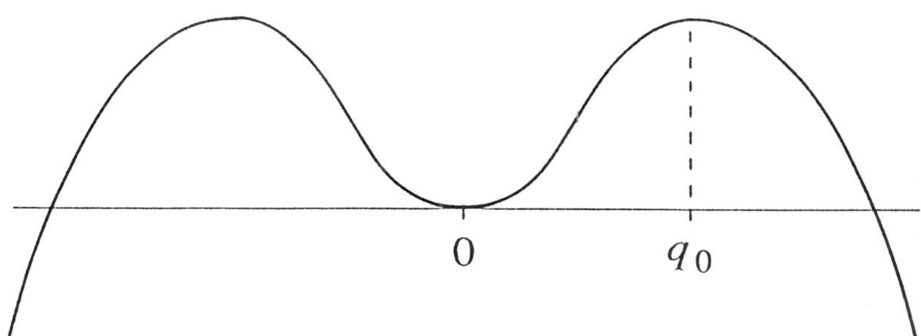

Fig. 6   Sketch of the effective potential for Eq. (53).

CONCLUDING REMARKS

It is useful at this point to make some observations that place the results that we have obtained for coupled linear fields in a more general context beyond the specific Hamiltonian model that we have analyzed. Some of these results are generic in nature and are therefore expected to occur in real systems in which there is weak nonlinear coupling between otherwise linear fields, even if the actual system is not precisely described by the (quadratic) Takeno Hamiltonian.

First we note that although we have specifically searched for soliton solutions (among all the possible solutions of the dynamical equations), we have actually found expressions for the dispersion relation and energy, $\tilde{\omega}(k)$ and $\bar{E}(k)$, that allow us to argue intuitively the behavior of the system even when solitons are only partially involved or not at all involved in the dynamical processes. In other words, the solutions in a sense transcend the approximations made to obtain them. Thus, for example, we used the dispersion and energy relations to reach rather generic conclusions about probable paths of thermalization, energy localization, consequences of the Benjamin-Feir instability and possible bond-breaking.

It is generic to linear fields coupled by weak nonlinearities that there is a relatively small red shift for all parameter values. For the quadratic Takeno Hamiltonian the maximum red shift of $\tilde{\omega}(k)$ away from $\omega(k)$ occurs at $k = 0$ and is of magnitude $\omega(0)/\sqrt{2}$. Similarly, the maximum energy lowering of $\bar{E}(k)$ relative to $E(k)$ is generically small, being at most 5% (again at $k = 0$) for our system. A third generic feature is the energy curve crossing at k^* related to the Benjamin-Feir instability. Thus, we would expect that long-wavelength plane waves in weakly coupled systems are unstable and transform to nonlinear wave trains of lower energy. An important consequence of this instability in *some* systems is the possible further instability that may actually lead to the *breaking of bonds*, as in fact occurs in the quadratic Takeno Hamiltonian. Although this effect is less generic than the others (e.g. it would not be expected in the general Takeno model), it nevertheless appears under rather general circumstances.

ACKNOWLEDGEMENT

This work was supported in part the U.S. Defense Advanced Research Projects Agency under ARPA order 7048, and by the Naval Research Laboratory under Grant No. N00014-88-K-2003.

REFERENCES

1. A. S. Davydov and N. I. Kislukha, Phys. Status Solidi **59**, 465 (1973); Zh. Eksp. Teor. Fiz. **71**, 1090 (1976) [Sov. Phys. - JETP **44**, 571 (1976)].

2. H. Fröhlich, H. Pelzer and S. Zieman, Philos. Mag. **41**, 221 (1950); Proc. R. Soc. London Ser. A **215**, 291 (1952); Adv. Phys. **3**, 325 (1954).

3. S. Takeno, Prog. Theor. Phys. **71**, 395 (1984).

4. S. Takeno, Prog. Theor. Phys. **73**, 853 (1985).

5. S. Takeno, Prog. Theor. Phys. **75**, 1 (1985).

6. X. Wang, D. W. Brown and K. Lindenberg, Phys. Rev. B **39**, 5366 (1989).

7. A. Migliori, P. Maxton, A. M. Clogston, E. Zirngiebl and M. Lowe, Phys. Rev. B **38**, 13464 (1988).

8. A. S. Davydov, Zh. Eksp. Teor. Fiz. **78**, 789 (1980) [Sov. Phys. - JETP **51**, 397 (1980)].

9. A. S. Davydov, Usp. Fiz. Nauk **138**, 603 (1982) [Sov. Phys. - Usp. **25**, 898 (1982)].

10. X. Wang, D. W. Brown, K. Lindenberg and B. J. West, Phys. Rev. A **37**, 3557 (1988).

11. D. W. Brown, Phys. Rev. A **37**, 5010 (1988).

12. K. Lindenberg, D. W. Brown and X. Wang, in *Far from Equilibrium Phase Transitions*, Lecture Notes in Physics Vol. 319, ed. L. Garrido (Springer, Berlin, 1988).

13. Z. Ivic and D. W. Brown, Phys. Rev. Letters **63**, 426 (1989).

14. H. Bolterauer and R. D. Henkel, Phys. Scripta **T13**, 314 (1986).

15. M. J. Skrinjar, D. V. Kapor and S. D. Stojanovic, Phys. Rev. A **38**, 6402 (1988).

16. Q. Zhang, V. Romero-Rochin and R. Silbey, Phys. Rev. A **38**, 6409 (1988).

17. X. Wang, D. W. Brown and K. Lindenberg, Phys. Rev. Letters **62**, 1706 (1989).

18. T. B. Benjamin and J. F. Feir, J. Fluid Mech. **27**, 417 (1967).

19. A. C. Newell, *Solitons in Mathematics and Physics* (Society for Industrial and Applied Mathematics, Philadelphia, 1985).

20. G. B. Whitham, *Linear and Nonlinear Waves* (Wiley, New York, 1974).

5

WHEN IS A SOLITON?

David W. Brown[a], Katja Lindenberg[a,b], and Xidi Wang[c]

[a] Institute for Nonlinear Science, R-002
[b] Department of Chemistry, B-040, and
[c] Department of Physics, B-019
University of California at San Diego, La Jolla, CA 92093 U.S.A.

> *"We [conclude] that classical concepts cannot be regarded as limiting forms of quantum concepts, but must instead be combined with limiting forms of quantum concepts in such a way that, in a complete description, each complements the other."*
>
> *--David Bohm* [1]

The title of this article is a paraphrase of another by the high-energy physicist Sidney Drell [2]. The latter article asks the question "When is a particle?" and deals with the changing standards of proof which lead us today to accept as "real" some fundamental particles (quarks) which we may never observe. Our situation is somewhat similar in that "hard evidence" of the existence of Davydov solitons remains elusive despite many years of effort expended in their study. As has been revealed during the discussions at this meeting, a part of this elusiveness does not arise from genuine problems of physics but from problems of communication between workers in the field. Among the latter is a considerable variance in the accepted meanings of the central terms "soliton" and "Davydov soliton" themselves. While most of us have a working knowledge of what the latter term means, the lack of a precise definition has been a root cause of some of the softness in the concept of the Davydov soliton. The physics of the underlying physical problem is, of course, completely indifferent to such linguistic difficulties which are purely of human origin; it is well, therefore, not to imbue them with undue importance. However, as the technical portion of this paper addresses a rather broad spectrum of behaviors open to an exciton in a deformable solid, we shall find it necessary to impose some precision on the terms to be used. In our general discussion, we will try use more general terms; when we encounter more distinct and special structures, we will try to preserve the distinctions in our language. It is inevitable that our terminology will conflict with that accepted by some segments of our audience; however, we hope that that will not prevent our message from being understood.

Our first and most important general term is not soliton, but *polaron*. By polaron we mean quite generally any quasiparticle excitation of a solid consisting of a conserved quantum of a massive field which maintains persistent and nontrivial correlations with the deformation or polarization quanta of the solid. It is essential to the quasiparticle concept that the interactions responsible for persistent correlations between zeroth-order fields should be essentially removed by incorporating the correlation property into the internal structure of the essentially free quasiparticle.

What then is a soliton? In classical mechanics, solitons are particle-like solutions of completely integrable nonlinear partial differential equations. After the fashion of our time, we may say that quantum solitons are quantum quasiparticles which are connected with classical solitons in a meaningful correspondence limit [3]. It is quite possible, even likely, that the polaron to which we address our attention may qualify as such a quantum soliton. To the extent that this is true, we may be able to use the correspondence principle to assist us in our study of the polaron, as a tether which connects the classical and quantum natures of the problem. However, it must be kept firmly in mind that in almost all cases the tether is too short, and we fail to arrive at a completely quantum mechanical solution. When this happens we must take care to discern which properties of our solution faithfully describe the quantum quasiparticle and which are the classical residue of a semiclassical method.

We shall argue that the entity which has come to be called the Davydov soliton represents a limiting form of the polaron in the standard adiabatic limit. In this respect, there is little useful distinction to be drawn between the Davydov soliton and the adiabatic large polaron which has been the subject of extensive study in the solid-state literature for many years [5-10].

For the sake of definiteness, we introduce the Fröhlich Hamiltonian [6] which is the traditional starting point for the study of the general transport problem

$$H = \sum_m E_m a_m^+ a_m - \sum_{mn} J_{mn} a_m^+ a_n + \sum_q \hbar \omega_q b_q^+ b_q + \sum_{qm} \hbar \omega_q (\chi_m^q b_q^+ + \chi_m^{q\,*} b_q) a_m^+ a_m \; . \qquad (1)$$

(Notation here and throughout this paper is that of Ref. 11.) In this general form any coupling geometry and any bare band structure can be accommodated. While most of our results can be obtained for this general case, most of our specific results are obtained for the translationally-invariant acoustic chain model with a "contact" interaction, which results from choosing the phonon dispersion relation ω_q and the dimensionless coupling function χ_n^q to have the forms

$$\omega_q = \omega_B \sin(\tfrac{1}{2}|ql|) \; , \qquad \chi_m^q = \chi^q e^{-iqR_m} = \frac{-2i\,\chi \sin(ql)}{\sqrt{2NM\hbar \omega_q^3}} e^{-iqR_m} \; . \qquad (2)$$

Throughout its evolution in interaction with the host medium, the exciton is constituted always as a single quantum of electronic energy. (For clarity, we avoid here the added complications which arise from considering the exciton to approximate a single quantum of vibrational energy; see Lindenberg, et al., this volume.) In this sense, the exciton is permanently far removed from a correspondence limit, and maintains at all times a highly quantum mechanical character. This does not prevent excitons from emulating classical particles, particularly at high temperatures where stochasticity dominates both kinds of systems; however, it does affect the correspondence between excitons and classical fields. Since it is the latter which plays the central role in the theory of solitons, one must take great care in identifying excitons with solitons.

Assessment of the quantum or classical nature of the lattice dynamics presents quite a different situation. Since phonon number is not conserved, the level of excitation of a particular phonon mode is influenced both by the temperature and by the interaction of phonons with the exciton. These two influences produce different effects; in this paper we focus on the influence of the exciton-phonon interaction. Due to this interaction, the solid contains at all times a gas of virtual phonons correlated with the exciton (the "phonon cloud"). The number of phonons of the mode q excited by the exciton-phonon interaction is proportional to the square of the exciton-phonon coupling constant $|\chi^q|^2$. Depending on the strength and q-dependence of the coupling constant, it is possible for the lattice dynamics to have significant and simultaneous contributions from a variety of phonon number states, including states of high phonon number. It is thus natural to expect the dynamics of the lattice to have a *semi*classical character. To the extent that the exciton-driven lattice dynamics is classical, the reaction of the exciton dynamics to this classical behavior closes a feedback loop which can give rise to nonlinear dynamics.

The degree to which the exciton-phonon interaction can induce classical motion in the lattice is clearly dependent upon the strength of the exciton-phonon coupling; thus, a prerequisite for significant anharmonicity is that the exciton-phonon coupling should be sufficiently strong. However, since the emergence of nonlinear dynamics in this problem requires a closed feedback loop, there is a second condition that the exciton must be sufficiently mobile to react to the classical motion; this condition requires that the exciton motion be sufficiently adiabatic in the standard sense. These latter characteristics of the system are quantifiable, so it is reasonable to expect that a quantitative understanding of the classical and quantum natures of exciton transport might be possible.

It is commonly said that Davydov's equations for exciton transport are classical. This statement means nothing more or less than that the equations of motion are the same as the Heisenberg equations of motion for the elementary exciton and phonon operators, regarding the operators as c-numbers. The wave nature of the exciton has not been lost, however, and neither have certain minimum uncertainty properties of the phonon states, so the finality with which the statement is often made can be misleading.

Davydov's derivation is predicated on the assumption that a good choice of a trial state for the system is the product state [4]

$$|D_2(t)> \equiv \sum_m \alpha_m(t)\, a_m^+ \exp\left\{ \sum_q [\beta_q(t)b_q^+ - \beta_q^*(t)b_q] \right\} |0> .$$
(3)

A second assumption, that the c-number amplitudes in this trial state vector evolve according to the Hamilton equations

$$i\hbar \dot{\alpha}_n(t) = \frac{\partial <H>}{\partial \alpha_n^*(t)} = E\alpha_n(t) - \sum_m J_{nm}\alpha_m(t) + \sum_q \chi_n^q \hbar\omega_q [\beta_q^*(t) + \beta_{-q}(t)]\alpha_n(t)$$
(4a)

$$i\hbar \dot{\beta}_q(t) = \frac{\partial <H>}{\partial \beta_q^*(t)} = i\hbar\omega_q \beta_q(t) + \hbar\omega_q \sum_m \chi_m^q |\alpha_m(t)|^2$$
(4b)

allows one to obtain the well-known Davydov equations. The remarkable result, of course, is that these coupled evolution equations cause correlations to develop between exciton and phonon amplitudes which give rise to soliton solutions.

A number of analyses spread over the last few years have made clear that the first of these two assumptions is the more limiting one [12]. Since the very early days of polaron theory, it has been known that an essential criterion for the validity of such a product state is the adiabatic condition noted above [9]. Thus, it should be possible to observe the breakdown of the D_2 Ansatz state and the "classical" equations based on it, by looking to the diabatic ("anadiabatic") limit.

In our problem, the diabatic limit can be reached by allowing the exciton transfer integrals J_{mn} to vanish. In this case, we are fortunate to have exact solutions of the fully linear and fully quantum mechanical problem. Among these exact solutions are the small-polaron states

$$|\Psi(t)> \equiv \sum_m \Psi_m(t) A_m^+ |0> = \sum_m \Psi_m(t) a_m^+ \exp\left\{ \sum_q [\chi_m^q b_q^+ - \chi_m^{q\,*} b_q] \right\} |0> ,$$
(5)

where χ_m^q is the same exciton-phonon coupling constant appearing in the Hamiltonian. Since there is no exciton motion in this limit, the time dependence of $\Psi_m(t)$ is a simple phase precession at the frequency Ω_m, where

$$\hbar\Omega_m = E_m - \sum_q |\chi_m^q|^2 \hbar\omega_q .$$
(6)

The exact small polaron state differs from Davydov's Ansatz state in two ways: first, the phonon coherent state amplitudes in the small polaron state are time independent, and second, the small polaron state contains not one phonon coherent state, but many--one per lattice site--each bearing an explicit correlation with a distinct lattice site. The first distinction is *not* merely due to the fact that the exciton probability distribution is time independent in this limit; were this the case, Ω_m would retain a dependence upon the now time-independent exciton probability distribution [12]. Instead, Ω_m in the small polaron state is totally independent of the exciton probability distribution, i.e., is state independent, being completely determined by the Hamiltonian parameters only. Whatever the origin of this first distinction (we suggest it is a quantum effect), it points to a fundamental departure in the character of the quantum state from that of the Davydov D_2 Ansatz state. The second distinction is clearly a quantum effect, since it points to the inadequacy of a single coherent state in representing the state of given phonon normal mode.

Davydov later generalized his Ansatz state in such a way as to *formally* include the small polaron states [13]. This proposal consisted of replacing the single coherent state amplitude $\beta_q(t)$ with a set of N amplitudes $\{\beta_{qn}(t)\}$, each bearing an explicit correlation with the exciton at a distinct site of the lattice.

$$|D_1(t)> \equiv \sum_m \alpha_m(t)\, a_m^+ \exp\left\{ \sum_q [\beta_{qm}(t) b_q^+ - \beta_{qm}^*(t) b_q] \right\} |0> . \tag{7}$$

Davydov's original treatment of the D_1 Ansatz state considered the $\alpha_m(t)$ and $\beta_{qm}(t)$ to be dynamical variables to which Hamilton's equations of motion could be applied. It is at this point where we must depart from Davydov's theory. From the perspective of the variational principle, the more general D_1 Ansatz state should be superior to both the D_2 Ansatz state *and* the small polaron state as a tool for the study of the general quantum mechanical transport problem; it is now well known, however, that this expectation is not fulfilled if the Hamilton equation method is used [12, 14, 15]. To improve upon Davydov's theory of exciton transport, it is necessary to improve upon the methods used to obtain evolution equations from the Ansatz state.

Without repeating detailed analyses existing in the literature [16, 17], it is easy to grasp why such a generalization should be necessary. Since Hamilton's equations describe the evolution of classical dynamical variables, it is understandable that equations (4) for the D_2 Ansatz state should yield reasonable results. However, the generalization from N lattice variables $\{\beta_q(t)\}$ to N^2 lattice variables $\{\beta_{qn}(t)\}$ does not simply add $N(N-1)$ classical degrees of freedom -- the added variables describe *non*classical motion. While it is not necessary that the time development of these nonclassical variables be inordinately complex, it is unreasonable to expect this time development to be given by classical equations of motion.

To determine the optimal equations of motion we apply the time-dependent variational principle [18-20]

$$\delta \int_{t_1}^{t_2} dt <\psi(t) | i\hbar \frac{d}{dt} - H | \psi(t)> = 0 , \tag{8}$$

recently used by Skrinjar, *et al.* [16] and Zhang, *et al.* [17] to obtain optimized D_1 equations. Were (8) to be applied to a general state vector free from artificial constraints, (8) would be equivalent to the Schrödinger equation. However, when we choose a necessarily limited class of trial states, we implicitly constrain the variation so that the system of evolution equations which result may not enjoy full agreement with the Schrödinger equation despite the consistency of the method with quantum mechanics. Clear measures of the accuracy of these optimized equations do not yet exist, though it is known from formal considerations that their solutions deviate from the true quantum solutions in the general case [21]. It is important, therefore, to understand how (8), *as an approximation scheme*, minimizes the divergence of trial solutions from exact, but unknown quantum mechanical solutions at every instant of time.

Consider the differential change $(i\hbar)^{-1} H |\psi(t)> dt$ which occurs when the trial state vector is propagated forward according to the *exact* quantum dynamics from time t to time $t + dt$. While this differential can be computed exactly, the differential equations we derive cannot reproduce this differential change precisely, owing to the limited form of our trial state. What the time-dependent variational principle does for us is insure that the differential change delivered by the approximate evolution equations produces a minimal deviation from the exact quantum state at the considered instant of time.

This minimization is at once local and global. It is local in that for t_2 close to t_1, the minimization is essentially a purely differential condition as noted above; however, such a differential condition, even though imposed uniformly at every instant of time, does not guarantee that the errors accumulated in propagating for a long interval of time remain small. That is, it is possible that the exact and approximate evolutions which follow from a particular initial condition may diverge significantly at long times despite the local accuracy obtained in this way. The minimization is global, however, in that for t_2 and t_1 which are well separated, it is the accumulated deviation which is minimized. Since the variation is carried out for t_2 and t_1 which are arbitrary, the result is a compromise between these two minimizations.

In the quantum Hilbert space, as in classical phase space, a cluster of initial states sweep out a set of non-crossing trajectories which may be viewed as a flow. Approximate trajectories will, in general, intersect the exact trajectories, evolving away from the particular quantum trajectory attached to a particular initial condition; however, the time-dependent variational principle causes the angle of intersection between the exact and approximate trajectories (a local property) to be minimized at every point along the approximate trajectory (a global property). In this way the variational principle produces approximate trajectories which "go with the flow" as faithfully as the limited form of the trial state allows.

It is easy to see that approximate trajectories should remain close to the true quantum trajectories in those regions of Hilbert space in which our choice of trial states is fortuitously good; however, it is natural to expect that in certain regions of Hilbert space the quantum evolution may be particularly incompatible with the approximate evolution. This incompatibility should be manifested in the appearance of a substantial component of the approximate evolution transverse to the quantum flow. This has the consequence that approximate trajectories should be repelled from regions of Hilbert space in which agreement is poor. Thus, we can visualize our approximate evolution as a flow in Hilbert space which generally flows around or moves rapidly through certain repelling regions and tends to eddy or dwell in certain attracting regions. The best solutions are those which dwell the longest in the attracting regions; i.e., those with least average transverse intersection with the true quantum trajectories.

The Davydov equations (4), despite their complexity, are conceptually simple and intuitively appealing. The optimized D_1 equations [16, 17], while doubtless superior, so redouble this complexity that in practical terms they have not yet taken us very far toward an improved understanding of the basic problem. It would be useful to establish an intermediate picture which improves on Davydov's D_2 equations in perhaps a less comprehensive way, but which retains their relative simplicity.

To motivate such an intermediate picture, we return to the observation that Davydov equations can be obtained by regarding the lattice (only) as a system of classical oscillators; one might, therefore, consider the smallness of Planck's constant as a formal indicator of one's proximity to the correspondence limit and hence to their accuracy. Since Planck's constant is not a physical variable, such considerations can be misleading if not handled with care; however, proceeding with care, we can obtain some useful information and be alerted to some difficulties with the conventional wisdom. Our basic observation is that the number of quanta in a particular mode of vibration is given (approximately) by

$$number\ of\ quanta \propto \frac{1}{\hbar}\ F\,[\ energy\ in\ the\ excited\ state\ ]\,. \tag{9}$$

For a given energy, the number of quanta in a mode is inversely related to Planck's constant, so that the correspondence limit can be formally implemented by allowing $\hbar$ to vanish at a fixed energy. The "classical" nature of Davydov's equations thus requires that each normal mode of vibration participating in the evolution be highly excited. This poses an immediate problem since we cannot conclude this from Davydov's equations with any generality; indeed, one can compute the lattice energy in the soliton state and find a number of physically meaningful regimes in which the relevant count of vibrational quanta is quite low. One such regime is the weak coupling regime; another is the diabatic or small-J regime. We focus on the latter.

In the diabatic limit, as noted above, we have exact solutions which represent the quasiparticles known as small polarons. When the small polaron effective mass is large but finite, the same unitary transformation that diagonalizes the $J = 0$ Hamiltonian [22] is commonly and reasonably assumed to identify good basis states for use in perturbation theory [23]. In the small polaron basis the transfer integrals J_{mn} are renormalized to smaller values $\bar{J}_{mn}$, reflecting the lower mobility of the correlated entity consisting of the exciton and the small-polaron lattice deformation or "phonon cloud". What is significant for our present discussion is that this renormalization goes roughly as[†]

$$\bar{J}_{mn} = J_{mn}\ \exp\left\{ -C < number\ of\ quanta\ in\ the\ phonon\ cloud > \right\} \tag{10}$$

where C is a "classical" constant. The classical or quantum nature of the lattice vibrations comprising the phonon cloud thus shows up in the narrowing of the quasiparticle energy band. If we try to apply a correspondence principle here, we would conclude that the quasiparticle energy band would have to be greatly reduced in a classical regime. This appears to be in contradiction to our previous discussion of

---

[†] It is important to note that the number of quanta referred to here is not $\sum_q |\chi^q|^2$, which is divergent, but $4\sum_q |\chi^q|^2 \sin^2(\tfrac{1}{2}ql)$, which is convergent.

the Davydov equations, where one finds that the assumption of classical lattice dynamics involves no explicit renormalization of the quasiparticle energy band (cf. (4a)).

This is a clear example of the tension which exists between the concept of the small polaron and the concept of the Davydov soliton. The difficulty in this case appears to lie in the compromise between the classical and quantum natures of the vibrations engaged in the correlations defining the polaron; the small polaron concept is not clearly applicable in such cases where the exciton-phonon correlation is an essentially classical phenomenon, and the Davydov soliton appears similarly unsuited to the complementary quantum phenomenon.

To separate that which is most quantum mechanical in our treatment of the D_1 state from that which is most classical, one might proceed as follows. First, we observe that there must exist N functionals of the N^2 amplitudes $\{\beta_{qn}(t)\}$ which *do* evolve very much like classical variables; these are the complex mode amplitudes one obtains by forming the expectation value of the annihilation operator

$$<b_q>(t) = <D_1(t)|\,b_q\,|D_1(t)> = \sum_m \beta_{qm}(t)\,|\alpha_m(t)|^2 \, . \tag{11}$$

The time development of these quantities, like that of all quantum expectation values corresponding to classical generalized coordinates, is subject to Ehrenfest's theorem [24], and thus can be expected to approximate classical motion. On the other hand, the N^2 quantities $\Delta\beta_{qm}(t)$ defined by

$$\Delta\beta_{qm}(t) = \beta_{qm}(t) - <b_q>(t) \tag{12}$$

describe the spatial and temporal fluctuations of the quantum trajectory about this classical motion. A brief examination of (11) and (12) shows that in any reasonably general case the quantum fluctuations $\Delta\beta_{qm}(t)$ must be both time and space dependent. To complete the separation, we can apply the variational principle (8) subject to the constraint that the N functionals $<b_q>(t)$ satisfy Hamilton's equations. The effect of the constraint is to reduce the number of variables describing quantum fluctuations from N^2 to $N(N-1)$. The advantage of such an approach is to allow one to develop an approximate description of the quantum evolution while preserving the character of the classical motion.

Recently, Brown and Ivic [25] have developed a simpler approximation scheme which, while more limited in scope, is directed to the same end. In the place of the quantum expectation value (11), an arithmetic average is used to define a semiclassical variable $\tilde{\beta}_q(t)$

$$\tilde{\beta}_q(t) = \frac{1}{N}\sum_m \beta_{qm}(t) \, . \tag{13}$$

Next one assumes that the time development of the corresponding quantum variables can be neglected, such that we can define N^2 quantities $\tilde{\beta}_{qm}$ by the relation[†]

$$\tilde{\beta}_{qm} = \beta_{qm}(t) - \tilde{\beta}_q(t) \tag{14}$$

in analogy with (12). This is consistent with studies of the diabatic limit, in which only the spatial dependence of $\tilde{\beta}_{qm}$ is of crucial importance. Since the "true" classical variables are given by

$$<b_q>(t) = \sum_m \tilde{\beta}_{qm}\,|\alpha_m(t)|^2 + \tilde{\beta}_q(t) \, , \tag{15}$$

it is clear that this approximate decomposition is equivalent to the assumption that the "true" quantum fluctuations $\Delta\beta_{qn}(t)$ take the form

$$\Delta\beta_{qn}(t) = \tilde{\beta}_{qn} - \sum_m \tilde{\beta}_{qm}\,|\alpha_m(t)|^2 \, . \tag{16}$$

[†] The constraint which implements the reduction from N^2 to $N(N-1)$ independent quantum variables is already implicit in (14) since $\sum_m \tilde{\beta}_{qm} = 0$ for all q.

The special choice of a time-independent quantum variable $\tilde{\beta}_{qm}$ is thus consistent with the observation above that the "true" quantum fluctuations must be time and space dependent in a general case. Thus, while more limited in scope and based on "convenient" variables, the approach of Brown and Ivic addresses the same issue as the more complex albeit more desirable approach noted above, that of achieving a tangible separation of the classical and quantum attributes of the system evolution.

Having motivated our choice of variables, it is well to define in a compact and formally precise way the variational trial state to be used. To define these states we denote with tildes quantities defined in a basis of mixed exciton-phonon states; e.g. [22],

$$\tilde{a}_m \equiv U a_m U^+ = a_m \exp\left\{ \sum_q [\tilde{\beta}_{qm} b_q^+ - \tilde{\beta}_{qm}^* b_q] \right\} , \qquad \tilde{b}_q \equiv U b_q U^+ = b_q + \sum_q \tilde{\beta}_{qm} a_m^+ a_m , \qquad (17)$$

$$U \equiv \exp\left\{ -\sum_{qm} [\tilde{\beta}_{qm} b_q^+ - \tilde{\beta}_{qm}^* b_q] a_m^+ a_m \right\} . \qquad (18)$$

In terms of these operators we define the trial state vector

$$|\tilde{D}(t)> \equiv \sum_m \tilde{\alpha}_m(t) \, \tilde{a}_m^+ \exp\left\{ \sum_q [\tilde{\beta}_q(t) \tilde{b}_q^+ - \tilde{\beta}_q^*(t) \tilde{b}_q] \right\} |0> . \qquad (19)$$

By transforming back into the basis of bare states, our $\tilde{D}$ states can be put in the form of D_1 states with the identifications

$$\alpha_n(t) = \tilde{\alpha}_n(t) \exp\left\{ -\frac{1}{2} \sum_q [\tilde{\beta}_{qn}^* \tilde{\beta}_q(t) - \tilde{\beta}_{qn} \tilde{\beta}_q^*(t)] \right\} , \qquad \beta_{qn}(t) = -\tilde{\beta}_{qn} + \tilde{\beta}_q(t) . \qquad (20)$$

Like the general D_1 state, the $\tilde{D}$ states contain as special cases all D_2 states (set $\tilde{\beta}_{qn} = 0$) and the exact $J_{mn} = 0$ small polaron states (set $\tilde{\beta}_q(t) = 0$, $\tilde{\beta}_{qn} = \chi_n^q$). In this respect the $\tilde{D}$ states have the basic properties desired of a unifying theory of polaron and soliton dynamics. Of course, the Hamiltonian is not diagonal in these states when $J_{mn} \neq 0$; however, the interaction between dressed excitons ($\tilde{a}$, $\tilde{a}^+$) and dressed phonons ($\tilde{b}$, $\tilde{b}^+$) is generally weaker than the interaction between bare excitons (a, a^+) and bare phonons (b, b^+). Because of this weaker interaction, a factored dressed state such as (19) affords a more plausible description of the microscopic dynamics than does a factored bare state such as (3) in those parameter regimes where quantum effects are important.

Thus, applying (8) we obtain the formal evolution equations

$$i\hbar \dot{\tilde{\alpha}}_n(t) + i\hbar \sum_q [\tilde{\beta}_{qn} \dot{\tilde{\beta}}_q^*(t) - \tilde{\beta}_{qn}^* \dot{\tilde{\beta}}_q(t)] \tilde{\alpha}_n(t) = \frac{\partial <H>}{\partial \tilde{\alpha}_n^*(t)} , \qquad (21a)$$

$$i\hbar \dot{\tilde{\beta}}_q(t) - i\hbar \sum_m \tilde{\beta}_{qm} \frac{d}{dt} |\tilde{\alpha}_m(t)|^2 = \frac{\partial <H>}{\partial \tilde{\beta}_q^*(t)} , \qquad (21b)$$

in which the energy functional $<H>$ is the expectation value of H in the $\tilde{D}$ state. This already offers a significant simplification over the original D_1 equations obtained by Zhang $et\ al.$ and Skrinjar $et\ al.$; however, in order to further simplify our analysis, we assume a particular form for the dressing parameters [26-28]

$$\tilde{\beta}_{qn} = \delta \chi_n^q . \qquad (22)$$

This actually involves two assumptions at once. The first assumption is that the spatial variation of $\tilde{\beta}_{qn}$ affects only the phase of the dressing parameters; the second is that the degree of dressing, expressed as a fraction of the small polaron value, is the same for each mode [26-28]. Both assumptions are motivated by the demands of the diabatic limit, in which such a form must hold. The only constraint provided by the adiabatic limit is that the $\tilde{\beta}_{qn}$ should vanish. The assumption (22) meets these conditions in the simplest possible way; the limit $\delta = 1$ allows (19) to include the highly quantum mechanical small polaron states, and when $\delta = 0$, (19) reduces to the semiclassical D_2 Ansatz state. As a cautionary note we observe that, even in this simple approximation, recovery of small polarons and other states having significant quantum character is *nontrivial*, since dialing up an interesting value of δ does not transparently determine the magnitude, time dependence, and implicit spatial structure of the classical lattice variables $\{\tilde{\beta}_q(t)\}$. As will be seen presently, the intuitive exercise of extrapolating from the well-known diabatic and adiabatic limits can be misleading.

This simplification streamlines calculations significantly and allows the energy functional $<H>$ to be expressed in the more convenient form

$$<H> = \sum_m [E - \delta(2 - \delta)E_b]|\tilde{\alpha}_m(t)|^2 - \sum_m \tilde{J}\tilde{\alpha}_m^*(t)[\tilde{\alpha}_{m+1}(t) + \tilde{\alpha}_{m-1}(t)]$$

$$+ \sum_q \hbar\omega_q |\tilde{\beta}_q(t)|^2 + (1 - \delta)\sum_{ql} \hbar\omega_q [\chi_m^q\tilde{\beta}_q(t)^* + \chi_m^{q\,*}\tilde{\beta}_q(t)]|\tilde{\alpha}_m(t)|^2 , \tag{23}$$

$$E_b = \sum_q |\chi^q|^2\hbar\omega_q , \quad \tilde{J} = Je^{-\delta^2 S} , \quad S = 4\sum_q |\chi^q|^2\sin^2(\tfrac{1}{2}ql) . \tag{24}$$

in which E_b is the small polaron binding energy and $\tilde{J}$ is the renormalized transfer matrix element *partially* reduced from the bare value J by interactions with phonons. (It is important to note that the small polaron binding energy (24) and the soliton binding energy from Davydov's theory differ significantly in origin and value; for the acoustic chain model we consider, the latter is given in terms of the former as $E_b^2/3J$.) Integrating (21b) and using the result to eliminate the explicit phonon variables $\tilde{\beta}_q(t)$, we obtain an integro-differential equation for the dressed-exciton probability amplitudes in the form

$$i\hbar\dot{\tilde{\alpha}}_n(t) = [E - \delta(2 - \delta)E_b]\tilde{\alpha}_n(t) - \tilde{J}[\tilde{\alpha}_{n+1}(t) + \tilde{\alpha}_{n-1}(t)]$$

$$- (1 - \delta)^2\sum_m K_{mn}(0)|\tilde{\alpha}_m(t)|^2\tilde{\alpha}_n(t) + (1 - \delta)\sum_m K_{mn}(t)|\tilde{\alpha}_m(0)|^2\tilde{\alpha}_n(t)$$

$$+ \int_0^t d\tau\sum_m K_{mn}(t - \tau)\frac{d}{d\tau}|\tilde{\alpha}_m(\tau)|^2\tilde{\alpha}_n(t) + \tilde{f}_n(t)\tilde{\alpha}_n(t) , \tag{25}$$

where

$$K_{mn}(t) = 2\sum_q \chi_m^q\chi_n^{q\,*}\hbar\omega_q\cos\omega_q t , \quad K_{mn}(0) = \frac{\chi^2}{w}(\delta_{mn+1} + 2\delta_{mn} + \delta_{mn-1}) , \tag{26}$$

$$\tilde{f}_n(t) = \sum_q \hbar\omega_q (\chi_n^q\tilde{\beta}_q^*(0)e^{i\omega_q t} + \chi_n^{q\,*}\tilde{\beta}_q(0)e^{-i\omega_q t}) . \tag{27}$$

This is the fundamental result of our time-dependent analysis, which, subject only to the simplifying restriction (22), is the general and exact consequence of applying the variational principle (8) to the trial state (19). This equation reflects a dynamic balancing of the system's soliton and small polaron characteristics. The detail of this balancing is perhaps non-intuitive, since in (22) χ^q is scaled by δ, while in (25) χ^q appears scaled by $\sqrt{\delta(2 - \delta)}$, δ, $(1 - \delta)$, $\sqrt{(1 - \delta)}$, and unity. The sense in which there is balance at all can be seen by considering the case of a static, preformed soliton in the strong-coupling regime, for which (25) reads

$$i\hbar\dot{\tilde{\alpha}}_n(t) \approx \left[E - \delta(2 - \delta)E_b - (1 - \delta)^2\sum_m K_{mn}(0)|\tilde{\alpha}_m(t)|^2 \right] \tilde{\alpha}_n(t) . \tag{28}$$

(This equation is approximate, since to be strictly valid it is necessary that $\delta = 1$.) Two energy-lowering mechanisms are clearly at work: the tendency of the system to form small polarons makes a linear, time-independent contribution proportional to the small polaron binding energy E_b; the tendency of the system to form solitons makes a nonlinear and generally time-dependent contribution which depends on the manner in which the excitation is distributed in space. The balance between these two tendencies is perhaps most clearly reflected in the sum rule

$$\delta(2 - \delta) + (1 - \delta)^2 = 1 , \tag{29}$$

which holds between the coefficients of the small polaron binding energy E_b and the standard cubic nonlinearity. The message of crucial importance here is that the two limiting natures of the general polaron state are not independent, but coexist in a *quantitative* balance.

If we specialize to the *large polaron limit* by setting $\delta = 0$, the small polaron binding energy E_b drops out of the equation and $\bar{J}$ reverts to the bare value J, exactly recovering Davydov's D_2 theory in the form given by Wang, *et al.* [11]. This was to be expected, of course, since our $\bar{D}$ Ansatz state reduces to Davydov's D_2 Ansatz state in this limit; however, if we specialize to the *small polaron limit* by setting $\delta = 1$, we obtain an altogether new equation of motion [29]

$$i\hbar\dot{\bar{\alpha}}_n(t) = (E - E_b)\bar{\alpha}_n(t) - \bar{J}[\bar{\alpha}_{n+1}(t) + \bar{\alpha}_{n-1}(t)]$$

$$+ \int_0^t d\tau \sum_m K_{mn}(t - \tau)\frac{d}{d\tau}|\bar{\alpha}_m(\tau)|^2\bar{\alpha}_n(t) + \bar{f}_n(t)\bar{\alpha}_n(t) , \tag{30}$$

in which $\bar{J}$ assumes the fully-reduced small-polaron value. This nonlinear evolution equation has a number of interesting properties. If we consider the special solution

$$\bar{\alpha}_n(t) = \bar{u}(k)e^{i(kR_n - [\bar{E}(k)/\hbar]t)} , \qquad \bar{E}(k) = (E - E_b) - 2Je^{-S}\cos kl , \tag{31}$$

with $\bar{\beta}_q(0) = 0$ (zero temperature), the equations of motion (30) automatically linearize, and we find that the Bloch states of the small polaron energy band are exact stationary solutions of the approximate evolution equation, confirming the agreement of our results with small polaron theory. On the other hand, it is clear from the form of (30) that *nonstationary* solutions of (30) must have nonlinear character, implying that polaron wave packets cannot propagate in the simple dispersive manner characteristic of energy band theory *even in the small polaron limit*.

Passing to the continuum limit, seeking special solutions having the D'Alembert property $|\alpha(y,\tau)| = |\alpha(y - v\tau)|$, and making a special choice of $\bar{\beta}_q(0)$ (in accordance with the corresponding discussion in [11]), we find a nonlinear Schrödinger equation

$$i\hbar\dot{\bar{\alpha}}(x,t) = -\frac{\hbar^2}{2\bar{m}}\frac{\partial^2}{\partial x^2}\bar{\alpha}(x,t) + \bar{E}(0)\bar{\alpha}(x,t) - \bar{G}(v)|\bar{\alpha}(x,t)|^2\bar{\alpha}(x,t) , \tag{32}$$

$$\bar{m} = \frac{\hbar^2}{2Jl^2}e^{\delta^2 S} , \qquad \bar{E}(0) = E - \delta(2 - \delta)E_b - 2Je^{-\delta^2 S} , \tag{33}$$

in which $\bar{m}$ is the linear effective mass and $\bar{E}(0)$ the bottom of the linear energy band. $\bar{G}(v)$ plays the same role as the velocity-dependent nonlinearity parameter in Davydov's theory, but generally has the reduced value

$$\bar{G}(v) = \frac{4\chi^2 l}{w}\left[(1 - \delta)^2 + \frac{v^2}{v_a^2 - v^2} \right] . \tag{34}$$

When $\delta = 0$, (32) is clearly Davydov's nonlinear Schrödinger equation and its soliton solution is the Davydov soliton. When $\delta \neq 0$, (32) is *not* Davydov's nonlinear Schrödinger equation; however, the different underlying physical picture is revealed only through a renormalization of the coefficients in this equation, allowing well-known results to be simply adjusted to our present purpose (See, e.g., [30]). Thus, we find soliton solutions of the form

$$\alpha(x,t) = \left[\frac{\kappa}{2}\right]^{\frac{1}{2}} \frac{e^{-i[\tilde{E}(0)/\hbar]t} \, e^{i(kx - \omega t)}}{\cosh[\kappa(x - \upsilon t)]} \, , \tag{35}$$

where

$$\hbar k \equiv \tilde{m}\upsilon \, , \quad \hbar\kappa \equiv \frac{\tilde{m}\tilde{G}(\upsilon)}{2\hbar} \, , \quad \hbar\omega \equiv -\frac{\tilde{m}\tilde{G}(\upsilon)^2}{8\hbar^2} + \tfrac{1}{2}\tilde{m}\upsilon^2 \, . \tag{36}$$

In keeping with the distinctions which can be drawn between the small and large polaron regimes, this nonlinear Schrödinger equation and its soliton solution in the small polaron limit ($\delta = 1$) differ from corresponding results of Davydov's theory ($\delta = 0$) in a number of ways. In the small polaron limit, the plane wave solutions of (32) are not bare exciton Bloch states as in Davydov's theory, but are instead small polaron Bloch states which already reflect an increased effective mass ($\tilde{m} > m$) and energy lowering ($\tilde{E}(0) < E(0)$) which results from the dressing of the bare exciton. In compensation for this renormalization of the exciton band structure, the nonlinear term in (32) is substantially weakened relative to the nonlinearity in Davydov's theory; indeed the linearity of the zero-velocity equation implies the *absence* of a static soliton solution. Slow soliton solutions of (32) exist for all values of δ; however, slow solitons in Davydov's theory have a limiting, finite size (given by the static soliton solution) while slow soliton solutions of (32) in the small polaron limit broaden into plane waves (small polaron Bloch states) as the group velocity goes to zero. Slow solitons in Davydov's theory are *lower* in energy than the plane wave solutions of the same nonlinear Schrödinger equation (excitons), and are separated from these plane waves by a finite energy gap. On the other hand, slow soliton solutions of (32) in the small polaron limit are *higher* in energy than the plane wave solutions of the same nonlinear Schrödinger equation (small polarons) and deform continuously into these plane waves with decreasing velocity without an energy gap. Finally, the soliton effective mass (cf. (37)) in Davydov's theory differs from (is greater than) the effective mass of the plane wave solutions (excitons), while the soliton solutions of (32) in the small polaron limit have the same effective mass as the plane wave solutions (small polarons).

When δ takes on values intermediate between 0 and 1, state characteristics intermediate between those of the small polaron and Davydov soliton are found.

The soliton mass can be given in the form

$$\tilde{m}_{sol} = \tilde{m}\left[1 + \frac{1}{6}\left[\frac{\tilde{G}(0)^2}{\hbar^2\upsilon_a^2}\right]\right] = m\, e^{\delta^2 S}\left[1 + (1 - \delta)^4\frac{1}{6}\left[\frac{G(0)^2}{\hbar^2\upsilon_a^2}\right]\right] \, . \tag{37}$$

The nonlinear correction to the linear mass ($\tilde{m}$) is strongly dependent on the dressing fraction; for example, $\delta = 1/2, 2/3, 9/10$ correspond to reductions of the nonlinear mass correction by factors of 16, 81, and 10,000, respectively. Since, in quantum mechanics, the effective mass of a quasiparticle is its most essential characteristic, this points to potentially serious problems with a number prevalent assumptions about the relevance of anharmonicity to the behavior of the polaron.

The width (in units of the lattice constant) of a static soliton according to (36) is given by

$$\lambda = (\kappa l)^{-1} = \frac{J}{E_b}\frac{e^{-\delta^2 S}}{(1 - \delta)^2} \, . \tag{38}$$

This presents us with another interesting example of how failing to observe the balance between the polaron's competing natures can lead to erroneous conclusions. Davydov's result is $\lambda = J/E_b$. If one incorporates a renormalized energy bandwidth into the nonlinear Schrödinger equation *without* making a compensating modification of the cubic term, one arrives at the result $\lambda \propto \exp(-\delta^2 S)$, leading to the conclusion that increasing the dressing fraction increases the efficacy of the nonlinear terms in localizing the excitation. The balanced conclusion, however, is quite the opposite. Provided that S is not too large,[†] increasing the dressing fraction weakens nonlinearity, resulting in a more delocalized excitation.

[†] When S exceeds a critical value, $\lambda(\delta)$ develops structure which complicates this discussion, but does not alter the conclusion.

Thus far we have treated δ as a free parameter which we are at liberty to manipulate; however, surely, for a given set of system parameters, there should be an optimal choice for δ corresponding to a minimum energy state. For such a choice of δ the equations of motion (25) describe the evolution of the system near this minimal state. Estimations of the optimal dressing fraction based on mean-field energy minimization arguments have been previously considered in the context of linear theory by Toyozawa [26], Yarkony and Silbey [27], and Venzl and Fischer [31] and in the context of Davydov's theory by Venzl and Fischer [31], Alexander and Krumhansl [32] and Sataric *et al.* [28]. Our present calculations generalize these analyses by considering the full nonlinear dependence of the system energy on exciton probability amplitudes, and by identifying the optimal states with solutions of the equations of motion (25).

Before discussing the general dependence of the dressing fraction on the important system parameters, we note that by means of a common mean-field analysis, it is possible to account for some of the effects which finite temperatures have on the dynamics; for our present purposes, the principal consequence of this approach is that the parameter S acquires a temperature dependence

$$S \to S(T) = 4\sum_q |\chi^q|^2 \sin^2(\tfrac{1}{2}ql)\coth(\hbar\omega_q/2k_B T) \ . \tag{39}$$

In what follows it is convenient to define the dimensionless quantity

$$B(T) = \frac{2J}{E_b}S(T) = \frac{8}{3\pi}\frac{2J}{\hbar\omega_B}\frac{S(T)}{S(0)} \ . \tag{40}$$

Since $B(T)$ is independent of the exciton-phonon coupling constant χ and $S(T)$ is independent of the resonance integral J, it is convenient to consider variations in $B(T)$ to be variations in J relative to a fixed phonon bandwidth $\hbar\omega_B$, and variations in $S(T)$ to be variations in the polaron binding energy E_b relative to $\hbar\omega_B$. Thus $B(T)$ is a measure of adiabaticity, and $S(T)$ is a measure of coupling strength.†

In terms of the dimensionless control parameters $B(T)$ and $S(T)$, the result of the linear theory of Toyozawa [26] and Yarkony and Silbey [27] can be represented

$$\delta = \frac{1}{1 + B(T)e^{-\delta^2 S(T)}} \quad provided \ that \quad \delta^2(1-\delta) \leq \frac{1}{2S(T)} \ . \tag{41}$$

This scaling relation admits a phase diagram exhibiting characteristics of a first-order phase transition [26]. The dotted curves in Figure 1 show the dependence of δ on $S(T)$ for various values of $B(T)$ as given by the "linear" formula (41). The formula analogous to (41) which results from minimizing the total average energy including nonlinear corrections is quite cumbersome and so we do not display it here; however, representative results have been determined numerically and are indicated by the solid curves in Figure 1. Clearly, the nonlinear corrections to the "linear" result based on (41) are quite small. Our calculations thus support the conventional picture of self-trapping as a robust one, with good quantitative agreement over a large region of the phase diagram.

Combining the two relations (41) with the width function (38), one can show that the transition point $S^*[B(T)]$, determined largely by adiabaticity, corresponds to a soliton width which is of the order of one lattice constant. The soliton width increases both as the coupling strength is decreased from S^* below the transition *or* increased from S^* above the transition. The *critical point* (using the approximate formula (41), $B_c = \tfrac{1}{2}e^{3/2}$, $S_c = 27/8$, $\delta_c = 2/3$) identifies a line separating an adiabatic $(B(T) > B_c)$ from a diabatic $(B(T) < B_c)$ regime. The self-trapping transition occurs only in the adiabatic regime; in the diabatic regime, soliton widths are typically significantly larger than a lattice constant for all coupling strengths.

At this point it is well to stand back and view our problem from some distance in order not to stumble over our own jargon. It is of relatively minor significance that we can find, at every point of

---

† While these temperature dependences are the result of the usual mean-field approach, one can find clear discrepancies between the detail of this temperature dependence and the results of Quantum Monte Carlo simulation. The interested reader may wish to compare these mean-field estimates with QMC results in Fig. 5 of Wang *et al.*, this volume. This discrepancy will be the subject of future study.

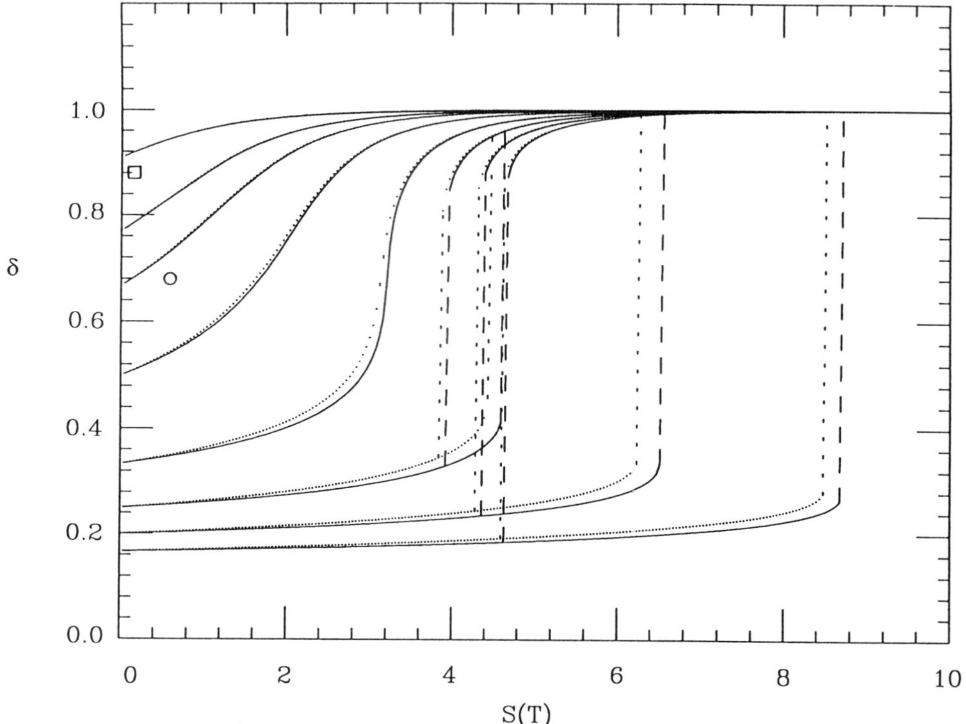

Fig. 1 Dressing fraction / Parameter space. Dotted lines ($\cdots\cdots$), $\delta[B(T),S(T)]$ resulting from the formula (41) which neglects nonlinear corrections to the energy. Solid lines (———), $\delta[B(T),S(T)]$ which results from including nonlinear corrections to the average energy. Each dotted/solid pair of curves corresponds to a fixed value of $B(T)$; from top to bottom: $B(T) = 0.1, 0.3, 0.5, 1.0, 2.0, 3.0, 4.0, 5.0$. Short dash (– – – –) and long dash (— — — —) lines connect the upper and lower branches of the linear and nonlinear $\delta[B(T),S(T)]$ curves respectively, forming hysteresis loops. The points marked $\square$ and $\bigcirc$ indicate the dressing fraction predicted for the α-helix at $T = 0K$ and $T = 300K$ respectively, using the standard parameters in Table 1.

this phase diagram, soliton solutions of a nonlinear Schrödinger equation which help us characterize the states of the quantum system. Even in the limit $\delta = 0$, the finite effective mass we find for the soliton surely signals the eventual delocalization of any soliton-like initial condition in the true quantum system. What is of greater importance are the estimates of intrinsic quasiparticle properties such as the polaron effective mass which we are able to determine with the assistance of soliton solutions. The theory of small polarons gives us an effective mass which increases exponentially with coupling strength as measured by S (set $\delta = 1$ in (37)), while Davydov's theory gives us an effective mass which increases only quadratically with S (set $\delta = 0$ in (37)). The unified theory of Brown and Ivic gives us a polaron effective mass with the exotic structure sketched in Figure 2. The mass catastrophe which appears here is just another manifestation of the self-trapping transition; a somewhat modified folding of the mass surface is implicit in the earlier linear approaches of Toyozawa [26] and Yarkony and Silbey [27]. The nonlinear corrections to such linear estimations of the polaron effective mass are evident primarily on the lower sheet of the mass surface, where the linear approach would lead to a recovery of the free exciton effective mass in the adiabatic limit; the corrected mass remains enhanced relative to that of the free exciton in the adiabatic limit.

This illustration makes it particularly evident that the different names we give to the polaron in different parameter regimes do not identify different quasiparticles, but refer instead to the different theoretical patches which we attempt to quilt together into a comprehensive understanding of *one*

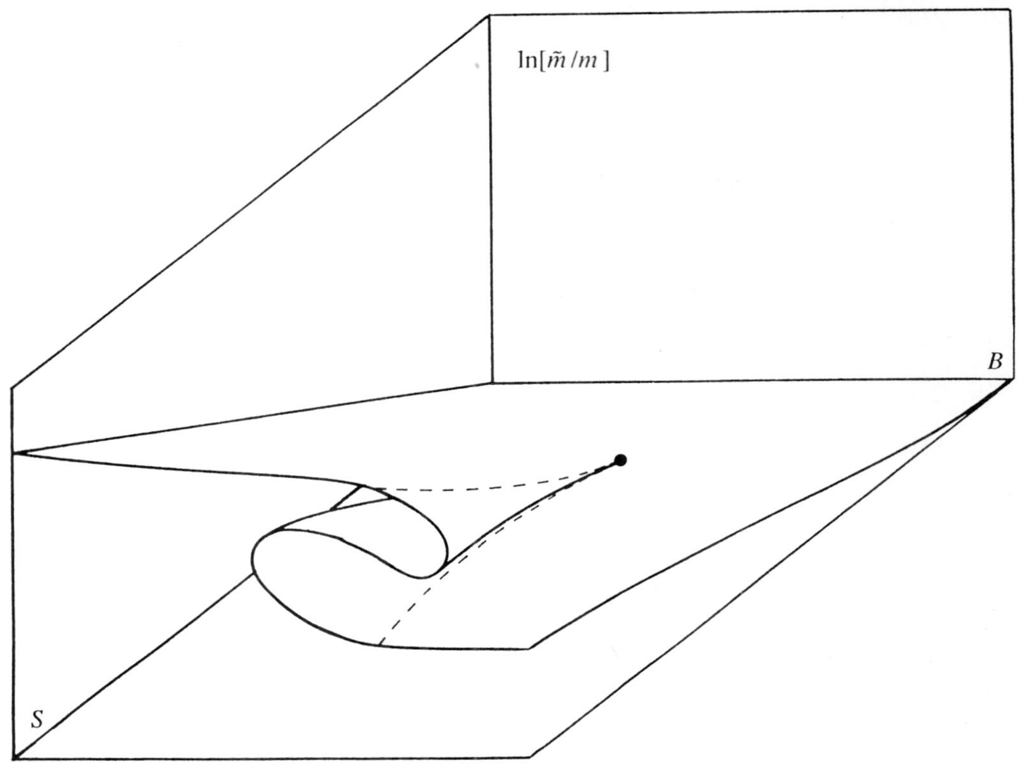

Fig. 2   Polaron effective mass (schematic).

quasiparticle. The situation is quite analogous to the problem of analytic continuation: Many analytic functions with singularities in the finite complex plane require for their complete specification a multiplicity of power series representations, each valid in a different region of the complex plane; analytic continuation is proven by showing the equality of these different representations along qualifying segments of boundary between the contiguous or overlapping representations. Such disputes as arise from time to time in polaron theory are occasionally due to a blurring of the boundaries between the varied theoretical descriptions which appear to be applicable to a particular circumstance. The theory of Brown and Ivic does not transcend these difficulties by any means, being hindered by its approximations like every other serious theory; however, the theory provides, as a by-product, a set of natural boundaries in parameter space which may be useful in clarifying the polaron lexicon (See Figure 3).

The first natural boundary is the line $B = B_c$ which, as noted above, separates the adiabatic regime $(B > B_c)$ from the diabatic regime $(B < B_c)$. In order to have a convenient name for this curve, we shall call it the "critical adiabat". Since either the adiabaticity parameter or its reciprocal may be used as an expansion parameter, this line should mark a real division between two classes of perturbation theory.

Specification of the second natural boundary is a bit more involved. Consider variations of S or δ subject to the constraint that the adiabaticity parameter B is held fixed. On such an "adiabat", the width of the soliton is infinite in both the weak and strong coupling limits and takes its minimum value at some intermediate value of coupling. Our second natural boundary is the locus of points in parameter space identifying these minimum-width states. On the weak-coupling side of the critical point, this locus of points is a simple curve which terminates at the critical point.† On the strong-coupling side of the critical point, this locus of points is the two-valued "coexistence curve" which

---

† Actually, the curve described crosses the coexistence curve slightly away from the critical point. At present, we are assuming that this misfit is due to the approximate nature of our analysis.

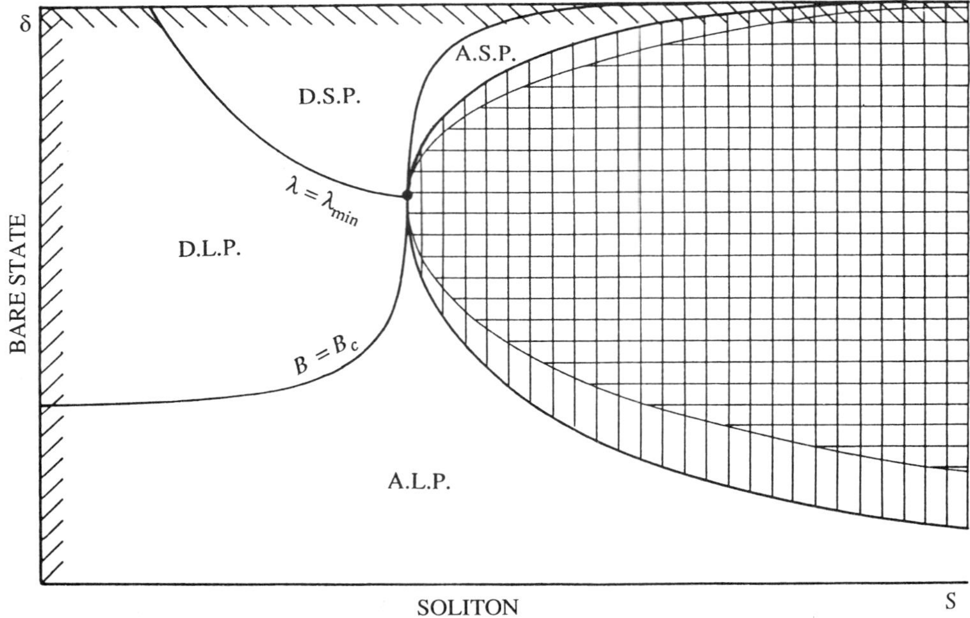

Fig. 3   Polaron lexicon: $\delta = 1$, *strict* diabatic limit = "Self-Trapped State". $\delta = 0$, *strict* adiabatic limit = "Soliton". $S = 0$, decoupled particle and phonon systems = "Bare State". $\lambda = \lambda_{min}$, locus of minimum width states = transition curve. $B = B_c$ = critical adiabat. Cross-hatched (###) region, unstable states. Semi-hatched (|||) region, metastable states. Backslashed (\) region, states self-trapped by virtue of disorder-induced localization. Slashed (///) region, states lacking classical particle-phonon correlations.

bounds the domain of metastable and unstable states which appear as part of the self-trapping transition. The union of the simple weak-coupling curve and the "coexistence curve" we shall call simply the "transition curve". It is this transition curve which we suggest marks the natural boundary between small polaron states and large polaron states.

The small polaron and large polaron domains are both cleaved by the critical adiabat, so we can identify four types of polaron states having at least qualitatively distinct characteristics: the diabatic small polaron, the adiabatic small polaron, the diabatic large polaron, and the adiabatic large polaron. We can also identify three kinds of *limiting* states: In the weak coupling limit, excitons and phonons are no longer interacting, so we recover the bare exciton states. In the adiabatic limit, we recover the Davydov soliton, and in the diabatic limit we recover the exact small polaron states; we call the latter self-trapped states in order to maintain a distinction between these *immobile* excitations and the mobile small polarons. It may appear that there should be a fourth limiting state, that which would be found in the strong coupling limit; it is easy to see, however, that increasing coupling strength along any adiabat inevitably results in a transition to a small polaron state, and that all small polaron states converge upon the self-trapped states noted above as coupling strength increases without bound.

It is important to bear in mind the fact that certain physical effects missing from our analysis will have a tangible impact on this map of parameter space; two of these are of particular importance: First, we must remember that our analytical method fails to completely account for all of the quantum fluctuations which affect the polaron state. That we find static soliton solutions at all is one indication of this. Schüttler and Holstein [33] have argued, persuasively, that adiabatic large polarons should cease to exist below some threshold coupling strength as the binding energies of the phonons comprising the phonon cloud fall below their zero-point energies [our interpretation of their argument]. We have not determined the boundary which this implies for our parameter space map; instead we have indicated with slashes (////) a fuzzy region in which exciton-phonon correlations should be substantially reduced from that expected of the large polaron. This can be viewed as a penetration of the limiting bare states into the finite-coupling regime; it is in this regime where traditional energy band theory should be expected to apply. Second, our entire analysis has presumed the absence of disorder. The most profound consequence of disorder in quantum transport theory is the phenomenon of Anderson localization. Disorder-induced localization occurs when the magnitude of the disorder measured by a variance of the appropriate energies is sufficiently large relative to the average tunneling matrix element, or, conversely, when the average tunneling matrix element is sufficiently small by the same measure. Since disorder is unavoidable in a real system, we must expect that in every real system there is some minimum tunneling matrix element below which otherwise mobile excitations become localized. In the study of polarons, we need to be concerned not only with the magnitude of the bare matrix element J, but also with magnitude of the dressed matrix element $\tilde{J}$. As above, we have not determined the boundary which these considerations imply for our parameter space map; instead we have indicated with backslashes (\\\\) a fuzzy region in which polaron states should be localized, and, absent incoherent transport mechanisms, immobile. This can be viewed as a penetration of the limiting self-trapped states into the finite-J regime; it is in this regime where polaron hopping theories should be expected to apply.

It must be noted that the mean-field characterization of soliton states which has been the basis of the above discussion must be severely limited when thermal fluctuations are sufficiently strong. Even in perfectly crystalline solids, thermal fluctuations present random fields capable of localizing excitations. The soliton widths we quote must therefore be viewed as upper bounds, maximum widths which are subject to erosion by coherence-degrading mechanisms. This observation is is most important above the self-trapping transition, since the adiabatic small polaron states suggested by our analysis are easy prey for localization via disorder mechanisms.

Davydov's theory of envelope solitons is widely discussed as a model of energy transport in biological systems. A specific form of this model has come to be viewed as a test bed for the viability of the soliton concept in biological applications. The standard model is formulated in terms of the translationally-invariant acoustic chain Hamiltonian (1) with the system parameters estimated from known properties of the α-helix. While some variability exists in estimates of these parameters, the most widely accepted values [34] are those shown in Table 1. (For convenience of the reader, we have assembled in Table 2 a number of related quantities relevant to the present discussion.)

The central question in this area of study is whether nonlinearities in the dynamics of energy transport are sufficiently strong to organize energy into coherent and stable nondispersing excitations which may serve as energy carriers in bioenergetic processes. A quantity which plays a central role in these considerations is the soliton width in the continuum approximation. In Davydov's theory, the number of sites spanned by a static soliton is obtained from the $\delta \to 0$ limit of (37). Using the α-helix parameters from Table 1, this number is less than unity. The obvious conclusion is that Davydov solitons in the α-helix would have to be confined to one, or at most a few, lattice sites, a conclusion which is supported by analytical studies including discreteness corrections [11] and by numerical studies of Davydov's theory in its discrete-lattice formulation [35-37]. However, straightforward calculation of the dressing fraction δ (the "linear" formula (41) is adequate) shows the optimal state to be approximately 88% dressed at $T = 0K$. In terms of our parameter space map, this places the α-helix deep in the diabatic large polaron regime, where, despite sharing some qualitative attributes with the adiabatic large polaron, the quasiparticle is quite unlike the Davydov soliton in some important quantitative respects. For example, this substantial degree of dressing has the immediate consequence that the strength of static nonlinearities (in the nonlinear Schrödinger equation, for example) must be reduced relative to Davydov's theory by a factor of 70 at $T = 0K$. Since the exciton-phonon coupling is weak ($S(0) \approx 0.04 S_c$), this weakening of nonlinearity is not significantly compensated for by reductions in the renormalized transfer matrix element $\tilde{J}$, which amount to only 11% at $T = 0K$.

We can compare our mean-field characterizations to the the recent quantum Monte Carlo (QMC) results of Wang $et\ al.$ [38] obtained for exactly the same model using the same system parameters from Table 1. Wang $et\ al.$ found evidence of structures 2–3 sites wide at $T = 0K$ and found essentially complete localization on a single site above $T = 11K$. In order to clearly understand this comparison, it is crucial that we examine the $same$ quantity computed in the QMC simulation. Wang $et\ al.$ computed the thermal equilibrium expectation value of the operator

$$\hat{C}_l = \sum_n \chi(\hat{Q}_{n+l+1} - \hat{Q}_{n+l-1})a_n^+ a_n \ . \tag{42}$$

Table 1 First panel, system parameters for the α-helix as given in [34]; second panel, derived quantities which appear in equations of motion; third panel, relevant energy scales in common units; fourth panel, ratios of intrinsic energy scales.

quantity	symbol	value	unit
transfer integral	J	7.8	cm^{-1}
lattice constant	l	5.4	$Å$
lattice mass	M	114	m_p
stiffness coefficient	w	13	$N\ m^{-1}$
coupling constant	χ	6.2×10^{-11}	N
free exciton mass	m	7.359×10^{-2}	m_p
nonlinearity parameter	$G(0)$	3.987×10^{-2}	$eV\ Å$
speed of sound	υ_a	4.459	km/s
transfer integral	J	7.8	cm^{-1}
small polaron binding energy	E_b	14.89	cm^{-1}
acoustic phonon bandwidth	$\hbar\omega_B$	87.67	cm^{-1}
room temperature	$300K$	208.5	cm^{-1}
	$\dfrac{E_b}{\hbar\omega_B}$	0.1698	
	$\dfrac{2J}{\hbar\omega_B}$	0.1779	
	$\dfrac{2J}{E_b}$	1.0477	

(Units here are chosen for notational convenience; we note that this diagnostic operator is expressed in terms of *bare* exciton and phonon operators.) Computing the thermal equilibrium expectation value of $\hat{C}_l$ in a quasi-static approximation (cf. [25] for details), we find the result

$$<\hat{C}_l>_T \approx -2\sum_q |\chi^q|^2 e^{-iqR_l} \hbar\omega_q \left[\delta + (1-\delta)<|P^q|^2>_T \right] \tag{43a}$$

$$= -\delta K_{l0}(0) - (1-\delta)\sum_{mn} K_{mn}(0) <|\alpha_m|^2|\alpha_{n+l}|^2>_T \ , \tag{43b}$$

in which P^q is the Fourier transform of the exciton probability distribution

$$P^q = \sum_n e^{-iqR_n} |\bar{\alpha}_n|^2 \ . \tag{44}$$

The observed deformation function thus decomposes into a "small-polaron part" (set $\delta = 1$) having weight δ and a "soliton part" (set $\delta = 0$) having weight $(1-\delta)$. The small-polaron part is wholly independent of exciton probability amplitudes, while the soliton part provides an image of the average shape of exciton probability distribution in space.

We wish to consider the form taken by this function under a number of different assumptions. First we consider the limit $\delta \to 1$ where clearly one finds

$$\lim_{\delta \to 1} <C_l>_T = -K_{l0}(0) \ . \tag{45}$$

Table 2   Upper panel, ratios of thermal to intrinsic energy scales; lower panel, quantities related to the determination of the dressing fraction and soliton width.

quantity	$T = 0$	$T = 300K$
$\dfrac{k_B T}{\hbar \omega_B}$	0	2.378
$\dfrac{k_B T}{E_b}$	0	14.00
$\dfrac{k_B T}{J}$	0	26.73
$S(T)$	0.1441	0.5848
$B(T)$	0.1510	0.6126
δ	0.8810	0.6791
$e^{-\delta^2 S(T)}$	0.8942	0.7636
λ	33.08	6.365

From previous discussions of the system parameter space, we know this limit is compatible with a great variety states and temperatures, particularly in the strong-coupling regime. We note in particular that we obtain this result without any knowledge of the actual shape of the polaron wave function, e.g., whether the latter might be localized or extended in space. To examine other cases, however, we must make some specific assumptions about the shape (and statistical properties) of the wave function amplitudes α_n.

At high temperatures, one might expect the polaron to evolve essentially classically, perhaps being highly localized in space and remaining highly localized while hopping (perhaps incoherently) from site to site. If we detail the time dependence of this evolution by prescribing a integer-valued function $r(t)$, then it is easy to see from (43b) that

$$|\alpha_m|^2 = \delta_{m\,r(t)} \implies <C_l>_T = -K_{l0}(0), \tag{46}$$

which is exactly the same result which obtains for the $\delta \to 1$ limit. Thus, complete localization is verified to be a sufficient condition for recovering the small-polaron deformation.

To obtain the last result we did not need to specify the form or the statistical properties of the classical trajectory $r(t)$, although at high temperatures it is natural to expect $r(t)$ to be a tracing of a random walk. Even for such an incoherent evolution in real space, however, this latter example presents a situation which is in fact quite coherent if viewed from k-space:

$$\alpha^k(t) = \frac{1}{\sqrt{N}} \sum_m e^{-ikm} \left[\delta_{m\,r(t)} e^{-i\theta_m(t)} \right] = \frac{1}{\sqrt{N}} e^{-ikr(t)} e^{-i\phi(t)}. \tag{47}$$

Since the function $r(t)$ is common to every Fourier component, the incoherence seen in real space requires all Fourier components to evolve in lock-step. While this can be consistent with the requirements of an infinite-temperature equilibrium density matrix,

$$<\alpha^k(t)\alpha^{k'}(t)^*>_{T=\infty} = \frac{1}{N} \delta_{kk'}, \tag{48}$$

it hardly represents our more common expectations regarding the statistical properties of Fourier amplitudes; viz., that phases of different Fourier components at high temperature should be uncorrelated. This latter situation corresponds not to a localized excitation, but to a tortuously fragmented spatial distribution without a persistently meaningful centroid. Our deformation function can be calculated for this case since one can show that under these conditions[†]

$$<|P^q|^2>_T \approx \delta_{q0} + N^{-1}(1 - \delta_{q0}) \ . \tag{49}$$

Since the $q = 0$ term does not contribute to the deformation function, this leads to the result

$$<C_l>_T \approx - [\delta + N^{-1}(1 - \delta)] \, K_{l0}(0) \ . \tag{50}$$

The soliton part of the deformation function is $O\{N^{-1}\}$ and therefore is negligible in a relative sense all but the most extreme cases, and is negligible in an absolute sense in a macroscopic system. Thus, we must conclude that were Davydov's theory ($\delta = 0$) applicable, we should expect to see a completely featureless deformation function when the real space probability amplitudes are distributed essentially randomly, as found by Lomdahl and Kerr [35] in their dynamical simulation of the $300K$ regime. The QMC simulations of Wang *et al.* do *not* find such a featureless correlation at high temperatures, suggesting that the Davydov theory fails to describe the true correlation properties in this regime. Better agreement between Davydov's theory and QMC simulations can be found at lower temperatures and in other, more favorable, parameter regimes (cf. Wang *et al.*, this volume); however, wherever we have made a test of the ideas presented in this paper against results of Quantum Monte Carlo simulation, we have found that improvement over Davydov's theory can be obtained by incorporating some degree of dressing according to the proposal of Brown and Ivic [25].

The compelling message of this work is this: While the Davydov soliton ($\delta = 0$) and the self-trapped state ($\delta = 1$) are useful as limiting concepts describing limiting natures of the polaron, the general polaron states presented to us by nature are not always neatly organized around one or the other of these two extremes; logical consistency and experimental (QMC) evidence both point to the need for a balanced theoretical approach.

In concluding, we note that for systems near the adiabatic limit, arguments based on thermal equilibrium, however reasonable, may be of limited relevance to real non-equilibrium processes such as solitons are supposed to mediate. For example, Schüttler and Holstein [33] have calculated the cross section presented by the adiabatic large polaron for the scattering of thermal phonons, and have found the polaron to be essentially transparent to phonons in this regime. Such transparency to phonons should signal anomalously long lifetimes against thermal decay, in general support of the motivating idea that solitons may play a role as energy carriers in bioenergetic processes. One must beware, however, of the role adiabaticity plays in such arguments. We note that in a recent analysis of self-trapping dynamics in a simple dynamical system model mimicking polaron behavior, Brown [39] showed in a natural way how damping which vanishes in the diabatic and adiabatic limits can be of considerable importance in the intermediate regime between these extremes. This suggests that the transparency to phonons found by Schüttler and Holstein may be unique to the adiabatic limit, so that one should beware of assuming this property to hold in a general case. Clearly, further study of the general case is needed.

ACKNOWLEDGEMENT

The authors are happy to acknowledge many fruitful discussions with Dr. Zoran Ivic during his visit to the Institute for Nonlinear Science. This work was supported in part by the U.S. Defense Advanced Research Projects Agency under ARPA Order 7048.

[†] Eq. (49) corrects an error in the original paper of Brown and Ivic [25] where this average was claimed to be equal to unity at high temperatures. On the basis of this erroneous estimate, Brown and Ivic concluded that the small-polaron deformation function should be found at high temperatures regardless of the value of δ, *including* the $\delta = 0$ limit which is Davydov's theory. Our conclusion here somewhat different from that stated in [25].

REFERENCES

1. David Bohm, *Quantum Theory*, (Prentice-Hall, New York, 1951).

2. Sidney Drell, Physics Today **31** No. 6, 23 (1978).

3. T. D. Lee, *Particle Physics and Introduction to Field Theory*, Contemporary Concepts in Physics, Vol. 1, edited by H. Feshbach, N. Bloembergen, L. Kadanoff, M. Ruderman, S. B. Treiman and H. Primakoff (Harwood Academic Publishers, New York, 1988).

4. A. S. Davydov, Phys. Stat. Sol. **36**, 211 (1969); A. S. Davydov and N. I. Kislukha, Phys. Stat. Sol. **59**, 465 (1973); Zh. Eksp. Teor. Fiz. **71**, 1090 (1976) [Sov. Phys.-JETP **44**, 571 (1976)]; A. S. Davydov, Phys. Scr. **20**, 387 (1979).

5. L. D. Landau, Phys. Zeit. Sowjetunion 3, 664 (1933); S. I. Pekar, Zh. Eksp. Theor. Fiz. **16**, 335 (1946); L. D. Landau and S. I. Pekar, Zh. Eksp. Teor. Fiz. **18**, 419 (1948).

6. H. Fröhlich, H. Pelzer and S. Zieman, Phil. Mag. **41**, 221 (1950); H. Fröhlich, Proc. R. Soc. London Ser. A **215**, 291 (1952); Adv. Phys. 3, 325 (1954).

7. I. Pekar, *Untersuchungen Uber die Electronentheorie der Kristalle*, (Akademie Verlag, Berlin, 1954).

8. T. Holstein, Ann. Phys. (N.Y.) **8**, 326 (1959).

9. Emmanuel I. Rashba, in *Excitons*, edited by E. I. Rashba and M. D. Sturge (North-Holland, Amsterdam, 1982).

10. M. Ueta, H. Kanzaki, K. Kobayashi, Y. Toyozawa and E. Hanamura, *Excitonic Processes in Solids*, Springer Series in Solid-State Sciences, Vol. 60 (Springer-Verlag, Berlin, 1986).

11. Xidi Wang, David W. Brown, Katja Lindenberg and Bruce J. West, Phys. Rev. A **37**, 3557 (1988).

12. David W. Brown, Katja Lindenberg and Bruce J. West, Phys. Rev. A **33**, 4104 (1986); David W. Brown, Bruce J. West and Katja Lindenberg, Phys. Rev. A **33**, 4110 (1986).

13. A. S. Davydov, Zh. Eksp. Teor. Fiz. **51**, 789 (1980) [Sov. Phys.-JETP **78**, 397 (1980)].

14 M. J. Skrinjar, D. V. Kapor and S. D. Stojanovic, Phys. Rev. A **37**, 639 (1988).

15 David W. Brown, Bruce J. West and Katja Lindenberg, Phys. Rev. A **37**, 642 (1988).

16. M. J. Skrinjar, D. V. Kapor and S. D. Stojanovic, Phys. Rev. A. **38**, 6402 (1988).

17. Q. Zhang, V. Romero-Rochin and R. Silbey, Phys. Rev. A **38**, 6409 (1988).

18. P. Langhoff, S. T. Epstein and M. Karplus, Rev. Mod. Phys. A **44**, 602 (1972).

19 J. Frenkel, *Wave Mechanics, Advanced General Theory* (Clarendon Press, Oxford, 1974).

20. Peter Kramer and Marcos Saraceno, *Geometry of the Time-Dependent Variational Principle in Quantum Mechanics*, Lecture Notes in Physics, Vol. 140, edited by J. Ehlers, K. Hepp, R. Kippenhahn, H. A. Weidenmüller and J. Zittartz (Springer-Verlag, New York, 1981).

21. David W. Brown, Phys. Rev. A **37**, 5010 (1988).

22. I. G. Lang and Yu. A. Firsov Zh. Eksp. Teor. Fiz. **43**, 1834 (1962).

23. T. Holstein, Ann. Phys. (N.Y.) **8**, 343 (1959).

24 P. Ehrenfest, Z. Phys. **45**, 455 (1927).

25. David W. Brown and Zoran Ivic, Phys. Rev. B **40**, 9876 (1989-I).

26. Yutaka Toyozawa, Prog. Theor. Phys. **26**, 29 (1961); Yutaka Toyozawa, in *Organic Molecular Aggregates*, edited by P. Reineker, H. Haken and H. C. Wolf (Springer-Verlag, Berlin, 1983).

27. David Yarkony and Robert Silbey, J. Chem. Phys. **65**, 1042 (1976); **67**, 5818 (1977).

28. M. Sataric, Z. Ivic, Z. Shemsedini and R. Zakula, J. Molec. Elec. **4**, 223 (1988).

29. Zoran Ivic and David W. Brown, Phys. Rev. Lett. **63**, 426 (1989).

30. A. S. Davydov, *Solitons in Molecular Systems* (Reidel Publishing Co., Boston 1985).

31. Gerd Venzl and Sighart F. Fischer, J. Chem. Phys. **81**, 6090 (1984); Phys. Rev. B **32**, 6437 (1985).

32. D. M. Alexander, Phys. Rev. Lett. **54**, 138 (1985); D. M. Alexander and J. A. Krumhansl, Phys. Rev. B **33**, 7172 (1986).

33. H.-B. Schüttler and T. Holstein, Ann. Phys. **166**, 93 (1986).

34. A. C. Scott, Philos. Trans. R. Soc. London Ser. A **315**, 423 (1985).

35. P. S. Lomdahl and W. C. Kerr, Phys. Rev. Lett. **55**, 1235 (1985).

36. Albert F. Lawrence, James C. McDaniel, David B. Chang, Brian M. Pierce and Robert R. Birge, Phys. Rev. A **33**, 1188 (1986).

37. L. Cruzeiro, J. Halding, P. L. Christiansen, O. Skovgaard, and A. C. Scott, Phys. Rev. A **37**, 880 (1988).

38. Xidi Wang, David W. Brown, Katja Lindenberg, Phys. Rev. Lett. **62**, 1796 (1989).

39. David W. Brown, Phys. Rev. B **39**, 8122 (1989-II).

6

QUANTUM MONTE CARLO SIMULATIONS OF THE DAVYDOV MODEL

Xidi Wang[a], David W. Brown[b] and Katja Lindenberg[b,c]

[a] Department of Physics, B-019
[b] Institute for Nonlinear Science, R-002 and
[c] Department of Chemistry, B-040
University of California at San Diego, La Jolla, CA 92093 U.S.A.

ABSTRACT

Through an application of the quantum Monte Carlo technique, we investigate the thermal equilibrium properties of the one-dimensional model proposed by Davydov for the description of energy transport processes in the α-helix. The calculations in this paper are free from uncontrollable approximations. The deformation of the lattice about a single (mobile) excitation is computed at a number of temperatures for a variety of coupling strengths. Broad and smooth coherent localized quasi-particle units are observed at low temperatures for some parameters of the system. For the "standard" α-helix data, the quasi-particle is embedded in strong fluctuations and is very localized. At temperatures greater than a few Kelvins, the quasi-particle attains its most localized form. We also considered scenarios in which several excitations are present in the system simultaneously; some preliminary results for the density-density correlation are calculated. The structure of polaron clusters is found, and their implication for biological systems is discussed.

WHY QUANTUM MONTE CARLO CALCULATIONS?

The proposal by Davydov and Kislukha [1,2] that self trapping of vibrational energy in the form of solitons to may facilitate energy transfer in biological systems is quite appealing. An a extensive amount of work has been done analytically [3-5], numerically [6-12] and experimentally [13-15] to investigate of the feasibility of this mechanism in biological contexts. The self trapping of vibrational energy is interesting in its own right, of course, whether or not it proves to be related to biological processes.

The standard model Hamiltonian for the description of the α-helix has the form of a Fröhlich Hamiltonian

$$\hat{H} = \sum_i E a_i^+ a_i - J \sum_i (a_{i+1}^+ a_i + a_i^+ a_{i+1})$$

$$+ \sum_i \left[\frac{\hat{P}_i^2}{2M} + \frac{w}{2}(\hat{Q}_{i+1} - \hat{Q}_i)^2 \right] + \chi \sum_i (\hat{Q}_{i+1} - \hat{Q}_{i-1}) a_i^+ a_i \tag{1}$$

where a_i^+, a_i are the creation and annihilation operators of C=O bond excitations on the ith site of the

Davydov's Soliton Revisited, Edited by P.L. Christiansen and A.C. Scott
Plenum Press, New York, 1990

α-helix, and $\hat{P}_i$, $\hat{Q}_i$ are the momentum and position operators of the center of mass of the ith unit of the polypeptide backbone.

The validity of the Davydov Ansätze [16-20] and the stability of solitons at physiological temperatures [5,9-12] have been a focus of much attention recently. The exact solution of the general model (1) so far is unknown. The main stream of numerical studies [9-11] of (1) is to solve/integrate the differential equations obtained from the Davydov Ansätze. In the past, because of the approximate nature of the Ansätze [16-18], different conclusions have reached [9-11].

Since it can be immediately noticed the there are problems with the semiclassical Ansatz machinery, some believe that the results obtained from the Ansatz treatments can be misleading or deviate substantially from the true solution for Hamiltonian (1), particularly because quantum aspects may play an important role in some respects.

The fact that the mass M of the polypeptide group is fairly large can easily lead us to regard the lattice as a system of classical oscillators, without paying attention to the existence of *quantum fluctuations*. To see how strong the quantum fluctuations may be, we can consider the simple case when the excitations (C=O bond vibron excitations) and the phonons are decoupled. Using the normal-mode representation for the acoustic phonons, one can easily find that the correlation function of the lattice coordinates in the free ground state is given by

$$<0|\hat{Q}_m(t)\hat{Q}_n(t')|0> = \sum_q \frac{\hbar}{2MN\omega_q} e^{i[(m-n)ql - \omega_q(t-t')]} \tag{2}$$

where $\omega_q = 2\sqrt{w/M} \sin(|ql|/2)$ is the qth normal mode frequency. In particular, the *auto*correlation function is $<0|\hat{Q}_n^2(t)|0> = \sum_q \hbar/2MN\omega_q$, so that the normal mode with frequency ω_q makes the contribution to $<0|\hat{Q}_n^2|0>$ by a amount of $\hbar/2NM\omega_q$. For the case of an infinite acoustic chain, when $q \to 0$, this contribution is proportional to q^{-1} and therefore is divergent; for a finite chain, the contribution of each mode is finite except the translation mode having $q = 0$. Thus the divergent contribution of $q \to 0$ mode (the *massless* Goldstone mode) indicates that large quantum fluctuations may be present in the system. There is no reason that the same fluctuation should become dramatically small when we turn on the excitation-phonon interactions. Such quantum effects are difficult to describe in terms of classical or semiclassical theories.

The Hamiltonian (1) has the property of translational symmetry, i.e. (1) does not change if we make the replacement of $i \to i + j$ where j is an integer. In the Ansatz treatments [3-5,8,21] of the problem, the soliton solutions one obtains are localized in space, such that the translational symmetry of the system is broken. Because solitons or other localized states at neighboring sites interact, a band of delocalized states forms [22,23]. This will happen for any finite J, so generally, initially localized one-excitation state disperses, eventually broadening out to become uniformly distributed along the one-dimensional chain at infinite time. The speed of the dispersion depends on the effective mass (curvature) of the band. This quantum behavior is in contrast to the corresponding classical situation in which the localized states may persist forever, even though the Hamiltonian is translational invariant.

The translational invariance of the exact eigenstates does not rule out the possibility of strong local correlations between the excitations and the lattice. A simple analogy to this situation is the following: suppose that we have one electron and one proton in a box. The exact ground state of the total system is uniform throughout the box, but if we make a measurement to determine the position of the proton, we will also detect an electron near it. Due to this strong correlation property shared by the electron and proton, the interesting entity in this case is the hydrogen atom. As we will see, a similar situation arises in the study of the Davydov model; the measured position of the excitation is aways correlated with a lattice deformation, and this dressed entity behaves like a stable particle which we hereafter call a *quasi-particle*. Care must be taken in the interpretation of this quasi-particle's stability and lifetime. The quasi-particle may disperse (at low temperatures) or diffuse (at high temperatures) resulting at long times in site-occupation probabilities more or less uniform throughout the chain. This does not imply that the correlated unit has been destroyed [8-11], the strongly correlated unit (like hydrogen atom) may persist, since the quasi-particle after dispersion (wave nature) is still a particle (particle nature); that is, if we make a measurement we will detect a full excitation (because the original Hamiltonian conserves particle number).

There are also difficulties in generalizing the Ansatz treatment to the case where more then one excitation is involved. One interesting case, for example, is Davydov's original proposal that ATP molecule couples two excitations of C=O bond energy into the α-helix chain. To date there has been no reliable calculation for this situation.

Clearly more reliable studies of the Davydov Hamiltonian are needed. While there are no known exact *analytical* quantum solution for the general Hamiltonian (1), a *numerical* approach, namely, Quantum Monte Carlo (QMC) simulation, is able to perform specific calculations essentially exactly without employing any uncontrollable approximations. By using the exact path integral approach developed by Feynman, we can quantize the system by considering a large set of *c-number* path configurations. In this way, calculations for quantum mechanical systems are reduced to corresponding classical problems with an extra dimension [24] (usually called Trotter dimension or the quantum dimension). We shall see this explicitly in our derivation of the partition function of the Davydov model below.

As proposed by Suzuki [25], the path configurations of the quantum system can be sampled by Monte Carlo techniques. The weight with which each configuration appears in the ensemble is regraded as the same as it appears in the partition function of the system. In terms of the canonical ensemble, the partition function is given by

$$Z = \mathrm{Tr}\,(e^{-\beta \hat{H}})\,.\tag{3}$$

The expectation value of any observable $\hat{O}$ can be calculated by averaging over members of the ensemble

$$<\hat{O}>_{ensemble} = \frac{\mathrm{Tr}\,(\hat{O}e^{-\beta \hat{H}})}{\mathrm{Tr}\,(e^{-\beta \hat{H}})}\,.\tag{4}$$

The computational problem reduces to determining the weight with which various quantities enter this average; QMC determines these weights by appropriate sampling of the randomly generated ensemble members. The operator $e^{-\beta \hat{H}}$ filters out the low lying states; in the limit of zero temperature ($\beta \to \infty$), only the ground state is filtered out. (Comparing $e^{-\beta \hat{H}}$ with the usual propagator $e^{-i\hat{H}t}$, it is clear why β is usually called an *imaginary time*.) Formally this procedure is exact, involving no uncontrollable approximations. However, as a practical matter, only the ground state information is easy to obtain this way; information about the higher excited states becomes increasingly difficult to obtain.

In our system, we generally have one or a few excitations present in the C=O bonds. These excitations are regarded as bosonic particles. Since the total number of the excitations is conserved, by keeping the number of excitation *world lines* (paths in the path integral representation) fixed, we can carry out simulations for multiple-excitation systems, just as in 1-D fermion systems [27]. To be more specific, let us derive the functional integral representation of the partition function Z of the Davydov model in a form which can address multiple-excitation systems as well as the usual one-excitation systems.

PARTITION FUNCTION AND CHECKERBOARD DECOMPOSITION

We quantize the system (1) in terms of the classical variables via Feynman's path-integral representation. As Feynman showed, propagators of quantum mechanical systems can be calculated by summing exponentials of appropriate actions evaluated along all possible classical paths. As in the standard path-integral technique, we discretize the imaginary time β into L intervals ($\Delta\tau = \beta/L$), inserting at each division a complete set of states $|1>,|3>,|5>,...,|j>,...,|2L-1>$ where j is the label for the jth imaginary time division or "cut" (the reason for choosing odd j will be seen later). The partition function of the system is then

$$Z = \sum_{1,3,\cdots,2L-1} <1|e^{-\Delta\tau \hat{H}}|2L-1> \cdots <5|e^{-\Delta\tau \hat{H}}|3><3|e^{-\Delta\tau \hat{H}}|1>;\tag{5}$$

here the summation is over all the possible configurations of the complete sets of basis states. For the content of this paper, we choose $|j>$ to be direct products of phonon and exciton basis states. Thus, $|j> \equiv |\{n_{i,j}\}> \otimes |\{Q_{ij}\}>$. We use the abbreviations

$$|\{n_{i,j}\}> \equiv |n_{1,j}> \otimes |n_{2,j}> \otimes \cdots \otimes |n_{i,j}> \otimes \cdots \otimes |n_{N,j}>\tag{6a}$$

$$|\{Q_{i,j}\}> \equiv |Q_{1,j}> \otimes |Q_{2,j}> \otimes \cdots \otimes |Q_{i,j}> \otimes \cdots \otimes |Q_{N,j}>\tag{6b}$$

where $|n_{ij}\rangle$ is an eigenstate of the number operator $\hat{n}_i \equiv a_i^+ a_i$, at the jth imaginary time division; the eigenvalue of $\hat{n}_i$ at this cut is n_{ij}. $|Q_{i,j}\rangle$ is an eigenstate of the lattice position operator $\hat{Q}_i$ at the jth imaginary time division; the eigenvalue of $\hat{Q}_i$ at this cut is Q_{ij}. In order to be able to calculate the matrix elements in equation (5), we employ the *checkerboard decomposition* technique as in [27]. That is, we separate the Hamiltonian into two parts according to the odd or even site dependence of each term as follows:

$$\hat{H} = \hat{H}_2\{\hat{P}\} + \hat{H}_1\{\hat{Q}\} \,, \tag{7}$$

where $\hat{H}_1\{\hat{Q}\}$ contains all terms involving $\hat{Q}$'s only, as well as those terms involving odd-site excitation operators; $\hat{H}_2\{P\}$ contains all terms involving $\hat{P}$'s only, as well as those terms involving even-site excitation operators.

$$\hat{H}_1\{\hat{Q}\} \equiv \sum_{i\ odd} \left\{ \frac{w}{2}[(\hat{Q}_{i+1} - \hat{Q}_i)^2 + (\hat{Q}_{i+2} - \hat{Q}_{i+1})^2] \right\} \,,$$

$$+ \sum_{i\ odd} \left\{ \chi\left[(\hat{Q}_{i+1} - \hat{Q}_{i-1})a_i^+ a_i + (\hat{Q}_{i+2} - \hat{Q}_i)a_{i+1}^+ a_{i+1} \right] - J(a_{i+1}^+ a_i + a_i^+ a_{i+1}) \right\} \,, \tag{8a}$$

$$H_2\{\hat{P}\} \equiv \sum_{i\ even} \left\{ \frac{\hat{P}_i^2}{2M} + \frac{\hat{P}_{i+1}^2}{2M} \right\} - \sum_{i\ even} \left\{ J(a_{i+1}^+ a_i + a_i^+ a_{i+1}) \right\} \,. \tag{8b}$$

With this decomposition, each term of (8a) in curly braces ("$\{ \cdots \}$") commutes with each other such term in (8a). Similar commutation properties hold in (8b). The motivation for arranging the Hamiltonian in this way becomes clear when we apply the Trotter formula

$$e^{-\Delta\tau\hat{H}} = e^{-\Delta\tau\hat{H}_2\{\hat{P}\}} e^{-\Delta\tau\hat{H}_1\{\hat{Q}\}} [1 + O(\Delta\tau^2)] \,. \tag{9}$$

This is actually the only place where the discretization approximation for the Trotter dimension is used; as $\Delta\tau \to 0$ the approximation recovers the exact results.[†]

We now insert complete sets of basis states between the $\hat{H}_1$ and $\hat{H}_2$ exponentials, labeling with the even numbers. The partition function is given by

$$Z \approx \sum_{1,2\cdots 2L} \langle 1|e^{-\Delta\tau\hat{H}_2\{\hat{P}\}}|2L\rangle\langle 2L|e^{-\Delta\tau\hat{H}_1\{\hat{Q}\}}|2L-1\rangle \cdots$$

$$\cdots \langle 3|e^{-\Delta\tau\hat{H}_2\{\hat{P}\}}|2\rangle\langle 2|e^{-\Delta\tau\hat{H}_1\{\hat{Q}\}}|1\rangle \,. \tag{10}$$

Because all the $\{ \cdots \}$ within (8a) and (8b) commute with each other, we can further decompose the exponentials of $\hat{H}_2\{\hat{P}\}$ and $\hat{H}_1\{\hat{Q}\}$ into products of more wieldable terms

$$e^{-\Delta\tau\hat{H}_2\{\hat{P}\}} = e^{-\frac{\Delta\tau}{2M}\sum_i \hat{P}_i^2} \prod_{i\ even} e^{\Delta\tau J(a_{i+1}^+ a_i + a_i^+ a_{i+1})} \,, \tag{11a}$$

$$e^{-\Delta\tau\hat{H}_1\{\hat{Q}\}} = e^{-\Delta\tau\sum_i \frac{w}{2}(\hat{Q}_{i+1} - \hat{Q}_i)^2} \prod_{i\ odd} e^{-\Delta\tau\left\{ \chi\left[(\hat{Q}_{i+1} - \hat{Q}_{i-1})a_i^+ a_i + (\hat{Q}_{i+2} - \hat{Q}_i)a_{i+1}^+ a_{i+1} \right] - J(a_{i+1}^+ a_i + a_i^+ a_{i+1}) \right\}} \,. \tag{11b}$$

[†] It is easy to obtain better approximations in higher orders of $\Delta\tau$ by using high-order Trotter formulae, but in practice these are not found to be very useful [28,29].

To calculate the matrix elements of (11a) we insert a complete set of lattice momentum eigenstates between the phonon and excitation operators in (11a) and carry out the summations over momentum eigenstates by simple Gaussian integrations. The $\hat{H}_2$ matrix elements are then given by

$$<j+2|e^{-\Delta\tau \hat{H}_2\{\hat{P}\}}|j+1> = \tag{12}$$

$$\left[\frac{M}{2\pi\Delta\tau}\right]^N e^{-\Delta\tau\sum_i \frac{1}{2}M\left[\frac{Q_{i,j+2}-Q_{i,j+1}}{\Delta\tau}\right]^2} <\{n_{i,j+2}\}|\prod_{i\ even} e^{\Delta\tau J(a_{i+1}^+ a_i + a_i^+ a_{i+1})}|\{n_{i,j+1}\}> ,$$

where N is the total number of sites. The matrix elements of $H_1\{\hat{Q}\}$ can be evaluated in a similar way using lattice position eigenstates, resulting in

$$<j+1|e^{-\Delta\tau H_1\{\hat{Q}\}}|j> = \prod_i \delta(Q_{i,j+1}-Q_{i,j})\ e^{-\Delta\tau \frac{w}{2}\sum_i (Q_{i+1,j}-Q_{i,j})^2} \tag{13}$$

$$\times <\{n_{i,j+1}\}|\prod_{i\ odd} e^{-\Delta\tau\left\{\chi\left[(Q_{i+1,j}-Q_{i-1,j})a_i^+ a_i+(Q_{i+2,j}-Q_{i,j})a_{i+1}^+ a_i\right] -J(a_{i+1}^+ a_i + a_i^+ a_{i+1})\right\}}|\{n_{i,j}\}> .$$

Thanks to the δ functions in (13), on substituting terms like (12) and (13) back in equation (10), we can easily carry out the integration over the coordinates of the lattice $Q_{i,j}$. The partition function then becomes

$$Z = \sum_{\{Q_{i,j},n_{i,j}\}} \left[\frac{M}{2\pi\Delta\tau}\right]^{NL} \prod_{j=1,3...2L-1}\left\{\prod_i e^{-\Delta\tau\left[\frac{M}{2}\left[\frac{Q_{i,j+2}-Q_{i,j}}{\Delta\tau}\right]^2 + \frac{w}{2}(Q_{i+1,j}-Q_{i,j})^2\right]}\right.$$

$$\prod_{i\ even}\left[\left[<n_{i+1,j+2}|\otimes<n_{i,j+2}|\right] e^{\Delta\tau J(a_{i+1}^+ a_i + a_i^+ a_{i+1})}\left[|n_{i,j+2}>\otimes|n_{i+1,j+2}>\right]\right]$$

$$\prod_{i\ odd}\left[\left[<n_{i+1,j+1}|\otimes<n_{i,j+1}|\right] e^{-\Delta\tau\left\{\chi\left[(Q_{i+1,j}-Q_{i-1,j})a_i^+ a_i+(Q_{i+2,j}-Q_i)a_{i+1}^+ a_{i+1}\right] -J(a_{i+1}^+ a_i + a_i^+ a_{i+1})\right\}}\right.$$

$$\left.\left.\left[|n_{i,j}>\otimes|n_{i+1,j}>\right]\right]\right\} \tag{14}$$

This is the basic formula upon which our Monte Carlo simulations are based. Pictorially, two sheets of data ($n_{i,j}$ and $Q_{i,j}$) interact according to a checkerboard pattern as in Figure 1.

The number of the exciton world lines is given by the number of the excitations $N_{ex} = \Sigma_i n_{i,j}$. The number of the phonon world lines is the same as the number the lattice sites. Note from equation (14) that we need to carry out extensive summations of large products of cumbersome matrix element factors, over many possible world line configurations. It appears that a horrendous volume of work has to be done to obtain any useful result; however, thanks to the very efficient sampling algorithm we use (see section 4), no global quantities have to be calculated allowing the task of performing the summation to be handled without exhausting computer resources.

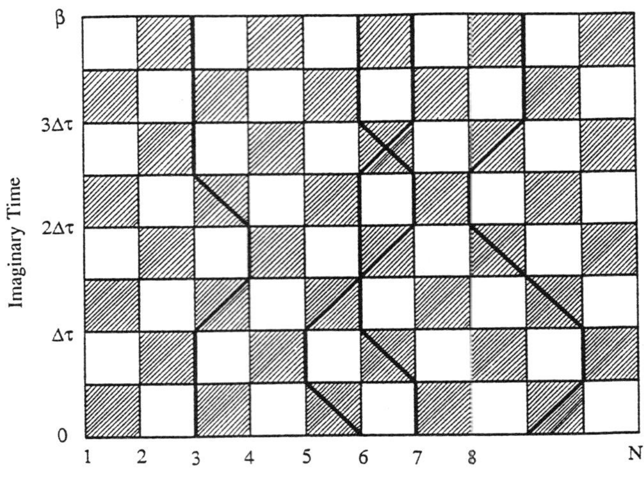

Imaginary Time (vertical axis): β, $3\Delta\tau$, $2\Delta\tau$, $\Delta\tau$, 0

Lattice Sites (horizontal axis): 1 2 3 4 5 6 7 8 ... N

Lattice Sites

Fig. 1 World line configurations of the excitation and lattice. One possible world line configuration of a four-excitation system is drawn. $n_{i,j}$ and $Q_{i,j}$, reside on a checkerboard where the horizontal direction represents the sites of the real space lattice and the vertical direction represents the imaginary time direction (Trotter direction). $n_{i,j}$ occupy those intersections with odd j; $Q_{i,j}$ occupy intersections with both odd and even j. Periodic boundary conditions apply in both the real space and imaginary time directions. World lines traverse the imaginary time interval from 0 to β, with each cut in this direction representing one member of the ensemble; propagation in the imaginary time direction changes from one member of the ensemble to another. The shaded boxes represent the dimer cell where the hopping in the imaginary time is allowed by the dimerization procedure as in (8a and 8b); hopping acrossing the unshaded boxes is not allowed. Notice that for bosonic excitations, the world lines are allowed to cross, touch or lie on top of each other. This aspect differs from the fermionic world line configuration as in [27].

DIMER MATRIX ELEMENTS

To speed up the computer simulations, we tabulate in advance the formulae needed to calculate the matrix elements appeared in (14). The essential part of finding general matrix elements is the evaluation of the typical matrix element

$$<m',n' \mid e^{-\Delta\tau[E_1 a_1^\dagger a_1 + E_2 a_2^\dagger a_2 - J(a_1^\dagger a_2 + a_2^\dagger a_1)]} \mid m,n>, \tag{15}$$

where $E_1 \equiv \chi(Q_{i+1,j} - Q_{i-1,j})$, $E_2 \equiv \chi(Q_{i+2,j} - Q_{i,j})$ for odd j, and $E_1 = E_2 = 0$ for even j. The product states $|m,n> \equiv |m> \otimes |n>$ are constrained such that $m+n = N_d$ here N_d is the total number of world lines passing through this dimer (note $N_d \leq N_{ex}$). To determine this matrix element, we first find the real time matrix element for the corresponding Schödinger equation. The wave function at time t is given by $|\psi(t)> = e^{-iHt} |\psi(0)>$ and the corresponding matrix element of (15) is then $<m',n' \mid e^{-i\hat{H}_d t} \mid m,n>$ under the mapping $\Delta\tau \rightarrow it$, where the dimer Hamiltonian is defined as

$$\hat{H}_d \equiv E_1 a_1^+ a_1 + E_2 a_2^+ a_2 - J(a_1^+ a_2 + a_2^+ a_1). \tag{16}$$

Generally, we have to solve a linear dimer problem, but with the added wrinkle of having to contend with the presence of multiple excitations. We rewrite $\hat{H}_d$ in matrix form and make a unitary transformation **u** to diagonalize this matrix

88

$$\hat{H}_d = \mathbf{a}^{+}\cdot\mathbf{H}\cdot\mathbf{a} = (\mathbf{u}\cdot\mathbf{a})^{+}\cdot(\mathbf{u}\cdot\mathbf{H}\cdot\mathbf{u}^{+})\cdot(\mathbf{u}\cdot\mathbf{a}) = \mathbf{b}^{+}\cdot\Lambda\cdot\mathbf{b} \tag{17}$$

where we have defined *operator* vectors $\mathbf{a}$ and $\mathbf{b}$ in terms of the operators a_1, a_2 and the operators b_1, b_2; also we have defined the *scalar* matrices $\mathbf{H}$ and Λ by

$$\mathbf{a} \equiv \begin{bmatrix} a_1 \\ a_2 \end{bmatrix}, \quad \mathbf{H} = \begin{bmatrix} E_1 & -J \\ -J & E_2 \end{bmatrix}, \quad \mathbf{b} \equiv \begin{bmatrix} b_1 \\ b_2 \end{bmatrix}, \quad \Lambda = \begin{bmatrix} \lambda_1 & 0 \\ 0 & \lambda_2 \end{bmatrix} \tag{18}$$

λ_1, λ_2 are the eigenvalues of the matrix $\mathbf{H}$. In the Heisenberg representation, the propagation of an operator through an imaginary time interval $\Delta\tau$ can be expressed as $\mathbf{a}(\Delta\tau) = e^{-\Delta\tau H_d}\mathbf{a}(0) = \mathbf{u}^{+}\cdot e^{-\Delta\tau\Lambda}\cdot\mathbf{u}\cdot\mathbf{a}(0)$. Thus we can follow the propagation across the checkerboard using tabulated values of the propagator matrix elements $e_{\mu\nu}(\Delta\tau) \equiv <\mu|\mathbf{u}^{+}e^{-\Delta\tau\Lambda}\mathbf{u}|\nu>$.

If we assume at the initial imaginary time the wave function of the dimer is $|\Psi_{m,n}(0)> = (m!\,n!)^{-1/2}|\,m,n>$, where $|\,m,n>$ is the dimer state with m excitations on the left and n excitations on the right, from (16)-(18) we can see that the state at imaginary time $\Delta\tau$ is then given by

$$|\Psi_{m,n}(\Delta\tau)> = (m!n!)^{-1/2}[e_{11}(\Delta\tau)a_1^{+}+e_{12}(\Delta\tau)a_2^{+}]^m[e_{21}(\Delta\tau)a_1^{+}+e_{22}(\Delta\tau)a_2^{+}]^n|0,0> . \tag{19}$$

The matrix element in (15) is thus the overlap of the wave function at time 0 with that at time $\Delta\tau$. With the help of (15) and (19) we find

$$<\Psi_{m',n'}(0)|\Psi_{m,n}(\Delta\tau)> = \left[\frac{m'!(N_d-m')!}{m!(N_d-m)!}\right]^{\frac{1}{2}} \sum_s C_m^s C_{N_d-m}^{m'-s} e_{11}^s e_{22}^{N-m-m'+s} e_{12}^{m-s} e_{21}^{m'-s} , \tag{20}$$

where the summation is over the interval $Max(0, m+m'-N_d) \leq s \leq Min(m, m')$ and $C_m^n = m!/n!(m-n)!$ is binomial coefficient. Physically (20) represents the transition amplitude of the world line number (m,n) to the world line number (m',n') in the dimer cell.

SIMULATION ALGORITHMS

Our canonical ensemble partition function can be put in the form

$$Z = \sum_{\{Q_{i,j},n_{i,j}\}} W\{Q_{i,j},n_{i,j}\} \tag{21}$$

(neglecting a normalization factor which never appears in our simulation). Here $W\{n_{i,j},Q_{i,j}\}$ is the product of the many matrix element factors which appeared in (14). The two sheets of data $n_{i,j}$ and $Q_{i,j}$ determine the configuration of the excitation and lattice world lines on the checkerboard. Each configuration represents a member of the ensemble. The weight $W\{n_{i,j},Q_{i,j}\}$ is a positive-definite number which is chosen to be the weight of the occurrence for this member of the ensemble. Our task now is to sample world line configurations according to this weight so that the expectation value of any observable can be calculated by averaging the observable over the sample. Were it necessary in every case to calculate $W\{n_{i,j},Q_{i,j}\}$ in its entirety, an enormous amount of computer time would be needed and little of practical value could be accomplished. Fortunately, there exist some algorithms that allow such global calculations to be avoided. Two important algorithms are often used to sample the weight $W\{n_{i,j},Q_{i,j}\}$ -- the *heat bath algorithm* and the *Metropolis algorithm*, (cf. [26,27]).

To sample the excitation world line configuration $n_{i,j}$, we use the heat bath algorithm [26,27]. We first make a local trial move of the world line from $\{n_{i,j}\}$ to $\{n'_{i,j}\}$. We then accept the trial move with probability $P_{excitation} = R_{excitation}/(1 + R_{excitation})$, where $R_{excitation}$ is the ratio of the weight $W\{n'_{i,j},Q_{i,j}\}$ after the trial move to the weight $W\{n_{i,j},Q_{i,j}\}$ before the trial move. $R_{excitation}$ is a local quantity because the contributions to $W\{n_{i,j},Q_{i,j}\}$ from unmoved parts of the world line cancel. To insure that the number of the excitation world lines is always equal to the fixed number of excitations, we need to conserve the number of the world lines (i.e. disallowing world lines which branch, begin or end in the middle of the checkerboard). Also we need to insure that our trial moves can reach every possible world line configuration so that we can not be trapped in only part of the world line configuration space. By deforming the world line sequentially along each world line as in [27], all the possible configurations can be reached (see Figure 2).

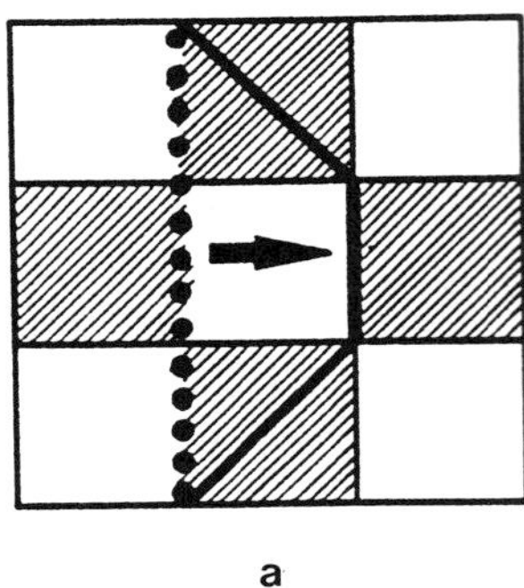

a

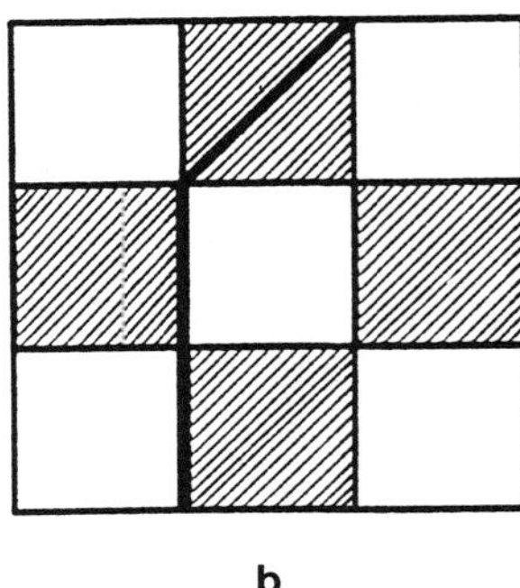

b

Fig. 2a A trial move for a one-excitation world line. The dotted line is the world line before the trial move and the solid line, after the trial move. The matrix elements of the four shaded boxes (dimers) in Figure 2a are affected due to this trial move; these matrix elements are then calculated after and before this trial move, and ratio of these matrix elements are then used to determine the possibility of accepting or rejecting this trial move according heat bath algorithm.

Fig. 2b A trial move for lattice. When a trial move of the lattice $Q_{i,j} \to Q_{i,j}+\Delta$ is made, in addition to the matrix element change in the pure phonon part as shown in (14), there is also a matrix element change due to interaction between the excitations and phonons. Only two shaded boxes (left and right of the trial move) are affected due to the matrix element changes; these two changes are then calculated after and before this trial move, and ratio of these matrix elements is the used to determine the possibility of accepting or rejecting this trial move according to the Metropolis algorithm.

We then look into the matrix element formulae, carry out the calculation of the ratio R, and accept or reject this configuration as required by the heat bath algorithm. If this trial move is accepted, then this world line configuration is used to calculate the average of observables; if this trial move is rejected, we simply disregard this configuration.

To sample the world line configuration of the lattice $\{Q_{i,j}\}$, we use the Metropolis algorithm [26,27,30], since this algorithm is better suited to the continuous nature of the lattice coordinates. We make a local trial move, $\{Q_{i,j}\} \to \{Q'_{i,j}\}$ and calculate the ratio of the matrix element $R_{lattice} = W\{n_{i,j}, Q'_{i,j}\}/W\{n_{i,j}, Q_{i,j}\}$ with the formulae provided in the last section, and we accept the trial move with the probability $P_{lattice} = Min\{1, R_{lattice}\}$ [26] etc.. As when sampling the excitation part, we make trial moves sequentially along each world line of the lattice, so that the entire sheet of lattice $\{Q_{i,j}\}$ can be sampled. Generally the local move $Q_{i,j} \to Q'_{i,j}$ changes the action of both the excitation and lattice parts (i.e. the trace of part in (14) and phonon kinetic energy part in (14)). As can be seen in Figure 2b, two shaded boxes in the vicinity of the local trial move are involved in the calculation of the action changes.

We proceed by updating the data sheets $\{n_{i,j}\}$ and $\{Q_{i,j}\}$ alternatively and call one updating cycle a "sweep". To initialize the simulations, a number (typically a few thousand) of warm-up sweeps are made and discarded. After this warm-up procedure, we commence the measurement of observables.

90

Because of sampling errors (limited sampling of an infinite configuration set) and discretization errors (limited number of imaginary time "cuts"), we need to control the error in our calculation carefully. To reduce sampling errors, we generally increase the number of sweeps contributing to our measurement until the observable we are measuring is essentially unchanged with any further increase in the number of the sweeps. Discretization error is generally second order or higher in $\Delta\tau$. In principle as $\Delta\tau \to 0$ (increase the number of cuts), we should recover the exact limit; however, to do this we would have to increase the number of divisions L in the Trotter dimension. This increases the size of our two dimension data arrays, $n_{i,j}$ and $Q_{i,j}$, causing the computational time to be increased, so practically we must strike a compromise between these two factors. Given a finite amount of computer resources, there appears to be an optimum value for $\Delta\tau$, usually we choose it to be in the range of 0.05–0.2. In our simulations, decreasing this value further has not shown any appreciable improvement in accuracy.

There are many ways to check the correctness of our code. First of all, when the excitation and the phonon are decoupled, we can solve the problem exactly, allowing us compare our result against exact calculations. For the decoupled phonon part we can compare our measured distribution of the lattice positions with the exact ground state wave function as in [30]. For the decoupled excitation part, we have the case of free excitation motion for which we can check our measurement of the effective mass against the mass of the free excitation. When the interaction between the excitations and the phonon is turned on in the $J = 0$ case, we can check the lattice deformation in our simulation against the results of exact analytical calculations for the fully-coupled system. Although, the foregoing quantities cannot be determined exactly in the general case, there is one quantity can be calculated analytically exactly at any temperature, i.e. the total lattice contraction [34]. In all of the above cases, our simulation have excellent agreements with the known results.

ONE-EXCITATION SIMULATIONS

It is convenient to simulate in terms of dimensionless data which are scaled such that the amplitude of the quantum fluctuations of the lattice is of the order of unity, analogous to the quantizing of the simple harmonic oscillator. The lattice positions are scaled such that $\hat{Q}_n \to (\hbar/M\,\Omega)^{1/2}\hat{Q}_n$ and the corresponding momenta are scaled such that $\hat{P}_n \to (\hbar M\,\Omega)^{1/2}\hat{P}_n$. We then scale the energy by transfer matrix element J. The transfer matrix element is thus equal to unity in the scaled Hamiltonian, and the commutation relation of $\hat{P}_m$ and $\hat{Q}_n$, has the form of $[\hat{Q}_n , \hat{P}_m] = \delta_{mn}$. We thus have four independent parameter left in our simulation: χ, w, M and β. Though there are some uncertainties about what constitutes the best choice of parameters for the simulation of the α-helix proteins, the "standard data" from [19] are used for our simulations. The "standard" data and the scaled data are given in Table 1.

Table 1 "Standard" data for the α-helix [19] and the dimensionless data used in simulation. The scaling is chosen such that $J = \hbar = 1$ and $[Q_m , P_n] = \delta_{mn}$. The "scaling" column gives the scaled data in terms of the original dimensioned parameters, and the result is shown in the last column.

parameter	value	unit	scaling	scaled data
J	7.8	cm^{-1}	1	1.00
l	5.4	Å	$\hbar^{1/2}M^{-1/4}w^{-1/4}$	66.1
M	114	m_p	$\hbar^{-1}M^{1/2}w^{-1/2}J$	0.178
w	13	$N\,m^{-1}$	$\hbar M^{-1/2}w^{1/2}J^{-1}$	5.60
χ	6.2×10^{-11}	N	$\hbar^{1/2}J^{-1}M^{-1/4}w^{-1/4}\chi$	3.27

We typically use 24-site chains, so we expec' encounter large, but finite quantum fluctuations. When there is only one excitation present in the system, we have only one excitation world line. A typical low-temperature world line configuration for a single excitation is shown in Figure 3, where it can be seen that the quantum fluctuations are quite large. The lattice deformation is embedded in these fluctuations; however, it is not easy to visualize this deformation because the r.m.s. amplitude of the random quantum fluctuations is significantly larger than the coherent lattice deformation.

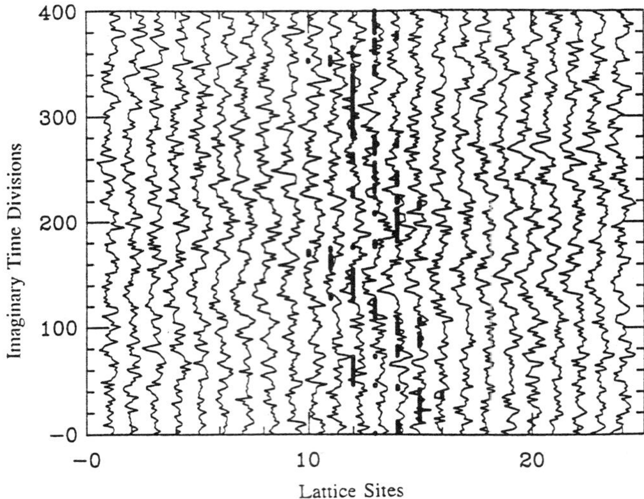

Fig. 3 A typical world line configuration for α-helix parameters at $0.27K$. In this simulation, β has been subdivided into 400 cuts. Vertices of the polygonal arcs represent the value at each cut of the deviation $Q_n - R_n$ of each lattice mass from its free-lattice equilibrium position position R_n. (The size of these deviations has been magnified by a factor of 30 for ease of viewing.) Similarly, open circles (O) represent the location of the excitation at each cut. There may appear to the reader to be more than one excitation present at a particular cut despite the fact that every cut contains exactly one excitation. This illusion is due to occasional rapid (in β) oscillations of the excitation between adjacent sites, which are difficult for the eye to resolve due to the high density of cuts. Careful comparison of an occupied and an unoccupied region shows how the weak, average deformation (see Figure 4) is realized amidst a sea of intrinsic quantum noise.

In our simulation, because of the translational invariance of the Hamiltonian, the probability of the finding the excitation at a particular site is the same for any site in the crystal. The average lattice deformation is thus distributed uniformly over the lattice at any temperature. However, for each individual measurement, we expect the contraction of the lattice to be centered around the measured position of the excitation, instead of being uniformly distributed. This local correlation property can be revealed by the excitation-lattice correlation function.

$$C_i = <\sum_j n_{m,j}(Q_{m+i+1,j} - Q_{m+i-1,j})>_{ensemble} \cdot \tag{22}$$

Algorithmically, we first measure the position of the excitation, and then measure the configuration of the lattice relative to this excitation, and average the configurations over ensemble members sampled. For the "standard" α-helix data given in Table 1, we have carried out such measurements at several temperatures, the results of which are displayed in Figure 4.

For all the temperatures we measured, we found the lattice deformation to be quite localized; however, for temperatures below $2.8K$, the correlation radius clearly extends beyond the nearest neighbor sites. Because of the quantum tunneling of the excitation among neighboring sites, the excitation world line wanders randomly from one site to another while propagating in imaginary time. (The world line of a classical particle is straight.) The extreme excursion of the world line away from its centroid[†] is in principle bounded only by the size of the system; however, the effective width of

---

[†] The location of the centroid varies from measurement to measurement, and is found with equal probability at every point of the lattice, reflecting the translational invariance of the Hamiltonian.

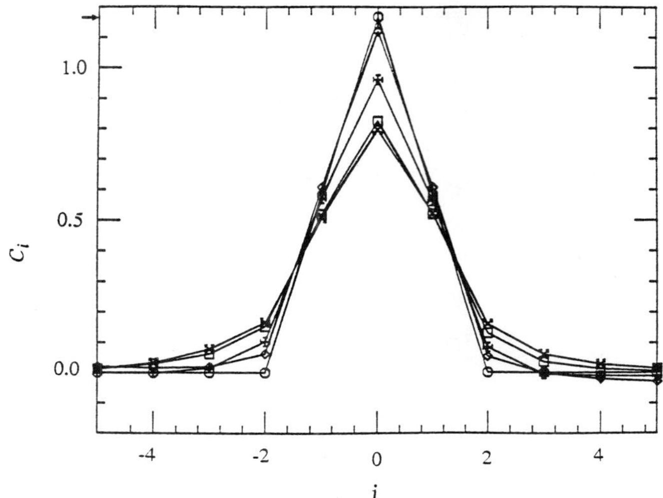

Fig. 4 Simulation results for the deformation function C_i (cf. (5.7)) using the α-helix parameters of Table 1: a) (×) $T = 0.27K$, b) (□) $T = 0.7K$, c) (+) $T = 2.8K$, d) (◇) $T = 11.2K$, e) (○) $T = \infty$. The latter data are obtained by examining the QMC algorithm and observing that the only possible result of a QMC survey at infinite temperature is that displayed. The vertical axis is in simulation units. The arrow (→) indicates the value of C_0 at infinite temperature, 9.5×10^{-2}Å or 1.76% of the lattice constant. Only 11 lattice sites are shown; however, every QMC run was made on a lattice of at least 24 sites. Solid lines through the data are provided to aid the eye only; no theoretical fit has been performed to obtain these curves.

this meander in the quantum dimension typically spans a modest number of sites, roughly three in the low-temperature example shown in Figure 3. Quantum tunneling influences the lattice deformation in such a way that the observed correlation radius is typically somewhat greater than the range of the exciton-phonon interaction, but somewhat less than the effective width of the world line meander in the quantum dimension. As the temperature increases, quantum tunneling effects are suppressed by increasingly important thermal effects, with the result that the excitation evolves in an increasingly classical manner. In QMC simulation, this shows up in the straightening out of the excitation world line with increasing temperature and in the attendant decrease in the effective width of the meander of the world line in the quantum dimension.[††] The net result is that the correlation radius decreases with increasing temperature; at temperatures above a few Kelvins, the correlation function attains its narrowest limit.

For different coupling strengths χ, the picture looks qualitatively the same as in Figures 3 and 4. Were the excitation completely localized on one site, the maximum deformation (C_0 in (5.7)) would increase linearly with χ [32]; deviations from this behavior must be a consequence of quantum tunneling processes.

For a one-excitation system, the density-density correlation function

$$D_{i,j} \equiv \langle n_{m,n}\, n_{m+i,n+j} \rangle_{ensemble} \tag{23}$$

contains useful information about the excitation world line propagation in the imaginary time. Because of the translational invariance of the system, after ensemble averaging, the density-density

---

[††] Computationally, as the temperature increases, β decreases, and the possibility of excitation world line hopping to the neighbor by crossing the shaded box decreases. This can be seen from (20) since the matrix element for the one excitation world line to crossing the shaded box is roughly proportional to $\Delta\tau$ (= β/L).

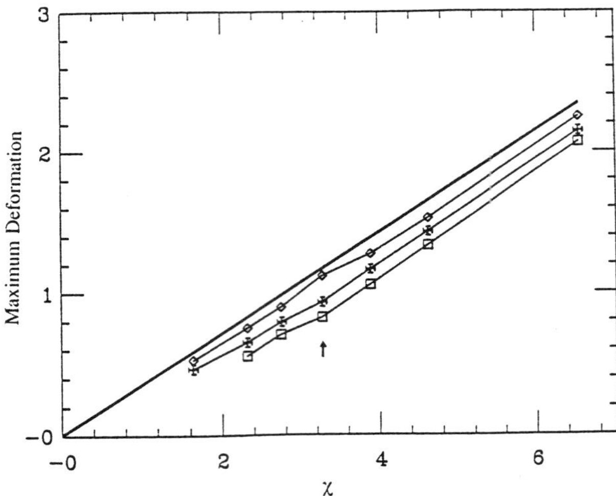

Fig. 5 The maximum depth of the deformation (C_0 in (22)) at different temperatures for different χ's. The maximum depth shows an apparent linear dependence on χ (at least for large χ), signaling the very localized nature of the quasi-particle, the thick solid line is the exact results for the case $J=0$ obtained from the analytical calculation. The arrow indicates the χ from Table 1.

correlation is independent of m and n. The the effective mass m^* of the quasi-particle can be found by the plotting density-density correlation function $D_{i,j}$ for fixed j such that

$$[-\ln(D_{i,L})]^{\frac{1}{2}} = \left[\frac{2m^*}{\beta}\right]^{\frac{1}{2}}(R_{i+m} - R_m) \tag{24}$$

where R_n is the nth site equilibrium position when excitation and lattice are decoupled, and we have chosen j to be at the position of $\beta/2$. For a single free particle, the above plot yields a straight line (in the continuum limit). The effective mass m^* can be derived from the slope of this line and can be verified to be in agreement with the bare mass of the free excitation. Should this plot still yield a straight line when the coupling between the excitation and phonon is turned on, we then expect the the effective mass of the quasi-particle is determined by slope of the line.

By increasing the hopping integral J and decreasing the coupling strength χ, broad and smooth quasi-particle profiles can be obtained. In Figure 6a, we show a smooth soliton-like profile obtained for hypothetical data. In Figure 6b, we show density-density plot for the same data used to obtain Figure 6a. The effective mass obtained this way differs from the effective mass which would be obtained from Davydov's for these parameters. Our preliminary results seem to be in better agreement with the result obtained using partial dressing technique of Brown and Ivic [31] (See also Brown *et al.*, this volume).

MULTIPLE-EXCITATION SIMULATIONS

Davydov's original idea for energy transport in the α-helix was that lysis of ATP might couple two quanta of C=O vibrational excitations into the α-helix chain. So far there has been no reliable multiple-excitation theory. In the case of the $J=0$ limit, the Hamiltonian can be diagonalized by making the usual small polaron transformation [32]. If we denote the small polaron operator by A_i^+, such that $a_i^+ \rightarrow A_i^+$; and denote the dressed phonon operator by B_q, such that $b_q \rightarrow B_q$ where b_q is the

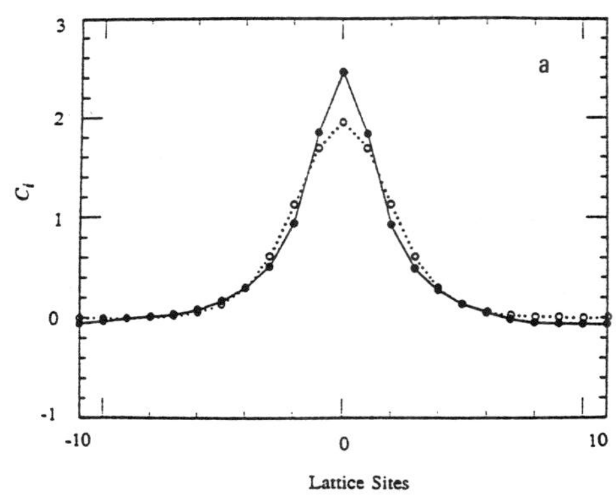
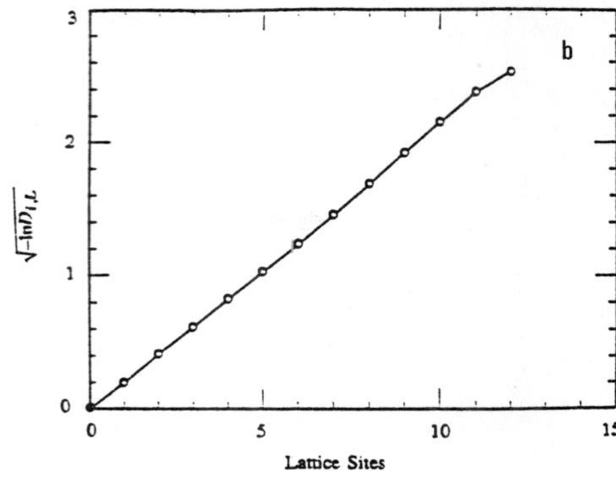

Fig. 6a Comparison of QMC results with Davydov's theory for the hypothetical data $\chi = 1.0$, $J = 6.0$, $M = 0.4$, $w = 0.4$ and $l = 1.0$. Solid line is the simulation result. The temperature of the simulation is very low, $k_B T = 0.1 E_b^D = 0.024 E_b^P$, where E_b^D is the binding energy from Davydov coherent state Ansatz [3,4], and E_b^P is the polaron binding energy [31,21]. The dotted line is the lattice deformation plotting from the D_2 Ansatz theory [21].

Fig. 6b The logarithmic plot of the density-density correlation for the same data as used in Figure 6a. The nearly perfect straight line indicates the quasi-particle behaves like a free particle with a well defined effective mass m^* which can be found from the slope of the line by the relation (5.9); for this particular data, $m^* = 0.61$.

bare acoustic phonon operator of the qth normal mode, the Hamiltonian (1) in this new basis takes the form

$$\hat{H}_{sm.pol.} = \sum_n [E - \frac{1}{2}K(0)]A_n^+ A_n + \sum_q \hbar\omega_q B_q^+ B_q - \frac{1}{2}\sum_{mn} K_{mn}(0)A_m^+ A_n^+ A_m A_n , \qquad (25)$$

where

$$K_{mn}(0) = \frac{\chi^2}{w}(\delta_{m,n-1} + 2\delta_{m,n} + \delta_{m,n+1}) \qquad (26)$$

gives the polaron-polaron interaction strength, and $K(0) \equiv K_{00}(0)$, (note that $K_{mn}(t)$ is the same convolution kernel as appeared in [21,32]). The Hamiltonian in the new basis decomposes into small polaron and dressed phonon parts. Here we are only interested in the polaron part. It is easy to verify that when each excitation is localized (but not necessarily on the same site), we find the (unnormalized) eigenfunctions $|E\{v\}> = \prod_i A_i^{+v_i} | 0 >$ where v_i is the number of excitations on the ith site. The eigenvalues of the small polaron part of Hamiltonian (25) are given by

$$E_{J=0}\{v\} = N[E - \frac{1}{2}K(0)] - \frac{1}{2}\sum_{m,n} K_{mn}(0)v_m(v_n - \delta_{mn}) \qquad (27)$$

The last term (27) arises from the two-body interaction term in the Hamiltonian (25), which gives rise to an attractive force among the small polarons. From the results of our one-excitation simulations, we know the shape of the "soliton" presented in the system is in fact very close to that of the small polaron. We thus expect that the polaron-polaron attractive interaction in (6.1-6.3) should afford a reasonable description of the multi-particle aspects of the "α-helix" system (Table 1). It is interesting

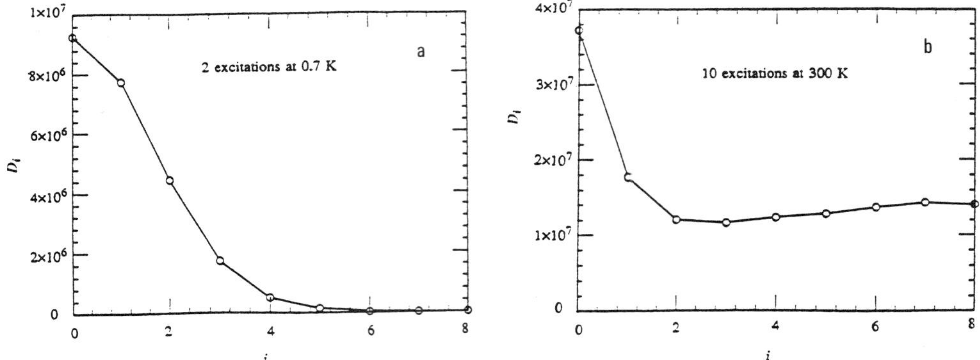

Fig. 7a The unnormalized density-density autocorrelation function $D_{i,0}$, for a 16-site lattice with two excitations present. The temperature is $k_B T = 0.033\chi^2/w = 0.033E_b^P$. The vertical axis represents the number of states measured in the simulation. The correlation function decays to essentially zero indicates that for the majority of our measured configurations, the two excitations are bound together.

Fig. 7b Similar to Figure 7a but for the 10 excitations at the temperature of $k_B T = 14.0\chi^2/w = 14.0E_b^P$, for the "standard" α-helix data from Table 1; this corresponds to $300K$.

that using the data from Table 1 for $N = 2$, the polaron-polaron interaction supports a *bound state* of a pair of polarons.

Fortunately, we can implement multiple-excitation QMC for a fixed number world lines without having to limit ourselves to the $J = 0$ case. Our QMC simulation for the multiple-excitation case is still in the preliminary stage, the complete scanning of the parameter space being yet to be done; however, a few interesting phenomena are already evident. For of the two-excitation case, because of the existence of the bound state noted above, at low temperatures, the world lines of the two excitations tend to lie close to each other, usually occupy nearest neighbor lattice sites; moreover, since we are simulating *identical bosonic* particles, the excitations tend to orbit each other in the Trotter dimension, exchanging positions rapidly in imaginary time. On the checkerboard, this means that the two excitation world lines tend to *cross* each other in the shaded box (cf. (20)).

For multiple-excitation systems, the density-density auto correlation $D_{i,0}$ contains information about the clustering of the excitations. When the excitations tend to cluster, $D_{i,0}$ decays essentially to zero over a distance approximating the radius of the cluster; otherwise, when the excitations move relatively freely with little correlation, $D_{i,0}$ rapidly approaches a non-negligible constant value related to the density of free excitations. At low temperatures, we find that two quasi-particles tend to form a bound-pair excitation as can be seen from Figure 7a. At higher temperatures, these pair excitations appear to dissociate. At physiological temperatures, even with as many as 10 excitations simultaneously present in the system (which according to (27) allows for very large binding energies) excitation clusters still appear to dissociate (see Figure 7b).

CONCLUSION

We have carried out simulations for the Davydov model at finite temperatures in thermal equilibrium by means of the QMC technique. No uncontrollable approximation has been made in carrying out our calculations. We expect numerical errors arising from the discretization of imaginary time and

96

finite sampling to be within a few percent. Our results represent the execution of a *Gendankenexperiment*, and thus stand quite independently of how different theories might be brought to bear to explain them.

The states we have sampled have translational symmetry, due to the symmetry of the Hamiltonian; however, these states can be viewed as superpositions of localized objects. At very low temperatures, and in a certain "hypothetical" parameter region far from the "α-helix" parameters in Table 1, we find the rather broad and smooth quasi-particle profiles which qualitatively resemble the Davydov soliton. Typically, these quasi-particles are embedded in large fluctuations. In general, quantum fluctuations dominate at low temperatures, and at high temperatures, the thermal fluctuations dominate.

For the "standard" α-helix data, we have carried out simulations over a wide range of temperatures. The quasi-particle profile appears to be rather narrow at any temperature, however, as the temperature is increased, the width of the quasi-particle monotonically decreases, such that narrowing is essentially complete at temperatures above a few Kelvins. The combined amplitude of thermal and quantum fluctuations is generally large relative to the coherent lattice deformation at any temperature; however, despite the tempestuousness of the lattice dynamics, the net contraction of the lattice is found to be temperature independent, as expected [34].

In our multiple-excitation simulations we find that clusters of quasi-particles will form at low temperatures, but that these clusters tend to dissociate at high temperatures; for the "standard" α-helix data, even a 10-excitation cluster will dissociate at $300K$. Because of the quantum symmetry obeyed by identical boson excitations, quasi-particles in a cluster tend to interchange positions rapidly in imaginary time.

Since our calculation is for thermal equilibrium only, only those dynamical properties of the system which persist at infinite times can be observed; these include such properties as the quasi-particle effective mass and the exciton-phonon correlation radius. In the sense of [8-11,33], this is the regime which sets in after the soliton's "death". Since we only calculate the *imaginary time* evolution in our QMC simulation, the *real time* dynamics of the initial-value problem is still unknown. However, because of the localized nature of the quasi-particle at high temperature, it is hard to imagine that the quasi-particle would undergo a smooth and coherent dynamical motion. It is more natural, for example, to expect diffusive motion instead. Taken together, these results raise serious questions about the practical utility of the soliton concept in biological systems; however, further real-time studies may be required to provide more definitive conclusions about quasi-particle dynamics.

ACKNOWLEDGEMENT

This work was supported in part by the National Science Foundation under grant no. DMR 86-19650-A01, and by the Center for Nonlinear Studies at Los Alamos under INCOR grant no. CNLS 89-425.

REFERENCES

1.   A. S. Davydov and N. I. Kislukha, Phys. Status Solidi **59**, 465 (1973); Zh. Eksp. Teor. Fiz. **71**, 1090 (1976) [Sov. Phys. JETP **44**, 571 (1976)].

2.   A. S. Davydov, Physica Scr. **20**, 387 (1979); Usp Fiz. Nauk. **138**, 603 (1982) [Sov. Phys. Usp. **25**, 898 (1982)]; Zh. Eksp. Teor. Fiz. **51**, 789 (1980) [Sov. Phys. JETP **78**, 397 (1980)].

3.   A. S. Davydov, *Solitons in Molecular Systems*, (Reidel, Boston, 1985).

4.   A. S. Davydov, in *Solitons*, edited by S. E. Trullinger, V. E. Zakharov, and V. I. Pokrovsky (North-Holland, New York, 1986).

5.   A. S. Davydov, Sov. Phys. JETP **51**(2), 397 (1980).

6.   A. C. Scott, Phys. Rev. A **26**, 587 (1982); **27**, 2767 (1982).

7.   J. C. Eilbeck, P. S. Lomdahl and A. C. Scott, Phys. Rev. B **30**, 4703 (1984).

8.   P. S. Lomdahl and W. C. Kerr, Phys. Rev. Lett. **55**, 1235 (1985).

9.   Albert F. Lawrence, James C. McDaniel, David B. Chang, Brian M. Pierce and Robert R. Birge, Phys. Rev. A **33**, 1188 (1986).

10. P. S. Lomdahl and W. C. Kerr, in *Physics of Many Particle Systems*, edited by A. S. Davydov (Naukova Dumka, Kiev, 1988).

11. L. Cruzeiro, J. Halding, P. L. Christiansen, O. Skovgaard, A. C. Scott, Phys. Rev. A **37**, 880 (1988).

12 Xidi Wang, David W. Brown and Katja Lindenberg, Phys. Rev. Lett. **62**, 1769 (1989).

13. G. Careri, U. Buontempo, F. Carta, E. Gratton and A. C. Scott, Phys. Rev. Lett. **51**, 304 (1983); G. Careri, U. Buontempo, F. Galluzi, A. C. Scott, E. Gratton and E. Shyamsunder, Phys. Rev. B **30**, 4689 (1984); A. C. Scott, E. Gratton, E. Shyamsunder and G. Careri, Phys. Rev. B **32**, 5551 (1985).

14. R. Knox, D. Magde, L. Hancock, and D. Wuttke, Bull. Am. Phys. Soc. **32**, 799 (1987).

15 A. C. Scott, I. J. Bigio and C. T. Johnston, Phys. Rev. B **39**, 12883 (1989).

16. David W. Brown, Katja Lindenberg and Bruce J. West, Phys. Rev. A **33**, 4014 (1986).

17. David W. Brown, Katja Lindenberg and Bruce J. West, Phys. Rev. A **33**, 4110 (1986).

18. David W. Brown, Katja Lindenberg, and Bruce J. West, Phys. Rev. Lett. **57**, 2341 (1986); 3124 (1986); Phys. Rev. B **35**, 6169 (1987).

19. A. C. Scott, Phil. Trans. R. Soc. A **315**, 423 (1985).

20 W. C. Kerr and P. S. Lomdahl, Phys. Rev. B **35**, 3629 (1987).

21. Xidi Wang, David W. Brown, Katja Lindenberg and Bruce J. West, Phys. Rev. A **37**, 3557 (1988).

22 R. Rajaraman, in *Solitons and Instantons* (North-Holland, New York, 1948).

23 Gerd Venzl and Sighart F. Fischer, J. Chem. Phys. **81**, 6090 (1984); Phys. Rev. B **32**, 6437 (1985).

24. A. M. Polyakov, *Gauge Fields and Strings* (Harwood Academic Publishers, New York, 1987).

25. M. Suzuki, S. Miyashita and A. Kuroda, Prog. Theor. Phys. **58**, 1377 (1977).

26. J. W. Negele and H. Orland in *Quantum Many-Particle Systems*, (Addison-Wesley, Reading, Mass., 1988).

27. J. E. Hirsch, R. L. Sugar, D. J. Scalapino and R. Blankenbecker, Phys. Rev. B **26**, 5033 (1982).

28. R. M. Fye, Phys. Rev. B **33**, 6271 (1986).

29. M. Suzuki, Commun. Math. Phys. **51**, 183 (1976).

30. M. Creutz and B. Freedman, Ann. Phys. **132**, 427 (1981).

31. David W. Brown and Zoran Ivic, Phys. Rev. B **40**, 9876 (1989-I).

32. Xidi Wang, David W. Brown and Katja Lindenberg, Phys. Rev. B **39**, 5366 (1989).

33. J. P. Cottingham and J. W. Schweitzer, Phys. Rev. Lett. **62**, 1792 (1989).

34. Xidi Wang, Doctoral Thesis (University of California, San Diego, 1989).

7

QUANTUM EFFECTS ON THE DAVYDOV SOLITON

H. Bolterauer

Institut für Theoretische Physik, Justus-Liebig-
Universität Giessen, D-6300 Giessen, FRG

Abstract

A simple method is introduced to calculate the quantum lifetime of a Davydov soliton.
We calculate the quantum lifetime for the usual parameter set in the Hamiltonian.
The influence of the quantum number and of a prior excited lattice on this lifetime is
shown. An important result is that the Davydov ansatz is exact for a classical lattice.
For this case, the soliton is stable against quantum fluctuations.

1. Introduction

1973 Davydov [1-3] proposed a model for energy transport in quasi one dimensional
biological systems. The basic idea for this model is, that the transport is done due
to separated energy packages, the so called Davydov solitons, which can freely travel
through the system. Solitons are well known for some classical systems which have to
be translational invariant and nonlinear. In these systems solitons have a remarkable
stability and therefore seem well suited to transport energy. The Davydov model, on
the other hand, is a quantum system and the soliton should be a one quantum state.

Davydovs treatment starts with a special ansatz for the soliton wave function

$$| \Psi_S >= \sum_n f_n(t)b_n^+ \mid 0 > U(t)|\varphi > \tag{1.1a}$$

$$U(t) = \exp\{\frac{1}{i\hbar} \sum_n (Q_n(t)p_n - P_n(t)q_n)\} \tag{1.1b}$$

which is symmetry breaking. His method to derive equations for the parameters f_n,
Q_n and P_n seem to be somewhat special and there was some questioning about this
method [4,5], but we showed that we can also use the well known time dependent
variational principle to derive the same equations [6].

Davydov's Soliton Revisited, Edited by P.L. Christiansen and A.C. Scott
Plenum Press, New York, 1990

The question remains, if the ansatz (1.1) is good enough [4] or to ask in a more principle manner: How can a soliton solution, which means a localized wavefunction, remain stable for a quantum system which is translational invariant [7-9]. Indeed, the right understanding of the consequences of translational invariance, seem crucial to us:

> Translational invariance is necessary for a freely moving soliton.

> On the other hand, it is an additional symmetry of the Hamiltonian, which means that stationary states must also be eigenstates of a translational operator. However, a non moving stable soliton should be a stationary state.

Using this translational symmetry we showed in [7], that stationary states must be Bloch states. Therfore they are not localized [8,9]. Only in the limiting case of a vanishing band width, (which means a degenerate set of states), allows the construction of localized stable states. If we consider a classical lattice we just have this case and we therefore see a stable Davydov soliton. For a non classical lattice a localized state will never remain stable. Davydov [3] argued that the soliton should remain stable because of a so called topological stability. Indeed the soliton solution has special boundary conditions: The distortion of the lattice on the lefthand side is different from the right hand side. This prevents a decay in exzitons. But it would not prevent a decay of the Davydov soliton in Bloch states which are themselves constructed out of solitons. It was this idea that we used to calculate the lifetime τ of the soliton, with the result that the soliton should decay in about 10^{-13} s. In this paper we again present a calculation of τ but without using any model of the decay.

In the following we use

$$H_S = H_L + H_O + H_I$$

$$H_L = \sum_{n=1}^{L} \left(\frac{p_n^2}{2M} + 1/2K(q_n - q_{n-1})^2 \right)$$

$$H_O = \sum_{n=1}^{L} \left(\epsilon\, b_n^+ b_n - J\, b_n^+ (b_{n+1} + b_{n-1}) \right)$$

$$H_I = \chi \sum_{n=1}^{L} b_n^+ b_n (q_n - q_{n-1}) \tag{1.2}$$

as Hamiltonian for the system.

2. The uncertaincy of energy and the quantum lifetime

In this paper we consider a nonmoving Davydov soliton, which means the operator X, describing the position of the soliton

$$X = \sum_n b_n b_n^+ n \tag{2.1}$$

has a time independent expectation value. If the soliton should be stable it must be an eigenstate of the system. If it is not an eigenstate we calculate the lifetime in the following way:

We start with the wellknown unequality for two operators A and B.

$$(\Delta A)^2(\Delta B)^2 \geq \left| \frac{1}{2i} < [A, B] > \right|^2 \tag{2.2}$$

For B we now take the Hamiltonian H. A should be an operator which describes the typical form of the soliton, for example, it could be

$$A = (X - < X >)^2 \tag{2.3}$$

which describes the broadness. We then immediately get

$$\Delta E \Delta A \geq \frac{\hbar}{2} \left| \frac{d}{dt} < A > \right|$$

or

$$\Delta E \left| \frac{\Delta A}{\frac{d}{dt} < A >} \right| \geq \frac{\hbar}{2} \tag{2.4}$$

Tl.e quantity τ

$$\tau = \frac{\Delta A}{\frac{d}{dt} < A >} \tag{2.5}$$

is the characteristic time where $< A >$ is essentially changing , therefore τ is the lifetime of the soliton.

3. The calculation of ΔE

We start with the Davydov ansatz

$$| \Psi_S >= \sum_n f_n(t) b_n^+ | 0 > U(t)|\varphi > \tag{3.1a}$$

$$U(t) = \exp \left\{ \frac{1}{i\hbar} \sum_n (Q_n(t)p_n - P_n(t)q_n) \right\} \tag{3.1b}$$

Here $|0 >$ is the vacuum state for the oscillators (H_O) and $|\varphi >$ is an eigenstate of H_L. Since we are considering a non moving soliton, we can use the usual variational principle to calculate the parameters f_n, P_n and X_n. Using the Hamiltonian (1.2) this directly gives

$$-\frac{\chi^2}{K}|f_n|^2 f_n - J(f_{n+1} + f_{n-1}) = E_f f_n \tag{3.2}$$

with

$$X_n - X_{n-1} = -\frac{\chi}{K}|f_n|^2$$
$$P_n = 0 \tag{3.3}$$

Let us assume that we have a solution of (3.2). We now directly calculate

$$< H^2 > = < \Psi_S|H^2|\Psi_S >$$
$$= \sum_m f_m < \varphi|U^+\{H_L + \epsilon + \chi(q_m - q_{m-1}) - J(A^+ + A^-)\}^2 U|\varphi > f_m \tag{3.4}$$

with

$$A^+ f_m = f_{m+1}; \quad A f_m = f_{m-1} \tag{3.5}$$

Shifting the unitary operator U to the left side and using

$$F(q_m)U = U F(q_m + Q_m) \tag{3.6}$$

for an arbitrary function $F(q_m)$ yields

$$< H^2 > = \sum_m f_m < \varphi|\{H_L + \epsilon + \chi(q_m - q_{m-1}) - J(A^+ + A^-)$$
$$+\chi(Q_m - Q_{m-1}) + \frac{1}{2}K\sum_n[(Q_n - Q_{n-1})^2 + 2(q_n - q_{n-1})(Q_n - Q_{n-1})]\}^2|\varphi > f_m \tag{3.7}$$

We know that for all stationary states of the lattice, linear and cubic terms in q and p will vanish.

$$< q_m > = < q_m^2 q_n > = < q_n p_m^2 > = 0$$

Using this fact, together with (3.3),(3.2) and $\sum_n |f_n|^2 = 1$ results in

$$< H^2 > =$$
$$\sum_m f_m < \varphi|\left\{H_L + \epsilon + \frac{1}{2}\frac{\chi^2}{K}\sum_n f_n^4 - \frac{\chi^2}{K}|f_m|^2 - J(A^+ + A^-)\right\}^2|\varphi > f_m$$
$$+\chi^2 \sum_m f_m^2 < \varphi|\left\{(q_m - q_{m-1}) - \sum_n(q_n - q_{n-1})|f_n|^2\right\}^2|\varphi >$$
$$= \sum_m f_m < \varphi|\left\{H_L + \epsilon + \frac{1}{2}\frac{\chi^2}{K}\sum_n f_n^4 + E_f\right\}^2|\varphi > f_m$$
$$+\chi^2 \sum_m f_m^2 < \varphi|\left\{(q_m - q_{m-1}) - \sum_n(q_n - q_{n-1})|f_n|^2\right\}^2|\varphi > \tag{3.8}$$

A similar calculation yields for $< H >$

$$< H >= \sum_m f_m < \varphi | \{ H_L + \epsilon + \frac{1}{2} \frac{\chi^2}{K} \sum_n f_n^4 + E_f \} | \varphi > f_m \qquad (3.9)$$

Using (3.5) and (3.2) we get

$$< H^2 >=< H >^2 + \chi^2 \sum_m f_m^2 < \varphi | \{ (q_m - q_{m-1}) - \sum_n (q_n - q_{n-1}) | f_n |^2 \}^2 | \varphi >$$
$$(3.10)$$

and

$$\Delta E^2 = \chi^2 \sum_m f_m^2 < \varphi | \{ (q_m - q_{m-1}) - \sum_n (q_n - q_{n-1}) | f_n |^2 \}^2 | \varphi > \qquad (3.11)$$

Since $\sum_m |f_m|^2 = 1$ we finally arrive at a relatively simple expression for the uncertaincy ΔE, which determines the lifetime of the soliton.

$$\Delta E^2 = \chi^2 < \varphi | (q_m - q_{m-1})^2 | \varphi >$$
$$-\chi^2 \sum_{m,n} f_m^2 f_n^2 < \varphi | (q_m - q_{m-1})(q_n - q_{n-1}) | \varphi > \qquad (3.12)$$

Before going on further, let us first ask when $\Delta E = 0$ which means the soliton remains stable. We see three possibilities:

a. No coupling between the two subsystems ($\chi = 0$).
   We have no real localized soliton. The soliton solution in this limiting case is nothing else than an exziton which is an eigenstate of H_O (see [7]).

b. The lattice is classical which means $< q_n^2 >=< q_n q_m >=< q_n >= 0$. In this case the ansatz (1.1) is exact and the soliton equations (3.2) and (3.3) can be obtained directly [7].

c. The form of the soliton is extremly narrow

$$f_m^2 = \delta_{m,m_0} \qquad (3.13)$$

where m_0 is the position of the soliton. But (3.13) is a solution of (3.2) only for the limiting case of $\chi \to \infty$.

One of these conditions is necessary for the stability of the Davydov soliton for T=0. Condition b. seems to be the most realistic possibility.

If we generalize the Davydov soliton allowing a large quantum number N

$$N = \sum_n b_n^+ b_n$$

we have a completely new effect which can stabilize the soliton: The effect of Bose condensation.

Instead of (1.1a) we then use the ansatz

$$| \Psi_S >= \frac{1}{\sqrt{N!}} \left\{ \sum_n f_n(t) b_n^+ \right\}^N | 0 > U(t) |\varphi > \quad ; \sum_n |f_n|^2 = 1 \qquad (3.14)$$

Again using the variational principle we get, as a generalisation of the case $N = 2$ which was discussed in [10], the following:

$$-\frac{N\chi^2}{K} |f_n|^2 f_n - J(f_{n+1} + f_{n-1}) = E_f f_n \qquad (3.15)$$

with

$$X_n - X_{n-1} = -\frac{N\chi}{K} |f_n|^2$$
$$P_n = 0 \qquad (3.16)$$

These are the equations (3.2) and (3.3) but now with an effective coupling parameter $\sqrt{N}\chi$ instead of χ. If we calculate ΔE we have the same effect arriving at

$$\Delta E^2 = N^2 \chi^2 < \varphi|(q_m - q_{m-1})^2|\varphi >$$
$$- N^2 \chi^2 \sum_{m,n} f_m^2 f_n^2 < \varphi|(q_m - q_{m-1})(q_n - q_{n-1})|\varphi > \qquad (3.17)$$

For $N \to \infty$ both ΔE^2 and E are proportional to N^2. Therfore we have

$$\frac{\Delta E}{E} = 0 \qquad (3.18)$$

which means that (3.14) *is a macroskopic quasiclassical state comparable with the ground state of a ferromagnet,* which also remains stable even if it does not have the symmetry of the Hamiltonian.

Similar arguments hold for the Fröhlich theory, where the so called Fröhlich Bose condensation can be seen as a macroscopic soliton in momentum space [11].

Let us now try to calculate ΔE for the parameters usually used for the Davydov model. For the following we assume that $|\varphi >= |\varphi_0 >$ where $|\varphi_0 >$ is the groundstate of the unperturbed phonon system

Since $< q_m q_n >$ is only dependent on the difference $|m - n|$ we have

$$\Delta E^2 = \chi^2 < \varphi_0|(q_m - q_{m-1})^2|\varphi_0 > (1 - \sum_n |f_n|^4)$$
$$- 2\chi^2 \sum_{n=1}^{\infty} < \varphi_0|(q_n - q_{n-1})(q_0 - q_{-1})|\varphi_0 > \sum_m |f_{n+m}|^2 |f_m|^2 \qquad (3.19)$$

To calculate the remaining quantities we use the well known transformations

$$q_n = \sqrt{\frac{1}{L}} \sum_\lambda e^{in\lambda} q_\lambda = \sqrt{\frac{1}{L}} \sum_\lambda e^{-in\lambda} q_\lambda^+$$

$$q_\lambda = \sqrt{\frac{\hbar}{2M\omega_\lambda}} (a_\lambda + a_{-\lambda}^+) \tag{3.20}$$

which diagonalize H_L.

$$H_L = \sum_\lambda \hbar\omega_\lambda \left(\frac{1}{2} + a_\lambda^+ a_\lambda\right) \tag{3.21}$$

Here λ is the phonon wavenumber and ω_λ the corresponding frequency.

$$\lambda = \frac{2\pi}{L}l; \quad l = 0, \pm1, \pm2, ...; \quad \omega_\lambda = 2\sqrt{\frac{K}{M}} |sin(\tfrac{1}{2}\lambda)| \tag{3.22}$$

Using

$$< \varphi_0 | q_\lambda \, q_{\lambda'}^+ | \varphi_0 > = \delta_{\lambda,\lambda'} \frac{\hbar}{2M\omega_\lambda} \tag{3.23}$$

we have

$$< \varphi_0 | (q_{n_1} - q_{n_1-1})(q_{n_2} - q_{n_2-1}) | \varphi_0 > = \frac{1}{L} \sum_\lambda e^{i(n_1-n_2)\lambda} < q_\lambda q_\lambda^+ > \, 4sin^2(\tfrac{\lambda}{2})$$

$$= \frac{1}{L} \sum_\lambda e^{i(n_1-n_2)\lambda} \frac{\hbar}{\sqrt{KM}} |sin(\tfrac{\lambda}{2})| = \frac{\hbar}{L\sqrt{KM}} \sum_\lambda cos\lambda(n_1 - n_2) \, |sin(\tfrac{\lambda}{2})| \tag{3.24}$$

For large L we can replace the sum by an integral

$$\sum_\lambda \approx \frac{L}{2\pi} \int_{-\pi}^{\pi} d\lambda$$

we finally have

$$< \varphi_0 | (q_{n_1} - q_{n_1-1})(q_{n_2} - q_{n_2-1}) | \varphi_0 > = \frac{\hbar}{2\pi\sqrt{KM}} \int_{-\pi}^{\pi} d\lambda \, cos\lambda(n_1 - n_2) \, |sin(\tfrac{\lambda}{2})|$$

$$= -\frac{2\hbar}{\pi\sqrt{KM}} \frac{1}{(n_1 - n_2)^2 - 1} \tag{3.25}$$

Here we used

$$\int_0^{\frac{\pi}{2}} cos(2lx)sin(x) = \frac{1}{1 - (2l)^2}$$

This now gives the final expression for ΔE

$$\Delta E^2 = A \left\{ 1 - \sum_m |f_m|^4 - 2 \sum_{n=1}^{\infty} \frac{1}{1-(2n)^2} \sum_m |f_{n+m}|^2 \, |f_m|^2 \right\}$$

$$A = \frac{2\chi^2 \hbar}{\pi \sqrt{KM}} \tag{3.26}$$

We calculated this expression using the following parameters

$$M = 114.2 \, m_{prot}$$
$$K = 19.8 \, \frac{J}{m^2}$$
$$\epsilon = 3.28 \; 10^{-20} \, J \tag{3.27}$$
$$J = 1.55 \; 10^{-22} \, J$$
$$\chi = 6.2 \;\; 10^{-11} \, N$$

Since the result is not sensitive to the exact form of f_n (only the prefactor A is important), we used the continuous form of f_n

$$f_n = \sqrt{B} \, sech(2Bn); \quad B = \frac{\chi^2}{8KJ} \tag{3.28}$$

The result is

$$A = 1.3 \; 10^{-43}$$
$$\Delta E^2 = 0.8 \, A = 1.07 \; 10^{-43} \tag{3.29}$$

or

$$\Delta E = 3.2 \; 10^{-22}$$
$$\tau = 3.2 \; 10^{-13} \tag{3.30}$$

showing the same order of magnitude as in a previous calculation [7], where an inequality was used to give a rough estimate of ΔE.

In our opinion τ is much to short to allow the Davydov soliton to transport energy remaining localized.

We want to point out that this quantum lifetime would increase when we use an excited state for $|\varphi >$ instead of the phonon groundstate $|\varphi_0 >$. Then, instead of (3.23) we have to use

$$< \varphi | q_\lambda q_{\lambda'}^+ | \varphi > = \delta_{\lambda,\lambda'} \left(n_\lambda + \frac{1}{2} \right) \frac{\hbar}{M\omega_\lambda} \tag{3.31}$$

where n_λ is the number of excited phonons with wavenumber λ.

References

1. Davydov, A.S., J. Theor. Biol. 38 (1973) 559.

2. Davydov, A.S. and Kislukha, N.I., Phys. Stat. Sol. B59 (1973).

3. See Davydov, A.S., Solitons in Molecular Systems, Riedel, Dordrecht (1985) for bibliographies.

4. Brown, D.W., Lindenberg, K. and West, B.J., Phys. Rev A33, 4110 (1986)

5. Brown, D.W., Lindenberg, K. and West, B.J., Phys. Rev A33, 4114 (1986)

6. Bolterauer, H. and Henkel, R.D., Phys. Scripta T13 (1986) 314

7. Bolterauer, H. and Opper, M. "The Quantum lifetime of the Davydov soliton"; submitted to Physica Scripta November 1988

8. Venzl, G. and Fischer, F., J. Chem. Phys. 81, 6090 (1984)

9. Venzl, G. and Fischer, F., J. Phys. Rev. B32, 6437 (1985)

10. Bolterauer, H., Henkel, R. D. and Opper, M., "Resonant and Quasiclassical Excitations of Solitons in the Alpha-Helix", in Structure, Coherence and Chaos in Dynamical Systems, edited by P. L. Christiansen and R. D. Parmentier, Manchester University Press 1986.

11. Bolterauer, H. and Tuszyński, J. A., "Fröhlichs Condensation in a Biological Membrane Vieved as a Davydov Soliton", Journal of Biological Physics, 17, 41 (1989)

8

DAVYDOV ANSATZ AND PROPER SOLUTIONS OF SCHRÖDINGER EQUATION FOR FRÖHLICH HAMILTONIAN

M. Škrinjar, D. Kapor and S. Stojanović

Institute of Physics, Faculty of Sciences
University of Novi Sad, Dr I. Đuričića 4
YU – 21 000 Novi Sad, Yugoslavia

1. INTRODUCTION

Theory of Davydov solitons in molecular systems is rather well established nowadays[1], but certain problems dealing with the quantum-mechanical foundations of the theory still remain, especially those concerning the applicability of various equations of motion, both quantum and classical.

The basic ingredients of the theory are the Hamiltonian describing the system and trial functions enabling us to go over from operators to c-numbers. The essential interactions in the system are described by Fröhlich's Hamiltonian[2]:

$$H = \Delta \sum_{n} B_{n}^{+} B_{n} - I \sum_{n} \left(B_{n}^{+} B_{n+1} + B_{n+1}^{+} B_{n} \right) + \sum_{n,q} \hbar\omega_{q} B_{n}^{+} B_{n} \left[\chi_{n}^{q} b_{q}^{+} + \chi_{n}^{q\,*} b_{q} \right] +$$

$$+ \sum_{q} \hbar\omega_{q} b_{q}^{+} b_{q} \qquad (1.1)$$

Here, B_{n}^{+} (B_{n}) are boson creation (annihilation) operators for quanta of intramolecular vibrations at site n with energy Δ; b_{q}^{+} (b_{q}) create (annihilate) acoustic phonons with the energy $\hbar\omega_{q}$; I is the intersite transfer energy produced by dipole-dipole interactions.

The coupling constant χ_{n}^{q} satisfies: $\chi_{n}^{*q} = \chi_{n}^{-q}$ and in the nearest-neighbour approximation for the ordered chain, it can be written as

Davydov's Soliton Revisited, Edited by P.L. Christiansen and A.C. Scott
Plenum Press, New York, 1990

$$\chi_n^q = -2\chi i \sqrt{\frac{\hbar}{2MN\omega_q}} \; \frac{\sin qa}{\hbar\omega_q} \; e^{-iqna} = \chi_q \, e^{-iqna} \qquad (1.2)$$

where a is the lattice constant, M is the mass of the molecule and N the number of molecules in the chain. Nonlinear coupling term $(\sim\chi)$ arises from the modulation of the one-site energy by the molecular displacements.

There exist two trial functions, titled also Davydov Ansatz, with common notation:

$$|D_1(t)\rangle = \sum_n \psi_n(t) B_n^+ |0\rangle_{ex} \, |\beta_n(t)\rangle \qquad (1.3a)$$

$$|\beta_n(t)\rangle = \exp\left[-\sum_q \left(\beta_{nq}^*(t) b_q - \beta_{nq}(t) b_q^+\right)\right] |0\rangle_{ph} \qquad (1.3b)$$

and

$$|D_2(t)\rangle = |\varphi\rangle |\beta\rangle \qquad (1.4a)$$

$$|\varphi\rangle = \sum_n \psi_n(t) B_n^+ |0\rangle_{ex} \qquad (1.4b)$$

$$|\beta\rangle = \prod_q |\beta_q\rangle = \prod_n |\beta_n\rangle = e^{-S} |0\rangle_{ph} =$$

$$= \exp\left[\sum_q \left(\beta_q(t) b_q^+ - \beta_q^*(t) b_q\right)\right] |0\rangle_{ph} \qquad (1.4c)$$

It follows from the principles of Quantum Mechanics[3] that simple trial function in the form of tensor product of exciton and phonon function can not be the solution of Schrödinger equation (SE) for the Hamiltonian with interaction (1.1). This was the reason that recently much more attention has been paid to the trial function $|D_1\rangle$ [4,5,6,7,8,9] and most of the limitations for its application are known now. On the other hand, most of the original work was based on $|D_2\rangle$ - Ansatz[1], so we return to it and try to establish quantum-mechanical foundations for its application and the limits of its validity.

The structure of the work is as follows. In Sec.2 we review classical and quantum equations of motion under the assumption that SE is valid for $|D_2\rangle$. Its validity will be studied in more detail in Sec.3, while the concluding part will offer some open questions.

2. SCHRÖDINGER EQUATION AND CLASSICAL EQUATIONS OF MOTION

First, of all, we must notice that the function $|D_2\rangle$ should be normalized, and the normalizing condition will be used in the form

$$\sum_n |\psi_n|^2 = C \tag{2.1}$$

because this relation represents a constraint in the variational calculations.

Initially, we shall look at the Lagrangian formalism. The strict formulation [10,11] is based on the expression

$$L = \frac{i\hbar}{2} \langle D_2 | \overset{\leftrightarrow}{\frac{\partial}{\partial t}} | D_2 \rangle - \langle D_2 | H | D_2 \rangle \tag{2.2}$$

giving

$$L = \frac{i\hbar}{2} \sum_n \left[\dot{\psi}_n \psi_n^* - \dot{\psi}_n^* \psi_n \right] + \frac{i\hbar}{2} \sum_q \left[\dot{\beta}_q \beta_q^* - \dot{\beta}_q^* \beta_q \right] C -$$

$$- \Delta \sum_n |\psi_n|^2 + I \sum_n \left[\psi_n^* \psi_{n+1} + \psi_{n+1}^* \psi_n \right] - C \sum_q \hbar\omega_q |\beta_q|^2 -$$

$$- \sum_{n,q} \hbar\omega_q |\psi_n|^2 \left[\chi_n^q \beta_q^* + \chi_n^{q*} \beta_q \right] \tag{2.3}$$

Once again, let us remember that the equation (2.1) represents a constraint, so we have to perform the variation in the following way:[9]

$$\delta \int \left[L - \lambda \left(\sum_n |\psi_n|^2 \right) \right] dt = 0 \tag{2.4}$$

where λ is the (undetermined) Lagrangian multiplier.

This leads to the following set of equations:

$$Ci\hbar\dot{\beta}_q = C\hbar\omega_q \beta_q + \sum_{n'} \hbar\omega_q |\psi_{n'}|^2 \chi_n^q, \tag{2.5a}$$

$$-Ci\hbar\dot{\beta}_q^* = C\hbar\omega_q \beta_q^* + \sum_{n'} \hbar\omega_q |\psi_{n'}|^2 \chi_n^{q*}, \tag{2.5b}$$

$$i\hbar\dot{\psi}_n = \left[\Delta + \sum_q \hbar\omega_q \left(|\beta_q|^2 + \chi_n^q \beta_q^* + \chi_n^{q*} \beta_q \right) - \right.$$

$$\left. - \frac{i\hbar}{2} \sum_q \left[\dot{\beta}_q \beta_q^* - \dot{\beta}_q^* \beta_q \right] + \lambda \right] \psi_n - I \left[\psi_{n+1} + \psi_{n-1} \right] \tag{2.6}$$

Setting C=1 in (2.5), the system (2.5), (2.6) represents the system of Euler-Lagrangian equations (ELE) for the system with the Lagrangian (2.3) and the constraint (2.1).

Let us now look at the Heisenberg equation of motion (HE) for the operator b_q:

$$i\hbar \dot{b}_q = \hbar\omega_q b_q + \sum_n \hbar\omega_q \chi_n^q B_n^+ B_n \tag{2.7}$$

Averaging it over $|D_2\rangle$ we obtain

$$i\hbar \dot{\beta}_q C = \hbar\omega_q \beta_q C + \sum_n \hbar\omega_q |\psi_n|^2 \chi_n^q \tag{2.8}$$

(Two comments are necessary here. First, Heisenberg equation for B_n does not lead to any relevant result, because when averaged over single--particle function $|\varphi\rangle$ it vanishes. Second, transition from (2.7) to (2.8) implicitely assumes the validity of SE for $|D_2\rangle$.)

Finally, let us write down SE for $|D_2\rangle$:

$$i\hbar \frac{\partial |D_2\rangle}{\partial t} = H|D_2\rangle \tag{2.9}$$

This is an assumption, leading to the following equation:

$$i\hbar \sum_n \dot{\psi}_n B_n^+ |0\rangle_{ex} |\beta\rangle + \frac{i\hbar}{2} \sum_{n,q} \psi_n \left[\dot{\beta}_q \beta_q^* - \dot{\beta}_q^* \beta_q \right] B_n^+ (0)_{ex} |\beta\rangle +$$

$$+ i\hbar \, e^{-S} \sum_{n,q} \psi_n \dot{\beta}_q B_n^+ |0\rangle_{ex} b_q^+ |0\rangle_{ph} =$$

$$= \Delta \sum_n \psi_n B_n^+ |0\rangle_{ex} |\beta\rangle - I \sum_n \left[\psi_{n+1} + \psi_{n-1} \right] B_n^+ |0\rangle_{ex} |\beta\rangle +$$

$$+ \sum_{n,q} \hbar\omega_q \psi_n \left[\chi_n^q \beta_q^* + \chi_n^{*q} \beta_q + |\beta_q|^2 \right] B_n^+ |0\rangle_{ex} |\beta\rangle +$$

$$+ e^{-S} \sum_{n,q} \hbar\omega_q \psi_n \left[\beta_q + \chi_n^q \right] B_n^+ |0\rangle_{ex} b_q^+ |0\rangle_{ph} \tag{2.10}$$

We are going to project (2.10) onto several relevant directions.

a) projection onto the direction $B_n^+ |0\rangle_{ex} |\beta\rangle$ leads to

$$i\hbar \dot{\psi}_n = \left[\Delta - \frac{i\hbar}{2} \sum_q \left(\dot{\beta}_q \beta_q^* - \dot{\beta}_q^* \beta_q \right) + \sum_q \hbar\omega_q \left(|\beta_q|^2 + \chi_n^q \beta_q^* + \chi_n^{*q} \beta_q \right) \right] \psi_n -$$

$$- I \left[\psi_{n+1} + \psi_{n-1} \right] \tag{2.11a}$$

b) projecting (2.10) onto $\sum_n \psi_n B_n^+ |0\rangle_{ex} e^{-S} b_q^+ |0\rangle_{ph}$ gives

$$C i\hbar \dot{\beta}_q = \hbar\omega_q \beta_q C + \sum_n \hbar\omega_q |\psi_n|^2 \chi_n^q \tag{2.11b}$$

c) projection onto the direction $B_n^+|0\rangle_{ex}\ e^{-S}\ b_q^+|0\rangle_{ph}$ gives

$$i\hbar\dot\beta_q = \hbar\omega_q\beta_q + \hbar\omega_q\chi_n^q \tag{2.11c}$$

This last relation is obviously meaningless, since β_q is site independent.

On the other hand, by multiplying (2.11c) with $|\psi_n|^2$ and summing over n, leads us again to (2.11b). This means that (2.11b) is a consequence of (2.11c), but that they should be both fulfiled. In fact, substituting (2.11 a,c) into SE leads to an identity. We are now the position to conclude that $|D_2\rangle$ does not satisfy SE, except and only in the case $\chi_n^q = 0$, because that is the only case when (2.11c) can be fulfiled [7].

Next, we notice that ELE agree with (2.11a,b) only in the case $\lambda=0$. This is reasonable because the variational calculation with the Lagrangian L leads to SE only for $\lambda=0$ ($\lambda\neq0$ in fact changes only the phase).

Finally, let us note that if we suppose that β_q and ψ_n follow the dynamics determined by ELE, than the averaged SE is valid.

$$\langle D_2|i\hbar\frac{\partial}{\partial t}|D_2\rangle = \langle D_2|H|D_2\rangle \tag{2.12}$$

Let us now look at the particular case when C=1 is apriori fixed ($\delta C=0$). The Lagrangian becomes

$$L = \frac{i\hbar}{2}\sum_n\left(\dot\psi_n\psi_n^* - \psi_n^*\dot\psi_n\right) + \frac{i\hbar}{2}\sum_q\left(\dot\beta_q\beta_q^* - \beta_q\dot\beta_q^*\right) - \langle D_2|H|D_2\rangle \tag{2.13}$$

and the minimalization is performed for $\lambda=0$. The condition

$$\delta\int L dt = 0 \tag{2.14}$$

leads to the following set of equations

$$i\hbar\dot\beta_q = \hbar\omega_q\beta_q + \sum_n\hbar\omega_q|\psi_n|^2\chi_n^q \tag{2.15a}$$

$$i\hbar\dot\psi_n = \Delta\psi_n - I\left[\psi_{n+1} + \psi_{n-1}\right] + \psi_n\sum_q\hbar\omega_q\left(\chi_n^q\beta_q^* + \chi_n^{q*}\beta_q\right) \tag{2.15b}$$

which is the set of Davydov's equations.

It is important to notice that eq. (2.15b) is not consistentent with any of previously derived equations, including averaged SE.

3. THE EQUATION OF MOTION FOR $|D_2(t)\rangle$

The main conclusion of the previous section can be put in the very simple form

$$|D_2(t)\rangle \neq e^{-\frac{i}{\hbar}Ht} |D_2(0)\rangle \tag{3.1}$$

for $\chi_n^q \neq 0$. We shall now look for the equation of time evolution of $|D_2(t)\rangle$ by substituting ELE (2.11a,b) (with C=1) into the expression for the time derivative of $|D_2\rangle$. This implies that we shall suppose that the dynamics of $\psi_n(t)$ and $\beta_q(t)$ is governed by ELE.

$$i\hbar \frac{\partial |D_2\rangle}{\partial t} = \sum_n i\hbar \dot{\psi}_n B_n^+ |0\rangle_{ex} B\rangle + \frac{i\hbar}{2} \sum_{n,q} \psi_n \left[\dot{\beta}_q \beta_q^* - \dot{\beta}_q^* \beta_q \right] \times$$

$$\times B_n^+ |0\rangle_{ex} |\beta\rangle + i\hbar\, e^{-S} \sum_{n,q} \psi_n \dot{\beta}_q B_n^+ |0\rangle_{ex} b_q^+ |0\rangle_{ph} \equiv$$

$$\equiv \Delta \sum_n \psi_n B_n^+ |0\rangle_{ex} |\beta\rangle + \sum_{n,q} \psi_n \hbar\omega_q \left[|\beta_q|^2 + \chi_n^q \beta_q^* + \chi_n^q \beta_q \right] B_n^+ |0\rangle_{ex} |\beta\rangle -$$

$$- I \sum_n \left[\psi_{n+1} + \psi_{n-1} \right] B_n^+ |0\rangle_{ex} |\beta\rangle + e^{-S} \sum_{n,q} \hbar\omega_q \psi_n |\psi_n|^2 \chi_n^q B_n^+ |0\rangle_{ex} b_q^+ |0\rangle_{ph} +$$

$$+ e^{-S} \sum_{n,q} \hbar\omega_q \psi_n \chi_n^q B_n^+ |0\rangle_{ex} b_q^+ |0\rangle_{ph} \tag{3.2}$$

This leads to

$$i\hbar \frac{\partial |D_2\rangle}{\partial t} = H |D_2\rangle + |\delta(t)\rangle \tag{3.3}$$

where

$$|\delta(t)\rangle = e^{-S} \sum_{n,q} \hbar\omega_q \psi_n \left[\gamma_q - \chi_n^q \right] B_n^+ |0\rangle_{ex} b_q^+ |0\rangle_{ph} \tag{3.4}$$

and

$$\gamma_q = \sum_{n'} \chi_{n'}^q |\psi_{n'}|^2 \tag{3.5}$$

For the sake of simplicity, we can put $|\delta(t)\rangle$ in the following form

$$|\delta(t)\rangle = \left[- H_{int} + v \right] |D_2(t)\rangle \tag{3.6}$$

where

$$v = \sum_{n,q} \hbar\omega_q B_n^+ B_n \left[\gamma_q b_q^+ + \gamma_q^* b_q \right] + \sum_n \Delta B_n^+ B_n \tag{3.7}$$

and

$$\Delta_n = \delta_n - \sum_n \hbar\omega_q \left[\beta_q^+ \gamma_q + \gamma_q^* \beta_q \right] \tag{3.8a}$$

$$\delta_n = \sum_q \hbar\omega_q \left[\chi_n^q \beta_q^* + \chi_n^{q*} \beta_q \right] \tag{3.8b}$$

We see that time derivative of $|D_2\rangle$ can be written in the form

$$i\hbar \frac{\partial}{\partial t} |D_2\rangle = \bar{H} |D_2\rangle \tag{3.9}$$

where the effective Hamiltonian

$$\bar{H} = H_{ex} + H_{ph} + v \tag{3.10}$$

can be written as

$$\bar{H} = \sum_n (\Delta + \Delta_n) B_n^+ B_n - I \sum_n B_n^+ \left[B_{n+1} + B_{n-1} \right] + \sum_q \hbar\omega_q b_q^+ b_q +$$
$$+ \sum_{n,q} \hbar\omega_q B_n^+ B_n \left[\gamma_q b_q^+ + \gamma_q^* b_q \right] \tag{3.11}$$

The explicit expression for (3.9) is

$$i\hbar \sum_n \dot{\psi}_n B_n^+ |0\rangle_{ex} |\beta\rangle + \frac{i\hbar}{2} \sum_{n,q} \psi_n \left[\dot{\beta}_q \beta_q^* - \dot{\beta}_q^* \beta_q \right] B_n^+ |0\rangle_{ex} |\beta\rangle +$$
$$+ i\hbar\, e^{-S} \sum_{n,q} \psi_n \dot{\beta}_q B_n^+ |0\rangle_{ex} b_q^+ |0\rangle_{ph} =$$
$$= \sum_n \psi_n \left[\Delta + \sum_q \hbar\omega_q \left[|\beta_q|^2 + \chi_n^q \beta_q^* + \chi_n^{q*} \beta_q \right] \right] B_n^+ |0\rangle_{ex} |\beta\rangle +$$
$$+ e^{-S} \sum_q \hbar\omega_q \left[\beta_q + \gamma_q \right] b_q^+ |0\rangle_{ph} |\varphi\rangle \tag{3.12}$$

It can be easily seen that the equation (3.12) can be projected only onto two directions: $B_n^+ |0\rangle_{ex} |\beta\rangle$ and $e^{-S} b_q^+ |0\rangle_{ph} |\varphi\rangle$, which give directly ELE.

Let us now introduce the following Hamiltonian:

$$H_D = H_{ex} + H_{ph} + \sum_{n,q} \hbar\omega_q B_n^+ B_n \left[\chi_n^q \beta_q^* + \chi_n^{q*} \beta_q \right] +$$
$$+ \sum_{n,q} \hbar\omega_q \left[|\psi_n|^2 \left[\chi_n^q b_q^+ + \chi_n^{q*} b_q \right] - \sum_{n,q} \hbar\omega_q |\psi_n|^2 \left[\chi_n^q \beta_q^* + \chi_n^{q*} \beta_q \right] \equiv$$

$$\equiv \sum_n (\Delta + \delta_n)\, B_n^{+} B_n \; - \; I \sum_n B_n^{+}\left[B_{n+1} + B_{n-1} \right] \; + \; \sum_q \hbar\omega_q\, b_q^{+} b_q \; +$$

$$+ \sum_q \hbar\omega_q \left[\gamma_q\, b_q^{+} + \gamma_q^{*}\, b_q \right] - \sum_q \hbar\omega_q \left[\gamma_q\, \beta_q^{*} + \gamma_q^{*}\, \beta_q \right] \tag{3.13}$$

One can calculate

$$H_D \,|D_2\rangle = \sum_n \psi_n \left[\Delta + \sum_q \hbar\omega_q \left(|\beta_q|^2 + \chi_n^{q} \beta_q^{*} + \chi_n^{q\,*} \beta_q \right) \right] B_n^{+} \,|0\rangle_{ex}\, |\beta\rangle \; +$$

$$+ \, e^{-S} \sum_q \hbar\omega_q \left(\beta_q + \gamma_q \right) b_q^{+} \,|0\rangle_{ph}\, |\varphi\rangle$$

which means that

$$\overline{H}\,|D_2\rangle \equiv H_D\,|D_2\rangle \tag{3.14}$$

and our results can be put in the following form:

$$i\hbar\,\frac{\partial}{\partial t}\,|D_2\rangle = H_D\,|D_2\rangle = H\,|D_2\rangle + |\delta\rangle \tag{3.15a}$$

$$\langle D_2 | \delta \rangle = 0 \tag{3.15b}$$

We can now ask the essential questions: if the dynamics of the function $|D_2(t)\rangle$ is determined by the time dependence of $\psi_n(t)$ and $\beta_q(t)$ which satisfy ELE, what is the relation between

$$|D_2(t)\rangle = T\, e^{\frac{1}{i\hbar}\int_0^t H_D(t')\,dt'} \,|D_2(0)\rangle \tag{3.16}$$

and the (unknown) function $|\psi(t)\rangle$, which is solution of SE for the Fröchlich's Hamiltonian H (1.1):

$$i\hbar\,\frac{\partial |\psi(t)\rangle}{\partial t} = H\,|\psi(t)\rangle \tag{3.17a}$$

or

$$|\psi(t)\rangle = e^{\frac{1}{i\hbar} Ht} \,|\psi(0)\rangle \tag{3.17b}$$

if they both satisfy the same initial condition

$$|D_2(0)\rangle = |\psi(0)\rangle \quad ? \tag{3.18}$$

Furthermore, what are the consequences of the calculations of various physical quantities with $|D_2\rangle$ instead of $|\psi\rangle$.

$$\langle D_2(t)\,|\,F(B,b,0)\,|\,D_2(t)\rangle \overset{?}{=} \langle \psi(t)\,|\,F(B,b,0)\,|\,\psi(t)\rangle = \langle \psi(0)\,|\,F_H(B,b,t)\,|\,\psi(0)\rangle$$

Equation of motion for the average values in the state $|D_2\rangle$ is

$$i\hbar \frac{d}{dt} \langle D_2(t)|F|D_2(t)\rangle = \langle D_2|[F,H_D]|D_2\rangle \tag{3.19a}$$

or

$$i\hbar \frac{d}{dt} \langle D_2(t)|F|D_2(t)\rangle = \langle D_2|[F,H]|D_2\rangle + \langle D_2|F|\delta\rangle + \langle \delta|F|D_2\rangle \tag{3.19b}$$

The exact average value satisfies the equation

$$i\hbar \frac{d}{dt} \langle \psi(t)|F|\psi(t)\rangle = i\hbar \frac{d}{dt} \langle \psi(0)|F_H|\psi(0)\rangle = \langle \psi(0)|\,F,H\,|\psi(0)\rangle \tag{3.20}$$

Let us look at two important examples.

1. $F = H$

One can see that although H and H_D not commute, the following is true:

$$\langle D_2|[H,H_D]|D_2\rangle = 0 \tag{3.21}$$

implying that

$$i\hbar \frac{d}{dt} \langle D_2|H|D_2\rangle = 0 \tag{3.22}$$

The following reasoning

$$\langle D_2(t)|H|D_2(t)\rangle = \langle D_2(0)|H|D_2(0)\rangle = \langle \psi(0)|H|\psi(0)\rangle = \langle \psi(t)|H|\psi(t)\rangle$$

shows that the quantity

$$\mathcal{H} = \langle D_2(t)|H|D_2(t)\rangle = \langle \psi(t)|H|\psi(t)\rangle \tag{3.23}$$

is the same when calculated with both functions.

2. $F = \sum_n B_n^+ B_n$

One can easily show that $\left[\sum_n B_n^+ B_n, H_D\right] = 0$, meaning that

$$\langle D_2|\sum_n B_n^+ B_n|D_2\rangle = \sum_n |\psi_n|^2 = \text{const} \quad \text{and following the above reasoning}$$

$$\langle D_2|\sum_n B_n^+ B_n|D_2\rangle = \langle \psi|\sum_n B_n^+ B_n|\psi\rangle \tag{3.24}$$

Thus we conclude that the constants of motion $\langle H\rangle$ and $\langle \sum_n B_n^+ B_n\rangle$ are also constants of motion for $|D_2\rangle$ and due to the same initial condition (3.18) they retain the same initial value. One can show the same also for the total momentum of the system

$$P = \sum_k \hbar k B_k^+ B_k + \sum_q \hbar q b_q^+ b_q \tag{3.25}$$

4. CONCLUDING REMARKS

We have seen that although very simple, $|D_2\rangle$ trial function describes well the constants of motion of the system. One can also construct an effective differential equation, which substitutes SE. In our opinion, this means that while working within the range of applications discussed previously, $|D_2\rangle$ is quite suitable, under the assumption that dynamics of the amplitudes is determined by ELE.

The problems arise, if one wishes to describe the energy transfer. The results obtained for the average values of $\frac{d}{dt} B_n^+ B_n$ and $\frac{d}{dt} B_n^+ B_{n+1}$, calculated with $|D_2\rangle$ and $|\psi\rangle$ are completely different, as can be estimated from the study of the limiting cases ($I=0$ and $\chi=0$). Further work in this direction is necessary and we think that one should be very cautious when studying kinetics with $|D_2\rangle$ - Ansatz.

REFERENCES

1. A.S. Davydov, *Solitons in Molecular Systems (Solitony v molekulyarnyh sistemah)*, Naukova dumka, Kiev (1988) (in Russian).

2. H. Fröhlich, Electrons in Lattice Fields, *Adv. Phys.* **3**, 325 (1954).

3. A. Messiah, *Quantum mechanics*, Wiley, New York (1962).

4. A.S. Davydov, Dvizhenie solitona v odnomernoi molekulyarnoi reshetke s uchetom teplovih kolebanii, *Zh. Eksp. Teor. Fiz.* **78**, 789 (1980) *Sov. phys. - JETP* **51**, 397 (1980)].

5. a) D.W. Brown, K. Lindenberg and B.J. West, Applicability of Hamilton's equations in the quantum soliton problem, *Phys. Rev.* B **33**, 4104 (1986).

   b) D.W. Brown, B.J. West and K. Lindenberg, Davydov solitons: New results at variance with standard derivations, *Phys. Rev.* B **33**, 4110 (1986).

6. W.C. Kerr and P.S. Lomdahl, Quantum-mechanical derivation of the equations of motion for Davydov solitons, *Phys. Rev.* **35**, 3629 (1987)

7. D.W. Brown, Balancing the Schrödinger equation with Davydov Ansätze, *Phys. Rev.* A **37**, 5010 (1988).

8. M.J. Škrinjar, D.V. Kapor and S. Stojanovic, Classical and quantum approach to Davydov's soliton theory, *Phys Rev.* B **38**, 6402 (1988).

9. Q. Zhang, V. Romera-Rochin and R. Silbey, Variational approach to Davydov soliton, *Phys. Rev,* **B 38**, 6409 (1988).

10. J.R. Klauder, The Action Option and a Feynman Quantization of Spinor Fields in Terms of Ordinary C-Numbers, *Ann. of Phys.* **11**, 123 (1960).

11. L.R. Mead and N. Papanicolaou, Holstein-Primakoff theory for many-body systems, *Phys. Rev.* B **28**, 1633 (1983).

Comment 1 by Kapor: ELE lead to the following corrected equation:

$$i\hbar\dot{\psi}_n = \left[\Delta - \tfrac{1}{2}\sum_{n,q}\hbar\omega_q|\psi_n|^2(\chi_n^q\beta_q^* + \chi_n^{q*}\beta_q)\right]\psi_n - I(\psi_{n+1} - \psi_{n-1}) + $$
$$+ \psi_n\sum_q \hbar\omega_q(\chi_n^q\beta_q^* + \chi_n^{q*}\beta_q). \tag{1}$$

The correction term (second in the square bracket) has the form $-\tfrac{1}{2}\mathcal{H}_{int}\psi_n$. Looking for Davydov's type of solution:

$$\beta_q(t) = \frac{\hbar\omega_q\exp(-iqvt)}{\hbar vq - \hbar\omega_q}\sum_n |\psi_n(0)|^2\chi_n^q \tag{2}$$

leads to the following integro-differential equation (in the continuum approximation)

$$i\hbar\frac{\partial\psi}{\partial t} = (\Delta - 2I)\psi - Ia^2\frac{\partial^2\psi}{\partial x^2} - G(v)|\psi|^2\psi + \frac{G(v)}{2}\psi\frac{1}{\alpha}\int |\psi|^4\, dx. \tag{3}$$

The general solution of this equation is not known at present, but Davydov solution, if introduced, satisfies it, because the integral turns into a phase factor. This implies that Davydov's soliton in fact satisfies ELE, but we do not know if there exists some other solution of the same form. It is worth mentioning that all necessary elements for the formulation of the above approximation were derived already in Reference[6].

Comment 2 by Kapor: Problem of the energy transfer $\langle\sum_n B_n^\dagger B_n\rangle = ?$ In the solitonic approximation

$$|D_2\rangle \approx \psi_0(x,t) = \phi(x - vt)\exp(ikx - i\omega_0 t) \tag{4}$$

$$\mathcal{E}_n^s = \langle D_2(t)|B_n^\dagger B_n|D_2(t)\rangle = \phi^2(x - vt). \tag{5}$$

It can be also derived in momentum space:

$$\psi_n = \frac{1}{N}\sum_q (t)\exp(iqna) \qquad \varphi_q(t) = \phi(q)\exp\left[-iqvt - i(\omega_0 - kv)t\right]$$

$$\phi(q) = \frac{\pi}{\sqrt{g(v)}}\frac{1}{\cosh\left[\frac{\pi a}{g(v)})(q - k)\right]} \qquad g(v) = \frac{4\chi^2 ma^2}{\hbar^2\kappa(1 - s^2)} \tag{6}$$

We have

$$
\begin{aligned}
\mathcal{E}_n^s &= \frac{1}{N^2} \sum_{q_1 q_2} \langle D_2(t)|B_{q_1}^\dagger B_{q_2}|D_2(t)\rangle \exp[-i(q_1 - q_2)na] \\
&= \frac{1}{N} \sum_{q_1} \phi(q_1) \exp[-iq_1(na - vt)]\frac{1}{N} \sum_{q_2} \phi(q_2) \exp[-iq_2(na - vt)] \\
&\Rightarrow \phi^2(x - vt).
\end{aligned}
\tag{7}
$$

Let us compare it with exact calculation $(|D_2(0)\rangle = |\psi(0)\rangle)$:

$$
\begin{aligned}
\mathcal{E}_n^\phi &= \langle D_2(0)|B_n^\dagger(t)B_n(t)|D_2(0)\rangle \\
&= \frac{1}{N^2} \sum_{k_1 k_2} \sum_{q_1 q_2} \phi(k_1)\phi(k_2) \exp[-i(q_1 - q_2)na] \times \\
&\quad \times \langle 0|B_{k_1}(0)B_{q_1}^\dagger(t)B_{q_2}(t)B_{k_2}^\dagger(0)|0\rangle.
\end{aligned}
\tag{8}
$$

This expression can give the value $\phi^2(x - vt)$ only if we suppose the following form:

$$
B_q(t) = B_q(0) \exp[-i(w_0 + qv)t].
\tag{9}
$$

Unfortunately, one can show that such form of the solution is in contradiction with the Heisenberg equations of motion, so the results for the energy transport should not be treated using Davydov's solution.

Comment 3 by Kapor: It is necessary to stress that there exists a simple explanation why ELE equation for $\dot{\beta}_q$ is equal to Hamilton's equation and averaged Heisenberg equation, but no such thing could be said about equation for $\dot{\psi}_n$. The reason is that β_q is an average value of an observable and ψ_n is not. This also explains why in the case of $|D_1\rangle$ ansatz, even the equations of β_{qn} disagree.

9

UNITARY TRANSFORMATION AND "DECOUPLING" OF EXCITONS AND PHONONS IN ACN

D. Kapor, M. Škrinjar and S. Stojanović

Institute of Physics, Faculty of Sciences
University of Novi Sad
Dr I. Đuričića 4, YU 21 000 Novi Sad, Yugoslavia

1. INTRODUCTION

Crystalline acetanilide (ACN) is the subject of numerous theoretical and experimental work due to two remarkable features [1,2]:

a) amide groups display bond distances similar to those occuring in polypeptides and

b) the appearance of the unconventional amide I band at 1650 cm^{-1}.

Experimental studies on pure and deuterated samples using various techniques are in progress [3,4] and the theoretical explanations range from Davydov soliton theory [5] up to the polaron theory. The most recent study [6] presents a polaron approach based on the analogy with F-center theory, where the new excitations are just an intermediate step toward Davydov solitons. This last study inspired us to treat the problem as a typical one of coupled exciton-phonon system. Our aim is to demonstrate the capabilities of unitary transformation method because, one can often encounter its incorrect application in the literature. The results will be applied to the study of some optical properties of ACN.

2. UNITARY TRANSFORMATION OF EXCITON-PHONON HAMILTONIAN

Following Scott et. al. [6] we assume that the system is described by the following Hamiltonian:

$$H = H_0 + JV \tag{2.1}$$

where

Davydov's Soliton Revisited, Edited by P.L. Christiansen and A.C. Scott
Plenum Press, New York, 1990

$$H_0 = \sum_{n=1} \mathcal{H}_0{}_n \tag{2.2a}$$

$$\mathcal{H}_0{}_n = \Omega_0 B_n^+ B_n + \sum_{j=1}^{M} \left[\omega_j b_{nj}^+ b_{nj} + \chi_j \left(b_{nj}^+ + b_{nj} \right) B_n^+ B_n \right] \qquad (\hbar=1) \tag{2.2b}$$

B_n^+ create the amide-I vibrational excitation (excitons or vibrons) with the energy Ω_0, ω_j is the frequency of the j-th optical phonon branch and χ_j is corresponding coupling energy. We consider the chain of R sites, and suppose that there appear M phonon modes. J is the nearest-neighbour coupling energy and

$$V = \sum_{n=1}^{R} \left[B_n^+ B_{n+1} + B_n^+ B_{n+1} \right] \tag{2.2c}$$

Periodic boundary conditions are assumed.

Essential assumption is that CO vibrational energy is localized at a single ACN molecule, so, in the lowest approximation we can set J=0. We are looking for the unitary transformation which can decouple excitons and phonons in H_0. For this reason we introduce the unitary operator

$$U = \exp \left[-\sum_{j',n'} \frac{\chi_{j'}}{\omega_{j'}} \left(b_{n'j'}^+ - b_{n'j'} \right) B_n^+ B_n \right] \tag{2.3}$$

which leads to the following transformed operators:

$$\tilde{b}_{nj} = U^+ b_{nj} U = b_{nj} - \frac{\chi_j}{\omega_j} B_n^+ B_n \tag{2.4a}$$

$$\tilde{B}_n = B_n \exp \left[-\sum_j \frac{\chi_j}{\omega_j} \left(b_{nj}^+ - b_{nj} \right) \right] \tag{2.4b}$$

The Hamiltonian transformed with U

$$\tilde{H} = U^+ H U = \tilde{H}_0 + J\tilde{V} \tag{2.5}$$

expressed in terms of initial (old) operators has the form

$$\tilde{H}_0 = \sum_n \tilde{\mathcal{H}}_0{}_n \tag{2.6a}$$

$$\tilde{\mathcal{H}}_0{}_n = \left(\Omega_0 - \gamma \right) B_n^+ B_n - \gamma B_n^+ B_n^+ B_n B_n + \sum_j \omega_j b_{nj}^+ b_{nj} \tag{2.6b}$$

with

$$\gamma = \sum_{j=1}^{M} \frac{\chi_j^2}{\omega_j} \tag{2.7}$$

and the experimental estimate [2] for ACN is $\gamma \simeq 27.4 \ cm^{-1}$.

$$\tilde{V} = \sum_{n} \left[T_{n,n+1} B_n^+ B_{n+1} + h.c \right] \tag{2.8}$$

where

$$T_{n,n+1} = \exp \left[\sum_{j} \frac{\chi_j}{\omega_j} \left(b_{nj}^+ - b_{n+1,j}^+ - h.c. \right) \right] \tag{2.9}$$

It is important to remember that in the same time the wave function is also transformed in the following way

$$|\tilde{\psi}\rangle = U^+ |\psi\rangle \qquad |\psi\rangle = U|\tilde{\psi}\rangle \tag{2.10}$$

Let us now study the paricular, case J=0. The Hamiltonian (2.6) is completely decoupled and in the following we shall suppress the site subscript.

The solution of the Schrödinger equation

$$\mathcal{H}_{0_n} |\tilde{\phi}(N, m_j)\rangle = E^O(N, m_j)|\tilde{\phi}(N, m_j)\rangle \tag{2.11}$$

is sought in the product form

$$|\tilde{\phi}(N, m_j)\rangle = |N\rangle \prod_{j=1}^{M} |m_j\rangle \qquad \text{where } B_n^+ B_n |N\rangle = N|N\rangle \text{ and} \tag{2.12}$$
$$b_j |m_j\rangle = \sqrt{m_j}|m_j - 1\rangle$$

with the result.

$$E^{(O)}(N, m_j) = \Omega_0 N - \gamma N^2 + \sum_{j} m_j \omega_j \tag{2.13}$$

We are also interested in the wave-function $|\phi\rangle$ related to $|\tilde{\phi}\rangle$ by (2.10).

$$|\phi(N, m_j)\rangle = \exp \left[-\sum_{j} \frac{\chi_j}{\omega_j} \left(b_j^+ - b_j \right) B^+ B \right] |\phi(N, m_j)\rangle = |N\rangle \prod_{j} |\phi_j\rangle \tag{2.14}$$

$$|\phi_j\rangle = \exp \left[-N \frac{\chi_j}{\omega_j} \left(b_j^+ - b_j \right) \right] |m_j\rangle =$$

$$= \exp \left[-N \frac{\chi_j}{\omega_j} \left(b_j^+ - b_j \right) \right] \frac{1}{\sqrt{m_j!}} b_j^{+ \, m_j} |0\rangle_{ph} =$$

$$= \frac{1}{\sqrt{m_j!}} e^{-[\ ]} b_j^{+ \, m_j} e^{[\ ]} \cdot e^{-[\ ]} |0\rangle_{ph} =$$

$$= \frac{1}{\sqrt{m_j!}} \left(b_j^+ + N \frac{\chi_j}{\omega_j} \right)^{m_j} \exp \left[-N \frac{\chi_j}{\omega_j} \left(b_j^+ - b_j \right) \right] |0\rangle_{ph}$$

Some operator algebra and series manipulation, leads to the following expression (where we had changed m_j into n_j, in order to get close to the notation of Ref.6):

$$|\phi_j\rangle = \exp\left[-N\frac{\chi_j}{\omega_j}\left(b_j^+ - b_j\right)\right]|n_j\rangle =$$

$$\sqrt{n_j!}\,\exp\left(-\frac{N^2\chi_j^2}{2\omega_j^2}\right)\sum_{m_j=0}^{\infty}\frac{\left(-\frac{N\chi_j}{\omega_j}\right)^{m_j-n_j}}{\sqrt{m_j!}}\sum_{m=0}^{n_j}(-1)^m\binom{n_j}{n_j-m}\frac{1}{m!}\left(N^2\frac{\chi_j^2}{\omega_j^2}\right)^m \qquad (2.15)$$

and using Gradshteyn-Ryzhik[7] definition of Laguerre polinomial, we obtain

$$|\phi_j\rangle = \sqrt{n_j!}\,\exp\left(-\frac{N^2\chi_j^2}{2\omega_j^2}\right)\sum_{m_j=0}^{\infty}L_{n_j}^{m_j-n_j}\left(-\frac{N^2\chi_j^2}{\omega_j^2}\right)\frac{\left(-\frac{N\chi_j}{\omega_j}\right)^{m_j-n_j}}{\sqrt{m_j!}} \qquad (2.16)$$

which is precisely the result of Scott et. al.[6]. In this way we have shown that unitary transformation approach can reproduce this result.

Furthermore, it is very easy to explain the way in which coherent phonon states appear in the system. In fact, if there exists even a single excitation in the system (N=1), then the lowest energy of the system appears for $n_j=0$ for all j. In this case

$$|\phi_j\rangle = \exp\left[-\frac{\chi_j}{\omega_j}\left(b_j^+ - b_j\right)\right]|0\rangle_{p,h} \qquad (2.17)$$

which is precisely coherent phonon state.

Let us demonstrate it. If the wave-function of the system is $|\psi_0\rangle = |\phi(1,0)\rangle$, the action of

$$\mathcal{H}_0 = \Omega_0 B^+B + \sum_j \chi_j B^+B\left(b_j^+ + b_j\right) + \sum_j \omega_j b_j^+ b_j \qquad (2.18)$$

gives

$$\overline{\varepsilon_0} = \langle\psi_0|\mathcal{H}_0|\psi_0\rangle = \Omega_0 \qquad (2.19)$$

while, for $|\psi_j\rangle = |1\rangle\prod_j|\phi_j\rangle$

$$\varepsilon_0 = \langle\psi_j|\mathcal{H}_0|\psi_j\rangle = \Omega_0 + \sum_j\left(-\frac{\chi_j}{\omega_j} - \frac{\chi_j}{\omega_j}\right)\chi_j + \sum_j\frac{\chi_j^2}{\omega_j} =$$

$$= \Omega_0 - \gamma < \overline{\varepsilon_0} \qquad (2.20)$$

What we encounter here, is so called "displaced vacuum state", because if we define the vacuum state for the operators $\tilde{b}_j$

$$\tilde{b}_j |\tilde{0}\rangle = 0 \qquad (2.21)$$

using the unitary transformation (2.3), we see that

$$b_j |\tilde{0}\rangle = b_j \exp\left[N \sum_j \frac{\chi_j}{\omega_j}\left(b_j^+ - b_j\right)\right] |0\rangle_{ph} = N \frac{\chi_j}{\omega_j}|\tilde{0}\rangle$$

So $|\tilde{0}\rangle$ is the coherent state for the "old" operators.

Thus, we can conclude the following: If one, or several vibronic (exciton) states are excited in the system, the phonon subsystem goes over to the coherent state $\prod_j |\phi_j\rangle$ which corresponds to the lower energy of the system (displaced vacuum for the new operators $\tilde{b}_j$).

3. ABSORPTION IN ACN

In order to study the absorption in ACN, we start with the standard expressions[8,9]. The dipole moment of the system is

$$\vec{P} = \frac{1}{V} \sum_{n,f} \left[\vec{d}_{nf} B_{nf}^+ + \vec{d}_{nf}^* B_{nf}\right] \qquad (3.1)$$

where $\vec{d}_{nf}$ denotes the dipole moment of the molecule at site n in the state "f". Vector potential of the incoming radiation is given by

$$\vec{A}(n,t) = -\frac{ic}{2\omega}\left[\vec{E}_0 e^{ikn-i\omega t} - \vec{E}_0^* e^{-ikn+i\omega t}\right] \qquad (3.2)$$

So the interaction is described by

$$H_{int} = -\frac{i}{c}\sum_{n,f} \omega_f \vec{A}(\vec{n},t)\left[\vec{d}_{nf} B_{nf}^+ - \vec{d}_{nf}^* B_{nf}\right] =$$

$$= -\frac{1}{2\omega}\sum_{n,f} \omega_f \left(\vec{E}_0 \vec{d}_{nf}\right)\left(B_{nf}^+ - B_{nf}\right)\left[e^{ikn-i\omega t} - c.c.\right] \qquad (3.3)$$

The polarization can be expressed in terms of dielectric constant in the following way:

$$\langle\vec{P}(\vec{r},t)\rangle = \frac{1}{2}\left[\frac{\varepsilon^+(\omega,k) - 1}{4\pi}\vec{E}_0 e^{ikr-i\omega t} + c.c.\right] \qquad (3.4)$$

We shall study the particular case of just one excited level f=1, and once again neglect the energy transfer (J=0). The Hamiltonian of the system is given by

$$H = H_0 + H_{int} \tag{3.5a}$$

where H_0 is given by (2.2) and H_{int} in this particular case becomes:

$$H_{int} = - \frac{\omega_f}{2\omega} \sum_n \left[B_n^+ - B_n \right] \left[\vec{E}_0 \vec{d}_n \right] \left[e^{ikn-i\omega t} - c.c. \right] \tag{3.5b}$$

Applying the unitary transformtion (2.3), we obtain

$$\tilde{H}^{\cdot} = H_{eq} = U^+ H U = \sum_n \tilde{\mathcal{H}}_{n0} + \sum_n \tilde{\mathcal{H}}_{n,int}(t) \tag{3.6a}$$

where $\tilde{\mathcal{H}}_{n0}$ is given by (2.6) and

$$\tilde{\mathcal{H}}_{n,int}(t) = - \frac{\omega_f}{2\omega} \left\{ B_n^+ \exp\left[\sum_j \frac{\chi_j}{\omega_j} \left(b_{n_j}^+ - b_{n_j} \right) \right] - h.c. \right\} \times$$

$$\times \left[\vec{E}_0 \cdot \vec{d}_n \right] \left[e^{ikn-i\omega t} - e^{-ikn+i\omega t} \right] \tag{3.6b}$$

We known the solution (2.13) of the eigenproblem of $\tilde{\mathcal{H}}_0$.

Let us suppose that there are no excitons (vibrons) in the system (N=0, $|N\rangle = |0\rangle$) and the crystal (phonon subsystem) is in thermodynamical equilibrium at the temperature $kT = \Theta = 1/\beta$. Then the statistical operator is given by

$$\rho_0 = |0\rangle\langle 0| \, e^{-\beta \tilde{\mathcal{H}}_{ph}} \tag{3.7}$$

or in the interaction picture

$$\rho_t = e^{-i\tilde{\mathcal{H}}_0 t} \, S(t,t_0) \, \rho_0 \tilde{S}^1(t,t_0) \, e^{i\tilde{\mathcal{H}}_0 t} \tag{3.8}$$

and the average value for any physical quantity F is

$$\langle F \rangle_t = \langle S^{-1}(t,t_0) \, F(t) \, S(t,t_0) \rangle_{\rho_0} \tag{3.9a}$$

where F(t) is given in the interaction picture

$$F(t) = e^{i\tilde{\mathcal{H}}_0 t} \, F_s(t) \, e^{-i\tilde{\mathcal{H}}_0 t} \tag{3.9b}$$

and

$$\langle A \rangle_{\rho_0} = Tr \left[\rho_{ph} \, \langle 0|A|0\rangle \right] \tag{3.9c}$$

Applying the approximation linear in

$$\tilde{w}(t) = e^{i\tilde{\mathcal{H}}_0 t} \tilde{\mathcal{H}}_{int} e^{-i\tilde{\mathcal{H}}_0 t} \tag{3.10}$$

we obtain the polarization per site:

$$\langle \tilde{\vec{\mathcal{P}}} \rangle \simeq \langle \tilde{\vec{\mathcal{P}}} \rangle_{\rho_0} + \frac{1}{i} \int_{t_0}^{t} dt' \langle \tilde{\vec{\mathcal{P}}}(t)\, \tilde{w}(t') - \tilde{w}(t')\tilde{\vec{\mathcal{P}}}(t) \rangle_{\rho_0} \tag{3.11}$$

where

$$\tilde{\vec{\mathcal{P}}}(t) = e^{i\tilde{\mathcal{H}}_0 t}\, \tilde{\vec{\mathcal{P}}}(0)\, e^{-i\tilde{\mathcal{H}}_0 t} =$$

$$= \frac{1}{a} \left\{ \vec{d}\, B^+(t)\, \exp\left[\sum_j \frac{\chi_j}{\omega_j} \left(b_j^+(t) - b_j(t) \right) \right] + h.c. \right\} \tag{3.12}$$

where the time-dependence of the operators $B(t)$ and $b_j(t)$ is governed by the equations of motion. In particular

$$i\frac{dB}{dt} = \left[B, \tilde{\mathcal{H}}_0 \right] = (\Omega_0 - \gamma)B - 2\gamma\, B^+ BB \tag{3.13}$$

and also

$$i\frac{d}{dt}\, B^+ B = 0 \tag{3.14}$$

A phase change

$$B(t) = A(t)\, e^{-i(\Omega_0 - \gamma)t} \tag{3.15}$$

eliminates the first term in (3.13) to give

$$i\frac{dA}{dt} = 2\gamma\, A^+ A\, A \tag{3.16a}$$

$$A(t) = 2i\gamma\, A^+ A \int_0^t A(t')dt' + B(0) \tag{3.16b}$$

(3.16) is solved by the interative procedure to give:

$$B_n(t) = \exp\left\{ -it\left[\Omega_0 - \gamma - 2\gamma\, B_n^+(0)B_n(0) \right] \right\} B_n(0) \tag{3.17}$$

and one can also show

$$b_{n_j}(t) = b_{n_j}(0)\, e^{-i\omega_j t} \tag{3.18}$$

Now one can use the Green function (GF) approach [8,9] in order to evaluate the polarization. The choice of ρ_0 (3.7) implies $\langle \tilde{\vec{\mathcal{P}}} \rangle_{\rho_0} = 0$. Some GF vanish, so we obtain the following result valid in the long wavelength approximation ($\vec{k}=0$), in the resonance region ($\omega_f = \Omega_0$):

$$\langle \tilde{\vec{P}} \rangle = \frac{i\Omega_0 \vec{d}(\vec{d}\vec{E}_0)}{2\omega a} e^{-i\omega t} \lim_{\eta \to 0} \int_0^\infty d\sigma \, e^{i\omega\tau - \eta\tau} \times$$

$$\times \left\{ - \langle\langle B_n(\tau) \exp\left[- \sum_j \frac{\chi_j}{\omega_j} \left[b_{n_j}^+(\tau) - b_{n_j}(\tau) \right] \right] \vec{B}_n^+(0) \times \right.$$

$$\times \exp\left[\sum_j \frac{\chi_j}{\omega_j} \left[b_{n_j}^+(0) - b_{n_j}(0) \right] \right] \rangle\rangle + \langle\langle B_n(0) \exp\left[-\sum_j \frac{\chi_j}{\omega_j} \times \right.$$

$$\left. \times \left[b_{n_j}^+(0) - b_{n_j}(0) \right] \right] \vec{B}_n^+(\tau) \exp\left[\sum_j \frac{\chi_j}{\omega_j} \left(b_{n_j}^+(\tau) - b_{n_j}(\tau) \right) \right] \rangle\rangle \right\} + c.c. \quad (3.19)$$

Here $\tau = t - t'$. The interaction is "turned on" adiabatically by the introduction of η term ($\tilde{w}(t' = \pm\infty) = 0$).

One can evalute exactly the averages appearing in (3.19).

$$\langle\langle \quad \rangle\rangle = \langle 0 | BB^+ | 0 \rangle \, \mathrm{Tr} \left\{ e^{-\Sigma} e^{+\Sigma} \rho_{ph} \right\}$$

$$\langle 0 | B(\tau)B^+(0) | 0 \rangle = e^{-i(\Omega_0 - \gamma)\tau}$$

$$\langle \exp\left[-\sum_j \frac{\chi_j}{\omega_j} \left[b_j^+(t) - b_j(t) \right] \right] \exp\left\{ \sum_j \frac{\chi_j}{\omega_j} \left[b_j^+(0) - b_j(0) \right] \right\} \rangle_{\rho_{ph}} =$$

$$= \prod_j \langle \exp\left\{ -\frac{\chi_j}{\omega_j} \left[b_j^+(0)e^{i\omega_j t} - b_j(0)e^{-i\omega_j t} \right] \right\} \exp\left\{ \frac{\chi_j}{\omega_j} \left[b_j^+(0) - b_j(0) \right] \right\} \rangle_{\rho_{ph}}$$

Using some operator identities[8] we arrive to:

$$\langle \quad \rangle = e^{-W(\Theta)} \exp\left\{ \sum_j \frac{\chi_j}{\omega_j^2} \left[\coth \frac{\omega_j}{2\Theta} \cos\omega_j\tau - i \sin\omega_j\tau \right] \right\}$$

where

$$W(\Theta) = \sum_j \frac{\chi_j^2}{\omega_j^2} \coth \frac{\omega_j}{2\Theta} \quad (3.20)$$

The final result is

$$\langle \tilde{\vec{\mathcal{P}}} \rangle = - \frac{i\Omega_0}{2\omega a} \, \vec{d} \, (\vec{d} \cdot \vec{E}_0) \, e^{-i\omega t} \, e^{-W(\Theta)} \times$$

$$\times \lim_{\eta \to 0} \int_0^\infty dt \left\{ \exp\left[i(\omega + \Omega_0 + \gamma)t - i \sum_j \left(\frac{\chi_j}{\omega_j}\right)^2 \sin\omega_j t - \eta t + \right. \right.$$

$$\left. + \sum_j \left(\frac{\chi_j}{\omega_j}\right)^2 \cos\omega_j t \right] - \exp\left[i(\omega + \Omega_0 - \gamma)t + i \sum_j \left(\frac{\chi_j}{\omega_j}\right)^2 \sin\omega_j t + \right.$$

$$\left. \left. + \sum_j \left(\frac{\chi_j}{\omega_j}\right)^2 \coth\frac{\omega_j}{2\Theta} \cos\omega_j t \right] \right\} + c.c. \tag{3.21}$$

This expression is an exact one for $J=0$ and follows the linear relation [3,4] between the polarization $\langle \tilde{\vec{\mathcal{P}}} \rangle$ and the electric field $\vec{E}_0$ in the tensorial form $(k=0)$, where

$$\varepsilon_{ij}^{\perp} = - i \, \frac{4\pi\Omega_0}{\omega a} \, d_i d_j \Gamma(\omega,\Theta) + \delta_{i,j} \tag{3.22}$$

where Γ denotes the term $e^{-W(\Theta)} \times$ the integral figuring in (3.21). This result is similar to the one obtained by Davydov [10].

If we accept the approximate expressions $\overline{\cos\omega_j \tau} = \overline{\sin\omega_j \tau} = 0$, in the vicinity of the resonance we obtain

$$\Gamma(\omega,\Theta) = e^{-W(\Theta)} \, \frac{1}{i(\omega - \Omega_0 + \gamma + i\eta)} \tag{3.23}$$

$$\varepsilon_{ij}^{\perp} = \delta_{ij} - \frac{4\pi\Omega_0}{\omega a} d_i d_j \, e^{-W(\Theta)} \, \frac{1}{\omega - \Omega_0 + \gamma + i\eta} \tag{3.24}$$

The absorbed energy $I(\omega,\Theta)$ is proportional to $\mathrm{Im}\{\varepsilon^{\perp}\}$ [9]:

$$I(\omega,\Theta) = - \frac{(\Omega_0 \vec{E}_0 \vec{d})^2}{2a\omega} \, \mathrm{Im}\{\Gamma(\omega,\Theta)\} \tag{3.25}$$

Let us study the example of the field directed along x-axis $(\vec{E}_0 \vec{d} = E_0 d_x)$. We obtain the absorption along x-axis:

$$I_x(\omega,\Theta) = \frac{\Omega_0^2 E_0^2}{8\pi} \, \mathrm{Im} \, \varepsilon_{xx}^{\perp} \tag{3.26a}$$

$$\mathrm{Im} \, \varepsilon_{xx}^{\perp} = \frac{\eta \, e^{-W(\Theta)}}{(\omega - \Omega_0 + \gamma)^2 + \eta^2} \tag{3.26b}$$

giving

$$I_x(\omega, \Omega) = \frac{\Omega_0^2 E_0^2}{8\pi} \frac{\eta \, e^{-W(\Theta)}}{(\omega - \Omega + \gamma)^2 + \eta^2} \qquad (3.26c)$$

Studing the above expressions we can conclude the following. In the approximation $J=0$, we can not determine the shape (the width) of exciton (vibron) line given by $\eta = \eta(\Theta)$. This approximation allows only to determine the shift of the line $\Omega_0 \to \Omega_0 - \gamma$. It is important to notice that even the exact solving of the integral $\Gamma(\Theta, \omega)$, would not allow the determination of the linewidth $\eta(\Theta)$, but only the line shift would be changed [10].

Since the expression (3.21) for the polarization is an exact one for $J=0$, the impossibility of determining $\eta(\Theta)$ follows from the fact that in the equivalent Hamiltonian "dressed" excitons (i.e. polarons) and phonons are decoupled, so there exists no mechanism by which the polaron line at $\omega = \Omega_0 - \gamma$ could be broadened. Namely, polarons ("dressed" excitons or vibrons) interact with phonons only for $J \neq 0$ and in that case one could (approximately) determine polaron line linewidth and give the interpretation of the integral intensity of the absorbed energy of 1650 cm^{-1} in ACN.

Approximate polaron line linewidth can be determined (in the first approximation) if we average the transformed Hamiltonian of energy transfer (2.8) over the translationally invariant single-particle exciton function

$$|\tilde{\psi}\rangle_{ex} = \frac{1}{\sqrt{R}} \sum_{k=1}^{R} e^{ikna} \tilde{B}_n^+ |\tilde{0}\rangle_{ex} \qquad (3.27)$$

and over the phonon ensamble:

$$\langle\langle J\tilde{V}\rangle\rangle = \mathrm{Tr}\left[\langle\tilde{\psi}|J\tilde{V}|\tilde{\psi}\rangle\rho_{ph}\right] = 2J \, e^{-W(\Theta)} \, \mathrm{coska} \qquad (3.28)$$

In this approximation, the polaron linewidth satisfies the inequality

$$\Delta E < 4J \, e^{-W(\Theta)} \qquad (3.29)$$

We conclude that the same factor $e^{-W(\Theta)}$ (which is also Franck-Condon factor [6]) which reduces the intensity of the absorbed energy of 1650 cm^{-1} polaron band, reduces also the width of the same line, and in this way increases the lifetime of the localized state.

Let us briefly summarize our results. By performing the unitary transformation of the Hamiltonian of the system of excitons interacting with optical phonons, we were able to solve the eigen-problem and determine the optical properties of the system, for J=0 in a rather simple way. The case J≠0 demands the application of the perturbation theory, and some qualitative estimates are presented here. Finally, let us note that the problem of interaction with acoustical phonons, leading to "standard" Davydov solitons, was not treated here.

REFERENCES

1. G. Careri, *Search for cooperative Phenomena in Hydrogen-Bonded Amide Structures* in *Cooperative Phenomena*, H. Haken and W. Wagner eds, Springer-Verlag, Berlin (1973).

2. G. Careri, U. Bountempo, F. Galluzzi, A.C. Scott, E. Gratton and E. Shyamsunder, Spectroscopic evidence for Davydov-like solitons in acetanilide, *Phys. Rev.* B **30** 4689 (1984).

3. G. Careri, E. Gratton and E. Shyamsunder, Fine structure of the amide I band in acetanilide, *Phys. Rev.* A **37** 4048 (1988).

4. M. Barthes, Optical anomalies in acetanilide: Davydov solitons, localised modes or Fermi resonance? Submitted to J. Mol. Liquids (special issue: Molecular aspects of systems of biological interest).

5. C. Eilbeck, P.S. Lomdahl and A.C. Scott, Soliton structure in crystalline acetanilide, *Phys. Rev.* B **30**, 4703 (1984).

6. A.C. Scott, I.J. Bigio and C.T. Johnston, Polarons in acetanilide, *Phys. Rev.* B **39** 12883 (1989).

7. I.S. Gradshteyn and and I.M. Ryzhik, *Table of Integrals, Series and Products,* Acad. Press, New York (1965).

8. A.S. Davydov, *Solid state Theory (Teoriya tverdogo tela)*, Nauka, Moskva (1976) (in Russian).

9. A.S. Davydov, *Theory of Molecular Excitons (Teoriya molekulyarnyh eksitonov)*, Nauka, Moskva (1968) (in Russian) Chap. IV.

10. A.S. Davydov, *Solitons in Molecular Systems (Solitony v molekulyarnyh sistemah)*, Naukova dumka, Kiev (1988) (in Russian) § **38**.

10

SOLITON GENERATION IN INFINITE AND HALF-INFINITE MOLECULAR CHAINS

Larisa Brizhik

Institute for Theoretical Physics

Academy of Sciences of the Ukrainian SSR, Kiev

Abstract

It is shown that the presence or the absence of a soliton excitation threshold and the threshold values depends on the character of the excitation distribution at the initial time. A study is made of the time evolution of an excitation given at the initial moment as a hyperbolic secant, a rectangular step, and a decreasing exponential, for an infinite chain. A mechanism for generating solitons by exciting the impurity molecules is considered.

1 Introduction

The question whether the excitation-phonon coupling constant is large enough to sustain a Davydov soliton (**DS**) at biological levels of energy in 1-D molecular systems is widely discussed in the literature [1,2]. In this connection Brizhik [3] and Brizhik and Davydov [4] have made an attempt to study the time evolution of the excitation distribution in an infinite molecular chain given at the initial time moment in different forms for different values of exciton-phonon coupling constant. The analogous problem for half-infinite chains was studied by Brizhik, Gaididei, A. Vakhnenko and V. Vakhnenko in Reference [5]. Here we present the main results on the problem.

2 Davydov Soliton in an Infinite Molecular Chain

The set of equations describing the collective excitation in the region of dipole vibrations of repetitive subunits (peptide groups) in 1-D system in the continuum approximation has the well-known form [6]

$$\left[i\hbar \frac{\partial}{\partial t} + J \frac{\partial^2}{\partial x^2} + \chi \rho(x,t) \right] \Psi(x,t) = 0$$

$$\left(\frac{\partial^2}{\partial t^2} - \frac{V_o^2}{a^2} \frac{\partial^2}{\partial x^2} \right) \rho(x,t) + \frac{\chi}{M} \frac{\partial^2}{\partial x^2} |\Psi(x,t)|^2 = 0. \tag{1}$$

In the case of excitation propagating with constant velocity $V < V_o$ along the chain, Equations (1) reduce to the nonlinear Schrödinger equation (**NSE**)

Davydov's Soliton Revisited, Edited by P.L. Christiansen and A.C. Scott
Plenum Press, New York, 1990

$$\left(i\frac{\partial}{\partial\tau} + \frac{\partial^2}{\partial x^2} + 2g|\Psi|^2\right)\Psi(x,\tau) = 0 \tag{2}$$

with dimensionless quantities

$$\tau = \frac{\hbar t}{2ma^2}, \qquad g = \frac{\chi^2}{2Jw(1-s^2)}, \qquad s = \frac{V}{V_o}. \tag{3}$$

The NSE (2) at any nonzero value of g and, consequently, nonzero value of the coupling parameter χ has a normalized solution in the form of a soliton

$$\Psi(x,t) = \frac{1}{2}\sqrt{g}\,\operatorname{sech}\left[\frac{g(x-4k\tau)}{2}\right]\exp\left[i(2kx - \omega\tau)\right] \tag{4}$$

with parameters

$$\omega = 4(k^2 - g^2/16), \qquad 2k = mVa/\hbar. \tag{5}$$

As it is well known, Zakharov and Shabat [7] have developed the inverse scattering method (ISM) for the NSE (2) which allows one to study the time evolution of the initial excitation, corresponding to an arbitrary function $\Psi(x,0)$ rapidly decreasing at infinity.

According to the ISM, one may put in correspondence to the NSE the linear scattering problem for the eigenvectors and eigenvalues

$$v = \begin{pmatrix} v_1 \\ v_2 \end{pmatrix}, \qquad \zeta = \xi + i\eta, \tag{6}$$

satisfying the set of equations

$$v_1' + i\zeta v_1 = q(x)v_2, \quad v_2' - i\zeta v_2 = -q^*(x)v_1 \tag{7}$$

in which

$$q(x) = i\sqrt{g}\,\Psi(x,0). \tag{8}$$

Solving the set of Equations (7) one can find the scattering data

$$\zeta_j, \quad c_j(\tau) = b(\zeta_j,\tau)\left(\left[\frac{\partial a(\zeta)}{\partial\zeta}\right]_{\zeta=\zeta_j}\right)^{-1}, \quad j = 1,2,\ldots N,$$
$$R(\xi,\tau) = b(\xi,\tau)/a(\xi,\tau) \tag{9}$$

which determine the function $q(x,\tau) = 2K(x,x)$ and consequently, $\Psi(x,\tau)$.

Here ζ_j are the values at which the scattering coefficient vanishes in the ζ upper half-plane, and $K(x,y)$ is the solution of the Gelfand-Levitan-Marchenko integral equation

$$K(x,y) = F^*(x+y,\tau) - \int_x^\infty ds \int_x^\infty dz\, F^*(s+y,\tau)\,F^*(s+z,\tau)\,K(x,z) \tag{10}$$

with

$$F(x,\tau) = \frac{1}{2\pi}\int_{-\infty}^\infty R(\xi,\tau)\exp(i\xi x)d\xi + \sum_{j=1}^N c_j(\tau)\exp(i\zeta_j x). \tag{11}$$

The real part ξ_j or ζ_j determines the velocity of the velocity of the jth soliton and its imaginary part η_j determines its amplitude.

The trivial case when an initial excitation is given in the form of a soliton hyperbolic secant

$$\Psi(x,0) = \frac{1}{2}\sqrt{g}\,\operatorname{sech}\left(\frac{gx}{2}\right)\exp(2ikx) \tag{12}$$

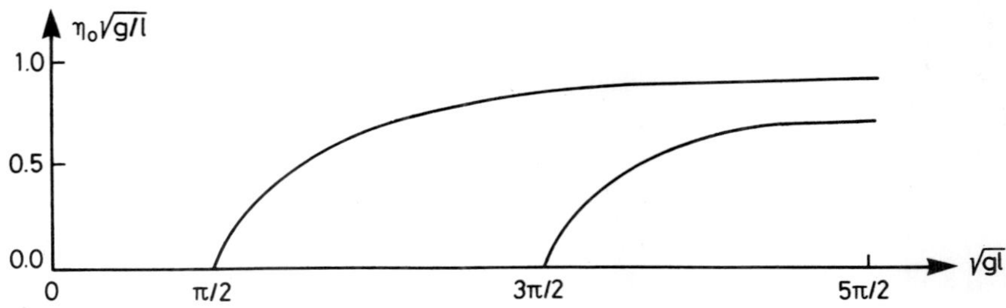

Figure 1. The $\eta_0\sqrt{g/\ell}$ dependence on $\sqrt{g/\ell}$.

is the case of a reflectionless potential with scattering data $b(\xi) = 0, \zeta_1 = -k + ig/4$ as follows from the solution of Equations (7). Solving next the Gelfand- Levitan-Marchenko Equation (10) for the case (12), one finds the wave function $\Psi(x,\tau)$. It has the form of a soliton impulse at any nonzero value of the nonlinearity parameter g in agreement with the general conception of solitons.

The next case we shall be concerned with is the case of a rectangular step

$$\Psi(x,0) = \begin{cases} \ell^{-1/2}\,\exp(2ikx), & \text{if } 0 \le x \le \ell, \\ 0, & \text{if } x < 0 \text{ or } x > \ell. \end{cases} \tag{13}$$

Solving Equations (7) one finds the scattering data and the equation for scattering coefficient zeros, which has the solution $\zeta_o = -k + i\eta_o$, where η_o satisfies the equation

$$\eta_o = -\sqrt{g/\ell - \eta_o^2}\,\operatorname{ctg}\left(\ell\sqrt{g/\ell - \eta_o^2}\right). \tag{14}$$

Figure 1 represents the value $\eta_o\sqrt{g/\ell}$ as a function of $\sqrt{g\ell}$. If $\sqrt{g\ell} < \pi/2$, Equation (14) has no solution. At $\pi/2 < \sqrt{g\ell} \le 3\pi/2$ it has a single solution, and with $\eta_o < \sqrt{g/\ell}$ and at $(2N-1)\pi/2 < \sqrt{g\ell} \le (2N+1)\pi/2$ it has N solutions, each of them corresponding to a localized (soliton) state with parameters $\eta_j \ne \eta_i < \sqrt{g/\ell}$, moving with the same constant velocity proportional to $\xi_i = -k$.

The solution of Gelfand-Levitan-Marchenko Equation (10) for $g\ell < \pi^2/4$ describes the propagation of the constant phase $(x - 1 - 4k\tau)^2/4\tau = const = \gamma^2/4$ along the x-axis according to the diffusion law $x - \ell - 4k\tau = \gamma\sqrt{\tau}$ with the amplitude independent of x and decreasing with time as $\tau^{-1/2}$, when the condition $x - 2\ell \ll \tau$ is satisfied. In the opposite case $x - 2\ell \gg \tau$ the solution to the Equation (7) has the same form but with an amplitude depending upon x and decreasing with time.

The solution to Equation (7) has quite another form in the case $g\ell > \pi^2/4$. Now $\Psi = \Psi_s(x,\tau) + \Psi_1(x,\tau)$, where $\Psi_s(x,\tau)$ describes the soliton

$$|\Psi_s(x,\tau)|^2 = 4\eta_o^2/g\cosh^2 z, \qquad z = 2\eta_o(x - x_o - 4k\tau) \tag{15}$$

with an amplitude $2\eta_o/\sqrt{g}$, width $1/2\eta_o$ and velocity $V = 2\hbar k/ma$, and $\Psi_1(x,\tau)$ describes the tail of the solution (15) decreasing with time.

Thus if the nonlinearity parameter is smaller than the critical value $g_{cr} = \pi^2/4\ell$ the initial rectangular step is transformed into a spreading and decreasing wave function. When the nonlinearity parameter exceeds this critical value, the soliton is formed: its width decreases as g increases. As $g \to g\pi^2/4\ell$, the major contribution to the wave function is provided by the soliton since in this case $\int |\Psi(x,\tau)|^2 dx \to 0.85$ and the role of a tail proves to be negligibly small. With the nonlinearity parameter $\pi^2(2N-1)^2 < 4g\ell \leq (2N+1)^2\pi^2$, N solitons are formed. However, the major contribution to the wave function is still provided by the soliton with a parameter $\eta_o \approx g/\ell$.

Thus we have shown that the threshold character of soliton generation in molecular chains is determined by the specific manner of their excitation, i.e. by the choice of the initial wave function. Analytical results for the initial excitation in the form of a rectangular step considered here explain well the computer results of the corresponding discrete problem [1].

3   DS in a Half-Infinite Molecular Chain

The consideration of the soliton generation problem in the first part of the paper disregarded the fact that DS is a topologically stable entity and, for this reason, it can be created or destroyed at the chain ends only. Hence there arise two questions: what new results one can expect with accounting for chain ends and in what way really can a DS be excited in a chain? One of the possible mechanisms of soliton creation was suggested in [5]: an excitation (it may be an electron, light quantum etc.) is captured by the impurity molecule at the end of the principal chain (it may be a chromophore molecule, for example) with the following fast excitation transfer onto the neighboring molecule of the chain with which the impurity is connected in a nonresonant way. (This is just the stage to form the initial condition for the soliton creation). Disposing different impurity molecules in the chain at different sites, one can change the localization of initial signal and its parameters, and, thus, govern the generation soliton parameters.

The redistribution of the initial excitation of the impurity molecule connected with the n_0th molecule of the chain along the chain is described by the system of equations

$$
i\hbar \frac{\partial}{\partial t}\rho_{n\,m} =
$$
$$
i\Gamma \delta_{n\,n_0}\delta_{n_0\,m}R +
$$
$$
+ J\left[(1-\delta_{n\,1})\rho_{n-1\,m} + \rho_{n+1\,m} - (1-\delta_{1\,m})\rho_{n\,m-1} - \rho_{n\,m+1}\right] +
$$
$$
+ \chi^2\left[\rho_{n+1\,n+1} + \rho_{n\,n} + (1-\delta_{n\,1})(\rho_{n\,n} + \rho_{n-1\,n-1}) - \rho_{m+1\,m+1} - \right.
$$
$$
\left. - (1-\delta_{1\,m})(\rho_{m\,m} + \rho_{m-1\,m-1})\right]\rho_{m\,m}/w, \quad n,m = 1,2,3,...,\infty,
$$

$$
\hbar\frac{\partial}{\partial t}R = -\Gamma R, \quad \rho_{n\,m}(t=0) = 0, \quad R(t=0) = 1. \tag{16}
$$

The longitudinal displacement β_n of the nth molecule from its equilibrium position is connected with the probability $\rho_{n\,n}$ of the excitation being localized on the molecule by

$$
\beta_n = \frac{\chi}{w}\sum_{m=n}^{\infty}(\rho_{m\,m} + \rho_{m+1\,m+1}), \quad n = 1,2,3,...,\infty, \tag{17}
$$

when the inertia of the elastic subsystem is small: $\hbar^2 w/4MJ^2 \ll 1$. Here, R is the probability of the excitation presence on impurity, Γ is the constant of nonresonance excitation transfer from the impurity molecule onto a neighboring one of the basic chain, M is the molecule mass, J is the resonance excitation transfer constant, w is the chain elasticity, and χ is the exciton-phonon coupling constant.

If the condition $\Gamma \gg J$ is fulfilled, the excitation is localized at the molecule nearest to the impurity at time moment to $(\hbar/\Gamma \ll t_0 \ll \hbar/J)$: $\rho_{n\,m}(t=0) \approx \delta_{n\,n_0}\delta_{n_0\,m}$. The inverse transfer of the excitation to the impurity is impossible $R(t \geq t_0) = 0$.

Introducing dimensionless time $\tau = J(t - t_0)/\hbar$ it is possible to obtain the equation for the probability amplitudes Ψ_n

$$i\hbar\frac{d\Psi_n}{d\tau} - \Psi_{n+1} - (1 - \delta_{n\,1})\Psi_{n-1} + 2\Psi_n - 2g\left[|\Psi_{n+1}|^2 + |\Psi_n|^2 + \right.$$
$$\left. +(1 - \delta_{n\,1})(|\Psi_n|^2 + |\Psi_{n-1}|^2)\right]\Psi_n = 0,$$

$$\Psi_n(\tau = 0) = \delta_{n\,n_0}, \quad \rho_{n\,m} = \Psi_n\Psi_m^*, \quad n = 1, 2, ..., \infty, \tag{18}$$

where the nonlinearity parameter g is determined in (3). Equations (18) for a half-infinite chain coincide with equations for an infinite chain

$$i\hbar\frac{d\Psi_n}{d\tau} - \Psi_{n+1} - \Psi_{n-1} + 2\Psi_n - 2g\left[|\Psi_{n+1}|^2 + |\Psi_{n-1}|^2 + \right.$$
$$\left. +(2 - \delta_{n\,1} - \delta_{-n\,1})|\Psi_n|^2\right]\Psi_n = 0,$$

$$\Psi_n(\tau = 0) = (\delta_{n\,n_0} - \delta_{-n\,n_0}), \quad -\infty < n < \infty \tag{19}$$

at $n \geq 1$, which give the NSE (2) on the continuum approximation.

Thus we have shown that the problem discussed is the same as that for an infinite chain but in the region $x > 0$ only by antisymmetric initial excitation. That is why in what follows we shall seek the solution to the NSE (2) with initial condition given in the form of antisymmetric rectangular step

$$\Psi(x,0) = \begin{cases} 0 & -\infty < x < -x_0 + \frac{1}{2} - \ell, \\ -\ell^{-1/2} & -x_0 + \frac{1}{2} - \ell < x < -x_0 + \frac{1}{2}, \\ 0 & -x_0 + \frac{1}{2} < x < x_0 - \frac{1}{2}, \\ \ell^{-1/2} & x_0 - \frac{1}{2} < x < x_0 - \frac{1}{2} + \ell, \\ 0 & x_0 - \frac{1}{2} + \ell < x < \infty, \end{cases} \tag{20}$$

within the ISM [7].

The equation for scattering coefficient zeroes follows from (7) with (20) [8]

$$\left[(z^2 + G^2)^{1/2}\,\mathrm{ctg}\,(z^2 + G^2)^{1/2} - iz\right]^2 + G^2\exp(2i\,\Delta z) = 0. \tag{21}$$

where

$$z = \zeta\ell = -u + iv, \quad G^2 = 2g\ell, \quad \Delta = (2x_0 - 1)/\ell. \tag{22}$$

The roots of this equation, if they do exist, enter in pairs

$$z_j \pm (\Delta, G) = \mp u_j(\Delta, G) + iv_j(\Delta, g), \quad (u_j, v_j > 0). \tag{23}$$

However, in a real physical situation, i.e. solitons in semi-infinite chain, only the roots with negative real part $z_j^+(\Delta, G)$ are of interest.

The analysis of (20) shows that at any nonzero Δ the soliton generation has a threshold with respect to G. Under the condition $G_{cr\,1}(\Delta) < G \leq G_{cr\,2}(\Delta)$, one soliton is created, two solitons appear at $G_{cr\,2} < G \leq G_{cr\,3}$, etc. The soliton velocities turn out to be bounded above: $u_j(\Delta, G) < u_{cr\,j}(\Delta)$ and each pair of values $G_{cr\,j}(\Delta)$ and $u_{cr\,j}(\Delta)(j = 1, 2)$ is defined from the equation

$$\sqrt{u_{cr}^2 + G_{cr}^2}\,\mathrm{ctg}\,\sqrt{u_{cr}^2 + G_{cr}^2} = iG_{cr}\,\exp(-iu_{cr}\Delta) - iu_{cr}. \tag{24}$$

It follows from (24) that $G_{cr\,j}(\Delta)$ and $u_{cr\,j}(\Delta)$ decrease monotonically with increasing Δ (see Figure 2) so that

$$G_{cr\,j}(0) = u_{cr\,j}(0) = (j - \frac{1}{2})\frac{\pi}{\sqrt{2}}$$

$$G_{cr\,j}(\Delta \gg 1) \approx (j - \tfrac{1}{2})\frac{\pi}{2},$$

$$u_{cr\,j}(\Delta \gg 1) \approx (j - \tfrac{1}{2})\frac{\pi}{\Delta}. \tag{25}$$

The dependence of $v_1(\Delta, G)$ on Δ is shown in Figure 3. The dependence of single soliton excitation efficiency

$$D(\Delta, G) = \frac{\int_0^\infty dx |\Psi_1^s(x, \tau \to \infty)|^2}{\int_0^\infty dx |\Psi(x, 0)|^2} = \frac{4\eta_1(\Delta, G)}{G^2} \tag{26}$$

on parameters G and Δ is nonmonotonous (Figures 4 and 5) as might be expected from energy balance conditions [5]. The soliton velocity decreases monotonically with increasing Δ (Figure 6) or G (Figure 7).

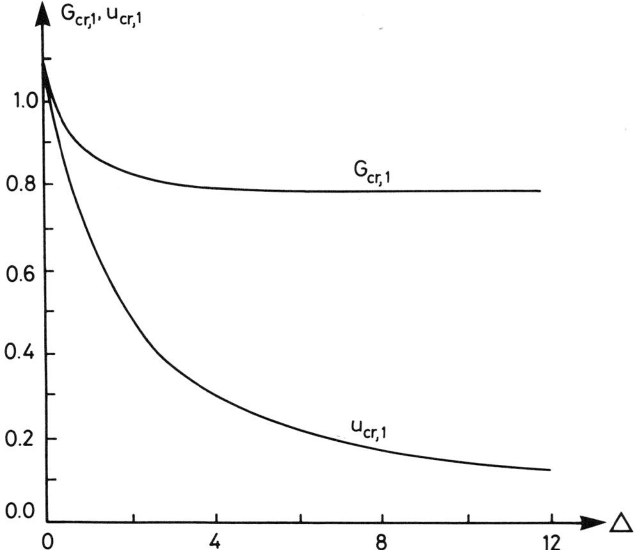

Figure 2. The dependence of $G_{cr,1}$ and $v_{cr,1}$ on Δ.

Thus the analysis of the time evolution of the initial excitation of a rectangular step in a semi-infinite chain indicates the possibility to govern the soliton generation efficiency and also soliton parameters. At the same time the threshold character of soliton generation is the result of a specific form of initial excitation distribution as was shown in Part 1.

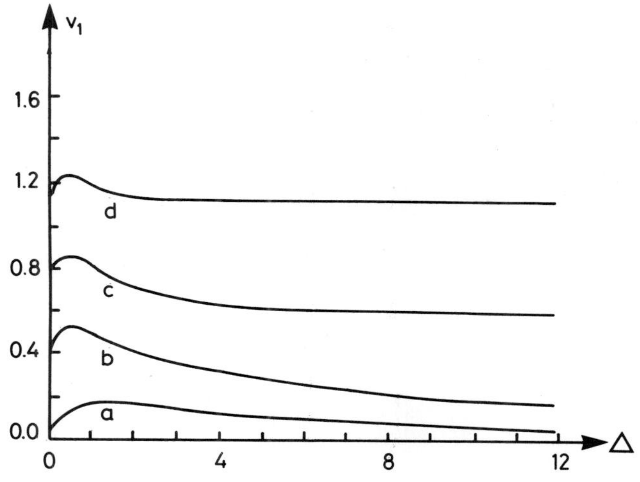

Figure 3. The value v_1 as a function of Δ: a) $G = \sqrt{2}\pi/4$,   b) $G = \pi/2$,   c) $G = 5\pi/8$, d) $G = 3\pi/4$.

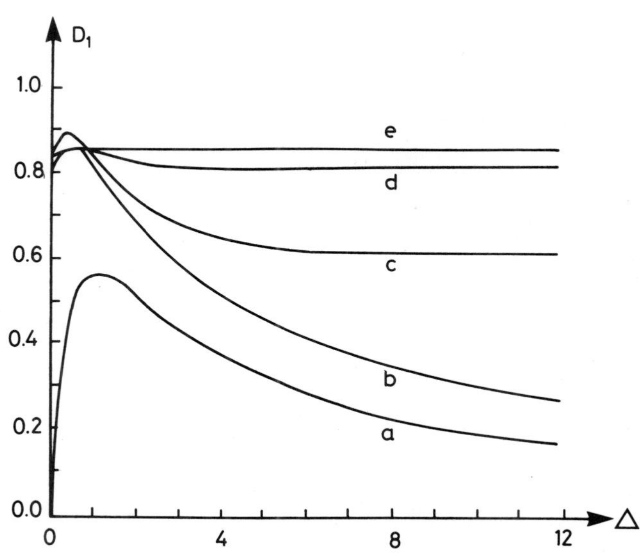

Figure 4. The dependence of D_1 on G: a) $\Delta = 0.0$,   b) $\Delta = 0.4$,   c) $\Delta = 1.0$, d) $\Delta = 2.0$,   e) $\Delta = 12.0$,   f) $\Delta \to \infty$.

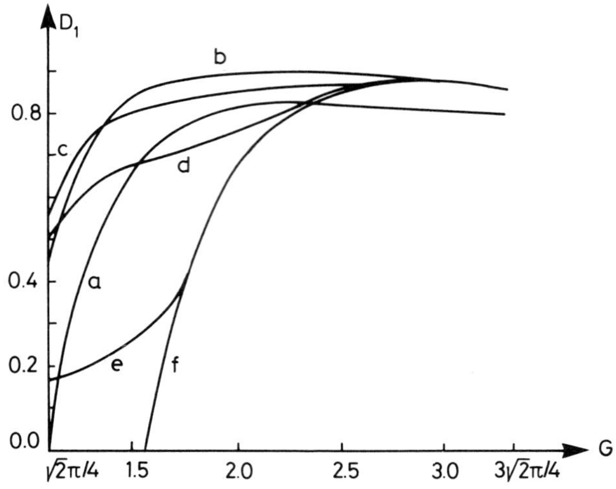

Figure 5. The depencence of D_1 on Δ: a) $G = \sqrt{2}\pi/4$,   b) $G = \pi/2$,   c) $G = 5\pi/8$, d) $G = 3\pi/4$,   e) $G = 3\sqrt{2}\pi/4$.

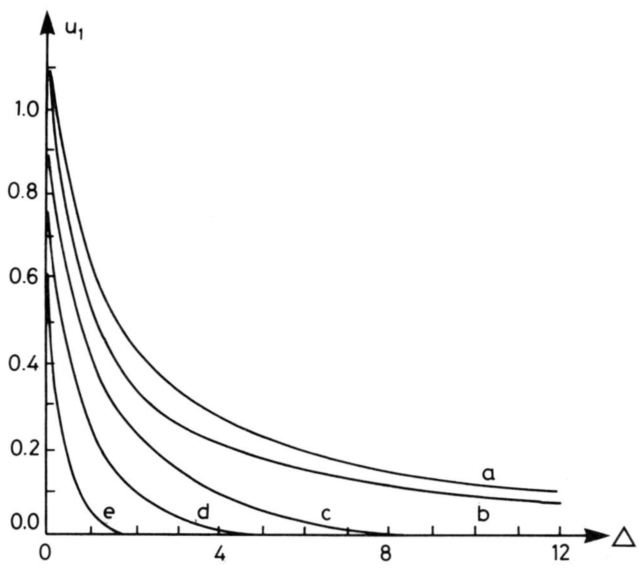

Figure 6. The dependence of u_1 on Δ: a) $G = \sqrt{2}\pi/4$,   b) $G = \pi/2$,   c) $G = 5\pi/8$, d) $G = 3\pi/4$,   e) $G = 3\sqrt{2}\pi/4$.

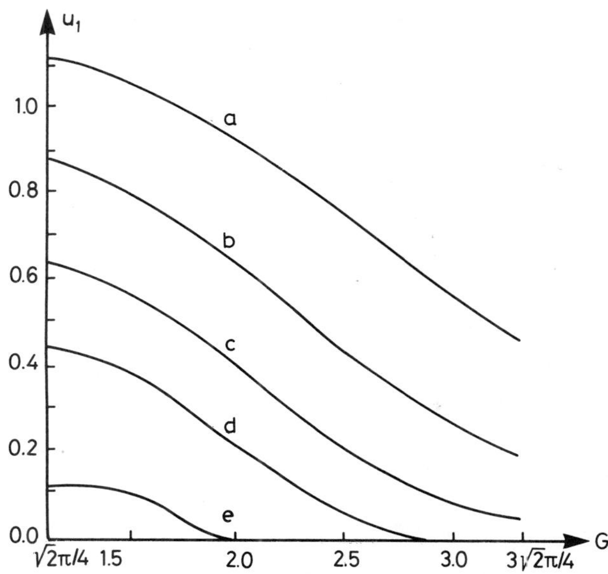

Figure 7. The dependence of u_1 on G: a) $\Delta = 0.0$,   b) $\Delta = 0.4$,   c) $\Delta = 1.0$, d) $\Delta = 2.0$,   e) $\Delta = 12.0$.

References

[1] J.M. Hyman, D.W. McLaughlin and A.C. Scott, On Davydov's alpha-helix solitons, **Physica D 3**; 23 (1981).

[2] J.C. Eilbeck, Davydov soliton, 16mm mute film available from Swift Film Productions, 1 Wood Road, Wimbledon, London SW DHN, Great Britain.

[3] L.S. Brizhik, The time evolution of the nonlinear Schrödinger equation solutions, Prepr. ITP-81-134R, Kiev (1981).

[4] L.S. Brizhik and A.S. Davydov, Soliton excitations in one-dimensional molecular systems, **phys. stat. sol. (b) 115**; 615 (1983).

[5] L.S. Brizhik, Yu.B. Gaididei, A.A. Vakhnenko, and V.A. Vakhnenko, Soliton generation in semi-infinite molecular chains, **phys. stat. sol. (b) 146**; 605 (1988).

[6] A.S. Davydov, **Solitons in molecular systems**, D. Reidel Pub. Co., Dordrecht, Boston, Lancaster (1987).

[7] V.E. Zakharov and A.B. Shabat, The exact theory of two-dimensional focusing and one-dimensional automodulation of waves in nonlinear systems, **Zh. Eksper. Teor. Fiz.** (Russ) **61**; 118 (1971).

[8] L.S. Brizhik, "The excitation and interactions of solitons, including extra fields", Cand. Thesis, Kiev (1984).

11

SOLITON DYNAMICS IN THE EILBECK-LOMDAHL-SCOTT MODEL FOR HYDROGEN-BONDED POLYPEPTIDES

Alexander V. Savin[a] and Alexander V. Zolotaryuk[b]

[a]Institute for Physico-Technical Problems
119034 Moscow, USSR
[b]Institute for Theoretical Physics, Academy of
Sciences of the Ukrainian SSR, 252130 Kiev, USSR

INTRODUCTION

The transfer of vibrational energy along quasi-one-dimensional molecular systems such as chains of hydrogen-bonded peptide groups (PG's) by means of self-trapped states (solitary waves or solitons) was first suggested by Davydov and Kislukha[1] in order to explain how the energy released by hydrolysis of adenosine triphosphate can be localized and moved along proteins providing important biological processes[2,3]. The soliton formation in this model[1,3] is due to the coupling of the high-frequency intramolecular $C = O$ stretch mode (the amide-I excitation, with frequency about 1665 cm^{-1}) in PG's and the acoustic mode (the intermolecular relative displacement field) of PG's with associated side groups through the dependence of the amide-I energy on the distances to neighbouring left and right molecules (PG's). After the numerous theoretical studies[3] on this acoustic-mode-coupled soliton theory, the experimental results performed by Careri and coworkers[4] for crystalline acetanilide (ACN) became very important for the question of the existence of self-trapped localized states in quasi-one-dimensional molecular systems since the material ACN contains chains of hydrogen-bonded PG's similar to protein molecules. In this material there are four optical low-frequency modes with which the high-frequency intramolecular amide-I excitation can be coupled. Using a Davydov-type theory Eilbeck, Lomdahl, and Scott[5] have constructed a model which attributes the observed anomalous sideband (at 1650 cm^{-1}) of the amide-I absorption peak (at 1665 cm^{-1}) to a localized self-trapped state formed by coupling the amide-I and optical phonon modes. Later on this soliton theory had been extended to include the original acoustic phonon coupling[6] and altered by Takeno[7] using his oscillator-lattice formalism.

In this paper we study by numerical simulations the interaction of two mobile solitons and the dynamical stability of bisolitons in the ELS-model. For simplicity we restrict ourselves, as in the case of Takeno[7], to the consideration of one optical mode. In this case, one-particle (when only one quantum of the amide-I excitation is present in the chain) equations of motion resemble those of the Holstein model[8] for electron self-trapping. In the presence of any definite number of amide-I quanta they can be generalized to the similar form[3,9-12] as in the original acoustic-phonon-coupled Davydov theory. Therefore the soliton dynamics should be investigated in the framework of many-particle equations of motion. For some purposes to be clarified below we also use a slightly generalized version of the model including a positive dispersion term for low-frequency optical phonons (as in the Holstein model[8,13]) and an anharmonicity for these phonons[14].

EQUATIONS OF MOTION

Consider a finite or infinite one-dimensional lattice with spacing a_o of identical molecules (PG's) with a high-frequency intramolecular (the amide-I excitation) mode. The Hamiltonian for the ELS-model (resticted to one low-frequency optical mode) is given by the sum of three operators:

$$H = H_{ex} + H_{ph} + H_{int} \tag{1}$$

where

$$H_{ex} = \sum_n \left[E_o B_n^+ B_n - J(B_n^+ B_{n+1} + B_{n+1}^+ B_n) \right], \tag{1a}$$

$$H_{ph} = \sum_n : \left[(\hat{p}_n^2/2M) + U(\hat{y}_n) - L\hat{y}_n\hat{y}_{n+1} \right] : , \tag{1b}$$

$$H_{int} = \sum_n \chi \hat{y}_n B_n^+ B_n . \tag{1c}$$

Here B_n^+ and B_n are boson creation and annihilation operators for quanta of the intramolecular vibrational excitation with energy E_o at the n-th site, J is the intersite transfer energy produced by dipole-dipole interactions, $\hat{y}_n$ and $\hat{p}_n$ are the low-frequency optical mode position and momentum operators at the n-th site, respectively, satisfying the canonical commutation relations

$$\left[\hat{y}_n, \hat{p}_{n'}\right] = i\hbar\delta_{nn'} , \left[\hat{y}_n, \hat{y}_{n'}\right] = 0 = \left[\hat{p}_n, \hat{p}_{n'}\right] ; \tag{2}$$

M is the reduced mass and the coupling constant χ arises from dependence of the on-site amide-I energy on the intramolecular displacements. For some analytical techniques to be described below the optical phonon part of the Hamiltonian (1b) is normal-ordered and may contain the dispersion term[13] with L, $0 \leq L < K/2$. The function $U(y)$ is supposed to have minimum at $y = 0$ with $U(0) = 0$ and $U''(0) = K > 0$ (a prime denotes differentiation). The operator-valued function $: U(\hat{y}_n):$ is defined as a Taylor expansion in the Wick powers

$$: \hat{y}_n^m: = \sum_{k=0}^{[m/2]} \frac{m!}{k!(m-2k)!} \left(- \frac{1}{2} \left\langle 0 \left| \hat{y}_n^2 \right| 0 \right\rangle \right)^k \hat{y}_n^{m-2k} . \tag{3}$$

This formula follows from the commutation relations (2) and a combinational argument. The other two terms in (1b) are defined in a similar way:

$$: \hat{p}_n^2: = \hat{p}_n^2 - \left\langle 0 \left| \hat{p}_n^2 \right| 0 \right\rangle , \tag{4a}$$

$$: \hat{y}_n \hat{y}_{n+1}: = \hat{y}_n \hat{y}_{n+1} - \left\langle 0 \left| \hat{y}_n \hat{y}_{n+1} \right| 0 \right\rangle . \tag{4b}$$

Let us consider the general case when in the chain there is an arbitrary conserved number N of amide-I excitation quanta. To describe approximately the quantum evolution of this collective excitation we proceed in the standard way using the generalized Davydov ansatz[15,16,10-12]

$$\left| \Psi_N(t) \right\rangle = (N!)^{-1/2} \sum_{n_1, \ldots, n_N} a_{n_1 \ldots n_N}(t) B_{n_1}^+ \ldots B_{n_N}^+ e^{-A(t)} \left| 0 \right\rangle \tag{5}$$

where $\left| 0 \right\rangle$ is the bare vacuum state,

$$A(t) = \frac{i}{\hbar} \sum_n \left[y_n(t) \hat{p}_n - p_n(t) \hat{y}_n \right] , \tag{5a}$$

and the approximation of the quantum evolution has been introduced by the factored form of (5). There is a series of papers[17] devoted to the problem of validity of this type of approximation. The probability amplitude $a_{n_1 \ldots n_N}(t)$ is supposed to be symmetric with respect to any permutation of $n_1, \ldots, n_N$ and is normalized as

$$\left\langle \Psi_N(t) \middle| \Psi_N(t) \right\rangle = \sum_{n_1, \ldots, n_N} \left| a_{n_1 \ldots n_N}(t) \right|^2 = 1 , \tag{6}$$

so that for all times there are N quanta in the state (5):

$$\left\langle \Psi_N(t) \middle| \sum_n B_n^+ B_n \middle| \Psi_N(t) \right\rangle = N . \tag{7}$$

If there are no additional restrictions for $a_{n_1 \ldots n_N}(t)$ any number $(0,1,\ldots,N)$ of quanta can be found at each lattice site (Bose statistics). In another interesting case no more than one quantum can be created at any site (a two-level system). In this case $a_{n_1 \ldots n_N}(t)$'s vanish if some of the integers $n_1, \ldots, n_N$ coincide (Pauli statistics).

The classical equations of motion for $y_n(t)$, $p_n(t)$, and $a_{n_1 \ldots n_N}(t)$ can be established in different ways resulting in the same form where the functions $a_{n_1 \ldots n_N}(t)$ may differ by a phase factor[18]. But in any case the formula

$$\langle \Psi_N(t) \mid :\hat{y}_n^m: \mid \Psi_N(t) \rangle = y_n^m(t) \tag{8}$$

is useful which can be established by using the relations (2),(3),

$$\langle 0 \mid \hat{y}_n^{2k} \mid 0 \rangle = \frac{(2k)!}{k!} \left(\frac{1}{2} \langle 0 \mid \hat{y}_n^2 \mid 0 \rangle \right)^k, \quad \langle 0 \mid \hat{y}_n^{2k+1} \mid 0 \rangle = 0, \tag{9}$$

and the Weyl identity. From (4) and (8) we immediately obtain

$$\langle \Psi_N \mid :\hat{p}_n^2: \mid \Psi_N \rangle = p_n^2, \quad \langle \Psi_N \mid :\hat{y}_n \hat{y}_{n+1}: \mid \Psi_N \rangle = y_n y_{n+1}, \tag{10a}$$

$$\langle \Psi_N \mid : U(\hat{y}_n): \mid \Psi_N \rangle = U(y_n). \tag{10b}$$

To derive the aforementioned equations we start from the Lagrangian

$$\mathcal{L} = \langle \Psi_N(t) \mid i\hbar \frac{\partial}{\partial t} - H \mid \Psi_N(t) \rangle \tag{11}$$

with the Hamiltonian (1). The differentiation in (11) can be performed in the same way as it was done by Kerr and Lomdahl[18]. Therefore

$$\mathcal{L} = i\hbar \sum_{n_1, \ldots, n_N} a^*_{n_1 \ldots n_N} \dot{a}_{n_1 \ldots n_N} + \frac{1}{2} \sum_n (y_n \dot{p}_n - \dot{y}_n p_n) -$$

$$- \sum_{n_1, \ldots, n_N} a^*_{n_1 \ldots n_N} \left(NE_o + \chi \sum_{j=1}^{N} y_{n_j} \right) a_{n_1 \ldots n_N} -$$

$$- J \sum_{j=1}^{N} (a_{n_1 \ldots n_j - 1 \ldots n_N} + a_{n_1 \ldots n_j + 1 \ldots n_N}) -$$

$$- \sum_n \left[(p_n^2 / 2M) + U(y_n) - L y_n y_{n+1} \right] \tag{12}$$

and the Euler-Lagrange equations which follow from (12) have the form

$$i\hbar \dot{a}_{n_1 \ldots n_N} = NE_o a_{n_1 \ldots n_N} - J \sum_{j=1}^{N} (a_{n_1 \ldots n_j - 1 \ldots n_N} +$$

$$+ a_{n_1 \ldots n_j+1 \ldots n_N}) + \chi \left(\sum_{j=1}^{N} y_{n_j} \right) a_{n_1 \ldots n_N} , \qquad (13a)$$

$$M\ddot{y}_n = L (y_{n-1} + y_{n+1}) - U'(y_n) - $$

$$- N \chi \sum_{n_1, \ldots, n_{N-1}} \left| a_{n_1 \ldots n_{N-1} n} \right|^2 . \qquad (13b)$$

The equations of this type have been derived previous-ly[10-12] for the acoustic-phonon-coupled Davydov theory. Note that some additional relations such as, for example,

$$\left[\sum_{n'} : U(\hat{y}_{n'}) : , \; \hat{p}_n \right] = i\hbar : U'(\hat{y}_n): \qquad (14)$$

should be used in derivation of Eqs.(13) if to follow Kerr and Lomdahl[18].

In the following it is convenient to use the dimension-less description introducing a new lattice spacing l_o and time T by the relations

$$l_o = \hbar c_o/a_o |J|, \; T = c_o l_o t/a_o , \; c_o^2 = K a_o^2/M. \qquad (15)$$

The dimensionless lattice fields are introduced as

$$\phi_{n_1 \ldots n_N}(T) = \prod_{j=1}^{N} (\pm 1)^{n_j} C^{1/2} a_{n_1 \ldots n_N}(t) \exp\left\{ \frac{i}{\hbar} \left[N(E_o - 2|J|)t \right] \right\} \qquad (16a)$$

where $\; C = \chi^2 a_o^2/l_o^2 |J| M(c_o^2 - 2 v_o^2) , \; v_o^2 = L a_o^2/M < c_o^2/2,$ and

$$q_n(T) = (\chi/l_o^2 |J|) y_n(t) . \qquad (16b)$$

Then in these new variables we can rewrite the Lagrangian (12), Eqs.(13), and also write the corresponding Hamiltonian function. We have

$$i \frac{d\phi_{n_1 \ldots n_N}}{dT} = l_o^{-2} \left[2N \phi_{n_1 \ldots n_N} - \sum_{j=1}^{N} (\phi_{n_1 \ldots n_j-1 \ldots n_N} + \right.$$

$$\left. + \phi_{n_1 \ldots n_j+1 \ldots n_N}) \right] + \left(\sum_{j=1}^{N} q_{n_j} \right) \phi_{n_1 \ldots n_N} , \qquad (17a)$$

$$l_o^2 \frac{d^2 q_n}{dT^2} = s_o^2 (q_{n+1} - 2q_n + q_{n-1}) - $$

$$- (1 - 2s_o^2) \left(\frac{dV}{dq_n} + N \sum_{n_1, \ldots, n_{N-1}} \left| \phi_{n_1 \ldots n_{N-1} n} \right|^2 \right) \qquad (17b)$$

where $V(q) = (1 - 2s_o^2)^{-1}(\chi^2 a_o^2 U/l_o^4 J^2 M c_o^2 - s_o^2 q^2)$, $s_o = v_o/c_o$, so that $V''(0) = 1$. The probability amplitude $\phi_{n_1 \ldots n_N}$ is normalized as

$$\sum_{n_1, \ldots, n_N} \left| \phi_{n_1 \ldots n_N}(T) \right|^2 = C . \tag{18}$$

REDUCTION TO A ONE-SOLITON PROBLEM

Eqs.(17) with the constraint (18) describe the nonlinear dynamics of N amide-I excitation quanta interacting with intramolecular low-frequency displacements q_n's . In the following two particular cases these equations can be reduced to a one-quantum (one-particle) problem. The first case is a sufficiently dilute "gas" of one-particle localized states when the N-particle probability amplitude can be represented as

$$\phi_{n_1 \ldots n_N}(T) = (N!)^{-\frac{1}{2}} C^{\frac{1-N}{2}} \sum_{P} \phi_{n_{P_1}}^{(1)}(T) \ldots \phi_{n_{P_N}}^{(N)}(T) . \tag{19}$$

Here the sum is taken over all permutations of $n_1, \ldots, n_N$ and the one-particle amplitudes $\phi_n^{(i)}(T)$'s are orthogonal and normalized as

$$\sum_{n} \phi_n^{(i)*}(T) \phi_n^{(j)}(T) = C \delta_{ij} . \tag{20}$$

In the other case all N quanta can be localized at one lattice site so that the N-particle amplitude can be chosen in the direct product form

$$\phi_{n_1 \ldots n_N}(T) = N^{-\frac{N}{2}} C^{\frac{1-N}{2}} \prod_{j=1}^{N} \phi_{n_j}(T) \tag{21}$$

where $\phi_n(T)$ is normalized by

$$\sum_{n} \left| \phi_n(T) \right|^2 = NC . \tag{22}$$

The both representations of the N-particle wave function given by (19) and (21) allow us to reduce Eqs.(17) to the one-particle evolution equations of the same form, i.e. to Eqs. (17) with $N = 1$. The only difference is the normalization (compare (20) and (22)).

Now we pass to treating one-soliton problem with $N \geq 1$ in (22). Substituting the ansatz

$$\phi_n(T) = \varphi_n(T) \exp\left\{ i\left[nkl_o - (4l_o^{-2}\sin^2\frac{kl_o}{2} + \lambda)T \right] \right\} \tag{23}$$

into Eqs.(17) with $N = 1$ and (22) with any $N \geq 1$ we obtain the set of equations

$$\frac{d\varphi_n}{dT} = l_o^{-2}\sin(kl_o)(\varphi_{n-1} - \varphi_{n+1}) \,, \tag{24a}$$

$$l_o^{-2}\cos(kl_o)(2\varphi_n - \varphi_{n-1} - \varphi_{n+1}) + q_n\varphi_n = \lambda\varphi_n \,, \tag{24b}$$

$$l_o^2\frac{d^2q_n}{dT^2} = s_o^2(q_{n+1} - 2q_n + q_{n-1}) - (1 - 2s_o^2)(\frac{dV}{dq_n} + \varphi_n^2) \,, \tag{24c}$$

$$\sum_n \varphi_n^2 = NC \,. \tag{25}$$

Here k is the wave number and λ is the spectral parameter to be determined. To get travelling-wave solutions of a stationary profile with a dimensionless velocity $s = v/c_o$ in the continuum limit we approximately substitute $\varphi_n(T)$ by $\varphi(\xi)$ and $q_n(T)$ by $q(\xi)$. Then Eq. (24a) gives the relation between velocity s and wave number k : $s = 2l_o^{-1}\sin(kl_o)$ while Eqs. (24b,c) and (25) are transformed to

$$-\cos(kl_o)\,\varphi'' + q\varphi = \lambda\varphi \,, \tag{26a}$$

$$l_o^2(s^2 - s_o^2)q'' + (1 - 2s_o^2)(dV/dq + \varphi^2) = 0 \,, \tag{26b}$$

$$l_o^{-1}\int \varphi^2(\xi)d\xi = NC \,, \qquad \xi = nl_o - sT \,. \tag{27}$$

For $s = s_o$ Eqs. (26) can be easily decoupled and integrated[19,20] resulting in an implicit soliton solution normalized by (27). In the harmonic limit $(V = q^2/2)$ this solution is transformed to the simple explicit form:

$$\varphi = (Cl_o\mu/2)^{1/2}\mathrm{sech}(\mu\xi), \quad q = -2\cos(kl_o)\mathrm{sech}^2(\mu\xi),$$

$$\lambda = -\cos(kl_o)\mu^2 \,, \tag{28}$$

where $\mu = Cl_o/4$ and $\cos(kl_o) = (1 - l_o^2s_o^2/4)^{1/2}$. This solution becomes stationary in the limit $s_o = 0$ when dispersion L disappears, but numerical simulations[14] have also proved the existence of mobile stable solitons if $L = 0$. The introduction of dispersion $L > 0$ into the model does not yield any new physics, but it is useful for the purpose to have a mobile soliton solution in an analytical simple form.

Another way for this purpose is the introduction into the model of a cubic anharmonicity:

$$V(q) = q^2/2 + \gamma q^3/3 \,. \tag{29}$$

Using the relations (28) as an ansatz from Eqs. (26) we find (if $s_o = 0$) that $\mu > 0$ satisfies the following cubic equation:

$$12\mu l_o s_1^2(4\mu^2 l_o^2 s_1^2 + 1) + C\gamma = 0 \tag{30}$$

where s_1 is a root of the equation

$$\cos(kl_o) = (1 - l_o^2 s^2/4)^{1/2} = -3l_o^2 \gamma^{-1} s^2 \; , \qquad \gamma < 0 \; . \qquad (31)$$

Therefore, any cubic anharmonicity with $\gamma < 0$ (see (29)) does not change the topology of the "solitonic part" $(q < 0)$ of the potential $V(q)$ and allows us to present a soliton profile with some velocity $s_1 > 0$ in an explicit analytical form. In the limit $\gamma \longrightarrow 0$ and $s_1 \longrightarrow 0$, $\lim(\gamma s_1^{-2}) = -3l_o^2$, Eq.(30) gives $\mu = Cl_o/4$, i.e. the solution (28).

Let us compare the total energy of N one-particle solitons with that of an N-particle soliton. For simplicity take immobile solitons when $\gamma = 0$ and $s_o = 0$. Simple calculations give $E = -C|J|l_o^4 N/48$ for N one-particle solitons and $E = -C|J|l_o^4 N^3/48$ for an N-particle soliton. Therefore, an N-particle soliton state is more favourable and the ratio of these energies increases as N^2.

NUMERICAL SIMULATION OF THE TWO-SOLITON DYNAMICS

In this section we present the numerical results for the two-quantum $(N = 2)$ dynamics in the cyclic chain of 50 molecules. The parameter values have been chosen to be in the correspondence with the previous studies[4-7]: $a_o = 5\text{Å}$, $M = m_p$ (m_p is the proton mass), $J = 4 \text{ cm}^{-1}$, $K = 13 \text{ N/m}$, $\chi = 2.5 \times 10^{-11} \text{N}$, so that $l_o = 11.73$, $C = 4.4 \times 10^{-3}$, $c_o = 4.4 \times 10^3 \text{ m/s}$. Next we require without loss of generality that $L = 0$ and $V = q^2/2$.

We use a standard fourth order Runge-Kutta scheme to integrate Eqs.(17). The integration step $h = .2$ ensures the conservation of the total probability (18) accurate to six decimal places and the total energy to five places. The initial conditions for Eqs.(17) were chosen by using continual one-soliton solutions of Eqs.(26) and the relations (19)-(22) for $N = 2$.

For finding two-component soliton solutions of Eqs.(26) we use the variational approach starting from the one-particle Lagrangian related to Eqs.(26). As a result we obtain the extremum problem:

$$\int \left(\cos(kl_o)\varphi'^2 - \frac{1}{2} l_o^2 s^2 q'^2 + \frac{1}{2} q^2 + \varphi^2 q \right) d\xi \longrightarrow \text{extr} \qquad (32)$$

where we put

$$s_o = 0 \; , \quad V = q^2/2 \; . \qquad (33)$$

To solve this problem with taking into account the condition (27) different sets of functions $\varphi(\xi)$ and $q(\xi)$ can be used. Previously[14] we have used a three-parameter family of the trial functions. But in the case (33) Eq.(26b) permits to reduce the number of variational parameters to one. The following choice of trial functions:

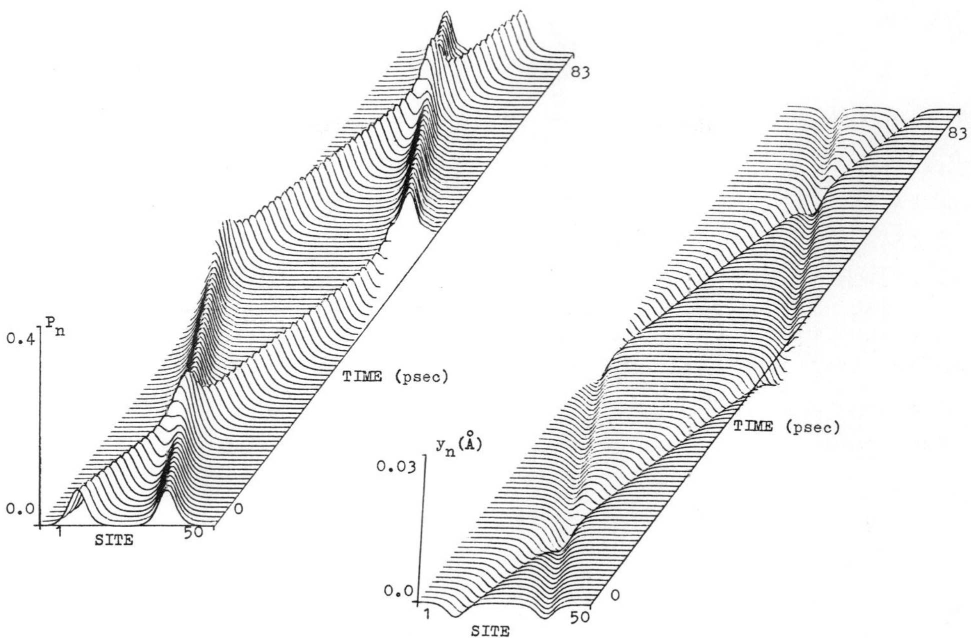

Fig. 1. Head-on elastic collision of two one-particle
solitons for the Bose and Pauli statistics':
Incident velocities: $s_1 = -s_2 = 0.015$.

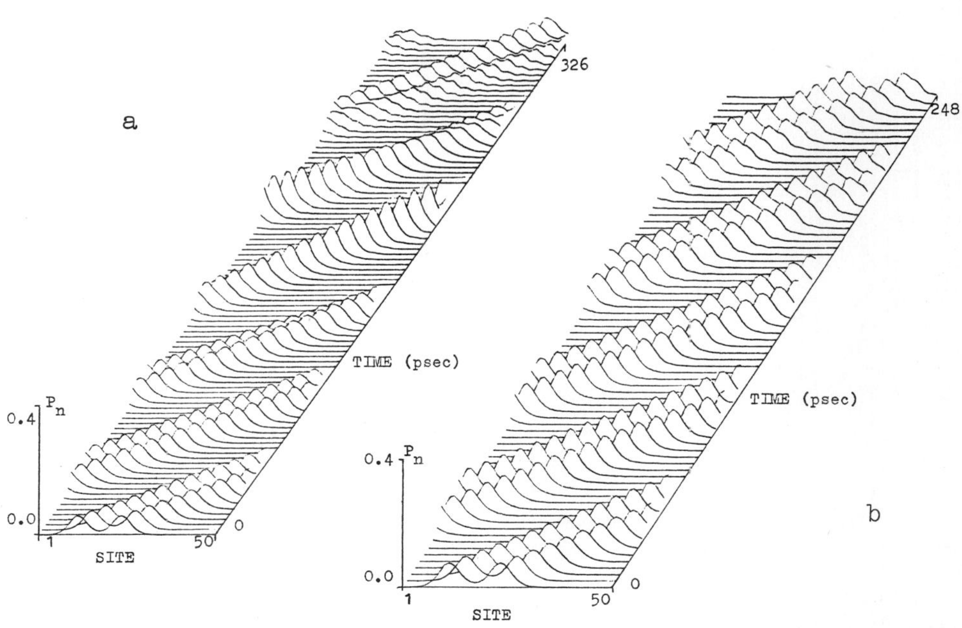

Fig. 2. Uniform motion of the two types of bisolitons
formed by a partial intersection of single-
quantum soliton profiles travelling with velocities $s_1 = s_2 = 0.015$.

$$\varphi^2 = (3NCr^3/s)\left[(1 + 4r^2)/6r^2 - \mathrm{sech}^2(r\xi/l_0 s)\right]\mathrm{sech}^2(r\xi/l_0 s),$$

$$\tag{34a}$$

$$q = - (NCr/2s)\,\mathrm{sech}^2(r\xi/l_0 s),\tag{34b}$$

where r is a single variational parameter from the interval $0 < r < 2^{-1/2}$, satisfies Eqs.(26b) and (27) in the case (33) and allows us to calculate the integral in (32) explicitly. The resulting expression is a rather complicated function of r which was studied numerically. It appears to have only one maximum. In (34) $N = 1$ if one-particle solitons are initially located at the chain ends and $N = 2$ if two quanta are located at one site (a bisoliton state). In order to have the continuum limit in the latter case the coupling constant was chosen to be smaller: $\chi = 1.5 \times 10^{-11} N$. In the intermediate case when one-soliton profiles are intersected the value for N is between 1 and 2 and it was calculated numerically according to the condition (6).

For graphical representation of the two-dimensional lattice field $a_{n_1 n_2}$ (or $\phi_{n_1 n_2}$) we have introduced the quantity $P_n = \sum_{n_1} |a_{n_1 n}|^2$ which means the probability to find no less than one quantum at the n-th molecule.

Using different combinations of initial conditions produced by the one-particle problem we can study various interaction properties by numerical simulations. The head-on collision of two one-particle solitons was shown to be purely elastic. Fig.1 illustrates the case of equal incident velocities. One-particle solitons run through each other keeping their profiles and velocities after collision. This effect does not depend on the type of statistics (Bose or Pauli). The solitons do not interact and can travel with the same velocity if they are sufficiently distant from each other. Their interaction was observed only when their profiles were partially intersecting. In this case they effectively attract each other aspiring to create a more stable bisoliton state. In the case of the Bose statistics this attraction results in a mutual wobble with some frequency (see Fig.2a). A bisoliton with an oscillating profile is formed, center of which travels with a constant velocity. In the other case (see Fig.2b) the effective attraction of two single-quantum solitons and their rarefaction caused by the Pauli statistics are balanced leading to the formation of a bisoliton with a stationary two-humped profile.

CONCLUSIONS

Thus, we have numerically studied in this paper the dynamics of two quanta of the amide-I excitation coupled with intramolecular low-frequency displacements of atoms in hydrogen-bounded polypeptides by using many-particle equations of motion for the Eilbeck-Lomdahl-Scott model restricted to one optical mode. Some analytical work has been also done in order to have appropriate initial conditions for studying this dynamics.

It has been shown that the head-on collision of two single-quantum amide-I solitons is purely elastic. The solitons pass through each other retaining their profiles and velocities but they interact only when their profiles are intersecting. This interaction is attractive and for sufficiently small relative velocities of these solitons a two-quanta soliton (bisoliton) can be formed. The final profile for the excitation under this interaction essentially depends on the type of quantum statistics. In the case of the Bose statistics (two quanta are allowed to be at one lattice site) the wobble of two single-quantum solitons passing through each other with some low frequency has been observed while the interdiction of two amide-I quantum to be at one site (the Pauli statistics) leads to the creation of a dynamically stable bisoliton state with a stationary two-humped profile, Initially prepared one-humped bisoliton state keeps its profile in the case of the Bose statistics and breaks in the chain with the Pauli statistics. The interaction of a single soliton with an exciton has been also studied numerically in the framework of the two-particle equations. It was shown that the soliton is dynamically stable while the exciton is unstable.

REFERENCES

1. A. S. Davydov and N. I. Kislukha, Solitary excitations in one-dimensional molecular chains, Phys. Status Solidi B 59:465 (1973).
2. A. S. Davydov, "Biology and Quantum Mechanics", Pergamon, Oxford (1982).
3. A. S. Davydov, "Solitons in Molecular Systems", Reidel, Boston (1985), and references therein.
4. G. Careri, U. Buontempo, F. Galluzzi, A. C. Scott, E. Gratton, and E. Shyamsunder, Spectroscopic evidence for Davydov-like solitons in acetanilide, Phys.Rev.B 30:4689 (1984).
5. J. C. Eilbeck, P. S. Lomdahl, and A. C. Scott, Soliton structure in crystalline acetanilide, Phys.Rev.B 30:4703 (1984).
6. P. S. Lomdahl and W. C. Kerr, Finite temperature effects on models of hydrogen-bonded polypeptides, in: "Physics of Many Particle Systems", A. S. Davydov, ed., Naukova Dumka, Kiev, N12 (1987).
7. S. Takeno, Vibron solitons and soliton-induced infrared spectra of crystalline acetanilide, Prog.Theor.Phys. 75:1 (1986).
8. T. Holstein, Studies of polaron motion. Part I. The molecular crystal model, Ann. Phys. 8:325 (1959).
9. L. S. Brizhik and A. S. Davydov, Pairing of electrosolitons in soft molecular chains, Fiz. Nizk. Temp. (Soviet Low Temp. Phys.) 10:748 (1984).
10. D. Lj. Mirjanić, M. M. Marinković, G. Knežević, and B. S. Tošić, Two-particle solitary waves, Phys. Status Solidi B 121:589 (1984).
11. Lj. Ristovski, G. S. Davidović-Ristovski, and V.Ristić, Bisolitons in a molecular polymer chain, Phys. Status Solidi B 136:615 (1986).
12. A. V. Zolotaryuk, Many-particle Davydov solitons, in: "Physics of Many Particle Systems", A. S. Davydov, ed., Naukova Dumka, Kiev, N13 (1988).
13. A. S. Davydov and V. Z. Enol'skii, Motion of an excess electron in a molecular chain when interaction with optical phonons is taken into account, Zh. Eksp. Teor. Fiz.

(Sov. Phys. - JETP) 79:1888 (1980).
14. A. V. Zolotaryuk and A. V. Savin, Solitons in molecular chains with intramolecular nonlinear interactions, Physica D, in press.
15. A. S. Davydov and A. A. Eremko, Radiative lifetime of solitons in molecular crystals, Ukr. Fiz. Zh. (Ukrainian Physical Journal) 22: 881 (1977).
16. A. S. Davydov, Solitons in molecular systems, Phys. Scripta 20:387 (1979).
17. X. Wang, D. W. Brown, and K. Lindenberg, Vibron solitons, Phys. Rev. B 39:5366 (1989); see also related references therein.
18. W. C. Kerr and P. S. Lomdahl, Quantum-mechanical derivation of the equations of motion for Davydov solitons, Phys. Rev. B 35:3629 (1987).
19. A. S. Davydov and A. V. Zolotaryuk, Solitons in molecular systems with nonlinear nearest-neighbour interactions, Phys.Lett.A 94:49 (1983).
20. A. S. Davydov and A. V. Zolotaryuk, Electrons and excitons in nonlinear molecular chains, Phys. Scripta 28:249 (1983).

12

INFLUENCE OF DAVYDOV SPLITTING ON SOLITONS IN ALPHA-HELIX

Lj. Mašković, B.S. Tošić and M.J. Škrinjar

Institute of Physics, Faculty of Sciences, University of
Novi Sad, Dr I. Đuričića 4, YU 21 000 Novi Sad, Yugoslavia

INTRODUCTION

Our aim is to study the influence of Davydov splitting onto the properties of solitons in α-proteins. We are going to demonstrate how different approaches lead to nearly same soliton properties.

We start with the standard analysis of soliton theory as proposed by Davydov [1,2,3,4]. Section 2 contains a different approach. We shall use the perturbational method and compare the results with those obtained by the standard analysis. Two cases will be discussed: the first one when 3 peptide chains lie in parallel planes and the second one when they are distributed on a helicoid. We shall present in Sec.3. also a more simplified approach valid in the case when phonon displacements are independent on α, leading to the same results.

1. REVIEW OF THE STANDARD THEORY OF SOLITONS IN α-PROTEINS

We start by presenting the standard analysis of soliton states in α-proteins, in order to indicate its essential features. Later on, we shall present different approaches avoiding some inconsistencies, but being approximate.

Following references [5,6,7], we approximate the spiral (helix) by three chains of peptide groups, mutually interacting through hydrogen bonds. The system of elementary excitations consists of intramolecular vibrations of peptide groups (vibrons) and vibrations of the peptide groups as whole (phonons), with vibron-phonon interaction included.

The Hamiltonian of the system has the form [8]:

Davydov's Soliton Revisited, Edited by P.L. Christiansen and A.C. Scott
Plenum Press, New York, 1990

$$H = H_v + H_p + H_{vp} \qquad\qquad (1.1)$$

where

$$H_v = \Delta \sum_{n\alpha} B^+_{n\alpha} B_{n\alpha} - J \sum_{n\alpha} \left[B^+_{n\alpha} B_{n+1,\alpha} + B^+_{n-1,\alpha} B_{n\alpha} \right] +$$

$$+ L \sum_{n\alpha} \left[B^+_{n\alpha} B_{n,\alpha+1} + B^+_{n,\alpha+1} B_{n\alpha} \right] \; ; \quad \alpha=1,2,3 \qquad\qquad (1.2)$$

is the Hamiltonian of the vibron subsystem;

$$H_p = \frac{1}{2} \sum_{n\alpha} \left[\frac{1}{M} P^2_{n\alpha} + \mathcal{H} \left(u_{n\alpha} - u_{n-1,\alpha} \right)^2 \right] \; ; \qquad \alpha=1,2,3 \qquad\qquad (1.3)$$

is the Hamiltonian of the phonon subsystem and

$$H_{vp} = \chi_1 \sum_{n\alpha} B^+_{n\alpha} B_{n\alpha} \left[u_{n+1,\alpha} - u_{n-1,\alpha} \right] + \chi_2 \sum_{n\alpha} \left[B^+_{n\alpha} B_{n+1,\alpha} + \right.$$

$$\left. + B^+_{n+1,\alpha} B_{n\alpha} \right] \left[u_{n\alpha} - u_{n-1,\alpha} \right] \; ; \qquad \alpha=1,2,3 \qquad\qquad (1.4)$$

is the Hamiltonian of vibron- phonon interaction.

Notation is as follows: $\Delta = \varepsilon + D_0$, $\varepsilon \sim 0.25$ eV is the threshold of vibronic energy of the peptide group, D_0 describes the interaction between peptide groups within the same chain, J characterizes excitation transfer in the chain, L is the interaction between chains, χ_1 and χ_2 are vibron-phonon coupling constants u – are the displacements of the whole peptide groups, $\mathcal{H}$ is the force constants in the chain, M is the mass of the peptide group and $B^+_{n\alpha}$ create vibron excitations at the site n of α-th chain.

Soliton states in the system described by the Hamiltonian (1.1) is analyzed in terms of the wave-function.

$$|\psi(t)\rangle = \sum_{n\alpha} \tilde{\varphi}_{n\alpha}(t) \, e^{S(t)} B^+_{n\alpha} |0\rangle \qquad\qquad (1.5)$$

$$\sum_{n\alpha} |\tilde{\varphi}_{n\alpha}(t)|^2 = 1$$

$$S(t) = - \frac{i}{\hbar} \sum_{n\alpha} \left[\beta_{n\alpha}(t) \, p_{n\alpha} - \pi_{n\alpha}(t) \, U_{n\alpha} \right]$$

Using Hamilton's formalism, i.e. looking for the minimum of the functional $\langle\psi(t)|H| \psi(t)\rangle$, one obtains the equationof motion for the amplitudes $\tilde{\varphi}_{n\alpha}$. Averaging the equation of motion for the phonon operators $P_{n\alpha}$ and $U_{n\alpha}$ over the states $|\psi(t)\rangle$, we obtain the equation of motion for the amplitudes $\beta_{n\alpha}$. Combining them in the continuum approximation, we arrive to:

$$\left[i\hbar \frac{\partial}{\partial t} - (\varepsilon_0 + W - 2J) + J \frac{\partial^2}{\partial \xi^2} + \frac{4\chi^2}{Mv_0^2(1-\nu_k^2)} |\tilde{\varphi}_\alpha(\xi,t)|^2 \right] \tilde{\varphi}_\alpha(x,t) =$$

$$= L \left[\tilde{\varphi}_{\alpha+1}(x,t) + \tilde{\varphi}_{\alpha-1}(x,t) \right] \qquad\qquad (1.6)$$

$$\chi = \chi_1 + \chi_2 \; ; \quad \xi = x - v_k t \; ; \quad v_k = \frac{2d_0^2 J}{\hbar} k$$

$$W = \frac{1}{d_0} \int\limits_{-\infty}^{+\infty} \left[M \left(\frac{\partial \beta}{\partial t}\right)^2 + \mathcal{H} \, d_0^2 \left(\frac{\partial \beta}{\partial \xi}\right)^2 \right] d\xi \; , \qquad \alpha = 1,2,3$$

$$\nu_k = \frac{v_k}{v_0} \; ; \qquad v_0^2 = d_0^2 \frac{\mathcal{H}}{M}$$

The solution of (1.6) is looked for in the form

$$\tilde{\varphi}_\alpha(x,t) = A_\alpha \phi(\xi) \, e^{ikx - i\omega t} \; , \qquad \overset{*}{A}_\alpha = A_\alpha \; ; \quad \overset{*}{\phi} = \phi \qquad\qquad (1.7)$$

with the normalizing condition:

$$\int\limits_{-\infty}^{+\infty} d\xi \phi^2(\xi) = 1 \; ; \qquad \sum_\alpha A_\alpha^2 = 1$$

Substituting (1.7) into (1.6), we obtain a new system of equations

$$d_0^2 J A_\alpha \frac{d^2\phi}{d\xi^2} = \left[E_0 - E \right] A_\alpha \phi + L \left(A_{\alpha+1} + A_{\alpha-1} \right) \phi - G |A_\alpha|^2 A_\alpha \phi^3 \qquad (1.8)$$

$$E_0 = W + \Delta - 2J + d_0^2 k^2 J$$

$$G = \frac{4\chi^2}{Mv_0^2(1-\nu_k)}$$

The normalized solution of (1.8) is

$$\phi(\xi) = \sqrt{\frac{d_0 \mu}{2}} \; \frac{1}{\mathrm{ch}\mu\xi} \qquad\qquad (1.9)$$

Combination of (1.8) and (1.9), leads to the system of equations for the determination of the coefficients A_α and energies E. This system allows for various solutions, including Davidov's [3] symmetric ($A_1 = A_2 = A_3 = \frac{1}{\sqrt{3}}$) and antisymmetric ($A_1 = -A_3 = \frac{1}{\sqrt{2}}$, $A_2 = 0$) solitons, as well as many others. The detailed analysis will be given is Chap.3.

2. PERTURBATION TREATMENT IN THE SOLITON THEORY IN α-PROTEINS

In order to avoid inconsistencies of the standard analysis, we shall apply the perturbational approach to calculate soliton energies and wave-functions.

The wave-function must be expressed in terms of the operators creating real vibron excitations. The operators describing these excitations are obtained in the diagonalization procedure of the vibronic Hamiltonian (1.2). The states created by the operators $\hat{B}^+_{n\alpha}$ were triply degenerate with the energy $E = \Delta - 2J \cos k\, d_0$. After the diagonalization, this energy level splits into three levels with the energies: $E_1 = \Delta - 2J \cos k\, d_0$, $E_{2,3} = E_1 \pm L\sqrt{2}$. The excitations with these energies are created by some new Bose-operators $\hat{b}_{n\alpha}$ and the function $|\psi(t)\rangle$ must be expressed in terms of these operators.

We now proceed to the diagonalization procedure. This first step is transition to momentum space.

$$B_{n\alpha} = \frac{1}{\sqrt{N}} \sum_{k} B_{k\alpha}\, e^{i k n d_0} \tag{2.1}$$

giving

$$H_v = \sum_{k\alpha} \left[\Delta - 2J \cos d_0 k \right] B^+_{k\alpha} B_{k\alpha} + L \left(B^+_{k\alpha} B_{k,\alpha+1} + B^+_{k,\alpha+1} B_{k\alpha} \right) \tag{2.2}$$

Now, we perform the transformation to the operators $\hat{b}$:

$$B_{k\alpha} = \sum_{s} \Theta_{\alpha s}\, b_{ks} \tag{2.3}$$

The unitarity conditions of the transformation are

$$\sum_{s} \Theta_{\alpha s} \Theta_{\alpha' s} = \delta_{\alpha\alpha'} \; ; \qquad \sum_{\alpha} \Theta_{\alpha s} \Theta_{\alpha s'} = \delta_{ss'}$$

The transformation (2.3) leads to the following form of the vibronic Hamiltonian

$$H_v = \sum_{sk} E_s(k)\, b^+_{ks} b_{ks} \tag{2.4}$$

Now, we discuss two particular cases:

a) all three chains of peptide groups lie in the paralel planes $(L_{12} = L_{13} = L \; ; \; L_{23} = 0)$. Following levels are obtained

$$E_1 = \Delta - 2J \cos d_0 k$$

$$E_2 = \Delta - 2J \cos d_0 k + L \sqrt{2} \qquad (2.5)$$

$$E_3 = \Delta - 2J \cos d_0 k - L \sqrt{2}$$

The coefficients can be chosen in the following way:

$$\Theta_{11} = \frac{1}{\sqrt{2}} \; ; \quad \Theta_{12} = \frac{1}{2} \; ; \quad \Theta_{13} = \frac{1}{2}$$

$$\Theta_{21} = 0 \; ; \quad \Theta_{22} = \frac{1}{\sqrt{2}} \; ; \quad \Theta_{23} = - \frac{1}{\sqrt{2}} \qquad (2.6)$$

$$\Theta_{31} = - \frac{1}{\sqrt{2}} \; ; \quad \Theta_{32} = \frac{1}{2} \; ; \quad \Theta_{33} = \frac{1}{2}$$

b) Chains are helicoidally distributed $(L_{12} = L_{13} = L_{23} = L)$. We obtain only two levels, one of them doubly degenerate:

$$E_1 = \Delta - 2J \cos d_0 k + 2L$$

$$E_2 = E_3 = \Delta - 2J \cos d_0 k - L \qquad (2.7)$$

The coefficients in this case become:

$$\Theta_{11} = \frac{1}{\sqrt{3}} \; ; \quad \Theta_{21} = \frac{1}{\sqrt{3}} \; ; \quad \Theta_{31} = \frac{1}{\sqrt{3}}$$

$$\Theta_{12} = \frac{1}{\sqrt{3}} \; ; \quad \Theta_{22} = \frac{1}{\sqrt{3}} e^{-\frac{2\pi i}{3}} \; ; \quad \Theta_{32} = \frac{1}{\sqrt{3}} e^{\frac{2\pi i}{3}} \qquad (2.8)$$

$$\Theta_{13} = \frac{1}{\sqrt{3}} \; ; \quad \Theta_{23} = \frac{1}{\sqrt{3}} e^{\frac{2\pi i}{3}} \; ; \quad \Theta_{33} = \frac{1}{\sqrt{3}} e^{-\frac{2\pi i}{3}}$$

The Hamiltonian (2.4) written in direct space in the nearest-neighbour approximation takes the form:

$$H_v = \sum_{n\alpha} \Delta_\alpha b_{n\alpha}^+ b_{n\alpha} - 2J \sum_{n\alpha} b_{n\alpha}^+ \left(b_{n+1,\alpha} + b_{n-1,\alpha} \right) \qquad (2.9)$$

a) $\Delta_1 = \Delta \qquad \Delta_{2,3} = \Delta \pm L \sqrt{2}$

b) $\Delta_1 = \Delta + 2L \qquad \Delta_2 = \Delta_3 = \Delta - L$

Total Hamiltonian, expressed in terms of new operators has the form

$$H = H_0 + H_1 + H_2 \qquad (2.10)$$

where

$$H_0 = \sum_{n\alpha} \left[\Delta_\alpha b_{n\alpha}^+ b_{n\alpha} - 2J b_{n\alpha}^+ \left(b_{n+1,\alpha} + b_{n-1,\alpha} \right) \right] + \chi \sum_{n\alpha} \Theta_{\alpha\alpha}^2 b_{n\alpha}^+ b_{n\alpha} \left[u_{n+1,\alpha} + \right.$$

$$u_{n-1,\alpha}\Big] + \frac{1}{2}\sum_{n\alpha}\left[\frac{1}{M}p_{n\alpha}^2 + \mathcal{H}\left(u_{n\alpha} - u_{n-1,\alpha}\right)^2\right] \qquad \alpha = 1,2,3 \qquad (2.11)$$

$$H_1 = \chi\sum_{n\alpha\alpha'}\Theta_{\alpha'\alpha}^2\, b_{n\alpha}^+ b_{n\alpha}\left(u_{n+1,\alpha'} - u_{n-1,\alpha'}\right) \qquad \alpha' \neq \alpha \qquad (2.12)$$

$$H_2 = \chi\sum_{n\alpha\alpha'\gamma}\Theta_{\alpha'\alpha}\Theta_{\alpha'\gamma}\, b_{n\alpha}^+ b_{n\gamma}\left(u_{n+1,\alpha'} - u_{n-1,\alpha'}\right) \qquad \alpha \neq \gamma \qquad (2.13)$$

$$\chi_1 = \chi\;;\quad \chi_2 = 0$$

We see that in H_0 the variables corresponding to different indices are decoupled, so we shall use it as zero - order Hamiltonian for the perturbational calculation of the contributions of $\hat{H}_1$ and $\hat{H}_2$.

The unperturbed wave-function is sought in the following form

$$|\psi^{(0)}(t)\rangle = \sum_{n\alpha}\varphi_{n\alpha}(t)\, e^{S(t)}\, b_{n\alpha}^+ |0\rangle \qquad (2.14)$$

$$\sum_{n\alpha}|\varphi_{n\alpha}(t)|^2 = 1$$

Applying the same approach as in standard analysus to the functional $\langle\psi^0(t)|\hat{H}_0|\psi^{(0)}(t)\rangle$, and performing the continnum transition, we obtain the system of equations

$$i\hbar\frac{\partial\varphi_\alpha}{\partial t} = (\Delta_\alpha - 2J)\,\varphi_\alpha - d_0^2 J\frac{\partial^2\varphi_\alpha}{\partial x^2} + 2\chi\varphi_\alpha\Theta_{\alpha\alpha}^2\frac{\partial\beta_\alpha}{\partial x} \qquad (2.15)$$

The equations of motion for phonon-displacement give:

$$\frac{\partial^2\beta_\alpha}{\partial t^2} - v_0^2\frac{\partial^2\beta_\alpha}{\partial x^2} = \frac{2\chi}{M}\Theta_{\alpha\alpha}^2\frac{\partial}{\partial x}|\varphi_\alpha|^2 \qquad (2.16)$$

The solutions of this system are

$$E_\alpha^{(0)}(k) = \Delta_\alpha - 2J + Jd_0^2 k^2 - \Theta_{\alpha\alpha}^8 f(k) \qquad \alpha = 1,2,3 \qquad (2.17)$$

$$f(k) = \frac{\chi^4}{JM^2 v_0^4 (1 - v_k^2)^2}\left[1 - \frac{2}{3}\frac{1+v_k^2}{1-v_k^2}\right]$$

$$|\psi^{(0)}(t)\rangle = \frac{1}{d_0}\int_{-\infty}^{+\infty}d\xi\,\varphi_\alpha(x,t)b_\alpha^+(\xi)|0\rangle \qquad (2.18)$$

$$\varphi(\alpha) = \frac{d_0\sqrt{\Omega}}{2}\Theta_{\alpha\alpha}^2\frac{e^{\,ikx - \dfrac{iE_\alpha^{(0)}(k)}{\hbar}t}}{\mathrm{ch}\,\dfrac{d_0\Omega}{2}\Theta_{\alpha\alpha}^4\,\xi}$$

$$\Omega = \frac{2\chi^2}{d_0^2 \, JM(1 - \nu_k^2)\nu_0^2}$$

The expressions for $\hat{H}_1$ and $\hat{H}_2$ in the continnum approximation are:

$$H_1 = - \frac{d_0 \chi^2 \Omega}{Mv_0^2(1-\nu_k^2)} \sum_{\alpha\alpha'} \Theta^6_{\alpha'\alpha'} \Theta^2_{\alpha'\alpha} \int_{-\infty}^{+\infty} d\xi \; \frac{b_\alpha^+(\xi)b_\alpha(\xi)}{ch^2 \frac{d_0\Omega}{2} \Theta^4_{\alpha'\alpha'}\xi} \qquad \alpha' \neq \alpha \qquad (2.19)$$

$$H_2 = - \frac{d_0 \chi^2 \Omega}{Mv_0^2(1-\nu_k^2)} \sum_{\alpha\alpha'\alpha''} \Theta^6_{\alpha'\alpha'} \Theta_{\alpha'\alpha} \Theta_{\alpha'\alpha''} \int_{-\infty}^{+\infty} d\xi \; \frac{b_\alpha^+(\xi)b_{\alpha''}(\xi)}{ch^2 \frac{d_0\Omega}{2} \Theta^4_{\alpha'\alpha'}\xi} \qquad \alpha'' \neq \alpha \qquad (2.20)$$

The Hamiltonian $\hat{H}_1$, being diagonal in α, contributes only in the first-order theory in χ^2, while $\hat{H}_2$ leads to the corrections in second-order ($\sim\chi^4$). Explicit expressions for the corrections $\delta E_\alpha^{(i)}$ ($i=1,2$) are:

$$\delta E_\alpha^{(1)}(k) = \langle\psi_\alpha^{(0)}(0)|H_1|\psi_\alpha^{(0)}(0)\rangle = - \frac{d_0^3 \chi^2 \Omega^2}{4Mv_0^2(1-\nu_k^2)} \Theta^4_{\alpha\alpha} \sum_{\alpha'} \Theta^6_{\alpha'\alpha'} \Theta^2_{\alpha'\alpha} I^{(1)}_{\alpha\alpha'} \qquad (2.21)$$

$$\alpha' \neq \alpha$$

$$I^{(1)}_{\alpha\alpha'} = \int_{-\infty}^{+\infty} \frac{d\xi}{ch^2 \frac{d_0\Omega}{2} \Theta^4_{\alpha\alpha}\xi \; ch^2 \frac{d_0\Omega}{2} \Theta^4_{\alpha'\alpha'}\xi}$$

$$\delta E_\alpha^{(2)}(k) = \sum_{\alpha''} \frac{v^2_{\alpha\alpha''}}{E_\alpha^{(0)}(k) - E_{\alpha''}^{(0)}(k)} \quad ; \qquad \alpha'' \neq \alpha \qquad (2.22)$$

$$v_{\alpha\alpha''} = \langle\psi_\alpha^{(0)}(0)|H_2|\psi_{\alpha''}^{(0)}(0)\rangle = - \frac{d_0^3 \chi^2 \Omega^2}{4Mv_0^2(1-\nu_k^2)} \sum_{\alpha'} \Theta^6_{\alpha'\alpha'} \Theta_{\alpha'\alpha} \Theta_{\alpha'\alpha''} \Theta^2_{\alpha''\alpha''} I^{(2)}_{\alpha,\alpha',\alpha''}$$

$$\alpha'' \neq \alpha$$

$$I^{(2)}_{\alpha,\alpha',\alpha''} = \int_{-\infty}^{+\infty} \frac{d\xi}{ch^2 \frac{d_0\Omega}{2} \Theta^4_{\alpha'\alpha'}\xi \; ch \frac{d_0\Omega}{2} \Theta^4_{\alpha\alpha}\xi \; ch \frac{d_0\Omega}{2} \Theta^4_{\alpha''\alpha''}\xi}$$

In this way we obtain the expressions for the energy up to χ^4

$$E_\alpha(k) = E_\alpha^{(0)}(k) + \delta E_\alpha^{(1)}(k) + \delta E_\alpha^{(2)}(k) \qquad (2.23)$$

and the wave function correct up to order χ^2

$$|\psi_\alpha^{(1)}(t)\rangle = |\psi_\alpha^{(0)}(t)\rangle + \sum_{\alpha''} \frac{v_{\alpha\alpha''}}{E_\alpha^{(0)}(k) - E_{\alpha''}^{(0)}(k)} |\psi_{\alpha''}^{(0)}(t)\rangle \qquad \alpha' \neq \alpha \qquad (2.24)$$

Let us look for the more explicit expressions for our two particular cases:

a) $L_{12} = L_{13} = L \qquad L_{23} = 0$

$$E_1(k) = \Delta - 2J + Jd_0^2 k^2 - 2\eta_k g_k - g_k j_{14}^{22} - 16g_k^2 \left(\frac{R_{12}^2}{L\sqrt{2}} - \frac{R_{13}^2}{L\sqrt{2} - \frac{15}{8}\eta_k g_k} \right) \qquad (2.25a)$$

$$E_2(k) = \Delta + L\sqrt{2} - 2J + Jd_0^2 k^2 - 2\eta_k g_k -$$

$$- g_k \left(j_{11}^{22} + \frac{1}{2} j_{14}^{22} \right) + 16g_k^2 \left[\frac{R_{12}^2}{L\sqrt{2}} - \frac{R_{23}^2}{L\sqrt{2} - \frac{15}{8}\eta_k g_k} \right] \qquad (2.25b)$$

$$E_3(k) = \Delta - L\sqrt{2} - 2J + Jd_0^2 k^2 - \frac{1}{8}\eta_k g_k - 3g_k j_{14}^{22} -$$

$$16 g_k^2 \left[\frac{R_{13}^2}{L\sqrt{2} - \frac{15}{8}\eta_k g_k} + \frac{R_{23}^2}{2L\sqrt{2} - \frac{15}{8}\eta_k g_k} \right] \qquad (2.25c)$$

with the following notation

$$\eta_k = 1 - \frac{2}{3} \frac{1+\nu_k^2}{1-\nu_k^2} \quad ; \qquad g_k = \frac{1}{32} \frac{\chi^4}{JM^2 v_0^4 (1-\nu_k^2)^2}$$

$$R_{12} = R_{21} = \frac{1}{\sqrt{2}} \left(j_{11}^{22} - \frac{1}{2} j_{14}^{22} \right)$$

$$R_{13} = R_{31} = \sqrt{2} \left(j_{14}^{13} - \frac{1}{8} j_{14}^{31} \right) \qquad (2.26)$$

$$R_{23} = R_{32} = j_{14}^{13} - \frac{1}{8} j_{14}^{31}$$

$$j_{ab}^{mn} = \int_{-\infty}^{+\infty} \frac{dx}{ch^m ax \, ch^n bx}$$

b) the case $L_{12} = L_{13} = L_{23} = L$

$$E_1(k) = \Delta + 2L - 2J + Jd_0^2 k^2 - \frac{1}{3^4} f_k - \frac{4}{3^5} \frac{d_0 \chi^2 \Omega}{Mv_0^2 (1-\nu_k^2)} + \frac{8 \cdot 16^2 \cdot 24^2}{3^{11}} \frac{g_k^2}{L} \qquad (2.27a)$$

$$E_{2,3}(k) = \Delta - L - 2J + Jd_0^2 k^2 - \frac{1}{3^4} f_k - \frac{4}{3^5} \frac{d_0 \chi^2 \Omega}{Mv_0^4 (1-\nu_k^2)} \pm \frac{8 \cdot 16^2 \cdot 24^2}{3^{11}} \frac{g_k^2}{L} \qquad (2.27b)$$

The treatment of realistic vibronic states as zero-order approximation and the application of perturbation method leads to splitting of doubly degenerate level in the case of helix.

Our analysis has shown that in both studied cases there appear three types of excitations. Two types have properties close to Davydov's symmetric and antisymmetric solitons, while soliton properties of the third type are weakly expressed.

3. SOLITON PROPERTIES FOR NEGLECTED DISTORSION

The particular case when molecular displacements do not depend on the chain (α), i.e. $u_{n\alpha} \to u_n$ for $\alpha=1,2,3$ represents an extremely simple problem if exciton amplitudes $\tilde{\varphi}$ (1.5) are transformed in the same way as the operators:

$$\tilde{\varphi}_{n\alpha} = \sum_s \Theta_{\alpha s} \, \varphi_{n\alpha} \tag{3.1}$$

We shall treat only the case $L_{12}= L_{13}= L_{23}= L$.

Since

$$\sum_\alpha \Theta_{\alpha s} \Theta_{\alpha s'} = \delta_{ss'}$$

the equation for the amplitudes $\varphi_{n\alpha}$ and $\tilde{\varphi}_{n\alpha}$ are equivalent yet using $\varphi_{n\alpha}$ leads to the simplier expressions.

$$i\hbar\dot{\varphi}_{ns} = \Delta_s \varphi_{ns} - J\left(\varphi_{n+1,s} + \varphi_{n-1,s}\right) + \chi\left(u_{n+1} - u_{n-1}\right)\varphi_{ns} \tag{3.2}$$

Summing over α the expression $\dot{\pi}_{n\alpha} = M\ddot{\beta}_{n\alpha}$, we obtain

$$3\dot{\pi}_n = 3M\ddot{\beta}_n = 3\mathcal{H}\left(\beta_{n+1} - 2\beta_n + \beta_{n-1}\right) + \chi\sum_s\left(|\varphi_{n+1,s}|^2 - |\varphi_{n-1,s}|^2\right) \tag{3.3}$$

The system of equations of motion takes the following form in the continuum approximation

$$i\hbar\dot{\varphi}_s(x,t) = (\Delta_s - 2J)\,\varphi_s - Jd_0^2\,\frac{\partial^2\varphi_s}{\partial x^2} + 2d_0\chi u_x\varphi_s \tag{3.4a}$$

$$3M\ddot{\beta}(x,t) - 3\mathcal{H}d_0^2\,\frac{\partial^2\beta}{\partial x^2} = 2\chi\,d_0\sum_s\frac{\partial|\varphi_s|^2}{\partial x} \tag{3.4b}$$

Let us suppose that $|\varphi_s|^2 = |\varphi_s|^2(x-vt)$. We have

$$\frac{\partial u}{\partial x} = - \frac{2\chi d_0}{3Mv_0^2(1-\nu_k^2)} \sum_s |\varphi_s|^2 \tag{3.5}$$

In order to "decouple" the equations for φ_{ns}, we shall introduce a new function

$$\varphi_{ns} = e^{-\frac{i}{\hbar}(\Delta_s - 2J)t} \Psi_{ns} \tag{3.6}$$

to obtain

$$i\hbar\dot{\Psi}_{ns} = - Jd_0^2 \frac{\partial^2 \Psi_{ns}}{\partial x^2} + 2\chi d_0 \frac{\partial u}{\partial x} \Psi_{ns} \tag{3.7}$$

Next change is

$$\Psi_{ns}(x,t) = a_s \phi(x,t) \tag{3.8}$$

where a_s can be complex. The equation (3.5) becomes

$$\frac{\partial u}{\partial x} = - \frac{2\chi d_0 |\phi|^2}{3Mv_0^2(1-\nu_k^2)} \sum_s |a_s|^2 \tag{3.9}$$

One of the possibilities for the normalization of the constant a_s is

$$\sum_s |a_s|^2 = 3 \tag{3.10}$$

so

$$\frac{\partial u}{\partial x} = \frac{-2\chi d_0 |\phi|^2}{Mv_0^2(1-\nu_k^2)} \tag{3.11}$$

If we substitute (3.8) and (3.11) into (3.7), we arrive to the nonlinear Schrödinger equation for $\phi(x,t)$:

$$i\hbar\dot{\phi} + Jd_0^2 \frac{\partial^2 \phi}{\partial x^2} + \frac{4\chi^2 d_0^2}{Mv_0^2(1-\nu_k^2)} |\phi|^2 \phi = 0 \tag{3.12}$$

Solitonic solution can be looked for in the form

$$\phi(x,t) = A(x-vt)\, e^{ikx - i\omega t} \tag{3.13}$$

$$A(\xi) = \frac{A_0}{ch\,\alpha\xi} \; ; \quad \xi = x-vt$$

and we obtain

$$\hbar\omega = Jd_0^2 k^2 - Jd_0^2 \alpha^2 \tag{3.14}$$

$$GA_0^2 = 2Jd_0^2 \alpha^2 \tag{3.15}$$

$$G = \frac{4\chi^2 d_0^2}{Mv_0^2(1-v_k^2)} = 2J\, d_0^2 \alpha^2 \tag{3.16}$$

$$J d_0^2 k^2 = \frac{\hbar^2 k^2}{2m}$$

and the normalizing condition becomes

$$\sum_{ns} |\varphi_{ns}|^2 = 1 \;\Rightarrow\; \sum_s \frac{1}{d_0} \int_{-\infty}^{+\infty} |\varphi_s|^2 = 1$$

$$\sum_s |a_s|^2 \frac{1}{d_0} \int_{-\infty}^{+\infty} A^2 \, dx = 1 \tag{3.17}$$

which gives (with (3.10)):

$$\frac{1}{d_0} \int_{-\infty}^{+\infty} A^2 \, dx = \frac{1}{3} \tag{3.18}$$

so that

$$\frac{2A_0^2}{\alpha d_0} = \frac{1}{3} \tag{3.19}$$

combining (3.15) and (3.18) we obtain

$$\alpha = \frac{G}{12 J d_0} \;;\qquad A_0^2 = \frac{1}{6}\, \alpha\, d_0 = \frac{G}{72 J} \tag{3.20}$$

(as compared to Davydov's $\alpha_D = \dfrac{G}{4 J d_0}$).

Notice that (3.17) allows a different normalization for a_s, without any substantial influence, so the condition (3.10) is chosen in order to arrive to Davydov's NSE (3.12).

Soliton dimension is also simply determined:

$$l_0 = \frac{2}{\alpha} \quad (l_0 = 3 l_D, \quad l_D - \text{Davydov's length})$$

$$\hbar\omega = J\, d_0^2 k^2 - \frac{G^2}{144 J} \tag{3.21}$$

The final solution of (3.4) is

$$\varphi_s(x,t) = a_s A(x-vt)\, e^{ikx - i\omega_s t} \tag{3.22}$$

where

$$\omega_s = \Delta_s - 2J + \hbar\omega \tag{3.23}$$

The wave function depends on a_s:

$$|\psi(a_1, a_2, a_3)\rangle = \sum_n a_s A(x-vt)\, e^{ikx - i\omega_s t}\, e^{S(t)}\, b_{ns}^+ |0\rangle \tag{3.24}$$

Introducing

$$\tilde{\varphi}_{n\alpha} = \sum_{s} \Theta_{\alpha s} \varphi_{ns}$$

(3.24) can be put in the form

$$|\psi(a_1,a_2,a_3)\rangle = \sum_{n\alpha} \tilde{\varphi}_{n\alpha} B^+_{n\alpha} e^{S(t)} |0\rangle \qquad (3.25)$$

where the amplitudes in the continuum approximation (taking the coefficients of b) case from the previous section) are:

$$\tilde{\varphi}_1(x,t) = \frac{1}{\sqrt{3}} A(x-vt) e^{ikx} \left[a_1 e^{-i\omega_1 t} + (a_2 + a_3) e^{-i\omega_2 t} \right]$$

$$\tilde{\varphi}_2(x,t) = \frac{1}{\sqrt{3}} A(x-vt) e^{ikx} \left[a_1 e^{-i\omega_1 t} + e^{-i\omega_2 t} \left(a_2 e^{\frac{2\pi i}{3}} + a_3 e^{-\frac{2\pi i}{3}} \right) \right]$$

$$\tilde{\varphi}_3(x,t) = \frac{1}{\sqrt{3}} A(x-vt) e^{ikx} \left[a_1 e^{-i\omega_1 t} + e^{-i\omega_2 t} \left(a_2 e^{-\frac{2\pi i}{3}} + a_3 e^{\frac{2\pi i}{3}} \right) \right] \qquad (3.26)$$

Various choices of parameters a_s lead to a series of solitonic solutions, as can be illustrated in the following examples

1) $a_2 = a_3 = 0$, $\qquad a_1 = \sqrt{3}$ $\qquad\qquad$ symmetric soliton

$$\tilde{\varphi}_1(x,t) = \tilde{\varphi}_2(x,t) = \tilde{\varphi}_3(x,t) = A(x-vt) e^{ikx-i\omega_1 t}$$

2) $a_1 = 0$, $\quad a_2 = \sqrt{3}$, $\quad a_3 = 0$ $\qquad$ antisymmetric soliton

$$\tilde{\varphi}_1(x,t) + \tilde{\varphi}_2(x,t) + \tilde{\varphi}_3(x,t) = 0$$

$$\tilde{\varphi}_1(x,t) = A(x-vt) e^{ikx-i\omega_2 t}$$

$$\tilde{\varphi}_2(x,t) = \tilde{\varphi}_1(x,t) e^{\frac{2\pi i}{3}}$$

$$\tilde{\varphi}_3(x,t) = \tilde{\varphi}_1(x,t) e^{-\frac{2\pi i}{3}}$$

3) $a_1 = 0$, $\quad a_2 = -a_3 = \frac{\sqrt{3}}{2}$ $\qquad$ antisymmetric soliton

$$\tilde{\varphi}_1(x,t) = 0$$

$$\tilde{\varphi}_2(x,t) = \frac{i}{2} A(x-vt) e^{ikx-i\omega_2 t}$$

$$\tilde{\varphi}_3(x,t) = -\tilde{\varphi}_2(x,t)$$

Total energy of the system, for $u_{n\alpha} = u_n$ is

$$E(a_1, a_2, a_3) = \frac{1}{3} \sum_s |a_s|^2 (\Delta_s - 2J) + \frac{\hbar^2 k^2}{2m} - \frac{G^2}{4 \cdot 108\ J} \frac{1 - 5\nu_k^2}{1 - \nu_k^2} \qquad (3.27)$$

$$\Delta_1 = \Delta + 2L \ ; \qquad \Delta_2 = \Delta_3 = \Delta - L$$

$$\text{for} \quad \sum_s |a_s|^2 = 3 \ , \quad G = \frac{4\chi^2 d_0^2}{M v_0^2 (1 - \nu_k^2)} \ ,$$

$$E = (\Delta - 2J) + \frac{\hbar^2 k^2}{2m} + \frac{1}{3}\left[2L|a_1|^2 - L\left(|a_2|^2 + |a_3|^2 \right)\right] -$$

$$- \frac{G^2}{4 \cdot 108\ J} \frac{1 - 5\nu_k^2}{1 - \nu_k^2} \qquad (3.28)$$

For the solitons with same velocity, the lowest energy belongs to the solitons with $a_1 = 0$, and the highest for $a_2 = a_3 = 0$, and solutions given by the general expressions (3.26) have energies lying between:

$$E_{min} = \Delta - 2J - L + \frac{\hbar^2 k^2}{2m} - \frac{G^2}{4 \cdot 108\ J} \frac{1 - 5\nu_k^2}{1 - \nu_k^2}$$

$$E_{max} = \Delta - 2J + 2L + \frac{\hbar^2 k^2}{2m} - \frac{G^2}{4 \cdot 108\ J} \frac{1 - 5\nu_k^2}{1 - \nu_k^2}$$

$$E_{max} - E_{min} = 3L \qquad (3.29)$$

We should notice that final solutions are rather close to Davydov's ones, but with amplitudes which depend on $\xi = x - vt$ and t as well, giving Scott's interchannel oscilations [6].

5. CONCLUDING REMARKS

The aim of our analysis of solitons in α-proteins was to avoid some inconsistencies appearing in the standard treatment.

We applied perturbational approach with zero-order approximation representing the Hamiltonian of true vibronic states. Two cases were analysed: a) all three chains lying in the paralel planes and b) chains distributed over a helix. In the first case, we obtain three energy leves even in zero-order, while in the second case, one doubly degenerate level splits into two. In this way, we obtain three types of excitations in both cases. Two types have properties similar to Davydov's symmetric and antisymmetric solitons, while soliton character of the third type is weakly expressed.

We have also studied the simplified case of no distorsion. Final expressions for energy and soliton solutions, allow us to reproduce all previously published results and obtain a whole spectrum of new solutions.

REFERENCES

1. A.S. Davydov and N.I. Kislukha, Solitary excitons in one-dimensional chain, *Phys. Stat. Sol. (b)* **59** 465 (1973).

2. A.S. Davydov, Solitons in molecular system, *Phys. Scripta* **20** 387 (1979).

3. A.S. Davydov, Solitons in one-dimensional chains, *Phys. Stat. Sol. (b)* **75** 735 (1976).

4. A.S. Davydov, *Biology and Quantum Mechanics* (*Biologiya i kvantovaya mehanika*), Naukova dumka, Kiev (1979) (in Russian).

5. A.S. Davydov, A.A. Eremko and A.I. Sergienko, Solitons in α-helix protein molecules (Solitoni v α-spiralnyh belkovih molekulah), *Ukr. Fiz. Zhur.* **23**, 983 (1978) (in Russian).

6. A.C. Scott, Launching a Davydov soliton: Soliton analysis, *Phys. Scripta* **29** 279 (1984).

7. S. Takeno, Vibron Solitons and Coherent Polarization in a Exactly Tractable Oscillator - Lattice System, *Prog. Theor. Phys.* **73** 853 (1985).

8. A.S. Davydov, Solitons in quasi-one-dimensional structures (Solitony v kvaziodnomernih molekulyarnyh strukturah, *Usp. Fiz. Nauk* **183**, 603 (1982) (in Russian).

9. A.S. Davydov, *Solitons in Molecular Systems* (*Solitony v molekulyarnyh sistemah*), Naukova dumka, Kiev (1988) in (Russian).

13

INTERACTION OF AN EXTRA ELECTRON WITH OPTICAL PHONONS IN LONG MOLECULAR CHAINS AND IONIC CRYSTALS

V.Z. Enol'skii

Institute of Metal Physics

Vernadsky str. 36, Kiev-I42, 252680, USSR

I. INTRODUCTION

After Davydov pioneer paper [I] on the energy transfer in biological systems the attention of investigators was drawn to different problems of electron-phonon interaction in molecular chains [2,3] . The effect of acoustic phonons with the dispersion law

$$\Omega(k) = k V_{ac} \qquad (\text{I.I})$$

on the motion of an extra electron (and exciton) in a one dimensional molecular chain was studied by Davydov [4-6] . It was shown that the stable motion of electron (exciton) with velocities less than a constant group velocity V_{ac} of a longtudinal sound is accompanied by a local chain deformation, and the motion of this collective deformation is described by a solitary wave which does not change its form and velocity. This wave, called as soliton, can travel only with the speed less than the sound velocity V_{ac}.

The following questions arise: i) do self-trapped electron states exist in the case of the electron interaction with optical phonons possesing the dispersion

$$\Omega^2(k) = \Omega_0^2 + k^2 V_0^2, \qquad (\text{I.2})$$

where V_0 is the minimal phase velocity of optical phonon? and ii) is there in this case limitation on the speed of the soliton wave?

The answer on the first question is known: we refer to Holstein polaron in the case of one-dimensional chain and short-range interaction and to Pekar polaron in the case of three dimensional ionic crystal and long-range interaction (see, for instance [7]). The answer on the second question presents the object of the present paper. We consider here a long one-dimensional molecular chain and a three-dimensional crystal and a self-trapped electron state in them. It appears, that in the case of optical phonons there is a limitation on the speed of the self-trapped electron - the motion is possible at the velocities $V < V_0$. This question was considered

Davydov's Soliton Revisited, Edited by P.L. Christiansen and A.C. Scott
Plenum Press, New York, 1990

by Davydov with the author in [8-IO] .

2. BASIC EQUATIONS DESCRIBING THE ELECTRON MOTION IN A ONE-DIMENSIONAL MOLECULAR CHAIN

The ground state energy and the electron wave functions in the field of an isolated electric dipole are calculated in the paper of Turner and Anderson [II] . The overlap of the electron wave function in the chain of molecules (with the period a) leads to the collectivization of electron states. The states of an extra electron involved in this chain from a donor in the approximation of effective electron mass are characterized by Hamiltonian

$$H_{el} = \frac{1}{a}\int \Psi^*(x,t)\left[\mathcal{E}_0 - \frac{\hbar^2}{2m}\frac{\partial^2}{\partial x^2}\right]\Psi(x,t)\,dx, \qquad (2.I)$$

where $\mathcal{E}_0$ is the energy of the bottom of conduction band. The quantity $|\Psi(x,t)|^2 dx/a$, normalized by the condition

$$\frac{1}{a}\int|\Psi(x,t)|^2 dx = 1, \qquad (2.2)$$

characterizes the probability of distribution of an electron on the segment dx/a . The integration in Eqs. (2.I), (2.2) and in all the following expressions (without a special remark) is also carried out in infinite limits.

Let $u_n(t)$ be the relative displacement of atom in a molecule n which determines the branch of longitudinal optical vibrations. In the continuum approximation it is written in the form

$$H_{int} = \frac{\chi}{a}\int|\Psi(x,t)|^2 u(x,t)\,dx, \qquad (2.3)$$

where x/a at $x = na$ determines the number of the branch point. χ is the interaction parameter which has an energy per length dimension.

If Ω_0 is the frequency of intramolecular vibrations corresponding to the replacements u_n then the optical phonons in the chain with the dispersion (I.2) are discribed by the Hamiltonian

$$H_{ph} = \frac{M}{2a}\int\left[(\partial u/\partial t)^2 + \Omega_0^2 u^2 + v_0^2(\partial u/\partial x)^2\right]dx, \qquad (2.4)$$

where M is the reduced mass of atoms responsible for the intramolecular vibration.

Considering the expression

$$H = H_{el} + H_{ph} + H_{int} \qquad (2.5)$$

as the hamiltonian of the electron interacting with optical vibrations and taking into account that it depends on generalized coordinates $\Psi(x,t)$, $u(x,t)$ and the generalized momenta $i\hbar\Psi^*(x,t)$, $M\,\partial u(x,t)/\partial t$ we obtain the equation of motion

$$\frac{\partial^2 u}{\partial t^2} + \Omega_0^2 u - V_0^2 \frac{\partial^2 u}{\partial x^2} + \frac{\chi}{M}|\Psi|^2 = 0, \qquad (2.6)$$

$$i\hbar\frac{\partial\Psi}{\partial t} - \mathcal{E}_0\Psi + \frac{\hbar^2}{2m}\frac{\partial^2\Psi}{\partial x^2} - \chi u\Psi = 0. \qquad (2.7)$$

Owing to the translational symmetry of the system we should find the solutions of Eqs. (2.6) and (2.7) in the form of excitations travelling along the chain with a constant velocity V. For this purpose we introduce using the equalities

$$u(x,t) = u(\xi), \quad \Psi(x,t) = \exp[i(kx - \omega t)]\,\varphi(\xi) \qquad (2.8)$$

real functions $u(\xi)$ and $\varphi(\xi)$ depending on the dimensionless variable $\xi = (x - x_0 - Vt)/a$. In this case from the equation (2.6) there follows

$$u(\xi) = -\frac{\chi}{M\Omega_0^2}\int d\xi_1\, w(\xi-\xi_1)\,\varphi^2(\xi_1), \qquad (2.9)$$

where

$$w(\xi) = \frac{1}{2\pi}\int\frac{\exp(iq\xi)\,dq}{1+\epsilon(q+i\epsilon)^2} =$$

$$= \begin{cases} (1/2\sqrt{\epsilon})\,\exp(-|\xi|/\sqrt{\epsilon}), & \text{if } \epsilon > 0, \qquad (2.10a) \\[2mm] \delta(\xi) & \text{if } \epsilon = 0, \qquad (2.10b) \\[2mm] (\theta(-\xi)/\sqrt{-\epsilon})\sin(|\xi|/\sqrt{\epsilon}), & \text{if } \epsilon < 0, \qquad (2.10c) \end{cases}$$

where $\epsilon = (V_0^2 - V^2)/a^2\Omega_0^2$, be the step function: $\theta(\xi) = 1$, if $\xi > 0$ and $\theta(\xi) = 0$, if $\xi < 0$. Then substituting the expressions (2.8) and (2.9) into Eq. (2.7) we find, at the values

$$k = mV/\hbar, \quad \hbar\omega = \Lambda + \mathcal{E}_0 + (1/2)mV^2$$

the integrodifferential equation

$$\left[\frac{\hbar^2}{2ma^2}\frac{d^2}{d\xi^2} + \Lambda + \frac{\chi^2}{M\Omega_0^2}\int d\xi_1\, W(\xi-\xi_1)\varphi^2(\xi_1)\right]\varphi(\xi) = 0, \qquad (2.\text{II})$$

which determines the value Λ and the function $\varphi(\xi)$ normalized by the condition

$$\int \varphi^2(\xi)\,d\xi = 1. \qquad (2.\text{I2})$$

The localized solutions of Eq. (2.II) satisfying the normalization condition (2.I2) will be called solitons.

Knowing Λ and $\varphi(\xi)$ and using the expressions (2.9), we may calculate the intramolecular displacements $u(\xi)$ following the electron motion and two integrals of motion, namely the total energy $E(V)$ and the momentum $P(V)$ of the excitation travelling with the velocity V. Taking into account the equality $\partial u/\partial t = -(V/a)\,du/d\xi$, these integrals of motion are determined by the expressions

$$E(V) = \mathcal{E}_0 + \frac{mV^2}{2} + \int \varphi(\xi)\left[\chi u(\xi)\varphi(\xi) - \frac{\hbar^2}{2ma^2}\frac{d^2\varphi(\xi)}{d\xi^2}\right]d\xi +$$

$$+ \frac{M\Omega_0^2}{2}\int\left[\frac{V^2+V_0^2}{a^2\Omega_0^2}\left(\frac{du(\xi)}{d\xi}\right)^2 + u^2(\xi)\right]d\xi, \qquad (2.\text{I3})$$

$$P(V) = V\left[m + \frac{M}{a^2}\int\left(\frac{du(\xi)}{d\xi}\right)^2 d\xi\right]. \qquad (2.\text{I4})$$

It follows from the Eq. (2.9), (2.I0) that the intramolecular displacement $u(\xi)$ has at infinity the oscillatory tail at velocities $V > V_0$. Therefore the integrals (2.I3) and (2.I4) diverge at speeds $V > V_0$ and soliton does not exist in this case.

3. APPROXIMATION OF SMALL VELOCITIES OF SOLITON

In molecular lattices the inequality $V_0^2 \ll a^2\Omega_0^2$ is usually satisfied, therefore at small velocities of soliton, when $V \sim V_0$ the parameter $|\epsilon|$ is considerably less than unity. In this case the kernel $W(\xi)$ of Eq. (2.II) is represented in the form of the approximate equality

$$W(\xi) = \delta(\xi) + \epsilon\frac{d^2}{d\xi^2}\delta(\xi) \qquad (3.\text{I})$$

and the integrodifferential equation (2.II) is reduced to the stationary modified nonlinear Schrödinger equation considered, in particular in [I3,I4]

$$\frac{d^2\varphi}{d\xi^2} + 4\alpha\varphi^3 + 4\alpha\epsilon\varphi\frac{d^2\varphi^2}{d\xi^2} - \lambda\varphi = 0 \qquad (3.2)$$

with values of dimensionless parameters $\alpha = ma^2\mathcal{X}^2/2\hbar^2\Omega_0^2 M$, $\lambda^2 = -2ma^2\Lambda/\hbar^2$.

Multiplying the Eq. (3.2) by $d\varphi/d\xi$ and integrating we obtain

$$(d\varphi/d\xi)^2 = \lambda^2\varphi^2 B/A, \qquad (3.3)$$

where $A = 1 + 8\alpha\epsilon\varphi^2$, $B = 1 - (2\alpha/\lambda^2)\varphi^2$. There follows from the Eq. (3.3) that

$$\xi = (1/2\lambda)\ln(|\sqrt{A} - \sqrt{B}|/(\sqrt{A} + \sqrt{B})) + \sqrt{|\epsilon|}\,C, \qquad (3.4)$$

where $C = 2\,\mathrm{arctg}\sqrt{A/4\epsilon\lambda^2 B}$, if $\epsilon \geqslant 0$, and $C = \ln|\sqrt{A} - \sqrt{4|\epsilon|\lambda^2 B}| - \ln(\sqrt{A} + \sqrt{4|\epsilon|\lambda^2 B})$, if $\epsilon < 0$.

When $\epsilon = 0$ the Eq. (3.2) is reduced to the stationary nonlinear Schrödinger equation. In this case from Eqs. (3.4), (2.12) there follows

$$\varphi(\xi) = \sqrt{\alpha/2}\,\mathrm{sech}(\alpha\xi), \quad \Lambda = -\hbar^2\alpha^2/2ma^2 \qquad (3.5)$$

At $\epsilon \neq 0$, satisfying the inequality $8\alpha\epsilon\varphi^2 \ll 1$, Eq. (3.4) yields the approximate expression

$$\xi = (1/2\lambda)(\ln((1-\sqrt{B})/(1+\sqrt{B})) - 4\lambda^2\epsilon\sqrt{B}. \qquad (3.6)$$

Solving Eq. (3.6) relative to φ^2 with an accuracy to the terms of the order of $(\epsilon\lambda^2)^2$ we find

$$\varphi^2 = (\lambda^2/2\alpha)\,\mathrm{sech}^2(\lambda\xi)(1 + 4\epsilon\lambda^2\tanh^2(\lambda\xi)).$$

The normalization condition results in $\lambda = \pm[1 - (4/3)\epsilon\alpha^2]$. We find for the function $u(\xi)$,

$$u(\xi) = -(\mathcal{X}\lambda^2/2\alpha M\Omega_0^2)\,\mathrm{sech}^2(\lambda\xi)(1 + 2\epsilon\lambda^2(4 - 5\,\mathrm{sech}^2(\lambda\xi))).$$

Then using Eqs. (2.13), (2.14), we may calculate the energy and momentum of the soliton, i.e. electron "worn by deformation" travelling with a constant velocity V. With an accuracy to the terms of the order of ϵ^2 they are determined by the expressions

$$E(V) = E(0) + (1/2)m_{sol}V^2, \quad P(V) = m_{sol}V, \qquad (3.7)$$

where m_{sol} is the effective soliton mass,

$$m_{sol} = m(1 + 8\hbar^2\alpha^4/15\,a^4 m^2 \Omega_0^2),$$

and $E(0)$ is the energy of an soliton at rest,

$$E(0) = -\hbar^2\alpha^2/6ma^2\left[1 - 8\alpha^2 V_0^2/5a^2\Omega_0^2\right] + \mathcal{E}_0$$

It is important that the energy $E(0)$ is separated from of the bottom the conductivity band by the gap.

4. GENERALIZATION OF PEKAR'S POLARON ADIABATIC THEORY

Autolocalized clusters arising in ionic crystals when the extra electron field interacts nonlinearly with the field of inertia polarization are called Pekar's polaron [I4,I5]. Their motion is possible only when their velocity does not exceed the lowest of group velocities in each of such fields. The group velocity V_g of an extra electron is proportional to its conductivity bandwidth and proves to be comparatively large. The group velocity V_0 of the polarization field motion is determined by the space dispersion of optical phonons. The inequality $V_0 < V_g$ is satisfied in ionic crystals. Therefore, V_0 is the limiting velocity for polarons. If there is no positive space dispersion, the field of optical phonons is fixed. That is why the polarons arising in such medium are fixed. If there exists negative dispersion, the electron motion is not related to that of the polarization field.

We study here Pekar polaron model [I5] taking into account the inertia and positive dispersion of optical phonons of the polarizazation field.

Let us assume that in the continuum approximation the ionic crystal polarization field is characterized by the dispersion law (I.2) where $k = (k_1, k_2, k_3)$ is vector. The extra electron in an isotropic ion crystal near conductivity band bottom (energy $\mathcal{E}_0$) is characterized by charge e and the effective mass m^*. The field of inertia polarization of the crystal is characterized by the effective dielectric permeability $\tilde{\varepsilon}$ [I5].

The state of the electron interacting with the polarization field is described by the extra electron wave function $\Psi(\bar{r},t)$, which is normalized by the condition

$$\int |\Psi(\bar{r},t)|^2 d^3r = 1, \tag{4.I}$$

and the polarization vector $\bar{P}(\bar{r},t)$. The Hamiltonian has the form

$$H = \int d^3r\left[\Psi^*\left(\mathcal{E}_0 - \frac{\hbar^2}{2m^*}\nabla_{\bar{r}}^2\right)\Psi + \right.$$

$$\left. + \frac{2\pi\tilde{\varepsilon}}{\Omega_0^2}\left(\Omega_0^2\bar{P}^2 + (\partial\bar{P}/\partial t)^2 - \right.\right. \tag{4.2}$$

$$-V_0^2 \bar{P} \nabla_{\bar{r}}^2 \bar{P}) - \bar{P}\bar{D}\Big],$$

where the vector

$$\bar{D} = -e\nabla_{\bar{r}} \int |\Psi(\bar{r}',t)|^2 \frac{d^3r}{|\bar{r}-\bar{r}'|} \tag{4.3}$$

determines the induction of the electric field. Expression (4.2) yields the following equations:

$$\Big[i\hbar\frac{\partial}{\partial t} - \mathcal{E}_0 + \frac{\hbar^2}{2m^*}\nabla_{\bar{r}} - e\varphi\Big]\Psi = 0, \tag{4.4}$$

$$\Big[\frac{\partial^2}{\partial t^2} + \Omega_0^2 - V_0^2\nabla_{\bar{r}}^2\Big]\varphi = -\frac{e\Omega_0^2}{\tilde{\varepsilon}} \int |\Psi(\bar{r}',t)|^2 \frac{d^3r'}{|\bar{r}-\bar{r}'|}, \tag{4.5}$$

where $\varphi(\bar{r},t)$ is the polarization vector potential,

$$\nabla_{\bar{r}}\varphi = 4\pi\bar{P}. \tag{4.6}$$

We define the polaron, moving with constant velocity V, as a soliton solution to Eqs. (4.4), (4.5). Since the solitons are formed due to the interaction between the two fields they can be called twocomponent solitons. We shall seek for a soliton solution to (4.4),(4.5) in the form

$$\varphi(\bar{r},t) = \varphi(\bar{\rho}), \quad \Psi(\bar{r},t) = a^{-3/2}\Phi(\bar{\rho})\,exp(i(\bar{k}\bar{r}-\omega t)), \tag{4.7}$$

where $\varphi(\bar{\rho})$ and $\Phi(\bar{\rho})$ are smooth, rapidly decreasing at $\bar{\rho} = (\xi,\eta,\zeta) \to \infty$, and vanishing only at infinity, real functions of the dimensionless vector $\bar{\rho} = (x/a, y/a, (z-z_0-Vt)/a)$, a is the lattice constant, $|\bar{k}| = m^*V/\hbar$ be the system energy. The Green's function $G(\bar{g})$ for (2.5) is determined by the expression

$$G(\bar{\rho}) = \frac{1}{(2\pi)^3} \int \frac{exp(i\bar{q}\bar{\rho})d^3q}{1 - \sigma^2(q_z+i\varepsilon)^2 + \sigma_0^2\bar{q}^2} \tag{4.8}$$

where the following notations are used: $\sigma_0^2 = V_c^2/a^2\Omega_0^2$, $\sigma^2 = V^2/a^2\Omega_c^2$. The Green function (4.8) takes into account the nonlocality of the interaction between the electron and the polarization field which is due to phonon dispersion ($\sigma_0 \neq 0$) and also the time retardation due to the electron motion ($\sigma \neq 0$).

If the conditions

$$0 \leq s^2 = \sigma^2/\sigma_0 \leq 1 \tag{4.9}$$

are satisfied, the Green's function (4.8) takes the form

$$G_1(\bar{\rho}) = \frac{\exp[-\sqrt{\zeta^2+(1-s^2)(\xi^2+\eta^2)}/\sigma_0\sqrt{1-s^2}]}{4\pi\sigma_0\sqrt{\zeta^2+(1-s^2)(\xi^2+\eta^2)}} \; . \qquad (4.10)$$

For $s^2 > 1$ it has the value

$$G_2(\bar{\rho}) = \frac{\cos[\sqrt{\zeta^2-(s^2-1)(\xi^2+\eta^2)}/\sigma_0\sqrt{s^2-1}]}{2\pi\sigma_0\sqrt{\zeta^2-(s^2-1)(\xi^2+\eta^2)}} \; , \qquad (4.11)$$

if the inequalities, $\zeta < 0$, $\zeta^2 > (s^2-1)(\xi^2+\eta^2)$ are satisfied and equals zero if these inequalities are not satisfied.
    Solutions to (4.5) obtained by means of Green's function
G_2 determine the potential $\varphi(\bar{\rho})$ oscilating with constant
amplitude at $\bar{\rho} \to \infty$. Therefore, the solutions satisfying
the normalization condition (4.1) can be expressed only through
the Green's function G_1 . In other words, the solutions localized in space are possible only at velocities less or equal
to V_0 . In the absence of optical phonon dispersion ($V_0 = 0$)
the excitation is stationary. Using the Green function (4.10)
the solution to (4.5) with the account Eq. (4.7), can be written as

$$\varphi(\bar{\rho}) = -\frac{e}{\tilde{\varepsilon}a}\int\frac{G_1(\bar{\rho}-\bar{\rho}_2)\,\Phi^2(\bar{\rho}_1)\,d^3\rho_1 d^3\rho_2}{|\bar{\rho}_2-\bar{\rho}_1|} \qquad (4.12)$$

Substituting this value into Eq. (4.4) we obtain the integro-
differential equation for the enveloping $\Phi(\bar{\rho})$ of the function (4.7),

$$\left[\frac{\hbar^2}{2m^*a^2}\nabla_{\bar{\rho}}^2 + \Lambda + \frac{e^2}{\tilde{\varepsilon}a}\int\frac{G_1(\bar{\rho}-\bar{\rho}_2)\,\Phi^2(\bar{\rho}_1)\,d^3\bar{\rho}_1 d^3\bar{\rho}_2}{|\bar{\rho}_2-\bar{\rho}_1|}\right]\Psi(\bar{\rho})=0, \qquad (4.13)$$

where Λ is the electron energy with respect to the conduction band bottom in a potential well $\varphi(\bar{\rho})$ moving with velocity V . The solution of Eq. (4.13) can be calculated with
the help of a variational method by minimizing the functional

$$J(\Phi) = -\frac{\hbar^2}{2m^*a^2}\int d^3\bar{\rho}\,\Phi(\bar{\rho})\left[\nabla_{\bar{\rho}}\Phi(\bar{\rho}) + \right. \qquad (4.14)$$

$$\left. + \gamma\Phi(\bar{\rho})\int\frac{G_1(\bar{\rho}-\bar{\rho}_2)\,\Phi^2(\bar{\rho}_1)\,d^3\bar{\rho}_1 d^3\bar{\rho}_2}{|\bar{\rho}_1-\bar{\rho}_2|}\right] \; ,$$

where $\nabla_{\bar{\rho}}^2 = \partial^2/\partial\xi^2 + \partial^2/\partial\eta^2 + \partial^2/\partial\zeta^2$ and $\gamma = e m^* a/\varepsilon \hbar^2$.

With no dispersion and fixed polaron ($\sigma_0 = \sigma = 0$) the Green's function $G(\bar{\rho} - \bar{\rho}_1)$ is reduced to the delta function $\delta(\bar{\rho} - \bar{\rho}_1)$. In this case the functional (4.14) is transformed into that of Pekar [15] and the zero approximation functional $J(\phi)$ can be minimized by a direct variational method. Since the electron motion is accompanied by a polarization of cylindrical symmetry it is possible to use the following normalized function as test one:

$$\phi(\bar{\rho}) = (2/\pi)^{3/4} \alpha^{1/2} \beta \, exp[-\alpha\zeta^2 - \beta(\xi^2 + \eta^2)], \qquad (4.15)$$

which is dependent on two parameters α and β (see Ref. [7]).

5. ENERGY AND WAVE FUNCTION OF A SLOWLY MOVING POLARON

With the continuum approximation applied, this theory allows one to study only excitations with the localization region exceeding considerably the lattice constant. In this case the optical phonon dispersion is small and the inequality $\sigma_0^2 \ll 1$ is fulfilled. Since the space-localized solitons can arise in the ionic crystal only at velocities V less than V_0 , then there always holds the inequality

$$\sigma^2 \leqslant \sigma_0^2 \ll 1. \qquad (5.1)$$

For characteristic ionic crystals $a\Omega_0 \approx 2.5 \times 10^4$ cm/s. Hence, the inequality (5.1) is well satisfied. Therefore, in calculating the function (4.15) by means of the functional (4.14) one can use the approximate Green's function

$$G_1(\bar{\rho}) = [1 + \sigma_0^2(\partial^2/\partial\xi^2 + \partial^2/\partial\eta^2) - \sigma^2\partial^2/\partial\zeta^2]\delta(\bar{\rho}) \qquad (5.2)$$

It has been shown in [9] that in this case the square of the wave function $\phi(\bar{\rho})$ that characterizes the spatial quasi-particle distribution is determined by the expressions

$$\phi^2(\bar{\rho}) = (2/\pi)^{3/2} \alpha \beta^2 \, exp[-2\beta^2(\bar{\rho}^2 + (12/7)\alpha_0\sigma^2\zeta^2)]$$

with the values $\beta = \alpha_0 + 6\alpha_0^3[(2/7)\sigma^2 - \sigma_0^2]$, $\alpha = \beta[1 + (12/7)\times \alpha_0\sigma^2]^{1/2}$, $\alpha_0 = \gamma/3\pi^{1/2}$. Thus for the fixed polaron ($\sigma = 0$) the quasi-particle spatial distribution probability has spherical symmetry. The increased polaron velocity results in an intensified localization. The increased dispersion σ_0 reduces the localization region. When the polaron velocity is different from zero the constant value $\phi^2(\bar{\rho})$ are located on the surface having the shape of a flattened rotation ellipsoid with the axis directed along the polarization velocity.

Now making use of Eq. (4.3) one can calculate the polarization field induction vector through which the principal values characterizing the slowly moving polaron are expressed. So, e.g. the soliton energy $E(V)$ is determined by the Eq.

(3.7) in which the energy of a soliton at the rest
has the value

$$E(0) = \mathcal{E}_0 - \frac{a^3}{8\pi\tilde{\mathcal{E}}} \int \left[\bar{D}(\bar{\rho})^2 + \sigma_0^2 (\nabla_{\bar{\rho}} \bar{D})^2 \right] d^3\rho \approx$$

$$\approx \mathcal{E}_0 - 0.053\, E_a\, (m^*/m\tilde{\mathcal{E}}\,)(1 - 0.142\, \gamma^2 \sigma_0^2\,) \tag{5.3}$$

and its effective mass is

$$m_{eff} = m_{sol} = m^* + \frac{a}{4\pi\tilde{\mathcal{E}}\Omega_0^2} \int d^3\rho \left[(\partial\bar{D}/\partial\zeta)^2 - \right.$$

$$\left. - 2\sigma_0 (\partial/\partial\zeta \nabla_\rho \bar{D})^2 + 3\sigma^2 (\partial^2\bar{D}/\partial\zeta^2)^2 \right] = \tag{5.4}$$

$$= m^* + \frac{4e^2\gamma^2}{9\pi^2 a^2 \Omega_0^2} \left[1 + (\gamma^2/5\pi)((181/21)\sigma^2 - 18\sigma_0^2) \right],$$

where m is the free electron mass, $E_a = me^4/\hbar^2$ the
atomic energy unit.

It should be noted that without dispersion and with
extremely small velocity ($\sigma \to 0$) Eqs. (5.3), (5.4) coin-
cides with the equations obtained heuristically in Landau
and Pekar's paper [16].

6. CONCLUSION

Thus the localized solitons can travel in long molecular chains and ionic crystals only when there exists positive
dispersion with velocities less or equal to V_0 . To carry
out a successive calculation of the polaron effective mass
at very small velocities it is necessary to take into account
the small dispersion. Otherwise the self-trapped electron is
immovable.

If quasi-particles move with velocities $V > V_0$, spati-
cally localized solutions such as (4.7) do not exist. Using
the variational method, the authors have obtained in [9] solutions oscillating at infinity. Consequently at such velocities the localization is possible only within a short time
interval since in its motion the electron is detached from
the polarization field.

Analogous conclusions can be made considering the extra
electron motion in a one-dimensional molecular chain when
it is autolocalized due to the short-range interaction with
optical phonons [6].

So at velocities $V > V_0$, nonlocalized states described
by plane waves are possible in long molecular chains and
ionic crystals. In this case the function $\Phi(\bar{\rho})$ and the
polarization field are independent of $\bar{\rho}$ and distributed
uniformly with negligibly small density in the crystal. The
electron motion in such state is described by wave packets.
It is accompanied by a small "cloud" of virtual phonons con-
stantly changing each other.

REFERENCES

I. A. S. Davydov, Biology and Quantum Mechanics, Pergamon, Oxford (1982).

2. A. S. Scott, Dynamics of Davydov soliton, *Phys. Rev. A*, 26:678 (1982).

3. A. S. Scott, The vibrational structure of Davydov soliton, *Phys. Scr.*, 25:651 (1982).

4. A. S. Davydov, The effect of electron-phonon interaction on the electron motion in one-dimensional molecular system, *Teor. Mat. Fis.*, 40:408 (1979) (in Russian).

5. A. S. Davydov, The soliton motion in one-dimensional molecular chain with regard of thermal oscillations, *Zn. Exper. Teor. Fiz.*, 78:789 (1980) (in Russian).

6. A. S. Davydov, Solitons, Bioenergetics and the mechanisms of muscle contraction, *Intern. J. Quant. Chem.*, 16:5 (1979).

7. J. Appel, Polarons, *Sol. St. Phys.*, 21:1 (1968).

8. A. S. Davydov, V. Z. Enol'skii, The theory of motion of an extra electron in a molecular chain with allowance for interaction with optical phonons, *Zn. Exper. Teor. Fiz.*, 79:1888 (1980).

9. A. S. Davydov, V. Z. Enol'skii, Translation-invariant theory of strong particle-field coupling, *Zn. Exper. Teor. Fiz.*, 81:1088 (1981).

10. A. S. Davydov, V. Z. Enol'skii, On the question of effective mass for Pekar polaron, *Zn. Exper. Teor. Fiz.*, 94:177 (1988).

11. I. E. Turner, V. E. Anderson, Ground state energy eigenvalues and eigenfunctions for an electron in electron-dipole field, *Phys. Rev.* 174:81 (1968).

12. A. Nakamura Damping and modificatiom of exciton solitary waves, *J. Phys. Soc. Jap.* 42:1824 (1977).

13. I. V. Simenog, On the asymptotics of stationary nonlinear Schrödinger equation, *Teor. Mat. Fiz.*, 30:3 (1977).

14. A. G. Litvak, A. M. Sergeev, On the one-dimensional collapse of plasma waves, *Lett. Zn. Exper. Teor. Fiz.*, 27:549 (1978).

15. L. D. Landau, On the electron motion in crystal lattice, *Phys. Zs. Sowiet.*, 3:664 (1933).

16. L. D. Landau, S. I. Pekar, Polaron effective mass, *Zn. Exper. Teor. Fiz.*, 18:419 (1948).

14

SELF-TRAPPING IN A MOLECULAR CHAIN WITH SUBSTRATE POTENTIAL

A.V. Zolotaryuk[1],   St. Pnevmatikos[2,3],   A.V. Savin[4]

[1] Institute for Theoretical Physics, UkrSSR Academy of Sciences 252130 Kiev, USSR
[2] Research Center of Crete, P.O. Box 1527, 711 10 Heraklio Crete, Greece
[3] University of the Aegean, 83200 Karlovassi, Samos, Greece
[4] Institute for Physico-Technical Problems, 119034 Moscow, USSR

ABSTRACT

We introduce here an improved version of the well known Davydov model for α-helix proteins and other hydrogen-bonded molecular chains, where the coupling of the chain with its atomic environment is taken into account via the introduction of an on-site harmonic potential for each molecule. For some standard sets of values for the physical parameters, the self-trapping mechanism occurs and two-component soliton excitations are generated in the molecular chain. The first component is the well known pulse excitonic (or electronic) soliton while the second component for the molecular vibrations is quite different. Due to the one-minimum on-site potential the vibrational component becomes a localized wave with zero asymptotic values. In the contrary of the initial Davydov model low energy exciton periodic solutions are also possible in this model. Both solitons and excitons are obtained here using a steepest descent numerical minimization technique. Their dynamics and stability are studied by numerical simulations of the discrete initial equations.

INTRODUCTION

A fundamental problem in biochemistry is to understand how metabolic energy is stored and transported in biological molecules[1]. For instance, the mechanism for transport along proteins of the free energy ($\approx$ 0.42 eV or 10 kcal/mol or 3350 cm^{-1}) released by hydrolysis of adenosine triphosphate (ATP) into adenosine diphosphate (ADP) attracted considerable attention. In the mid-1970's Davydov and Kislukha[2] suggested that energy transport in proteins and other quasi-one-dimensional molecular chains could occur by a soliton mechanism. The main physical idea is that a coupling between intramolecular vibrations (amide-I, e.g. CO stretching, vibrations) and molecular displacements of peptide groups (PGs) lead to a self-trapped long lived state of two component soliton type which transports energy along the chain ($\approx$ 100 PGs). This concept has been elaborated in subsequent papers by Davydov and Scott and co-workers[3,4]

Davydov's Soliton Revisited, Edited by P.L. Christiansen and A.C. Scott
Plenum Press, New York, 1990

Around 1970 Careri noted that the peptide bond angles and lengths in crystalline acetanilide $(CH_3CONHC_6H_5)_x$ or ACN are almost identical to those in natural proteins. In ACN, which is an organic solid, there are chains of hydrogen bonded peptide groups running through the crystal in a manner quite similar to the three chains of the α-helix protein. Careri began a systematic spectral study of ACN and he soon found an "unconventional" amide-I absorption line at 1650 cm^{-1} that is red shifted from the conventional peak by about 15 wave-numbers[5]. Careri's observation is considered today as an experimental confirmation of the Davydov soliton in ACN[6].

The Davydov Hamiltonian is practically a coupled electron-phonon Hamiltonian of the Fröhlich type which is the same as that used for the polaron problem. In this approach the molecular chain does not interact with its atomic environment. However, this is never true. Each hydrogen-bonded molecular chain in the α-helix protein or in ACN is tightly attached to a three dimensional complex skeleton and in general the PGs have well defined equilibrium positions around of which the vibrations take place. The simplest way to describe the interaction of the molecular chain with such an atomic environment is to introduce in the original Davydov Hamiltonian one sequence of harmonic on-site potentials which associate to each PG of the chain one equilibrium position.

Our aim in this paper is to show that this slight but very reasonable modification to the Davydov Hamiltonian produces important changes on the original model.

EQUATIONS OF MOTION

Let us consider a schematic representation of the hydrogen-bonded polypeptide chain in the α-helix protein or in the ACN crystal (see fig.1):

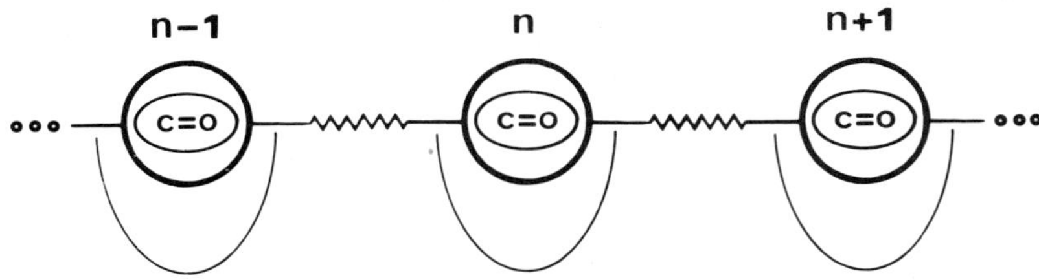

Fig.1   Schematic representation of a polypeptide molecular chain with on-site potentials.

In such a representation we focus our attention on the main degrees of freedom of the real system which in the case of fig.1 are:

a) The high-frequency intramolecular CO stretching vibrations coupled between each other through a quantum mechanical dipole-dipole interaction.

b) The low-frequency molecular vibrations of the peptide group of atoms (seen as one molecule) coupled with their first neighbors through classical harmonic pair potentials (which model the hydrogen bonds).

c) The influence of the atomic environment seen as an one-minimum on-site potential for each molecular peptide group.

According to the standard Davydov theory, the Lagrangian of such a system can be written as follows:

$$L\{a_n, a_n; u_n, u_n\} = \sum_n \left\{ i\hbar a_n{}^* da_n/dt + J a_n{}^*(a_{n-1} + a_{n+1}) + \frac{1}{2} M(du_n/dt)^2 - \right.$$

$$\left. - \frac{1}{2} M(v_o/a_o)^2 (u_{n+1} - u_n)^2 - [E_o + \chi (u_{n+1} - u_{n-1})] a_n{}^* a_n - \frac{1}{2} K u_n{}^2 \right\}$$

(1)

where M is the molecular mass of the entire PG, a_o the intermolecular spacing at equilibrium, $M(v_o/a_o)^2$ the intermolecular coupling constant, E_o is the energy of an amide-I quantum, J the intersite transfer energy produced by dipole-dipole interactions, u_n is the displacement of the n-th molecular group from its equilibrium and a_n is the probability to find the amide-I quantum at the site n, normalized as

$$\sum_n |a_n(t)|^2 = 1 \qquad (2)$$

The nonlinear coupling constant χ arises from modulation of the on-site amide-I energy by the molecular displacements. In the Lagrangian (1) we introduced the last term, with $K>0$, which is the on-site harmonic potential.

The corresponding Euler-Lagrange equations of motion for n=1,2,...,N which follow from (1) have the form :

$$i\hbar da_n/dt = [E_o + \chi(u_{n+1} - u_{n-1})]a_n - J(a_{n+1} + a_{n-1}) \qquad (3a)$$

$$Md^2u_n/dt^2 = M(v_o/a_o)^2 (u_{n+1} - 2u_n + u_{n-1}) + \chi(|a_{n+1}|^2 - |a_{n-1}|^2) - Ku_n \qquad (3b)$$

In the case where the coupling constant χ vanishes then longitudinal sound waves would be governed by the equations:

$$Md^2u_n/dt^2 = M(v_o/a_o)^2 (u_{n+1} - 2u_n + u_{n-1}) - Ku_n \qquad (4b)$$

These equations have plane wave vibrational solutions with the following dispersion relation:

$$\Omega^2 = \Omega_o^2 + 4\,\omega_o^2\,\sin^2(ka_o/2) \qquad (5b)$$

where $\Omega_o=(K/M)^{1/2}$, $\omega_o=v_o/a_o$ and $-\pi/a_o \leq k \leq \pi/a_o$ is the first Brillouin zone for the wave-vector k. In the same time, the amide-I vibrations would be governed by the following Schrödinger equation:

$$i\hbar\,da_n/dt = E_o a_n - J(a_{n+1} + a_{n-1}) \qquad (4a)$$

This equation is now decoupled by the equation (4b) and has excitonic plane wave solutions with the following dispersion relation:

$$\hbar\omega = E_o - 2J + 4\,J\,\sin^2(ka_o/2) \qquad (5a)$$

The excitonic frequency band $\omega(k)$ given by the dispersion relation (5a) is much higher than the vibrational frequency band $\Omega(k)$ given by the dispersion relation (5b) because $E_o-2J>>\Omega_o$. These last constants define the frequency gap that both bands present below their lower values. The characteristic velocity v_o is not anymore the long wavelength sound velocity, as it is when $\Omega_o \to 0$.

In the following it is convenient to use a dimensionless description introducing a new lattice spacing l_o and a new time τ by the relations :

$$l_o = \hbar v_o / a_o |J| \quad , \qquad \tau = l_o v_o t / a_o \tag{6}$$

and also the following dimensionless variables:

$$\Phi_n(\tau) = (\pm 1)^n \, \alpha^{1/2} \, a_n(t) \, \exp\left[\frac{i}{\hbar} (E_o - 2|J|)t \right] \tag{7a}$$

where

$$\alpha = (2a_o \chi / l_o v_o)^2 / M|J|$$

and the sign "+" ("-") is considered when $J > 0$ ($J < 0$), and

$$q_n(\tau) = (2\chi / l_o |J|) \, u_n(t) \tag{7b}$$

Using now (6) and (7), we can write equations (3) in the following dimensionless form:

$$i \frac{d\Phi_n}{d\tau} = l_o^{-2} (2\Phi_n - \Phi_{n-1} - \Phi_{n+1}) + \frac{1}{2l_o} (q_{n+1} - q_{n-1}) \, \Phi_n \tag{8a}$$

$$\frac{d^2 q_n}{d\tau^2} = l_o^{-2} (q_{n+1} - 2q_n + q_{n-1}) + \frac{1}{2l_o} (|\Phi_{n+1}|^2 - |\Phi_{n-1}|^2) - A q_n \tag{8b}$$

where

$$A = (a_o / l_o v_o)^2 \ K/M$$

and the probability function $\Phi_n(\tau)$ is now normalized as follows:

$$\sum |\Phi_n(\tau)|^2 = \alpha \tag{9}$$

Note that the transformation (7a) takes out from the carrier wave the high frequency oscillations and rescales $\Phi_n(\tau)$ to a slow varying regime.

NUMERICAL CALCULATION OF INITIAL CONDITIONS

In order to proceed with a numerical study of the dynamics of equations (8) we need appropriate initial conditions, which we obtain in this section using numerical techniques as well. For this purpose we proceed with a higher degree of approximation by introducing the following ansatz:

$$\Phi_n(\tau) = \varphi_n(\tau) e^{i\theta_n(\tau)} = \varphi_n(\tau) \, \exp\{i[nkl_o - (4l_o^{-2}\sin^2(kl_o/2) + \lambda)\tau]\} \tag{10}$$

where λ is a parameter to be determined. Using (10) the equations (8) and (9) become :

$$\frac{d\varphi_n}{d\tau} = l_o^{-2} \sin(kl_o) \, (\varphi_{n-1} - \varphi_{n+1}) \tag{11a}$$

184

$$l_o^{-2}\cos kl_o\ (2\varphi_n - \varphi_{n-1} - \varphi_{n+1}) + \frac{1}{2l_o}\ (q_{n+1} - q_{n-1})\ \varphi_n = \lambda\varphi_n \tag{11b}$$

$$\frac{d^2 q_n}{d\tau^2} = l_o^{-2}\ (q_{n+1} - 2q_n + q_{n-1}) + \frac{1}{2l_o}\ (\varphi_{n+1}{}^2 - \varphi_{n-1}{}^2) - Aq_n \tag{11c}$$

and

$$\sum_n \varphi_n^2(\tau) = \alpha \tag{12}$$

The standard way to solve such a system of discrete equations is to go to the continuum limit. In this case, looking for travelling wave solutions, we introduce a new space variable $\xi = x - s\tau$, with $x = nl_o$ and $s = v/v_o$ the dimensionless wave velocity, where the wave solutions look stationary. Then equation (11a) defines a relation between the velocity s and the wave vector k :

$$s = 2l_o^{-1}\sin(kl_o)\ , \qquad 0 \le k \le \pi/l_o \tag{13}$$

while the two other equations become:

$$-\cos(kl_o)\varphi_{\xi\xi} + q_\xi\varphi = \lambda\varphi \tag{14a}$$

$$(s^2-1)q_{\xi\xi} + Aq = (\varphi^2)_\xi \tag{14b}$$

where indices here denote total derivatives.
However, the system (14) cannot be solved analytically. Therefore, only numerical approaches can be used. For this purpose, we go back to the Lagrangian (1) and rewrite it in the following dimensionless form:

$$L = \frac{|J|}{\alpha}\ l_o^2 \sum_n \left[i\Phi_n^* \frac{d\Phi_n}{d\tau} + l_o^{-2}\ \Phi_n^*\ (\Phi_{n+1} - 2\Phi_n + \Phi_{n-1}) + \right.$$

$$\left. + \frac{1}{2}\left(\frac{dq_n}{d\tau}\right)^2 - \frac{1}{2}\left(\frac{q_{n+1}-q_{n-1}}{l_o}\right)^2 - \frac{1}{2}\ Aq_n^2 - \frac{q_{n+1}-q_{n-1}}{2l_o}\ \Phi_n\Phi_n^* \right] \tag{15}$$

From here one can derive the Hamiltonian by introducing the appropriate conjugate momenta. Substituting in the Lagrangian (15) the ansatz (10) and using the stationarity relations for the travelling wave solutions in the continuum limit, from where:

$$\frac{dq_n}{d\tau} = - s\ \frac{dq_n}{dx} = - s\ \frac{q_{n+1}-q_n}{l_o} \tag{16}$$

we obtain the following form for the Lagrangian:

$$L[\varphi_n, d\varphi_n/d\tau; \theta_n, d\theta_n/d\tau; q_n, dq_n/d\tau] = \frac{|J|}{\alpha}\ l_o^2 \sum_n \left[\lambda\varphi_n^2 + 2l_o^{-2}\cos kl_o\cdot\varphi_n(\varphi_{n+1}-\varphi_n) \right.$$

$$\left. - \frac{1}{2l_o}\ (q_{n+1} - q_{n-1})\varphi_n^2 - \frac{1}{2}\ Aq_n^2 + \frac{1}{2}\ (s^2-1)\left(\frac{q_{n+1}-q_n}{l_o}\right)^2 \right] \tag{17}$$

where the dependence on θ_n and $d\theta_n/d\tau$ is given implicity since

$$\theta_{n+1} - \theta_n = kl_o \quad\text{and}\quad d\theta_n/d\tau = -[4l_o^{-2}\sin^2(kl_o/2) + \lambda] \tag{18}$$

Now, we can write the corresponding "Hamiltonian" for the Lagrangian (17), using the following definitions for the conjugate momenta:

$$\alpha_n = L/(d\varphi_n/d\tau) = 0, \quad \beta_n = L/(d\theta_n/d\tau) = -|J|\, l_o^2 \varphi_n^2/\alpha \ , \quad \gamma_n = L/(dq_n/d\tau) \tag{19}$$

So, the "Hamiltonian" becomes:

$$H_o = \frac{|J|}{\alpha}\, l_o^2 \sum_n \left\{ 2l_o^{-2}\varphi_n\,[\varphi_n - \cos(kl_o)\,\varphi_{n+1}] + \frac{q_{n+1}-q_{n-1}}{2l_o}\,\varphi_n^2 + \right. \tag{20}$$

$$\left. + \frac{1}{2}\,(1-s^2)\,\left(\frac{q_{n+1}-q_n}{l_o}\right)^2 + \frac{1}{2}\,Aq_n^2 \right\}$$

Note that, for static solutions ($s=0$) the expression (20) is exactly the Hamiltonian for the discrete system. This means that, in order to obtain numerically from (20) static solutions we don't need the continuum limit consequence (16).

In the general case ($s\neq0$), imposing periodic boundary conditions and selecting appropriate initial values for $\{\varphi_n^{\,0},q_n^{\,0}\}_{n=1}^{N}$, one can obtain dynamical initial conditions by solving the following minimization problem:

$$\sum_{n=1}^{N}\left[-2l_o^{-2}\cos(kl_o)\varphi_n\varphi_{n+1} + \frac{q_{n+1}-q_{n-1}}{2l_o}\varphi_n^2 + \frac{1}{2}(1-s^2)\left(\frac{q_{n+1}-q_n}{l_o}\right)^2 + \frac{1}{2}Aq_n^2\right] \to \min \tag{21}$$

with

$$\sum_{n=1}^{N}\varphi_n^2 = \alpha \tag{22}$$

The periodic boundary conditions for the molecular chain impose the following constraint on the available velocity values:

$$kl_o = 2\pi m/N \quad\text{and}\quad s=2l_o^{-1}\sin(2\pi m/N) \quad\text{with}\quad m=0,1,..., N-1$$

<u>Physical quantities to be measured</u>

Let $\{\varphi_n^{\,0},\ q_n^{\,0}\}_{n=1}^{N}$ be a smoothly dependent on n solution of the minimization problem (21). Then the initial data for the system of equation (8) are

$$\Phi_n|_{\tau=0} = \varphi_n^{\,0}e^{iknl_o} \quad , \quad q_n|_{\tau=0} = q_n^{\,0}$$

$$\left.\frac{dq_n}{d\tau}\right|_{\tau=0} = -s\,\frac{\overset{o}{q}_{n+1} - \overset{o}{q}_{n-1}}{2l_o} \tag{23}$$

Since we use here the central derivative the total energy (20) should be redefined as follows:

$$E = \frac{|J|}{\alpha}\, l_o^2 \sum_{n=1}^{N} \Big[-2l_o^{-2}\,\cos(kl_o)\,\overset{o}{\varphi}_n\cdot\overset{o}{\varphi}_{n+1} + \frac{\overset{o}{q}_{n+1}-\overset{o}{q}_{n-1}}{2l_o}\,\overset{o}{\varphi}_n{}^2 +$$

$$\tag{24}$$

$$+\ \frac{1}{2}\Big(\frac{\overset{o}{q}_{n+1}-\overset{o}{q}_n}{l_o}\Big)^2 + \frac{1}{2}\,s^2\Big(\frac{\overset{o}{q}_{n+1}-\overset{o}{q}_{n-1}}{2l_o}\Big)^2 + \frac{1}{2}\,A\overset{o}{q}_n{}^2 \Big]$$

Besides the energy (24), the following two parameters related to the solutions of the problem (21) should be defined: **the degree of localization**[7]

$$L = \sum_{n=1}^{N} \overset{o}{P}_n{}^2 \quad \text{with} \quad \overset{o}{P}_n = \overset{o}{\varphi}_n{}^2/\alpha \tag{25}$$

and **the mean square width** (only for localized solutions) :

$$D = 1+2\Big[\sum_{n=1}^{N} \overset{o}{P}_n\,(m-n)^2\Big]^{1/2} \quad \text{with} \quad m = \sum_{n=1}^{N} n\,\overset{o}{P}_n{} \tag{26}$$

where m is the center of the localization region and the difference is taken in modN. For example:

when	$\overset{o}{\varphi}_{N/2} = \alpha^{1/2}$		$L=1$
		then	$D=1$
	$\overset{o}{\varphi}_n = 0$ if $n \neq N/2$		$m=N/2$
when	$\overset{o}{\varphi}_{N/2} = \overset{o}{\varphi}_{N/2+1}$		$L=1/2$
		then	$D=2$
	$\overset{o}{\varphi}_n = 0$ if $n\neq N/2,\ N/2+1$		$m=(N+1)/2$

<u>Initial points for the minimization scheme</u>

The minimization problem (21) has been solved numerically using the steepest descent minimization method with a variable step. It was observed that the type of the solution (soliton, exciton, ...) depends on the choice of the initial configuration $\{\varphi_n,q_n\}_{n=1}^{N}$ for the minimization process. The following two initial sets of points have been considered:

α) **s-points** (strongly localized state) :

$$\varphi_{N/2} = \alpha^{1/2}, \quad \varphi_n = 0 \ \text{if} \ n\neq N/2 \ , \quad q_n = 0 \ \text{for} \ n=1,2,\,...,\,N \tag{27a}$$

β) **e-points** (quasi-extended state) :

$$\varphi_n = (\alpha/2N)^{1/2} \ \text{if} \ n=1,2,\,...,\,N/4,\ 3N/4+1,\ 3N/4+2,\,...,\,N$$

$$\varphi_n = (3\alpha/2N)^{1/2} \ \text{if} \ n=N/4+1,\ N/4+2,\,...,\,3N/4 \tag{27b}$$

$$q_n = 0 \qquad \text{for } n=1,2, \ldots, N.$$

The s-points are closed to the soliton state for the φ_n field. When the minimization process starts (descents) from these points then an exciton solution appears only if the soliton state cannot exist. The e-points are between an exciton and a soliton state having rather an extended nature. When the minimization process starts from e-points then a soliton solution can be obtained only if the exciton state cannot exist. In the former case one can assert the exciton state is stable and the soliton unstable. In the last case, the opposite can be asserted. Finally, we may have cases where both the soliton and the exciton states can appear from the minimization scheme starting by s and e points respectively. In this case, we assert that the two states can coexist. Note that the solutions obtained by the minimization scheme are exact only when s=0 (static solutions). For s≠0, we considered some continuum limit type approximations and therefore the stability of these solutions should be checked by numerical simulations of the equations of motion (8) with the so generated initial conditions $\{\Phi_n, q_n\}_{n=1}^{N}$.

Parameter values

We have used the following standard parameter values : $E_0=0.22$eV (in fact, in our dimensionless description this parameter does not appear), $a_0=4.5$Å, $M=114.2\ m_p$ (m_p is the proton mass), $J=1.55\times10^{-22}$ Joules, so that the maximum group velocity for excitons to be $v_e=2\hbar^{-1}a_0 J=1.32\times10^3$ m/s. For the string constant K_a which is related to the characteristic phonon velocity v_0 by the relation $v_0=(K_a/M)^{1/2}a_0$, and the coupling constant χ we have chosen two sets of values:

"Hard" chain $(v_e<v_0)$		"Soft" chain $(v_e>v_0)$
$K_a = 13$N/m		$K_a = 0.13$ N/m
$\chi = 4^{\times}10^{-11}$ N	therefore	$\chi = 0.4^{\times}10^{-11}$ N
$v_0 = 3.7^{\times}10^3$ m/s		$v_0 = 3.7^{\times}10^2$ m/s
$l_0 = 5.6,\ \alpha=0.1,\ 41.4$		$l_0 = 0.56,\ \alpha=10.1,\ \beta=41.4$
$v_e/v_0 = 0.356$		$v_e/v_0 = 3.56$

Initial conditions

Using now the steepest descent numerical scheme for the minimization problem (21) and keeping K (the on-site potential coefficient) as a free parameter we have produced the following initial conditions:

a) "Hard" chain : Starting with s-points we obtained static (s=0) soliton numerical solutions for $K\leq3.0$ N/m. Starting with e-points we obtained static soliton solutions for $K<1.0$ N/m and static exciton solutions for $K\geq1.0$ N/m.

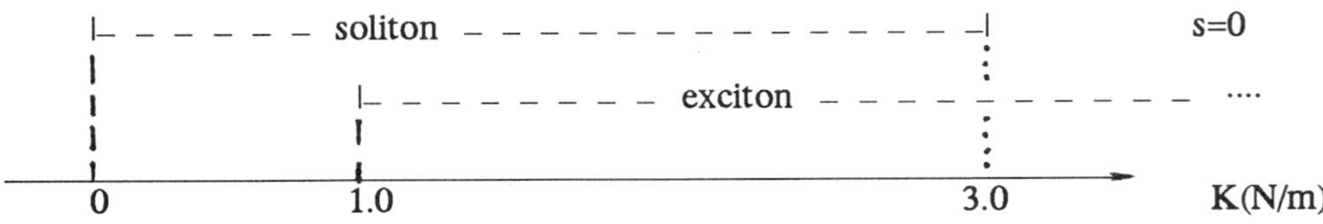

These results do not depend on the chain length (for sufficiently long chains). We note that for 1.0 N/m $\leq$ 3.0 N/m both soliton and exciton modes are possible. At K=0 only the soliton mode emerges (self-trapping) in agreement with the standard Davydov theory[2,3]. Increasing K the amplitude of the soliton (PG displacements) decreases and the soliton width increases. For K>3.0 N/m the soliton solutions are not anymore possible in this chain. In figure 2, we plot the soliton (E_{sol}) and the exciton (E_{ex}) energy versus the parameter K.

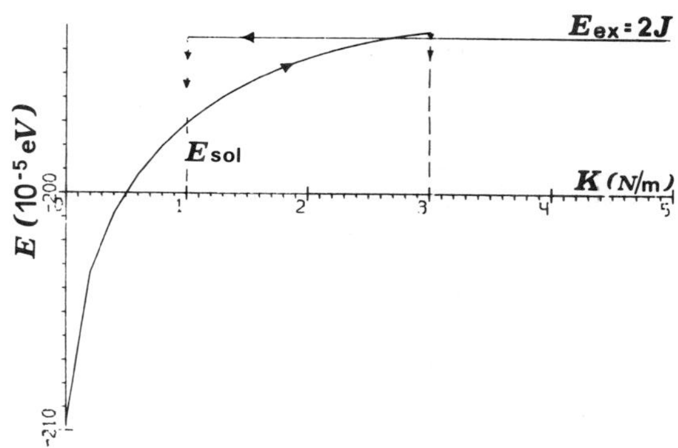

Fig.2   *The soliton and exciton energy versus K for s=0.*

These results can also be obtained by using a variational method[8].

Similar results hold also for s≠0. For example, in the case of s=0.25, starting with s-points, we obtained soliton solutions for K≤5.6N/m and exciton solutions for K>5.6N/m. Starting with e-points we obtained soliton solutions for K<1.5N/m and exciton solutions for K≥1.5N/m. Therefore, for s≠0, there is also a window in the K values where solitons and excitons exist simultaneously. In figure 3, we plot the initial conditions in the $\varphi_n(0)$ and $q_n(0)$ fields for s=0.25 and K=2.0N/m. Due to the presence of the on site potential the soliton in the $q_n(0)$ variable has zero asymptotic values. When K=0, this soliton obtains the standard kink form predicted by the Davydov theory[2-4].

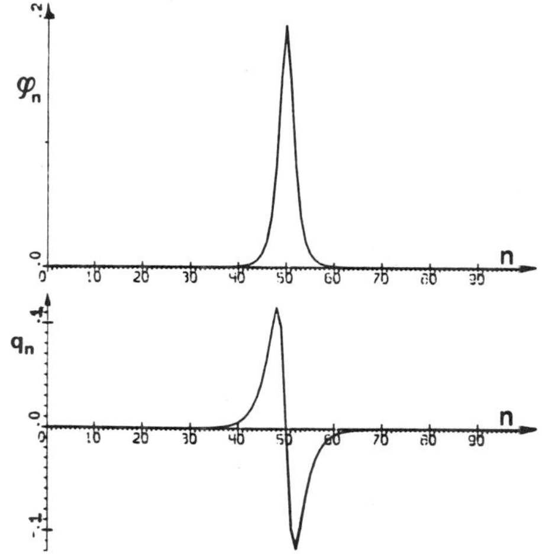

Fig.3   *Soliton initial condition in the $\varphi_n(0)$ and $q_n(0)$ variables for K=2.0 N/m and s=0.25. Here, we have E_{sol}=-1.411x10^{-3}eV<E_{ex}, L=0.248 and D=3.48.*

If now, we keep K fixed and we vary s then we again generate different solutions for different velocities. Similarly, we find windows in the s values where solitons and excitons coexist.

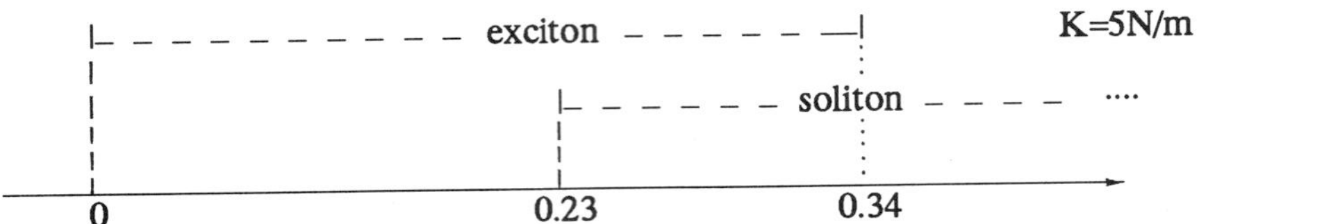

The soliton state is stable only for velocities $s>s_0(K)$ where the critical velocity $s_0(K)$ increases with increasing K. For K=0 we have $s_0(0)=0$, as is expected from the standard Davydov theory. The soliton width increases with decreasing velocity and becomes infinite at s_0. For $s<s_0(K)$ only exciton states are stable.

b) "Soft" chain : Similar results are also obtained for a soft chain. But here we may also have "supersonic" soliton states (s>1). However, these solitons are very narrow and their stability should be affected by discreteness effects.

NUMERICAL SIMULATION OF THE DYNAMICS OF EXCITATIONS

The dynamical stability of these excitations is studied numerically by integrating with a Runge-Kutta scheme the equations of motion (8) with periodic boundary conditions. The accuracy of the numerical simulation is controlled through the conservation of energy $E(\Phi_n, q_n)$ of the molecular chain and the conservation of the normalisation condition (9). The total energy here is given by the formula:

$$E = \frac{|J|}{\alpha} l_0^2 \sum_{n=1}^{N} \left[\frac{1}{2} \left(\frac{dq_n}{d\tau}\right)^2 + \frac{1}{2l_0^2} (q_{n+1} - q_n)^2 + \frac{A}{2} q_n^2 - \right.$$

$$\left. - \frac{2}{l_0^2} \Phi_n^* \Phi_{n+1} + \frac{1}{2l_0} (q_{n+1} - q_{n-1}) \Phi_n^* \Phi_n \right] \tag{28}$$

In the equations of motion (8), we also may add a term $(\Gamma/l_0)dq_n/d\tau$ in order to take into account damping mechanisms when that is needed.

a) "Hard chain" : Let us consider the case K=5N/m and introduce in the chain one soliton initial condition with velocity s=0.25. In fig.4a, we plot in a three-dimensional diagram the time evolution of the $P_n=|\Phi_n|^2/\alpha$ component of the solution, which represents the probability to find the excitation at the n-th PG group (we plot only the envelope line). In fig.4b, we distinguish the initial (dashed lines) and final (full lines) snapshot of the simulation and we plot the probability envelope P_n together with the carrier wave $r_n=(Re\Phi_n)^2/\alpha$, the displacement q_n and its derivative q_n'. The last snapshot is obtained for $\tau=11200$ and the soliton still conserve its initial face in both fields with a slightly modified velocity (s=0.23). The wave is propagated over 462 cells.

190

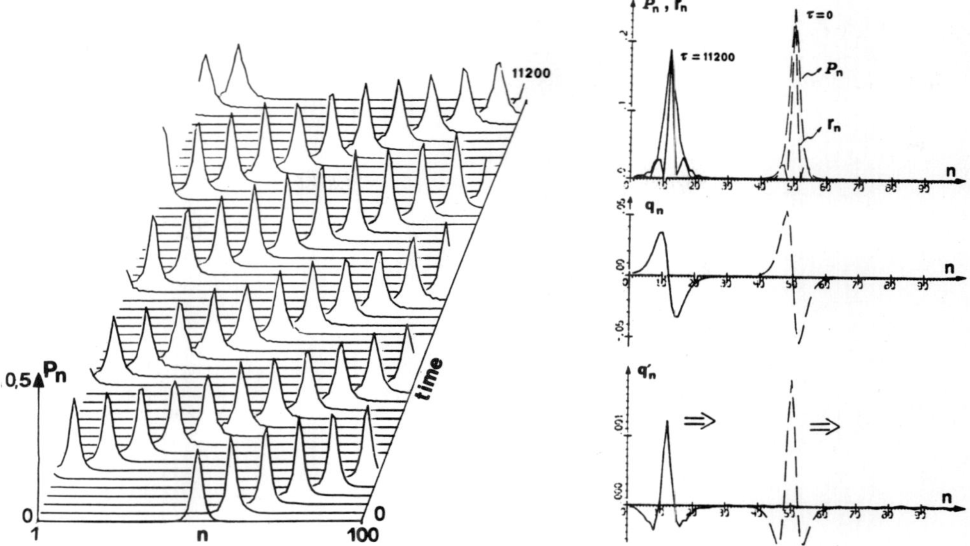

Fig.4   *Soliton propagation (s=0.25) in a "hard" chain with $\Gamma=0$, $K=5N/m$ with periodic boundary condition. a) Time evolution of the probability amplitude. b) Initial and last time snapshots of both variables P_n, q_n and the derivative q_n'.*

In order to verify the stability of both excitations against small perturbations, we introduce random small amplitude displacements of PG groups into the initial data. So, the initial condition (23) should be rewritten as follows:

$$\Phi_n\big|_{\tau=0} = \varphi_n^{\,o}\, e^{iknl_o} \qquad , \qquad q_n\big|_{\tau=0} = q_n^{\,o} + Q\xi_n$$

$$\frac{dq_n}{d\tau}\bigg|_{\tau=0} = -s\,\frac{q_{n+1}^{\,o} - q_{n-1}^{\,o}}{2l_o} \qquad , \qquad n=1,2,\,...,\,N$$

(29)

where ξ_n is a uniformly distributed random quantity from $[-1,1]$ and $Q = \max\limits_{n=1,...,N} |q_n^{\,o}|$, for the soliton initial condition, and as follows:

$$\Phi_n\big|_{\tau=0} = (\alpha/N)^{1/2}\, e^{inkl_o} \qquad , \qquad q_n\big|_{\tau=0} = Q\xi_n$$

$$\frac{dq_n}{d\tau}\bigg|_{\tau=0} = 0 \, , \qquad n = 1,2,\,...,\,N$$

(30)

where the values kl_o are taken from the set $2\pi m/N$, $m=0,1,2,...,$ $N/2$ in order that the equation $s=0.23=(2/l_o)\sin kl_o$ to be satisfied as exactly as possible, for the exciton initial condition. The numerical simulation (see fig.5) has shown that both excitations are stable against small oscillations in a "hard" chain with $K=5N/m$, $\Gamma=0$ and initial velocity $s=0.23$.

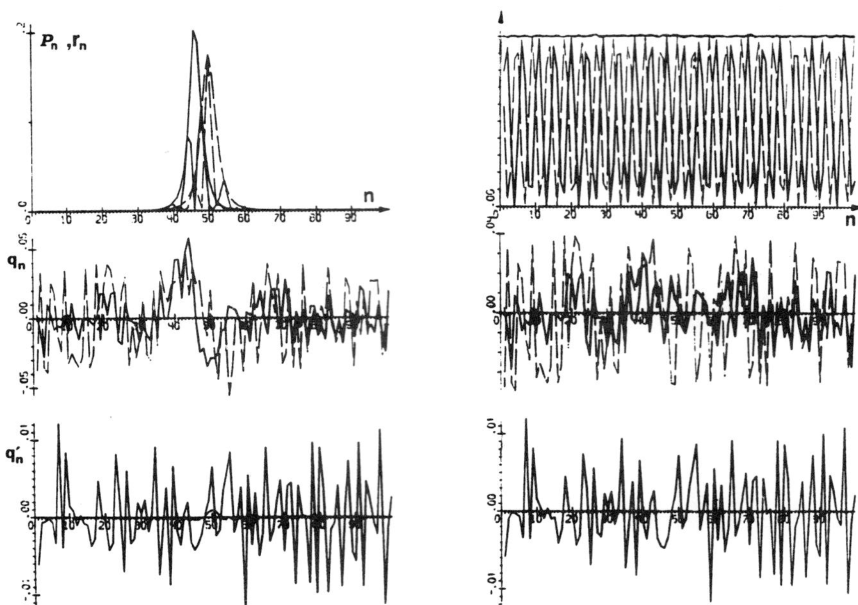

Fig.5 *Stability of the soliton (a) and exciton (b) initial conditions (dashed lines). The last snapshots (full lines) are obtained for $\tau=11200$.*

Now, let us consider a dissipative molecular chain ($\Gamma=0.2$) and study the stability of a finite life time soliton state. For K=5N/m the critical value of velocity for which the soliton state stops to exist (see generation of initial conditions for $s\neq0$ in the previous section) is s(K)=0.23, while for K=2.5N/m this critical velocity is s(K)=0. In general the damping acts on the soliton propagation by decaying its amplitude and velocity and increasing its width. Starting with an initial velocity s=0.25, we observe that (see fig.6) in the first lattice (K=5N/m) the continuous decay of the soliton velocity due to the damping leads the wave to a dynamical situation where the soliton is not anymore stable (fig.6a). Therefore, the soliton initial state disperses. In the second lattice the soliton state is always stable for any $s\neq0$, so the decay is very smooth and does not involve any destruction of the soliton (fig.6b).

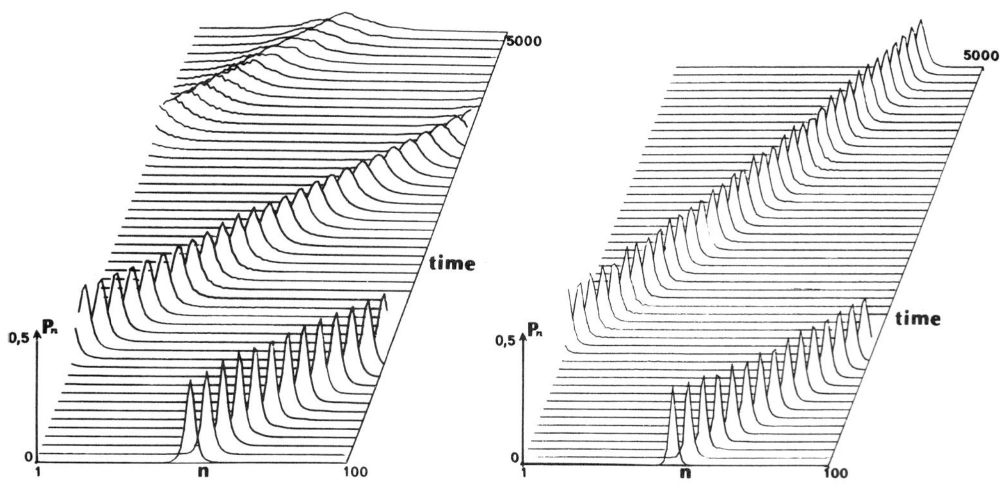

Fig.6 *Damping effects ($\Gamma=0.2$) in a "hard" chain with K=5N/m (a) and K=2.5N/m (b).*

b) "Soft" chain : In this molecular chain we may have both subsonic and supersonic soliton states. However, the numerical simulations showed that only the subsonic solitons are stable (see fig.7). The subsonic class of initial conditions behave like the soliton states in the hard chain (qualitatively), while the supersonic initial conditions decay to the more stable exciton states.

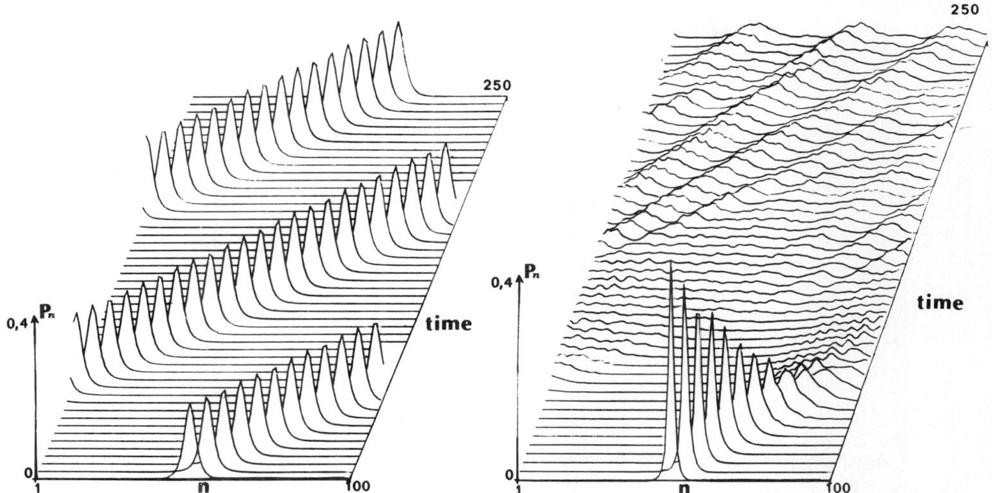

Fig.7   Soliton time evolution in the "soft" chain. a) $K=0.037N/m$, $\Gamma=0$, initial velocity $s=0.5$ and final velocity $s'=0.49$. b) $K=0.105N/m$, $\Gamma=0$, $s=1.01$.

Therefore, the minimization of the expression (21), starting with appropriate initial points, seems to be able to give us stable soliton and exciton dynamical waves which may even coexist in the same molecular chain.

DISCUSSION AND CONCLUSIONS

We have considered in this paper the standard Davydov discrete Hamiltonian with a slight but important modification: the molecular chain is considered to interact with its atomic environment through the influence on each PG of a frozen on-site harmonic potential which introduces one fixed equilibrium position to each molecule. When the coefficient of this on-site potential goes to zero ($K\to 0$) then our chain goes to the standard Davydov chain. However, when $K\neq 0$ some important modification to the Davydov conclusions[2-4] appear.

i) For $K>0$ the displacement field is a localized wave with zero asymptotic values (see fig.3) while in the Davydov theory this field is a kink. In a real molecular system one kink excitation should displace the half of the PGs which means that in an infinite chain this excitation requires an infinite activation energy since each molecular group should have a well defined equilibrium position. In our model this discrepancy disappears since the PG soliton component has zero asymptotic values.

ii) In the standard Davydov Hamiltonian only the soliton state is stable, so the molecular chain has no other choice than to create soliton excitations after any initial condition of some appropriate amplitude (self-trapping). Our model supports both exciton and soliton excitation. For sufficiently large values of the ratio K/K_a only exciton states are stable. For small values of this ratio only soliton modes are stable (self-trapping). While for intermediate values both solitons and excitons are stable. Which form is generated depends on the way we excite initially the chain.

iii) In sufficiently soft molecular chains supersonic soliton states are obtained. However, the stability of these supersonic solitons is questionable.

The dynamical equations of motion for K>0 cannot be solved analytically even in the continuum limit. For this reason, we developed a very interesting numerical method based on a steepest descent minimization scheme and we obtained exact static and approximate dynamic numerical solutions of the exciton and soliton type for the discrete molecular system. The stability of these waves have been studied numerically.

The present approach consists of an improvement on the standard Davydov theory for the self-trapping in biological macromolecules (for instance α helix proteins). We believe that our solitons should be more stable in thermal fluctuations and maybe their lifetime is longer at physiological temperatures. The introduction of temperature to our Hamiltonian as well as some other improvements to out model will be the subject of a next paper.

REFERENCES

1. D. Green, Science 181, 583, (1973); Ann. N. Y. Acad. Sci. 227, 6 (1974).
2. A. S. Davydov and N. I. Kislukha, Phys. Stat. Sol. (b) 59, 465 (1973); A. S. Davydov, J. Theor. Biol. 38, 559 (1973); and "Biology and Quantum Mechanics" (Pergamon, New York 1982).
3. A. S. Davydov and N. I. Kislukha, Zh. Eksp. Teor. Fiz. 71, 1090 (1976) [Sov. Phys. JETP 44, 571 (1976)]; A. S. Davydov, Usp. Fiz. Nauk 138, 603 (1982) {Sov. Phys. Usp. 25, 898 (1982)].
4. J. M. Hyman, D. W. McLaughlin and A. C. Scott, Physica 3D, 23 (1981); A. C. Scott, Phys. Rev. A26, 578 (1982); and 27, 2767 (1983); and Phys. Scr. 25, 651 (1982); L. MacNeil and A. C. Scott, Phys. Scr. 29, 284 (1984).
5. G. Careri, in "Cooperative Phenomena", eds H. Haken and M. Wagner. (Springer-Verlag, Berlin p. 391 (1973)).
6. G. Careri, U. Buontempo, F. Carta, E. Gratton and A. C. Scott, Phys. Rev. Lett., 51, 304 (1983); G. Careri, U. Buontempo, F. Galluzzi, A.C. Scott, A. C. Gratton and E. Shyamsunder, Phys. Rev. B30, 4689 (1984).
7. V. A. Kuprievich, Physica 14D, 3, 395 (1985).
8. A. V. Zolotaryuk and A. V. Savin, unpublished (1989).

Section II

Exciton-Phonon Interactions

A lover's ear will hear the lowest sound.

William Shakespeare

In this section are presented calculations of exciton-phonon coupling (**EPC**) parameters using *ab initio* self consistent field (**SCF**) molecular orbital (**MO**) techniques. To put the problem in context, consider a "zero order" description of exciton dynamics:

$$i\hbar \dot{a}_n = E_n a_n - J_n(a_{n+1} - a_n) - J_{n-1}(a_n - a_{n-1}).\qquad(1)$$

In this equation, phonon interactions are ignored, $E_n/\hbar$ is the amide-I site frequency, and $\hbar/J_n$ is the transfer time for exciton probability between adjacent sites. The dynamic variables a_n $(n = 1, 2, \ldots, N)$ can be considered either the quantum mechanical probability amplitude at site n or the corresponding parameters in Davydov's trial wave function (see Chapter 2 by Kerr and Lomdahl for a detailed study of this problem).

Now assume, as a first order correction to Equation (1), that phonon interactions modulate the parameters E_n and J_n. If the phonon amplitude at site n is the length, R_n, of the hydrogen bond adjacent to the nth amide-I oscillator, there are at least three EPC mechanisms:

1. The change in site energy E_n resulting from a change in the adjacent hydrogen bond length, dE_n/dR_n. Kuprievich calls this parameter $\chi\prime$. Pierce and Østergård call it χ_1^+.

2. The change in E_n resulting from a change in the other (nonadjacent) neighboring hydrogen bond length, dE_n/dR_{n-1}. Kuprievich calls this parameter $\chi\prime\prime$. Pierce and Østergård call it χ_1^-.

3. The change in dipole dipole coupling energy J_n resulting from a change in the intermediate hydrogen bond length, dJ_n/dR_n. Kuprievich calls this parameter χ^{res}, while Pierce, Østergård and Davydov et al. [1] call it χ_2. Following a suggestion of Davydov [2], this parameter was calculated to be [3]

$$\chi^{res} = -\frac{3J_n}{R_n} \tag{2}$$

since $J_n \propto R_n^{-3}$. With $J_n = 7.8\text{cm}^{-1}$ and $R_n = 4.5\text{Å}$, $dJ_n/dR_n = 1.0$ piconewtons.

There are some experimental measurements of the sum $\chi\prime$ (in the notation of Kuprievich). The first of these

$$\chi\prime + \chi\prime\prime = +62 \quad \text{piconewtons} \tag{3}$$

was inferred by Careri from a comparison of amide-I frequencies and hydrogen bond lengths for a series of amide crystals [4]. The "+" sign in Equation (3) indicates that the amide-I frequency increases as the hydrogen bond becomes longer. A second experimental value

$$\chi\prime + \chi\prime\prime = +35 \quad \text{piconewtons} \tag{4}$$

can be inferred by comparing the amide-I frequencies in crystalline acetanilide (**ACN**) with the hydrogen bond lengths at two different temperatures [5,6].

Ab initio calculations of $\chi\prime$ and $\chi\prime\prime$ are difficult; the numerical computations are costly and inaccurate. Using sophisticated techniques of data analysis to improve accuracy, Kuprievich finds for a formamide dimer

$$|\chi\prime\prime| \ll |\chi\prime| \tag{5}$$

and $\chi\prime$ in the range

$$\chi\prime = -30 \quad \text{to} \quad -60 \quad \text{piconewtons.} \tag{6}$$

The magnitude of this result is in approximate agreement with the experimental values, but the sign is not.

Pierce obtains

$$\left.\begin{array}{l} \chi\prime = +26 \quad \text{piconewtons} \\ \chi\prime\prime = 0 \quad \text{piconewtons} \end{array}\right\} \tag{7}$$

but only by comparing results for two different conformations of the formamide dimer. For the same confirmation, he finds

$$\chi\prime = -7 \quad \text{piconewtons.} \tag{8}$$

Pierce also discusses corresponding values for the NH stretching (amide-A) mode.

Østergård presents some preliminary results of a research program which aims to resolve the differences noted above. He finds the calculated values of $\chi\prime$ to be quite sensitive to model system geometry, and this sensitivity compounds the difficulties in obtaining accurate *ab initio* results.

From a broader perspective, such sensitivity might be expected. As Careri has emphasized (see [4] and his chapter in this volume), hydrogen bonds with $R(\text{N} - \text{H} \cdots \text{O}) \approx 2.8\text{Å}$ are "intermediate" between "weak" bonds (dominated by electrostatic interactions) and "strong" (valence) bonds. In the intermediate range, the shape of the electron cloud will depend strongly on electrostatic potentials which, in turn, depend strongly upon the shape of the cloud. It is just this effect which is presumed to cause self-trapping of amide-I vibrational energy.

References

[1] A.S. Davydov, A.A. Eremko, and A.I. Sergienko, Ukr. Fiz, Zh. **23**, 983 (1978).

[2] A.S. Davydov, private communication, September, 1979.

[3] A.C. Scott, Phys. Rev. A **26**, 578 (1982).

[4] G. Careri, U. Buontempo, F. Galluzzi, A.C. Scott, E. Gratton and E. Shyamsunder, Phys. Rev. B **30**, 4689 (1984).

[5] A.C. Scott, unpublished notes, August 1987.

[6] H.J. Wasserman, R.R. Ryan and S.P. Layne, Acta Cryst. **41**, 783 (1983).

15

ON THE CALCULATIONS OF THE EXCITON-PHONON COUPLING PARAMETERS

IN THE THEORY OF DAVYDOV SOLITONS

Victor A. Kuprievich

Institute for Theoretical Physics
Academy of Sciences of the Ukrainian SSR
252130 Kiev, USSR

INTRODUCTION

Davydov soliton (DS) theory[2,3] considers collective amide-1 vibrations (excitons) in the chains of the hydrogen-bonded peptide groups in the peptide α-helix and their coupling with longitudinal deformations of the chains (acoustic phonons). The coupling leads to the H-bond length changes, thus resulting, under some conditions, in a self-trapping of the excitation . Due to the exciton-phonon coupling (EPC) the longitudinal α-helix deformations affect the exciton Hamiltonian, changing   the resonance parameters (which describe the hopping of the excitation between H-bonded peptide groups), and nonresonance (diagonal) parameters (which are the excitation energy of each peptide group). Accordingly, the EPC can be characterized by three parameters: the "resonance" parameter χ^{res} defined as the derivative of the hopping integral with the respect to the corresponding H-bond length and two "nonresonance" parameters, χ' and χ'', which are the derivatives of the amide-1 excitation energy E in the peptide group with the respect to its distances to the "right" and "left" groups in the chain. In other words, χ' and χ'' present the extra forces with which some peptide group acts, when excited, upon its right and left neighbours in the chain.

The computer simulations performed by Scott and co-workers[3-5] showed the EPC threshold in soliton formation estimated as 30 - 60 pica-Newtons (pN) in overall EPC magnitude. Consequently, the problem of estimating the coupling parameters in the real α-helix is of great importance for the DS theory.

The continuum treatment shows that the self-trapping ability is determined by the overall EPC magnitude, which involves all the three parameters in the conbination

$$\chi^{cont} = (\chi' + \chi'')/2 + \chi^{res} . \qquad (1)$$

As estimated in[4], $\chi^{res} \simeq 1$ pN that is negligible in comparison with the threshold magnitude, so the nonresonance parameters only are of interest.

For considering the χ^{cont} magnitude in the DS theory it is useful that the amide-1 vibration is highly characteristic, concentrating mostly in the C=O bond[6]. In the α-helix H-bonds are χormed between an oxygen atom of one peptyde group and hydrogen atom of the neighbour one:

$$...H-N-C=O...H-N-C=O...H-N-C=O...$$

Hence, it seems to be reasonable to consider that amide-1 vibration in some peptide group is perturbed mainly by that of the two neighbour groups which is adjacent to the oxygen atom, thus implying, qualitatively, the principal conribution of the 'right' parameter χ' to overall EPC magnitude. Moreover, the amide-1 localization property allows also to estimate χ', at least in the first approximation, within the framework of a simple model. The first results obtained by the author and Kudritskaya[7] in this way show that the nonresonance EPC is well superior to the resonance one and the calculated value $|\chi'| = (30 - 50)$ pN is comparable, by the order of magnitude, to the threshold. The later calculations[8,9] performed along similar line with some improvements yield χ' estimations which are generally much smaller by absolute values and mostly negative (the correct χ' sign in[7] is also negative that was left out of account there because of χ-sign symmetry of soliton dynamics).

On the other hand, available experimental evidence, basing on the relation between the H-bond length and amide-1 frequency in the crystals of acetanilide and a number of its derivatives, shows that χ' is positive with the overal magnitude of 100 to 120 pN [10]. Also, the positive χ' of about 30 pN can be obtained from the length – frequency relation for acetanilide crystal at different temperatures [11]. In the view of the striking discrepancy between the theoretical and experiomental data and the essential differences between the calculated EPC values the overall scheme of χ' calculations should be made more accurate.

When estimating χ' on the base of *ab initio* calculations[7-9], it is important to use proper analytic potential to fit the *ab initio* energies determined at different geometries. Up to now the fiting is carried out using the potential of the harmonic oscillator whose frequency depends on the H-bond length. Here, a more general scheme which allows to take explicitly into account anharmonicity effects is considered and some applications to χ' estimations are regarded.

PROCEDURE

According to the definition, the general procedure for evaluating χ (χ is either of χ' and χ'') can be outlined as involving the following stages.

Stage (i). The form of amide-1 vibration in the peptide group is determined.

Stage (ii). The adiabatical potential energy U as a function of the amide-1 vibration amplitude X in some peptide group of the chain and the length Y of adjacent H-bond is evaluated for chosen set of points (X_i, Y_i).

Stage (iii). The values $U(X_i, Y_i)$ are fitted by some analytical function V(X,Y). By solving, finally, the Schroedinger equation with the potential V(X,Y), where Y is regarded as external parameter, the amide-1 excitation energy E and its derivative with respect of Y are determined, thus obtaining the estimation of the corresponding χ.

The system under consideration is intricate so a sequence of approximations is helpful to get the final result. According to the simplest approach used in [7] the above procedure is specified as follows.

Stage (i). Basing on the localization property of the amide-1 vibration the normal mode is represented by the stretching vibration of the C=O group, so the C=O bond length is treated as X coordinate. It should be noticed that only the superior of the EPC parameters χ' can be estimated in this line.

Stage (ii). Treating adiabatically the nuclea motion in the electron nucleus system of the peptyde group chain, the potential energy function $U(X,Y)$ is just the total energy of the system, with nuclear coordinates considered to be external parameters. Due to the fact that *ab initio* calculations are highly costed (the required computer time rises as the fourth power with the size of the system), calculations of $U(X_i,Y_i)$ are carried out for the formamide dimer as a model of the minimal-size α-helix fragment containing two H-bonded peptide groups. Running *ab initio* calculations of this 48-electron system at the lowest relevant level, the Hartree – Fock – Roothaan method is used implemented on the minimal STO-3G basis set.

Stage (iii). The fit quadratic in X and linear in Y is applied, thus allowing to calculate χ' with the aid of simple expression which involves the excitation energy E, C=O bond force constant K and its derivative K_Y, the latter is evaluated through energy increments.

It is the above simplest procedure that has been implemented in [7] to obtain the first tentative estimations of χ'. The fit used requires the computations of $U(X,Y)$ only in six points, at three positions of X for each of two positions of Y. It is of importance to note that Xi and Yi positions should be choosen near equilibrium point, where $U(X,Y)$ reaches its minimum. Hence, extra calculations are required to find this point by the geometry optimization, at least partial, when experimental structural data are used as it is done in [7].

To improve the simplest approach, the larger basis set 4-31G is used in [8] for formamide dimer calculations and the optimization of the dimer geometry is performed for each of X and Y positions. In [9] the calculations are done also on the basis of the complete amide-1 normal mode together with its simplified stretcing C = O model. The *ab initio* computations on the minimal STO-3G level has been used both to carry out the normal mode analysis for formamide monomer and to evaluate $U(X_i,Y_i)$ for the dimer.

In [7-9] the simplest quadratic fit is used to evaluate K as a function of Y (the higher level fit is used in [8] only along Y coordinate). However, the extensive study [9] where the diferent geometries has been explored, both planar and nonplanar, shows that calculated χ are highly sensitive to the details of geometry and fitting scheme. Thus some systematical procedure for finding the minimum point and fitting calculated $U(X_i,Y_i)$ is desirable to take into account anharmonicity effects.

A genaral fitting scheme aimed at χ estimations can be proposed as outlined in the following way. Let $U(X_i,Y_i)$ be the potential energies calculated within the minimum area. The function $V(X,Y)$ fitting the $U(X_i,Y_i)$ can be generally written in the form

$$V(X,Y) = v(x,P_1,P_2,\ldots,P_N) + W(y), \quad x = X - \bar{X}, \quad y = Y - \bar{Y}, \qquad (2)$$

where $v(x,P_1,P_2,\ldots P_N)$ is the function of assumed analytical form with the minimum position at $x = 0$ which is independent of the variables $P_1, P_2, \ldots P_N$ completely specifying the function; these variables together with $\bar{X}$ are considered as functions of Y. The function $W(y)$ representing the H-bond potential energy together with the energy origin shift is also of given analytical form with the minimum at $y = 0$. The simplest possible $W(y)$ is quadratic

$$W(y) = W_0 + W_y y + W_{yy} y^2 / 2. \tag{3}$$

By solving the Schroedinger equation

$$(T + v)\phi_k(x) = \varepsilon_k \phi_k(x), \tag{4}$$

where T is the kinetic energy operator, the energy levels ε_k and amide-1 excitation energy $E = \varepsilon_1 - \varepsilon_0$ are obtained as functions of the external parameter Y. Thus, the value of χ is defined by the formula

$$\chi = \sum_{i=1}^{N} \frac{\partial E}{\partial P_i} \frac{dP_i}{dy} , \tag{5}$$

where the derivatives refer to the minimum point $x = y = 0$. Obviously, so far as the χ evaluations are concerned , one can take into account only the linear dependence of P_i on y

$$P_i = P_{io} + P_{iy} y \tag{6}$$

Accordingly, the complete set involves the following 2N+6 numerical parameters

$$\bar{X}_0, \ \bar{X}_y, \ P_{1o}, \ P_{2o}, \ldots P_{No}, \ P_{1y}, \ P_{2y}, \ldots P_{Ny}, \ W_0, \ W_y, \ K_y, \ \bar{Y}$$

which can be determined by the least-squares fit minimizing the function

$$Q(P_1, P_2, \ldots P_N) = \sum_i [U(X_i, Y_i) - V(X_i, Y_i)]^2. \tag{7}$$

As a result the coordinates of minimum position Xo, Yo as well as the parameters P_{iy} are obtained in the same run, thus allowing to evaluate χ by the final formula

$$\chi = \sum_{i=1}^{N} \frac{\partial E}{\partial P_i} P_{iy}. \tag{8}$$

The choice of X and Y positions where the potential energy should be calculated depends on the form of fitting function and the computation accuracy. Formally, this fitting scheme needs at least 2N+6 points (X_i, Y_i) forming, though not necessary, a rectangular lattice, either regular or irregular. This scheme enables us to easily control the fitting accuracy, further points can be directly involved into the calculations to improve it in a systematic manner or to smooth roundoff errors in computed values $U(X_i, Y_i)$.

NUMERICAL RESULTS OF DIFFERENT FITTINGS

We exemplify the above scheme using the functions of different kinds to fit the formamide dimer energies . In order to apply higher-order fitting functions which would take account of the anharmonicity we must consider energy sets with at lest four X positions each. Two sets of this

Table 1 Potential energies for formamide dimer
calculated in Ref. 9 by *ab initio* method at
different C=O length X and H-bond length Y.

i	Y_i	X_i	$U(X_i, Y_i)$
1	-0.1889725	-0.04	-364.678670
2	-0.1889725	-0.02	-364.337183
3	-0.1889725	0	-363.593025
4	-0.1889725	0.02	-362.472089
5	0	-0.04	-364.891457
6	0	-0.02	-364.614309
7	0	0	-363.938525
8	0	0.02	-362.890182

kind obtained by Hartree-Fock-Roothaan *ab initio* calculations on minimal STO-3G basis are available from and presented in Table 1. The atomic units are used throughout this study, milli-Hartree for energy ($4.35942 \cdot 10^{-21}$ Joule) and Bohr for length ($5.2916791 \cdot 10^{-11}$ m). For convenience, listed energies are substracted by -333000 milli-Hartree.

Since only two Y positions are presented, -0.1889725 and 0, the general scheme is restricted neglecting y^2 term in (3) and fixing Y_0 = -0.09448625, so resulted χ' refers to this middle Y_0 position. (our calculations have shown that the change of Y_0 within the range -1 to 1 Bohr produces relatively small effect on calculated χ'). From now on we omit the prime at χ' (which is only regarded in the following as the stretching C=O model is applied for the amide-1 vibration).

<u>Quadratic fit</u>. The simplest quadratic fitting potential is used initially aiming to examine the present technique and illustrate the manner of its application. The potential considered is written in the form

$$v(x) = Kx^2/2, \quad K = K_0 + K_y y. \qquad (9)$$

The excitation energy is expressed by the oscillator formula ($\hbar = 1$ in atomic units)

$$E = \sqrt{K/\mu}, \qquad (10)$$

where μ is reduced mass for amide-1 vibration. This yields the expression for χ

$$\chi = E K_y/2K_0 \qquad (11)$$

used in the preceeding calculations. Estimating χ we put $E = \bar{E}$, where $\bar{E}$ = 7.563541 is amide-1 excitation energy, which coresponds to the observed frequency 1660 cm^{-1}. To get the final result, the relation between force units is used: 1 milli-Hartree / Bohr^2 = 823.8268 pica-Newtons. To evaluate the parameters K_0 and K_y together with X_0, X_y, W_0 and W_y, a six-point set is sufficient. Choosing in different manners six-point sets and minimizing Q to find K_0 and K_y, we reproduced exactly the corresponding results of Ref.9; the whole eight-points set has also been exploited.

$\underline{\text{Fit with }} x^3 \underline{\text{anharmonicity}}$. To consider the effects of the anharmonicity, first the fitting potential with x^3 term is applied

$$v(x) = Kx^2/2 + Ax^3, \quad K = K_0 + K_y y, \quad A = A_0 + A_y y. \tag{12}$$

The corresponding excitation energy can be obtained from the perturbative expression for energy levels[12] and written as follows

$$E = E_0 + E_1, \quad E_1 = -15A^2 E_0^2/2K^3, \tag{13}$$

where E_0 is energy, expressed by (10). By differentiating E one can easily find

$$\chi = E_0 \, K_y/2K_0 + 2E_1 \, (A_y/A_0 - K_y/K_0), \tag{14}$$

where the energies E_0 and E_1 correspond to $Y=Y_0$. The energy E_0, involving unknown reduced mass μ, is determined from (13) by setting $E = \bar{E}$, that leads to

$$E_0 = 2\bar{E}/(1 + \sqrt{1 - 30A_0^2 \bar{E}/K_0^3}\,). \tag{15}$$

$\underline{\text{Fit with }} x^4 \underline{\text{anharmonicity}}$. The next fitting function involves x^4 term instead of x^3 one

$$v(x) = Kx^2/2 + Ax^4. \tag{16}$$

The calculations in this case are carried out along the same line as in the previous example using the expression for the excitation energy[12] and the related formulas corresponding to (13)-(15)

$$E = E_0 + E_1, \quad E_1 = 3A(E_0/K)^2, \tag{17}$$

$$\chi = E_0 \, K_y/2K_0 + E_2(A_y/A_0 - K_y/K_0), \tag{18}$$

$$E_0 = 2\bar{E}/(1 + \sqrt{1 + 12 \, A_0\bar{E}/K_0^2}\,). \tag{19}$$

$\underline{\text{Morse fit}}$. The last of the potentials used is of Morse form

$$v(x) = D(e^{-2bx} - 2e^{-bx} + 1). \tag{20}$$

In contrast with the two former cases, the exact solutions of the Schroedinger equation are known for this nonpolinomial potential[12]. Using them the following expressions can be obtained

$$E = 2Dt(1 - t), \quad t = b/\sqrt{2\mu D}, \tag{21}$$

$$\chi = 2D_0 t[(1 - 2t)b_y/b_0 + D_y/2D_0]. \tag{22}$$

The equality $E = \bar{E}$ is again useful to determine t

$$t = (1 - \sqrt{1 - 2\bar{E}/D_0}\,)/2. \tag{23}$$

Table 2 lists the principal parameters minimizing $Q(P_1, P_2, \ldots P_N)$ when each of the four functions $v(x)$ is used to fit the potential energy values presented in the Table 1; the calculated χ are also given. The other parameters are in the ranges: $X_0 = (-0.0462$ to $-0.0450)$, $X_y = (0.01484$ to $0.0166)$, $W_0 = (-364.806$ to $-364.398)$, and $W_y = (-1.050$ to $-1.025)$. The calculated reduced mass μ (in the units of the proton mass) ranges from

Table 2 Results of application of different fitting functions to
the energies listed in Table 1. Fitting parameters are
given in atomic units corresponding to the choice of
milli-Hartree for energy and Bohr for length.

fit	K_0	A_0	K_y	A_y	χ (pN)
harmonic	969.152		-54.598		-17.551
anharm. x^3	1081.718	-541.354	-99.232	-20.286	-31.138
anharm. x^4	1033.769	-3871.462	-73.976	-1862.248	-58.715
Morse *					-33.286

* D_0 = 457.353, b_0 = -1.0897, D_y = -49.994, b_y = -0.00186.

8.1 (x^4 fit) to 10.1 (x^3 and Morse fits), which can be compared with the
reduced mass of the C=O diatomics ($\simeq$ 7 proton mass).

DISCUSSION

Let us summarize the results of χ_0 calculations listed in the Table 2.
As expected fron the consideration in , the estimation of χ obtained on
the basis of quadratic fit is the smallest by absolute value and much
below the soliton threshold. When an anharmonicity is taken into account,
the calculated χ magnitude considerably increases. The results obtained
when the x^3 term is included and those obtained by means of Morse fit are
quite similar, which seems to be connected with the close relation between
these asymmetric fits.

An analysis shows that the x^3 term does not contribute much to the
calculated χ (the value of the second term on the right-hand side of (14)
is only -2.172 pN). However, x^3 presence in the fitting function leads to
the considerable increase of K_y, thus the magnitude of χ turns out to be
doubled. The symmetric potential involing x^4 anharmonicity even more
enlarges the calcualted magnitude of χ. In contrast to x^3 anharmonicity,
the x^4 term modifies Y-dependence of potential in such a way, that only a
half of the total e change with Y is due to the direct effect of K_y. The
other part of the change is caused by the Y-dependence of the x^4 term.

Although the magnitude of χ calculated with x^4 term is clearly close
to the threshold for soliton formation, the result, obviously, should be
considered as tentative. Further refinements are desirable along the
following lines.

(i) To improve *ab initio* energy calculations the extended basis set
of, say, double-dzeta quality, should be used; polarization basis
functions and electron correlation effects, at least in the H-bond area,
may prove to be important.

(ii) The accurate amide-1 normal mode is more preferable for χ
estimating than the simple C=O stretching model.

(iii) A more flexible fitting potential should be used, involving at
least both x^3 and x^4 anharmonicity terms. In this case, more (X_i, Y_i)
points with the computed *ab initio* energies are required. The extra
points, above their minimal number, are desirable to exam the potential

used. Eventually nonpolinomial potentials, like the flat-bottomed Reid potential[13] used in H-bond theory, may happen to be more satisfactory for the problem under consideration.

Note in conclusion that a special attention should be paid to the discrepancy between negative theoretical values of χ and positive experimental ones. The problem is not unusual for H-bond theory - a similar question arises when considering the changes in O-H stretching motion with H-bond formation[14]. The possibility still exists that both χ definitions are not directly connected and observed H-bond-length - frequency relation can not be completely explained by the consideration of the single H-bond but rather is a consequence of some cooperative effects involving the whole chain of H-bonded molecules. Anyhow, futher studies are needed to resolve the χ-sign problem for the reason of its importance not only for the Davydov soliton theory but also for the hydrogen bond theory on the whole.

ACKNOWLEDGMENTS

I am most grateful to Alwyn Scott for helpful discussions and to Niels Ostergard for the *ab initio* energies of the formamide dimer.

REFERENCES

1. A. S. Davydov, "Biology and quantum mechanics", Pergamon Press, Oxford (1982).

2. A. S. Davydov, "Solitons in molecular systems", Reidel Publishing Co., Dordrecht (1985).

3. J. M. Hyman, D. W. McLaughlin, and A. C. Scott, On Davydov's alpha-helix solitons, <u>Physica</u> 3D:23 (1981).

4. A. C. Scott, Dynamics of Davydov solitons, <u>Phys</u>. <u>Rev</u>.A 26:578 (1982).

5. L. MacNeil and A. C. Scott, Launching of Davydov soliton: II. Numerical studies, <u>Phys</u>. <u>Scripta</u> 29:284 (1984).

6. M. V. Volkenshtein, "Molecular biophysics" (Molekulyarnaya biofizika), Nauka, Moscow (1975).

7. V. A. Kuprievich and Z. G. Kudritskaya, Davydov solitons and determination of the exciton-phonon-interaction parameters <u>in</u>:"Modern problems of the solid state physics and biophysics" (Sovremennyje problemy fiziki twerdogo tela i biofiziki), V. G. Bar'yakhtar, ed., Naukova dumka, Kiev (1982).

8. B. M. Pierce, A. F. Lawrence, and D. B. Chang, A theoretical study of the interaction between amide-1 and hydrogen bond stretching vibrations in hydrogen-bonded polypeptides, <u>in</u>: "Spectroscopy of biological molecules", A. J. P. Alix, L. Bernard, and M. Manfait, eds., Wiley, New York (1985).

9. N. Ostergard, "Ab initio calculations for hydrogen bonds in relation to biomolecualr dynamics" (thesis), The Technical University of Denmark, Lyngby (1988).

10. G. Careri, U. Buontempo, F. Galluzzi, A.C. Scott, E. Gratton, and E. Shyamsunder, Spectroscopic evidence for Davydov-like solitons in acetanilide, Phys. Rev. B 30:4689 (1984).

11. A. C. Scott, private communication (1989).

12. L. D. Landau and E. M. Liftshitz, "Quantum mechanics", Pergamon, London and New York (1959) v. 1.

13. C. Reid, Semiempirical treatment of the hydrogen bond, J. Chem. Phys. 30:182 (1959).

14. N. D. Sokolov, Dynamics of a hydrogen bond in: "Hydrogen bond" (Wodorodnaya swyaz), Nauka, Moscow (1981).

16

QUANTUM CHEMICAL CALCULATIONS OF MOLECULAR PARAMETERS DEFINING DAVYDOV SOLITON DYNAMICS IN POLYPEPTIDES

Brian M. Pierce

Hughes Aircraft Company, Bldg. A-1, M/S 3C924
P.O. Box 9399, Long Beach, California, 90810-0399, USA

ABSTRACT

Ab-initio-SCF-MO theory is used to calculate the molecular parameters defining the dynamics of Davydov solitons arising from the excitation of either the amide-I or $v(NH)$ vibration in a hydrogen-bonded polypeptide or polyamide chain. Both the split-valence 4-31G and STO-3G atomic orbital basis sets are utilized in this study, and a hydrogen-bonded, linear formamide dimer is employed as a model of the chain. The theoretical analysis of the linear dimer consists of calculating (1) equilibrium geometries and electronic charge distributions, (2) vibrational normal modes, (3) adiabatic and non-adiabatic potential energy curves as a function of hydrogen bond length, $R(N\text{---}O)$, (4) electric dipole moment derivatives for the amide-I and $v(NH)$ modes as functions of $R(N\text{---}O)$, and (5) force constants and frequencies for the amide-I and $v(NH)$ modes as functions of $R(N\text{---}O)$. These calculations yield Davydov soliton parameters for the amide-I and $v(NH)$ modes that compare well with the relevant experimental data. The $v(NH)$ mode is calculated to be more strongly coupled to the hydrogen bond stretching vibration than is the amide-I mode. Theoretical treatments of the dynamics of Davydov solitons in polypeptides and polyamides should consider including the $v(NH)$ mode and other modes involving the vibration of the N-H bond, as well as the amide-I mode.

INTRODUCTION

One of the central issues of bioenergetics is how energy arising from chemical reactions is transferred within enzymes and other proteins. Green first proposed in 1973, that the amide-I vibration involving the stretching of the C=O bond in the peptide unit of a protein could be used for the storage and transport of such energy[1]. A linearized model of amide-I vibrational excitation predicts decay to thermal vibrations through transition dipole-transition dipole coupling within a few picoseconds. This lifetime seemed to eliminate propagation of an amide-I excitation as an intermediate step in biological energy transfer because most biochemical processes proceed on much longer time scales[1,2]. Davydov suggested in 1977 that a nonlinear coupling between the amide-I single quantum vibrational excitation and acoustical phonons in the hydrogen-bonded polypeptide chain of a protein could offset the effect of transition dipole-transition dipole dispersion, and thus provide a mechanism for the transfer of amide-I vibrational energy down the chain[3]. In other words, this vibrational exciton-acoustical phonon coupling was postulated to produce a dynamically stable amide-I vibrational excitation that propagates as a solitary wave (also known as a Davydov soliton)[3-6].

A representation of the structure of a hydrogen-bonded polypeptide chain segment is given in Fig. 1. This segment contains three peptide units labelled i-1, i, and i+1, which are linked in the longitudinal direction by hydrogen bonds (indicated by dashed lines). Covalently-bonded groups are shown branching off the ---H-N-C=O---H-N-C=O--- spine in the transverse direction. In the case of the α-helix structure found in many proteins, these groups are connected through an intervening number of related groups to form a covalently-bonded chain that spirals along the longitudinal axis[7].

The vibrations of a hydrogen-bonded polypeptide chain can be divided into two parts: low-frequency motions involving displacements of the peptide units, i.e., the acoustical or optical phonons, and higher frequency, normal mode vibrations of each peptide unit. The hydrogen bonds between the peptide units are

Davydov's Soliton Revisited, Edited by P.L. Christiansen and A.C. Scott
Plenum Press, New York, 1990

much weaker than the covalent bonds within a peptide. Thus, in mechanical models of hydrogen-bonded polypeptide chains, the hydrogen bonds are viewed as soft springs connecting the coupled collection of stiff springs and point masses that defines a peptide.

Although hydrogen bonds are weak, their vibrations can be coupled rather strongly to selected internal, normal mode vibrations of molecules joined by hydrogen bonds[8]. For example, consider the amide-I vibration of a peptide in a hydrogen-bonded polypeptide chain. Because (1) an important component of this mode is the stretching of the C=O bond, and (2) the oxygen atom participates in a hydrogen bond with the neighboring peptide (see Fig. 1), it is reasonable that changes in the hydrogen bond length resulting from the propagation of low-frequency longitudinal phonons along the chain will perturb the amide-I vibration. Conversely, the excitation of the amide-I vibration into its first excited vibrational state increases the equilibrium C=O bond length, which shortens the equilibrium length of the hydrogen bond. This shortening can then perturb the phonons. Couplings between high-frequency vibrational excitations (vibrational excitons) and low-frequency phonons are therefore possible in hydrogen-bonded molecular systems.

Experimental studies of the sensitivity of intramolecular, normal mode vibrational frequencies to hydrogen bond lengths in hydrogen-bonded molecular crystals have provided estimates of vibrational exciton-phonon couplings. In the case of the amide-I vibration, Careri, et. al.[9] plotted the amide-I frequency versus the N---O distance, or hydrogen bond length, for a variety of amide molecular crystals, performed a linear fit of the data, and obtained a coupling equal to +310 cm^{-1}/Å or +62 pN. Scott[10] estimated this coupling to be +175 cm^{-1}/Å or +35 pN using vibrational[9] and structural[11] data for the acetanilide molecular crystal at 113K and 300K. In the case of the stretching vibration of the N-H bond that participates in hydrogen bonding, i.e., the ν(NH) mode, Lautie', et. al.[12] plotted the ν(NH) frequency versus the N---O distance for a large number of crystals, and obtained a coupling equal to +1700 cm^{-1}/Å or +340 pN for N---O distances in the range of 2.8-3.0 Å. In the case of the N-H out-of-plane vibration, i.e., the γ(NH) mode, Bandekar and Zundel[13] studied the temperature dependence of the ν(NH) and γ(NH) vibrational frequencies in uracil and derivatives, and found the two modes to be of similar sensitivity. In conclusion, there are a number of intramolecular amide vibrations that are coupled to hydrogen bond stretching vibrations associated with phonons. Not surprisingly, these intramolecular vibrations involve the motions of atoms whose bonds are most directly affected by hydrogen bonding.

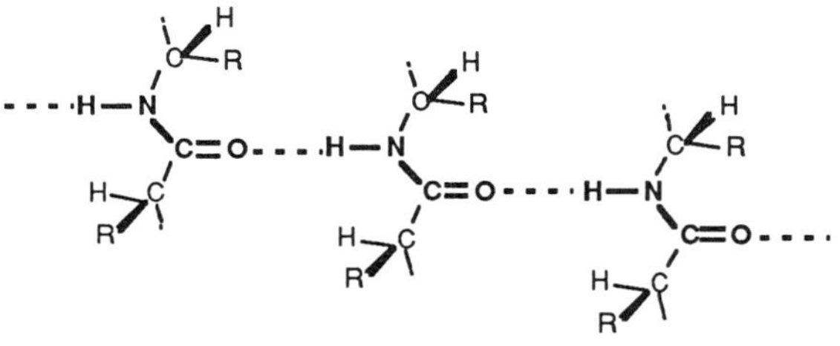

Figure 1.   The Bonding Structure of a Segment of a Hydrogen-Bonded Polypeptide Chain.

As stated in the opening paragraph to this section, the coupling between an amide-I vibrational exciton and a longitudinal acoustical phonon in a hydrogen-bonded polypeptide chain was postulated by Davydov[3] to produce a dynamically stable amide-I vibrational excitation that propagates as a solitary wave. The Hamiltonian that has been used to define the dynamics of Davydov solitons along the three hydrogen-bonded "spines" in a protein α-helix includes only the coupling between the amide-I vibrational excitation and longitudinal acoustical phonons that change the hydrogen bond lengths[5,6,14]. The salient molecular parameters in this Hamiltonian are as follows: the quantum energy of an amide-I vibration ($h\omega_I/2\pi$); transition dipole-transition dipole interaction energy, or longitudinal coupling, between an amide-I vibration on a particular peptide unit and those on neighboring ones (J); the hydrogen bond stretching force constant (w); the mass of the peptide unit (M); the non-resonance amide-I vibrational exciton-phonon coupling [$\chi_1(I) = (1/2\pi)\partial h\omega_I/\partial R(N---O)$, where R(N---O) is the hydrogen bond length]; and the resonance amide-I vibrational exciton-phonon coupling [$\chi_2(I) = \partial J/\partial R(N---O)$]. The $\chi_1(I)$ term is the change in the amide-I vibrational energy with respect to the longitudinal expansion and compression of the α-helix, while the $\chi_2(I)$ term is the change in longitudinal coupling between amide-I vibrations on neighboring peptide groups with respect to longitudinal changes in the α-helix. Scott[6,10] obtained estimates of all of these parameters using the results of experimental vibrational spectroscopic studies of polypeptides and the acetanilide molecular crystal: $\hbar\omega_I/2\pi$ = 1660 cm^{-1}, $|J|$ = 7.8 cm^{-1}, w = 19.5 N/m, M = 114.2 amu, $\chi_1(I)$ = +35 pN, and $\chi_2(I)$ = 1 pN. The $\chi_1(I)$ parameter is of particular interest because the pioneering numerical study of Davydov soliton dynamics in an α-helix by Scott[6] predicted a *threshold* for soliton formation at $\chi_1(I)$ = +35 pN. Subsequent theoretical studies of the effect of temperature on Davydov soliton dynamics also emphasize the importance of $\chi_1(I)$[14,15]. Therefore, the accurate determination of $\chi_1(I)$ is critical to the Davydov soliton model.

In this paper, $\chi_1(I)$ and other parameters in the Hamiltonian are calculated using the approximate quantum mechanical treatment of many-electron molecular systems provided by *ab-initio*-self-consistent-field (SCF) molecular orbital (MO) theory[16,17]. The advantage of this theoretical approach is that we can directly calculate how changes in the inter- and intramolecular coordinates associated with vibrational excitons and phonons result in different molecular electron density distributions, and hence different vibrational properties, e.g., force constants and harmonic frequencies, for the molecular system. An *ab-initio*-SCF-MO calculation of a hydrogen-bonded polypeptide chain would be a formidable task because of the large number of electrons. However, there exist simpler models of the chain that are computationally tractable. The simplest such model is the hydrogen-bonded linear formamide dimer. Bonding structures of three different conformers of the linear formamide dimer are shown in Fig. 2, along with the cyclic dimer and monomer.

Kuprievich and Kudritskaya[18,19] were the first to calculate $\chi_1(I)$ for the linear formamide dimer using *ab-initio*-SCF-MO theory. The STO-3G minimal atomic orbital basis set was employed in this effort, and the amide-I vibrational normal mode was approximated as only a C=O bond stretching vibration. The $\chi_1(I)$ coupling was calculated to range from -27 to -36 pN, depending on the values of the C=O bond stretching force constant (k_{CO}) used in the calculations. The *magnitudes* of these values of $\chi_1(I)$ are consistent with the above-cited experimental estimates and Scott's threshold for soliton formation[6], but the calculated *negative signs* of $\chi_1(I)$ are not in agreement with the experimentally-estimated positive sign. Similar *ab-initio*-SCF-MO calculations of $\chi_1(I)$ for the dimer using the STO-3G[14,20] and 4-31G[21] basis sets have also yielded negative values of $\chi_1(I)$: -2 to -3 pN[14], -7 pN[20,21]. Note that the magnitudes of these $\chi_1(I)$'s are less than the range calculated by Kuprievich and Kudritskaya[18,19]. Given the range of calculated magnitudes for $\chi_1(I)$ and the discrepancy between calculated and experimentally-estimated signs for $\chi_1(I)$, more extensive *ab-initio*-SCF-MO investigations of $\chi_1(I)$ and other couplings are needed to help resolve these issues.

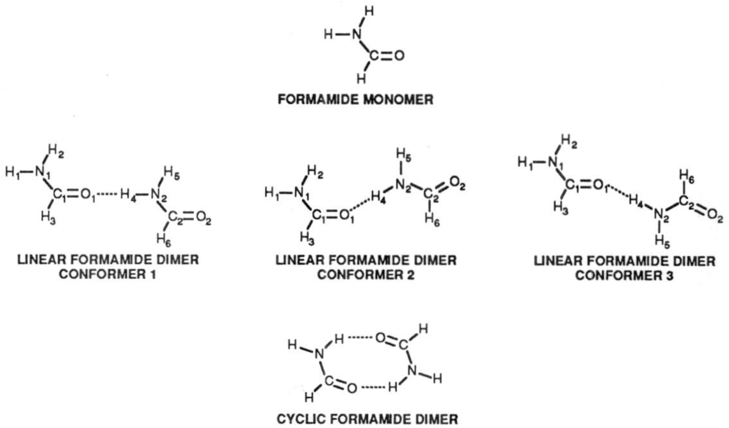

Figure 2. Bonding Structures of Three Different Conformers of the Linear Formamide Dimer, the Cyclic Dimer, and the Monomer.

The hydrogen-bonded, linear formamide dimer was employed as the model system in the present investigation of the molecular parameters defining Davydov soliton dynamics in polypeptides and polyamides. The split-valence 4-31G and STO-3G basis sets were used in this investigation, which comprised several steps. First, equilibrium geometries and electronic charge distributions for the formamide monomer and the

211

three conformers of the linear dimer were calculated. The second step comprised the computation of vibrational normal modes for the equilibrium geometry of Conformer 3 of the linear dimer. The third step consisted of calculating adiabatic and non-adiabatic potential energy curves as a function of R(N---O) for Conformer 1 of the linear dimer. These curves are important to determining w and the anharmonicity of the hydrogen bond stretching vibration. The fourth step involved the calculation of dipole moment derivatives for the amide-I and ν(NH) modes in Conformer 1 of the dimer as a function of R(N---O). These relationships are fundamental to the J and χ_2 terms for the amide-I and ν(NH) modes. The fifth and final step consisted of the calculation of the force constants and normal mode frequencies for the amide-I and ν(NH) modes in Conformer 1 of the dimer as a function of R(N---O). These relationships are basic to the χ_1 terms for the amide-I and ν(NH) modes.

THEORETICAL

This section consists of two parts. In the first part, the vibrational exciton-phonon interaction component of the Davydov Hamiltonian for a hydrogen-bonded polypeptide chain and the coupling parameters are presented. The second part discusses the use of the quantum mechanical *ab-initio*-SCF-MO theory to calculate the coupling parameters and molecular properties specific to the linear formamide dimer.

Vibrational Exciton-Phonon Interaction Hamiltonian

The Hamiltonian for the interaction of a vibrational exciton of normal mode α with longitudinal acoustical phonons in a discrete chain of N molecular units forming a regular one-dimensional lattice was derived by Kuprievich and Kudritskaya[18,19] in a form differing slightly from that reported by Davydov[4]:

$$H_{int}(\alpha) = \sum_{i=1}^{N} b_i^+(\alpha) b_i^-(\alpha) [\chi_1^+(\alpha)(u_{i+1} - u_i) + \chi_1^-(\alpha)(u_i - u_{i-1})]$$

$$+ \sum_{i=2}^{N} [b_i^+(\alpha) b_{i-1}^-(\alpha) + b_{i-1}^+(\alpha) b_i^-(\alpha)]\, \chi_2(\alpha)(u_i - u_{i-1}), \tag{1}$$

where $b_i^+(\alpha)$ and $b_i^-(\alpha)$ are the creation and annihilation operators of the vibrational excitation of mode α in unit i; u_i is the phonon-induced displacement of this unit from its equilibrium position; $\chi_1^+(\alpha)$ and $\chi_1^-(\alpha)$ are the non-resonance vibrational exciton-phonon coupling terms concerning the changes in the vibrational excitation energy of mode α in unit i with respect to phonon-induced changes in the distances between units i and i+1, and units i and i-1, respectively; $\chi_2(\alpha)$ is the resonance vibrational exciton-phonon coupling term concerning the change in interaction energy between excitations of mode α in units i and i-1 with respect to a phonon-induced change in the distance between these units. Note that the terms in the first sum with $\chi_1^-(\alpha)$ at i=1, and with $\chi_1^+(\alpha)$ at i=N, should be omitted because both terminal units have only one neighbor.

The explicit formulations of $\chi_1^+(\alpha)$, $\chi_1^-(\alpha)$, and $\chi_2(\alpha)$ for a dimeric group consisting of peptide units i and i+1 are as follows:[18,19]

$$\chi_1^+(\alpha) = (h/4\pi\omega_{\alpha,i})\partial(\partial^2 E_{i,i+1}/\partial Q_{\alpha,i}^2)/\partial R_{i,i+1}, \tag{2}$$

$$\chi_1^-(\alpha) = (h/4\pi\omega_{\alpha,i+1})\partial(\partial^2 E_{i,i+1}/\partial Q_{\alpha,i+1}^2)/\partial R_{i,i+1}, \tag{3}$$

and

$$\chi_2(\alpha) = (h/4\pi(\omega_{\alpha,i}\omega_{\alpha,i+1})^{1/2})\partial(\partial^2 E_{i,i+1}/\partial Q_{\alpha,i}\partial Q_{\alpha,i+1})/\partial R_{i,i+1}, \tag{4}$$

where $\omega_{a,j}$ is the mode α vibrational angular frequency for units j=i and i+1; $R_{i,i+1}$ is the distance between units i and i+1, taken to be equal to the hydrogen bond length between these units; $E_{i,i+1}$ is the interaction energy of units i and i+1, or total energy of the group in the Born-Oppenheimer approximation; and $Q_{\alpha,j}$ is the mass-weighted normal mode coordinate for the α mode in units j=i and i+1. Because the second derivatives of the total energy of the dimer with respect to its vibrational normal coordinates are proportional to the harmonic force constants for these modes, the coupling parameters in Eqs. (2) - (4) can be expressed as

$$\chi_1^+(\alpha) = (h/4\pi\omega_{\alpha,i}\mu_\alpha)\partial k_{\alpha,i}/\partial R_{i,i+1}, \tag{5}$$

$$\chi_1^-(\alpha) = (h/4\pi\omega_{\alpha,i+1}\mu_\alpha)\partial k_{\alpha,i+1}/\partial R_{i,i+1}, \tag{6}$$

and

$$\chi_2(\alpha) = \quad (h/4\pi(\omega_{\alpha,i}\omega_{\alpha,i+1})^{1/2}\mu_\alpha)\partial k_{\alpha,i;\alpha,i+1}/\partial R_{i,i+1}, \tag{7}$$

where μ_α is the reduced mass for mode α, $k_{\alpha,j}$ is the harmonic force constant for mode α in units $j=i$ and $i+1$, and $k_{\alpha,i;\alpha,i+1}$ is the harmonic interaction force constant between mode α in units i and $i+1$. Furthermore, because $k_{\alpha,i} = \mu_\alpha\omega_{\alpha,i}^2$, the expressions for $\chi_1^+(\alpha)$ and $\chi_1^-(\alpha)$ in Eqs. (5) and (6) become

$$\chi_1^+(\alpha) = \quad (1/2\pi)\partial h\omega_{\alpha,i}/\partial R_{i,i+1} \tag{8}$$

and

$$\chi_1^-(\alpha) = \quad (1/2\pi)\partial h\omega_{\alpha,i+1}/\partial R_{i,i+1}. \tag{9}$$

Note that the frequencies for mode α in Eqs. (8) and (9) are those of the hydrogen-bonded units in the dimeric group, not those of isolated units. Thus, the shift in frequency for an isolated unit resulting from complexation, i.e., the D term in Davydov exciton theory[22], is included in the frequencies in Eqs. (8) and (9).

The resonance vibrational exciton-phonon coupling term in Eq. (4) can be approximated[6] as the change in the transition dipole-transition dipole interaction energy for mode α in units i and $i+1$ with respect to the change in $R_{i,i+1}$:

$$\chi_2(\alpha) = \quad \partial J(\alpha)/\partial R_{i,i+1}. \tag{10}$$

The expression for $J(\alpha)$ is given by[23]

$$J(\alpha) = \quad (1/\varepsilon)|\Delta\mu_{\alpha,i}||\Delta\mu_{\alpha,i+1}|X_{i,i+1}(\alpha), \tag{11}$$

where ε is the dielectric constant, taken to be 1; $|\Delta\mu_{\alpha,j}|$ is the magnitude of the transition dipole moment for mode α in units $j=i$ and $i+1$; and $X_{i,i+1}(\alpha)$ is the geometrical factor

$$X_{i,i+1}(\alpha) = \quad [e_{\alpha,i}\cdot e_{a,i+1} - 3(e_{\alpha,i}\cdot e_{i,i+1})(e_{\alpha,i+1}\cdot e_{i,i+1})]/|R_{i,i+1}(\alpha)|^3, \tag{12}$$

$e_{\alpha,j}$ being the unit vector defining the direction of the transition dipole moment for mode α in units $j=i$ and $i+1$; and $e_{i,i+1}$ and $|R_{i,i+1}(\alpha)|$ being the direction and magnitude of the distance vector between the centers of the transition dipoles for mode α in the two units.

If the electric dipole moment of the dimeric group, μ, is expanded in terms of the normal coordinates, $Q_{\alpha,j}$, and harmonic oscillator wavefunctions are used to evaluate the quantum mechanical expression for $|\Delta\mu_{\alpha,j}| = |\langle 1_{\alpha,j}|\mu|0_{\alpha,j}\rangle|$, then[23]

$$|\Delta\mu_{\alpha,j}| = \quad (4.1058/\nu_\alpha^{1/2})|\partial\mu/\partial Q_{\alpha,j}|, \tag{13}$$

where $|\Delta\mu_{\alpha,j}|$ is in units of D; ν_α, in units of cm^{-1}, is the unperturbed frequency of mode α in a single isolated unit; and $\partial\mu/\partial Q_{\alpha,j}$ is in units of $D\cdot\text{\AA}^{-1}\cdot amu^{-1/2}$. The expression for $J(\alpha)$ in Eq. (11) then becomes[23]

$$J(\alpha) = \quad (84,862)/\nu_\alpha)|\partial\mu/\partial Q_{\alpha,i}||\partial\mu/\partial Q_{\alpha,i+1}|X_{i,i+1}(\alpha), \tag{14}$$

where $J(\alpha)$ is in units of cm^{-1}, and $X_{i,i+1}(\alpha)$ in $\text{\AA}^{-3}$. Note that the integrated infrared (IR) absorption intensity for mode α, A_α, is directly proportional to $|\partial\mu/\partial Q_{\alpha,i}|^2 \cong |\partial\mu/\partial Q_{\alpha,i}||\partial\mu/\partial Q_{\alpha,i+1}|$, and so $J(\alpha)$ can also be approximated as

$$J(\alpha) \cong \quad (a_1/\nu_\alpha)A_\alpha X_{i,i+1}(\alpha), \tag{15}$$

where a_1 is a constant.

By inserting the above approximation of $J(\alpha)$ in Eq. (14) into the expression for $\chi_2(\alpha)$ given in Eq. (10), one obtains

$$\chi_2(\alpha) \cong J(\alpha)[(1/I_{\alpha,i})(\partial I_{\alpha,i}/\partial R_{i,i+1}) + (1/I_{\alpha,i+1})(\partial I_{\alpha,i+1}/\partial R_{i,i+1}) - (3/|R_{i,i+1}(\alpha)|], \qquad (16)$$

where $\chi_2(\alpha)$ is in units of cm^{-1}/Å, and $I_{\alpha,j} = |\partial\mu/\partial Q_{\alpha,j}|$. It is assumed in Eq. (16) that the unit vectors, $e_{\alpha,j}$ and $e_{i,i+1}$, do not change as $R_{i,i+1}$ changes.

In summary, the two types of molecular parameters in the interaction Hamiltonian can be expressed in terms of vibrational spectroscopic properties of the ground electronic state for the system of interest. The non-resonance vibrational exciton-phonon coupling term for mode α is related to the *shift in the center frequency* of the IR absorption peak, or Raman scattering peak, for the mode as a result of a change in the hydrogen bond length; the resonance coupling term is related to a similarly induced *shift in the integrated intensity* of the IR absorption peak for mode α. Molecular parameters in the exciton and phonon Hamiltonians are also related to vibrational spectroscopic properties. In the exciton Hamiltonian, the quantum energy for the excitation of mode α and the longitudinal coupling terms are related to the center frequency of the IR absorption peak, or Raman scattering peak, for the mode and the integrated IR absorption intensity for mode α, respectively. In the phonon Hamiltonian, the force constant is related to the center frequency of the Brillouin scattering peak for the longitudinal acoustical phonon. The only other molecular parameter in the phonon Hamiltonian is the weight of the molecular unit, which is easily determined from the known atomic composition of the unit. Thus, a theoretical method that is successful in treating the vibrational spectroscopic properties of the ground electronic state for a hydrogen-bonded molecular system should be successful in calculating the salient molecular parameters in all three components of the Hamiltonian describing Davydov soliton dynamics in these systems.

In the case of a hydrogen-bonded peptide or amide system, there are several normal mode vibrations internal to the peptide or amide units that are coupled to longitudinal phonons involving vibrations of the hydrogen bonds between the units. As discussed in the Introduction, these normal modes comprise vibrations of the C=O and N-H covalent bonds on either side of the given hydrogen bond (see Fig. 1). Theoretical studies of Davydov solitons in hydrogen-bonded, peptide or amide chains have focussed on the solitons resulting from the coupling between only the amide-I (largely C=O bond stretch) vibrational exciton and longitudinal phonons[6,14,24]. The Hamiltonian for this simple case consists of an amide-I exciton component containing the amide-I excitation energy and transition dipole-transition dipole interaction energies, a longitudinal phonon component containing the phonon energies, and the interaction component given in Eq. (1) that includes the amide-I exciton-phonon coupling terms. More realistic Hamiltonians should consider including (1) the other vibrational excitons coupled to longitudinal phonons, and (2) phonon-mediated couplings between these intramolecular vibrations on the same and different molecular units, and with single and multi-quantum excitations.

Molecular Orbital Calculations

The mapping of the total energy and atomic charge distribution for a hydrogen-bonded chain of peptide or amide molecular units as a function of the positions of the nuclei in these units is fundamental to the determination of the molecular parameters important to the Davydov Hamiltonian. These surfaces can be calculated using an approximate quantum mechanical treatment of many-electron molecular systems. The treatment we selected for our studies is *ab-initio* self-consistent-field (SCF) molecular orbital (MO) theory[16,17]. This theory invokes the Born-Oppenheimer approximation to separate the motion of the nuclei in a molecule from the much quicker electronic motion, and thereby calculates molecular electronic wavefunctions and energies as functions of the positions of the nuclei. These wavefunctions are expressed in terms of molecular orbitals, which are composed of linear combinations of atomic orbitals.

The nature of the atomic orbital basis set employed in an *ab-initio*-SCF-MO calculation chiefly determines the size of the molecular system being treated and the accuracy of the calculated molecular properties. We used two types of basis sets in our calculations. The first and simplest one is termed an STO-3G minimal basis set, and it uses (1) one basis function for each atomic orbital in the inner and valence shells of an atom, (2) a Slater-type orbital (STO) to represent the radial component of the given basis function, and (3) a linear combination of three Gaussian (3G) functions to approximate the STO[16]. The second, more extensive basis set is termed a 4-31G split-valence basis set, and it consists of (1) one basis function for each atomic orbital in the inner shell of an atom, with each of these basis functions approximated by a linear combination of four Gaussian functions, and (2) two basis functions for each atomic orbital in the valence shell of an atom, with one basis function defining a contracted component of the atomic orbital and approximated by a linear combination of three Gaussian functions, and the other a diffuse component approximated by a single Gaussian function[16]. Both the STO-3G and 4-31G basis sets were used with standard parameters[16]. The 4-31G basis set is much more flexible than the STO-3G basis set, and is better suited to describe electronic charge distributions in polar and anisotropic molecules. The two basis sets are best used in calculations of relative energies, energy-minimized geometries, charge distributions and electric dipole moments[16]. Consequently, molecular vibrational spectroscopic properties and the parameters in the

Davydov Hamiltonian should also be treated reasonably well by the STO-3G and 4-31G basis sets. Absolute energies for molecular systems are not calculated well by these basis sets because of the "basis set superposition error"[25].

The calculation of energy and atomic charge distribution surfaces for a hydrogen-bonded peptide or amide chain would be a formidable task using *ab-initio*-SCF-MO theory, even with the STO-3G basis set. In order to make this task more tractable, we followed the example of Kuprievich and Kudritskaya[18,19] in using the hydrogen-bonded linear formamide dimer to model a dimeric group in the chain (see Fig. 2). We then restricted our attention to the potential energy surface defining inter- and intramolecular vibrations of the dimer about its calculated, energy-minimized (or equilibrium) geometry.

Several different *ab-initio*-SCF-MO calculations of the linear formamide dimer were performed in order to compute the molecular parameters important to the Davydov Hamiltonian. These calculations are listed below, along with a brief description of the method of calculation.

Equilibrium geometry. The energy of the dimer was minimized with respect to the Cartesian coordinates of all the atoms in the dimer. This calculation was accomplished by utilizing energy gradient and optimization routines included in the *ab-initio*-SCF-MO software package. The threshold for convergence of the optimization routine was when the magnitude of the largest Cartesian component of the calculated gradient for an atom dropped below 0.0005 Hartree/Bohr. The only constraint on the geometry optimization was that the dimer had to remain planar. This constraint was implemented because a planar structure corresponds closely to energy minima calculated for the dimer using the STO-3G and 4-31G basis sets, and a planar geometry reduced the computational time of the optimization calculation. It is noted that the 4-31G basis set is better than the STO-3G basis set for analyses of the geometries of amide molecules[26]. For example, the optimized geometry for the formamide monomer (see Fig. 2) was calculated to be exactly planar using the 4-31G basis set[26], in agreement with the measured geometry[27], while the STO-3G basis set yielded a slightly non-planar geometry[26].

Vibrational normal mode analysis. Vibrational normal mode analyses were performed for equilibrium geometries of the linear formamide dimer. The force constant matrix was determined by numerical differentiation of analytically-calculated energy gradients in the Cartesian coordinate system, with shifts of ± 0.01 Å for each Cartesian coordinate of each atom. Normal mode harmonic frequencies and atomic displacement coordinates were calculated by direct diagonalization of the mass-weighted force constant matrix. The normal mode vibrational analysis provides (1) the fundamental frequency of the intramolecular vibrational mode α in the exciton component of the Davydov Hamiltonian, ω_α, and (2) the non-mass-weighted, atomic displacement coordinates of this mode.

Intermolecular potential energy curve. Starting with the appropriate, equilibrium geometry, adiabatic and non-adiabatic potential energy curves were calculated for the linear formamide dimer as a function of hydrogen bond length, $R(N\text{---}O)$, or $R_{i,i+1}$ in Eqs. (2) - (16). The adiabatic curve was computed by fixing the positions of the oxygen and hydrogen atoms on the ends of the hydrogen bond for a given value of $R(N\text{---}O)$, and then allowing the other atoms in the dimer to find their energy minima (see above). The dimer was constrained to be planar in the calculation of the intermolecular potential energy curves. The curves were used in (1) the determination of the force constant of the hydrogen bond stretching vibration, w, that appears in the phonon component of the Davydov Hamiltonian, and (2) the investigation of the effect of hydrogen bonding on the intramolecular coordinates and charge distribution for each of the units in the dimer.

Electric dipole moment derivative. According to Eqs. (14) and (16), the magnitudes of the electric dipole moment derivatives for mode α in units i and i+1, $|\partial\mu/\partial Q_{\alpha,i}|$ and $|\partial\mu/\partial Q_{\alpha,i+1}|$, are important to the calculation of (1) the transition dipole-transition dipole interaction energy, $J(\alpha)$, in the exciton component of the Davydov Hamiltonian, and (2) the resonance exciton-phonon coupling term, $\chi_2(\alpha)$, in the interaction component. The $|\partial\mu/\partial Q_\alpha|$ was evaluated numerically at the calculated equilibrium value of $R(N\text{---}O)$ for the computation of $J(\alpha)$, and as a function of $R(N\text{---}O)$ for $\chi_2(\alpha)$.

Non-resonance exciton-phonon coupling. The most direct way to calculate the non-resonance vibrational exciton-phonon coupling terms for mode α, $\chi_1^+(\alpha)$ and $\chi_1^-(\alpha)$, is to perform vibrational normal mode calculations for similar conformers of the dimer with different values of $R(N\text{---}O)$, and then numerically evaluate the derivative expressions for $\chi_1^+(\alpha)$ and $\chi_1^-(\alpha)$ in Eqs. (8) and (9). This approach is very good at treating all intramolecular vibrational modes that are strongly coupled to phonon-induced changes in hydrogen bond length, but it is also very expensive in terms of computational time. Cruder, but cheaper, estimates of $\chi_1^+(\alpha)$ and $\chi_1^-(\alpha)$ for selected intramolecular modes can be obtained by (1) utilizing the fundamental expressions for $\chi_1^+(\alpha)$ and $\chi_1^-(\alpha)$ in Eqs. (2) and (3), (2) approximating the normal mode displacements by changes in only a few internal coordinates, and (3) numerically evaluating the derivatives in Eqs. (2) and (3).

This approach works well for the calculation of the couplings for the ν(NH) mode, $\chi_1^+[\nu(\text{NH})]$ and $\chi_1^-[\nu(\text{NH})]$, because this normal mode is completely defined by an N-H bond stretching motion (see below); the treatment of the couplings for the amide-I normal mode, $\chi_1^+(\text{I})$ and $\chi_1^-(\text{I})$, is more complicated because this normal mode is not completely defined by the C=O bond stretching vibration (see below). A final note concerning $\chi_1^+(\alpha)$ and $\chi_1^-(\alpha)$ for the linear formamide dimer is that $\chi_1^+(\alpha)$ refers to the intramolecular vibrational mode α localized in the formamide monomeric unit on the *left side* of the hydrogen bond in Fig. 2, and $\chi_1^-(\alpha)$ to the mode localized in the unit on the *right side*.

RESULTS AND DISCUSSION

Equilibrium Geometry

Calculated equilibrium geometries of the three conformers of the linear formamide dimer shown in Fig. 2 are given in Table 1, along with calculated equilibrium geometries for the formamide monomer, measured bond lengths and bond angles for crystalline formamide[28], and the observed geometry for the formamide monomer in the vapor phase[27]. The geometry of the isolated linear dimer is not known to have been measured. All calculated geometries in Table 1 were determined using the 4-31G basis set.

Table 1. Calculated Equilibrium Geometries of the Formamide Monomer and Hydrogen-Bonded Linear Dimers, Compared with Observed Structures of the Monomer in the Vapor Phase and Crystalline Formamide

		THEORY						EXPT.	
	Mon.[b]	Dimer						Mon.[d]	Crystal[e]
Internal		Linear 1[c]		Linear 2[c]		Linear 3[c]			
Coord.[a]		Left	Right	Left	Right	Left	Right		
r(CN)	1.346	1.338	1.338	1.339	1.339	1.338	1.339	1.352	1.30
r(CO)	1.216	1.221	1.222	1.221	1.221	1.221	1.221	1.219	1.255
r(CH)	1.081	1.079	1.081	1.079	1.081	1.080	1.082	1.098	
r(NH^t)	0.989	0.990	0.996	0.990	0.997	0.990	0.997	1.002	
r(NH^c)	0.993	0.993	0.992	0.993	0.992	0.993	0.992	1.002	
$\angle$(NCO)	124.9	124.4	125.4	124.7	125.2	124.5	125.3	124.7	121.5
$\angle$(NCH)	113.5	114.5	113.5	114.3	113.6	114.2	113.5	112.7	
$\angle$(HCO)	121.6	121.1	121.1	121.0	121.2	121.3	121.2		
$\angle$(CNH^t)	121.6	122.0	121.8	121.8	122.3	121.8	121.9	120.0	
$\angle$(CNH^c)	119.8	119.6	118.9	119.9	118.6	119.7	118.6	118.5	
$\angle$(H^{t}NH^c)	118.6	118.4	119.3	118.3	119.1	118.5	119.4		
r(NO*)		2.916		2.964		2.980			2.935
$\angle$(CNO*)		118.6		124.2		119.2			118.5
$\angle$(CON*)		176.8		150.9		143.9			120
$\angle$(H^{t}NO*)		3.3		1.8		2.7			
r(O*H^t)		1.922		1.968		1.985			

[a]Bond lengths in Å and bond angles in degrees. An atom with an * indicates an atom in the left molecular unit of the dimer (see Fig. 2). A t or c indicates the N-H hydrogen that is *trans* or *cis* to the C=O bond.
[b]Present work for the monomer.
[c]Present work for Conformers 1, 2 and 3 (see Fig. 2).
[d]Observed planar structure for the monomer in the vapor phase, Ref. [27].
[e]Observed bond lengths and bond angles for crystalline formamide, Ref. [28].

Calculations of the three conformers of the linear formamide dimer were performed in order to investigate the sensitivity of the internal coordinates, and hence the intramolecular vibrations, of the monomeric units to the C=O---H angle, $\angle$C=O---H. This issue was a concern because the anisotropic distribution of electron density on the oxygen atom in a C=O bond can result in a preferred $\angle$C=O---H for hydrogen-bonded amide or peptide systems[29]. The anisotropic distribution on the oxygen atom consists of so-called lone-pair electron density concentrated at ~±120° with respect to the C=O→ bond axis[29]. Thus, one of the three conformers has $\angle$C=O---H = ~180° (Conformer 1), and the other two have $\angle$C=O---H at ~120° (Conformer 2) and ~-120° (Conformer 3). The distinction between Conformers 2 and 3 is that the non-bonded repulsive interactions between the hydrogen atoms of the dimer are greater for Conformer 2, which results in an equilibrium hydrogen bond length for this conformer different from that of Conformer 3.

Table 2. Calculated Equilibrium Electronic Charge Distributions for the Formamide Monomer and Conformer 1 of the Hydrogen-Bonded Linear Dimer

PROPERTY	*MONOMER*	*LINEAR DIMER 1*
q (e)[a]		
H_1	+0.37	+0.38
H_4		+0.46
N_1	-0.90	-0.89
N_2		-0.95
C_1	+0.59	+0.62
C_2		+0.58
O_1	-0.61	-0.66
O_2		-0.64
H_2	+0.39	+0.39
H_5		+0.37
H_3	+0.17	+0.19
H_2		+0.16
$|\mu|$ (D)[b]	4.47	10.33

[a]Mulliken charge in units of electron charge for each atom in the molecule. See Fig. 2 for the atom numbering.
[b]Magnitude of the calculated electric dipole moment for the molecule in units of Debye.

The greatest changes in the geometry of the formamide monomer resulting from the formation of hydrogen-bonded linear dimers are calculated to occur in the bond lengths along the H-N-C=O spine of the monomer. As shown in Table 1, the C=O bond lengths for both monomeric units in all three conformers *increase* by an average of 0.005 Å; the length of the N-H bond of the right monomer that participates in the hydrogen bond *increases* by an average of 0.007 Å; the C-N bond lengths of both monomers *decrease* by an average of 0.008 Å; the bond lengths of the remaining bonds in the dimer undergo changes of ≤ 0.002 Å. The calculated average change for all bond angles is ≤ 1°.

The *directions* of the calculated changes in the C=O and C-N bond lengths are in agreement with those of the observed changes obtained by comparing the measured geometries of the formamide monomer[27] and crystal[28] given in Table 1; the *magnitudes* of the calculated and observed changes are different principally because the hydrogen bond network in the crystal[28] is more complex than that represented by the linear dimers. These findings are also consistent with those of other *ab-initio*-SCF-MO studies of the linear[30,31] and cyclic[32,33] formamide dimers. The calculated hydrogen bond lengths (r(NO*) in Table 1) for the three conformers of the linear dimer are in slightly better agreement with the measured value for the crystal than is the calculated value of the cyclic dimer equal to 2.878 Å[33]. As with the cyclic dimer[33], the hydrogen bonds for the three conformers are calculated not to be linear, i.e., $\angle$(H^{t}NO*) is not equal to 0°.

The calculated internal coordinates of the monomeric units in the three conformers of the linear formamide dimer are very similar. As a result, the sensitivity of the internal coordinates to $\angle$C=O---H is calculated to be low when using the 4-31G basis set. This low sensitivity is reflected in the relative energies calculated for the three conformers; Conformer 3 has the lowest energy of the three at -337.376494667 a.u., with Conformers 1 and 2 at 8 and 105 cm^{-1} above this energy. Thus, the calculations indicate that the potential energy curve for the linear dimer as a function of $\angle$C=O---H is very shallow. The potential energy curve as a function of hydrogen bond length [r(NO*) or R(N---O)] and the sensitivity of the internal coordinates to R(N---O) are discussed in a section below.

Table 3. Calculated Vibrational Frequencies for the Formamide Monomer and Conformer 3 of the Hydrogen-Bonded Linear Dimer, Compared with Observed Frequencies for the Monomer in the Vapor Phase and Crystalline Formamide

| THEORY | | | Mon.[b] | Mon.[c] | Cryst.[d] |
| Linear Dimer 3[a] | | | | | |
$\omega(cm^{-1})$	$\omega(cm^{-1})$	Assignment[e]	$\omega(cm^{-1})$	$\nu(cm^{-1})$	$\nu(cm^{-1})$
Intramolecular, in-plane					
3959	3917	ν(NH) antisym.	3964	3545m	
3824	3759	ν(NH) sym.	3826	3451m	
3263	3236	ν(CH)	3249	2852m	
1893	1882	ν(CO) + NH_2 sciss. + δ(CH)	1898	1734s	
1851	1825	NH_2 sciss. [δ(NH)]	1822	1572m	
1559	1555	δ(CH)	1561	1378w	
1425	1417	ν(CN) + NH_2 rock + δ(CH)	1391	1255m	
1221	1194	NH_2 rock + ν(CN) + δ(CH)	1178	1059w	1140m, 1133w
645	629	δ(NCO) + NH_2 rock	623	565w	657s, 635s
Intramolecular, out-of-plane					
1198	1196	CH out-of-plane [γ(CH)]	1188	1030w	1063s, 1047w
864	780	NH_2 wag. [γ(NH)]	745	289s	843m
743	618	τ(CN) [NH_2 twist]	584	602w	675m
Intermolecular[f]					
	137	in-plane			224s, 155m
	80	in-plane			171m
	27	in-plane			146s, 224w
	117	out-of-plane			233vs
	40	out-of-plane			212s, 198s
	14	out-of-plane			115m, 88w

[a]Present work.
[b]Monomer, Ref. [26].
[c]Monomer in the vapor phase, Ref. [35,36]. The letters w, m, s, and vs next to the frequencies indicate whether the infra-red absorption intensity for the mode is weak, medium, strong, or very strong, respectively.
[d]Crystal, Ref. [28]. See footnote c for significance of the letters.
[e]ν is bond stretching; δ is in-plane bond bending; γ is out-of-plane bond bending; sciss. is scissoring; rock is rocking; wag. is wagging; τ is torsional twisting; and twist is twisting.
[f]Note that the intermolecular frequencies less than 50 cm^{-1} should be considered very approximate.

The electronic charge distributions for the equilibrium geometries of the three conformers are calculated to be very similar when using the 4-31G basis set. The Mulliken atomic charges[16,34] and electric dipole moments calculated for the formamide monomer and Conformer 1 of the linear dimer are presented in Table 2. As expected, the greatest change in the equilibrium electronic charge distribution of the formamide monomer resulting from the formation of a hydrogen-bonded linear dimer is calculated to be the increased polarization of the $C_1=O_1$ and N_2-H_4 bonds involved in the hydrogen bond; the H_4 and C_1 atoms become more *positive* by 0.09 and 0.03 e, respectively, and the O_1 and N_2 atoms become more *negative* by 0.05 e (see Fig. 2 for the atom numbering). The changes in the charges on the other atoms in the monomer are $\leq$ 0.03 e. This polarization of the $C_1=O_1$ and N_2-H_4 bonds is manifested in the fact that the magnitude of the electric dipole moment for Conformer 1 is calculated to be more than two times that of the monomer.

The calculated effect of hydrogen bonding on the internal coordinates and electronic charge distribution of the formamide monomer suggests that a hydrogen bond stretching vibration in the linear dimer will be coupled most strongly to higher frequency vibrations involving the covalent bonds along the N_1-C_1=O_1---H_4-N_2-C_2=O_2 spine of the dimer. These couplings are investigated in greater detail in the following sections.

Vibrational Normal Mode Analysis

The calculated vibrational normal mode, harmonic frequencies for Conformer 3 of the linear formamide dimer are listed in Table 3, together with the calculated harmonic frequencies for the monomer[26], and measured fundamental frequencies for crystalline formamide[28] and the formamide monomer in the vapor phase[35,36]. The vibrational spectrum of the isolated linear dimer is not known to have been measured. All calculated frequencies in Table 3 were determined using the 4-31G basis set. The vibrational normal mode frequencies for Conformer 3 were calculated because this conformer has the lowest energy of the three investigated. In this paper, the vibrational normal mode analysis focuses on the intramolecular vibrations most strongly coupled to the hydrogen bond stretching vibration; a more complete analysis that includes a discussion of the intermolecular vibrations and calculations of IR absorptivities and Raman scattering intensities for the intra- and intermolecular modes is given in Ref. [37].

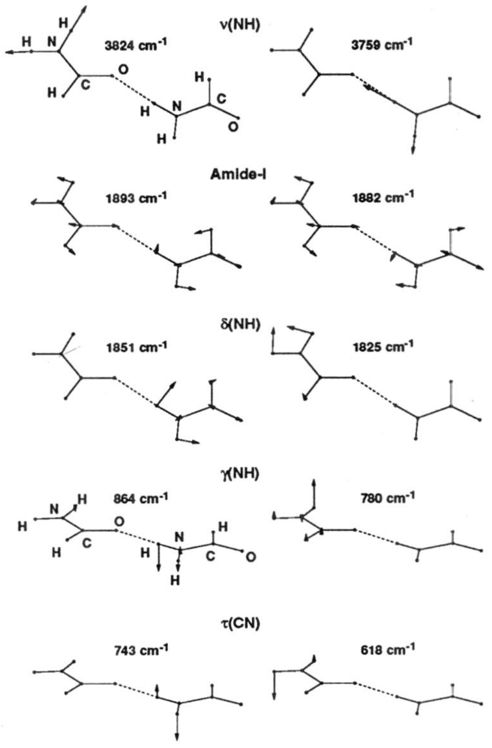

Figure 3.   Calculated Non-mass-weighted Atomic Displacement Coordinates for Selected Intramolecular Vibrational Normal Modes of Conformer 3 of the Linear Formamide Dimer.

The calculated and measured frequencies for the formamide monomer have been analyzed in detail by Sugawara, et. al.[26]. Although the calculated frequencies for the in-plane modes are 10-20% higher than the measured ones, the relative values of the calculated and measured frequencies are in very good agreement. In this regard, the 4-31G basis set is considered to be adequate to treat the in-plane modes[26]. As shown in Table 3, the out-of-plane modes are not treated as well by the 4-31G basis set. Sugawara, et. al.[26] needed to include the polarization functions for the carbon, nitrogen, and oxygen atoms, i.e., expand to a 4-31G* basis set, in order to adequately treat the out-of-plane modes. Only the 4-31G basis set was employed in our vibrational normal mode anaylsis of Conformer 3 of the linear formamide dimer in order to make the calculations more tractable. Consequently, the frequencies and atomic displacements calculated for the out-of-plane modes are less reliable than those for the in-plane modes.

The vibrational states of the monomer split into two components in Conformer 3 with one component localized on each monomeric unit. The calculated shifts in the frequencies of these components for a given mode with respect to the frequency of the related mode in the monomer indicate the sensitivity of this mode to hydrogen bonding. As revealed in Table 3, the greatest shifts are for the mode consisting of the torsional twisting of the C-N bond, or the τ(CN) mode, and the modes involving the out-of-plane bending, stretching, and in-plane bending of the N-H bond in the right monomeric unit of the dimer, i.e., the γ(NH), ν(NH), and δ(NH) modes, respectively. A smaller shift is calculated for the mode involving the stretching of the C=O bond in the left monomeric unit of the dimer, i.e., the ν(CO) or amide-I mode, and even smaller shifts are calculated for modes involving the stretching, in-plane bending, and out-of-plane bending of the C-H bond, i.e., the ν(CH), δ(CH), and γ(CH) modes, respectively. The directions and magnitudes of the frequency shifts calculated for the τ(CN) and γ(NH) modes are in agreement with the shifts indicated by the measured frequencies for crystalline formamide[28] and the monomer in the vapor phase[35,36] (see Table 3). Furthermore, the calculated directions and magnitudes of the shifts for the ν(NH) and γ(NH) modes agree with those measured by Rasanen[38] in his matrix IR spectroscopic study of association of formamide. Intramolecular vibrations involving the C-N, N-H, and C=O bonds of the appropriate monomeric unit in the dimer should therefore be the vibrations most strongly coupled to the intermolecular, hydrogen bond stretching vibration. This finding is consistent with the above comparison of the calculated equilibrium geometries of the formamide monomer and linear dimers.

The non-mass-weighted atomic displacement coordinates calculated for the τ(CN) mode, the γ(NH) mode, the symmetric ν(NH) mode, the δ(NH) mode, and the ν(CO) mode (amide-I) are presented in Fig. 3. In contrast to the modes involving the motion of the N-H bond, the amide-I mode is delocalized over the monomeric unit.

Intermolecular Potential Energy Curve

The adiabatic potential energy curve calculated as a function of hydrogen bond length, R(N---O), for Conformer 1 of the linear formamide dimer is presented in Fig. 4. This calculation was performed using the 4-31G basis set. The adiabatic curve, as well as the non-adiabatic one (not shown), are fit well by the Morse function

$$E[R(N\text{---}O)](cm^{-1}) = 1620[1 - e^{-1.79(R(N\text{---}O)(Å) - 2.942)}]^2. \tag{17}$$

The energies of the first five vibrational states of the Morse oscillator that describes the hydrogen bond stretching vibration for Conformer 1 are indicated by horizontal dashed lines in Fig. 4. The vertical dashed lines in Fig. 4 indicate the range of values of R(N---O) covered by the hydrogen bond stretching vibration at room temperature (kT $\cong$ 200 cm^{-1}). This range is from ~2.78 Å to ~3.20 Å.

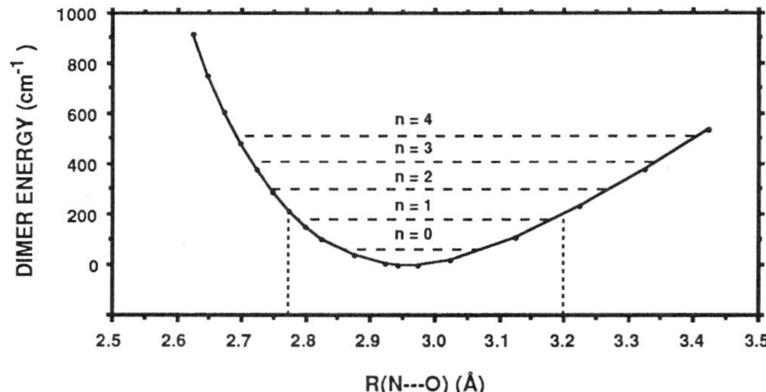

Figure 4.   Adiabatic Potential Energy Curve Calculated as a Function of R(N---O) for Conformer 1 of the Linear Formamide Dimer.

The potential energy curve in Fig. 4 is important to the phonon component of the Davydov Hamiltonian because it provides an estimate of the hydrogen bond stretching force constant (w), and it indicates the validity of the assumption of harmonic motion for a longitudinal acoustical phonon in a hydrogen-bonded polypeptide or polyamide[5,6,14]. The equilibrium w calculated for the curve in Fig. 4 is 20 N/m, which is in agreement with the value of 23 N/m calculated for the cyclic formamide dimer using the 4-31G basis set[33]. Both of these calculated w's are close to the value of 19.5 N/m used by Scott in his study of Davydov solitons in an α-helix[6].

Consistent with vibrational spectroscopic studies of hydrogen-bonded dimers in the vapor phase[8], the potential energy curve in Fig. 4 is anharmonic. This result implies that the assumption of harmonic motion for the longitudinal acoustical phonon is not a good one. *It would be worthwhile to investigate the sensitivity of Davydov soliton dynamics to the anharmonicity of longitudinal acoustical phonons in a hydrogen-bonded polypeptide or polyamide.*

In the course of calculating the adiabatic potential energy curve as a function of R(N---O) for Conformer 1, one also calculates the changes in internal coordinates and electronic charge distributions of the monomeric units in the dimer as a function of R(N---O). The calculated changes in bond length with respect to the equilibrium values at R(N---O) = 2.942 Å are plotted in Fig. 5(a); the calculated changes in atomic electron density with respect to the equilibrium values are plotted in Fig. 5(b).

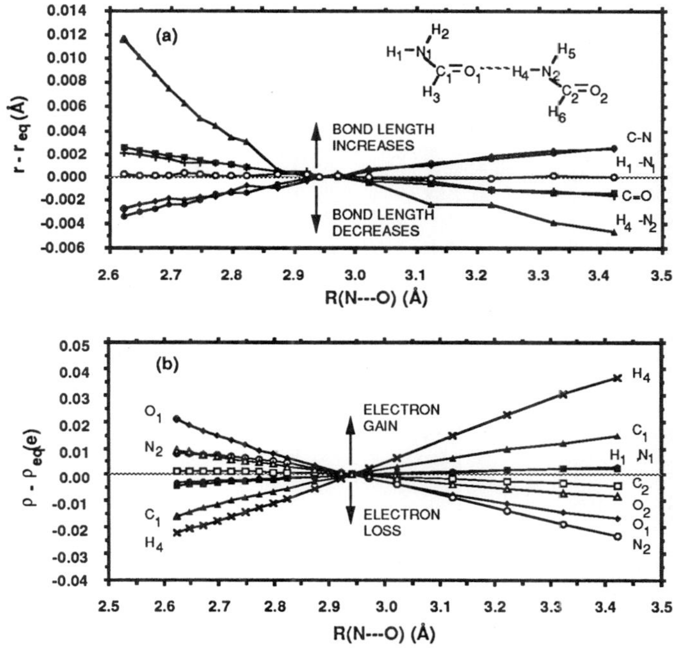

Figure 5. (a) Calculated Changes in Bond Lengths as a Function of R(N---O) for Conformer 1 of the Linear Formamide Dimer. (b) Calculated Changes in Atomic Charges as a Function of R(N---O) for Conformer 1 of the Linear Formamide Dimer.

The bond length changes for the N_1-H_2, C_1-H_3, N_2-H_5, and C_2-H_6 bonds are not shown in Fig. 5(a) because they were calculated to be unaffected by changes in R(N---O). The C-N bond in *both* monomeric units change by virtually the same amount as a function of R(N---O); the two curves defining these changes are denoted by the C-N label in Fig. 5(a). The C=O bond lengths in *both* monomeric units also change by virtually the same amount as a function of R(N---O); the two curves defining these changes are denoted by the C=O label in Fig. 5(a). Not surprisingly, the N_2-H_4 bond length changes much more with repsect to R(N---O) than does the N_1-H_1 bond length. The N_2-H_4 bond is the most sensitive to changes in R(N---O), with the C-N and C=O bonds following in decreasing sensitivity. The N_2-H_4 bond and C=O bonds *lengthen* with decreasing R(N---O), while the C-N bonds *shorten*. All of these results are consistent with the above comparison of calculated equilibrium geometries for the formamide monomer and linear dimers.

The changes in electron density on atoms H_2, H_3, H_5, and H_6 are not shown in Fig. 5(b) because they were calculated to be unaffected by changes in R(N---O). As discussed in the above analysis of the equilibrium electronic charge distributions in the formamide monomer and dimer, the charges on atoms C_1, O_1, H_4, and N_2 are the most affected by hydrogen bonding.

The curves in Figs. 5(a) and 5(b) support the statement made above in the section on equilibrium geometries and charges that a hydrogen bond stretching vibration in the linear dimer will be coupled most strongly to higher frequency vibrations involving the covalent bonds along the N_1-C_1=O---H_4-N_2-C_2=O_2 spine of the dimer.

Vibrational Exciton-Phonon Interactions

Resonance vibrational exciton-phonon coupling. According to Fig. 5(b), the atomic charge distributions for the C_1=O_1 and N_2-H_4 covalent bonds on either side of the hydrogen bond in the linear dimer are calculated to be the most sensitive to changes in R(N---O). The transition dipole moments, and hence the χ_2's and J's [see Eqs. (10)- (16)], for the mode invoving C_1=O_1 bond stretching [v(CO) or amide-I] and the N_2-H_4 bond stretching mode [v(NH)] should therefore be strongly coupled to the hydrogen bond stretching vibration. If one approximates the v(CO) and v(NH) modes as comprising only changes in the C_1=O_1 and N_2-H_4 bond lengths, then the transition dipole moments for these modes can be written as $(1/\mu_{CO})\partial\mu/\partial r_{CO}$ and $(1/\mu_{NH})\partial\mu/\partial r_{NH}$, where μ_{XY} is the reduced mass for the X-Y bond oscillator, μ is the electric dipole moment of the dimer, r_{CO} is the length of bond C_1=O_1, and r_{NH} is the legnth of bond N_2-H_4. It is clear from the calculated normal mode atomic displacements for the v(NH) mode in Fig. 3 that it can be approximated by a change in N-H bond length; however, it is not clear from the displacements for the amide-I mode in Fig. 3 that it can be approximated by a change in C=O bond length. In the case of calculating the *transition dipole moment* for the amide-I mode, the approximation is a good one because the dominant contribution to this transition dipole moment is from the change in the polar C=O bond[39]. In the case of calculating the non-resonance vibrational coupling for the amide-I mode (see below), one must be more careful in approximating the amide-I mode by a change in the C=O bond length.

The square of $\partial\mu/\partial r_{CO}$ for the C_1=O_1 bond was used to approximate the interaction between the transition dipole moments for the amide-I mode on the two monomeric units; in a similar fashion, the square of $\partial\mu/\partial r_{NH}$ for the N_2-H_4 bond was employed to approximate the interaction for the v(NH) mode. These approximations result in calculations of |J| and χ_2 for the modes that are upper limits on the actual values that would be calculated for the dimer. The $(\partial\mu/\partial r_{CO})^2$ and $(\partial\mu/\partial r_{NH})^2$ were evaluated numerically for Conformer 1 of the linear formamide dimer using electric dipole moments calculated with the 4-31G basis set. These terms are plotted as functions of R(N---O) in Fig. 6, and are independent of the adaiabatic or non-adaiabatic nature of the charge distributions and geometries calculated as a function of R(N---O). It is seen in Fig. 6 that $(\partial\mu/\partial r_{CO})^2 > (\partial\mu/\partial r_{NH})^2$ for the range of R(N---O)'s investigated, but $(\partial\mu/\partial r_{NH})^2$ is more sensitive to changes in R(N---O).

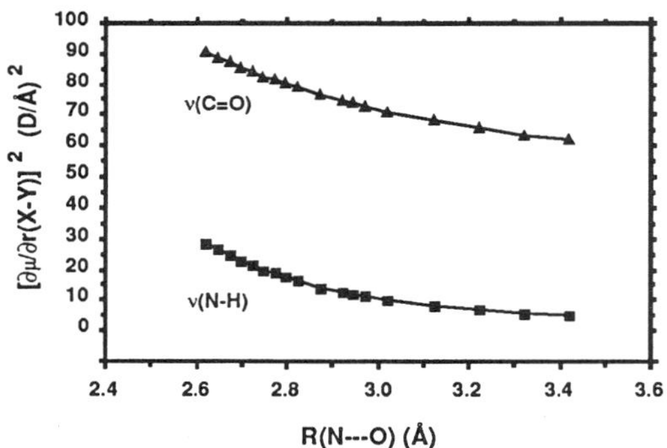

Figure 6.   Calculated Electric Dipole Moment Derivatives for the v(CO) and v(NH) Modes as Functions of R(N---O) for Conformer 1 of the Linear Formamide Dimer.

The upper limits on $|J|$ calculated for the amide-I mode, $|J(I)|$, are 11, 9, and 7 cm^{-1} for $R(N\text{---}O)$ equal to 2.722, 2.916, 3.113 Å. The upper limits on χ_2 calculated for the amide-I mode, $\chi_2(I)$, at these values of $R(N\text{---}O)$ are 3, 2, and 1 pN. The upper limits on $|J|$ calculated for the $\nu(NH)$ mode, $|J[\nu(NH)]|$, are 11, 6, and 3 cm^{-1} for $R(N\text{---}O)$ equal to 2.722, 2.916, 3.113 Å. The upper limits on χ_2 calculated for the $\nu(NH)$ mode, $\chi_2[\nu(NH)]$, at these values of $R(N\text{---}O)$ are 8, 3, and 0.5 pN. Our calculated values of $|J(I)|$ and $\chi_2(I)$ are in good agreement with the values of 7.8 cm^{-1} and 1 pN used by Scott in his study[6]. The calculated magnitudes of $|J(I)|$ and $|J[\nu(NH)]|$, and hence $\chi_2(I)$ and $\chi_2[\nu(NH)]$, are on a similar scale even though $(\partial\mu/\partial r_{CO})^2 > (\partial\mu/\partial r_{NH})^2$ because the reduced mass for the C=O bond is greater than that for the N-H bond. This finding is in agreement with the similar IR intensities calculated for the amide-I and $\nu(NH)$ modes of the cyclic formamide dimer[33]. Because the magnitudes of the derivatives, $|\partial\mu/\partial r_{CO}|$ and $|\partial\mu/\partial r_{NH}|$, increase with decreasing $R(N\text{---}O)$, $|J(I)|$, $\chi_2(I)$, $|J[\nu(NH)]|$, and $\chi_2[\nu(NH)]$ are likewise dependent on $R(N\text{---}O)$. *The effect of this dependence on the dynamics of Davydov solitons has not been investigated, and it may be worthwhile to do so.* It is also noted that $|J[\nu(NH)]|$ and $\chi_2[\nu(NH)]$ are calculated to be more sensitive to changes in $R(N\text{---}O)$ than are $|J(I)|$ and $\chi_2(I)$. This finding, along with the calculated magnitudes of $J[\nu(NH)]$ and $\chi_2[\nu(NH)]$, suggests that *one may want to include the $\nu(NH)$ mode in theoretical studies of the dynamics of Davydov solitons in polypeptides and polyamides (see below).*

Non-resonance vibrational exciton-phonon coupling. If one approximates the amide-I and $\nu(NH)$ modes as comprising only changes in the $C_1=O_1$ and $N_2\text{-}H_4$ bond lengths of the linear formamide dimer, then $\chi_1^+(I)$ for the left monomeric unit can be written as $(h/4\pi\omega_I\mu_{CO})\partial k_{CO}/\partial R(N\text{---}O)$, and $\chi_1^-[\nu(NH)]$ for the right monomeric unit as $(h/4\pi\omega_{\nu(NH)}\mu_{NH})\partial k_{NH}/\partial R(N\text{---}O)$, where ω_j is the fundamental harmonic frequency for mode j, μ_{XY} is the reduced mass for the X-Y bond oscillator, k_{CO} is the harmonic force constant for the $C_1=O_1$ bond, and k_{NH} is the harmonic force constant for the $N_2\text{-}H_4$ bond [see Eqs. (5) and (6)]. The $\nu(NH)$ mode pictured in Fig. 3 for the dimer is clearly dominated by changes in the N-H bond lengths, while the amide-I mode pictured in Fig. 3 is composed of more than just changes in the C=O bond length, e.g., there is NH$_2$ in-plane scissoring and C-H in-plane bond bending. As a result, the approximation for $\chi_1^-[\nu(NH)]$ stated above should be better than that for $\chi_1^+(I)$. The validity of these approximations is discussed below. Also, remember in the following discussion that $\chi_1^+(\alpha)$ refers to the intramolecular vibrational mode α localized in the formamide monomeric unit on the *left side* of the hydrogen bond in Fig. 2, and $\chi_1^-(\alpha)$ to the mode localized in the unit on the *right side*.

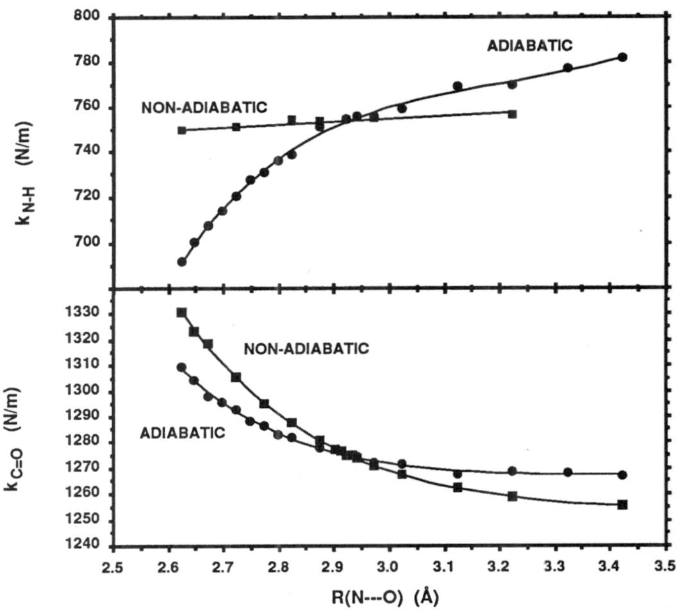

Figure 7. Calculated Stretching Force Constants for the $C_1=O_1$ and $N_2\text{-}H_4$ Bonds as a Function of $R(N\text{---}O)$ for Conformer 1 of the Linear Formamide Dimer.

The force constants, $k_{CO} = \partial^2 E_{dimer}/\partial r_{CO}{}^2$ and $k_{NH} = \partial^2 E_{dimer}/\partial r_{NH}{}^2$, associated with the adiabatic and non-adiabatic intermolecular potential energy curves were evaluated numerically for Conformer 1 of the linear formamide dimer using total dimer energies calculated with the 4-31G basis set. These force constants are plotted as functions of R(N---O) in Fig. 7. The distinction between the adiabatic and non-adiabatic force constants in Fig. 7 is seen to be very significant. The adiabatic force constants are considered to be more realistic principally because $\omega_{v(NH)}$ is observed to be more sensitive to R(N--O) than ω_I[8]. As expected from the inverse dependence of $r(N_2\text{-}H_4)$ on R(N---O) shown in Fig. 5(a), the adiabatic k_{NH} *decreases* with decreasing R(N---O) in Fig. 7. However, the adiabatic k_{CO} *increases* with decreasing R(N--O) in Fig. 7, despite the inverse dependence of $r(C_1\!=\!O_1)$ on R(N---O) shown in Fig. 5(a).

As shown in Fig. 8, the inverse relationship between the adiabatic and non-adiabatic k_{NH}'s and R(N---O) leads to *positive* values for $\chi_1{}^-[v(NH)]$, which are in agreement with experimental data[12]; the direct relationship between the adiabatic and non-adiabatic k_{CO}'s and R(N---O) results in *negative* values for $\chi_1{}^+(I)$, which are in disagreement with the positive ones estimated using vibrational spectroscopic data[9,10]. These findings are consistent with other *ab-initio*-SCF-MO studies that used only the N-H and C=O force constants to calculate $\chi_1{}^-[v(NH)]$[40] and $\chi_1{}^+(I)$[14,18-21], respectively. Thus, while the k_{NH} approximation of $\chi_1{}^-[v(NH)]$ is definitely valid, the k_{CO} approximation of $\chi_1{}^+(I)$ is very suspect.

There are several ways of improving upon the k_{CO} approximation of $\chi_1(I)$. One way is to calculate the shift in the amide-I *normal mode frequency*, $h\omega_I/2\pi$, in response to a change in R(N---O), instead of only calculating the shift in k_{CO}. This improvement is significant because, as stated above, the amide-I normal mode vibration is not completely composed of a C=O bond stretch, and experimental studies always measure normal mode vibrations. Another way is to use a more realistic model of a hydrogen-bonded polyamide like a linear formamide trimer, or a linear formamide-N-methylacetamide-formamide chain. The reason for this improvement is that studies of delocalized modes like the amide-I generally require a trimer in which both the CO and NH bonds of the center amide molecule participate in hydrogen bonds[40]. A third way is to increase the sophistication of the *ab-initio*-SCF-MO procedure, for example, utilizing more extensive atomic orbital basis sets like 6-31G* or 6-31G**[16,17]. A fourth way suggested and discussed by Kuprievich in his contribution to these Proceedings is to perform a more extensive mapping of the potential energy surface for the dimer as a function of the appropriate intra- and intermolecular vibrational coordinates, and then use an anharmonic potential energy curve for the amide-I vibration to calculate $\chi_1(I)$.

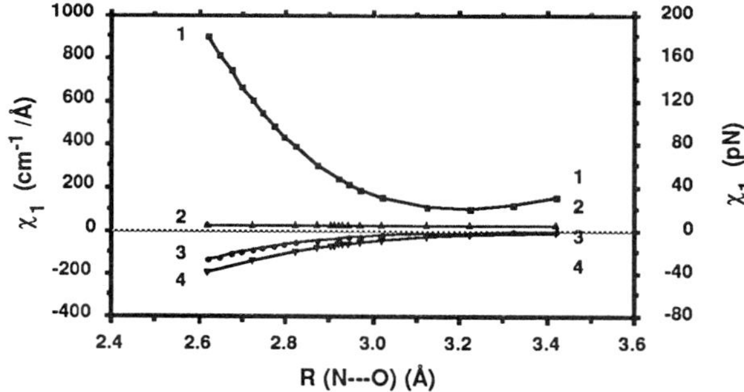

Figure 8. Calculated Non-resonance Vibrational Exciton-Phonon Coupling Terms for the $C_1\!=\!O_1$ and $N_2\text{-}H_4$ Bond Stretching Vibrations as a Function of R(N---O) for Conformer 1 of the Linear Formamide Dimer. Curve 1 is Adiabatic $\chi_1{}^-[v(NH)]$; Curve 2 is Non-adiabatic $\chi_1{}^-[v(NH)]$, Curve 3 is Adiabatic $\chi_1{}^+(I)$; and Curve 4 is Non-adiabatic $\chi_1{}^+(I)$.

In this paper, we investigated the improvement concerning the shift in the amide-I normal mode frequency by performimg vibrational normal mode analyses for equilibrium geometries of Conformers 2 and 3 of the linear dimer, which have different hydrogen bond lengths. The STO-3G basis set was used in these calculations. In the case of Conformer 2, the R(N---O) was calculated to be 2.822 Å, the ω_I and $\omega_{v(NH)}$ for the left monomeric unit were 2098 cm^{-1} and 4063 cm^{-1}, and ω_I and $\omega_{v(NH)}$ for the right monomeric unit were 2111 cm^{-1} and 3984 cm^{-1}. In the case of Conformer 3, the R(N---O) was calculated to be 2.799 Å, the ω_I and $\omega_{v(NH)}$ for the left monomeric unit were 2095 cm^{-1} and 4064 cm^{-1}, and ω_I and $\omega_{v(NH)}$ for the right monomeric were 2111 cm^{-1} and 3979 cm^{-1}. These results for Conformers 2 and 3 were then used to numerically evaluate the derivatives defining χ_1^+ and χ_1^- in Eqs. (8) and (9). The calculated values of $\chi_1^+(I)$ and $\chi_1^-[v(NH)]$, as well as $\chi_1^-(I)$ and $\chi_1^+[v(NH)]$, are given in Table 4, together with the calculated values of $\chi_1^+(I)$ and $\chi_1^-[v(NH)]$ determined using the force constant approximation, and experimentally estimated values of $\chi_1^+(I)$ and $\chi_1^-[v(NH)]$. The salient results of the STO-3G vibrational normal calculations are (1) a positive value of $\chi_1^+(I)$ is calculated that has a magnitude close to those of the experimentally-estimated ones[9,10], and (2) the value of $\chi_1^-[v(NH)]$ is in very good agreement with those calculated using the k_{NH} approximation, thereby providing more support for this approximation. It is disconcerting that there is such a large difference between the calculated and experimentally-estimated magnitudes of $\chi_1^-[v(NH)]$[12]. One reason for this difference is that spectroscopic data for a wide variety of hydrogen-bonded compounds were used to generate this empirical estimate, and if only hydrogen-bonded amides are considered, then the magnitude of $\chi_1^-[v(NH)]$ is much smaller[40].

Table 4. Calculated Non-resonance Vibrational Exciton-Phonon Coupling Terms for the Linear Formamide Dimer, Compared with Measured Values for Selected Hydrogen-bonded Molecular Systems

| | χ_1 (pN)[a] | | | |
| | amide-I | | $v(NH)$ | |
METHOD	χ_1^+	χ_1^-	χ_1^+	χ_1^-
Theory				
Ab-initio STO-3G: $k_{C=O}$[b]	-27	-	-	-
Ab-initio STO-3G: $k_{C=O}$[c]	-2	-	-	-
Ab-initio 4-31G: k_{X-Y}[d]	-7	-	-	+48
Ab-initio STO-3G: k_{N-H}[e]	-	-	-	+52
Ab-initio STO-3G: ω_{X-Y}[f]	+26	0	-5	+43
Experiment				
Selected molecular systems[g]	+62			
Acetanilide at 113 and 300K[h]	+35			
Selected molecular systems[i]				+339

[a]Non-resonance vibrational exciton-phonon coupling terms for the amide-I and $v(NH)$ modes (see Table 3 and Fig. 3). See text for the definition of χ_1^+ and χ_1^- terms.
[b]Ref. [19]. Determined by calculating the shift in $k_{C=O}$ with respect to R(N---O).
[c]Ref. [14]. Determined as in footnote b.
[d]Present work. Determined by calculating the shift in k_{X-Y} with respect to R(N---O).
[e]Present work. Determined by using the k_{N-H}'s calculated as a function of R(N---O) in Ref. [40].
[f]Present work. Determined by calculating the shift in normal mode frequency with respect to R(N---O).
[g]Ref. [9].
[h]Ref. [10], using the results in Refs. [9] and [11].
[i]Ref. [12].

It is encouraging that the initial disagreement between calculated[14,18-21] and experimental[9,10] values of $\chi_1^+(I)$ can be resolved by a more complete treatment of the amide-I normal mode. Additional theoretical studies are needed to confirm this result, particularly calculations employing a more sophisticated basis set like the split-valence 4-31G. Also, as stated above, it is important to calculate $\chi_1^+(I)$ for a more realistic model of a polyamide chain like the formamide-N-methylacetamide-formamide linear trimer. The value of $\chi_1{}^+(I)$ calculated for this model will likely be greater than than for the dimer because hydrogen bond strengths have been found to increase with increasing chain length[8,41,42].

Although one must be careful in utilizing the force constant approximation to calculate $\chi_1{}^+(I)$ and $\chi_1{}^-[\nu(NH)]$, the curves of $\chi_1{}^+(I)$ and $\chi_1{}^-[\nu(NH)]$ in Fig. 8 indicate that these non-resonance vibrational exciton-phonon couplings are very dependent on R(N---O). *The effect of this dependence on the dynamics of Davydov solitons in polypeptides and polyamides has not been investigated, and it may be worthwhile to do so.* Furthermore, the magnitude of $\chi_1{}^-[\nu(NH)]$ and its sensitivity to R(N---O) are greater than that for $\chi_1{}^+(I)$. This finding supports the suggestion made above that *one may want to include the $\nu(NH)$ mode in theoretical studies of the dynamics of Davydov solitons in polypeptides and polyamides.* Other intramolecular modes that one may want to include in these studies are the $\delta(NH)$, $\gamma(NH)$, and $\tau(CN)$ modes shown in Fig. 3. The $\gamma(NH)$, and $\tau(CN)$ modes would be of particular interest because of (1) the large difference in frequencies between the monomeric and dimeric modes, and (2) the large integrated IR absorption intensities for these modes. Furthermore, the $\gamma(NH)$ mode has been measured to be just as sensitive to changes in R(N---O) as the $\nu(NH)$ mode[13].

SUMMARY AND CONCLUSIONS

Ab-initio-SCF-MO theory was used to calculate the molecular parameters defining the dynamics of Davydov solitons arising from the excitation of either the amide-I or $\nu(NH)$ vibration in a hydrogen-bonded polypeptide or polyamide chain. Both the split-valence 4-31G and STO-3G basis sets were utilized in this study, and a hydrogen-bonded, linear formamide dimer was employed as a model of the chain. The theoretical analysis of the linear dimer consisted of several steps. First, equilibrium geometries and electronic charge distributions for the formamide monomer and the three conformers of the linear dimer were calculated. The second step comprised the computation of vibrational normal modes for the equilibrium geometry of Conformer 3 of the linear dimer. The third step consisted of calculating adiabatic and non-adiabatic potential energy curves as a function of R(N---O) for Conformer 1 of the linear dimer. These curves are important to determining w and the anharmonicity of the hydrogen bond stretching vibration. The fourth step involved the calculation of electric dipole moment derivatives for the amide-I and $\nu(NH)$ modes in Conformer 1 of the dimer as a function of R(N---O). These relationships are fundamental to the J and χ_2 terms for the amide-I and $\nu(NH)$ modes. The fifth and final step consisted of the calculation of the force constants and normal mode frequencies for the amide-I and $\nu(NH)$ modes in Conformer 1 of the dimer as a function of R(N---O). These relationships are basic to the χ_1 terms for the amide-I and $\nu(NH)$ modes.

The calculated parameters for the amide-I and $\nu(NH)$ modes are summarized in Table 5, together with those for the amide-I mode determined by Scott[6,10]. There is good agreement between Scott's[6,10] parameters and our calculated ones. This result is consistent with the success of *ab-initio*-SCF-MO theory in treating the vibrational spectroscopic properties of molecules in the ground electronic state.

The calculated effect of hydrogen bonding on the internal coordinates and electronic charge distribution of the formamide monomer suggests that a hydrogen bond stretching vibration in the linear dimer will be coupled most strongly to higher frequency vibrations involving the covalent bonds along the N_1-C_1=O_1---H_4-N_2-C_2=O_2 spine of the dimer. This finding is supported by a comparison of the vibrational normal modes calculated for the monomer and dimer. The specific vibrations of interest in regard to vibrational exciton-phonon coupling are the $\nu(NH)$, $\gamma(NH)$, $\delta(NH)$ and $\tau(CN)$ modes, as well as the amide-I mode.

As shown in Table 5, the $\nu(NH)$ mode is calculated to be more strongly coupled to the hydrogen bond stretching vibration than is the amide-I mode. Because the $\gamma(NH)$ mode is measured to be just as sensitive to changes in R(N---O) as the $\nu(NH)$ mode[13], the vibrational exciton-phonon coupling for the $\gamma(NH)$ mode should be comparable to that for the $\nu(NH)$ mode. Thus, more realistic Hamiltonians that describe the dynamics of Davydov solitons in polypeptides and polyamides should consider including (1) additional vibrational excitons coupled to longitudinal phonons, and (2) phonon-mediated couplings between these intramolecular vibrations on the same and different molecular units, and with single and multi-quantum excitations. It is noted that if the $\nu(NH)$ mode is included in the Davydov soliton Hamiltonian, then it may be necessary to also account for the Fermi resonance between this mode and the $\delta(NH)$ mode (amide-II mode)[43].

Other modifications in the Davydov soliton Hamiltonian suggested by the *ab-initio*-SCF-MO calculations for the linear formamide dimer are as follows. First, the anharmonic potential energy curve for hydrogen bond stretching indicates that it would be worthwhile to investigate the sensitivity of Davydov soliton dynamics to the anharmonicity of the longitudinal acoustical phonons in a hydrogen-bonded polypeptide or polyamide. Second, the dependence of ω_α, $J(\alpha)$, $\chi_1^+(\alpha)$, $\chi_1^-(\alpha)$ and $\chi_2(\alpha)$ on R(N---O) for a given normal mode α should be considered. Molecular dynamics simulations[44] may also want to account for the R(N---O) dependence of these terms.

Table 5.   Molecular Parameters for the Davydov Hamiltonian

Parameter	PRESENT WORK[a]		SCOTT[b]
	Amide-I	ν(NH)	Amide-I
$h\omega/2\pi$ (cm^{-1})	1882[c]	3917[c], 3759[c]	1660
$|J|$ (cm^{-1})	9[d]	6[d]	7.8
χ_1^+ (pN)	-7[d], +26[e]	-5[e]	+35
χ_1^- (pN)	0[e]	+48[d], +43[e]	+35
χ_2 (pN)	+2[d]	+3[d]	+1
w (N/m)	20[d]	20[d]	19.5

[a]Parameters for the amide-I and ν(NH) modes calculated using *ab-initio*-SCF-MO theory.
[b]Refs. [6] and [10]. These parameters were extracted from experimental vibrational and structural studies of polypeptides and the acetanilide molecular crystal.
[c]Calculated from a vibrational normal mode analysis of Conformer 3 of the formamide dimer using the 4-31G basis set. The two ν(NH) frequencies are for the anti-symmetric and symmetric modes in the right monomeric unit of the dimer.
[d]Calculated for the equilibrium geometry of Conformer 1 of the formamide dimer using the 4-31G basis set.
[e]Calculated from vibrational normal mode analyses for Conformers 2 and 3 of the formamide dimer using the STO-3G basis set.

Additional *ab-initio*-SCF-MO calculations should be performed for more realistic models of polypeptides and polyamides, e.g., the formamide-N-methylacetamide-formamide linear trimer, in order to better define the parameters for the amide-I and ν(NH) modes, and those of other high-frequency modes, coupled to low-frequency deformations of the hydrogen bond. These calculations should also investigate the sensitivity of the parameters to the sophistication of the atomic orbital basis set.

ACKNOWLEDGMENTS

It is a pleasure to acknowledge the inspiration provided by Prof. A.C. Scott and the many discussions with him, Dr. D.B. Chang, Prof. D. Christensen, Prof. P.L. Christiansen, Prof. O. Faurskov-Nielsen, Dr. F. Fillaux, Dr. V.A. Kuprievich, Dr. A.F. Lawrence and Mr. N. Østergård. Prof. O. Faurskov-Nielsen, Prof. P.L. Christiansen, and the Danish National Science Foundation are thanked for their support of work conducted at the University of Copenhagen and the Laboratory of Applied Mathematical Physics at the Danish Technical University.

REFERENCES

1.   D.E. Green, *Science* 181:583 (1973).
2.   D.E. Green, *Ann. N.Y. Acad. Sci.* 227:6 (1974).
3.   A.S. Davydov, *Studia Biophysica* 62:1 (1977).
4.   A.S. Davydov, "Biology and Quantum Mechanics," Pergamon, New York (1982); *Sov. Phys. Usp.* 25:898 (1983).
5.   A.S. Davydov, A.A. Eremko, and A.I. Sergienko, *Ukr. Fiz. Zh.* (Russ. Ed.) 23:983 (1978).
6.   A.C. Scott, *Phys. Rev. A* 26:578 (1982).
7.   L. Pauling, "The Nature of the Chemical Bond," Cornell University, Ithaca (1960); R.E. Dickerson and I. Geis, "The Structure and Action of Proteins," Benjamin/Cummings, Menlo Park (1969).
8.   C. Sandorfy, *in*: "Hydrogen Bonds", edited by P. Schuster, Springer, New York (1984).
9.   G. Careri, U. Buontempo, F. Galluzzi, A.C. Scott, E. Gratton, and E. Shyamsunder, *Phys. Rev. B* 30:4689 (1984).
10.   A.C. Scott, private communication (1987).
11.   H.J. Wasserman, R.R. Ryan, and S.P. Layne, *Acta Cryst.* C41:783 (1983).

12. A. Lautie', F. Froment, and A. Novak, Spectroscopy Lett. 9:289 (1976).

13. J. Bandekar and G. Zundel, *Spectrochim. Acta, Part A* 38:815 (1982).

14. A.F. Lawrence, J.C. McDaniel, D.B. Chang, B.M. Pierce, and R.R. Birge, *Phys. Rev. A* 33:1188 (1986).

15. See the papers in these Proceedings concerning temperature effects on Davydov solitons.

16. W.J. Hehre, L. Radom, P.v.R. Schleyer, and J.A. Pople, "*Ab-Initio* Molecular Orbital Theory," Wiley, New York (1986).

17. H.F. Schaefer, "The Electronic Structure of Atoms and Molecules. A Survey of Rigorous Results," Addison-Wesley, New York (1977).

18. V.A. Kuprievich and Z.G. Kudritskaya, *Acad. Sci. Ukranian SSR, Inst. for Theor. Phys.* Preprint ITP-82-63E (1982).

19. V.A. Kuprievich and Z.G. Kudritskaya, *Acad. Sci. Ukranian SSR, Inst. for Theor. Phys.* Preprint ITP-82-64E (1982).

20. N. Østergård, Master's Thesis, The Technical University of Denmark, Lyngby, Denmark, 1988.

21. B.M. Pierce, A.F. Lawrence, and D.B. Chang, *in*:"Spectroscopy of Biological Molecules," edited by A.J.P. Alix, L. Bernard, and M. Manfait, Wiley, New York (1985).

22. A.S. Davydov, "The Theory of Molecular Excitons", McGraw-Hill, New York (1962).

23. T.C. Cheam and S. Krimm, Chem. Phys. Lett. 107:613 (1984).

24. See the papers in these Proceedings concerning theoretical studies of Davydov solitons in hydrogen-bonded, peptide or amide systems.

25. J. Andzelm, M. Klobukowski, E. Radzio-Andzelm, *J. Comput. Chem.* 5:146 (1984).

26. Y. Sugawara, Y. Hamada, A.Y. Hirakawa, M. Tsuboi, S. Kato, and K. Morokuma, *Chem. Phys. Lett.* 50:105 (1980).

27. E. Hirota, R. Sugisuki, C.J. Nielsen, and G.O. Sorensen, *J. Chem. Phys.* 49:251 (1974).

28. K. Itoh and T. Shimanouchi, *J. Molec. Spectros.* 42:86 (1972).

29. R. Taylor, O. Kennard, and W. Versichel, *J. Am. Chem. Soc.* 105:5761 (1983).

30. M. Dreyfus and A. Pullman, *Theo. Chim. Acta* 19: 20 (1970).

31. P. Hobza, F. Mulder, and C. Sandorfy, *J. Am. Chem. Soc.* 104:925 (1982).

32. T. Ottersen, *J. Molec. Struct.* 26:365 (1975).

33. M.J. Wojcik, A.Y. Hirakawa, M. Tsuboi, S. Kato, and K. Morokuma, *Chem. Phys. Lett.* 100:523 (1983).

34. A.E. Reed, R.B. Weinstock, and F. Weinhold, *J. Chem. Phys.* 83:735 (1985).

35. J.C. Evans, *J. Chem. Phys.* 22:1228 (1954); 31:1435 (1959).

36. S.T. King, *J. Phys Chem.* 75:405 (1971).

37. B.M. Pierce, N. Østergård, P.L. Christiansen, and O. Faurskov-Nielsen, to be published.

38. M. Rasanen, *J. Molec. Struct.* 102:235 (1983).

39. T.C. Cheam and S. Krimm, *Chem. Phys. Lett.* 107:613 (1984).

40. T.C. Cheam and S. Krimm, *J. Molec. Struct.* 146:175 (1986).

41. J.F. Hinton and R.D. Harpool, *J. Am. Chem. Soc.* 99:349 (1977).

42. A. Pullman, H. Berthod, C. Giessner-Prette, J.F. Hinton, D. Harpool, *J. Am. Chem. Soc.* 100:3991 (1978).

43. A.T. Tu, "Raman Spectroscopy in Biology: Principles and Applications", Wiley, New York, 1982.

44. See contribution by A. Giansanti in these Proceedings.

17

ON *AB INITIO* ESTIMATIONS OF THE NONLINEARITY PARAMETERS

IN THE DAVYDOV MODEL

Niels Østergård

Laboratory for Applied Mathematical Physics
The Technical University of Denmark, Bldg. 303
DK–2800 Lyngby, Denmark

1. Introduction

In the Davydov soliton model of α–helix dynamics discussed in this book the parameters can be obtained from experimental data, but for the nonlinearity parameters (χ's) these experimental estimates are of a comparatively indirect and uncertain nature. Since some analytic and numeric studies show that soliton propagation in the discrete chain only is possible for $|\chi|$ within certain limits it is of interest to look for alternative ways of estimating χ's.

Three χ's will be discussed here: The forward coupling χ_1^+ and the backward coupling χ_1^- which are the derivatives of the excitation energy for the amide–I mode in a peptide group with respect to the intermolecular distance to its nearest neighbours to the right and to the left respectively along the hydrogen bonded spine, and the cross term coupling χ_2 which is the derivative of the coupling energy between the amide–I modes in two such adjacent groups with respect to their distance. Of these three χ's, χ_1^+ can be expected to be the most significant, followed by χ_1^-. In various studies of the Davydov model the assumptions $\chi_1^- = 0$ or $\chi_1^- = \chi_1^+$ have been used, usually with $\chi_2 = 0$ and the remaining parameter denoted χ ($\chi = \chi_1^+ + \chi_1^-$ or $\chi = (\chi_1^+ + \chi_1^-)/2$).

In 1982 Kuprievich and Kudritskaya in Kiev first calculated χ_1^+ using *ab initio* theory [1]. Since, this idea has been followed up by Pierce et al. [2–4] and by the author [5]. Also, all three groups contribute to the present book [6,7]. The numerical *ab initio* approach is based directly on quantum mechanics making no use of empirical data. Thus it is in a sense unbiased, but it leaves a number of decisions to be made in the exact procedure resulting in varying signs and magnitudes of the χ values obtained.

First the model system must be chosen; in the literature only the formamide dimer *in vacuo* has been investigated. This is the smallest conceivable system with two hydrogen bonded peptide groups. Next one must decide on a level of approximation in the *ab initio* calculations. The Hartree–Fock level (HF) with a minimal basis set (STO–3G) or a split valence basis set (4–31G) has been explored. The heart of *ab initio* calculations is the approximation of the electronic ground state wave function for frozen coordinates of the nuclei by minimization of the energy within a certain wave function ansatz. HF means that this ansatz is in the form of a Slater determinant; each molecular orbital entering into the determinant is expanded and optimized in a specific incomplete basis set [8]. The choice of model system and *ab initio* level has been governed by computing resources. Also in this study the formamide dimer will be used and the *ab initio* level is HF/4–31G. A third choice to be made is the conformation or geometry of the model system. Previous studies have focused on planar conformations which gives a

Davydov's Soliton Revisited, Edited by P.L. Christiansen and A.C. Scott
Plenum Press, New York, 1990

saving in computer time. In the present study the two formamide molecules are forced into a geometry that approximates the relative positions of hydrogen bonded peptide groups in α–helix proteins. This was also attempted in ref. [5] but the resulting geometry had a hydrogen bond that was about 1 Å too long ($R_{N-H\cdots O} = 3.8$ Å).

The χ_1 parameters are derivatives of the amide–I excitation energy E_0 with respect to the intermolecular distance R. In most calculations the amide–I mode has been approximated by a C=O bond stretch, and the intermolecular distance has been taken to be the hydrogen bondlength $R_{N-H\cdots O}$. These approximations are avoided in the present study by performing normal mode analyses to compute the excitation energies, and by letting R measure displacement parallel to the helix axis. In addition to χ_1^+ the parameters χ_1^- and χ_2 are also calculated in the present work by fitting the two calculated amide–I normal modes in the dimer to the result of a simple degenerate perturbation analysis.

The paper is structured as follows: Section 2 discusses normal mode analysis of molecules and molecular systems in general and section 3 presents the degenerate perturbation analysis for dimers. Next follows a section where the Davydov Hamiltonian is derived, thereby establishing a direct connection between this model and the data calculated in the *ab initio* approach. The α–helix–like model system geometry is described in section 5 and the results of the normal mode analyses are given in section 6. As suggested by Kuprievich [6] the effects of intrinsic amide–I anharmonicity on the excitation energy E_0 is included by fitting calculated conformational energies to a quartic potential function in section 7. Finally some supplementary calculations are presented in section 8 and the results are discussed in section 9.

2. Normal Mode Analysis of Molecules

Normal modes or normal vibrations in molecules are nothing but eigenmodes of the linearized dynamics of the nuclei, but to fix some notation and terminology it will be briefly described in this section. In the case of a molecule with a peptide group one of the normal modes is the amide–I mode, dominated by the C=O stretching vibration.

In the Born–Oppenheimer approximation the nuclei move in a potential defined by the Coulomb repulsions and attractions between the electrons in the ground state and the nuclei. If $\underline{r} = (x_1, y_1, z_1, \cdots, x_N, y_N, z_N)$ is the 3N vector of Cartesian displacement coordinates for the N nuclei the potential energy is, Taylor expanded to the 2nd order:

$$V = V_0 + \frac{1}{2}\underline{r}^t \underline{\underline{F}}\,\underline{r} \tag{2-1}$$

where $\underline{\underline{F}}$ is the 3N x 3N real symmetric force constant matrix. $\underline{r} = \underline{0}$ is taken to be the equilibrium configuration; thus there is no gradient term. The kinetic energy is:

$$T = \frac{1}{2}\underline{\dot{r}}^t \underline{\underline{M}}\,\underline{\dot{r}}$$

where $\underline{\underline{M}} = \text{diag}(m_1, m_1, m_1, \cdots, m_N, m_N, m_N)$ is the matrix of atomic masses. A harmonic solution $\underline{r}(t) = \underline{r}_0 \exp(i\omega t)$ to the dynamic equations $-\underline{\underline{F}}\,\underline{r} = \underline{\underline{M}}\,\underline{\ddot{r}}$ is most easily obtained by transforming to mass weighted coordinates $\underline{x} = \underline{\underline{M}}^{1/2}\underline{r}_0$ and introducing the mass weighted force constant matrix (MWFCM) $\underline{\underline{A}} = \underline{\underline{M}}^{-1/2}\underline{\underline{F}}\,\underline{\underline{M}}^{-1/2}$. Then the eigenmodes are found by solving the eigenvalue equation

$$\underline{\underline{A}}\,\underline{x} = \lambda \underline{x} \tag{2-2}$$

where $\lambda = \omega^2$. Using the 3N orthonormal solution vectors $\underline{x}_i$ as a basis for the normal mode coordinates q, the potential and kinetic energies are simultaneously diagonalized:

$$V = V_0 + \tfrac{1}{2} q^t \underline{\underline{\Lambda}} q \qquad , \qquad \underline{\underline{\Lambda}} = \mathrm{diag}(\lambda_1, \cdots, \lambda_{3N}) \; ;$$

$$T = \tfrac{1}{2} \dot{q}^t \underline{\underline{I}} \dot{q} \qquad , \qquad \underline{\underline{I}} = \mathrm{diag}(1, \cdots, 1) \quad .[1]$$

In the present context the *ab initio* calculations give the potential energy as a function of the coordinates of the nuclei. To perform a normal mode analysis two steps are carried out: (i): Minimize the potential energy to determine the equilibrium geometry. (ii): Calculate the 2nd derivatives making up the matrix $\underline{\underline{F}}$ in eq. (2–1) and solve the eigenvalue equation (2–2). In the present work the *ab initio* program Gaussian 86 by Pople et al. [9] was used; it has built–in an algorithm to calculate so–called analytic 2nd derivatives which makes it more efficient than programs using numerical 2nd derivatives calculated by varying each coordinate slightly. The program was run at the Amdahl VP1100 vector computer at the UNI·C computing center.

3. Normal Mode Analysis of a Dimer

A degenerate perturbation analysis.

Bringing two identical molecules (monomers) into close proximity (forming a dimer) they will perturb each other. If the monomer has the MWFCM $\underline{\underline{A}}$ the dimer will have

$$\underline{\underline{A}}_2 = \begin{pmatrix} \underline{\underline{A}} & \\ & \underline{\underline{A}} \end{pmatrix} + \epsilon \begin{pmatrix} \underline{\underline{D}} & \underline{\underline{C}} \\ \underline{\underline{C}}^t & \underline{\underline{E}} \end{pmatrix} = \begin{pmatrix} \underline{\underline{A}} + \epsilon \underline{\underline{D}} & \epsilon \underline{\underline{C}} \\ \epsilon \underline{\underline{C}}^t & \underline{\underline{A}} + \epsilon \underline{\underline{E}} \end{pmatrix}$$

where ϵ is a small perturbation parameter that will be set equal to 1 in the end. Let $\underline{x}$ be a non–degenerate eigenvector of $\underline{\underline{A}}$ with eigenvalue λ. Then $\underline{\underline{A}}_2$ will have two eigenvectors of the form

$$\underline{x}_2 = \begin{pmatrix} w_1 \underline{x} \\ w_2 \underline{x} \end{pmatrix} + O(\epsilon)$$

with eigenvalues

$$\lambda_2 = \lambda + \epsilon \lambda^{(1)} + O(\epsilon^2) \; .$$

Let $c = \underline{x}^t \underline{\underline{C}} \underline{x}$, $d = \underline{x}^t \underline{\underline{D}} \underline{x}$ and $e = \underline{x}^t \underline{\underline{E}} \underline{x}$. From the 1st order equation:

$$\begin{pmatrix} d & c \\ c & e \end{pmatrix} \begin{pmatrix} w_1 \\ w_2 \end{pmatrix} = \lambda^{(1)} \begin{pmatrix} w_1 \\ w_2 \end{pmatrix}$$

---

[1] When the normal modes are printed by the *ab initio* program they are normalized differently so that the matrix in the expression for T is $\mathrm{diag}(\mu_1, \cdots, \mu_{3N})$ where the μ_i are the reduced masses.

or, neglecting $O(\epsilon^2)$ in λ_2 and letting $\epsilon = 1$,

$$\begin{pmatrix} \lambda+d & c \\ c & \lambda+e \end{pmatrix} \begin{pmatrix} w_1 \\ w_2 \end{pmatrix} = \lambda_2 \begin{pmatrix} w_1 \\ w_2 \end{pmatrix} . \tag{3-1}$$

This is the eigenvalue equation for a system with two degrees of freedom. The two solutions are given the superscripts "+" and "−", respectively:

$$\lambda^+ = \lambda + \frac{d+e}{2} + R ,$$

$$\lambda^- = \lambda + \frac{d+e}{2} - R ,$$

$$\varphi^+ = \arctan \frac{w_2^+}{w_1^+} = \arctan \frac{\frac{e-d}{2}+R}{c} ,$$

$$\varphi^- = \arctan \frac{w_2^-}{w_1^-} = \varphi^+ + \frac{\pi}{2} , \tag{3-2}$$

where

$$R = \sqrt{\left(\frac{e-d}{2}\right)^2 + c^2} \geq 0 .$$

If the perturbation analysis to the lowest order presented above is valid, a normal mode analysis of the dimer should for each mode in the monomer show two modes that were orthogonal linear combinations of that mode in each molecule with frequencies $\omega = \sqrt{\lambda}$ and angles $\varphi = \arctan(w_2/w_1)$ given by eqs. (3-2). Limiting cases of such orthogonal linear combinations are: (i): Two localized modes $(w_1,w_2) = (1,0)$ and $(0,1)$ ($\varphi = 0$ and $\pi/2$) respectively. This is a trivial case of Anderson localization [10] and arises if $|c| << |(e-d)/2|$. (ii): A symmetric and an antisymmetric mode $2^{-1/2}(1,1)$ and $2^{-1/2}(1,-1)$ ($\varphi = \pi/4$ and $-\pi/4$) which results if $|c| >> |(e-d)/2|$. If the dimer is symmetric in the sense that the two molecules are symmetrically equivalent it follows that $e = d$ so that case (ii) is realized. This is not the case for formamide dimers, except for the cyclic dimer which is not treated in the present paper.

Given λ^+, λ^- and φ^+ as the result of a normal mode analysis of a dimer, one may recover $\lambda + d$, $\lambda + e$ and c by inverting eqs. (3-2):

$$\lambda + d = \frac{\lambda^+ + \lambda^-}{2} - \frac{\tan\varphi - \cot\varphi}{\tan\varphi + \cot\varphi} \frac{\lambda^+ - \lambda^-}{2}$$

$$\lambda + e = \frac{\lambda^+ + \lambda^-}{2} + \frac{\tan\varphi - \cot\varphi}{\tan\varphi + \cot\varphi} \frac{\lambda^+ - \lambda^-}{2}$$

$$c = \frac{\lambda^+ - \lambda^-}{\tan\varphi + \cot\varphi} . \tag{3-3}$$

These numbers constitute the MWFCM in eq. (3-1).

4. Derivation of the Davydov Hamiltonian

In the previous sections it was outlined how to compute a mass weighted force constant matrix (MWFCM) for e.g. the amide–I modes in a formamide dimer

$$\underline{\underline{A}}_2 = \begin{bmatrix} \lambda_1 & \\ & \lambda_2 \end{bmatrix} + \begin{bmatrix} d_1 & c_1 \\ c_1 & e_1 \end{bmatrix} . \tag{4–1}$$

Formally each λ is indexed with the number of the site (the peptide group) but since they all are the eigenvalue in the free molecule they are actually identical, $\lambda_1 = \lambda_2 = \lambda$. [2] The subscripts on the c's, d's and e's indicate the number of the hydrogen bond causing these perturbations; hydrogen bond number n connects peptide groups n and n+1.

Given the data in eq. (4–1) the best guess at a MWFCM for a trimer is

$$\underline{\underline{A}}_3 = \begin{pmatrix} \lambda_1 & & \\ & \lambda_2 & \\ & & \lambda_3 \end{pmatrix} + \begin{pmatrix} d_1 & c_1 & \\ c_1 & e_1 & \\ & & 0 \end{pmatrix} + \begin{pmatrix} 0 & & \\ & d_2 & c_2 \\ & c_2 & e_2 \end{pmatrix} = \begin{pmatrix} \lambda_1 + d_1 & c_1 & \\ c_1 & \lambda_2 + d_2 + e_1 & c_2 \\ & c_2 & \lambda_3 + e_2 \end{pmatrix}$$

and for a polymer chain (ignoring end effects):

$$\underline{\underline{A}}_N = \begin{pmatrix} \ddots & & & & \\ & \lambda_{n-1} + d_{n-1} + e_{n-2} & c_{n-1} & & \\ & c_{n-1} & \lambda_n + d_n + e_{n-1} & c_n & \\ & & c_n & \lambda_{n+1} + d_{n+1} + e_n & \\ & & & & \ddots \end{pmatrix}$$

This MWFCM describes a vibron system. The classical Hamiltonian is

$$H_{vib} = T + V$$

$$T = \tfrac{1}{2}\dot{\underline{q}}^t \underline{\underline{I}}\,\dot{\underline{q}} = \tfrac{1}{2}\sum_n \dot{q}_n^2$$

$$V = \tfrac{1}{2}\underline{q}^t \underline{\underline{A}}_N \underline{q} = \tfrac{1}{2}\sum_n \left[c_{n-1}q_{n-1} + (\lambda + d_n + e_{n-1})q_n + c_n q_{n+1} \right] q_n \tag{4–2}$$

where $\underline{q}$ is the vector of amide–I coordinates q_n for the individual sites n. Quantization is attained by the substitutions

$$q_n \rightarrow \hat{q}_n = q_n \; ,$$

$$\dot{q}_n \rightarrow \hat{p}_n = -i\hbar \frac{\partial}{\partial q_n} \; ,$$

$$H_{vib} \rightarrow \hat{H}_{vib}$$

---

[2] In a real α–helix varying side–chains along the chain may give varying λ's.

and so—called 2nd quantization is attained by changing to the representation with creation and annihilation operators

$$\hat{b}_n^\dagger = (2\hbar\omega_n)^{-1/2}\,(\omega_n\hat{q}_n - i\hat{p}_n)\ ,$$

$$\hat{b}_n = (2\hbar\omega_n)^{-1/2}\,(\omega_n\hat{q}_n + i\hat{p}_n)\ ,$$

where ω_n is chosen to be the squareroot of the diagonal element of $\underline{\underline{A}}_N$:

$$\omega_n = \sqrt{\lambda + d_n + e_{n-1}}\ . \tag{4-3}$$

This yields for eq. (4–2):

$$\hat{H}_{vib} = \sum_n \left(E_n(\hat{b}_n^\dagger\hat{b}_n + \tfrac{1}{2}) - J_n(\hat{b}_n^\dagger\hat{b}_{n+1} + \hat{b}_n\hat{b}_{n+1}^\dagger + \hat{b}_n^\dagger\hat{b}_{n+1}^\dagger + \hat{b}_n\hat{b}_{n+1}) \right) \tag{4-4}$$

where

$$E_n = \hbar\omega_n = \hbar\sqrt{\lambda + d_n + e_{n-1}}\ ,$$

$$-J_n = \frac{\hbar c_n}{\sqrt{\omega_n\omega_{n+1}}} \tag{4-5}$$

Now the phonon system is added:

$$\hat{H}_{ph} = \sum_n \left(\frac{\hat{P}_n^{\,2}}{2M} + \frac{W}{2}\,(\hat{U}_{n+1} - \hat{U}_n)^2 \right)$$

where $\hat{U}_n$ and $\hat{P}_n$ are the displacement and momentum operators for peptide group n in the phonon chain. M is the mass and W the spring constant. The quantity

$$R_n = U_{n+1} - U_n \tag{4-6}$$

is the deviation from equilibrium of the intermolecular distance between groups n and n+1 which corresponds to hydrogen bond n.

Assuming that c_n, d_n and e_n are functions of the phonon system variable R_n but independent of R_m, $m \neq n$, the vibron system parameters $E_n = E_n(R_{n-1}, R_n)$ and $J_n = J_n(R_n)$ will be functions of the R's as indicated, see eqs. (4–5). The dependence of J_n on the R's through $\sqrt{\omega_n\omega_{n+1}}$ is neglected here; if $\partial c/\partial R$, $\partial d/\partial R$ and $\partial e/\partial R$ are taken to be all of the same order, this is valid to the order c/λ. Taylor expanding E_n and J_n to the 1st order

$$E_n = E_0 + \chi_1^- R_{n-1} + \chi_1^+ R_n$$

$$-J_n = -J_0 + \chi_2 R_n \qquad (4\text{--}7)$$

one arrives at the Takeno Hamiltonian [11] (plus a zero point energy):

$$\hat{H}_{vib} + \hat{H}_{ph} = \sum_n \left[E_0 + \chi_1^-(\hat{U}_n - \hat{U}_{n-1}) + \chi_1^+(\hat{U}_{n+1} - \hat{U}_n) \right] (\hat{b}_n^\dagger \hat{b}_n + \tfrac{1}{2})$$

$$+ \sum_n \left[-J_0 + \chi_2(\hat{U}_{n+1} - \hat{U}_n) \right] \left[\hat{b}_n^\dagger \hat{b}_{n+1} + \hat{b}_n \hat{b}_{n+1}^\dagger + \hat{b}_n^\dagger \hat{b}_{n+1}^\dagger + \hat{b}_n \hat{b}_{n+1} \right]$$

$$+ \sum_n \left[\frac{\hat{P}_n^2}{2M} + \frac{W}{2}(\hat{U}_{n+1} - \hat{U}_n)^2 \right] . \qquad (4\text{--}8)$$

Note that in this treatment the vibron–phonon interaction $\hat{H}_{int}$ is a part of $\hat{H}_{vib}$. Finally to obtain the Davydov Hamiltonian one has to discard the non–resonant non–number–conserving coupling term $\hat{b}_n^\dagger \hat{b}_{n+1}^\dagger + \hat{b}_n \hat{b}_{n+1}$ that arose in the vibron Hamiltonian eq. (4–4); the corresponding diagonal term $\hat{b}_n^\dagger \hat{b}_n^\dagger + \hat{b}_n \hat{b}_n$ was here avoided by the choice of ω_n given in eq. (4–3).

Comparing the Taylor expansions of E_n and $-J_n$ (eqs. (4–7)) with their definitions (eqs. (4–5)) one finds

$$E_0 = \hbar\omega = \hbar\sqrt{\lambda + d + e}$$

$$-J_0 = \frac{\hbar c}{\omega} \qquad (4\text{--}9)$$

and

$$\chi_1^+ = \frac{\partial E_n}{\partial R_n} = \frac{\hbar}{2\omega_n} \frac{\partial d_n}{\partial R_n} = \frac{\hbar}{2\omega} \frac{\partial d}{\partial R} ,$$

$$\chi_1^- = \frac{\hbar}{2\omega} \frac{\partial e}{\partial R} ,$$

$$\chi_2 = \frac{\hbar}{2\omega} \frac{\partial c}{\partial R} . \qquad (4\text{--}10)$$

By performing *ab initio* normal mode analyses of the monomer and of the dimer at different values of R (eq. (4–6)) one may calculate all the parameters in the Hamiltonian eq. (4–8) except the peptide group mass M. First, from the analysis of the monomer one obtains λ and from an analysis of the dimer at R = 0 one obtains λ^+, λ^- and φ^+ for the amide–I mode, and from these one calculates $\lambda + d$, $\lambda + e$ and c (eqs. (3–3)). Then one can calculate E_0 and J_0 (eqs. (4–9)). Repeating the dimer analysis at a different value of R one can calculate $\partial d/\partial R$, $\partial e/\partial R$ and $\partial c/\partial R$ by numerical differentiation ($\Delta d/\Delta R$ etc). This also gives χ_1^+, χ_1^- and χ_2 (eqs. (4–10)). Finally, computing the conformational energy V at a third value of R one can calculate the force constant W

$$W = \frac{\partial^2 V}{\partial R^2} \qquad (4\text{--}11)$$

by numerical differentiation $\Delta^2 V/\Delta R^2$.

5. The α–helix–like Dimer Geometry

To make the formamide dimer model system as realistic as possible the two molecules are forced into a geometry that approximates two hydrogen bonded peptide groups in an α–helix. The planar peptide group conformation given by Corey and Pauling [12] and the dihedral angles at the C_α's $(\phi, \psi) = (132^O, 123^O)$ given in Dickerson and Geis [13] are used to construct a helix. This helix has 3.647 residues per turn and a rise along the axis of 5.503 Å per turn. The hydrogen bondlength is $R_{N-H\cdots O} = 2.890$ Å which is in the range 2.79 ± 0.12 Å given in ref. [12] determined from a variety of crystals. The 4 atoms $C=O\cdots H-N$ are roughly co–linear with the angles $\alpha_{O\cdots H-N} = 171^O$ within the range 180 ± 20° [12] and $\alpha_{C=O\cdots H} = 165^O$, but the two amide planes are rotated 64° relative to each other around this line. The planes are parallel with the helix axis within 0.5^O, and the $C=O\cdots N-H$ line is also roughly parallel with the axis.

To obtain a formamide dimer given two hydrogen bonded peptide groups in this helix the α–carbons are replaced by hydrogens labelled H_α (fig. 1) reducing the bondlengths to reasonable values but preserving the bond directions. However, in order to do the *ab initio* normal mode analysis the geometry must be in a minimum of the potential calculated with *ab initio*. Obviously the present geometry is far from an equilibrium for the formamide dimer, but freezing the positions of the H_α's in the optimization and in the normal mode analysis this problem may be overcome.

Freezing the H_α's in the normal mode analysis will of course affect the normal modes. To investigate this, similar calculations were performed for the formamide monomer. With free H_α's HF/4–31G *ab initio* calculations give an amide–I mode at 1900 cm^{-1}, compared to the measured IR frequency 1740 cm^{-1} in the vapour phase [14]. Note that HF calculations give systematically too high frequencies. In the calculated mode 82% of the kinetic energy is in the $C=O$ stretch, 7% in the NH_2 scissoring, and 7% in CH_α bending. Actually the NH_2 scissoring is rather an NH_α bend leaving only 0.5% for the NH_1 bend (see fig. 1b for atom labelling). Freezing the H_α's in the monomer the CH_α and NH_α bends are ruled out. The amide–I mode is found at 1897 cm^{-1} and has now 94% $C=O$ stretch and 4% NH_1 bend. A better model of a protein peptide group would be N–methyl–acetamide (NMA) where the C_α's are replaced by CH_3 groups rather than merely H atoms. Again CH_α and NH_α bends are ruled out. Here, the HF/4–31G amide–I mode is at 1876 cm^{-1} with 88% $C=O$ stretch, 6% NH_1 bend, and 2% $C_\alpha H$ bends. To conclude, the amide–I mode in formamide with frozen H_α's is at least as good an approximation to the mode in NMA, and thus presumably in polypeptides, as is the amide–I mode with free H_α's.

Returning to the dimer, the geometry was optimized with fixed H_α's. This caused the peptide planes to rotate slightly around their $H_\alpha - H_\alpha$ axes lengthening the hydrogen bond by 0.017 Å to $R_{N-H\cdots O} = 2.907$ Å. To calculate numerical derivatives $\Delta/\Delta R$ and $\Delta^2/\Delta R^2$ geometries with different intermolecular distances R are needed. These were obtained by displacing the H_α's in molecule 2 the length R in the direction of the helix axis followed by re–optimization with fixed H_α's (R is taken to be equal to zero in the previously described geometry). $R = -0.2, +0.2$ and $+0.4$ Å were

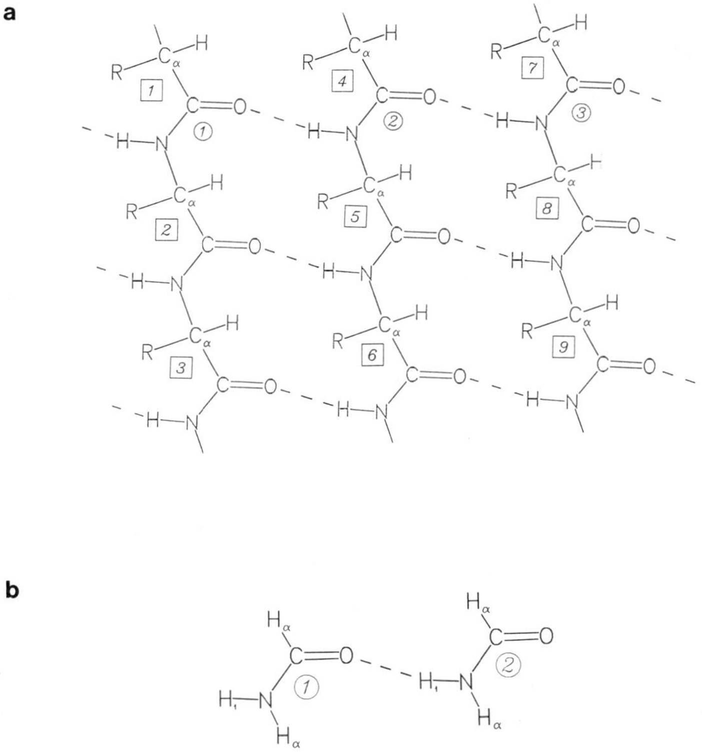

Figure 1. **a)** A cylindrical projection of a segment of an α–helix polypeptide chain with hydrogen bonds $(---)$, α–carbons (C_α) and side chains (R) indicated. The peptide groups C_α–NH–C=O–C_α are planar and the 4 atoms bonded to each C_α are in a tetrahedral arrangement with R and H pointing out from the helix. The helix axis is roughly parallel to the hydrogen bonds. Group numbers along the backbone are shown in boxes; numbers along one of the three spines are shown in circles. **b)** The formamide dimer used as a model for hydrogen bonded peptide groups in the α–helix. The C_α's have been replaced by hydrogens labelled H_α.

Table 1. Data for the three optimized α–helix–like formamide dimers.

R/Å	$R_{N-H\cdots O}$/Å	$R_{C=O_1}$/Å	$R_{C=O_2}$/Å	$E_{HF/4-31G}$/hartree
0.0	2.907	1.220272	1.221261	-337.375608999
0.2	3.078	1.220897	1.220751	-337.375409686
0.4	3.268	1.220865	1.220173	-337.374307664

Notes: $R_{C=O} = 1.215750$ Å in the optimized monomer.
1 hartree $= 4.35975\cdot10^{-18}$ J is the atomic unit of energy.

Table 2. Calculated amide–I modes.

R/Å	mode	$\bar{\nu}$/cm^{-1}	$\lambda/10^{30}$s^{-2}	φ/$^{\circ}$	non–amide–I
0.0	"+"	1886.92	0.126331	-33.98	0.25%
	"–"	1873.58	0.124551	$-33.95+90$	0.85%
0.2	"+"	1883.21	0.125835	-46.55	0.32%
	"–"	1872.11	0.124355	$-46.53+90$	0.54%
0.4	"+"	1882.94	0.125798	-53.57	0.29%
	"–"	1872.59	0.124418	$-53.56+90$	0.42%

Notes: $\bar{\nu} = \omega/2\pi c$; $\lambda = \omega^2$; $\varphi = \arctan(w_2/w_1)$ (see eq. (3–2)). The last column gives the percentage of the kinetic energy that didn't match the amide–I mode calculated for the monomer.

used, but in the optimization with $R = -0.2$ Å the rotation of the peptide planes in order to lengthen the short hydrogen bond was unacceptable. At $R = +0.2$ and $+0.4$ Å they rotated slightly in order to shorten the bond. Only the data with $R = 0$, $+0.2$ and $+0.4$ Å were used; in these cases the peptide planes rotated less than 3°. The resulting hydrogen bondlengths, C=O bondlengths and HF/4–31G energies are given in table 1.

6. Results from Normal Mode Analyses

In each of the three optimized formamide dimer geometries *ab initio* normal mode analyses were performed with frozen H_α's. The resulting amide–I modes are given in table 2. For the perturbation analysis developed in section 3 to be valid one must require $\varphi^- \approx \varphi^+ + 90^{\circ}$ and that the two modes are almost entirely amide I. The table shows an excellent agreement. Based on these data eqs. (3–3) give $\lambda + d$, $\lambda + e$ and c, see table 3a. The amide–I mode in the monomer found at $\bar{\nu} = 1897$ cm^{-1} gives $\lambda = 0.127692 \cdot 10^{30}$s^{-2}; using this value and the data (1) in table 3a one finds by eqs. (4–9)

$$E_0 = 3.703 \cdot 10^{-20}J \sim \bar{\nu}_0 = 1864 \text{ cm}^{-1} \; ,$$
$$-J_0 = -2.48 \cdot 10^{-22}J \sim \quad -12.5 \text{ cm}^{-1}$$

From the data in table 3a the numerical derivatives in table 3b may be computed. Using eqs. (4–10) and the values extrapolated to $R = 0$ we find:

238

Table 3. Calculated force constants and their derivatives.

a.	R/Å	$\lambda+d$	$\lambda+e$	c
		$10^{30}\,s^{-2}$	$10^{30}\,s^{-2}$	$10^{30}s^{-2}$
(1)	0.0	0.125775	0.125107	−0.000825
(2)	0.2	0.125055	0.125135	−0.000739
(3)	0.4	0.124905	0.125311	−0.000659

b.	R/Å	$\Delta d/\Delta R$	$\Delta e/\Delta R$	$\Delta c/\Delta R$
		$10^{36}s^{-2}m^{-1}$	$10^{36}s^{-2}m^{-1}$	$10^{36}s^{-2}m^{-1}$
(1)and(2)	0.1	−36.0	1.4	4.3
(2)and(3)	0.3	−7.5	8.8	4.0
extrapol.	0.0	−50.2	−2.3	4.4

$$\chi_1^+ = -7.55 \text{ pN},$$
$$\chi_1^- = -0.35 \text{ pN},$$
$$\chi_2 = 0.67 \text{ pN}.$$

The last Davydov Hamiltonian parameter that can be calculated is the spring constant W, eq. (4–11). Using the energies given in table 1 the result is

$$W = 0.02257 \text{ hartree}/\text{Å}^2 = 9.8 \text{ N/m}.$$

This can be compared to the value 20 N/m calculated by Pierce [7] at the same *ab initio* level but for a planar geometry, and to the values 13 N/m and 19.5 N/m used in various Davydov soliton studies.

7. Effects of Intrinsic Anharmonicity

Guided by the work of Kuprievich presented in this book [6] an attempt along the same lines has been made to include the effects of intrinsic amide–I anharmonicity on the excitation energy E_0 and its derivatives χ_1^+ and χ_1^-. Following Landau and Lifshitz [15] an oscillator with a Hamiltonian

$$\hat{H} = \tfrac{1}{2}\hat{p}^2 + \tfrac{1}{2}k\hat{q}^2 + k\alpha\hat{q}^3 + k\beta\hat{q}^4$$

will have the approximate 1st excitation energy

$$E_0 = \hbar\sqrt{k} + \hbar^2(-\frac{15}{2}\alpha^2 + 3\beta) \tag{7–1}$$

and thus

$$\frac{\partial E_0}{\partial R} = \frac{\hbar}{2\sqrt{k}}\frac{\partial k}{\partial R} + \hbar^2(-15\alpha\frac{\partial \alpha}{\partial R} + 3\frac{\partial \beta}{\partial R}) \tag{7–2}$$

To estimate α, β, $\partial\alpha/\partial R$ and $\partial\beta/\partial R$, HF/4–31G energies were calculated for geometries close to the optimized formamide dimers with R = 0, 0.2, and 0.4 Å, viz adding ζ_i times the amide–I eigenvector to the coordinates of molecule i, $\zeta_i = 0, \pm 0.05$ and ± 0.10 Å, i

Table 4. Parameters in the polynomial fit to the potential energy.

a.	$k/10^{30}s^{-2}$	$\alpha/10^{22}m^{-1}kg^{-1/2}$	$\beta/10^{44}m^{-2}kg^{-1}$
molecule 1	0.125763	−7.5461	93.259
molecule 2	0.125100	−7.0862	87.638
cross term	−0.000824	−	−

b.	$\frac{\partial k}{\partial R}/10^{36}s^{-2}m^{-1}$	$\frac{\partial \alpha}{\partial R}/10^{30}m^{-2}kg^{-1/2}$	$\frac{\partial \beta}{\partial R}/10^{54}m^{-3}kg^{-1}$
molecule 1	−49.9	8.2	−2.87
molecule 2	−2.1	−111.2	9.89
cross term	4.0	−	−

Table 5. Calculation of E_0 and χ_1^+, χ_1^- with contributions from intrinsic anharmonicity.

a.$(10^{-20}J)$	$\hbar\sqrt{k}$	$-\frac{15}{2}\hbar^2\alpha^2$	$+3\hbar^2\beta$	E_0
molecule 1	3.7398	−0.0475	+0.0311	3.7234
molecule 2	3.7300	−0.0335	+0.0292	3.7257

b.(pN)	$\frac{\hbar}{2\sqrt{k}}\frac{\partial k}{\partial R}$	$-15\hbar^2\alpha\frac{\partial \alpha}{\partial R}$	$+3\hbar^2\frac{\partial \beta}{\partial R}$	χ_1
molecule 1	−7.42	+0.10	−0.10	−7.42
molecule 2	−0.31	−1.31	+0.33	−1.29

= 1 and 2. This gives a total of 3 x 5 x 5 = 75 energies that were fitted with a polynomial to the 2nd order in R and to the 4th order in (ζ_1,ζ_2), giving 39 terms of order up to 6 (e.g. $R^2 \cdot \zeta_1^3 \cdot \zeta_2$), not counting terms linear in ζ_i that were omitted since $\zeta_1 = \zeta_2 = 0$ defines the geometry optimized with fixed R. The fit had an RMS error of $5 \cdot 10^{-6}$ hartree; the 75 energies fitted spanned a range of 0.045 hartree. Generally speaking the terms cubic or quartic in (ζ_1,ζ_2) involving *both* variables were insignificant. Thus the effect of anharmonicity on the off–diagonal parameters J_0 and χ_2 may be neglected. The relevant parameters from the fit are given in table 4. The first column corresponds to the $\lambda+d$, $\lambda+e$, c, and their derivatives, calculated by fitting the normal mode analysis to the perturbation analysis and presented in table 3 (compare to the data at R = O).

The resulting excitation energy E_0 by eq. (7–1) and χ_1^+ and χ_1^+ by eq. (7–2) are calculated in table 5. In the present calculation the effect of intrinsic anharmonicity is seen not to be as dramatic as in the data studied by Kuprievich [6].

8. Some Supplementary Calculations

In order to investigate individually the effects of some of the features of the calculations presented in the previous sections, a series of simpler calculations have been carried out. Except for the deviations explicitly noted, each calculation consists of 6 HF/4–31G energy calculations, viz. with R = 0 and 0.2 Å and with $\zeta_1 = 0$ and ± 0.05 Å ($\zeta_2 = 0$). This allows the calculation of $k_1 = \lambda+d = \partial^2 V/\partial\zeta_1^2$, at each R, followed by a calculation of $\chi_1^+ = (\hbar/2\omega)\partial k_1/\partial R$.

Case 1. Based on the 6 relevant points out of the 75 points used in the previous section, $k_1 = 0.126620 \cdot 10^{30}s^{-2}$ and $\chi_1^+ = -6.43$ pN. This is fairly close to the value χ_1^+

	J_0/cm^{-1}	χ_2/pN
ab initio (present work)	12.5	0.67
transition dipole [16]	7.8	1

$= -7.42$ pN obtained in the previous section, and is taken as a reference point for the following calculations.

Case 2. Calculating the energies of the same 6 geometries at the HF/STO–3G level, i.e. with a minimal basis set in the *ab initio* calculation, $k_1 = 0.159396 \cdot 10^{30}s^{-2}$ and $\chi_1^+ = -5.03$ pN. The agreement is reasonably good.

Case 3. Returning to HF/4–31G level but using the points with R = 0.2 and 0.4 Å : $\chi_1^+ = -1.12$ pN. This shows a strong dependence of χ on the intermolecular distance.

Case 4. Using R = 0 and 0.2 Å but omitting the geometry optimization of the dimer (each molecule is now in the geometry optimized for the monomer): $k_1 = 0.129306 \cdot 10^{30}s^{-2}$ and $\chi_1^+ = -1.63$ pN. This χ is significantly lower than the value in case 1. Most of this difference may be explained in the following way: Due to the intrinsic anharmonicity of the amide–I mode, and particularly of the C=O force constant, the calculated harmonic force constant is a function of the C=O bondlength. In case 1 this bondlength was allowed to vary when varying the hydrogen bondlength; see table 1 $(R_{C=O_1})$. All else equal, the variation of the C=O bondlength with the hydrogen bondlength combined with the C=O anharmonicity will give a variation of the C=O force constant with hydrogen bondlength, and thus a non–zero χ. This seems to be the main source for the χ–value in case 1; in case 4 this mechanism is ruled out by fixing $R_{C=O} = 1.215750$ Å independently of the hydrogen bondlength. Thus, a smaller χ value results.

Case 5. In a planar conformation with the same hydrogen bondlength but C=O$\cdots$N–H co–linear and again without optimization of the dimer geometry, $k_1 = 0.129325 \cdot 10^{30}s^{-2}$ and $\chi_1^+ = -1.32$ pN, quite similar to the previous case.

Case 6. Returning to the geometry from case 1 but approximating the amide–I mode with a C=O stretch (displacing both atoms by the same amount but in opposite directions): $k_1 = 0.132526 \cdot 10^{30}s^{-2}$, $\chi_1^+ = -6.6$ pN. The mass–weighted force constant is a little higher than for the full amide–I mode but the χ value is very similar.

9. Discussion

It is a major problem with *ab initio* calculations that there are no error bounds. Thus it is only by comparison to experiments that one can learn what level of approximation is sufficient for a particular property, and little is known about properties like χ's that essentially are 3rd derivatives of the energy surface (since force constants are 2nd derivatives). We find reasonable agreement between the 4–31G and the STO–3G basis sets (cases 1 and 2 in section 8). The results must converge when increasing basis sets. However, if the geometry used in the HF/STO–3G calculation had been optimized at the HF/STO–3G level rather than at the HF/4–31G level, the result might have been different. So far no calculations have been made to show the effect of expanding the ansatz to multideterminant electronic wavefunctions (i.e. using e.g. MP2 rather than HF).

A support for the validity of the present calculations can be found in the values of J_0 and χ_2 calculated in section 6, which are in good agreement with the transition dipole interaction values [16], see table 6. Note that ref. [16] assumes that the cross coupling is due to transition dipole interaction *only*. This is not the case for the *ab initio* values.

The sensitivity of the χ's on the model system geometry is disconcerting. Cases 3 and 4 in section 8 show by comparison to case 1 the strong dependence on hydrogen bondlength and C=O bondlength respectively. Case 5 may indicate that the angular arrangement of the two peptide groups is less important; however as case 4 this case ignores the variation of the C=O bondlength with hydrogen bond geometry. If the angular arrangement in fact *is* important, the discrepancy between the present result $\chi_1^+ = -7.4$ pN and the result $\chi_1^+ = +28$ pN from ref. [4] may be explained by the fact that ref. [4] obtains χ_1^+ by combining data calculated at different angles.

Compared to the experimental estimates of $\chi = \chi_1^+ + \chi_1^- = +62$ pN [17] and $+35$ pN [18,19], most *ab initio* estimates till now have been small and often negative. It seems that this is the result to be obtained from *ab initio* calculations at the HF level with the formamide dimer as model system. The outstanding questions are: (i): What will the result be using more accurate *ab initio* methods and/or larger model systems? (ii): Do the calcultions actually approach the same parameter that the experiments measure, and if not, which one is relevant to the Davydov model?

Acknowledgments

The present contribution owes its existence to many inspiring discussions with Victor Kuprievich and Brian M. Pierce during the NATO workshop and the MIDIT/ESF study centre, and to the support of my thesis advisors Peter Leth Christiansen, Alwyn C. Scott, Ole Faurskov Nielsen, and Sten Rettrup. The financial support from the Danish Technical Research Council under grant No. 16–4307.E and from EEC Science Programme under grant No. (89 100079/JU1) is acknowledged.

References

[1] V.A. Kuprievich and Z. Kudritskaya, Exciton description of interacting vibrations in a molecular chain, ITP–82–62E; On the interaction of vibrational excitons with acoustic phonons in a molecular chain, ITP–82–63E; Numerical evaluation of the exciton–phonon interaction parameters in the theory of Davydov solitons in polypeptide chains, ITP–82–64E, Institute for Theoretical Physics, Kiev (1982).

[2] B.M. Pierce, A.F. Lawrence, and D.B. Chang, A theoretical study of the interaction between amide–I and hydrogen bond stretching vibrations in hydrogen–bonded polypeptides, in "Spectroscopy of Biological Molecules", John Wiley & sons (1985) 87–89.

[3] A.F. Lawrence, J.C. McDaniel, D.B. Chang, B.M. Pierce, and R.M. Birge, Dynamics of the Davydov model in α–helix proteins: Effects of the coupling parameter and temperature, Phys. Rev. A33 (1986) 1188–1201.

[4] B.M. Pierce, N. Østergård, O.F. Nielsen, and P.L. Christiansen, *Ab initio* molecular orbital calculations of vibrational spectroscopic properties of the linear formamide dimer, in: A. Bertoluzza, C. Fagano, P. Monti (eds.): "Spectroscopy of Biological Molecules — State of the Art", Società editrice Esculapio, Rimini (1989).

[5] N. Østergård and P.L. Christiansen, "*Ab initio* calculations for hydrogen bonds in relation to biomolecular dynamics", Master's Thesis, Laboratory for Applied Mathematical Physics, The Technical University of Denmark, Lyngby (1988).

[6] V.A. Kuprievich, On the calculations of the exciton–phonon coupling parameters in the theory of Davydov solitons, in this volume.

[7] B.M. Pierce, in this volume.

[8] W.J. Hehre, L. Radom, P.v.R. Schleyer, and J.A. Pople, "*Ab initio* molecular orbital theory", John Wiley & sons (1986).

[9] M.J. Frisch, J.S. Binkley, H.B. Schlegel, K. Raghavachari, C.F. Melius, R.L. Martin, J.J.P. Stewart, F.W. Bobrowicz, C.M. Rohlfing, L.R. Kahn, D.J. Defrees, R. Seeger, R.A. Whiteside, D.J. Fox, E.M. Fleuder, and J.A. Pople, "Gaussian 86", Carnegie–Mellon Quantum Chemistry Publishing Unit, Pittsburgh PA (1984).

[10]   P.W. Anderson, Absence of diffusion in certain random lattices, Phys. Rev. 109 (1958) 1492–1505.

[11]   S. Takeno, Vibron solitons in one–dimensional molecular crystals, Prog. Theor. Phys. 71 (1984) 395–398.

[12]   R.B. Corey and L. Pauling, Fundamental dimensions of polypeptide chains, Proc. R. Soc. Lond. B141 (1953) 10–20.

[13]   R.E. Dickerson and I. Geis, "The structure and action of proteins", Benjamin/Cummings (1969).

[14]   J.C. Evans, Infrared spectrum and thermodynamic functions of formamide, J. Chem. Phys. 22 (1954) 1228–1234.

[15]   L.D. Landau and E.M. Lifshitz, "Quantum mechanics", Pergamon (1958).

[16]   A.C. Scott, Dynamics of Davydov solitons, Phys. Rev. A26 (1982) 578–595 and A27 (1983) 2767.

[17]   G. Careri, U. Buontempo, F. Galluzzi, A.C. Scott, E. Gratton, and E. Shyamsunder, Spectroscopic evidence for Davydov–like solitons in acetanilide, Phys. Rev. B30 (1984) 4689–4702.

[18]   H.J. Wasserman, R.R. Ryan, and S.P. Layne, Structure of acetanilide (C_8H_9NO) at 113K, Acta Cryst. C41 (1985) 783–785.

[19]   A.C. Scott, private communication (1987).

Section III

Temperature Stability

The report of my death was an exaggeration.
Mark Twain

The most contentious subject discussed during the workshop was the temperature stability of Davydov's soliton at physiological temperature (310K). There are two reasons for the difficulty in resolving this issue: i) Misunderstandings about the relevant parameter values, and   ii) Lack of an exact theory. The selection of appropriate parameter values in the Fröhlich Hamiltonian was discussed in our introductory remarks for Section I. These considerations are even more important in the present section.

There are three contributions in this section that find the solution to be unstable at 310K. The first of these is a numerical study by Lomdahl and Kerr which was previously published and is brought up-to-date in the present chapter [1]. The chapter by Förner and Ladik also reports numerical computations that are similar (but not identical) to those of Lomdahl and Kerr (see also Reference [2]). Finally the chapter by Schweitzer and Cottingham, also reported previously [3], describes an analytical study based on perturbation theory. This analysis permits them to calculate the soliton decay rate (Γ) caused by scattering into extended (non-soliton) states. All of these works approximate the alpha-helix with parameters of the "single channel model" listed in Table 1 of our introductory remarks to Section I. Thus the conclusions drawn must be reconsidered in the context of the "three channel model" parameters which are also listed in Table 1.

The numerical results of Förner and Ladik are more useful than those of Lomdahl and

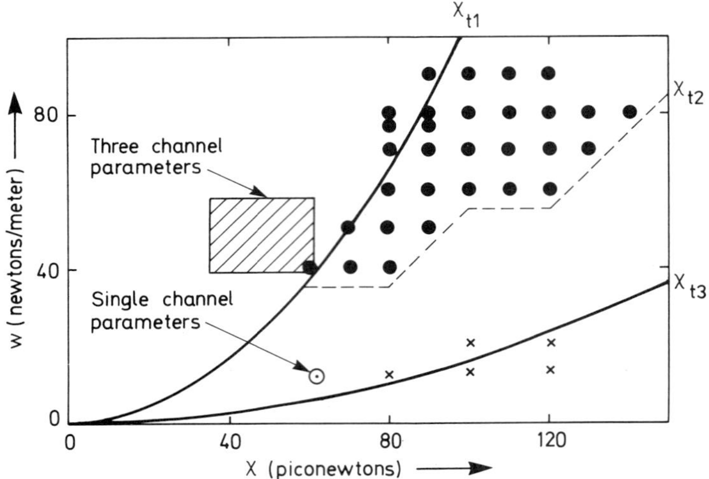

Figure 1. The $\chi - w$ parameter space of soliton propagation at 300K. See text for details.

Kerr because they explore a much wider range of parameter values. To appreciate this, refer to Figure 1 which is a modified version of Förner and Ladik's Figure 6. Here the black circles ($\bullet$) indicate parameters for which traveling solitons were observed and the crosses ($\times$) indicate parameters for stationary, self-trapped states. All of the results reported in the chapter by Lomdahl and Kerr were at the point ($\odot$) indicated as "single channel parameters." For $w \geq 40$ newtons/meter, Förner and Ladik indicate three threshold values of the exciton-phonon coupling parameter: χ_{t1}, χ_{t2}, and χ_{t3} (see Figure 1). It is helpful to consider these threshold values in some detail.

To proceed, notice that Förner and Ladik base their analysis upon the interaction energy operator indicated in Equation (2) of our introductory remarks to Section I. Thus for stationary (or slowly moving) self-trapped states, their dynamical equation is

$$(i\hbar\frac{d}{dt} - E_0)a_n + J(a_{n+1} + a_{n-1}) + \frac{\chi^2}{w}|a_n|^2 a_n = 0. \tag{1}$$

Measuring time in units of $\hbar/J$ and introducing the gauge transformation

$$a_n = \phi_n \, \exp[-it(E_0 - 2J)/J]$$

reduces Equation (1) to

$$i\dot{\phi}_n + \phi_{n+1} - 2\phi_n + \phi_{n-1} + \frac{\chi^2}{Jw}|\phi_n|^2 \phi_n = 0. \tag{2}$$

For $Jw/\chi^2 \gg 1$, an approximate, stationary solution of Equation (2) is

$$\phi_n \doteq \sqrt{\alpha/8} \, \text{sech} \, (\alpha n/4) \exp(it\alpha^2/16) \tag{3}$$

where $\alpha \equiv \chi^2/Jw$.

The highest threshold value of the exciton-phonon coupling parameter (χ_{t3}) indicates where in parameter space the self-trapped state does not propagate. It is evident that Equation (3) ceases to be a useful approximation for $\alpha > 4$ or

$$\chi_{t3} \doteq \sqrt{4Jw}. \tag{4}$$

This relation is indicated on Figure 1 in approximate agreement with the numerical results of Förner and Ladik.

The lowest threshold value of the exciton-phonon coupling parameter (χ_{t1}) indicates where the soliton forms. This question has been treated by Brizhik in Chapter 10 of these proceedings and in References [4]. For the initial condition which concentrates all of the amide-I vibrational energy at a single lattice point (this corresponds to the calculations of Förner and Ladik), the inequality $\chi^2/Jw \geq \pi^2/16$ must be satisfied for soliton formation [3]. Thus we expect

$$\chi_{t1} \doteq \sqrt{\pi^2 Jw/16}. \tag{5}$$

From Figure 1 we see that this condition is again in agreement with the numerical results of Förner and Ladik.

It is important to be aware that the value of χ_{t1} depends upon the initial conditions. Thus, as Brizhik has shown (see Chapter 10), if the initial conditions have the form of a soliton state, the corresponding threshold value for χ will be zero. In the context of Figure 1, the values of χ_{t1} might lie anywhere between those indicated and the left hand axis. We note further that one might expect an initial transition into a "quasimode" (in the sense defined by Arnol'd [5]) with the hyperbolic secant form of a soliton state.

Finally we turn to the intermediate threshold level, χ_{t2}. Förner and Ladik state that for $\chi > \chi_{t2}$ the exciton-phonon coupling "is large enough to allow the thermal fluctuations of the lattice to destroy the soliton." We do not have a simple derivation of the value of χ_{t2} corresponding to Equations (4) and (5). Understanding how it depends upon dipole-dipole coupling (J), temperature (T), and initial conditions should be an important objective of future research.

We are now in a position to consider which alpha-helix parameters should lead to formation of a long-lived soliton state. From Figure (1), the "single channel parameters" clearly do not. The "three channel parameters" appear at first glance to have only a minimal overlap at

$w = 40$ newtons/meter, and
$\chi = 60$ piconewtons.

However if we consider the fact that χ_{t1} will move to the left as the initial conditions become more "soliton-like," the three channel parameters fall well within the region where long-lived soliton states are expected.

The analytical results obtained by Schweitzer and Cottingham seem to be in approximate agreement with the numerical results of Förner and Ladik. For the single channel parameters, they find the decay time of a soliton-like initial state to be 2.6 picoseconds at 300K [6]. This is also in agreement with Lomdahl and Kerr. But with

$w = 52$ newtons/meter
$\chi = 62$ piconewtons
$J = 3.9 \times 10^{-23}$ joules

they find the lifetime of a soliton like initial state to be 78 picoseconds at 300K [6].

There are some differences of opinion about how long soliton-like initial conditions must survive before one can conclude stability. Förner and Ladik (see their Figure 5) judge a soliton to be stable after about 80 picoseconds. Schweitzer and Cottingham require about 500 picoseconds. We have been satisfied with less [7]. The point is: how do any of us decide? Perhaps we physical scientists should limit ourselves to the task of determining relevant facts and let the biological scientists decide whether and how the described soliton behavior might be useful.

We return now to the second reason for difficulty in finding agreement on soliton temperature stability: lack of an exact theory. This point is emphasized by Cruzeiro-Hansson in her contribution.

To gain perspective, let us briefly review some of the main features of each chapter. Davydov was the first to consider the effects of finite temperature on soliton propagation [8,9] and his chapter is a brief summary of that work. His approach is to introduce a new trial wave function $|\Phi\rangle$, defined in his Equations (10) and (11), in which the phonon part of the wave function depends explicitly upon both the site number (n) and the phonon mode number (q). In contrast to his standard trial wave function (which depends only upon n or q), this new trial function permits phase mixing between the exciton and phonon systems.[1] Starting with the total Hamiltonian operator, $\hat{H}$, the functional $\mathcal{H} = \langle \Psi | \hat{H} | \Psi \rangle$ is minimized under the constraint $\langle \Psi | \Psi \rangle = 1$ and thermally averaged. In the course of this analysis, it emerges that the dipole-dipole interaction energy should be multiplied by a Debye-Waller factor (DWF) of the form

$$\text{DWF} = \exp(-W). \tag{6}$$

Since at high temperature W is proportional to temperature, the DWF $\to 0$ as $T \to \infty$. This effect is physically reasonable because thermalized phonon states are expected to detune the exciton modes. Davydov's result is that the soliton width is inversely proportional to $(1 - T/T_0)$ where

$$T_0 = \frac{\pi}{2} \frac{\hbar}{k_B} \sqrt{\frac{w}{M}}. \tag{7}$$

At $T = T_0$, the soliton becomes infinitely wide and therefore identical to an extended state. For the three channel parameter values in Table 1 of the introductory remarks to Section I, T_0 lies between 100K and 120K.

The numerical analysis of Lomdahl and Kerr is based upon Davydov's standard wave function. Their approach is to append fluctuation and dissipation terms to the dynamic equations for phonon displacements [1]. Förner and Ladik also use the standard trial wave function, but they bring the phonon lattice to thermal equilibrium before introducing a soliton-like initial state [2]. The analytical approach of Cottingham and Schweitzer [3] is based upon perturbation theory and the standard trial wave function of Davydov.

The chapter by Satarić et al. also reviews and expands upon previously reported work [10]. Using Davydov's new (thermal) wave function and thermal averaging, they obtain approximate analytical expressions for the temperature dependence of soliton parameters including the soliton decay rate (Γ).

Bolterauer's contribution can be considered an extension of his Chapter 7 in Section I. After a discussion of the way thermal stability of a soliton depends upon the nature of the heat bath, he describes a thermodynamic variational principle which can be used to optimize a density operator ansatz. Contrary to the result of Davydov – see Equation (7) – he finds the soliton to be unstable at finite temperature against decay into delocalized states even

---

[1]In the terminology of Brown, Lindenberg and Wang, the standard wave function is called the "D_2 ansatz" while the new (thermal) one is called the "D_1 ansatz." See Chapter 5 for details.

when the lattice is assumed to be classical. This work should be carefully compared with the quantum Monte Carlo results presented by Wang et al. in Chapter 6. Again he calculates a thermal lifetime for a soliton that seems surprisingly short. This lifetime (.083 picoseconds) is a factor of twelve smaller than the value obtained with the exciton-phonon coupling (χ) assumed to be zero. Also the corresponding uncertainty in energy (1.2×10^{-21} joules) is twice as large as the entire exciton band width ($4J = 6.2 \times 10^{-22}$ joules).

Finally, Cruzeiro-Hansson et al. summarize previously published work which follows Davydov's thermal treatment [7], but they avoid any further assumptions or approximations by proceeding numerically rather than analytically. One novel feature in this analysis is a more realistic representation of the interaction energy operator $\hat{H}_{int}$. To see why this is necessary, consider a single channel (see Figure 1 of the introductory remarks to Section I) with the atomic sequence

$$\cdots \; \mathrm{H} - \mathrm{N} - \mathrm{C} \overset{\chi\prime}{=} \mathrm{O} \cdots \overset{\chi\prime\prime}{\mathrm{H} - \mathrm{N} - \mathrm{C}} = \mathrm{O} \; \cdots$$

Using the terminology of Kuprievich (see Chapter 11), $\chi\prime$ and $\chi\prime\prime$ are respectively the derivatives of the amide-I energies with respect to the lengths of the "right" and "left" hydrogen bonds. Thus

$$\hat{H}_{int} = \chi\prime \sum_n (\hat{u}_{n+1} - \hat{u}_n)\, \hat{B}_n^\dagger \hat{B}_n + \chi\prime\prime \sum_n (\hat{u}_n - \hat{u}_{n-1})\, \hat{B}_n^\dagger \hat{B}_n. \tag{8}$$

Writing phonon displacement operators ($\hat{u}_n$) in terms of phonon creation (annihilation) operators $\hat{b}_q^\dagger (\hat{b}_q)$, one recovers Davydov's Equation (5), but with his Equation (6) changed to

$$F(q) = \left[\frac{\hbar}{2M\omega_q}\right]^{1/2} \left[(\chi\prime - \chi\prime\prime)\,(\cos qa - 1) + i\,(\chi\prime + \chi\prime\prime)\,\sin qa\right]. \tag{9}$$

With this formulation, it is straightforward to compare Davydov's assumption ($\chi\prime = \chi\prime\prime$) with the approximation ($\chi\prime\prime = 0$) that is suggested by the *ab initio* calculations of Kuprievich, and this is done in Reference [7]. This general effect of the second ($\chi\prime\prime = 0$) assumption is to increase the range of $\chi\prime$ over which propagating soliton behavior is observed.

At this point, one generalization about the papers in this section can be made. The works of Lomdahl and Kerr, Förner and Ladik, and Schweitzer and Cottingham are based upon Davydov's standard trial wave function and, therefore, do not include the effects of the Debye-Waller factor indicated in Equation (6). The other contributions start from the new (thermal) wave function and do include the DWF. One effect of the DWF factor is to reduce the dipole-dipole coupling energies. As this effect is considered, it is appropriate also to include several additional couplings as was found necessary in Reference [3]. Thus even the improved parameters of the three channel model (see Table 1 in the introductory remarks to Section I) should be viewed as but a pale reflection of the real alpha-helix. Including more dipole interactions should tend to spread out the soliton, while including the DWF should reduce the soliton size.

Clearly the question of thermal stability of a soliton-like initial state must be considered unfinished business by the theoretical community. Future studies should attempt to include the following features.

1. Realistic parameter values including several dipole-dipole interactions.

2. The Debye-Waller factor.

3. The more realistic interaction energy operator indicated in Equations (8) and (9).

4. The phonon structure of a real alpha-helix [11].

5. The Takeno Hamiltonian rather than the Fröhlich Hamiltonian (see Chapters 3 and 4 of Section I).

We close these introductory comments by making brief references to works on soliton temperature stability that were not represented at the workshop. The first of these by Lawrence et al. [12] appeared at about the same time as the original publication of Lomdahl and Kerr [1] and was based on the same idea of appending fluctuation and dissipation terms to the dynamical equations for phonon displacement parameters. Unfortunately, this study uses a very early representation of the alpha-helix [13]; thus comparison of the results with later work is difficult. More recently Kadantsev, Lupichov and Savin have used Davydov's thermal analysis as a basis for numerical studies of autolocalized states on a cyclic chain [14] and on a chain with free ends [15]. They conclude that "thermal vibrations not only do not prevent the soliton transport of energy but ...become its necessary condition."

In the face of all these theoretical uncertainties, it is evident that the question of temperature stability of Davydov's soliton must also be investigated by experiment.

References

[1] P.S. Lomdahl and W.C. Kerr, Phys. Rev. Lett. **55**, 1235 (1985).

[2] H. Motschmann, W. Förner and J. Ladik, J. Phys.: Condens. Matter **1**, 5083 (1989)

[3] J.P. Cottingham and J.W. Schweitzer, Phys. Rev. Lett. **62**, 1792 (1989).

[4] A.C. Scott, Phys. Rev. A **26**, 578 (1982) [Errata: ibid. **27**, 2767 (1983)]; Physica Scripta **29**, 279 (1984).

[5] V.I. Arnol'd, Funct. anal. applic. **6**, 94 (1972).

[6] J.W. Schweitzer, private communication.

[7] L. Cruzeiro, J. Halding, P.L. Christiansen, O. Skovgaard and A.C. Scott, Phys. Rev. A **37**, 880 (1988).

[8] A.S. Davydov, Zh. Eksp. Teor. Fiz. **78**, 789 (1980); [Sov. Phys. JETP **51**, 397 (1980)].

[9] A.S. Davydov, phys. stat. sol. (b) **138**, 559 (1986).

[10] M. Satarić, Z. Ivić, Z. Shemsedini, and R. Žakula, J. Molec. Electronics 4, 223 (1988).

[11] O.H. Olsen, M.R. Samuelsen, S.B. Petersen and L. Nørskov, Phys. Rev. A **38**, 5856 (1988).

[12] A.F. Lawrence, J. C. McDaniel, D.B. Chang, B.M. Pierce and R.R. Birge, Phys. Rev. A **33**, 1188 (1986).

[13] J.M. Hyman, D.W. McLaughlin and A.C. Scott, Physica **3D**, 23 (1981).

[14] V.N. Kadantsev, Lupichov and Savin, phys. stat. sol. (b) **143**, 569 (1987).

[15] V.N. Kadantsev, Lupichov and Savin, phys. stat. sol. (b) **147**, 155 (1988).

18

THE QUANTUM THEORY OF SOLITONS WITH THERMAL VIBRATION TAKEN INTO ACCOUNT

A.S. Davydov

Institute for Theoretical Physics

Academy of Sciences of the Ukrainian SSR, Kiev

1   Introduction

Many investigations of autolocalized states—solitons—in molecular systems have used the classical description of molecular displacements from equilibrium positions at $T = 0$. In those cases quantum fluctuations of the equilibrium positions and thermal vibrations of molecule about new equilibrium position were not taken into account.

Only the quantum theory of molecular displacements from equilibrium positions can give a consistent description of the role zero point and thermal vibrations play in the creation of solitons in a molecular chain. This theory has been developed by the author [1] some time ago. Now I shall give a very brief description of this theory. As a model, we use the one-dimensional chain of N periodically repeated neutral molecules at sites na maintained in contact with a thermostat at temperature $T \neq 0$.

2   The stationary states of a quasi-particle in a one-dimensional chain

Stationary states of one extra electron or of one intermolecular vibration in the chain in the short-range approximation are described by the Hamiltonian

$$H = H_0 + H_{ph} + H_{int}, \tag{1}$$

where

$$H_0 = J \sum_{n=1}^{N} \left[2A_n^\dagger A_n - \left(A_n^\dagger A_{n+1} + \text{h.c.} \right) \right] \tag{2}$$

is the energy operator counted from the energy band bottom of a free quasi-particle with effective mass

$$m = \frac{\hbar^2}{2a^2 J}. \tag{3}$$

Davydov's Soliton Revisited, Edited by P.L. Christiansen and A.C. Scott
Plenum Press, New York, 1990

For a quantum description, it is convenient to express of the displacement, u_n, at the nth molecule from its equilibrium position, na, through the operators of creation, $\tilde{b}_q^\dagger$, and annihilation, $\tilde{b}_q$, of phonons by the formula

$$u_n = \left(\frac{a\hbar}{2MNV_0}\right)^{1/2} \sum_q |\xi|^{-1/2} \left(\tilde{b}_q + \tilde{b}_{-q}^\dagger\right) \exp(in\xi), \quad \xi = aq. \tag{4}$$

Here M is the mass of the molecule, and V_0 is the velocity of long-wave sound. The wave number, q, runs through N discrete values.

The energy operator of the short-range deformational interaction of a quasi-particle with the displacements has, in a linear approximation, the form

$$H_{int} = N^{-1/2} \sum_{n,q} F(q) A_n^\dagger A_n (\tilde{b}_q + \tilde{b}_{-q}^\dagger) \exp(in\xi), \tag{5}$$

$$F(q) = F^*(-q) = i\chi a \left[\frac{\varepsilon(q)}{2MV_0^2}\right]^{1/2} |\xi|^{-1} \sin \xi. \tag{6}$$

Here χa is the energy of the deformation potential. The energy operator of the longitudinal deformation energy of the chain,

$$H_{ph} = \sum_q \varepsilon(q) \tilde{b}_{-q}^\dagger \tilde{b}_q \tag{7}$$

where

$$\varepsilon(q) = \hbar V_0 |q| \tag{8}$$

is the energy of a phonon with a wave number q. Stationary states of the chain are described by the average of the energy

$$\mathcal{H} = \langle \Psi | \mathcal{H} | \Psi \rangle, \quad \langle \Psi | \Psi \rangle = 1. \tag{9}$$

In this expression, the wave function $|\Psi\rangle$ is defined by

$$|\Psi\rangle = \sum_n \prod_q \varphi_n(t) S_{nq}(t) A_n^\dagger |...\nu_q...\rangle \tag{10}$$

where

$$S_{nq}(t) \equiv \exp\left[\tilde{\beta}_{qn}^* \tilde{b}_q - \tilde{\beta}_{qn} \tilde{b}_q^\dagger\right] \tag{11}$$

is the unitary displacement operator. The functions $\tilde{\beta}_{qn}(t)$ are modulated plane waves

$$\tilde{\beta}_{qn} = \beta_{qn}(t) \exp(-in\xi). \tag{12}$$

The action of the unitary operator S_{nq} upon the operator $\tilde{b}_q$ and $\tilde{b}_q^\dagger$ leads to their displacement by complex numbers $\tilde{\beta}_{qn}$ and $\tilde{\beta}_{qn}^*$ because

$$b_q \equiv S_{nq}^\dagger \tilde{b}_q S_{nq} = \tilde{b}_q - \tilde{\beta}_{qn}. \tag{13}$$

These equations show that the interaction of a quasi-particle with the chain results in vibration of the molecules about the new equilibrium positions $\tilde{\beta}_{qn}$. These vibrations are characterized by the new creation and annihilation operators $(b_q^\dagger, b_q)$ of real phonons.

In the chain in thermal equilibrium with the thermostat at temperature θ (in energy units), the statistical average of (9) and others is reduced to replacing the quantum number ν_q of real phonons by their averages

$$\langle \nu_q \rangle = \left[\exp\left(\frac{\varepsilon(q)}{\theta}\right) - 1\right]^{-1}. \tag{14}$$

After calculation, the functional energy (9) is transformed to

$$\mathcal{H} \;=\; \sum_{n,q}\Big\{ J\left[2|\phi_{qn}|^2 - \{(\exp(-W_n)\phi_{qn}^{*}\phi_{qn+1} + \text{c.c.}\}\right] -$$
$$- N^{-1/2}F(q)|\phi_{qn}|^2(\beta_{qn} + \beta_{-qn}^{*}) + \varepsilon(q)\left[\langle\nu_q\rangle + |\phi_{qn}|^2|\beta_{qn}|^2\right]\Big\} \tag{15}$$

where the Debye-Waller factor W_n is defined by

$$W_n \;=\; \sum_q |\beta_{qn}|^2(1 + 2\langle\nu_q\rangle)(1 - \cos\xi)$$
$$\approx \tfrac{1}{2}\sum_q |\beta_{qn}|^2\xi^2(1 + 2\langle\nu_q\rangle). \tag{16}$$

The energy functional (15) describes both nonlocalized (bands) and autolocalized states. The autolocalized states are stationary states in which the distances or the orientation of the molecules change in some finite region of the chain, i.e. there is a local violation of translational symmetry. The localization region may occupy any part of the chain. Consequently, general translational symmetry is preserved. Therefore, the stationary states are characterized by the energy and the total momentum related to the movement of the localization region along the chain with a constant velocity that depends on the value of the wave number k.

The stationary nonlocalized states are also characterized by a certain value of wave number k, but in these states the probability of distribution of a quasi-particle is independent of n, and so the motion of a quasi-particle is absent. By superposing stationary nonlocalized states we can form nonstationary states (having no strictly defined energy) in which the probability of finding the particle is different from zero in a finite region that moves along the chain with group velocity. The size of the localization region in this state, however, grows continuously with time, i.e. the wave packet "smears".

The autolocalized states arising under the short-range interaction are described by nonlinear differential equations. We shall call them—*solitons*, to distinguish them from the autolocalized states first introduced in 1933 by Landau [2] and elaborated by Pekar [3] when describing the electron motion in ionic crystals. The latter states were called *polarons*, because they are caused by the long-range (coulomb) interaction of electrons with the field of electric polarization of a crystal which is described by longitudinal phonons. The properties of polarons are defined by integro-differential equations.

3   Smooth Autolocalized State—Single Soliton

Now we investigate the case when the autolocalization occurs in finite region, $L > a$, of a long molecular chain [4]. Assume that the state of quasi-particle in such a chain is described by the function

$$\phi_n(t) = \Phi(n)\exp\left[i(kna - \omega t)\right] \tag{17}$$

with a fixed wave number, k, and the real amplitude $\Phi(n)$ different from zero only in some finite (not too small) region of the chain, i.e. at values of n that satisfy the condition

$$|\tilde{n} - n| \le N_0 \ll N. \tag{18}$$

In this case, the function $|\beta_{qn}|^2$ in the modulated wave (12) is also equal to zero beyond the region (18) and weakly dependent on n inside this region. Since inside the region (18) the function $|\beta_{qn}|^2$ is weakly dependent on n, we equate it to a constant value $|\beta_{q\tilde{n}}|^2$.

Near the site $\tilde{n}$ in the region in which the quasi-particle is mainly distributed, the energy functional (15) takes the form

$$\mathcal{H}_n = J\left[2|\phi_n|^2 - (\phi_n^*\phi_{n+1} + \text{c.c.})\exp(-W_n)\right] +$$
$$+ \sum_q \varepsilon(q)\left(\langle \nu_q\rangle + |\beta_{qn}|^2\right) -$$
$$- N^{-1/2}\sum_q F(q)|\phi_n|^2\left(\beta_{qn} + \beta_{-qn}^*\right), \tag{19}$$

where $n = \tilde{n}$. Since site $\tilde{n}$ can be in any part of the chain, henceforth we shall not indicate the tilde sign explicitly.

The amplitudes β_{qn} contained in (19) are weakly dependent on n in the excitation region, so we can use an approximate equality

$$\beta_{qn}\beta_{qn+1}^* \approx \beta_{qn}^*\beta_{qn+1} \approx |\beta_{qn}|^2. \tag{20}$$

We take expressions (16) and (19) into account to obtain the differential-difference equations

$$i\hbar\frac{\partial\phi_n}{\partial t} = J\left[2\phi_n - (\phi_{n+1} + \phi_{n-1})\exp(-W_n)\right] - N^{-1/2}\sum_q F(q)\phi_n(\beta_{qn} + \beta_{-qn}^*) \tag{21}$$

$$i\hbar\frac{\partial\beta_{qn}}{\partial t} = \varepsilon(q)\beta_{qn} - N^{-1/2}F^*(q)|\phi_n|^2 + J(\langle \nu_q\rangle + \tfrac{1}{2})\xi^2|\phi_n|^2\beta_{qn}\exp(-W_n). \tag{22}$$

The real phonons with wave numbers q correspond to molecular vibrations about new equilibrium positions with frequencies $|q|V_0$. The time changes in the amplitudes of displacements from equilibrium positions β_{qn} are defined by the velocity V of the movement of a quasi-particle along the chain,

$$i\hbar\frac{\partial\beta_{qn}}{\partial t} = |q|V\beta_{qn}. \tag{23}$$

Using (23), we find from (22) and its complex conjugate

$$\beta_{qn} = \frac{F^*(q)|\phi_n|^2}{N^{1/2}|\xi|\varepsilon_0\left(1 + \pi^{-1}\delta_n G(\xi) + s\right)} \tag{24}$$

where

$$s = V/V_0,$$
$$\delta_n = \frac{J}{\varepsilon_0}\pi|\phi_n|^2\exp(-W_n),$$

and

$$\varepsilon_0 = \hbar V_0/a \tag{25}$$

is the maximum phonon energy. In states in which the quasi-particle localization region considerably exceeds the distances between the molecules, $|\phi_n|^2 \ll 1$, and at low temperature $(\theta < \theta_D)$ Equations (21) and (22) become

$$i\hbar\frac{\partial\phi_n}{\partial t} = J\left[2\phi_n - (\phi_{n+1} + \phi_{n-1})\exp(-W_n)\right] + G(n)|\phi_n|^2\phi_n \tag{26}$$

in which the nonlinearity parameter $G(n)$ is

$$G = \frac{D}{(1-s^2)}, \quad D = \frac{\chi^2 a^2}{MV_0^2}, \tag{27}$$

and the Debye-Waller factor W_n is defined by

$$W_n = B_n\left[1 + 2\exp\left(\frac{\pi\varepsilon_0}{\theta}\right)\right]\exp(-W_n), \tag{28}$$

where

$$B_n \approx 7.43 \times 10^{-5} \left(\varepsilon_0 J^2\right)^{-1} \left(\frac{D}{1-s^2}\right)^4 \ll 1. \tag{29}$$

In the continuum approximation, (26) takes the form

$$\left(i\hbar\frac{\partial}{\partial t} - 2J\left[1 - \exp(-W)\right] + \frac{\hbar^2 \exp(-W)}{2m}\frac{\partial^2}{\partial x^2} + G|\phi|^2\right)\phi = 0. \tag{30}$$

Its solutions on the infinite chain can be written in the form

$$\phi(x,t) = \Phi(z)\exp\left[i(kz - \omega t)\right], \tag{31}$$

where $\Phi(z)$ is a smooth real function in the system of coordinates, z, moving with constant velocity V; thus

$$z = x - x_0 - Vt, \quad V = \frac{\hbar k}{m} < V_0. \tag{32}$$

It obeys the equation

$$\left[\Lambda + \frac{\hbar^2}{2\tilde{m}}\frac{d^2}{dz^2} + G\Phi^2(z)\right]\Phi(z) = 0, \tag{33}$$

with

$$\tilde{m} = m\exp(W). \tag{34}$$

A localized solution of (33), normalized on the infinite chain by the condition

$$a^{-1}\int \Phi^2(z)dz = 1, \tag{35}$$

is defined by the function

$$\Phi(z) = (\tfrac{1}{2}aQ)^{1/2}\operatorname{sech}(zQ) \tag{36}$$

with parameters

$$Q = \frac{G\exp(W)}{4aJ}, \quad \Lambda = -\frac{\chi^2 a^2 \exp(W)}{16J}. \tag{37}$$

The energy of the deformation of the chain in the region of the nth molecule is

$$E_{def} = \frac{D}{2a(1-s^2)}\int \Phi^4(z)dz = \frac{D^2 \exp(W)}{24\,J(1-s^2)}. \tag{38}$$

The energy of the quasi-particle in the potential field of the deformation well

$$U(z) = -G\Phi^2(z) = -\frac{D^2 \exp(W)}{8J(1-s^2)^2}\operatorname{sech}^2(zQ), \tag{39}$$

that moves with velocity V is defined by the expression

$$\hbar\omega = 2J\left[1 - \exp(-W)\right] + a^2(k^2 - Q^2)J\exp(-W). \tag{40}$$

The first term in (40) indicates a decrease in the resonance interaction caused by the fluctuations in intermolecular distances. The second term characterizes the energy gain by binding at the quasi-particle in the field (39).

To calculate the total energy $E(V)$, transferred by a moving soliton, we must add to (40) the energy spent to form the deformation. So we obtain

$$E(V) = \hbar\omega + E_{def} = 2J\left[1 - \exp(-W)\right] - \frac{D^2 \exp(W)}{48J(1-s^2)} + \tfrac{1}{2}mV^2\exp(W). \tag{41}$$

3.1 Smooth State at $2\theta < \varepsilon_0$ and Small Velocity of Solitons

At a temperature θ that is smaller than the maximum energy of phonons, ε_0, and small velocities ($s^2 \ll 1$), the soliton total energy (41) can be written as

$$E(V) = E(0) - \frac{D^2 \exp(W)}{48J} + \tfrac{1}{2} M_{sol} V^2, \quad \theta < \varepsilon_0 \tag{42}$$

where

$$E(0) = 2J \left[1 - \exp(-W)\right] \tag{43}$$

is the energy of a soliton at rest, and

$$M_{sol} = m \exp(W) \left[1 + \frac{a^6 \chi^4}{12\hbar^2 M^2 V_0^6}\right] \tag{44}$$

is its effective mass. The momentum transferred by a soliton is

$$P(V) = mV \exp(W) \left[1 + \frac{a^6 \chi^4}{12\hbar^2 M^2 V_0^6}\right]. \tag{45}$$

At zero temperature the function $W = 0$. With rising temperature, the gain of energy when the rest quasi-particle is bound with a deformation, is defined by

$$\Delta E = -\frac{D^2}{48J} + \frac{W(2J^2 - D^2)}{48J}, \quad W \ll 1. \tag{46}$$

If the inequality $2J^2 < D^2$ is fulfilled, increasing temperature ($\theta < \varepsilon_0$) causes increasing W and stabilizes the soliton. Its binding energy and effective mass increase, but in agreement with (36) the effective size ($1/Q$) decreases.

3.2 Smooth state at temperature $2\theta > \varepsilon_0$ and small velocity of solitons

At temperatures exceeding the maximum phonon energy ε_0, and small velocities ($s^2 \ll 1$), the nonlinear parameter G that enters in (30) takes the form

$$G \approx D \left[1 - \frac{2\theta}{\pi \varepsilon_0}\right]. \tag{47}$$

It decreases linearly with increasing temperature.

In the same approximation the Debye-Waller factor

$$W \approx \frac{D\theta}{\pi \varepsilon_0^2} \left[1 + \frac{2\theta}{\pi \varepsilon_0}\right] \tag{48}$$

increases. So, increasing temperature (at $2\theta < \varepsilon_0$) is accompanied by an increasing of the soliton size and a decreasing of its maximum amplitude.

References

[1] A.S. Davydov, Quantum theory of motion of a quasi-particle in a molecular chain with thermal vibrations taken into account, **phys. stat. sol. (b) 138**, 559, (1986).

[2] L. Landau, Über die Bewegung der Electronen in Kristallgitter, Phys. Z. Sowjetunion **3**, 664 (1933).

[3] C.I. Pekar, Autolocalization of an electron in crystals, Zh. Eksp. and Teor. Fiz. **16**, 335 (1946).

[4] A.S. Davydov, Solitony v Molekulyarnykh Sistemakh, Naukova Dumka, Second edition, 1987, Kiev; **Solitons in Molecular Systems**, D. Reidel Publishing Company, Dordrecht (1985).

19

DAVYDOV SOLITONS AT 300 KELVIN: THE FINAL SEARCH

P. S. Lomdahl

Theoretical Division
Los Alamos National Laboratory
Los Alamos, NM 87545

W. C. Kerr

Olin Physical Laboratory
Wake Forest University
Winston-Salem, NC 27109-7507

The original proposal by Professor A. S. Davydov of a soliton mechanism for localization and transport of energy along linear chain molecules provided the impetus for several research efforts which have explored the properties of these nonlinear entities in differing degrees of realism. The general conclusion from all of this work is that the nonlinear equations of motion which have been used to describe these systems have soliton-like solutions when they are solved in the deterministic limit. This limit corresponds to the absolute zero of temperature, because it ignores the influence of random thermal perturbations on the system. However, the questions of existence and importance of the Davydov soliton remain controversial when non-zero temperature effects are taken into account, because numerical simulations and theoretical calculations done by independent research groups have reached diametrically opposed conclusions.

Our 1985 paper[1] was the first to simulate thermal perturbations at biologically relevant temperature (300 K). Since publishing that paper, we have done simulations for collisions of phonon wave packets with Davydov solitons and have also taken into account the presence of multiple quanta of the high frequency oscillator field in the Davydov equations of motion. We present these results here. However, for the major question about the temperature effects on the Davydov soliton, our conclusions remain unchanged from Ref. 1. We feel that we can not improve upon what we said there, so we refer the reader to that paper for a detailed discussion.

As explained in our other contribution in this volume, the multi-quantum Davydov *Ansatz* is

$$| \psi(t) > = \left[\sum_n a_n(t) B_n^\dagger \right]^Q \exp\left[-\frac{i}{\hbar} \sum_j \left[\beta_j(t) p_j - \pi_j(t) u_j \right] \right] | 0 >. \tag{1}$$

We insert this into the time-dependent Schroedinger equation using the Froehlich Hamiltonian and obtain the following equations of motion for the functions appearing in (1):

$$m\ddot{\beta}_n = w\,(\beta_{n+1} - 2\beta_n + \beta_{n-1}) + Q\chi(|a_{n+1}|^2 - |a_{n-1}|^2), \tag{2a}$$

$$i\hbar\dot{a}_n = -J\,(a_{n+1} + a_{n-1}) + \chi(\beta_{n+1} - \beta_{n-1})a_n. \tag{2b}$$

To describe the interaction of the system with a thermal reservoir at temperature T, we have added a damping force and a noise force,

$$F_n = -m\Gamma\dot{\beta}_n + \eta_n(t), \tag{3}$$

to (2a) for the molecular displacements. We have taken the correlation function for the random force to be

$$<\eta_n(t)\eta_{n'}(0)> = 2m\Gamma k_B T \delta_{nn'}\delta(t) \tag{4}$$

(k_B is Boltzmann's constant). This extension converts (2a) to Langevin equations. The effect of the two terms in (3) is to bring the system to thermal equilibrium; we have verified numerically that over sufficiently long time intervals the mean kinetic energy satisfies

$$<\sum_n \frac{1}{2}m\dot{\beta}_n^2(t)> = \frac{1}{2}Nk_B T \tag{5}$$

($<\cdots>$ denotes time average). Eqs. (2) with (3) included still imply the conservation of the norm $<\psi(t)|\psi(t)>$. The number of high frequency quanta present is

$$<\psi(t)|\sum_n B_n^\dagger B_n|\psi(t)> = Q \tag{6}$$

Our equations involve the combination of (4) and (5), which are the classical fluctuation-dissipation relation, with (2), which are obtained quantum mechanically. The justification for doing this is that for parameter values near those appropriate for the α helix a quantum of the highest-frequency acoustic mode $\hbar\omega_{max}$ is around 100 K. If we solve the equations at 300 K, then the occupation numbers of *all* phonon modes are accurately given by the classical distribution, and in that situation (5) is valid. The use of (5) for temperatures below, say, 200 K would not be valid because of quantum corrections to the phonon occupation numbers, but such temperatures are not biologically relevant. This point was emphasized in our original paper.

Table I. Parameter values.

	Discrete	Continuum	α-helix
w (N/m)	5.0	13.0	13.0
J (cm^{-1})	20.0	31.2	7.8
χ (10^{-10} N)	0.75	0.48	0.62
χ^2/wJ	2.83	0.29	1.91
t_0 (10^{-13} s)	1.95	1.21	1.21

We have solved the set of stochastic differential equations in (2), (3), and (4) using techniques developed by Greenside and Helfand.[3] The solutions were done for a chain of 100 sites with periodic boundary conditions. The parameters used are given in Table I; in addition the mass was always taken to be 114 proton masses ($m = 1.904\times10^{-25}$ kg). The rate of spatial variation of the variables is determined by the ratio χ^2/wJ, which is given in the table. The quantity t_0 is $(m/w)^{1/2}$ and is the time unit used for the calculations. The only parameter not determined from other considerations is the damping rate Γ in (4). We have used the (dimensionless) value $\Gamma = 0.005$, which is chosen so that the lowest (non-zero) phonon on our 100 site chain is substantially underdamped.

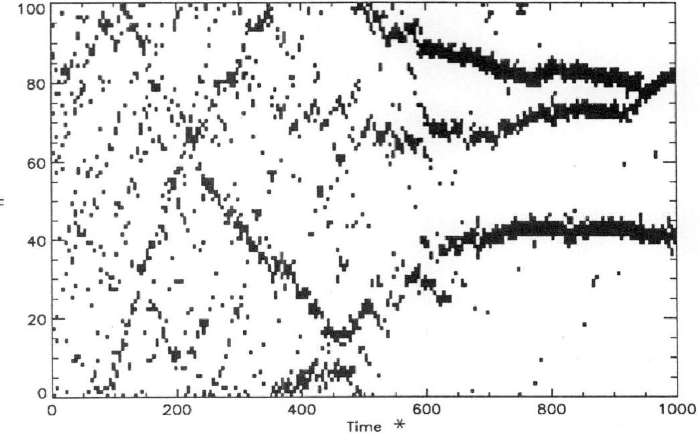

FIG. 1. Soliton detector for random initial conditions and $T = 0.01K$.

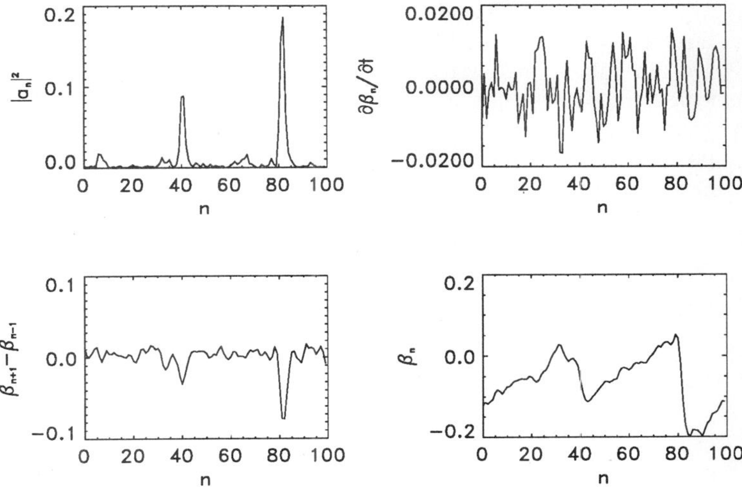

FIG. 2. Waveform graphs at time $t = 1000$ time units in Fig. 1.

This criterion is admittedly size dependent. However, since the self-trapping feature we want to study involves predominantly short wavelength phonons, we want to be sure that the damping force is not the major determinant of the evolution of those phonons. Furthermore, we have varied Γ from 0.0025 to 0.05 and have found no qualitative change in the results described here.

We present our results with certain diagnostics. One is "waveform" graphs: plots of $|a_n|^2$ and the discrete gradient $\beta_{n+1} - \beta_{n-1}$ as functions of n at a given t. A soliton is recognized as a maximum in $|a_n|^2$ and a minimum in $\beta_{n+1} - \beta_{n-1}$ occurring together. A second diagnostic is a "soliton detector": on the (t,n) plane, we mark those times and positions where both $|a_n|^2$ exceeds a certain level (chosen to be 0.02) and $\beta_{n+1} - \beta_{n-1}$ is negative. The temporal extent of the marked regions shows how long solitonlike entities can exist.

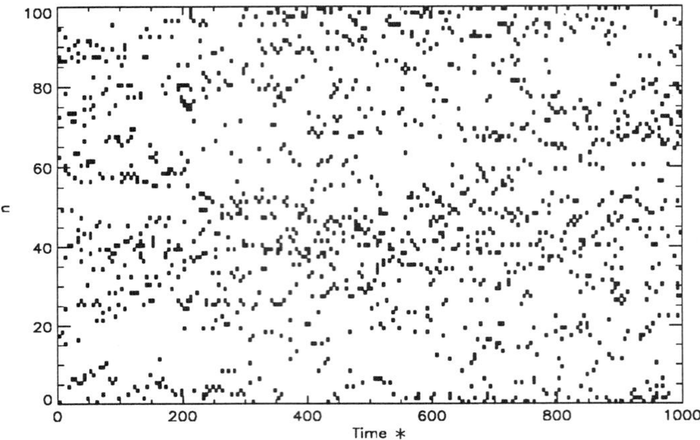

FIG. 3. Soliton detector at $T = 300$ K.

First we show an example which verifies that the equations (2) without thermal fluctuations do possess coherent, localized, propagating soliton solutions. Figure 1 shows the soliton-detector results at a very low temperature, $T = 0.02$ K, with the parameters labeled "discrete" in Table I and with random initial conditions. We see that several solitons are nucleated and move along the chain. The waveform graph in Fig. 2 shows the correlation of the maximum in $|a_n|^2$ and the minimum in $\beta_{n+1} - \beta_{n-1}$ which characterizes the soliton. This is clear evidence that solitons can form in this system at zero temperature. Figure 3 shows the effect of raising the temperature to 300 K with the other parameters remaining the same as for Figs. 1 and 2 and with random initial conditions. The nucleation and propagation processes that take place at low temperature are now seen not to occur. The random forces prevent the necessary correlations between the two fields from taking place.

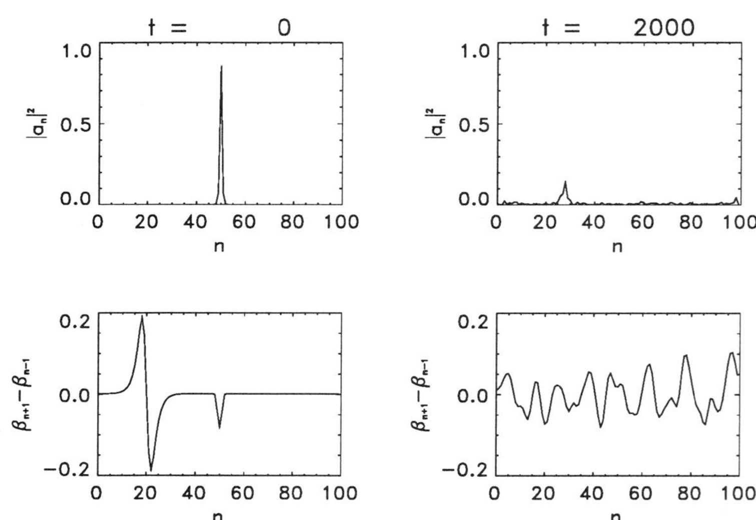

FIG. 4. Waveform graphs for collision of phonon wavepacket with Davydov soliton. Initial and final states are shown.

The previous figure shows that continual random displacements characteristic of thermal equilibrium prevent the soliton from forming when these displacements have a large enough magnitude. The following set of figures show a deterministic simulation (no random forces) in which a pre-existing Davydov soliton interacts repeatedly with a phonon wave-packet. Two different simulations were done. In the first the energy of the phonon wave-packet was 16 cm^{-1} which is approximately equal to the binding energy of the soliton (using α-helix parameters). For this case, the low energy *phonon wave-packet* is destroyed after about three collisions with the soliton. In the second simulation the *total energy* of the phonon wave-packet was approximately 425 cm^{-1}, which is equal to the average total energy *per particle* at $T = 300$ K. The initial condition for this case is shown in Fig. 4. The soliton detector picture (Fig. 5) shows 20 collisions with the wave-packet, out to $t = 2000$ time units. The black streak showing the soliton becomes more and more tenuous as these collisions proceed. The wave-form graphs for this case (Fig. 4) show that the *amplitude* of the soliton decreases from essentially 1.0 at $t = 0$ to about 0.15 at $t = 2000$. The normalization of the probability amplitudes is conserved, so these

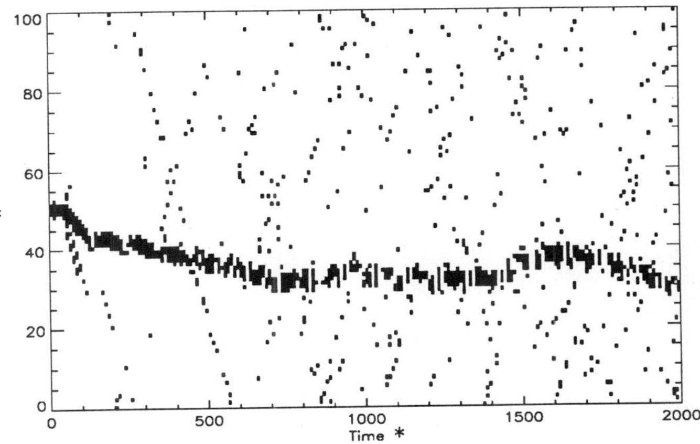

FIG. 5. Soliton detector for wavepacket-soliton collisions.

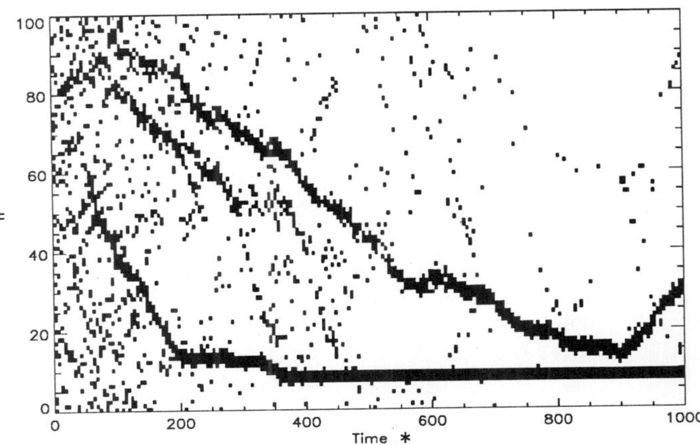

FIG. 6. Soliton detector of Davydov equations ($Q = 2$) with soliton initial condition and $T = 0.01$ K.

repeated collisions have distributed the initial sharply peaked distribution over the rest of the chain. The situation of thermal equilibrium at $T = 300$ K corresponds to repeated collisions with wave-packets with much higher energy than this value, so this simulation gives some insight into why the solitons are unstable at 300 K.

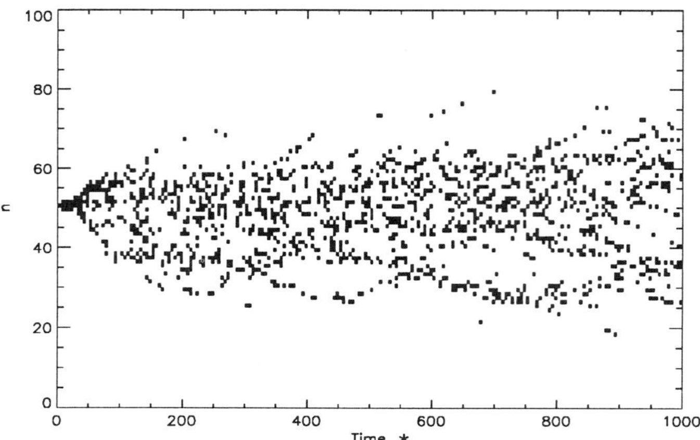

FIG. 7. Soliton detector of Davydov equations ($Q = 2$) with soliton initial condition and $T = 300$ K.

In our other contribution in this volume[2], we described a derivation of the Davydov equations which include multiple quanta of the high frequency oscillator system. We present here results for two quanta ($Q = 2$) for two temperatures $T = 0.01$ and 300 K. Fig. 6 shows a simulation at $T = 0.01$ K with random initial conditions. One sees that nucleation and propagation of solitons occurs here. (The fact that two solitons are nucleated in this run is coincidental and not related to the fact that the number of quanta used is $Q = 2$.) Figure 7 shows that raising the temperature to $T = 300$ K destroys the soliton present in the initial state. In Fig. 8 we show the case of $Q = 6$. It appears that increasing the number of quanta does not increase stability of the soliton.

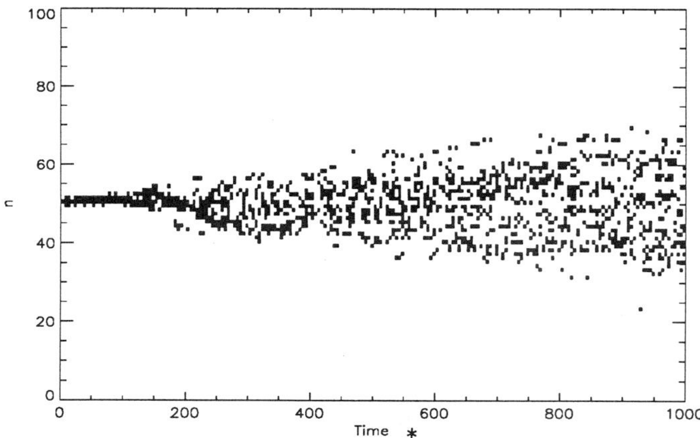

FIG. 8. Soliton detector of Davydov equations ($Q = 6$) with soliton initial condition and $T = 300$ K.

One purpose of this conference is to assess the current status of Davydov's proposal of the soliton mechanism for storage and transport of biological energy. We believe that recent developments, continuing the line initiated by our 1985 paper, conclusively show that the mechanism can not function as originally envisioned at the physiological temperatures which are of interest in biology. Our simulations are of course constrained by using the Davydov *Ansatz* and by introducing the temperature in a way which is valid only in the classical limit (which seems well satisfied for α-helix parameter values). The recent quantum-dynamical calculation of the soliton lifetime[4] and the quantum Monte Carlo (QMC) thermal equilibrium simulation of the *Hamiltonian*[5] are not constrained in these particular ways (although they have other limitations). (These calculations are also presented in this volume; see the contributions by Schweitzer *et al.* and by Wang *et al.* These two calculations have achieved an unusual synergism. The QMC calculations shows that solitons can not exist in an equilibrium system above 10 K, and the quantum-dynamic calculations shows that if they do form by some non-equilibrium mechanism, they last at most two picoseconds. Taken together these two calculations reach consistent conclusions covering both time and temperature variables and provide very strong evidence that the original soliton proposal does not work at biological temperature. The "crisis of bioenergetics" is still with us![6]

ACKNOWLEDGMENTS

The work at Los Alamos was performed under the auspices of the US DOE.

REFERENCES

1. P. S. Lomdahl and W. C. Kerr, Phys. Rev. Lett. **55**, 1235 (1985).
2. W. C. Kerr and P. S. Lomdahl, this volume.
3. H. S. Greenside and E. Helfand, Bell. Syst. Tech. J. **60**, 1927 (1981).
4. J. P. Cottingham and J. W. Schweitzer, Phys. Rev. Lett. **62**, 1792 (1989).
5. X. Wang, D. W. Brown and K. Lindenberg, Phys. Rev. Lett. **62**, 1796 (1989).
6. See Ann. N. Y. Acad. Sci. **227**, 108 (1974).

* **Editor's note:** The time units in Figures 1, 3, and 5–8 are t_0 from Table I. In all the figures presented here, this is 1.2×10^{-13} seconds (the "α-helix" parameters).

20

INFLUENCE OF HEAT BATH AND DISORDER ON DAVYDOV SOLITONS

Wolfgang Förner and János Ladik

Chair for Theoretical Chemistry and Laboratory of the
National Foundation for Cancer Research of the Friedrich-
Alexander-University Erlangen-Nürnberg, D-8520 Erlangen, FRG

ABSTRACT

With the help of the so-called $/D_2\rangle$ state vector ansatz the dynamics
of Davydov solitons is investigated. The soliton remains stable against
aperiodicity in the sequences of masses, spring constants, and coupling
constants. However, already a small aperiodicity in the sequence of dipole
interaction energies ($\rangle$ +2.5% $\overline{J}$) destroys the soliton. For values of the
parameters typical for polypeptide α-helices the soliton is stable up to
30K. For a temperature of T = 300K stable solitons can be found only for
spring constants $\overline{W} \geqslant$ 40N/m and coupling constants $\overline{X} \geqslant$ 50 pN. These
values of $\overline{W}$ are much larger than those usually accepted for proteins
($\overline{W}$ = 13 N/m).

I. INTRODUCTION

For the explanation of a wide variety of chemical and physical phenomena the introduction of non-linear forces turned out to be necessary,
e.g. a lattice bound by linear forces (harmonic forces) would have an
infinite heat conductivity /1/. In a channel near Edinburgh a non-disper-
sive localized wave packet has been observed as first example of a soliton /2/. Solitary solutions can occur only for non-linear wave equations,
since linear wave packets disperse rapidly. As some examples for the
introduction of soliton concepts in physics and chemistry let us mention
dynamics of ferro- and antiferromagnetic materials /3,4/, rotation around
carbon-carbon bonds in poly-ethylene /5/, phase changes in solids /6,7/,
dynamics of the sugar-phosphate backbone /8/ or the nucleotide bases /9/ in
DNA, and the spin-less charge transport in trans-polyacetylene /10/.
Many biological processes are associated with an energy transfer
through proteins, where this energy is released by hydrolysis of adenosine
triphosphate (ATP). The mechanism of this energy transport is not quite
clear /11/. As an alternative to electronic mechanisms /11/ one can assume
that the energy is stored as vibrational energy in the C = O stretching
mode (amide-I) of a polypeptide chain. Following Davydov's idea /12/ one
can take into account the coupling between the amide-I vibration and the
acoustic phonons in the lattice. Through this coupling non-linear terms
appear in the equations of motion, In this way the energy can be transported as a localized wave packet in solitary waves. Direct experimental
evidence for the existence of such solitons in proteins is still missing.
This is due to the complex structure of proteins, which makes such measure-

Davydov's Soliton Revisited, Edited by P.L. Christiansen and A.C. Scott
Plenum Press, New York, 1990

ments very difficult. However, in acetanilide crystals a substructure with
chains of hydrogen bonds similar to proteins is present. In low temperature
infrared and Raman spectra of these materials a new band in the amide-I
region appears. Up to now this band could only be explained with the help
of a model similar to the Davydov soliton concept in proteins /13/.

This paper deals with the dynamics of Davydov solitons in proteins.
In section II we discuss briefly the mathematical model for Davydov
solitons and the dependence of the dynamics obtained on the parameter
values. In section III the effects of aperiodicty in the chain are discussed. In section IV we investigate the influence of temperature on
Davydov solitons, while section V gives a summary of the results obtained.

II. MATHEMATICAL MODEL AND PARAMETER DEPENDENCES

The Hamiltonian used for this study is in the most simple form for
the system investigated by Davydov /12/, however, it is extended for the
possibility of disorder. Disorder is present in any protein due to the 20
different natural amino acids which form proteins. More sophisticated forms
of the Hamiltonian which incorporate more details of the protein structure
have lead qualitatively to the same results /14/.

$$
\hat{H} = \sum_n \left\{ E_o \hat{b}_n^+ \hat{b}_n - J_n (\hat{b}_n^+ \hat{b}_{n+1} + \hat{b}_{n+1}^+ \hat{b}_n) + \hat{p}_n^2/(2M_n) + (W_n/2)(\hat{q}_n - \hat{q}_{n-1})^2 \right.
$$
$$
\left. + X_n \hat{b}_n^+ \hat{b}_n (\hat{q}_n - \hat{q}_{n-1}) \right\} \tag{1}
$$

In equ. (1) $\hat{b}_n^+$ $(\hat{b}_n)$ are the usual boson creation (annihilation) operators
/15/ for the amide-I oscillators at sites n (see Fig. 1).

From infrared spectra the ground state energy of an isolated amide-I
oscillator can be deduced (E_o = 0.205 eV) /16/. Usually for all parameters
in equ. (1) site-independent mean values are used. The average value for
the dipole-dipole coupling between neighboring amide-I oscillators is
$\bar{J}$ = 0.967 meV /16/. The average spring constant of the hydrogen bonds
is taken usually to be $\bar{W}$ = 13 N/m /16/. In our preliminary paper /17/ we
used $\bar{W}$ = 76 N/m which is the spring constant for the hydrogen bonds in
the hydrogen carbonate dimer /18/. $\hat{p}_n$ is the momentum and $\hat{q}_n$ the positio.
operator of unit n. The average mass $\bar{M}$ is taken as that of myosin
($\bar{M}$ = 114 m_p; m_p = proton mass) /16/. The energy of the CO stretching
vibration in hydrogen bonds is a function of the length r of the hydrogen
bond ($E = E_o + \bar{X}r$) /19/. For $\bar{X}$ the experimental value is 62 pN /16/.
Ab initio calculations on formamide dimers usually lead to $\bar{X}$ = 30 - 50 pN
/20/.

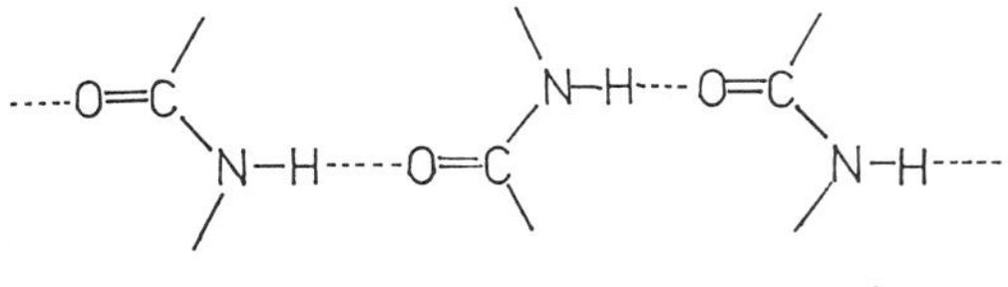

n-1   n   n+1

Fig. 1. Schematic picture of a hydrogen bonded channel in a protein

For the solution of the time dependent Schrödinger equation

$$\hat{H}|\psi\rangle = i\hbar\frac{\partial}{\partial t}|\psi\rangle \tag{2}$$

we use the displaced oscillator state ansatz (so called $|D_2\rangle$ - ansatz) of Davydov /12/

$$|\psi\rangle = \sum_n a_n'(t)\,\hat{b}_n^+\,\exp(-\hat{S}(t))\,|0\rangle \tag{3a}$$

$$\hat{S}(t) = \frac{-i}{\hbar}\sum_m (\hat{p}_m q_m(t) - \hat{q}_m p_m(t)) \tag{3b}$$

In this ansatz $q_n(t)$ is the expectation value of the position operator, $p_n(t)$ that of the momentum operator of the lattice unit n, $|0\rangle$ is the vacuum state, and $|a_n(t)|^2$ is the probability to find an amide I vibrational quantum at site n provided that $\sum_n |a_n|^2 = 1$.

However, in their recent work Brown et al. have shown /21,22/ that in the transportless case ($\bar{J} = 0$) the ansatz (3) leads to the correct time evolution of $q_n(t)$ but to an incorrect value of the phonon energy. The more sophisticated second ansatz ($|D_1\rangle$) of Davydov /12/ is shown to lead even to incorrect $q_n(t)$. Note that for $\bar{J} = 0$ the Hamiltonian can be solved exactly /21/.

Brown et al. /22/ have derived a non-linear density matrix equation of motion. However, this rather complicated theory was, at least to our knowledge only applied in linearized form to a two site system /23/. Using a time-dependent unitary transformation method on the Davydov Hamiltonian Mechtly and Shaw /24/ were able to derive equations of motion which are exact in the transportless limit ($\bar{J} = 0$). Their numerical simulations show qualitatively similar results as previous calculations /14/ using ansatz (3). However, soliton formation starts at a higher threshold value of $\bar{X}$ ($\bar{X}_{th} \approx 100$ pN) than in simulations using ansatz (3) ($\bar{X}_{th} \approx 40$ pN) /14/. Thus for qualitative studies one can use ansatz (3). However, one has to keep in mind its limitations. The equations of motion can be derived either by using the expectation value of $\hat{H}$ with (3) as classical Hamiltonian function /12,14/ or by quantum mechanical methods /25/. After the gauge transformation $a_n' = a_n \exp(-iE_0 t/\hbar)$ one obtains

$$i\hbar\,\dot{a}_n = -(J_n a_{n+1} + J_{n-1} a_{n-1}) + X_n(q_n - q_{n-1})a_n \tag{4a}$$

$$\dot{p}_n = W_{n+1}(q_{n+1} - q_n) - W_n(q_n - q_{n-1}) + X_{n+1}|a_{n+1}|^2 - X_n|a_n|^2 \tag{4b}$$

$$\dot{q}_n = p_n/M_n \tag{4c}$$

The complex equation (4a) was solved as a system of two coupled equations for the real and imaginary parts of a_n. The system of units eV for the energy, Å for lengths, and ps for the time proved to be suitable for numerical solution of (4). For each unit one has four coupled equations of first order in time. For this purpose a fourth order Runge-Kutta algorithm /26/ was used. With a time step size of 0.01 ps in the simulations described in this section and in section III the total energy was conserved up to $0.3 \cdot 10^{-5}$ eV (0.015 %). A possible imaginary part of the energy /17/ (which can occur due to numerical inaccuracies) was zero to an accuracy of $0.2 \cdot 10^{-17}$ eV. The norm $\sum_n |a_n(t)|^2 = 1$ was conserved up to $0.4 \cdot 10^{-6}$. Note that we use fixed chain ends and an inital excitation $a_n(o) = \delta_{n,N-1}$ (where N is the number of units, typically N = 200) was used. For the lattice $p_n(o) = q_n(o) = 0$ was applied in sections II and III.

For the purpose of comparison we give in Table 1 soliton (v) and sound (c)
velocities obtained for various average values of the parameters.

Table 1. Soliton velocities (v) and sound velocities (c) both in km/s
as functions of average parameter values in the Hamiltonian
(one parameter varied, all others kept at standard values:
$\overline{M}$ = 114 m_p, $\overline{J}$ = 0.967 meV, $\overline{X}$ = 62 pN, $\overline{W}$ = 13 N/m).

$\overline{M}/m_p$	v	c	$\overline{W}$(N/m)	v	c	$\overline{J}$(meV)[a]	v	$\overline{X}$(pN)[a]	v
10	0.79	12.73	5	0.00	2.29	0.0	0.00	0[c]	1.29
50	0.77	5.57	10	0.39	3.18	0.6	0.00	20[c]	1.29
100	0.74	3.87	15	0.85	4.05	0.8	0.46	30[b]	1.29
114	0.73	3.71	20	1.01	4.46	1.0	0.79	40	1.20
150	0.71	3.18	30	1.31	5.57	1.2	1.20	50	0.99
200	0.68	2.78	40[b]	1.41	6.36	1.4	1.31	60	0.78
			50[b]	1.46	8.10	1.6	1.56	70	0.45
			60[c]	1.49	8.91	1.8	1.78	80[d]	0.09
			70[c]	1.49	9.90	2.0	1.98	100	0.00
						2.5	2.41	120	0.00

a) $\overline{c}$ = 3.7 km/s ($\approx \sqrt{\overline{W}/\overline{M}}$ a, where a is the lattice constant of 4.5 Å);
b) c obtained numerically.
c) Slowly dispersive.
d) Fully dispersive.
   Irregular soliton propagation.

As to be expected /14/ the variation of $\overline{M}$ influences mostly the sound
velocity. For $\overline{W}$ we find a localized soliton for $\overline{W}$ < 10 N/m, a travelling
soliton for 10 N/m ≲ $\overline{W}$ < 40 N/m and dispersive behavior for larger values
of $\overline{W}$. If $\overline{J}$ < 0.8 meV we find a pinned soliton , for $\overline{J}$ > 2.5 meV
v approaches c and we find no more soliton. As found in earlier work also
/14,24/ for variation of $\overline{X}$ a travelling soliton is found for
40 pN < $\overline{X}$ < 80 pN. For larger values of $\overline{X}$ the soliton is pinned,
for smaller ones the system is dispersive.

III. DISORDER EFFECTS

To account for disorder we have used a random number generator with
equal probabilitiy of the different numbers within a given interval.
As reported in our preliminary study /17/, disorder in the mass sequence
destroys the soliton only for a large disorder starting from
(0.01 $\overline{M}$ ≲ M_n ≲ 50 $\overline{M}$). For (0.01 $\overline{M}$ ≲ M_n ≲ 10 $\overline{M}$) only the soliton velocity
is reduced to 0.59 km/s, the sound velocity to 2.12 km/s. In case of the
mass variation of natural amino acids (0.66 $\overline{M}$ ≲ M_n ≲ 1.79 $\overline{M}$) virtually
no change in the soliton dynamics is obtained, thus the average mass
approximation is certainly justified.
In Fig. 2 six examples of the evolution of $|a_k|^2$ as function of site
(k) and time are shown. As next step in addition to the natural degree
of disorder in the masses we have varied W_n. Up to a random variation of
± 20% $\overline{W}$ we find no change in the dynamics. For ± 30% $\overline{W}$ (Fig. 2a) the soliton velocity is somewhat reduced to 0.68 km/s. Finally, for ± 50% $\overline{W}$
(Fig. 2b) the excitation desperses slowly and the propagation is irregular. For variations in J_n alone or together with the natural mass variation the soliton is stable up to ± 5% $\overline{J}$. Thus the soliton is far more
sensitive against disorder in J_n than in the other parameters. If in
addition W_n is aperiodic the soliton is stable up to ± 10% $\overline{W}$ (Fig. 2c),
while at ± 20% $\overline{W}$ (Fig. 2d) slowly dispersive behavior shows up. Finally,

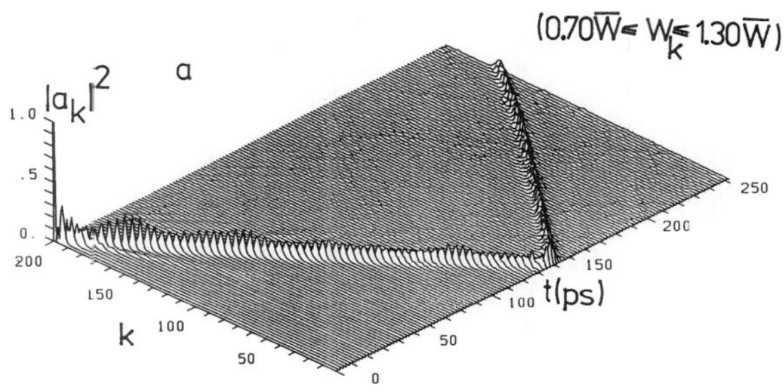

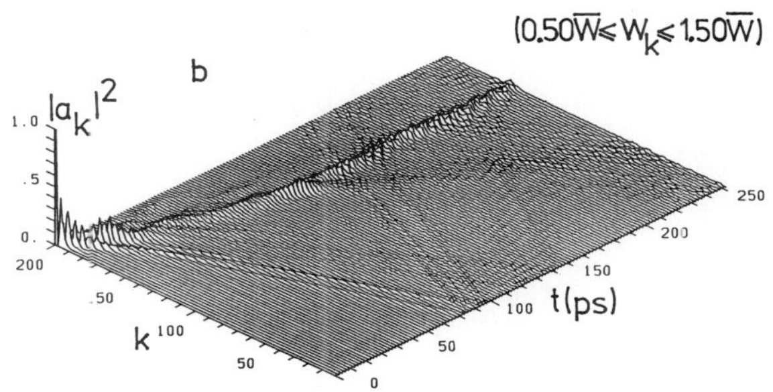

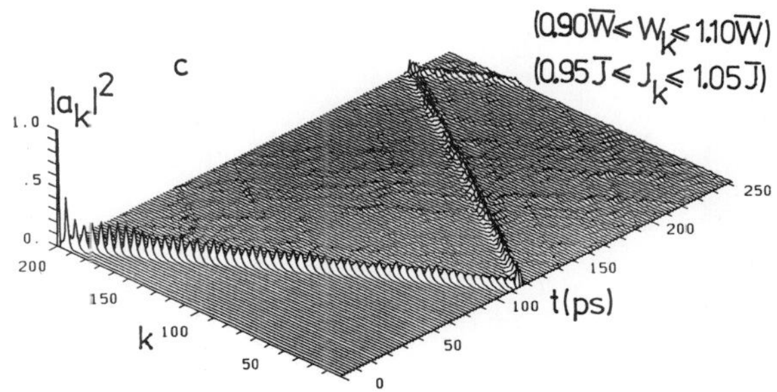

Fig. 2. (continued on next page)

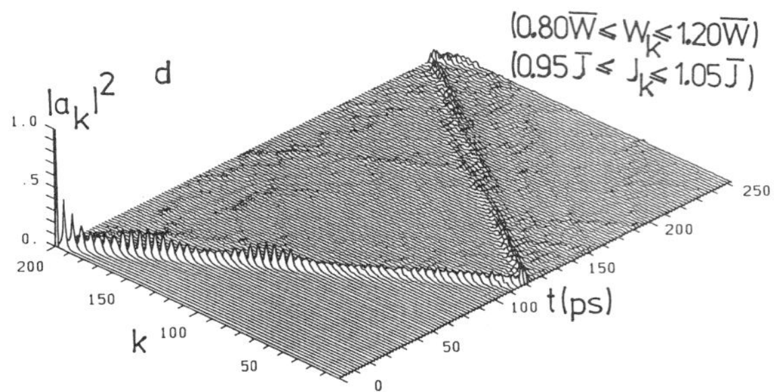

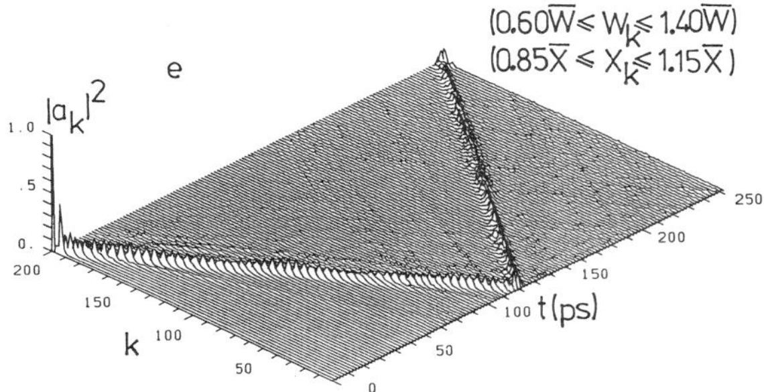

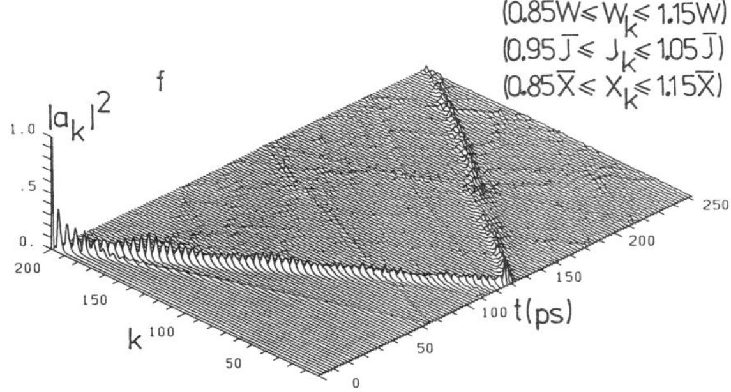

Fig. 2. $|a_k|^2$ as function of site (k) and time (t) for some typical examples of disorder (parameters not mentioned have standard values; for masses in all cases $(0.66\,\overline{M} \leqslant M_k \leqslant 1.79\,\overline{M})$).

if X_n alone is aperiodic or together with the natural mass variation, X_n can be varied up to $\pm$ 20% $\overline{X}$ without destruction of the soliton. However, if disorder also in W_n is introduced X_n can be varied up to $\pm$ 15% $\overline{X}$ and W_n up to $\pm$ 40% $\overline{W}$ (Fig. 2e). If finally, all four parameters are randomly varied the maximal possible disorder which still allows the existence of a soliton is $\pm$ 20% $\overline{W}$, $\pm$ 2.5% $\overline{J}$, and $\pm$ 10% $\overline{X}$. Fig. 2f shows a dispersive example: $\pm$ 15% $\overline{W}$, $\pm$ 5% $\overline{J}$, $\pm$ 15% $\overline{X}$.

For the maximal possible disorder we have calculated 10 different randomly chosen sequences to find out whether the soliton properties depend only on the magnitude of disorder or also on the individual sequences. As Table 2 shows the soliton velocity varies between 0.61 km/s and 0.80 km/s in this case.

Table 2. Soliton velocity (v) and sound velocity (c), both in km/s, for 10 different random sequences of parameters
($0.66\ \overline{M} \leqslant M_n \leqslant 1.79\ \overline{M}$; $0.90\ \overline{X} \leqslant X_n \leqslant 1.10\ \overline{X}$; $0.80\ \overline{W} \leqslant W_n \leqslant 1.20\ \overline{W}$; $0.975\ \overline{J} \leqslant J_n \leqslant 1.025\ \overline{J}$)

No.	1	2	3	4	5	6	7	8	9	10
v	0.74	0.71	0.74	0.71	0.80	0.67	0.70	0.61	0.77	0.77
c	3.56	3.43	3.43	3.56	3.43	3.56	3.56	3.43	3.43	3.43

In Fig. 3 we show as example the time evolution of $|a_k|^2$ for four of these sequences and for one example also the evolution of the lattice distortion.

However, the actual degree of disorder in proteins is unknown. The disorder in effective masses should be smaller than the mass interval of the natural amino acids since they are not free particles but covalently bound in the main polypeptide chain. Disorder in the other parameters should be mainly due to small influences of the side groups on the geometry of the main chain. Thus one can conclude that the naturally occurring disorder in the parameters should be smaller than the maximal disorder in which the soliton is stable. Only in case of J_n natural disorder may interfere with the soliton since the stability interval for J_n is rather small ($\pm$ 2.5% $\overline{J}$).

IV. TEMPERATURE EFFECTS

Since Davydov's result that solitons are stable at $T \approx 300K$ /27/, which he obtained via analytical considerations there has been a considerable discussion on this point in the literature. Halding and Lomdahl /28/ found stable pulses at T = 310K in classical molecular dynamics studies on peptide units moving in a Lennard-Jones potential. Lomdahl and Kerr /29/ and Lawrence et al. /30/ used the $|D_2\rangle$ ansatz together with a damping and a noise term to introduce temperature. They found stable solitons only for $T \approx 10K$. However, Bolterauer /31/ argued that their classical thermalization scheme should lead to incorrect results when applied to a quantum system. Bolterauer /31/ and other authors (see references in /31/) found solitons to be stable at $T \approx 300K$. Cruzeiro et al. /32/ derived a thermally averaged Hamiltonian and evolution equations using the $|D_1\rangle$ ansatz. They also found stable solitons at 300K. Very recently Cottingham and Schweitzer /33/ found a life time of only ≈ 1 ps for the soliton at 300K. They partially diagonalized the Davydov Hamiltonian. Using a thermal average for the phonon state they computed life times using first order perturbation theory (the soliton state is an exact eigenstate of the diagonizable part of the Hamiltonian). Wang et al. /34/ found solitons to be stable up to $\approx 7K$ with the help of quantum Monte Carlo simulations.

Since the Boltzmann factor for the excitation of CO oscillators at

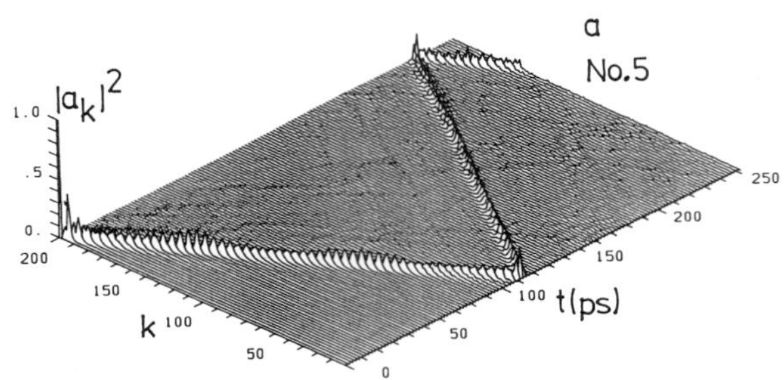

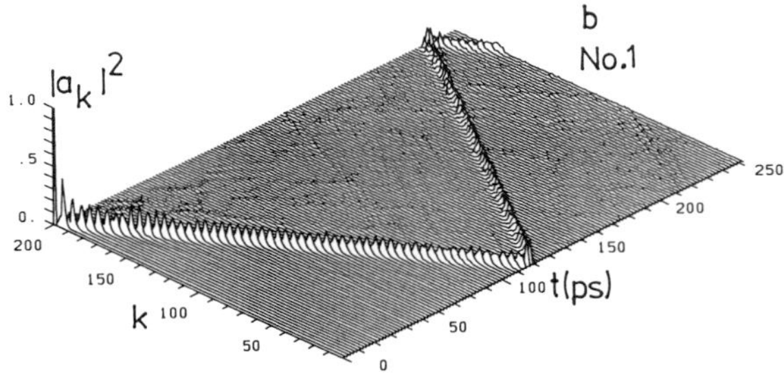

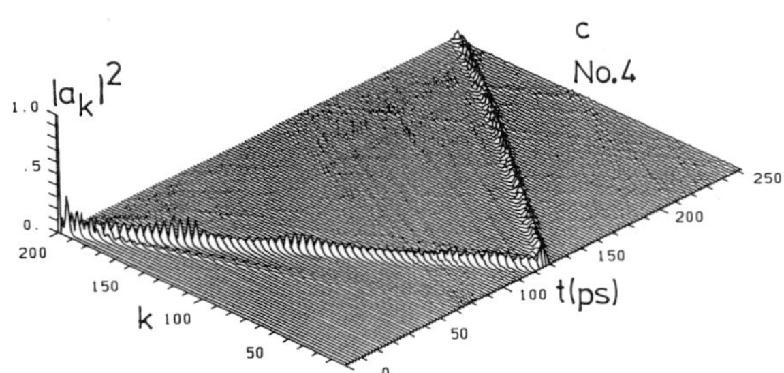

Fig. 3. (continued on next page)

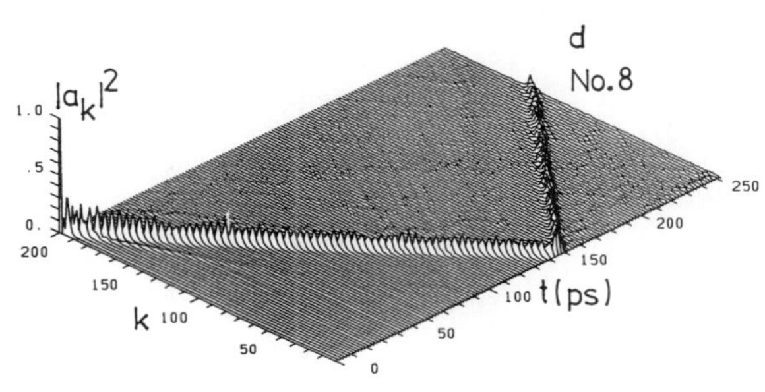

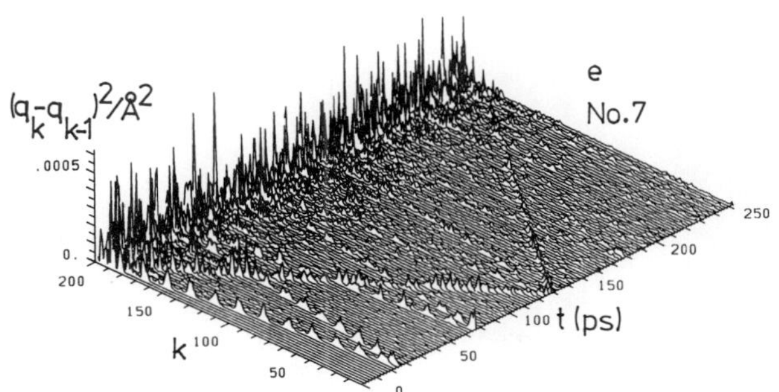

Fig. 3. $|a_k|^2$ as function of site (k) and time (t) for four of the aperiodic sequences given in Table 2 and $(q_k - q_{k-1})^2$ as function of site (k) and time (t) for one typical case.

300K is only $3 \cdot 10^{-4}$ one can assume that the heat bath affects the soliton
motion primarily via the lattice. We assume that the soliton moves fast
compared to thermal equilibration. Thus we introduce an energy Nk_BT
(where k_B is Boltzmann's constant) into the lattice prior to the soliton
start. Then the heat bath is switched off /31/. In our previous work /17/
we introduced an energy k_BT on each site as kinetic energy. The signs of
the momenta were determined randomly. In this work we first introduce an
additional potential term V' ($V' = W' \cdot (q_1^2 + q_N^2)$, $W' = 100\ \overline{W}$) to keep the
chain ends fixed. Then an energy of Nk_BT is distributed on the normal
modes using a Bose-Einstein distribution. The normal modes are excited
such that half of the energy is kinetic, the other half potential energy.
As in /17/ the decoupled lattice is allowed to relax through typically
120 ps (t_r). After that the soliton is started. In our previous work /17/ we
found the soliton to be completely unaffected by the heat bath for a
specific set of parameters ($\overline{W} = 76$ N/m, $\overline{X} = 90$ pN). This result remains
unchanged if the Bose-Einstein distribution is used. Also the use of
Maxwell-Boltzmann statistic did not change the results.

With a time step of 0.005 ps the equations of motion have been integrated over a time of 120 ps using again the Runge-Kutta method. Typically
the error in total energy remained smaller than 0.9% of its initial value,
the norm was conserved to an accuracy of 0.018‰, and the imaginary part
of the total energy was less than $0.3 \cdot 10^{-17}$ eV for T = 300K and N = 200.
The use of a time step $\tau/2$ only reduced the numerical errors but did not
change the results.

Fig. 4 shows the evolution of $/a_k(t)/^2$ using the generally accepted
parameter values for the α-helix for different temperatures. Obviously
the soliton is stable up to 30K. From 10K to 20K its velocity decreases
as expected. However, with onset of the dispersive behavior the velocity increases again (30K). At 40K the soliton is stable only for
40-50 ps, and disperses rapidly afterwards. Note, that if the dynamics at
40K would have been computed for ≈ 40 ps only, one would have reached the
wrong conclusion that the soliton might be stable.

Fig. 5 shows the evolution of $/a_k(t)/^2$ at T = 300K for different sets
of parameter values $\overline{W}$, $\overline{X}$. For the set appropriate for the α-helix the
excitation disperses slowly into the chain. With increasing $\overline{X}$ (100 pN) the
solitary wave is stabilized but pinned at the chain end. However, there
is still a slow dispersion present. Using $\overline{W} = 40$ N/m we find dispersive
character for $\overline{X} = 40$ pN. At $\overline{X} = 60 - 80$ pN a stable soliton travels
through the chain. For $\overline{X} = 60$ pN we show the results for two different
lattice relaxation times t_r to indicate their independence of t_r. If $\overline{X}$
is increased above 80 pN the soliton disperses again, however, at
$\overline{X} = 120$ pN a tendency to localization at the chain end starts to appear.

Finally in Fig. 6 our survey of the $\overline{W}$, $\overline{X}$ parameter space at 300K
is summarized. Obviously for a travelling soliton to be stable $\overline{W} \geq 40$ N/m
is required, which is well above the value for the α-helix (13 N/m).
For smaller values of $\overline{W}$ one needs too large fluctuations in q_k to accomo-
date $1/2\ Nk_BT$ as potential energy. Through the coupling ($\overline{X}$) these fluctuations destroy the soliton for $\overline{W} < 40$ N/m. For $\overline{W} \geq 40$ N/m we find three
threshold values for $\overline{X}$. If $\overline{X}$ is smaller than a value $\overline{X}_{t1}$ (≈ 60 pN for
$\overline{W} = 40$ N/m) the excitation disperses. This happens also for T = 0K.
Between $\overline{X}_{t1}$ and $\overline{X}_{t2}$ (≈ 80 pN for $\overline{W} = 40$ N/m) a travelling soliton appears
to be stable. For $\overline{X} > \overline{X}_{t2}$, however, dispersion shows up again. In this
region the coupling is large enough to allow the thermal fluctuations of
the lattice to destroy the soliton. Finally, for $\overline{X} > \overline{X}_{t3}$ (≈ 80 pN for
$\overline{W} = 13$ N/m) the soliton appears to be stabilized again, but it remains
localized at the chain ends. For $\overline{W} < 40$ N/m only $\overline{X}_{t3}$ exists. This
behavior for large $\overline{X}$ occurs also at T = 0K. It appears to be quite probable
that the instability region between $\overline{X}_{t2}$ and $\overline{X}_{t3}$ would vanish for larger
values of $\overline{W}$. However, the basic feature of Fig. 6 is that stable solitons
are found in the upper right region, while the parameters appropriate
for the α-helix are found in the lower left part.

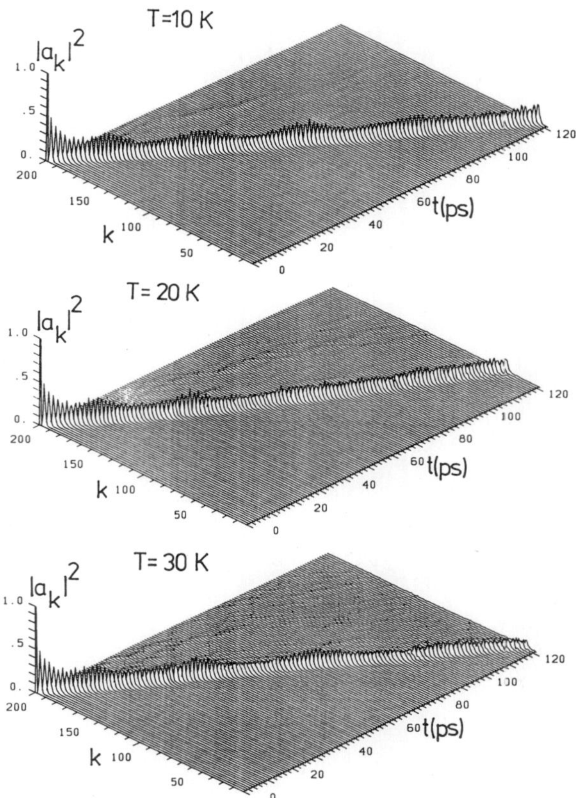

Fig. 4. (continued on next page)

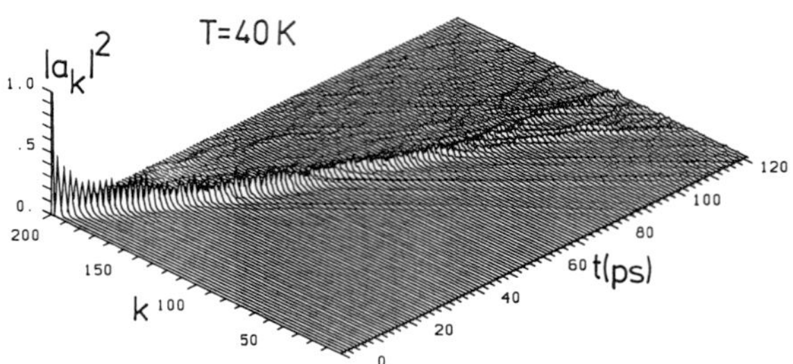

Fig. 4. $|a_k|^2$ as function of site (k) and time (t) for different temperatures ($\overline{M}$ = 114 m_p, $\overline{W}$ = 13 N/m, $\overline{X}$ = 62 pN, $\overline{J}$ = 0.967 meV).

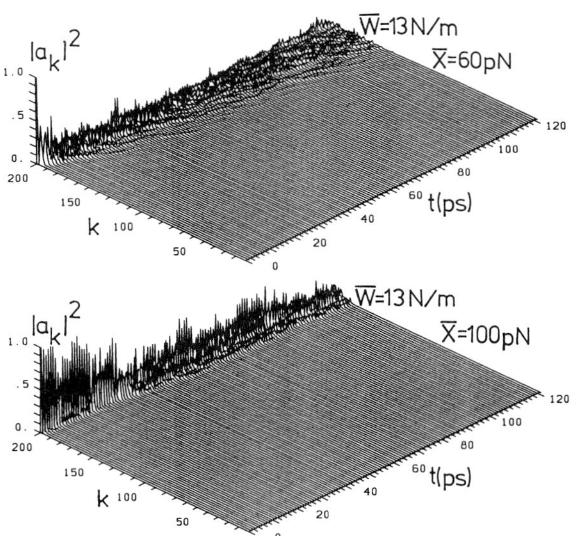

Fig. 5. (continued on next page)

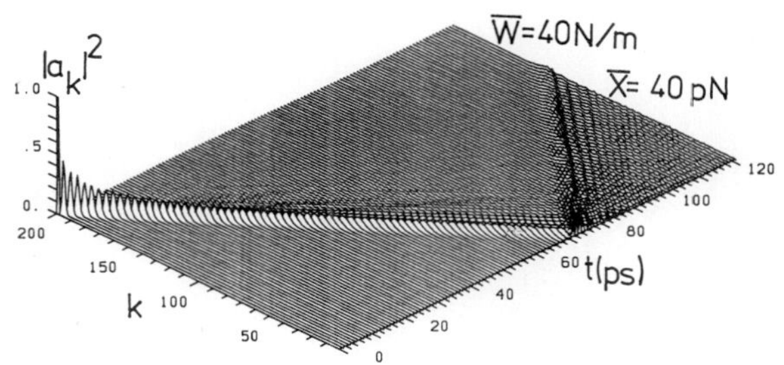

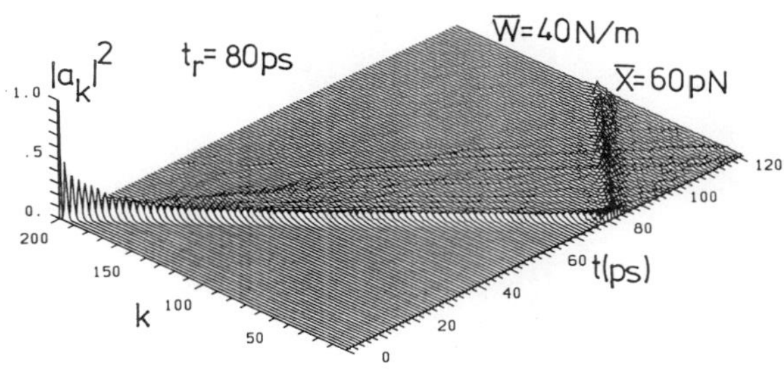

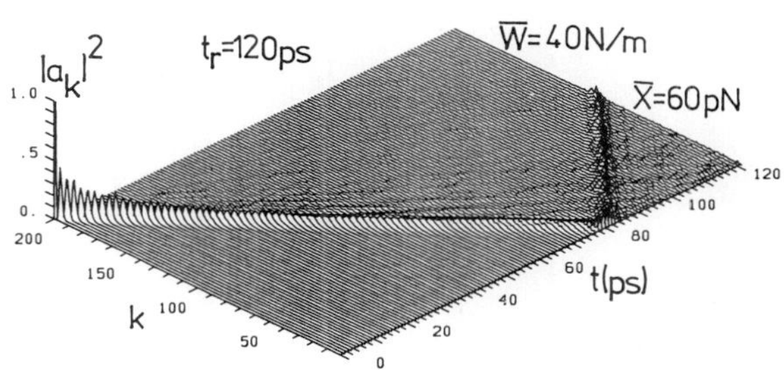

Fig. 5. (continued on next page)

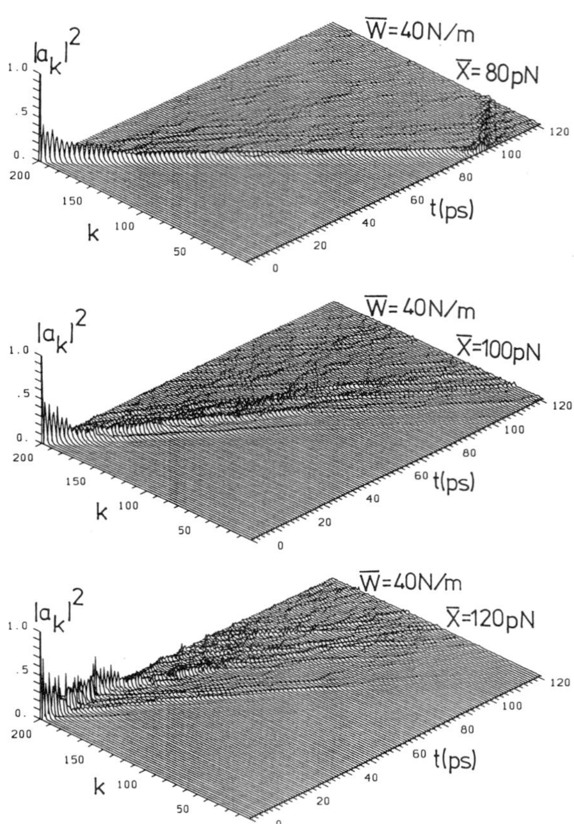

Fig. 5. Some examples of $|a_k|^2$ as function of site (k) and time (t) at temperature T = 300 K for different values of $\overline{W}$ and $\overline{X}$ (lattice relaxation time t_r = 120 ps if not otherwise mentioned).

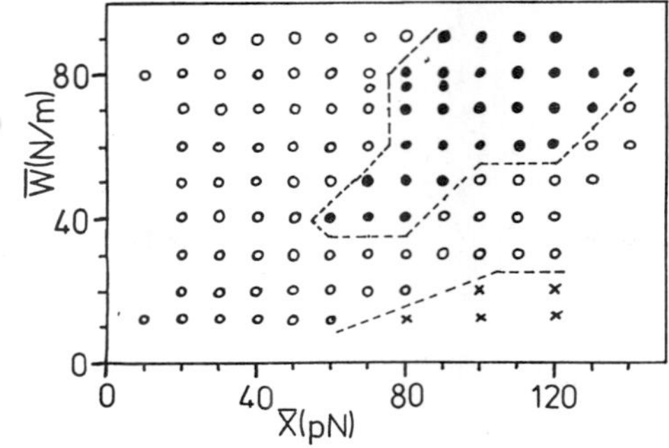

Fig. 6. Dependence of the dynamical properties of the Davydov system at
T = 300K on $\overline{W}$ and $\overline{X}$ (● travelling soliton, x localized, very
slowly dispersive excitation, o dispersive)

V. CONCLUSION

We found the Davydov soliton to be quite stable against aperiodicity
in the parameters, with the exception of the dipole coupling J_k. In this
case only for a disorder of up to $\pm$ 2.5% $\overline{J}$ remains the soliton stable.
The soliton velocity depends somewhat on the actual random sequence of the
parameter values.

Our results concerning temperature effects suggest that the Davydov
soliton is destroyed at 300K for the parameter values appropriate for an
α-helix. We have obtained only for much larger parameter values travelling
solitons. This finding disagrees with that of Cruzeiro et al. /32/. However,
in their work they followed the dynamics only for 20-30 ps, while as
mentioned above, there are cases where the soliton disperses only after
40-50 ps.

However, the value $\overline{W}$ = 13 N/m is taken from a study on crystalline
formamide /35/. Since in proteins any movement of a unit involves also
changes in the geometries of the covalent bonds between the unit and the
main chain, one may expect that in proteins the effective spring constants
of hydrogen bonds should be larger than the values for formamide.

ACKNOWLEDGMENT

We thank Professors A.C. Scott and P.L. Christiansen for making the
presentation and discussion of this work at the NATO Workshop on "Self-
trapping of vibrational energy in proteins" (1989) possible. This work was
supported by the "Deutsche Forschungsgemeinschaft" (Project-no. Ot 51/6)
for which the authors should like to express their sincere gratitude.

REFERENCES

/1/ E. Fermi, J. R. Pasta and S. M. Ulam, Los Alamos Report La-1940
     (1955).
/2/ J. Scott-Russell, Proc. Roy. Soc. Edinburgh, 319 (1844).
/3/ H. J. Mikeska, J. Phys. C 11: L29 (1978).
/4/ K. Maki, J. Low Temp. 41: 327 (1980).
/5/ K. J. Wahlstrand, J. Chem. Phys. 82:5247 (1985); K. J. Wahlstrand
     and P. G. Woylness, J. Chem. Phys. 82:3392 (1976).
/6/ S. Aubry, J. Chem. Phys. 64:3392 (1976).

/7/ M. A. Collins, A. Blumen, J. F. Currie and J. Ross, Phys. Rev. B19: 3630 (1979).

/8/ J. A. Krumhansl and D. M. Alexander, in "Structure and Dynamics: Nucleic Acids and Proteins", E. Clementi and R. H. Sarma, eds., Adenine Press, New York (2983) p. 61.

/9/ J. Ladik and J. Čižek, Int. J. Quant. Chem. 26:955 (1984); J. Ladik, in "Molecular Basis of Cancer", Part A, Alan Liss Comp. (1985) p. 343; D. Hofmann, W. Förner, and J. Ladik, Phys. Rev. A37: 4429 (1988); W. Förner, Phys. Rev. A38:939 (1988); W. Förner, Phys. Rev. A (submitted); W. Förner, P. Otto, J. Ladik, and F. Martino, Phys. Rev. A (submitted); D. Hofmann, W. Förner, and J. Ladik, J. Phys. C (submitted).

/10/ W. P. Su, J. R. Schrieffer, and A. J. Heeger, Phys. Rev. Lett. 42:1698 (1979); W. P. Su, Solid State Comm. 35:899 (1980); W. P. Su, J. R. Schrieffer, and A. J. Heeger, Phys. Rev. B22:2099 (1980); W. P. Su and J. R. Schrieffer, Proc. Natl. Acad. Sci. USA 77:5626 (1980); F. Guinea, Phys. Rev. B30:1884 (1984); A. R. Bishop, D. K. Campbell, P. S. Lomdahl, B. Horovitz, and S. R. Phillpot, Phys. Rev. Lett. 52:671 (1984); W. Förner, M. Seel, and J. Ladik, Solid State Comm. 57:463 (1986); J. Chem. Phys. 84:5910 (1986); A. Godzik, M. Seel, W. Förner, and J. Ladik, Solid State Comm. 60: 609 (1987); C.-M. Liegener, W. Förner, and J. Ladik, Solid State Comm. 61:203 (1987); S.R. Phillpot, D. Beariswyl, A.R. Bishop, and P.S. Lomdahl, Phys. Rev. B35:7533 (1987); S. Kivelson and D. E. Heim, Phys. Rev. B26:4278 (1982); A. J. Heeger and J. R. Schrieffer, Solid State Comm. 48:207 (1983); Z. Soos and S. Ramasesha, Phys. Rev. Lett. 51:2374 (1983); M. Sasai and H. Fukutome, Synth. Metals 9:295 (1984); W. P. Su, Phys. Rev. B34:2988 (1986); S. Kivelson and Wei-Kang Wu, Phys. Rev. B34:5423 (1986); C. L. Wang and F. Martino, Phys. Rev. B34:5540 (1986); W. Förner, C.L. Wang, F. Martino and J. Ladik, Phys. Rev. B37:4567 (1988); W. Förner, Solid State Comm. 63:941 (1987); R. Markus, W. Förner, and J. Ladik, Solid State Comm. 68:135 (1988); H. Orendi, W. Förner, and J. Ladik, Chem. Phys. Lett. 150:113 (1988);

/11/ A. Szent-Györgyi, Nature 148:157 (1941); Science 93:609 (1941); A.K. Bakhshi, P. Otto, J. Ladik and M. Seel, Chem. Phys. 20:683 (1986).

/12/ A.S. Davydov and N. I. Kislukha, Phys. Stat. Sol. B 59:465 (1973); A.S. Davydov, Phys. Scripta 20:387 (1979); A. S. Davydov, Soviet Phys. Usp. 25:898 (1982); A. S. Davydov, "Biology and Quantum Mechanics", Pergamon Press, Oxford (1982).

/13/ G. Careri, U. Buontempo, F. Galuzzi, A. C. Scott, E. Gratton, and E. Shyamsunder, Phys. Rev. B30:4689 (1984); J. C. Eilbeck, P. S. Lomdahl and A. C. Scott, Phys. Rev. B30:4703 (1984).

/14/ A. C. Scott, Phys. Rev. A26:578 (1982); A. C. Scott, in "Structure and Dynamics: Nucleic Acids and Proteins", E. Clementi and R. H. Sarma, eds., Adenine Press, New York (1983) p. 389; A. C. Scott, Physica Scripta 29:279 (1984); L. MacNeil and A. C. Scott, Physica Scripta 29:284 (1984); A. C. Scott, Phil. Trans. Roy. Soc. London A 315:423 (1985).

/15/ J. M. Ziman, "Elements of Advanced Quantum Theory", Cambridge University Press (1969).

/16/ N. A. Nevskaya and Y. N. Chirgadze, Biopolymers 15:637 (1976).

/17/ H. Motschmann, W. Förner, and J. Ladik, J. Phys. C 1:5083 (1989).

/18/ W. C. Hamilton and J. A. Ibers, "Hydrogen Bonding in Solids", W. A. Benjamin Press, New York (1968).

/19/ A. Novak, "Structure and Bonding", Vol. 18, Springer-Verlag, Berlin-New York (1974).

/20/ V. A. Kuprievich and V. E. Klymenko, Mol. Phys. 34:1287 (1977).

/21/ D.W. Brown, K. Lindenberg, and B.J. West, Phys. Rev. A 33:4104 (1986); D. W. Brown, B. J. West, and K. Lindenberg, Phys. Rev. A 33: 4110 (1986); D. W. Brown, Phys. Rev. A 37:5010 (1988).

/22/ D.W. Brown, K. Lindenberg, and B. J. West, Phys. Rev. B 35:6169 (1987).
/23/ D.W. Brown, K. Lindenberg, and B. J. West, Phys. Rev. B 37:2946 (1988).
/24/ B. Mechtly and P.B. Shaw, Phys. REv. B 38:3075 (1988).
/25/ W.C. Kerr and P. S. Lomdahl, Phys. Rev. B 35:3629 (1987).
/26/ J. Stiefel, "Einführung in die numerische Mathematik", Teubner Verlag, Stuttgart (1965).
/27/ A. S. Davydov, Zh. Eksp. Teor. Fiz. 78:789 (1980) Sov. Phys. JETP 51:397 (1980).
/28/ J. Halding and P.S. Lomdahl, Phys. Lett. A 124:37 (1987).
/29/ P. S. Lomdahl and W. C. Kerr, Phys. Rev. Lett. 55:1235 (1985).
/30/ A. F. Lawrence, J. C. McDaniel, D. B. Chang, B. M. Pierce, and R. R. Birge, Phys. Rev. A 33:1188 (1986).
/31/ H. Bolterauer, in "Structure, Coherence, and Chaos", Proc. MIDIT 1986 Workshop, Manchester University Press.
/32/ L. Cruzeiro, J. Halding, P. L. Christiansen, O. Skovgaard, and A. C. Scott, Phys. Rev. A 37:880 (1988).
/33/ J. P. Cottingsham and J. W. Schweitzer, Phys. Rev. Lett. 62:1792 (1989).
/34/ X. Wang, D. W. Brown, and K. Lindenberg, Phys. Rev. Lett. 62:1796 (1989).
/35/ K. Itoh and T. Shimanouchi, J. Mol. Spectr. 42:86 (1972).

21

PERTURBATION ESTIMATE OF THE LIFETIME OF THE DAVYDOV SOLITON AT 300 K

J. W. Schweitzer and J. P. Cottingham

Department of Physics and Astronomy
The University of Iowa
Iowa City, Iowa 52242-1479

INTRODUCTION

The soliton model proposed by Davydov[1] for energy transport in biological molecules is an important application of ideas from solid state theory to biological processes at the molecular level. It yields a compelling picture for the mechanism of energy transport in the α-helical protein molecule and consequently has been the subject of a large body of work.[2] However there has been controversy in recent years concerning whether the Davydov soliton is sufficiently stable at physiological temperatures to provide a viable explanation for energy transport. Since there seems to be wide agreement on the appropriate values for the parameters of the model for the α-helical protein application, it should be possible to resolve this question.

When we first addressed this question we felt that previous estimates of the stability of the Davydov soliton at finite temperatures were not convincing since they were based on either semi-classical treatments[3] or qualitative arguments.[4] We have recently reported[5] on a straightforward analytical quantum mechanical estimate of the finite temperature lifetime of the Davydov soliton-type excitation state having the form of a product of a single localized exciton and a coherent phonon state. Using the generally accepted values of the parameters, we found that the Davydov soliton in the presence of a thermal equilibrium distribution of phonons will decay to delocalized excitations at a rate which is much too fast for the soliton to travel the length of the protein molecule.

Our calculation made use of the formalism developed by Eremko, Gaididei, and Vakhnenko[6] where the Davydov soliton is represented by a single particle state. This single particle state is not an exact eigenstate of the model Hamiltonian which contains a term that mixes this localized state with delocalized states in the presence of the emission or absorption of a phonon. Lowest order perturbation theory was used to estimate the probability for transitions to delocalized states. The temperature dependence was introduced in a manner consistent with a slowly moving stable soliton where phonons relative to the locally deformed lattice are in thermal equilibrium. Although the final result is probably inconsistent with the assumption of a sufficiently stable soliton to allow for the thermal equilibrium of the phonons, the negative result is convincing evidence that the Davydov soliton of the form considered does not provide a mechanism for energy transport in the α-helical protein molecule, at least for parameter values in the range usually assumed.

In the present paper we discuss some details of our calculation that were not reported in the earlier paper.[5] We also consider a more general single soliton-type state that is represented by an n-particle state. Finally we report on how the values of the parameters would need to change in order to yield sufficiently stable solitons at biological temperatures.

MODEL HAMILTONIAN

The essential physics is assumed by Davydov to be governed by a Fröhlich-type Hamiltonian for a one-dimensional chain of molecular units on a lattice of N sites with lattice constant R and periodic boundary conditions.

$$
\begin{aligned}
H = &\sum_j \left[\epsilon_o B_j^+ B_j - J(B_j^+ B_{j+1} + B_{j+1}^+ B_j) \right] \\
&+ \frac{1}{2} \sum_j \left[p_j^2/m + w(u_j - u_{j+1})^2 \right] \\
&+ \chi \sum_j (u_{j+1} - u_{j-1}) B_j^+ B_j
\end{aligned}
\tag{1}
$$

where B_j^+ and B_j are the creation and annihilation operators for an intramolecular excitation on the molecular unit at the jth site and u_j and p_j are the displacement and momentum operators of the jth unit. Since one is interested in investigating the case where there is initially a soliton moving with a velocity v on the chain, it is convenient to do the analysis in a frame of reference where the soliton is at rest. Also in order to have simple analytical expressions we make the usual continuum approximation. Thus $\tilde{H} = H - vP$, where P is the total momentum, becomes

$$
\begin{aligned}
\tilde{H} = &\int_0^L dx \left[(\epsilon_o - 2J)\psi^+(x)\psi(x) + JR^2 \frac{\partial \psi^+}{\partial x} \frac{\partial \psi}{\partial x} \right. \\
&\left. - \frac{i\hbar v}{2} \left(\frac{\partial \psi^+}{\partial x} \psi(x) - \psi^+(x) \frac{\partial \psi}{\partial x} \right) \right] \\
&+ \sum_k \hbar(\omega_k - kv) a_k^+ a_k + \frac{1}{\sqrt{N}} \sum_k F(k)(a_{-k}^+ + a_k) \int_0^L dx\, e^{ikx} \psi^+(x)\psi(x)
\end{aligned}
\tag{2}
$$

where $\psi(x)$ is the field operator corresponding to B_j in the continuum limit and a_k^+ and a_k are the phonon creation and annihilation operators. Here $L = NR$, $-\pi \le kR \le \pi$, and

$$
\omega_k = (w/m)^{1/2} R|k| = v_a|k| \quad \text{and} \quad F(k) = i\chi(2\hbar/m\omega_k)^{1/2} kR \quad .
\tag{3}
$$

Following the procedure of Eremko et al.,[6] one introduces new phonon operators

$$
b_k = a_k - \frac{1}{\sqrt{N}} f_k
\tag{4}
$$

which describe phonons relative to a chain with a particular deformation. The vacuum state for the new phonons is then the coherent phonon state

$$
|\tilde{0}\rangle_{ph} = \exp\left[\frac{1}{\sqrt{N}} \sum_q (f_q\, a_q^+ - f_q^*\, a_q) \right] |0\rangle_{ph}
\tag{5}
$$

in terms of the phonons of the uniform lattice. The Hamiltonian $\tilde{H}$ is rewritten as

$$
\begin{aligned}
\tilde{H} = &\int_0^L dx\, \psi^+(x)(\epsilon_o - 2J + V(x) - JR^2 \frac{\partial^2}{\partial x^2} + i\hbar v \frac{\partial}{\partial x})\psi(x) \\
&+ W + \sum_k \hbar(\omega_k - kv) \left[b_k^+ b_k + \frac{1}{\sqrt{N}} (b_k^+ f_k + f_k^* b_k) \right] \\
&+ \frac{1}{\sqrt{N}} \sum_k F(k)(b_{-k}^+ + b_k) \int_0^L dx\, e^{ikx} \psi^+(x)\psi(x)
\end{aligned}
\tag{6}
$$

where

$$W = \frac{1}{N} \sum_k \hbar(\omega_k - kv)|f_k|^2 \tag{7}$$

$$V(x) = \frac{1}{N} \sum_k F(k)(f^*_{-k} + f_k)e^{ikx} \quad . \tag{8}$$

To describe the deformation corresponding to a single soliton in the subspace where

$$\int_o^L dx\,\psi^+(x)\psi(x) = n \quad , \tag{9}$$

i.e., a single soliton with excitation numbers equal to n, one chooses

$$V(x) = -2n^2\,J\mu^2\,\mathrm{sech}^2(n\mu x/R) \tag{10}$$

where

$$\mu = \frac{\chi^2}{wJ[1 - (v/v_a)^2]} \quad . \tag{11}$$

Since the continuum approximation requires spatial variations to be small on the scale of the lattice constant R, it is necessary for $n\mu$ to be not much greater than one. For given values of the parameters of the model this condition restricts n and v/v_a.

The operator $\tilde{H}$ is partially diagonalized by the transformation

$$\psi(x) = \sum_\alpha A_\alpha a_\alpha(x) \tag{12}$$

where

$$\left[-JR^2\frac{\partial^2}{\partial x^2} + i\hbar v\frac{\partial}{\partial x} + \epsilon_o - 2J + V(x)\right]a_\alpha(x) = E_\alpha a_\alpha(x) \quad . \tag{13}$$

For our choice of $V(x)$ there is only one bound state

$$a_s(x) = (n\mu/2R)^{1/2}\mathrm{sech}(n\mu x/R)\exp(i\hbar vx/2JR^2) \tag{14}$$

with energy

$$E_s = \epsilon_o - 2J - (\hbar^2 v^2/4JR^2) - n^2 J\mu^2 \tag{15}$$

and unbounded (delocalized) states

$$a_k(x) = \frac{1}{\sqrt{L}}\frac{kR + in\mu\tanh(n\mu x/R)}{[(n\mu)^2 + (kR)^2]^{1/2}}\exp(ikx + i\hbar vx/2JR^2) \tag{16}$$

with

$$E_k = \epsilon_o - 2J - (\hbar^2 v^2/4JR^2) + J(kR)^2 \quad . \tag{17}$$

Hence in this new representation

$$\begin{aligned}
\tilde{H} = {}& W + E_s A_s^+ A_s + \sum_k E_k A_k^+ A_k + \sum_q \hbar(\omega_q - qv)b_q^+ b_q \\
&+ \frac{1}{\sqrt{N}}\sum_q \hbar(\omega_q - qv)(b_q^+ f_q + f_q^* b_q)(1 - \frac{1}{n}A_s^+ A_s) \\
&+ \frac{1}{\sqrt{N}}\sum_{k,k',q} X(k,k',q)(b_{-q}^+ + b_q)A_{k'}^+ A_k \\
&- \frac{1}{N}\sum_{k,q}\overline{X}(k,q)(b_{-q}^+ + b_q)(A_s^+ A_{-k} - A_k^+ A_s)
\end{aligned} \tag{18}$$

where

$$X(k, k', q) = F(q) \int_o^L dx \, e^{iqx} a_{k'}^*(x) a_k(x) \simeq X(k, k+q, q)\delta_{k',k+q} \tag{19}$$

$$\begin{aligned}
\overline{X}(k, q) &= F(q) \int_o^L dx \, e^{iqx} a_k^*(x) a_s(x) \\
&= \frac{\pi F(q)}{\sqrt{2n\mu}} \frac{(qR)\mathrm{sech}[\pi(k-q)R/2n\mu]}{[(n\mu)^2 + (kR)^2]^{1/2}} \quad .
\end{aligned} \tag{20}$$

Here f_q is determined by $V(x)$ and the condition $(\omega_q - qv)f_q = (\omega_q + qv)f_{-q}^*$ which is required to obtain the simple $1 - \frac{1}{n}A_s^+ A_s$ dependence in $\tilde{H}$. One finds

$$f_q = \frac{i\pi\chi}{w\mu[1 - (v/v_a)^2]}\Big(\frac{m}{2\hbar\omega_q}\Big)^{1/2}(\omega_q + qv)\mathrm{csch}(\pi q R/2n\mu) \tag{21}$$

and

$$W = \frac{2}{3}n^3 J\mu^2 \quad . \tag{22}$$

For this f_q the $|\tilde{0}\rangle_{ph}$ of (5) is just the coherent phonon state introduced by Davydov generalized to the case of arbitrary n. The n-type single soliton is described by the "n-particle" state vector

$$|n\rangle = \frac{1}{\sqrt{n!}}(A_s^+)^n |0\rangle_{ex}|\tilde{0}\rangle_{ph} \quad . \tag{23}$$

Note that this is not an exact eigenstate of $\tilde{H}$ owing to the presence of the term in $\tilde{H}$ with $A_k^+ A_s$. One can also construct a coherent state with average excitation number n,

$$|n\rangle_c = e^{n^{1/2}(A_s^+ - A_s)}|0\rangle_{ex}|\tilde{0}\rangle_{ph} \tag{24}$$

which corresponds to an alternate form of the Davydov soliton state. One notes that

$$\langle n|\tilde{H}|n\rangle = {}_c\langle n|\tilde{H}|n\rangle_c = W + nE_s = (\epsilon_o - 2J - \hbar^2 v^2/4JR^2)n - \frac{1}{3}J\mu^2 n^3 \tag{25}$$

which demonstrates that both forms correspond to the same semi-classical soliton with excitation number equal to n. Also there is no lower bound to the energy if n is unrestricted.

PERTURBATION THEORY

For the discussion of the decay of the soliton state it is convenient to divide $\tilde{H}$ into $H_o + V$ where

$$\begin{aligned}
H_o &= W + E_s A_s^+ A_s + \sum_k E_k A_k^+ A_k + \sum_q \hbar(\omega_q - qv)b_q^+ b_q \\
&\quad + \frac{1}{\sqrt{N}}\sum_q \hbar(\omega_q - qv)(b_q^+ f_q + f_q^* b_q)(1 - \frac{1}{n}A_s^+ A_s) \quad .
\end{aligned} \tag{26}$$

This choice for H_o is such that $|n\rangle$ is the ground state of H_o with energy $W + nE_s$ in the subspace of excitation number equal to n, i.e., $A_s^+ A_s + \sum_k A_k^+ A_k = n$. Equally important, in this subspace the eigenstates have the simple form

$$|n-m; k_1, k_2 \cdots k_m; \{n_q\}\rangle = \frac{1}{\sqrt{(n-m)!}}(A_s^+)^{n-m} A_{k_1}^+ \dots A_{k_m}^+ |0\rangle_{ex} \prod_q \frac{(\hat{c}_q^+)^{n_q}}{\sqrt{n_q!}}|\tilde{0}\rangle_{ph}^{n-m} \tag{27}$$

where

$$\hat{c}_q = b_q + \frac{m}{n}\frac{1}{\sqrt{N}}f_q = a_q - \frac{n-m}{n}\frac{1}{\sqrt{N}}f_q \tag{28}$$

with $\hat{c}_q|\tilde{0}\rangle_{ph}^{n-m} = 0$. The corresponding energy

$$E^{(0)}_{n-m;k_1,\ldots k_m;\{n_q\}} = [1 - (m/n)^2]W + (n-m)E_s + \sum_{i=1}^{m} E_{k_i} + \sum_q \hbar(\omega_q - qv)n_q \quad . \tag{29}$$

Thus for $m = 0$, the excited state is an n-type soliton plus phonons relative to a chain with the deformation corresponding to the n-type soliton. For $m = n$, the excitations are delocalized and the phonons are relative to a chain without any deformation. Furthermore, except for small k, the delocalized states approximate ordinary excitons. For intermediate m the interpretation is less straightforward. The $n - m$ state is not exactly a $(n - m)$-type soliton with m delocalized excitations since the state $a_s(x)$ is a bound state in the deformation appropriate for the n-type soliton. Nevertheless the eigenstates of H_o provide a reasonably accurate analogue of the picture resulting from the usual semi-classical treatments.

The decay of a single n-type soliton in lowest order perturbation theory is caused by the part of V,

$$\overline{V} = -\frac{1}{N} \sum_{k,q} \overline{X}(k,q)(b^+_{-q} + b_q)(A^+_s A_{-k} - A^+_k A_s) \quad , \tag{30}$$

which allows for transitions between the n and $n - 1$ states. The average transition rate Γ from initial n-type soliton states is given by

$$\Gamma = \lim_{t \to \infty} \frac{1}{\hbar^2} \frac{d}{dt} \sum_i P_i^{ph} \sum_k \sum_{\{n_q'\}} \left| \int_o^t dt' \langle n, \{n_q\}_i | \overline{V}(t') | n-1, k, \{n_q'\} \rangle \right|^2 \tag{31}$$

where

$$\overline{V}(t) = e^{iH_o t} \, \overline{V} \, e^{-iH_o t} \quad . \tag{32}$$

Here the sum over i indicates a sum over initial sets of phonon occupation numbers and P_i^{ph} denotes a probability distribution which we assume to be a thermal equilibrium distribution. Since

$$e^{-iH_o t}|n, \{n_q\}\rangle = \exp\left\{ -i(W + nE_s)t/\hbar - i\sum_q(\omega_q - qv)b^+_q b_q t \right\} |n, \{n_q\}\rangle \tag{33}$$

and

$$e^{-iH_o t}|n-1, k, \{n_q'\}\rangle$$
$$= \exp\left\{ -i\left[(1 - \frac{1}{n^2})W + (n-1)E_s + E_k\right] t/\hbar - i\sum_q(\omega_q - qv)c^+_q c_q t \right\} |n-1, k, \{n_q'\}\rangle \tag{34}$$

where

$$c_q = b_q + \frac{1}{n}\frac{1}{\sqrt{N}}f_q \quad , \tag{35}$$

it is easy to show that

$$\Gamma = \frac{1}{\hbar^2}\frac{1}{N^2} \sum_{k,q,q'} \overline{X}^*(k,q)\overline{X}(k,q') \, 2\,Re \int_o^\infty dt \, U(q,q',t)$$
$$\exp\left\{ -\frac{i}{\hbar}[J(kR)^2 + (n^2 - \frac{2}{3}n)J\mu^2]t \right\} \tag{36}$$

where

$$U(q,q',t) = \ll \exp\left[i\sum_k(\omega_k - kv)b^+_k b_k t\right] (b^+_q + b_{-q})$$
$$\exp\left[-i\sum_k(\omega_k - kv)c^+_k c_k t\right] (b^+_{-q'} + b_{q'}) \gg \tag{37}$$

with

$$\ll A \gg = Tr\left\{ A \exp[-\beta \sum_q \hbar(\omega_q - qv)b_q^+ b_q]\right\} / Z_{ph} \tag{38}$$

where

$$Z_{ph} = \prod_q \{1 - \exp[-\beta\hbar(\omega_q - qv)]\}^{-1} \quad . \tag{39}$$

This rather unusual expression for Γ occurs because the phonons in the final states are relative to a different deformation.

EVALUATION OF $U(q,q',t)$

The critical step in the calculation of the rate Γ is the analytical evaluation of $U(q,q',t)$. The trace involved can be calculated using the occupation number states, but this is a very tedious calculation that involves the summation of infinite series. The calculation becomes quite easy if one uses coherent states $|z\rangle$ defined by

$$b_q|z\rangle = z_q|z\rangle \quad . \tag{40}$$

with

$$\langle z|z'\rangle = \exp\left\{ \sum_q (z_q^* z_q' - \frac{1}{2}|z_q|^2 - \frac{1}{2}|z_q'|^2)\right\} \quad . \tag{41}$$

Then

$$U(q,q',t) = \frac{1}{Z_{ph}} \int d\mu(z) \int d\mu(z') \, (z_q'^* + z_{-q}')(z_{-q'}^* + z_{q'})$$

$$\langle z| \exp\left\{ \sum_k (\omega_k - kv)[-\beta\hbar + it]b_k^+ b_k\right\} |z'\rangle \tag{42}$$

$$\langle z'| \exp\left\{ -i\sum_k (\omega_k - kv)[b_k^+ b_k + \frac{1}{n}\frac{1}{\sqrt{N}}(b_k^+ f_k + f_k^* b_k) + \frac{1}{n^2}\frac{1}{N}|f_k|^2]t\right\} |z\rangle \quad .$$

Here the integration measure is defined as

$$d\mu(z) = \prod_q \frac{1}{\pi} dx_q dy_q \quad \text{with} \quad x_q + iy_q = z_q \quad . \tag{43}$$

Since one can show that

$$\exp\{u b_q^+ b_q\}|z_q\rangle = \exp\{\frac{1}{2}|z_q|^2(e^{u+u^*} - 1)\}|e^u z_q\rangle \quad , \tag{44}$$

it immediately follows that the first matrix element in (42) equals

$$\exp\left\{ -\sum_q \left(\frac{1}{2}|z_q|^2 + \frac{1}{2}|z_q'|^2 - z_q^* z_q' \exp[(\omega_q - qv)(-\beta\hbar + it)]\right)\right\} \quad . \tag{45}$$

The second matrix element in (42) can be formulated as a path integral that can be evaluated exactly. Using the general relationship between the matrix element and the path integral:

$$\langle z_q'| \exp\left\{ -i\omega[b_q^+ b_q + u^* b_q + b_q^+ u + u^* u]t\right\} |z_q\rangle$$

$$= \exp\left[-\frac{1}{2}(|z_q'|^2 + |z_q|^2) - i\omega|u|^2 t\right] \int\limits_{\substack{y(0)=z_q \\ y^*(t)=z_q'^*}} D(y^*, y) \exp[iS(y^*, y)] \tag{46}$$

where

$$S(y^*, y) = \int_o^t dt' \left\{ iy^*(t')\frac{dy}{dt'} - \omega[y^*(t')y(t') + u^*y(t') + y^*(t')u] \right\} - iz_q'^* y(t) \quad , \qquad (47)$$

one can evaluate the path integral by standard techniques. The result for (46) is

$$\exp\left\{ -\frac{1}{2}(|z_q'|^2 + |z_q|^2) + z_q'^* z_q e^{-i\omega t} - (1 - e^{-i\omega t})(z_q'^* u + u^* z_q + |u|^2) \right\} \quad . \qquad (48)$$

Hence substituting for the matrix elements in (42) one obtains

$$U(q,q',t) = \frac{e^{g_n(t)}}{Z_{ph}} \int d\mu(z) \int d\mu(z')(z_q'^* + z_{-q}')(z_{-q'}^* + z_{q'})$$

$$\exp\left\{ - \sum_k \left(|z_k|^2 + |z_k'|^2 - z_k^* z_k' \exp[(\omega_k - kv)(-\beta\hbar + it)] \right. \right.$$

$$\left. \left. - z_k'^* z_k \exp[-i(\omega_k - kv)t] + \frac{1}{n}\frac{1}{\sqrt{N}}(z_k'^* f_k + f_k^* z_k)(1 - \exp[-i(\omega_k - kv)t]) \right) \right\}$$

$$\tag{49}$$

where

$$g_n(t) = -\frac{1}{n^2}\frac{1}{N}\sum_q |f_q|^2(1 - \exp[-i(\omega_q - qv)t]) \quad . \qquad (50)$$

The z' and z integrations can be performed without much effort. For example, the contribution from the term with the $z_q'^* z_{q'}$ factor, which we denote by $U_{abs}(q,q',t)$ since it is associated with the absorption of a phonon, is found to be

$$U_{abs}(q,q',t) = \frac{\exp[i(\omega_q - qv)t + g_n(t) + \zeta_n(t)]}{\exp[\beta\hbar(\omega_q - qv)] - 1}\left\{ \delta_{qq'} \right.$$

$$\left. - \frac{\frac{1}{n^2}\frac{1}{N}f_q^* f_{q'}(\exp[-i(\omega_q - qv)t] - 1)(\exp[i(\omega_{q'} - q'v)t] - 1)}{\exp[\beta\hbar(\omega_{q'} - q'v)] - 1} \right\}$$

$$\tag{51}$$

where

$$\zeta_n(t) = -\frac{4}{n^2 N}\sum_q \frac{|f_q|^2 \sin^2[\frac{1}{2}(\omega_q - qv)t]}{\exp[\beta\hbar(\omega_q - qv)] - 1} \quad . \qquad (52)$$

One notes that the breaking of translational symmetry by the deformation leads to off-diagonal terms corresponding to the loss of wavevector conservation. However these terms are proportional to $\frac{1}{N}f_q^* f_{q'}$ which is negligible when either $|q|$ or $|q'|$ is large compared with $2n\mu/\pi R$ as can be seen in the definition of f_q in (21). Furthermore recall that $-\pi \le qR \le \pi$ and the continuum approximation required $n\mu$ to be not much greater than one. Hence assuming $n\mu \ll \frac{\pi^2}{2}$ the off-diagonal terms are negligible except for a small region at the center of the Brillouin zone. Since the small wavevector terms do not significantly contribute to Γ owing to the q-dependence of $\overline{X}(k,q)$, one can ignore the off-diagonal terms in $U_{abs}(q,q',t)$ in evaluating Γ. Likewise the emission part of $U(q,q',t)$, which comes from the $z_{-q}' z_{-q'}^*$ term, is

$$U_{em}(q,q',t) = \frac{\exp[-i(\omega_q + qv)t + g_n(t) + \zeta_n(t)]}{1 - \exp[-\beta\hbar(\omega_q + qv)]}\left\{ \delta_{qq'} + O(\frac{1}{N}f_{-q'}^* f_{-q}) \right\} \quad . \qquad (53)$$

The remaining terms in $U(q,q',t)$ are $O(\frac{1}{N}f_q^* f_{-q'}^*)$ and $O(\frac{1}{N}f_{-q} f_{q'})$ respectively.

RESULTS

Using the $U(q,q',t)$ determined in the previous section, we obtain for the decay rate

$$\Gamma = \frac{2}{\hbar^2}\frac{1}{N^2}\sum_{k,q}|\overline{X}(k,q)|^2 Re\int_o^\infty dt\left\{\exp\left[-i\left(J(kR)^2 + (n^2 - \frac{2}{3}n)J\mu^2\right)t/\hbar + g_n(t) + \zeta_n(t)\right]\right.$$
$$\left.\left[\frac{\exp[i(\omega_q - qv)t]}{\exp[\beta\hbar(\omega_q - qv)] - 1} + \frac{\exp[-i(\omega_q + qv)t]}{1 - \exp[-\beta\hbar(\omega_q + qv)]}\right]\right\}$$

$$(54)$$

with

$$|\overline{X}(k,q)|^2 = \frac{\pi^2}{2n\mu}|F(q)|^2\frac{(qR)^2 \operatorname{sech}^2[\pi(k-q)R/2n\mu]}{(n\mu)^2 + (kR)^2} \quad . \qquad (55)$$

This is an essentially exact analytical expression for the decay rate at any temperature within lowest order perturbation theory provided $n\mu \ll \pi^2/2$. Note that in the case where a phonon of wavevector q is absorbed or a phonon of wavevector $-q$ is emitted, the delocalized excitation produced need not have wavevector equal to q. Wavevector is only approximately conserved by the $\operatorname{sech}^2[\pi(k-q)R/2n\mu]$ term. This, of course, is a consequence of the breaking of the translational symmetry by the deformation. Also one does not find the usual energy conservation. The terms $g_n(t)$ and $\zeta_n(t)$ occur because the phonons in the initial and final states are defined relative to different deformations.

For the soliton velocity equal to zero we can obtain an analytical expression for $g_n(t)$ defined by (50).

$$\lim_{v\to 0}\ g_n(t) = -g_o[ix\psi'(1 + ix) + \psi(1 + ix) - \psi(1)] \qquad (56)$$

where ψ is the digamma function and $x = \frac{1}{2}n\omega_a t$ and where

$$g_o = \frac{2\chi^2}{\pi\hbar w}\sqrt{\frac{m}{w}} = \frac{2}{\pi}\frac{J\mu}{\hbar v_a/R} \quad \text{and} \quad \omega_a = \frac{2\mu}{\pi}\sqrt{\frac{w}{m}} \quad . \qquad (57)$$

Also for $v = 0$ and sufficiently high temperatures $(T > nT_o)$

$$\lim_{v\to 0}\ \zeta_n(t) \simeq \frac{2g_o}{n}\frac{T}{T_o}[1 - \frac{1}{2}n\pi\omega_a t\coth(\frac{1}{2}n\pi\omega_a t)] \qquad (58)$$

where

$$T_o = \hbar\omega_a/k_B \quad . \qquad (59)$$

Furthermore if $g_o \ll 1$ and $T > nT_o$, one can show that

$$\frac{1}{\pi\hbar}Re\int_o^\infty dt\exp\left\{-i\left[J(kR)^2 + (n^2 - \frac{2}{3}n)J\mu^2 \mp \hbar\omega_q\right]t/\hbar + g_n(t) + \zeta_n(t)\right\}$$

$$\simeq \frac{\hbar\gamma/\pi}{\hbar^2\gamma^2 + [J(kR)^2 + (n^2 - \frac{2}{3}n)J\mu^2 \mp \hbar\omega_q]^2} \quad , \quad g_oT/T_o \ll 1$$

or

$$\simeq (4\pi\hbar^2 D^2)^{-1/2}\exp\left\{-\frac{[J(kR)^2 + n^2 J\mu^2 \mp \hbar\omega_q]^2}{4\hbar^2 D^2}\right\} \quad , \quad g_oT/T_o \gg 1 \qquad (60)$$

where

$$\hbar\gamma = \pi g_o k_B T \quad and \quad \hbar D = [\pi^2 n g_o k_B^2 T T_o/6]^{1/2} \quad . \qquad (61)$$

Thus the usual delta function in energy that one associates with perturbation theory is replaced by a Lorentzian or a Gaussian.

The generally accepted values of the parameters for the application to the α-helical protein molecule are

$$m = 114 m_p = 1.9 \times 10^{-25} kg \quad , \quad J = 1.55 \times 10^{-22} \, \text{Joules}$$

$$\chi = 6.2 \times 10^{-11} N \qquad , \qquad w = 13 \ N/m$$

and $v \leq 0.2 v_a$. For these values,

$$\mu = 1.9 \quad , \quad g_o = 0.215 \quad , \quad T_o = 77^\circ K \quad , \quad \text{and} \quad g_o T/T_o = 0.84$$

at $T = 300^\circ K$. Since the continuum approximation requires $n\mu \ll \pi^2/2$, only $n = 1$ adequately satisfies this requirement owing to the large value of μ. Also neither (63) nor (64) applies, although either will yield the correct order of magnitude for the decay rate with these parameters.

We have numerically evaluated the decay rate directly from (54) for $n = 1$ without approximating the integral or the functions $g_n(t)$ and $\zeta_n(t)$. The wavevectors were restricted to the Brillouin zone. This cut-off is consistent with the original lattice. However the continuum approximation completely neglects dispersion. We included dispersion by replacing the definitions of ω_q and $F(q)$ in (5) by

$$\omega_q = 2\sqrt{\frac{w}{m}}|\sin(\frac{1}{2}qR)| \quad \text{and} \quad F(q) = i\chi(\frac{2\hbar}{m\omega_q})^{1/2}\sin qR \quad . \tag{62}$$

The effect on Γ is to yield a somewhat smaller rate. At $T = 300^\circ K$ the calculated value for the lifetime defined as $1/\Gamma$ is $3.2 \times 10^{-13} s$. Since $\sqrt{m/w}$, the time to travel one lattice spacing at the speed of sound, is equal to $1.2 \times 10^{-13} s$, the soliton is not a viable mechanism for energy transport at biological temperatures if the parameters are correctly estimated. It is interesting to note that this value for the lifetime from a fully quantum mechanical perturbation treatment agrees well with estimates from the semi-classical numerical treatment of Lomdahl and Kerr.[3] We have also evaluated the lifetime as a function of temperature. The lifetime increases rapidly with decreasing temperature below T_o. For example, at $10^\circ K$ the lifetime is $1.1 \times 10^{-10} s$ which is sufficiently long for the soliton excitation to be a well-defined elementary excitation capable of energy transfer. This result is in qualitative agreement with the quantum Monte Carlo thermal equilibrium simulation[7] that found solitons only below $7^\circ K$ with this set of parameters.

It can be seen from (54) that the lifetime in units of the time $\sqrt{m/w}$ depends on J, m, χ, w, and T only as a function of μ, g_o, and T/T_o. At a given temperature the lifetime increases as μ and T_o increase and g_o decreases from the generally accepted values. Also it increases as n increases but not significantly for values such that $n < \pi^2/2\mu$. For $\mu = 1.9$ one finds that $g_o T/T_o < 0.02$ is required for viable solitons ($\Gamma^{-1}/\sqrt{m/w} > 5 \times 10^2$). This upper bound on $g_o T/T_o$ is almost two orders of magnitude smaller than its value of 0.84 at $T = 300^\circ K$ based on the accepted values for the parameters. A larger μ would permit a larger $g_o T/T_o$, however $\mu \simeq 2$ is about as large as possible if the soliton is to extend over several lattice sites. Hence very large changes in the parameter values would be required for the soliton mechanism for energy transport to work at biological temperatures. In particular, J needs to be about an order of magnitude smaller and w an order of magnitude larger since g_o/T_o is proportional to J/w for fixed μ.

The arguments[4] that the transition from soliton to delocalized excitation is forbidden by the Franck-Condon principle in a fully quantum mechanical treatment are obviously not valid since our result is essentially exact. The Franck-Condon type factor $|_{ph} < 0|\tilde{0} >_{ph}|^2 = \exp[-\frac{1}{N}\sum_{q\neq 0}|f_q|^2]$, which is extremely small since $\frac{1}{N}\sum_{q\neq 0}|f_q|^2 \to \infty$ as $N \to \infty$ owing to the small q dependence of $|f_q|^2$, does not appear in the calculation of the lifetime. The related factors $\exp[g(t)]$ and $\exp[\zeta(t)]$ that do appear are finite since in both $g(t)$ and $\zeta(t)$ the $|f_q|^2$ is multiplied by a quantity that is proportional to q as $q \to 0$.

REFERENCES

1. A. S. Davydov and N. I. Kislukha, <u>Phys. Stat. Sol. B</u> 57:465 (1973); A. S. Davydov, J. Theor. Biol. 38:559 (1973).

2. For Reviews: A. S. Davydov, <u>Sov. Phys. Usp.</u> 25:898 (1982); "Biology and Quantum Mechanics," Pergamon, New York (1982); "Solitons in Molecular Systems," D. Reidel, Dodrecht (1985).

3. P. S. Lomdahl and W. C. Kerr, <u>Phys. Rev. Lett.</u> 55:1235 (1985); A. F. Lawrence, J. C. McDaniel, D. B. Chang, B. M. Pierce, and R. R. Birge, <u>Phys. Rev. A</u> 33:1188 (1986).

4. A. S. Davydov, <u>Sov. Phys. JETP</u> 51:397 (1980); A. C. Scott <u>in</u> "Energy Transfer Dynamics," T. W. Barrett and H. A. Pohl, eds., Springer-Verlag, Berlin (1987), p. 167.

5. J. P. Cottingham and J. W. Schweitzer, <u>Phys. Rev. Lett.</u> 62:1792 (1989).

6. A. A. Eremko, Yu. B. Gaididei and A. A. Vakhnenko, <u>Phys. Stat. Sol. B</u> 127:703 (1985).

7. X. Wang, D. W. Brown and K. Lindenberg, <u>Phys. Rev. Lett.</u> 62:1796 (1989).

22

THE TEMPERATURE DEPENDENCE OF EXCITON-PHONON COUPLING IN THE CONTEXT OF DAVYDOV'S MODEL; THE DYNAMIC DAMPING OF SOLITON

M. Satarić (a), Z. Ivić (b) and R. Žakula (b)

a) Faculty of Technical Sciences, Novi Sad
   Yugoslavia
b) Institute of Nuclear Sciences "Boris Kidrič"
   Laboratory of Theoretical Physics, Belgrade
   Yugoslavia

ABSTRACT

The temperature-dependent behaviour of localized excitons in a linear molecular α-helix chain model is investigated. After transforming Davydov's original Hamiltonian for the α-helix soliton to include Bogoliubov's variational theorem, the temperature dependence of the "threshold" of solitonic solution is obtained. We also investigated the damping of a soliton under the influence of the thermal bath. The approach is based on the consequent microscopic treatment in context of quantum Langevin equation.

I INTRODUCTION

Davydov's solitons are robust localized waves that couple molecular amide-I vibrations to longitudinal sound in a quasi-
-one-dimensional α-helix protein chain. Davydov's model is popular in the context of theoretical explanations of different biological phenomena, for example the molecular treatment of animal muscle contraction |1| and the function of DNA |2|. In addition, this model is a cornerstone for the examination of measurements of infrared apsorption and Raman scattering on crystalline acetanilide, in a similar manner to laser-Raman spectra measured on metabolically active cells |3|.

It is pointed out that the formation of Davydov's soliton (DS) strongly depends on the various physical parameters. Furthermore we emphasize that a few questions need to be resolved before DS model can be applied to proteins in the living matter.

In the first place is the influence of thermal bath and of the strength of the exciton-phonon coupling (EPC) on the stability of DS. Davydov examined the soliton in a molecular chain which is in contact with a thermal bath |4|. His main result is that thermalization decreases the effective EPC and increases the soliton mass. We examined similar problem for

Davydov's Soliton Revisited, Edited by P.L. Christiansen and A.C. Scott
Plenum Press, New York, 1990

the case of biexciton soliton in a molecular chain $|5|$.

Lomdahl and Kerr $|6|$ proposed a model that included the interaction of DS with a thermal bath. The dissipation effects are taken into account phenomenologically by introducing damping coefficient and random forces into dynamic equations. They found that random forces corresponding to 10 K are sufficient to destroy DS. On the other hand, Lawrence et al. $|7|$ estimated (on the basis of similar treatment) that the upper limit of the existence of DS is approximatly 240 K. However, Bolterauer $|8|$ showed that the classical recipe of thermalization is inadequate for a phonon system, which is essentially a quantum system.

The main aim of our investigation was to employ a consequent quantum mechanical finite-temperature approach, attempting to answer whether it is realistic to expect DS to appear at biologically relevant temperatures. In first part of our work an exactly tractable model convenient to study the explicite temperature dependence of EPC is presented. Similar treatment related with experimental confirmation of DS existence in crystalline acetanilide is proposed by Alexander $|9|$.

Second part is dedicated to study of dynamical behaviour of DS in the presence of damping forces and noise forces. Instead of mentioned classical approach, here the quantum Langevin equation is used.

II   THE MODEL HAMILTONIAN

Let us begin by considering the model of molecular chain in which the collective excitations can propagate along the chain transferring the amide-I exciton energy $\Delta \sim 0,205$ eV. As usual we took only the most important effects into account.

The corresponding Hamiltonian in second quantized form is given by

$$H = \Delta \sum_n B_n^+ B_n - J \sum_n B_n^+ (B_{n+1} + B_{n-1}) + \sum_q \hbar\omega_q (a_q^+ a_q + 1/2) +$$

$$+ N^{-1/2} \sum_{n,q} F_n(q) B_n^+ B_n (a_q + a_{-q}^+) \qquad (2.1)$$

where the interaction between the amide-I excitation and the displacements of the peptide groups has the form

$$F_n(q) = 2i\chi \left(\frac{\hbar}{2M\omega_q}\right)^{1/2} \sin(R_o q) \exp(in R_o q) \quad i = \sqrt{-1} \qquad (2.1.a)$$

Here J is the accustomed resonance transfer integral between adjacent peptide groups separated by the distance R_o; χ is the nonlinear EPC arising from the fact that the magnitude of the exciton energy Δ depends on the momentary distance between adjacent peptide groups; ω_q is frequency of normal mode of wave vector q, corresponding to the free harmonic phonon Hamiltonian; at last, the operators B_n and B_n^+ destroy and create an amide-I exciton on the n-th peptide group, while a_q and a_q^+ destroy and create the phonons.

Crucial for the soliton solution is the following product ansatz originally given by Davydov |1|

$$|\psi(t)> = \sum_n \psi_n(t)B_n^+|0>_{ex} \prod_q |\alpha_q(t)> \tag{2.2}$$

which is a superposition of single exciton states $\psi_n(t)$ and coherent phonon states $|\alpha_q(t)>$ defined by the property that

$$a_q|\alpha_q(t)> = \alpha_q(t)|\alpha_q(t)> \tag{2.2.a}$$

for all of the phonon annihilation operators a_q. The amplitudes $\psi_n(t)$ and $\alpha_q(t)$ are treated as generalized coordinates with corresponding generalized momenta $i\hbar\psi_n^*, i\hbar\alpha_q^*$. The equations of motion for these generalized dynamic variables supposed to be classical Hamilton equations in which the expectation value $<H> = <\psi(t)|H|\psi(t)>$ of the quantum Hamiltonian (2.1) plays the role of Hamilton function as follows

$$i\hbar\dot{\psi}_n(t) = \frac{\partial<H>}{\partial\psi_n^*(t)} \quad ; \quad i\hbar\dot{\alpha}_q(t) = \frac{\partial<H>}{\partial\alpha_q^*(t)} \tag{2.3}$$

The basic question is whether the product ansatz (2.2) for this system is good enough. Bolterauer and Opper |10| recently showed that only in the limit of a purely classical phonon lattice the ansatz is exactly valid. Brown and co-workers |11,12| also demonstrated that the Hamilton equation propagation of the ansatz state vector (2.2) is not a solution of the Schrödinger equation.

From our opinion, Davydov's ansatz is based on the time-independent variational principle, which is one of many examples of Hartree's method. Having in mind that the base of Davydov's theory is the supposition of strong exciton-phonon interaction, then it is reasonable to take the trial state in Pekar's form |13|

$$\left.\begin{aligned}|\tilde{\psi}(t)> &= \exp(-i\frac{E}{\hbar}t)\,|\psi> \\[4pt] |\psi> &= \sum_k \psi_k B_k^+|0>_{ex}\prod_q|\alpha_q>\end{aligned}\right\} \tag{2.4}$$

by using the Fourier transform of exciton population

$$B_n^+ = N^{1/2}\sum_k B_k^+\exp(-inR_ok) \tag{2.4.a}$$

The variational parameters ψ_k and α_q are now time-indenpendent. The equations of motion for variational parameters can be formed by minimization of functional

$$F = \langle\psi|(H-\gamma P-\lambda)|\psi\rangle \quad , \tag{2.5}$$

where γ and λ are Lagrangean multipliers. The functional (2.5) must be optimized under the following restrictions; first, the amplitude $|\psi\rangle$ is subjected to the normalization condition $\langle\psi|\psi\rangle = 1$, second, the total momentum of the system is an integral of motion

$$P = \sum_k \hbar k B_k^+ B_k + \sum_q \hbar q a_q^+ a_q \tag{2.5.a}$$

Performing the minimization of eq.(2.5) in the long-wave continuum limit and seeking D'Alembert solution ($|\psi_{(x,t)}| = |\psi(x-vt)|$), after eliminating the phonon variable, the well-known nonlinear Schrödinger equation has the shape

$$(\Delta-2J+w-\lambda)\psi-JR_o^2\psi_{xx}+i\hbar\gamma\psi_x-G|\psi|^2\psi = 0 \tag{2.6}$$

where we introduced the set of denotations for phonon energy and for parameter of cubic nonlinearity as follows

$$w = \sum_q \hbar\omega_q(|\alpha_q|^2+1/2) \; ; \quad G = \frac{4\chi^2}{\kappa(1-s^2)} \; ; \quad s = \frac{v}{v_o} \tag{2.6.a}$$

v represents the solitonic velocity while v_o is the longitudinal sound velocity; κ is the chain elasticity constant.

From this procedure one can be generally infered that variational treatment reproduces all Davydov's results and can support soliton transport (at least at zero temperature).

III   THE TEMPERATURE DEPENDENCE OF EPC IN THE CONTEXT OF SOLITONIC FORMATION

In order to include the role of quantum fluctuations of molecular equilibrium positions just as well the thermal oscilations around the new equilibrium molecular positions, we now transform the Hamiltonian (2.1) by a unitary transformation $|14|$

$$U = N^{-1/2} \sum_{n,q} Y_n(q)(\hbar\omega_q)^{-1}B_n^+B_n(a_q-a_{-q}^+) \tag{3.1}$$

where the corresponding, as yet undetermined, variational parameter (VP) has the attribute $Y_n(-q) = Y_n^*(q)$. In the majority of papers dealing with such method the restriction of type $Y_n(q) = F_n(q)$ has been taken into account for all temperatures. However, Yarkony and Silbey $|15|$ are established that mentioned equality is only correct at high enough temperatures.

Starting from the fact that the main aim of our paper is to find the explicit thermal dependence of EPC, it is convenient to define the VP as follows

$$Y_n(q) = 2iy(T)(\frac{\hbar}{2M\omega_q})^{1/2} \sin(R_0 q)\exp(inR_0 q) \ , \qquad (3.2)$$

where $y(T)$ is a new VP which will be obtained afterwards. Performing the transformation $\tilde{H} = \exp(U)H\exp(-U)$, we found the new form of Hamiltonian (2.1)

$$\tilde{H} = E\sum_n B_n^+ B_n - J\sum_n \theta_n^+ \theta_{n+1} B_n^+ B_{n+1} + h.c. - 2A(T)\sum_n \left[(B_n^+)^2 B_n^2 + B_n^+ B_{n+1}^+ B_{n+1} B_n \right] +$$

$$+ \sum_q \hbar\omega_q (a_q^+ a_q + 1/2) + N^{-1/2} \cdot \sum_{n,q} G_n(q) B_n^+ B_n (a_q + a_{-q}^+) \ , \qquad (3.3)$$

where

$$\theta_n = \exp\{ N^{-1/2} \sum_q Y_n(q)(\hbar\omega_q)^{-1} \cdot (a_q - a_{-q}^+) \} \qquad (3.3.a)$$

is the operator destroying the virtual phonon which takes place in creation of "the phonon cloud" surronding the exciton at site n; the shifted energy of a such "dressed" exciton is $E = \Delta - 4A(T)$, where

$$A(T) = \kappa^{-1}\left[\chi y(T) - \frac{1}{2} y^2(T) \right] \qquad (3.3.b)$$

represents the effective dynamical exciton-exciton interaction; the effective nonlinear EPC in such framework is defined by

$$G_n(q) = 2i\left[\chi - y(T) \right](\frac{\hbar}{2M\omega_q})^{1/2} \cdot \sin(R_0 q)\exp(inR_0 q) \qquad (3.3.c)$$

We now infer that transformed Hamiltonian (3.3) describes the system of so-called "partially dressed" excitons interacting with lattice vibrations (the real phonons), and, in the other hand, mutually interacting through the exchange of virtual phonons.

Further development of this problem will have the perturbative character. The transformed Hamiltonian $\tilde{H}$ may be rewritten as

$$\tilde{H} = H_m + (\tilde{H} - H_m) \ , \qquad (3.4)$$

where the first term on the right hand side H_m represents the so-called "model Hamiltonian" (MH), while the second term $(\tilde{H} - H_m)$ may be treated as perturbation. In order to involve the thermalization it is convenient to choose the MH in the following manner; As first, for the sake of simplicity, we transform $\tilde{H}$ into the inverse space of exciton operators using mentioned Fourier transform (2.4.a), $\tilde{H}(n,q) \rightarrow \tilde{H}(k,q)$. Then, we make the partially averaging of $\tilde{H}(k,q)$ in such a way that only "dressed" excitonic part of Hamiltonian is being averaged with respect to the phonon basis, while the remaining part of $\tilde{H}(k,q)$ is unchanged. In this way we obtained MH in the form

$$H_m = \sum_k E(k) B_k^+ B_k - N^{-1/2} \sum_{k,q,\rho} v(k) B_{k+\rho}^+ B_{q-k}^+ B_q B_\rho + \sum_q \hbar\omega_q a_q^+ a_q$$

$$+ N^{-1/2} \sum_{k,q} G(q) B_{k+q}^+ B_k (a_q + a_{-q}^+) \tag{3.5}$$

The new energy parameters are defined as follows

$$E(k) = \Delta - 2A(T) - 2J\exp\left[-W(T)\right] \cdot \cos(R_o k)$$

$$v(k) = 4A(T)\cos^2\left(\frac{R_o k}{2}\right) \tag{3.5.a}$$

$$G(q) = 2i\left[\chi - y(T)\right]\left(\frac{\hbar}{2M\omega_q}\right)^{1/2} \cdot \sin(R_o q)$$

where $\exp\left[-W(T)\right]$ represents Debye-Waller's thermal factor which is the result of phonon averaging of "the cloud of virtual phonons" $<\theta_n^\pm \theta_{n+1}>_{ph} = \exp\left[-W(T)\right]$. This factor includes the contributions of quantum fluctuations caused by thermal vibrations of molecules about the new equilibrium positions.

More explicit shape of thermal factor is

$$W(T) = 4R_o^5 (\pi\hbar M v_o^3)^{-1} \cdot y^2(T) \int_o^{q_o} q\left(\overline{v}_q + 1/2\right) dq \; ; \; q_o = \frac{\pi}{R_o} \tag{3.5.b}$$

$$\overline{v}_q = \left[\exp(\beta\hbar\omega_q) - 1\right]^{-1} \; ; \; \beta = (k_B T)^{-1}$$

For full description of dynamics of system represented by MH it is necessary to find out $y(T)$. According to Bogoliubov's variational theorem $|15|$, the following inequality for free energy (FE) of the system is valid

$$\Phi \leq \Phi_o \equiv -\frac{1}{\beta}\ln\{Tr\left[\exp(-\beta H_m)\right]\} - <(H-H_m)>_o \tag{3.6}$$

where Φ represents the actual FE of the system (2.1), while Φ_o is the FE corresponding to the MH. The symbol $<...>_o$ denotes the averaging with respect to the canonical ensemble of noninteracting phonons and excitons.

For first instance, confining our attention on the qualitative analysis, we can neglect the dynamical interaction $v(k)$ in MH. Then the FE of the system can be expressed as a sum of FE of the noninteracting subsystems (excitonic and phononic).

$$\Phi_o = \sum_k E(k)\overline{n}\left[E(k)\right] - \frac{1}{\beta}\ln\{\overline{n}\left[E(k)\right]\} - \sum_q \hbar\omega_q \overline{v}_q - \frac{1}{\beta}\sum_q \ln(\overline{v}_q) \; , \tag{3.7}$$

where, besides introduced phonon population $\bar{v}_q$, we introduce
the average value of excitonic population $\bar{\eta}[E(k)]$

$$\bar{\eta}[E(k)] = \{\exp[\beta E(k)]-1\}^{-1} \qquad (3.7.a)$$

Because the phonon part of FE is independent of $y(T)$, we
can omit the two last terms in eq. (3.7) and examine only the
FE of "partially dressed" excitons . The value of $y(T)$ which
minimizes the FE is given by the conditions

$$\frac{\partial \Phi_o}{\partial y(T)} = 0 \quad ; \quad \frac{\partial^2 \Phi_o}{\partial y(T)^2} > 0 \qquad (3.8)$$

which reflect into two implicit relations

$$X^{-1} = 1+2\kappa JF(T)\exp\left[-X^2\chi^2 F(T)\right]$$

$$(3.8.a)$$

$$X^2(1-X) < \left[2\chi^2 F(T)\right]^{-1} \quad ; \quad X = \chi^{-1}y(T)$$

and where the temperature function $F(T)$ is expressed by the
integral

$$F(T) = 8MR_o(\alpha)^{-1}\int_o^{z_o} dz\cdot z\operatorname{cth}z \; ; \quad \left.\begin{array}{l} \sigma=\pi v_o\kappa^2\beta^2\hbar^3 \\ z_o=\pi v_o\hbar\beta(2R_o)^{-1} \end{array}\right\} (3.8.b)$$

In the limit of high temperatures, and on the other hand, for
low enough temperatures, the explicit value of the function
(3.8.b) may be written as follows

$$F(T) = \begin{cases} 4M\left[(\kappa\hbar)^2\beta\right]^{-1}\left[1+(\frac{\pi\hbar v_o\beta}{6R_o})^2\right] & ; \quad T \gg T_D \\[2ex] \pi R_o(v_o\hbar\kappa)^{-1}+\frac{2}{3}MR_o\pi\hbar\sigma^{-1} & ; \quad T\to 0 \end{cases} \qquad (3.8.c)$$

where $T_D = v_o\hbar(2R_o k_B)^{-1}$ is Debye's temperature of the system.
It is easy to show that the DS existence in a molecular chain
is very sensitive to variations of the relevant chain parameters. The parameters used in earlier numerical estimations
have fairly wide ranges of values $|16-19|$. Our intention is to
estimate the temperature dependence of DS formation by varying
the EPC in the rather wide range $(1-30)\cdot 10^{-11}N$. The results of
our numerical estimations are given in figs. 1. and 2. The
figs. 1. and 2. show how the ratio $X =\chi^{-1}y(T)$ depends on χ,
taking the temperature as parameter. The relevant parameters
for these function are $\kappa = 76Nm^{-1}$, $M = 1,9\cdot 10^{-25}kg$, $J =$
$= 1,55.10^{-22}J$. In this case the corresponding function has
the approximate shape

$$X = \{1+aT\left[1+(\frac{101}{T})^2\right]\}^{-1} \quad ; \quad a=2,39\cdot 10^{-3}K^{-1} \qquad (3.9)$$

For final discussion it is convenient to introduce the
effective EPC as $\chi_{eff.} = \chi-y(T)$. The first result of numerical

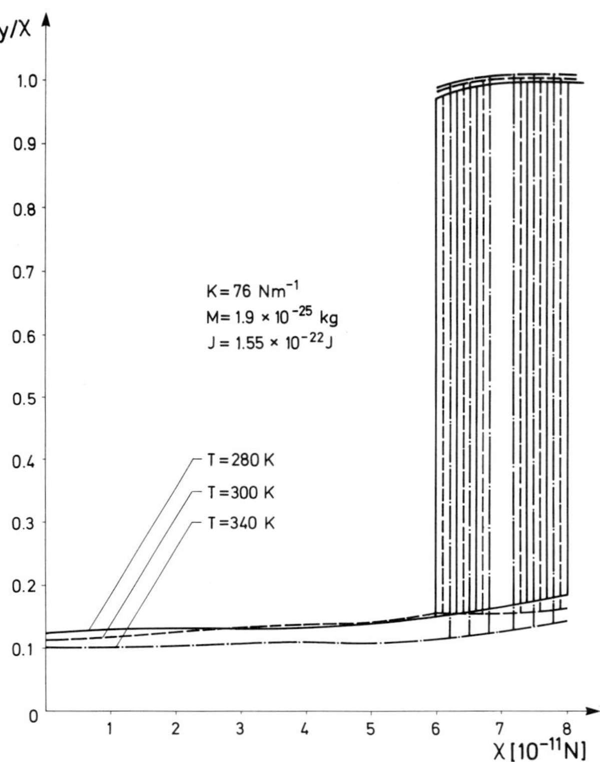

Figure 1. Temperature dependence of Davydov's soliton formation. See text for details.

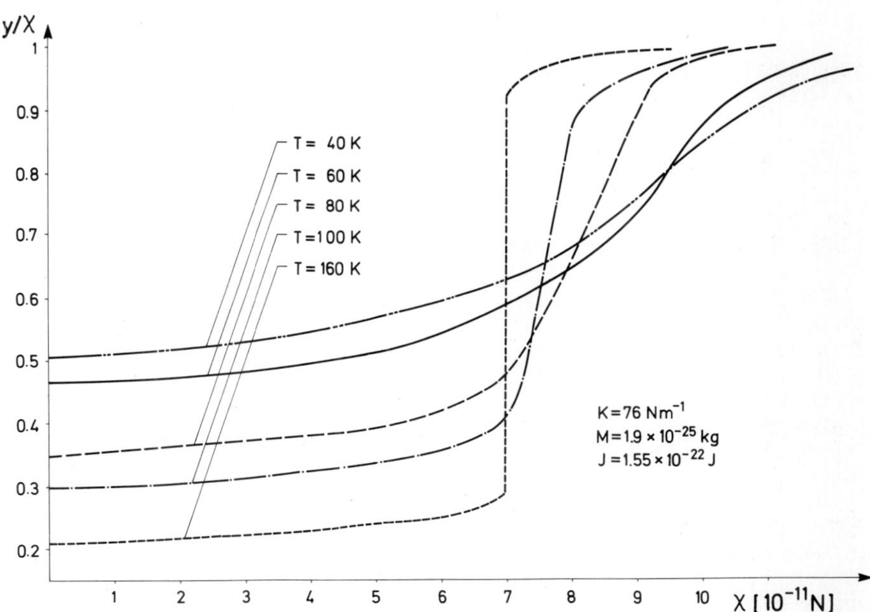

Figure 2.   Temperature dependence of Davydov's soliton formation.
See text for details.

analysis is that χ_{eff} decreases with increasing temperature, which is agreement with earlier predications.

For large enough values of χ the FE has two minima with respect to variations in y(T). This phenomenon appears even at biologically relevant temperatures if the EPC satisfies the inequality $\chi > 7,7 \cdot 10^{-11}$N. More precisely, in the region $(7,7-11) \cdot 10^{-11}$N, begining at a temperature of about 160 K, the two minima of FE change abruptly from a small value of y(T) to a large value. Physical insight in this circumstance reflects that under mentioned conditions one has a certain "phase transition" from the state of a "partially dressed" exciton to the state of a "fully dressed" exciton. For the first minimum χ_{eff} is still large enough that DS exists. For the second minimum $\chi_{eff} \simeq 0$, which is the consequence of the fact that the exciton-phonon mechanism of DS formation practically vanishes. However, it seems that then the exciton-exciton interaction can take over the role of DS formation $|20|$.

At last, we again emphasize that instead of a known region of solitonic existence $(3,4-6,2) \cdot 10^{-11}$N, for biologically relevant temperatures our treatment offers a new threshold or "solitonic window" in the range $(7,7-11) \cdot 10^{-11}$N.

IV   THE DYNAMIC DAMPING AND MODIFICATION OF DS AS A
       CONSEQUENCE OF THERMALIZATION

The problem of a quantum particle coupled to a quantum-mechanical heat bath is fundamental to many fields of physics; statistical mechanics, condensed matter, quantum optics etc. For such a system the description in terms of the quantum Langevin equation has a broad and general application $|21,22|$. We followed this method having in mind that it is a complete microscopic quantum description that can be characterized in a precise and general way.

The intrinsic point is that we examine the complete Hamiltonian consisting of three parts. Solitonic subsystem, described by model Hamiltonian without phonon part $H_S = H_m - H_{ph}$, interacts with thermal bath $H_B = H_{ph}$ through perturbative interaction $H_{SB} = (\tilde{H} - H_m)$.

Let us consider certain operator A(t) corresponding to solitonic subsystem in the Heisenberg's picture. The corresponding equation of motion is then

$$\dot{A}(t) = i\hbar^{-1}\left[H,A(t)\right] = iLA(t) \tag{4.1}$$

where the Liouville operator L is given as sum of three terms
$$L = \sum_j L_j,$$

$$L_j A(t) = \hbar^{-1}\left[H_j,A(t)\right] \; ; \quad j=S,B,SB \tag{4.1.a}$$

In order to transform (4.1) into the form of quantum Langevin equation it is necessary to separate the relaxation and the fluctuation part. This purpose can be achieved by using the technique of projection operator $\hat{P}$ defined as follows

$$\hat{P}A(t) = \text{Tr}(\hat{\rho}_B A(t)) = <A(t)>_B \quad , \qquad (4.2)$$

where $\hat{\rho}_B$ is the equilibrium statistical operator of thermal bath.

Applying the convenient transformations of eqs (4.1) and (4.2) we obtained $|23|$ that the amplitude of solitonic state is subjected under the law of dynamical evolution as follows (here $\psi_k(t)$ takes over the role of $A(t)$)

$$\dot{\psi}_k(t) = \exp(iLt)\cdot i(L_S + <L_{SB}>_B)\psi_k(0) +$$

$$+\exp(iLt)\int_0^\infty d\tau\{\exp(-iL_S\tau)<iL_{SB}\exp[i(L_S+L_B)\tau]\cdot iL_{SB}>_B\}\psi_k(0) +$$

$$+R_k(t) \qquad (4.3)$$

whence

$$R_k(t) = (1-\hat{P})\exp[i(L_S+L_B)t]\cdot iL_{SB}\psi_k(0) \qquad (4.3.a)$$

represents the fluctuation part of dynamical equation, and $\psi_k(0)$ is the solitonic amplitude at initial moment. The symbol $<...>_B$ represents the averaging defined by (4.2). Let us consider the dynamical variable $D(\alpha_q, \psi_k, h.c.)$ which depends on the variables of both interacting subsystems (phonon and soliton)

$$D = A(\alpha_q, \alpha_q^*)\cdot B(\psi_k, \psi_k^*) \qquad (4.4)$$

Then, the action of operator L_j is explicitly given by the next equation

$$L_j D = -i\hbar^{-1}\left\{\sum_q\left[\frac{\partial <H_j>}{\partial \alpha_q^*}\cdot\frac{\partial A}{\partial \alpha_q} - \frac{\partial <H_j>}{\partial \alpha_q}\cdot\frac{\partial A}{\partial \alpha_q^*}\right]B - \sum_k A\left[\frac{\partial <H_j>}{\partial \psi_k^*}\cdot\frac{\partial B}{\partial \psi_k} - \frac{\partial <H_j>}{\partial \psi_k}\frac{\partial B}{\partial \psi_k^*}\right]\right\}$$

$$(4.5)$$

The expectation values $<H_j>$ are given by averaging the corresponding Hamiltonian with respect to ansatz state (2.2). On the basis of fundamental relation (4.5), after performing the cumbersome calculations in eq(4.3), we finally have got

$$i\hbar\dot{\psi}_k(t) = [E(k)+\Delta E(k)]\psi_k(t) - GN^{-1}\sum_{q,\rho}\psi_{q+\rho}^*\psi_{k+q}\psi_\rho - \frac{i\hbar}{2}\Gamma_k + i\hbar R_k,$$

$$(4.6)$$

where the damping term Γ_k is given by the expression

$$\Gamma_k = \{2J\exp[-W(T)]\}^2 \frac{2\pi M}{\kappa \hbar^3} \sum_q |A(q)|^2 \Big[\bar{\nu}_q \delta(\varepsilon_{k+q}-\varepsilon_k+\hbar\omega_q)+$$

$$+(\bar{\nu}_q+1)\delta(\varepsilon_{k+q}-\varepsilon_k-\hbar\omega_q)\Big] \qquad (4.7)$$

while the exciton energy shift $\Delta E(k)$ has the value

$$\Delta E(k)=4\frac{MJ^2}{\kappa N}\exp[-2W(T)]\sum_q |A(q)|^2 \Big[\bar{\nu}_q \mathcal{P}(\frac{1}{\varepsilon_{k+q}-\varepsilon_k+\hbar\omega_q})+$$

$$+(\bar{\nu}_q+1)\mathcal{P}(\frac{1}{\varepsilon_{k+q}-\varepsilon_k-\hbar\omega_q})\Big] \qquad (4.8)$$

We emphasize that $A(q)$ is simply equal to $F(q)$ if $\chi_{eff} = 0$, i.e. if we have the system at high enough temperature. On the other hand, in the case where only one excitonic soliton is propagating, the equality $A(q) = G(q)$ is valid. The energy parameter ε_k is related with solitonic energy $E_S = \hbar\omega_S$ and with solitonic v as follows

$$\varepsilon_k = \hbar(kv-\omega_S) \qquad (4.9)$$

$\mathcal{P}(x)$ is the principal value of corresponding integral. At last, the fluctuation term $R_k(t)$ has the explicite form

$$R_k(t)=i\hbar^{-1}\frac{J}{N}\sum_{k'}\psi_{k'}\exp(i\omega_k t)\sum_{m,s}\Delta\theta_{ms}\exp\{i[(k'-k)mR_o+k'sR_o]\} \qquad (4.10)$$

where the virtual phonon fluctuations are

$$\Delta\theta_{ms} = \theta_m^+\theta_{m+s}-\langle\theta_m^+\theta_{m+s}\rangle_B \quad ; \quad s=\pm 1 \qquad (4.10.a)$$

Let us see in what manner the solitonic solution changes under the influence of thermal bath. First of all, it is clear that the equation (4.6) represents the Fourier transform of nonlinear Schrödinger equation (NSE), but now with the relaxation and fluctuation term. By using the accustomed recipe of transition from discret to the continuum description |24| we obtain the NSE

$$i\hbar\dot{\psi}(x,t)=\tilde{E}\psi(x,t)-JR_o^2\exp[-W(T)]\psi_{xx}-G|\psi|^2\psi-\frac{i\hbar}{2}\Gamma\psi(x,t)+$$

$$+R(x,t) \qquad (4.11)$$

where

$$\tilde{E} = \Delta-2J\exp[-W(T)]-2A(T)+\Delta E$$

$$(4.11.a)$$

$$R(x,t) = N^{-1/2}\sum_k R_k(t)\exp(ikx)$$

Inserting the explicit values of physical parameters appearing in new terms $\Gamma(x,t)$, ΔE and $R(x,t)$ we easily find that their range is much smaller in comparison with basic parameters of NSE, so that we can apply one of the convenient perturbative methods in final treatment.

Let us confine our attention, for the moment, to the case where the fluctuation term $R(x,t)$ can be neglected. In these instances all effects reflect in the change of solitonic amplitude which becomes the function of time $\psi_0 = \psi_0(t)$. The other parameters remain unchanged in this approximation.

By using the mean field method the solution of eq. (4.11) may look for in the form $\psi(x,t) = <\psi>+\delta\psi$. Inserting this solution into (4.11) and then averaging NSE with respect the phonon ensamble $<...>_B$, and having in mind that the condition $<\psi(x,t)> \gg \delta\psi$ is fulfilled, we get

$$i\hbar<\dot{\psi}>=\tilde{E}<\psi>-JR_0^2\exp\left[-W(T)\right]<\psi>_{xx}-G\left|<\psi>\right|^2<\psi>-\frac{i\hbar}{2}\Gamma<\psi> \qquad (4.12)$$

The solution of last equation has well-known bell-shaped form with time dependent amplitude

$$\psi_0(t) = \psi_0(0)\exp(-\Gamma t) \qquad (4.13)$$

This fact involves the change of solitonic domain of localization. The effects of decreasing of solitonic amplitude and of spreading of solitonic localization can be explained as a consequence of the processes of emission and apsorption of acoustic phonons during the solitonic motion.

REFERENCES

1. Davydov A.S., J. Theor. Biol. 38(1973) 359
2. Balanovski E. and Beaconsfield P., Phys. Rev. A 32 (1985) 3059
3. Lomdahl P.S., Mac Neil L., Scott A.C., Stoneham M.E. and Webb S.J., Phys. Lett. A 92(1982) 207
4. Davydov A.S., Sov. Phys. JETP. 78(1980) 789
5. Satarić M., Vujičić G. and Žakula R., Il Nuovo Cimento 63B(1981) 709
6. Lomdahl P.S. and Kerr W.C., Phys. Rev. Lett. 55 (1985) 1235
7. Lawrence A.F., McDaniel J.C., Chang D.B. and Pierce B.M., Phys. Rev. A 33(1986) 1188
8. Bolterauer H., Proceeding of the MIDIT 1986, workshop
9. Alexander D.M., Phys. Rev. Lett. 54, No2(1985) 138
10. Bolterauer H. and Opper M. "The quantum lifetime of the Davydov soliton", subjected for publication in Physica Scripta
11. Brown D.W., Lindenberg K. and West B.J., Phys. Rev. A, 33(1986) 4104
12. Brown D.W., West B.J. and Lindenberg K., Phys. Rev. A, 33(1986) 4110
13. Pekar S.I., Sov. Phys. JETP 16(1946) 341
14. Alexander D.M., Phys. Rev. Lett. 54, No2(1985) 138
15. Yarkony D. and Silbey R., J. Chem. Phys. 65(1976) 1042

16. MacNeil L. and Scott A.C., Physica Scripta 29(1984) 284
17. Careri G., Buontempo V., Karta F., Gratton E. and Scott A.C., Phys. Rev. Lett. 51(1983) 304
18. Eilbeck J., Lomdahl P.S. and Scott A.C., Phys. Rev. B 30(1984) 4703
19. Scott A.C., Phys. Lett. A 86(1981) 60
20. Satarić M., Ivić Z., Shemsedini Z. and Žakula R., Journal of Molecular Electronics 4(1988) 223
21. Valasek K., Zubarev D.N. and Kuzemski A.L., Sov. Phys. TMF 5(1970) 281
22. Shibata F. and Hashitsume N., J. Phys. Soc. Japan 44 (1978) 1435
23. Ivić Z., The Doctoral Thesis "The Dynamics and Transport Properties of Solitons in One-Dimensional Molecular Crystals in Contact with Thermal Bath", Belgrade (1988)
24. Satarić M., Ivić Z. and Žakula R., Physica Scripta 34(1986) 283.

Question by Kenkre: The question asked of you by Brown, concerning the unacceptable real decay of probability in your equation, is serious indeed since the total probability must be conserved even after your approximation. However, this problem, it would appear, can be fixed in the usual manner by the use of density matrix (rather than amplitude) equations in which only the off-diagonal elements in an appropriate representation suffer the decay at your rate Γ, the diagonal elements (the probabilities) being unaffected. Why don't you introduce this modification into your theory?

Reply: We must take into account (which follows from the procedure of derivation) that our ψ is not pure state because we used the averaging with respect to the phonon ensemble. On the other hand, in the similar approach, in the ferromagnetism, ψ is not Schrödinger's amplitude but the scalar field, and owing to this occasion we performed the mentioned treatment. Moreover, for the case of soliton as bound state of $N \gg 1$ excitons, the condition $\int |\psi|^2 dx = N$ is valid. The change of ψ reflects the nonconservation of the number of excitons in the cluster.

23

TEMPERATURE EFFECTS ON THE DAVYDOV SOLITON

H. Bolterauer

Institut für Theoretische Physik, Justus-Liebig-
Universität Giessen, D-6300 Giessen, FRG

Abstract

There are two effects which tend to delocalize the Davydov soliton. First, quantum
fluctuations which can also be seen as a zero point motion of the soliton position. Second, thermal fluctuations produced by the interaction with the surrounding solvent.
This solvent, which we now call heatbath, wants to force the system into thermal
equilibrium. We discuss under what circumstances a soliton in the Alpha-helix can
be seen as a thermodynamic equilibrium state. We show how the principle of a minimal free energy, formulated in a variational principle of thermodynamics, can be
used to optimize a given ansatz for the density operator. Thermal soliton theories
of Davydov and Krumhansl are discussed within this theory. Both, in principle, use
the same ansatz with no freedom to adjust for maximum entropy. In using a more
realistic ansatz, we always get the result that the soliton is delocalized. Finally, we
discuss how the strength of interaction, together with the internal structure of the
heatbath, determines the thermal lifetime of the soliton.

1. Introduction

We start with the model Hamiltonian for the Alpha-helix first invented by Davydov
[1-3]. This Hamiltonian

$$H_S = H_L + H_O + H_I$$

$$H_L = \sum_{n=1}^{L} (\frac{p_n^2}{2M} + 1/2K(q_n - q_{n-1})^2)$$

$$H_O = \sum_{n=1}^{L} (\epsilon\, b_n^+ b_n - J\, b_n^+(b_{n+1} + b_{n-1}))$$

$$H_I = \chi \sum_{n=1}^{L} b_n^+ b_n (q_n - q_{n-1}) \tag{1.1}$$

describes the two most important degrees of freedom in the alpha-helix with a non-linear interaction. Many authors [3-7] tried to improve this model and the necessary numerical calculations. Also, the question of stability was discussed, either of stability against thermal fluctuations [8-11,21] (with different results) or of quantum fluctuations [12], showing a finite quantum lifetime of the soliton. All of these calculations where founded on the product ansatz originated by Davydov

$$| \Psi_S >= \sum_n f_n(t) b_n^+ \ | \ 0 > U(t)\varphi_0(q)$$

$$U(t) = \exp \{ \frac{1}{i\hbar} \sum_n (Q_n(t)p_n - P_n(t)q_n) \} \tag{1.2}$$

The validity of this ansatz and the resulting equation of motion were questioned in [13] and in [18,19]. On the other hand, we where able to show that the validity of (1.2) depends on how classical the lattice (H_L) is. For a classical lattice the ansatz (1.2) is an exact one. The quantum lifetime of the Davydov soliton depends on the importance of quantum fluctuations in the lattice [12], which wants to delocalize the soliton [20,21]. In a further paper Davydov [14] improved this ansatz (1.2) to include correlations and at the same time the effect of a finite temperature. Numerical calculations on this improved theory followed, showing as a result that the Davydov soliton should be stable at room temperature [15]. Independently Alexander and Krumhansl [17] developed a theory to explain the temperature dependency of a soliton peak in the absorbtion spectra of ACN.

We will try to show that Davydov [14] and Alexander and Krumhansl [17] essentially developed the same theory, which means they used the same ansatz for the density operator ρ . Also we will discuss the limitation in this ansatz for ρ and we will show that an improved ρ would have a *delocalizing effect on the soliton.*

2. Basic Concepts

There are two basic questions which we try to answer in this paper:

1.) Is the Davydov soliton stable against thermal fluctuations?

2.) If we have no stability of the soliton, how long will such a localized state remain localized? Are we able to calculate the thermal life time without knowing the precise structure of the heatbath?

As fact number one, we have to consider that the Davydov soliton is an excited state of the system (1.1). If we forget about the conserved quantities for this system we can make the very general statement:
A macroscopic system (which means the thermodynamic limit is possible) goes to the thermal equilibrium by itself. For a small system the thermal equilibrium is

reached by virtue of the surrounding material,, the so called heatbath. Crucial for thermodynamics is, that this thermal equilibrium state can be defined without any further knowledge of the internal structure of the heatbath. There are only a few macroscopic quantities which we have to know, such as the temperature, the pressure ecetera. These quantities can be defined independently from the heatbath.

There is only one exception to this rule: If the considered system is exact integrable, then it would never go to thermal equilibrium by itself. For example an excited state of the Toda lattice would never thermalize, if the Toda lattice is not coupled to a heatbath. If the Toda lattice is coupled to a heatbath the whole system would go to thermal equilibrium.

If we consider the dynamics (time dependence) of an *excited state in a small system*, things are quite different. In this case, the knowledge of the internal structure of the heatbath is crucial. Together with the strength of interaction, it determines the time which the system needs to go from an excited nonequilibrium state to thermal equilibrium.

In our opinion, the Alpha-helix is not a macroscopic system. Furthermore, it is quasi one dimensional but surrounded by the three dimensional heatbath. Therefore, we think that it is necessary to make realistic assumptions about the interaction with the heatbath.

Let us now try to discuss different cases:

a.) No heatbath: In this case the time dependence of the density operator ρ is given by

$$\dot{\rho} = \frac{1}{i\hbar}[H, \rho] \tag{2.1}$$

If ρ is given by its diagonalized representation

$$\rho = \sum_n |\Psi_n > \rho_n < \Psi_n| \tag{2.2}$$

then every Ψ_n obeys the time dependent Schrödinger equation. In every state Ψ_n the soliton is formed. We essentially have quantum decay of the soliton.

b.) Coupling to a heatbath in a way that the number of quanta

$$N = \sum_n b_n^+ b_n \tag{2.3}$$

remains a conserved quantity. This is a very severe restriction. For room temperature T we have

$$T \ll \epsilon/k_B \approx 2400, \tag{2.4}$$

showing that a one quantum state is a highly excited state. At first, the assumption $\dot{N} = 0$ seems unrealistic. On the other hand, the so called topological stability of the soliton [3] could stabilize the quantum number. To our knowledge, all existing papers on the thermal stability of the Davydov soliton use this assumption. The consequence is, that a one quantum state like the Davydov soliton, remains a one quantum state. The question about the stability of the soliton is then a question if the soliton is itself a thermal equilibrium state or not.

c.) The coupling with the heatbath will not allow that N is conserved. As a consequence a one quantum soliton cannot be a thermal equilibrium state.

If we could show that the Davydov soliton is a thermal equilibrium state, then this statement can be made independently of any heatbath.

If, on the other hand, the Davydov soliton is not a thermal equilibrium state, then the consequence is, as before mentioned: The lifetime of such an excited state is determined by the coupling strength and by the internal structure of the heatbath. *Different assumptions about the heatbath give different answeres for the lifetime.* Although the case a.) seems not to be realistic, it is the simplest possible assumtion. Then the calculation of τ, the lifetime of the soliton, can be made relatively simple. A calculation along these lines using perturbation theory was done by Cottingham and Schweitzer [22].

For the following we will assume that N is conserved. To answer the question if the Davydov soliton is an equilibrium state, we first introduce a variational principle as a general method to make approximations for the thermal equilibrium state of a system. We will see how nicely existing theories of Davydov and Krummhansl fit into this approximation scheme. We will finally show that an improvement of these theories would delocalize the soliton. At the end we will present our own calculation of τ using the assumption that the coupling to the heatbath can be neglected.

3. The variational principle

Of course the possibility exists, that a soliton in the Alpha-helix could also polarize the nearest surrounding material, in the same manner as the lattice (H_L) is polarized. For the moment we do not want to include this effect in our consideration, which would mean that correlations between heatbath and system exist. The simplest assumption to describe a thermal equilibrium state is then a canonical ensemble. It is well known, that the principle of a minimal free energy F

$$F = E - TS = Min \tag{3.1}$$

can be used to calculate approximations for the density operator ρ in the canonical ensemble. Here E is the internal energy, T the temperature and S the entropy. The corresponding variational principle for ρ

$$\delta\, Tr\rho(H + kT \ln \rho - \lambda) = 0 \tag{3.2}$$

is, in the special case of a vanishing temperature T, nothing else than the normal variational principle of quantum mechanics

$$\delta < 0 \mid H - \lambda \mid 0 > = 0 \tag{3.3}$$

If we allow a free variation of ρ , the principle (1.2) directly results in

$$\rho = e^{-\frac{1}{kT}(H-\lambda-1)} \tag{3.4}$$

This describes the exact canonical ensemble

$$\rho = \frac{1}{Z} e^{-\frac{1}{kT}H}$$

since λ has to be adjusted in such a way that ρ is normalized.

We can immediately generalize (3.1) to another situation, where the heatbath and the system have more interaction. Again, we do not consider correlation effects, so that the total density operator is a product

$$\rho = \rho_S \, \rho_B$$

with

$$Tr_S \, \rho_S = 1$$

$$Tr_B \, \rho_B = 1$$

We asume that the interaction between system and heatbath takes place via a macroscopic "coordinate" X of the system. (If not, we would have correlations to consider)

$$H = H_S + H_B + H_1(X) \tag{3.5}$$

The whole system is now in thermal equilibrium and we can again use the variational principle (1.2). The result is:

$$\delta \, Tr_S \, \rho_S \, [H_S - kT \ln \rho_S + \; < H_1(X) >_B + \; < H_B >_B \; -kT < \ln \rho_B >_B] = 0$$

$$\delta \, Tr_B \, \rho_B \, [H_B - kT \ln \rho_B + \; < H_S + H_1 - kT \ln \rho_S >_S] = 0 \tag{3.6}$$

If $H_1(X)$ is a product

$$H_1 = -XP \tag{3.7}$$

we arrive at the usual "pressure term" as the effect of the heatbath.

$$\rho_S = \frac{1}{Z} \, e^{-\beta(H_S - PX)}$$

$$\beta = \frac{1}{k_B T} \tag{3.8}$$

To model the surrounding material (mostly water) of the alpha-helix as a real heatbath without any correlations does not seem to be very realistic. But we can, eventually, assume that the characteristic frequencies $J/\hbar$ and $\sqrt{K/m}$ for a one or two quantum state of the system are small in comparison to the typical frequencies of the surrounding material. The heatbath will arrive at equilibrium very fast. We can then try a thermodynamic adiabatic approach, which is similar to the Born-Oppenheimer approximation in quantum mechanics. (Compare the similar situation in molecules where the electrons are fast in comparison to the atomic nucleus). Again assuming (3.5) as the whole Hamiltonian, we use the ansatz

$$\rho = \rho_S \, \rho_B(X) \tag{3.9}$$

For the bath (which is fast), X has to be seen as a parameter. Therefore, we have

$$\delta \, Tr \left[\rho_B(H_B + H_1(X) - kT \ln \rho_B) \right] = 0 \tag{3.10}$$

to calculate ρ_B. For calculating ρ_S we again have

$$\delta \, Tr \left[\rho_S(H_S - kT \ln \rho_S + F_B(X)) \right] \tag{3.11}$$

This is nothing else than the variational principle (1.2) with the free energy of the heatbath

$$F_B(X) = < H_S + H_1(X) - kT \ln \rho_B >_B \tag{3.12}$$

as an additional potential for the system. We want to point out that the X - dependence of $F_B(X)$ is nontrivial, since the thermal average $<>_B$ is itself X - dependent.

For the following we assume that this potential can be included in H_S to give renormalized constants J^* (dielectric effects) and M^* (effective mass). We want to assume that this inclusion has been done. The result is an effective Hamiltonian, which we again call H_S. For this Hamiltonian we can again use (3.2) to calculate the one or two quantum states in thermal equilibrium characterized by ρ_S.

4. The one quantum state

The density operator for a conserved quantum number N has the following form

$$\rho = \frac{1}{Z} e^{-\beta P_N H_S P_N} \tag{4.1}$$

with P_N for N = 0, 1, 2 as

$$P_0 = | \, 0 > < 0 \, |$$

$$P_1 = \sum_n b_n^+ \, | \, 0 > < 0 \, | \, b_n$$

$$P_2 = \frac{1}{2} \sum_{n,m} b_n^+ b_m^+ \, | \, 0 > < 0 \, | \, b_m b_n \tag{4.2}$$

Let us first consider a noninteracting system $\chi = 0$. Then ρ_S must be the product

$$\rho_S = \rho_0 \, \rho_L \tag{4.3}$$

ρ_0 and ρ_L are very similar in structure since H_0 and H_L are both one particle operators, which can be diagonalized

$$H_L = \sum_\lambda \hbar\omega_\lambda (a_\lambda^+ a_\lambda + \frac{1}{2})$$

$$H_0 = \sum_\nu \hbar\omega_\nu b_\nu^+ b_\nu$$

$$\omega_\lambda = \sqrt{\frac{K}{M}} \sin\frac{\lambda}{2}$$

$$\omega_\nu = \epsilon - 2J\cos\nu \tag{4.4}$$

Here, we have used the well known transformations

$$b_n = \frac{1}{L} \sum_\nu e^{i\nu n} b_\nu$$

$$q_n = \sum_\lambda \sqrt{\frac{\hbar}{2ML\omega_\lambda}} (a_{-\lambda}^+ + a_\lambda) e^{i\lambda n}$$

$$p_n = -i \sum_\lambda \sqrt{\frac{\hbar M\omega_\lambda}{2L}} (a_\lambda - a_{-\lambda}^+) e^{i\lambda n} \tag{4.5}$$

λ is the phonon, ν the exziton wavenumber. The trivial zero quantum state is not considered here. The probability of finding special states in the one or two exziton band has the usual Boltzmann factor. For example, for the one quantum state we have

$$\rho_0 = \frac{1}{Z} \sum_\nu b_\nu^+ \mid 0 > e^{-\beta\hbar\omega_\nu} < 0 \mid b_\nu \tag{4.6}$$

We want to point out that this typical distribution is a result of a counterbalance between the energy and the entropy term in the variational principle (3.2).

Next, we consider the general case $\chi \neq 0$. We want to limit our discussion to the one quantum state. Of course ρ_S cannot be exactly calculated.

Let us first discuss the theories given by Davydov [14] and by Krumhansl [17]. Using our variational treatment (3.2) we can combine these two theories with the following ansatz for ρ

$$\rho_S = U\rho^0(f)U^+$$

$$U = \exp\left[\sum_{n,\lambda}(\beta^*_{\lambda n}a_\lambda - \beta_{\lambda n}a^+_\lambda)b^+_n b_n\right]$$

$$\rho^0(f) = \mid f><f \mid \rho^0_{ph}$$

$$\mid f> = \sum_n f_n b^+_n \mid 0> \tag{4.7}$$

The parameters in this ansatz are f and β . (4.7) is the formulation of the ansatz given by Krumhansl, using U_n instead of U

$$U\mid f> = U\sum_n f_n b^+_n \mid 0> = \sum_n U_n f_n b^+_n \mid 0>$$

$$U_n = \exp\left[\sum_\lambda(\beta^*_{\lambda n}a_\lambda - \beta_{\lambda n}a^+_\lambda)\right] \tag{4.8}$$

we get the formulation given by Davydov.

Since f is normalized to one

$$\sum_n \mid f_n \mid^2 = 1 \tag{4.9}$$

(3.2) simplifies to

$$\delta Tr\rho_S(H - kT\ln\rho_S) = 0 \tag{4.10}$$

To show the limitations of the ansatz (4.7), we now calculate the possible variation of the entropy term

$$-kT\,\rho_S\ln\rho_S \tag{4.11}$$

Since a unitary transformation does not change the trace, we have

$$Tr\rho_S\ln\rho_S = Tr\rho^0(f)\ln\rho^0(f) \tag{4.12}$$

However this expression vanishes since we used a "pure state" in the oscillatory part of the density operator.

$$\delta Tr\,\rho^0(f)\left[\ln\mid f><f\mid +\ln\rho^0_{ph}\right] \tag{4.13}$$

The fact that the variation of the entropy term vanishes, is certainly a severe limitation for this ansatz: The entropy term in our variational principle ((3.2) is responsible for the spreading of the energy in many different states, as we have seen for the noninteracting case. We remain with

$$\delta Tr\,\rho_S H = \delta Tr\,\rho^0(f)U^+HU \tag{4.14}$$

to calculate the parameters. (4.14) was used in both papers. To calculate the thermal energy one first has to calculate $U^+ H U$ which is easily done with the help of the following rules

$$U^+ a_\lambda U = a_\lambda - \sum_n \beta_{\lambda n} b_n^+ b_n$$

$$U^+ b_n U = b_n U_n$$

$$U^+ b_n^+ b_n U = b_n^+ b_n \tag{4.15}$$

The final result is

$$U^+ H U = \bar{H} = \bar{H}_L + \bar{H}_O + \bar{H}_I \tag{4.16}$$

with

$$\bar{H}_L = \sum_\lambda \hbar\omega_\lambda \left[\frac{1}{2} + (a_\lambda^+ - \sum_n \beta_{\lambda n}^* b_n^+ b_n)(a_\lambda - \sum_n \beta_{\lambda n} b_n^+ b_n) \right] \tag{4.17}$$

$$\bar{H}_O = \sum_n \epsilon b_n^+ b_n - J b_n^+ (b_{n+1} U_n^+ U_{n+1} + b_{n-1} U_n^+ U_{n-1}) \tag{4.18}$$

$$\bar{H}_I = \chi \sum_{n\lambda} F(\lambda) b_n^+ b_n \left[a_\lambda - \beta_{\lambda n} + a_\lambda^+ - \beta_{\lambda n}^* \right] e^{i\lambda n} \tag{4.19}$$

To calculate (4.19) we have used the transformations (4.5) to write the interaction Hamiltonian as

$$H_I = \chi \sum_{n=1}^{L} \sum_\lambda F(\lambda) b_n^+ b_n (a_\lambda + a_{-\lambda}^+) e^{i\lambda n}$$

$$F(\lambda) = \sqrt{\frac{\hbar}{2ML\omega_\lambda}} (1 - e^{-i\lambda}) \tag{4.20}$$

using the relation

$$Tr\, \rho_{ph}^0 U_n^+ U_{n+1} = \exp(W_{n\,n+1})$$

$$W_{n\,n+1} = \sum_\lambda [(n_\lambda + 1)\beta_{\lambda n}^* \beta_{\lambda n+1} + n_\lambda \beta_{\lambda n} \beta_{\lambda n+1}^* - (n_\lambda + \frac{1}{2})(|\beta_{\lambda n}|^2 + |\beta_{\lambda n+1}|^2)] \tag{4.21}$$

we arrive at the thermal expectation value of H.

$$Tr\rho_S H = Tr\rho^0(f)(\bar{H}_O + \bar{H}_L + \bar{H}_I)$$

$$Tr\, \rho^0(f)\bar{H}_O = \sum_n \left[\epsilon |f_n|^2 - J(f_n f_{n-1} e^{W_{n\,n+1}} + f_n f_{n+1} e^{W_{n\,n-1}}) \right]$$

$$Tr\, \rho^0(f)\bar{H}_L = \sum_\lambda \hbar\omega_\lambda\left(\frac{1}{2} + n_\lambda + \sum_n |\beta_{-\lambda n}|^2\right)$$

$$Tr\, \rho^0(f)\bar{H}_I = -\chi \sum_{n\lambda} F(\lambda)|f_n|^2 (\beta_{\lambda n} + \beta^*_{-\lambda n}) e^{i\lambda n} \tag{4.22}$$

Variation of this expression

$$\delta\, Tr \rho_S H = 0 \tag{4.23}$$

gives us the equations to calculate the parameters $\beta_{\lambda n}$, which was done in [15,16]. The result of these calculations is, that the Davydov soliton should be stable at room temperature.

5. Quantum and thermal Fluctuations

The theory described in the last chapter has two disadvantages:

    a) The ansatz (4.7) for the density operator ρ does not commute with $A\,Q$, where Q is the shifting operator in the lattice

$$Q^+ a_\nu Q = e^{i\nu} a_\nu \tag{5.1}$$

and A is the shifting operator in the oscillators.

$$A^+ b_\lambda A = e^{i\lambda} b_\lambda \tag{5.2}$$

    b) The ansatz does not allow that the entropy gets to its maximum.

Let us discuss these two points:

It is easy to show [12] that

$$[H_S, AQ] = 0 \tag{5.3}$$

holds. The eigenstates of H_S are not localized but must form Bloch states $|k>$. We discussed this fact in detail in [12]. The result is, that as long as the lattice has to be seen as non classical, the soliton wants to decay in the corresponding Bloch band. Therefore we will have a finite lifetime of the soliton even without taking the temperature effects into account. On the other hand, for a classical lattice the ansatz (1.2) is exact for $T = 0$ and the resulting soliton state is stable.

The question which remains is the following: Is it possible that the soliton remains stable if we have a classical lattice? Would not the interaction with the heatbath destabilize the soliton? This question can easily be answered. We again use the principle of minimal free energy (3.1). Let us assume we have a solution for the density operator using the ansatz (4.7) which shows a soliton at lattice site l. We will call this solution ρ_l. It has the free energy

$$F_l = Tr\{\rho_l(H + kTln\rho_l)\} = E_l - TS_l \qquad (5.4)$$

Since ρ_l is normalized to one, we can construct the following new delocalized density operator ρ.

$$\rho = \frac{1}{L}\sum_l^L \rho_l \qquad (5.5)$$

Let us compare the free energies for the two density operators. The free energy for the delocalized density operator is

$$F = Tr\{\rho(H + kTln\rho)\} = E - TS \qquad (5.6)$$

The two thermal energies are equal

$$E_l = E \qquad (5.7)$$

To calculate the entropy we use the fact that the eigenstates $|\Psi_{l,n}>$ of ρ_l

$$\rho_l = \sum_n |\Psi_{l,n} > \rho_{l,n} < \Psi_{l,n}| \qquad (5.8)$$

are orthogonal for different l

$$< \Psi_{l,n'}|\Psi_{l,n} >=< \Psi_{l,n}|(AQ)^{n'-n}|\Psi_{l,n} >= 0 \qquad (5.9)$$

This holds because the lattice is classical. Therefore we get

$$S = \frac{1}{L}\sum_l Tr\rho_l\, ln(\frac{1}{L}\rho_l) = S_l + k\, lnL \gg S_l \qquad (5.10)$$

showing an enormous increase of the entropy and an enormous decrease of the free energy of the delocalized state in comparison to the localized state.

$$F = F_l - kT\, lnL \ll F_l \qquad (5.11)$$

Even if the lattice is not classical we always have

$$S > S_l$$

We want to point out that the ansatz (5.5) is not a realistic approximation for the thermal equilibrium, since S has a wrong L - dependency. So we only showed that ρ and ρ_l both describe states far from thermal equilibrium. But we have seen that a localized soliton state is always a nonequilibrium state, which means that it has a finite lifetime.

6. Calculation of the lifetime

We do not know how the Alpha-helix is coupled to the heatbath. Therefor we ask now how a soliton state created in a thermalized lattice would develop in time, if we neglect the coupling to the heatbath. This gives a very simple calculation of τ for this case. We assume the Davydov soliton is given by the following density matrix:

$$\rho_S = U \rho^0(f) U^+$$

$$U = \exp \left\{ \frac{1}{i\hbar} \sum_n (Q_n(t) p_n - P_n(t) q_n) \right\}$$

$$\rho^0(f) = | f >< f | \rho^0_{ph}$$

$$| f > = \sum_n f_n b_n^+ | 0 > \tag{6.1}$$

This is ansatz (4.7), but with $\beta_{\lambda,n}$ independent of n and with

$$Q_n = \sum_\lambda \sqrt{\frac{\hbar}{2ML\omega_\lambda}} (\beta^*_{-\lambda} + \beta_\lambda) e^{i\lambda n}$$

$$P_n = -i \sum_\lambda \sqrt{\frac{\hbar M \omega_\lambda}{2L}} (\beta_\lambda - \beta^*_{-\lambda}) e^{i\lambda n} \tag{6.2}$$

At the same time it is the old Davydov ansatz (1.2) thermalized in the simplest possible way. To show this we write ρ in the following form

$$\rho = \sum_l |\Psi_{S,l} > \rho^0_{ph,l} < \Psi_{S,l}| \tag{6.3}$$

$|\Psi_{S,l} >$ has the form (1.2) but with the phonon groundstate $\varphi_0(q)$ replaced by the excited state $\varphi_l(q)$

$$|\Psi_{S,l} >= \sum_n f_n(t) b_n^+ | 0 > U(t)\varphi_l(q) \tag{6.3a}$$

$\rho^0_{ph,l}$ is just the probability to find the state $\varphi_l(q)$ in a lattice which is thermalized. Again we assume a nonmoving soliton. With the help of the variational principle we then obtain for the parameters the well known equations

$$-\frac{\chi^2}{K} |f_n|^2 f_n - J(f_{n+1} + f_{n-1}) = E_f f_n \tag{6.4}$$

with

$$Q_n - Q_{n-1} = -\frac{\chi}{K} |f_n|^2$$
$$P_n = 0 \tag{6.5}$$

We now assume that we have got a soliton solution of the equations (6.4) and (6.5). Every state $|\Psi_{S,l}>$ then has its own soliton. This should be the initial state $\rho(t=0)$. Let us now ask for the time development of $\rho(t)$. Since we have no coupling to a heatbath the time dependence of $\rho(t)$ is given by

$$\dot{\rho} = \frac{1}{i\hbar}[H,\rho] \tag{6.6}$$

An aquivalent formulation is the following: $\rho(t)$ remains in the form (6.3) but every state $|\Psi_{S,l}>$ develops in time according to the Schrödinger equation. The probability $\rho^0_{ph,l}$ to find the state $|\Psi_{S,l}>$ remains constant in time. Using the results of [23] we therefor arrive at

$$\tau \geq \frac{\hbar}{\Delta E} \tag{6.7}$$

with

$$\Delta E^2 = \chi^2 << \varphi_0|(q_m - q_{m-1})^2|\varphi_0 >> (1 - \sum_n |f_n|^4)$$
$$-2\chi^2 \sum_{n=1}^{\infty} << \varphi_0|(q_n - q_{n-1})(q_0 - q_{-1})|\varphi_0 >> \sum_m |f_{n+m}|^2|f_m|^2 \tag{6.8}$$

(6.8) is the same equation as $(3.19)^{*}$ but we have to replace the quantum averaging $< ..... >$ by thermal avaraging $<< .... >>$. Using the same transformation as in [23] we can again calculate

$$<< \varphi_0|(q_{n_1} - q_{n_1-1})(q_{n_2} - q_{n_2-1})|\varphi_0 >>= \frac{1}{L} \sum_{\lambda} e^{i(n_1-n_2)\lambda} << q_\lambda q_\lambda^+ >> 4sin^2(\frac{\lambda}{2}) \tag{6.9}$$

with

$$<< q_\lambda q_\lambda^+ >>= \frac{\hbar}{2M\omega_\lambda} coth\left(\frac{\hbar\omega_\lambda}{2k_BT}\right) \tag{6.10}$$

For $T = 0$ this gives

$$<< q_\lambda q_\lambda^+ >>= \frac{\hbar}{2M\omega_\lambda}$$

which we used in [23]. (6.10) also simplifies if

$$k_BT \gg \hbar\omega_\lambda \tag{6.11}$$

In this case we have

$$<< q_\lambda q_\lambda^+ >>= \frac{k_BT}{M\omega_\lambda^2} \tag{6.12}$$

This finally gives the very simple expression

---

*of Chapter 7

$$\Delta E^2 = \frac{\chi^2 k_B T}{K} \left\{ 1 - \sum_m |f_m|^4 \right\} \tag{6.13}$$

For $T = 300^0 K$ and the parameter set

$$
\begin{aligned}
M &= 114.2\, m_{prot} \\
K &= 19.8\ \frac{J}{m^2} \\
\epsilon &= 3.28\ \ 10^{-20} J \\
J &= 1.55\ \ 10^{-22} J \\
\chi &= 6.2\ \ \ 10^{-11} N
\end{aligned}
\tag{6.14}
$$

this yields

$$\Delta E = 1.2\ 10^{-21}$$

and the lifetime τ is calculated as

$$\tau = \frac{\hbar}{\Delta E} = 8.3\ 10^{-14}$$

This lifetime is shorter than in the case of $T = 0$. The reason is clear: The soliton is now feeling quantum and thermal fluctuations at the same time.

References

1. Davydov, A.S., J. Theor. Biol. 38 (1973) 559.

2. Davydov, A.S. and Kislukha, N.I., Phys. Stat. Sol. B59 (1973).

3. See Davydov, A.S., Solitons in Molecular Systems, Riedel, Dordrecht (1985) for bibliographies.

4. Scott, A.C., Phys. Rev. A26 (1982) 578.

5. Scott, A.C., Phys. Scripta 25 (1982) 651

6. MacNeil, L. and Scott, A.C. Phys. Scripta 29 (1984) 284

7. Bolterauer, H. and Henkel, R.D., Phys. Scripta T13 (1986) 314

8. Lomdahl, P.S. and Kerr, W.C., Phys. Rev. Lett. 55 (1985) 1235

9. Lawrence, A.F., McDaniel, J.C., Chang, D.B., Pierce, B. M. and Birge, R.R., Phys. Rev. A33 (1986) 1188

10. Scott, A.C., "On Davydov Solitons at 310 K", in Energy Transfer Dynamics edited by T. W. Barett and H. A. Pohl; Springer-Verlag 1987

11. Bolterauer, H., "Aspects of Quantum Mechanical Thermalisation in the Alpha-Helix", in Structure, Coherence and Chaos in Dynamical Systems, edited by P. L. Christiansen and R. D. Parmentier, Manchester University Press 1986.

12. Bolterauer, H. and Opper, M. "The Quantum lifetime of the Davydov soliton";
    submitted to Physica Scripta November 1988

13. Brown, B.W., Lindenberg, K. and West, B.J., Phys. Rev. A33 (1986) 4104

14. Davydov, A.S., Phys. stat. sol. B138 (1986) 559

15. Cruzeiro, L., Halding, J., Christiansen, P.L., Skovgaard, O. and Scott, A.C.,
    Phys. Rev. A37,880 (1988)

16. Kadantzev, V.N., Lupichov, L.N. and Savin, A.V., Phys. stat. sol.(b) 143, 569
    (1987) and Phys. stat. sol.(b) 147, 155 (1988)

17. Alexander, D.M. and Krumhansl, J.A., Phys. Rev. B33 (1986) 7172

18. Brown, D.W., Lindenberg, K. and West, B.J., Phys. Rev A33, 4110 (1986)

19. Brown, D.W., Lindenberg, K. and West, B.J., Phys. Rev A33, 4114 (1986)

20. Venzl, G. and Fischer, F., J. Chem. Phys. 81, 6090 (1984)

21. Venzl, G. and Fischer, F., J. Phys. Rev. B32, 6437 (1985)

22. Cottingham, J. P. and Schweitzer, J. W., Phys. Ref. Lett. 62, 1792 (1989)

23. Bolterauer, H., "Quantum effects on the Davydov soliton" this Volume

24

THERMAL STABILITY OF THE DAVYDOV SOLITON

L. Cruzeiro-Hansson, P.L. Christiansen* and A.C. Scott*

Dept. Crystallography, Birkbeck College, Malet St., London U.K.
*Lab. Applied Mathematical Physics, The Technical University of
Denmark, DK-2800 Lyngby, Denmark

INTRODUCTION

Although in the science of bioenergetics considerable progress has been made
in the past 20 years, the storage and transport of energy in biological systems is not
well understood. An answer to this problem was suggested in 1973 by Davydov, who
proposed that quantum units of peptide vibrational energy (in particular the amide-I
or C=O stretching vibration) might become "self-localized" through interactions with
lattice phonons[1,2]. Following his original suggestion many related studies have been
published by Davydov and his colleagues[3] and by others[4] on this "Davydov soliton".
The question of the thermal stability of the Davydov soliton at biological temperatures
has also been studied by Davydov, in a quantum mechanical framework[5] and by others
in classical simulations[6,7]. Here results obtained following Davydov's thermal treatment
without any approximations are presented and the discrepancies between the latter and
the classical simulations [6,7] are discussed. Finally, as a way to overcome the inconclu-
siveness of the latter approaches a search for exact dynamical solutions is proposed.

DAVYDOV'S THERMAL TREATMENT

The hamiltonian for a one dimensional molecular lattice with N sites is[1-3,5]

$$\hat{H} = \hat{H}_{ex} + \hat{H}_{ph} + \hat{H}_{int} \tag{1}$$

where $\hat{H}_{ex}$ is the unperturbed exciton operator, $\hat{H}_{ph}$ is the unperturbed phonon operator
and $\hat{H}_{int}$ is an operator describing the interaction between the intramolecular excitations
and the phonons. Thus,

$$\hat{H}_{ex} = \sum_{n=1}^{n} [\epsilon\, \hat{a}_n^\dagger \hat{a}_n - J(\hat{a}_n^\dagger \hat{a}_{n-1} + \hat{a}_n^\dagger \hat{a}_{n+1})] \tag{2}$$

where ϵ is the intramolecular excitation energy (only one excited state is considered), -J
is the dipole-dipole interaction energy and $\hat{a}_n^\dagger(\hat{a}_n)$ is the creation(annihilation) operator
for the intramolecular excitations. Furthermore,

Davydov's Soliton Revisited, Edited by P.L. Christiansen and A.C. Scott
Plenum Press, New York, 1990

$$\hat{H}_{ph} = \sum_q \hbar\Omega_q(\hat{b}_q^\dagger \hat{b}_q + 1/2) \tag{3}$$

where $\hbar\Omega_q$ is the energy of an acoustic phonon with wave number q, i.e.

$$\Omega_q = 2(\kappa/M)^{1/2} \,|\sin(qa/2)| \tag{4}$$

κ being the elasticity constant of the lattice, M the mass of the molecule in each site and a being the distance between sites. Finally,

$$\hat{H}_{int} = \sum_{q,n} F(q)\hat{a}_n^\dagger \hat{a}_n(\hat{b}_q + \hat{b}_{-q}^\dagger)e^{iqna} \tag{5}$$

where

$$F(q) = (\hbar/2MN\Omega_q)^{1/2}2i\chi \sin(qa) \tag{6}$$

χ being an anharmonic parameter related to the coupling between the intramolecular excitation and the displacement of the following bond in the chain. To facilitate comparisons with the results of Reference 6, we assume

$\epsilon = 1660$ cm^{-1}

J=7.8 cm^{-1}

κ=13 N/m

M=114 m$_p$

a=4.5 10^{-10} cm

The value of χ is still controversial and will be specified later. The general solution of Schrödinger's equation for the hamiltonian (1-6) is not known and as an approximation, Davydov has suggested the following ansatz:

$$|\psi_\nu(t)> = \sum_n \phi_n(t)\hat{a}_n^\dagger |0>_{ex} \hat{U}_n(t)|\nu> \tag{7}$$

where ϕ_n is the probability amplitude for an excitation in site n and $|\nu>$ is the phonon wave function

$$|\nu> = \prod_q (\nu_q!)^{-1/2}(\hat{b}_q^\dagger)^{\nu_q}|0>_{ph} \tag{8}$$

and $\hat{U}_n(t)$ is a unitary operator of the displacements

$$\hat{U}_n(t) = exp\left[\sum_q(\beta_{qn}^\star(t)\hat{b}_q - \beta_{qn}\hat{b}_{-q}^\dagger)\right] \tag{9}$$

We emphasize that (7-9) represents a mixed state in the phonon wave functions and that, since the displacement operator $\hat{U}_n(t)$ is site dependent, phase mixing between the exciton and the phonon systems *is* allowed. Davydov avoids dealing with an infinite dimension density matrix (derived from (7)) by calculating the thermally averaged hamiltonian H$_T$

$$H_T = \sum_\nu \rho_\nu H_{\nu\nu} \tag{10}$$

where

$$H_{\nu\nu} = <\psi_\nu(t)|\hat{H}|\psi_\nu(t)> \tag{11}$$

and

$$\rho_\nu = \frac{<\nu \mid exp(-\hat{H}_{ph}/kT) \mid \nu>}{\sum_\nu <\nu \mid exp(-\hat{H}_{ph}/kT) \mid \nu>} \tag{12}$$

k being the Boltzmann's constant and T the absolute temperature. The result is

$$H_T = \sum_n [\epsilon \mid \phi_n \mid^2 - J(\phi_n^\star \phi_{n-1} e^{W_{nn-1}} + \phi_n^\star \phi_{n+1} e^{W_{nn+1}}) - \tag{13}$$
$$- \mid \phi_n \mid^2 \sum_q F(q) e^{\imath qna}(\beta_{qn} + \beta_{-qn}^\star) +$$
$$+ \mid \phi_n \mid^2 \sum_q \hbar\Omega_q(\bar{\nu}_q + \mid \beta_{qn} \mid^2)] + 1/2 \sum_q \hbar\Omega_q$$

with

$$W_{nn\pm1} = -\sum_q [(\bar{\nu}_q + 1/2) \mid \beta_{qn} - \beta_{qn\pm1} \mid^2 + \frac{1}{2}(\beta_{qn}\beta_{qn\pm1}^\star - \beta_{qn}^\star\beta_{qn\pm1})] \tag{14}$$

We point out that the thermally averaged hamiltonian shows the temperature dependent reduction of the exciton bandwidth by the factor $e^{W_{nn\pm1}}$, a feature that does not occur in the classical simulations[6,7]. Since the formation of solitons is ultimately dependent on the ratio between J_{eff}/χ, being favoured by smaller values of this ratio, the quantum mechanical treatment can be expected to yield soliton solutions in conditions in which the classical simulations do not allow them. Considering that H_T is a classical hamiltonian in the variables ϕ_n, β_{qn} canonically conjugate to $\imath\hbar\phi_n^\star$ and $\imath\hbar\beta_{q,n}^\star$, the following dynamical equations are derived:

$$\imath\hbar\frac{d\phi_n}{dt} = \frac{\partial H_T}{\partial\phi_n^\star} = \epsilon\phi_n - J(\phi_{n-1} e^{W_{nn-1}} + \phi_{n+1} e^{W_{nn+1}}) - \tag{15}$$
$$-\phi_n \sum_q F(q) e^{\imath qna}(\beta_{qn} + \beta_{-qn}^\star) + \phi_n \sum_q [\hbar\Omega_q(\bar{\nu}_q + \mid \beta_{qn} \mid^2)]$$

$$\imath\hbar\frac{d\beta_{qn}}{dt} = \frac{\partial H_T}{\partial\beta_{qn}^\star} = -J[\phi_n^\star\phi_{n-1} e^{W_{nn-1}}[(\bar{\nu}_q + 1)\beta_{qn-1} - (\bar{\nu}_q + 1/2)\beta_{qn}] + \tag{16}$$
$$+\phi_{n+1}^\star\phi_n e^{W_{n+1n}}[\bar{\nu}_q\beta_{qn+1} - (\bar{\nu}_q + 1/2)\beta_{qn}] +$$
$$+\phi_n^\star\phi_{n+1} e^{W_{nn+1}}[(\bar{\nu}_q + 1)\beta_{qn+1} - (\bar{\nu}_q + 1/2)\beta_{qn}] +$$
$$+\phi_{n-1}^\star\phi_n e^{W_{n-1n}}[\bar{\nu}_q\beta_{qn-1} - (\bar{\nu}_q + 1/2)\beta_{qn}]] -$$
$$- \mid \phi_n \mid^2 F^\star(q) e^{-\imath qna} + \mid \phi_n \mid^2 \hbar\Omega_q\beta_{qn}$$

Numerical integration of (15-16) shows that there are soliton solutions above a threshold value of $\chi = 0.17 \ 10^{-10}$N. Figure 1 shows the results obtained when the initial condition is two sites excited in the middle of the chain (with N=51):

$$\mid \phi_{24}(0) \mid^2 = \mid \phi_{25}(0) \mid^2 = 0.5$$
$$\mid \beta_{qn}(0) \mid^2 = 0.0$$

Figure 1 shows that part of the initial excitation is lost to exciton waves (which do not induce any displacement), while the rest travels coherently along the chain, in soliton-like waves (with associated displacements).

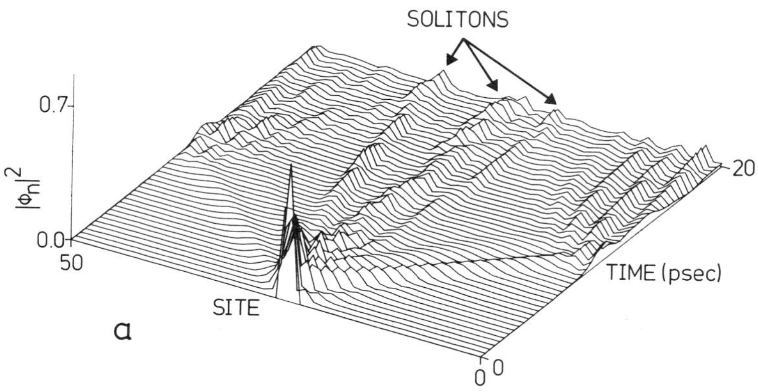

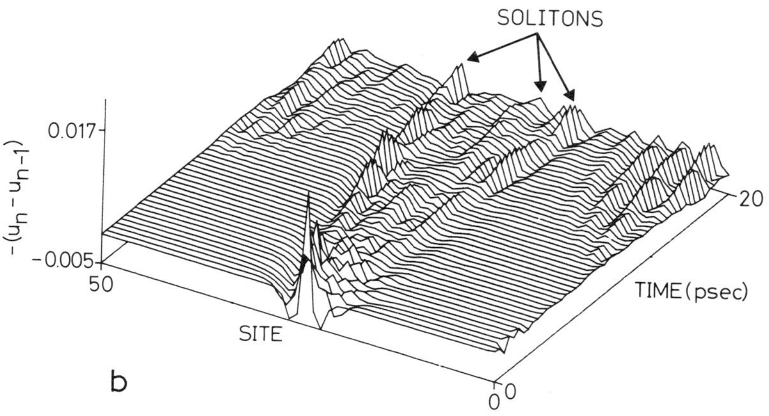

Figure 1. (a) Evolution of the probability of an excitation in site n, $|\phi_n|^2$ (n=0,...,50) and (b) the molecular displacement $-(u_n\text{-}u_{n-1})$ in Å. $\chi = 0.23 \times 10^{-10}$ N and T=310K. Two sites excited initially (see text).

Figure 2 shows that the fraction of the initial excitation that travels as a soliton-like wave is dependent on the initial condition. Indeed, if the initial condition is a sech pulse, i.e.

$$| \phi_n(0) |^2 \;\; = \;\; 0.56 \, sech^2[0.64 \, (n - 38)]$$
$$| \beta_{qn}(0) |^2 \;\; = \;\; 0.0$$

then the fraction of the energy that travels in a coherent fashion is increased.
The simulations do show scattering of the initial excitation energy into exciton, dispersive waves. This happens initially and also, although less dramatically, throughout the propagation. It is thus expected that the soliton-like waves have a finite lifetime. The question is however whether they last long enough, with sufficient energy to be useful for biological purposes. We point out that both initial conditions used correspond to bare excitons, i.e., the initial displacements were zero. Dressed initial conditions should lead to more stable soliton solutions. Morevoer, figure 2 already shows a soliton solution which survives for 30 psec. From these simulations, the conclusion is therefore that Davydov solitons can play a role in biological energy transport.

WHO IS RIGHT ?

The conclusion drawn in the previous section stands in contrast to those drawn from the classical simulations[6,7]. Lomdahl and Kerr[6] and Lawrence *et al.*[7] modelled the temperature effects with a generalized Langevin equation derived by adding damping and noise terms to the dynamical equations derived from an earlier Davydov ansatz for the wave function[1-3]. The noise was δ -correlated in space and time and it was verified that for sufficiently long times the mean kinetic energy of the lattice was the classical thermal energy. These classical simulations of thermal effects differ in several aspects from those presented above. (i) The ansatz for the wave function in the classical simulations is a product ansatz and therefore does not allow mixing of the excitation with the phonon states. The ansatz used by us (eqn(7)), with its site dependent displacement operator should be considered an improvement. (ii) Davydov's procedure of calculating a thermally averaged hamiltonian may underestimate the effect of thermal fluctuations, while the classical simulations, by using a Boltzmann distribution overestimate the latter effect (see also Scott[8] and Bolterauer[9]). (iii) The derivation of the equations of motion, in the case presented above assumes that ϕ_n , β_{qn} and $i\hbar\phi_n^{\star}$ and $i\hbar\beta_{qn}^{\star}$ are canonically conjugate variables, an assumption that has been recently criticized[10,11] .
The corresponding equations in the classical simulations, on the other hand, are derived in an exact quantum mechanical way. (iv) The classical simulations do not show the Debye-Waller factor that leads to a reduction of J (see eqns (13-14)), which the physics of the problem suggests should occur as temperature increases and which leads to a stabilization of the soliton solution.

A main difficulty consists in quantifying the degree of approximation involved in the different treatments (approximations (i), (ii) and (iii) in the quantum mechanical treatment and (i), (ii) and (iv) in the case of the classical simulations). The question at the head of this section is thus still an open one. Clearly more exact treatments are needed in order to decide whether the Davydov soliton is biologically viable. The next section presents some efforts in that direction.

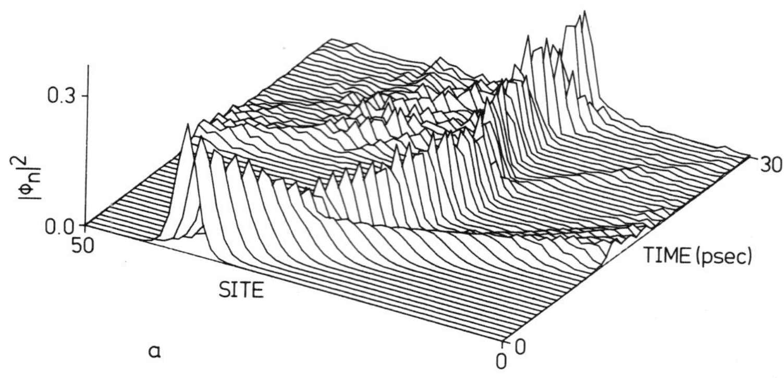

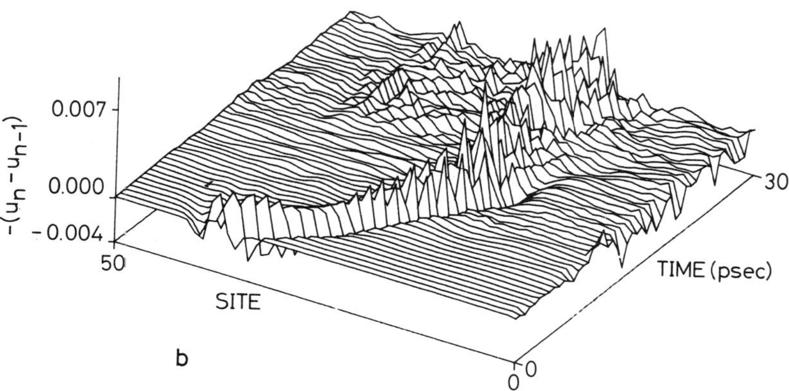

Figure 2. (a) Evolution of the probability of an excitation $| \phi_n |^2$ and (b) the molecular displacement $-(u_n-u_{n-1})$ in $\mathring{A}$ in site n (n=0,...,50). $\chi = 0.23{\times}10^{-10}$N, T=310K. Initial condition is a sech pulse (see text).

THE DAVYDOV DIMER

We use a transformation well-known from polaron theory[12]:

$$\hat{a}_n \rightarrow \hat{A}_n = \hat{R}^\dagger \hat{a}_n \hat{R} = \hat{a}_n exp[-\sum_q \frac{F(q)}{\hbar\Omega_q} e^{\imath qna}(\hat{b}_q - \hat{b}^\dagger_{-q})] \tag{17}$$

$$\hat{b}_q \rightarrow \hat{B}_q = \hat{R}^\dagger \hat{b}_q \hat{R} = \hat{b}_q + \sum_n \frac{F(-q)}{\hbar\Omega_q} e^{-\imath qna}\hat{a}^\dagger_n \hat{a}_n \tag{18}$$

with the unitary operator $\hat{R}$ defined by

$$\hat{R} = exp[-\sum_{q,n} \frac{F(q)}{\hbar\Omega_q} e^{\imath qna}(\hat{b}_q - \hat{b}^\dagger_{-q})\hat{a}^\dagger_n \hat{a}_n] \tag{19}$$

In terms of the new operators $\hat{A}_n$ and $\hat{B}_q$ the unperturbed exciton hamiltonian becomes

$$\hat{H}_{ex} = \sum_{n=1}^{N} [\epsilon\hat{A}^\dagger_n \hat{A}_n - J(\hat{A}^\dagger_n \hat{A}_{n-1}\hat{S}_{nn-1} + \hat{A}^\dagger_n \hat{A}_{n+1}\hat{S}_{nn+1})] \tag{20}$$

where

$$\hat{S}_{ll\pm 1} = exp[-\sum_q \frac{F(q)}{\hbar\Omega_q}(e^{\imath qla} - e^{\imath q(l\pm 1)a})(\hat{B}_q - \hat{B}^\dagger_{-q})] \tag{21}$$

The unperturbed phonon hamiltonian becomes

$$\hat{H}_{ph} = \sum_q \hbar\Omega_q(\hat{B}^\dagger_q \hat{B}_q + 1/2) - \sum_{q,m} F(q)e^{\imath qma}(\hat{B}_q + \hat{B}^\dagger_{-q})\hat{A}^\dagger_m \hat{A}_m + \tag{22}$$
$$+ \sum_{q,n,m} \frac{|F(q)|^2}{\hbar\Omega_q} e^{\imath q(n-m)a} \hat{A}^\dagger_n \hat{A}_n \hat{A}^\dagger_m \hat{A}_m$$

And the interaction hamiltonian becomes

$$\hat{H}_{int} = \sum_{q,n} F(q)e^{\imath qna}(\hat{B}_q + \hat{B}^\dagger_{-q})\hat{A}^\dagger_n \hat{A}_n - 2\sum_{q,n,m} \frac{|F(q)|^2}{\hbar\Omega_q} e^{\imath q(n-m)a} \hat{A}^\dagger_n \hat{A}_n \hat{A}^\dagger_m \hat{A}_m \tag{23}$$

The total hamiltonian in terms of the new operators is

$$\hat{H} = \sum_n [\Delta\hat{A}^\dagger_n \hat{A}_n - J(\hat{A}^\dagger_n \hat{A}_{n-1}\hat{S}_{nn-1} + \hat{A}^\dagger_n \hat{A}_{n+1}\hat{S}_{nn+1})] + \tag{24}$$
$$+ \sum_q \hbar\Omega_q(\hat{B}^\dagger_q \hat{B}_q + 1/2) + \sum_{q,n,m} \frac{|F(q)|^2}{\hbar\Omega_q} e^{\imath q(n-m)a} \hat{A}^\dagger_n \hat{A}^\dagger_m \hat{A}_n \hat{A}_m$$

where

$$\Delta = \epsilon - \sum_q \frac{|F(q)|^2}{\hbar\Omega_q} \tag{25}$$

The case $J = 0$, for which the hamiltonian (24) is diagonal, has been dealt by Brown *et al.*[13]. As in the latter work the focus here will also be on the first excited states for which the last term in (24) cancels. However, we shall consider the case $J \neq 0$ for a dimer (N=2). As an initial condition we take a localized excitation in a dressed lattice:

$$| \psi(0) >= \hat{A}_1^\dagger | 0 > \qquad (26)$$

And the idea is to calculate the solution at time t by acting with the evolution operator on this initial condition, i.e.

$$| \psi(t) >= e^{\lambda \hat{H}_D} | \psi(0) >= e^{\lambda \hat{H}_D} \hat{A}_1 \phi_1 | 0 >= \qquad (27)$$

$$(\hat{A}_1^\dagger + \lambda[\hat{H}_D, \hat{A}_1^\dagger] + \frac{\lambda^2}{2!}[\hat{H}_D,[\hat{H}_D,\hat{A}_1^\dagger]] + \frac{\lambda^3}{3!}[\hat{H}_D,[\hat{H}_D,[\hat{H}_D,\hat{A}_1^\dagger]]] + \cdots)e^{\lambda \hat{H}_D} \phi_1 | 0 >$$

where

$$\lambda = -\frac{\imath}{\hbar}t \qquad (28)$$

and $\hat{H}_D = \hat{H}$ (N=2). If the second last term in (24) is also neglected the series (27) leads to

$$| \psi(t) >= exp(-\frac{\imath}{\hbar}Et)[\cos(2J\frac{t}{\hbar})\hat{A}_1^\dagger - \imath \sin(2J\frac{t}{\hbar})\hat{A}_2^\dagger \hat{S}_{21}] | 0 > \qquad (29)$$

where

$$E = \epsilon - \sum_q (\frac{| F(q) |^2}{\hbar\Omega_q}) + 1/2 \sum_q \hbar\Omega_q \qquad (30)$$

This is an exact analytical solution of the Davydov dimer, for very short times and $T = 0K$. From (29) one finds:

$$| \phi_1(t) |^2 =< \psi(t) | \hat{A}_1^\dagger \hat{A}_1 | \psi(t) >=< \psi(t) | \hat{a}_1^\dagger \hat{a}_1 | \psi(t) >= \cos^2(2J\frac{t}{\hbar}) \qquad (31)$$

and

$$| \phi_2(t) |^2 =< \psi(t) | \hat{A}_2^\dagger \hat{A}_2 | \psi(t) >=< \psi(t) | \hat{a}_2^\dagger \hat{a}_2 | \psi(t) >= \sin^2(2J\frac{t}{\hbar}) \qquad (32)$$

That is, for very short times, the Davydov dimer behaves as two classical linear, coupled oscillators.

In terms of the bare operators $\hat{a}_n$, $\hat{b}_q$, the wave function (29) is

$$| \psi(t) >= exp[-\frac{\imath}{\hbar}Et] [\cos(2J\frac{t}{\hbar})\hat{a}_1^\dagger - \imath \sin(2J\frac{t}{\hbar})\hat{a}_2^\dagger] exp[\sum_q \frac{F(q)}{\hbar\Omega_q} e^{\imath qa}(\hat{b}_q - \hat{b}_{-q}^\dagger)] \qquad (33)$$

i.e. a product ansatz with $\beta_{qn}^\star = \frac{F(q)}{\hbar\Omega_q}e^{\imath qa}$. The corresponding displacement is

$$< \Psi(t)| \hat{u}_n | \Psi(t) >= -2 \sum_q \sqrt{\frac{\pi}{2MN\Omega_q}} e^{\imath qna} \frac{F(-q)}{\hbar\Omega_q} e^{-\imath qa} \qquad (34)$$

332

It is not dependent on t, i.e., it is essentially the initial displacement associated with the initial condition (26). Expression (34) is thus an indirect indication of the timescale in which the solution (29) and (33) are valid. A more precise indication can be obtained by considering the next terms in expansion (27). The full development of the latter expansion shows that terms of order λ^2 and onwards have been neglected. The wave function to order λ^2 is

$$| \psi(t) >= \quad exp(-\tfrac{\imath}{\hbar}Et)\{[\cos(2J\tfrac{t}{\hbar})\hat{A}_1^\dagger - \imath \sin(2J\tfrac{t}{\hbar})\hat{A}_2^\dagger \hat{S}_{21}] + \tag{35}$$
$$(-2J)\tfrac{t^2}{\hbar^2 2!}\{\textstyle\sum_q[F(q)(e^{\imath 2qa} - e^{\imath qa})(\hat{B}_q + \hat{B}_q^\dagger) + \tfrac{|F(q)(e^{\imath 2qa}-e^{\imath qa})|^2}{\hbar\Omega_q}]\hat{A}_2^\dagger \hat{S}_{21}\}\} \, | \, 0 >$$

To order λ^2 we still get

$$< \psi(t) \mid \hat{a}_1^\dagger \hat{a}_1 \mid \psi(t) >= \cos^2(2J\tfrac{t}{\hbar}) \tag{36}$$

and

$$< \psi(t) \mid \hat{a}_2^\dagger \hat{a}_2 \mid \psi(t) >= \ \sin^2(2J\tfrac{t}{\hbar}) + (\tfrac{t^2}{\hbar^2 2!}(-2J)^2)^2\{\textstyle\sum_q \mid L(q) \mid^2 + (\textstyle\sum_q \tfrac{|L(q)|^2}{\hbar\Omega_q})^2\} \tag{37}$$

where

$$L(q) = F(q)(e^{\imath 2qa} - e^{\imath qa}) \tag{38}$$

Using the parameters for the α-helix as specified above, we find that the second term in (37) can be neglected for times shorter than 10^{-13} sec. Therefore expression (32) and the approximation (29) hold for times shorter than 10^{-13} sec.
Summing the full expression (27) corresponds to obtaining a solution for any time t. In the absence of the latter, one possibility is to continue to add terms of higher order in time and thus obtain solutions for specific finite times.

THE GENERAL CASE

For any N, in the Heisenberg picture we get:

$$\imath\hbar\frac{d\hat{A}_1^\dagger}{dt} = \quad E\hat{A}_1^\dagger - J\hat{A}_2^\dagger\hat{S}_{21} - J\hat{A}_N^\dagger\hat{S}_{N1} \tag{39}$$
$$\imath\hbar\frac{d(\hat{A}_2^\dagger\hat{S}_{21})}{dt} = \quad E\hat{A}_2^\dagger\hat{S}_{21} - J\hat{A}_3^\dagger\hat{S}_{31} - J\hat{A}_1^\dagger + [\textstyle\sum_q \hbar\Omega_q\hat{B}_q^\dagger\hat{B}_q, \hat{A}_2^\dagger\hat{S}_{21}]$$
$$\vdots \qquad\qquad \vdots$$
$$\imath\hbar\frac{d(\hat{A}_N^\dagger\hat{S}_{N1})}{dt} = \ E\hat{A}_N^\dagger\hat{S}_{N1} - J\hat{A}_{N-1}^\dagger\hat{S}_{N-11} - J\hat{A}_1^\dagger + [\textstyle\sum_q \hbar\Omega_q\hat{B}_q^\dagger\hat{B}_q, \hat{A}_N^\dagger\hat{S}_{N1}]$$

The approximation that led to (29) and (33) corresponds to neglecting the commutators in the latter system of equations. In that case, although the variables are operators, the coeficients are real constants and thus we have the general rule that *the exact dynamical solution for very short times corresponds to solving the system of linear, coupled equations (39) with suitable initial conditions!*

THERMAL EFFECTS

When a solution for $T = 0K$ has been obtained, a solution for a finite temperature can be derived by considering the following initial condition:

$$| \Psi_\nu(0) > = \hat{A}_1^\dagger | \nu >$$ (40)

The solution for a time t is then

$$| \Psi_\nu(t) > = e^{-\frac{i}{\hbar}\hat{H}t} | \Psi_\nu(0) > = e^{-\frac{i}{\hbar}\hat{H}t}\hat{A}_1^\dagger | \nu >$$ (41)

and using the same development as in eq.(27) we get

$$| \Psi_\nu(t) > = (\hat{A}_1^\dagger + \lambda[\hat{H}, \hat{A}_1^\dagger] + \cdots)e^{-\frac{i}{\hbar}\hat{H}t} | \nu > = [solution\ for\ T = 0K]\, e^{-\frac{i}{\hbar}\hat{H}t} | \nu >$$ (42)

CONCLUSION

It is our view that the numerical simulations described in the first part of this article do not give clear answers to the question of the thermal stability of the Davydov soliton. Therefore, a method has been proposed to obtain exact solutions. This method bypasses the diagonalization of the Fröhlich Hamiltonian (1-6) by looking for exact *dynamical* solutions instead of eigenstates. As presented here it only provides answers for very short times. In fact, it constitutes a new program of research. We think that only when such a program has been carried out will it be possible to decide whether the Fröhlich Hamiltonian does predict the existence of soliton solutions at biological temperatures.

Acknowledgement

Stimulating discussions with P. A. Hansson are acknowledged.

References

1. A.S. Davydov, J. Theor. Biol. 38, 559 (1973).
2. A.S. Davydov and N.I. Kislukha, Phys. Status Solidi b 59 ,465 (1973).
3. See A.S. Davydov, Biology and Quantum Mechanics (Pergamon, New York, 1982); Usp. Fiz. Nauk. 138, 603 (1982)[Sov. Phys. – Usp. 25,898 (1982)], for extensive bibliographies.
4. A.C. Scott, Phys. Rev. A 26 , 578 (1982); Physica Scr. 29, 279 (1984); L. MacNeil and A.C. Scott, *ibid.* 29, 284 (1984).
5. A.S. Davydov, Zh. Eksp. Teor. Fiz. 78,789 (1980) [Sov. Phys.– JETP 51, 397 (1980)]; Institute of Theoretical Physics, Kiev, USSR, Report No. ITP-85-146E, 1985 (unpublished).
6. P.S. Lomdahl and W.C. Kerr, Phys. Rev. Lett. 55, 1235 (1985).
7. A.F. Lawrence, J.C. McDaniel, D.B. Chang, B.M. Pierce and R.R. Birge, Phys. Rev. A 33, 1188 (1986).
8. A.C. Scott, "On Davydov solitons at 310K", in Energy Transfer Dynamics, eds.

T.W. Barrett and H.A. Pohl, Springer-Verlag, Berlin-Tokyo, Chapt.15, 167 (1987).
9. H. Bolterauer, "Aspects of Quantum Mechanical Thermalization in the Alpha-helix", in Structure, Coherence and Chaos in Dynamical Systems, Manchester Univ Press, Manchester, Chapt. 55, 619 (1989).
10. D.W. Brown, K. Lindenberg and B.J. West, Phys. Rev. A **33**, 4104 (1986).
11. D.W. Brown, B.J. West and K. Lindenberg, Phys. Rev. A **33**, 4110 (1986).
12. T.D. Lee, F.E. Low and D. Pines, Phys. Rev. **90**, 297 (1953).
13. D.W. Brown, K. Lindenberg and B.J. West, J. Chem. Phys., **84** , 1574 (1986).

Question by Kenkre: What exactly is the dimensionless small parameter in orders of which you make the expansion in the second part of your talk to determine the evolution?

Reply: There is not a single parameter involved. The terms are of the form

$$\frac{\left(i\frac{t}{\hbar}\right)^n}{n!}\left[\ldots,\left[\hat{G}\ldots,\left[\hat{F}\ldots,\left[\hat{G},\ldots[\hat{F},\hat{A}_1^\dagger]\ldots\right]\right.\right.\right., \tag{1}$$

where n is the total number of commutators involved. Let there be m $\hat{G}$ operators and $n-m$ $\hat{F}$ operators. Then there will be m factors $\hbar R_q$ and $n-m$ $(-2J)$ terms. The dimensionless constant is of the form:

$$\left(\frac{t}{\hbar}\right)^n (2J)^{n-m}(\hbar R_q)^m. \tag{2}$$

Equation (1) also involves phonon operators, e.g.

$$\left[\hat{G},\left[\hat{F},\hat{A}_1^\dagger\right]\right] = -(2J)\left\{\sum_q\left[\frac{|F(q)(e^{iqa}-e^{i2qa})|^2}{\hbar R_q}+\right.\right.$$
$$\left.\left.+\hbar R_q\frac{F(q)}{\hbar R_q}(e^{i2qa}-e^{iqa})(\hat{B}_q+\hat{B}^\dagger_{-q})\right]\right\}\hat{A}_2^\dagger\hat{S}_2. \tag{3}$$

The commutator with two $\hat{G}$s involves a power of two of $(B_q+B^\dagger_{-q})$ with a factor of $\frac{1}{3!}$ from (1), a commutator with three involves a power of three with a factor of $\frac{1}{4!}$. So $\frac{1}{n!}$ factors compensate for the increase in the power of $(B_q+B^\dagger_{-q})$.

Section IV

Experimental Results

Now what I want is Facts.... Facts alone are wanted in life.
Mr. Gradgrind in *Charles Dickens's* **Hard Times**

In the early 1970's, Giorgio Careri began a novel experimental investigation of protein dynamics. His idea was to obviate many of the problems arising in the study of real protein by looking at hydrogen bonded crystals that might be regarded as "model proteins." The basic idea can be explained in the context of acetanilide ($CH_3CONHC_6H_5$) or ACN and the structure of a typical polypeptide chain in natural protein in Figure 1. The similarity of bond lengths and angles suggests that the dynamical behavior of ACN might provide clues to the corresponding behavior of natural protein.[1]

The comparison becomes more striking if we look at the structure of crystalline ACN shown in Figure 2 together with that of alpha-helix shown in Figure 1 of the introductory remarks to Section I. In both cases one sees clearly the hydrogen bonded peptide channel with the atomic structure

$$\cdots \; \mathbf{H - N - C = O} \; \cdots \; \mathbf{H - N - C = O} \; \cdots \; \mathbf{H - N - C = O} \; \cdots$$

that has been the starting point for many of the theories discussed in this volume.

---

[1] At present ACN is of no commercial value, but parahydroxyl ACN (or acetaminophen) is well known to tense North Americans as "Tylenol." ACN is oxidized in the body to acetaminophen, and in fact predates aspirin as a common analgesic. Early in the present century it was marketed under a variety of names including: "U-Re-Ka Headache Powders," and "Harper's CUFORHEDAKE BRANEFUDE." Being occasionally fatal, ACN gradually lost trade to aspirin. [1]

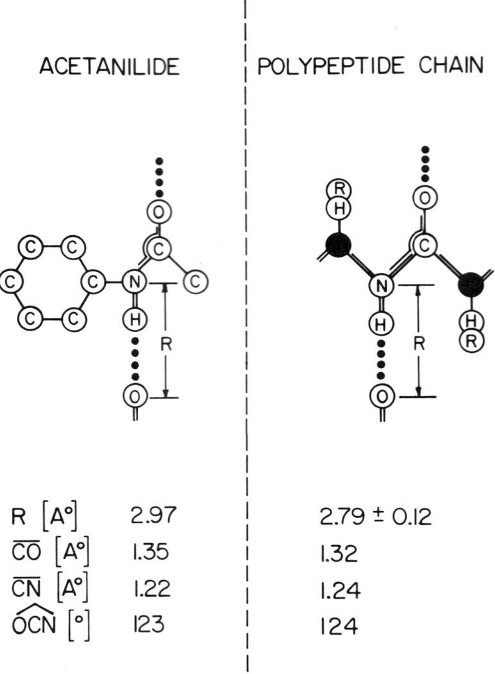

Figure 1. A comparison of bond lengths and angles in crystalline acetanilide and natural protein.

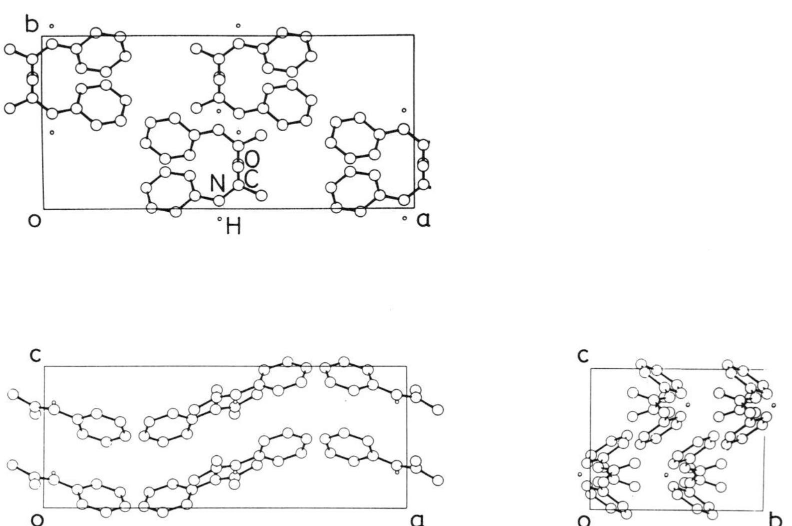

Figure 2. A unit cell of crystalline acetanilide.

Careri's intuition was soon rewarded by the discovery of an anomalous resonance at 1650 cm^{-1} in crystalline ACN [2], as shown in Figure 3. A decade of effort by Careri and his co-workers at the University of Rome established that this 1650 cm^{-1} peak was an amide-I (CO stretching) resonance which could not be conventionally assigned to a Fermi resonance, Davydov splitting, or a phase transition [3]. It is important to note that Careri's discovery and the subsequent experimental studies were made without any knowledge of Davydov's theory of self-trapped molecular vibrations. It was only in 1982 after he had exhausted all possibilities for conventional assignments that he considered dynamical self-trapping.

In Careri's chapter, the importance of the *intermediate* character of the hydrogen bond is emphasized. With reference to Figure 1, a change in the distance R alters the distribution of local charges and, thereby, the corresponding force constants. The resulting sensitivity of this effect may help to explain why the exciton-phonon coupling parameter

$$\chi = \frac{dE_0}{dR} \tag{1}$$

is so difficult to calculate using *ab initio* techniques (see Section II for further details). Careri also points out that the fine structure in the 1650 cm^{-1} and 1665 cm^{-1} peaks (see Figure 1) is not yet understood, but it may be related to a double well potential for the hydrogen bonding proton [4]. Finally he contributes some speculations concerning the role that a self-trapped state, similar to that observed in ACN, might play in enzymatic mechanisms.

Barthes describes measurements on several deuterated species of ACN using both incoherent neutron scattering (INS) and infrared absorption. One result is the observation of a new band centered at 750 cm^{-1} , which is assigned to an out of plane bending mode of the hydrogen bonding proton. The unusual width of this band suggests a double well potential for the proton.

Of particular importance with respect to the assignment of the 1650 cm^{-1} peak to self-trapping [3] are the observations of Barthes on the partially deuterated species: $CD_3CONHC_6D_5$. In this material, the hydrogen bonding proton is unchanged, but the 1650 cm^{-1} peak *disappears*. Furthermore a new anomalous peak (not found in undeuterated ACN) appears at 1480 cm^{-1}. These results suggest that the self-trapping mechanism in ACN may be much more complex than has hitherto been assumed.

The observations reported by Bigio et al. are based on Raman spectroscopy of carefully prepared ACN single crystals. These are probably the best available measurements of the integrated intensity versus temperature of the 1650 cm^{-1} peak. A small polaron theory (see also Chapter 9 by Kapor et al. in Section I) is used to show that this temperature dependence can be derived from a model in which twelve (or more) optical modes clustered around 70 cm^{-1} contribute to the self-trapping.

The chapter by Nielsen returns to Careri's original idea of using model proteins to learn about the behavior of natural protein [2]. He looks at simple amides in the liquid state (see his Table 1) in which the above mentioned hydrogen bonded peptide channels are expected to form. Using his "$R(\bar{\nu})$ representation" to improve resolution in the far-infrared Raman spectrum, Nielsen finds a characteristic band around 100 cm^{-1} which seems to involve a substantial transverse motion of the hydrogen bonding proton (see his Figure 3). The ubiquitous nature of this band confirms Careri's intuition [2,5] and the observations of Barthes concerning the importance of the hydrogen bonding proton as a source of anharmonicity in protein dynamics.

There are many model proteins for the experimental scientist to choose, and each choice requires a substantial commitment of time and effort. Both parachloro-ACN [3] and N-methylacetamide [6,7], for example, show evidence of anomalous amide-I bands similar to the

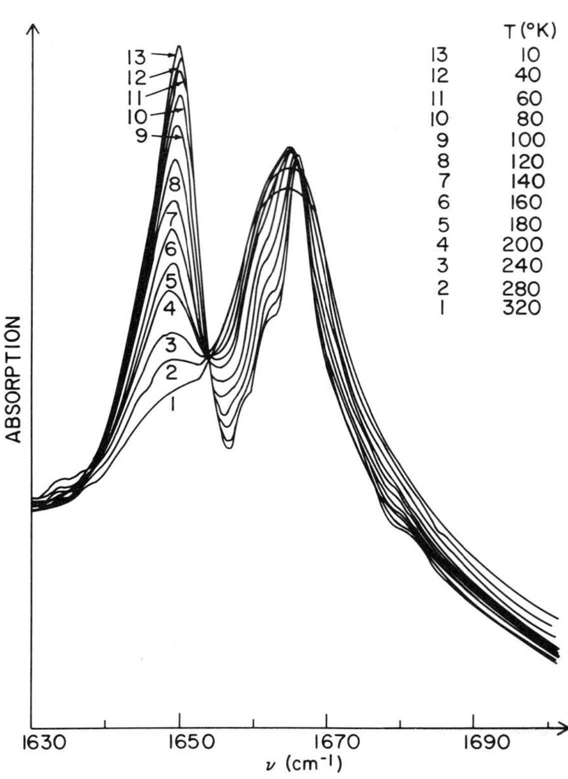

Figure 3. An infrared absorption spectrum of acetanilide in the amide-I region. Notice the anomalous peak at 1650 cm^{-1}.

1650 cm^{-1} band in ACN, but more work must be done before this can be concluded with confidence.

Migliori et al. have directed their attention toward yet another choice: ℓ-alanine (see their Figure 2). In making this selection they seem less motivated by the hope of modeling anharmonic effects in natural protein and more by a general interest in nonlinear energy transport processes, but the results may be biologically relevant (see Chapter 4 by Lindenberg et al.). This work concentrates upon the self-trapping of far infrared optical phonons by longitudinal acoustic phonons, and a simple mechanical model is suggested. Measurements of far infrared Raman intensities vs. temperature provide some evidence of localization. Coherent anti-stokes Raman spectroscopy is planned in the near future.

In the last chapter of this section, Knox et al. describe a pump-probe experiment being designed to observe the transport of vibrational energy along alpha-helix. Their aim is to provide a direct test of Davydov's original idea [8]. Considering the theoretical uncertainties presented in this volume, one must agree that such a test is needed.

Finally it seems appropriate to mention a recently published description of some transient-bleaching-recovery experiments on the 1650 cm^{-1} band of ACN (see Figure 3) using a free-electron laser source [9]. According to the authors, these experiments "indicate that the anomalous 1650 cm^{-1} band which appears on cooling of acetanilide crystals persists for at least several microseconds following rapid pulse heating." Such observations suggest that the 1650 cm^{-1} band is not evidence of a self-trapped vibrational state but "perhaps is due to multiple configurations for the amide group" related to a change in the position of the hydrogen bonding proton.

We were pleased to have Lewis Rothberg, one of the authors of Reference [9], at the workshop to present this research and to participate in informal discussions. During the course of these discussions, it appeared that the results of Reference [9] are at variance with some unpublished experiments carried out at the Los Alamos National Laboratory by Irving Bigio and Cliff Johnston on ACN using a pulsed CO laser source [10]. These experiments indicated that the 1650 cm^{-1} band changes its intensity on a microsecond time scale in a manner that is consistent with laser heating and the observations of Careri et al. [3].

Considering the disagreement between these two observations, it is important that both be repeated under carefully controlled conditions.

References

[1] C.C. Mann and M.L. Plummer, "The big headache," **The Atlantic Monthly** , October 1988.

[2] G.Careri, in **Cooperative phenomena** (H. Haken and M. Wagner, Eds.) Springer-Verlag, Berlin, Heidelberg and New York (1973) pp. 391-394.

[3] G. Careri, U. Buontempo, F. Galluzzi, A.C. Scott, E. Gratton and E. Shyamusunder, Phys. Rev. B **30**, 4689 (1984).

[4] G. Careri, E. Gratton and E. Shyamsunder, Phys. Rev. A **37**, 4048 (1988).

[5] G. Careri, U. Buontempo, F. Carta, E. Carta, E. Gratton and A.C. Scott, Phys. Rev. Lett. **51**, 304 (1983).

[6] I.J. Bigio, O. Faurskov Nielsen and C.T. Johnston in **Computer analysis for life science** (C. Kawabata and A.R. Bishop, Eds.) Omsha, Ltd, (1986).

[7] J. Halding Jensen, P.L. Christiansen, O. Skovgaard, O. Faurskov Nielsen and I.J. Bigio, Phys. Lett. A **117**, 123 (1986).

[8] A.S. Davydov, J. Theor. Biol. **38**, 559 (1973).

[9] W. Fann, L. Rothberg, M. Roberson, S. Benson, J. Madey, S. Etemad and R. Austin, Phys. Rev. Lett. **64**, 607 (1990).

[10] I.J. Bigio (private communication).

25

THE AMIDE-I BAND IN ACETANILIDE:
PHYSICAL PROPERTIES AND BIOCHEMICAL SUGGESTIONS

Giorgio Careri
Dipartimento di Fisica
Università di Roma I, 00185 Roma

Abstract

Measurements of infrared absorption and Raman scattering on crystalline acetanilide, a model system for proteins, are reviewed. A new band, close to the conventional amide-I band, was detected at low temperature. Equilibrium properties and spectroscopy data ruled out an explanation based on conventional assignments. A detailed analysis showed that a soliton model is in satisfactory agreement with the experimental data. The fine structure of this band, detected more recently, is discussed. Some possible biological implications are finally mentioned.

1   An infrared active soliton in a model crystal

Hydrogen bonding is widespread in biomacromolecules, presenting physical features that vary greatly from one case to another. In weak bonds this interaction can be treated as a problem in electrostatics involving a set of fixed and localized interacting charges. On the other hand, in the case of very strong hydrogen bonds, one is faced with a delocalized charge distribution to be treated according to valence theory. Between these two extremes, there are cases where the bonds are of intermediate strength. In these cases, one can model the complex state of affairs by assuming that the local charges depend upon their mutual distances and that these distances in turn depend upon the local charges. Thus in these intermediate cases one can visualize the microscopic source of the anharmonic coupling which gives rise to the nonlinear terms in the equations describing the dynamics of the system. This is the case of the N-H $\cdots$ O=C bond in proteins, for R(NO) distances close to 2.80 Å.

Acetanilide ($CH_3CONHC_6H_5$), or ACN, is an interesting solid because two close chains (spines) of nearly planar hydrogen-bonded amide groups run through the crystal, providing an interesting model for an array of hydrogen-bonded amides in one direction. Moreover, the bond distances in ACN are very close to those found in alpha-helices, where three similar spines are coiled along the helix axis. Since the physical properties of hydrogen-bonded systems are very sensitive to bond distances, we thought ACN would be a useful model system to be used in searching for new physical features of extended polypeptide chains and perhaps also proteins. We found that a new amide-I band appears at low temperature in crystalline ACN, red-shifted by 15 cm^{-1} from the primary amide-I band at 1665 cm^{-1} , and a large number of other experiments by Raman, X-rays, specific heat and isotopic substitution pointed out that the new band at 1650 cm^{-1} is characteristic of the amide group of ACN in crystal form. A detailed analysis by the usual exciton model cannot account for this new band.

Davydov's Soliton Revisited, Edited by P.L. Christiansen and A.C. Scott
Plenum Press, New York, 1990

Having excluded conventional explanations [1,2,3], we consider the possibility of assigning it to a collective excitation similar to the soliton proposed by Davydov for alpha-helix in proteins [4,5]. Davydov's soliton arises from a cooperative interaction between localized amide-I bond energy and lattice distortion. The bond energy acts through nonlinear coupling, as a source of lattice distortion. This lattice distortion reacts, again through nonlinear coupling, as a potential well to trap the bond energy and prevents its dispersion via dipole-dipole coupling effects. We followed the same theory with one important difference: for lattice distortion we substituted displacement of the hydrogen-bonded proton. The distinction is vital, because Davydov has shown that photon absorption by his intermolecular vibrational soliton is ruled out by the Franck-Condon principle. Here I shall limit myself to outlining the major points and presenting some conclusions. The main idea was that the effect of introducing localized amide-I energy could displace the ground state of the adjacent hydrogen-bonded proton. This displacement of the proton acts to trap the amide-I band energy and prevent its dispersion via dipole-dipole interaction effects. The combined excitation was proved to be a soliton, and we assigned the binding energy of this soliton to the experimentally observed red shift of 15 cm^{-1} from the conventional amide-I band to the unconventional amide-I band. A more recent theoretical work [6] has identified this unconventional band as a vibronic analog of a small Holstein polaron.

2   The fine structure of the amide-I band

The more recent theoretical work [6] suggests that the shift Δ of this polaron frequency from the unperturbed amide-I mode is due to the coupling with both acoustic and optical longitudinally polarized lattice modes. Moreover, acoustic coupling is found to contribute to about half of the total observed shift ($\Delta = 15$ cm^{-1}), and to dominate the temperature dependence of the absorption strength of the polaron peak. This last theoretical result prompted us to look at the fine structure of the unpolarized absorption spectrum in the amide-I region, where the absorption profile shows the presence of temperature dependent unresolved bands in the shoulder of the 1650 cm^{-1} peak (see Figure 3 of the introductory remarks to this section).

An analysis [7] of the 1665 cm^{-1} and 1650 cm^{-1} bands in terms of superposition of lorentzian components was performed on the same set of data already reported in Figure 3. At least two components were necessary to describe the shape of the 1650 cm^{-1} band in the entire temperature range investigated at 1645 and 1650 cm^{-1}. The decomposition of the band centered at 1665 cm^{-1} required also at least two components centered at 1662 and 1666 cm^{-1}. Polarization studies performed on thin crystals have shown that the subbands have the same polarization. Low temperature spectra of partially deuterated samples showed the presence of subbands at the same absorption frequencies found by the fitting procedure in the spectra of non-deuterated sample. The position of the center of the bands did not changed with temperature whereas the width and intensity were temperature dependent. We recognized that the total integrated intensity, i.e., the sum of the areas of the four component bands, remained fairly constant in the temperature range from 20 K to 300 K. Instead, the integrated intensity of the two lorentzian components in which we decomposed each band had different temperature behaviour. The component band at 1665 cm^{-1} and at 1645 cm^{-1} had similar, but complementary temperature behaviour, and the total area change due to the sum of these two bands was about 10 percent of the total area over all the temperature range. Instead, the integrated intensity of the two components bands at 1661 cm^{-1} and 1650 cm^{-1} had complementary behaviour. The sum of their integrated intensity was constant in the temperature range from 300 K to 20 K, but each of them had a strong temperature dependence. The integrated intensity of each of these two subbands changed about five-fold from 20 K to 300 K.

The use of a symmetric lorentzian shape forced the fitting algorithm to include a band at 1645 cm^{-1} to account for the nonsymmetry of the spectral shape. However, due to the small contribution of this band (less than 10 percent) to the total intensity and due to the small temperature variation, we have not included this band in the discussion. On the basis of our band decomposition and single crystal polarized spectra, we have observed that the temperature dependence of the integrated area of the band at 1666 cm^{-1} shows a change between 300 K and 100 K but then level off at lower temperature, while the band at 1662 cm^{-1} shows a much larger and continuous decrease of the integrated intensity as the temperature is lowered. According to the previous proposed explanation for the origin of the 1650 cm^{-1} band [2], the intensity of these two subbands centered at 1666 and 1662 cm^{-1} should have the same temperature dependence. The band analysis shows that this is not the case, therefore a modification of the proposed model [3] must be introduced to understand the behaviour of the ACN IR spectrum in the amide-I region.

One possibility that can qualitatively account for the existence of subbands is the existence of a double well structure, which could arise from the non-planarity of the amide group. Two equivalent positions for the N-H bond, corresponding to a deviation of about 0.1 radians from the planar geometry of the amide group have been proposed by Fillaux and deLozé [8] to explain the fine structure of the soft mode region shown in the Raman spectrum of N-methylacetamide. Thus one can propose that the asymmetry due to the double well potential could be the origin for the nonlinear coupling between the C=O vibration and the N-H bending needed to explain the amide-I anomalous behaviour. As a matter of fact, disregarding the weak 1645 cm^{-1} band, one could suppose that the normal amide-I band in ACN is split into two components with absorption at 1662 and 1666 cm^{-1}, because of tunnelling between the two wells, and that only the component at 1662 cm^{-1} could give rise to the anomalous band at 1650 cm^{-1}. This interpretation is based on the sum law for the 1650 cm^{-1} and 1662 cm^{-1} bands.

As a second possibility to account for the fine structure of the amide-I band in ACN, one can postulate that the band at 1662 cm^{-1} is due to a soliton which is stable at high temperature (an excited polaron state in the phonon field, visually a deformation of the trapped charge cloud). Then, still disregarding the weak band at 1645 cm^{-1} , the observed nearly constant intensity of the sum of the band at 1650 cm^{-1} and 1662 cm^{-1} could arise from the competition of the two soliton energy levels during the IR absorption. This model requires a treatment of the temperature dependence of the soliton band(s) different from the one initially proposed by Scott, et al. [2].

3 Biological implications

The work of ACN reported above has been motivated by a few relevant biological implications that we believe it would be appropriate to mention. Biochemical events involving the reactions or changes of state of a macromolecule may be supposed to occur one step at a time. Thus they may often be represented in terms of a network of closed pathways or loops. At equilibrium there will be no net circulation around the network, the conditions of detailed balance being fulfilled. A living system requires process, as in the form of a one-way circulation around the network. In the fully-developed organism, this state of affairs, which provides for the transduction of free energy from one chemical reaction to another, is made possible by the presence of highly developed polyfunctional enzymes. In an effort to provide a mechanism for this one-way circulation, we have recently introduced the concept of a trapped soliton as an energy packet created at one point of the network and liquidated at another [9]. We postulated the formation of a soliton trapped as a ligand on the protein matrix, for which the surrounding heat bath serves as a sink. This soliton is assumed to display physical features quite similar to those observed experimentally in ACN. All this follows from the intrinsic structure of the hydrogen-bonded polypeptide chain itself, because of its capability to let the vibrational soliton be stable or decay in different backbone conformations.

More recently, after the detection of soliton overtones in [3], we extended these considerations to the special conditions of prebiotic life [10]. Before nature developed chlorophyll, we suggest that there was a retention of a trapped soliton, which served as the driving mechanism for unidirectional circulation. To this end, we imagine that the trapped soliton resulted from absorption of solar IR radiation, a process which served the same function as the driving reaction that supplies chemical free energy under present, more complex conditions. We believe that the cyclical process considered above displays several attractive features in a prebiotic scenario. In the first place, this process mimics the present chlorophyll-assisted sun energy pathway although it is much simpler because it requires only one IR photon to be absorbed and stored as a trapped soliton. Moreover it depends on only one widespread chemical species, namely the amide group. It does not seem inconceivable that nature, at some state of evolution, might have developed a protoenzyme with IR properties similar to those exhibited by ACN at 90 K. Finally, and most importantly, this suggested outlook shows how the "biological arrow of time" can be traced back in a natural way to the general physical principle of symmetry breaking.

References

[1] Careri G., Buontempo U., Carta F., Gratton E. and Scott A.C. 1983. Phys. Rev. Lett. **51**, 304.

[2] Careri G., Buontempo U. Galluzzi F., Scott A.C., Gratton E. and Shymsunder E. 1984. Phys. Rev. B **30**, 4689.

[3] Scott A.C., Gratton E., Shymsunder E. and Careri G. 1985. Phys. Rev. B **32**, 5551.

[4] Davydov A.S. and Kishlukha N.I. 1973. phys. status solidi (b) **59**, 465.

[5] Davydov A.S. 1973. J. Theor. Biol. **38**, 559.

[6] Alexander D.M. and Krumhansl J.A. 1986. Phys. Rev. B **33**, 7172

[7] Careri G., Gratton E. and Shyamsunder E. 1988. Phys. Rev. A **37**, 4048.

[8] Fillaux F. and deLozé M. 1978. J. Raman Spectr. **7**, 224.

[9] Careri G. and Wyman J. 1984. Proc. Natl. Acad. Sci. U.S.A. **81**, 4386.

[10] Careri G. and Wyman J. 1985. Proc. Natl. Acad. Sci. U.S.A. **82**, 4115.

Question by Kenkre: An important message from your talk is that fluctuations introduced by the environment can be so important to the alpha-helix and produce such sensitive results that the consequent changes in the dielectric constant could very easily change the character of the soliton from propagation to pinned. I suppose that these fluctuation effects are negligible for a solid such as acetanilide for which Scott, you, and collaborators have provided a theoretical explanation of the spectra. For a solution, however, the effects could be substantial. Is it possible to design dynamic experiments to observe such change of soliton character?

Reply: You certainly understood my point quite well, but it is very difficult to study the amide-I band in solution because of the high water absorbance. The only possibility is to work on hydrated powders at low temperature, extending the work done by us in lysozyme powders at room temperature. In this last work, however, we did not detect an effect similar to the one reported in ACN.

26

INCOHERENT NEUTRON SCATTERING AND INFRA-RED MEASUREMENTS IN ACETANILIDE AND DERIVATIVES

Mariette Barthes*

G.D.P.C. – U.S.T.L.

34060 Montpellier cedex, France

Acetanilide ,(ACN), is an anharmonic orthorhombic crystal characterised by soft hydrogen- bonded chains of molecules. Anomalies in the I.R. and Raman mode spectra have been observed at low temperature . The theoretical interpretations of these unconventional effects involve   alternatively, Davydov-like solitons,localized "polaronic" modes,topological defects or Fermi resonance

In the present paper , first, incoherent neutron scattering measurements of the vibrational density of states, using samples with different specific deuterations ,are presented. These new data :
- allow to identify a very intense maximum, assigned to the N-H out-of-plane bending mode.
- display the specific behaviour of the methyl torsionnal modes , unobserved in optical investigations .
- confirm our previous inelastic neutron scattering studies, indicating no obvious anomalies in the range of frequency of the acoustic phonons .
- shows the existence of thermally activated quasi-elastic scattering above 100 K, assigned to the random diffusive motion of the methyl protons.

Secondly, new infra-red data are described . New absorption bands with an intensity showing an anomalous temperature dependence are observed . The assignment of the NH out-of-plane bending mode is confirmed,in agreement with neutron scattering data . A new, unconventional mode is observed also in $C_6D_5NHCOCD_3$ in the frequency region of the amide-II modes

These new results are discussed in the light of the recent theoretical models .

----

* In collaboration with R.ALMAIRAC,J.L.SAUVAJOL,J.MORET,G.PAGE (GDPC) ;
   R.CURRAT and A.J.DIANOUX (I.L.L.,Grenoble,France)

Davydov's Soliton Revisited, Edited by P.L. Christiansen and A.C. Scott
Plenum Press, New York, 1990

An important theoretical and experimental interest is being focused on acetanilide ($C_6H_5NHCOCH_3$, or ACN) , crystal displaying chains of hydrogen-bonded amide groups (>N-H···O=C<), with bond distances very close to those found in polypeptides. So ACN is considered as a model system for some biological molecules and proteins; it is also a potential candidate for the observation of Davydov solitons. The anomalous temperature dependence of some infrared and Raman modes of ACN [1-4], mainly observed in the region of the amide-I mode (or C=O stretching), has received several theoretical interpretations :

It has been assigned to the existence of a trapped Davydov soliton state[2,5], to vibron solitons [6], or to a vibronic analog of a small Holstein polaron [7,8,9] due to a coupling of the unperturbed amide-I mode with both acoustic and optic longitudinally polarized lattice phonons.This last model presents similarities to those of localised modes in ionic crystals containing impurities. The occurence of topological solitons has also been suggested [10]

These models have been questionned by an alternative explanation involving a temperature tuning of a Fermi resonance between the amide-I mode and a accidentally degenerated combination level [11]

To get more information about the nature of the non-linear excitations in this material, we have done some neutron scattering and optical investigations on ACN and some deuterated derivatives. Brillouin, and Raman scattering[12] will be published elsewhere Our coherent inelastic neutron scattering study [13] has revealed an important anharmonicity , but no observable anomaly on the dispersion and temperature dependence of the acoustic modes, in the fully deuterated derivative, in contradiction with some expectations. The sound velocities and the corresponding elastic constants, calculated from the slopes of the acoustic dispersion curves, reflect a strong anisotropy of the bonding between ACN molecules.

We present here new results of an incoherent neutron scattering (INS) study of four specifically deuterated species of ACN :

/1/ $C_6H_5NHCOCH_3$ (or fully hydrogenated ACN)

/2/ $C_6H_5NDCOCH_3$

/3/ $C_6D_5NHCOCD_3$

/4/ $C_6D_5NDCOCD_3$ (or fully deuterated ACN),

giving the temperature dependence of the phonon density of states and the quasi-elastic scattering. To precise some results of this study, a new investigation of infra-red absorption became necessary,and is presented in the second part of this paper.

1   INCOHERENT NEUTRON SCATTERING

MATERIALS AND INSTRUMENTATION

The synthesis of powdered samples of the four derivatives were achieved in our laboratory. The rate of deuteration of the benzene ring and of the methyl group is better than 99% and stable with time. The rate of deuteration of the proton of the amide group (-NHCO-) , controlled by IR absorption and Raman scattering , is better than 90%. Samples N°/2/ and /4/ were always handled in dry nitrogen or helium atmosphere, in order to prevent the H/D exchange on the H bond.

All measurements have been carried out at the Institut Laue-Langevin in Grenoble, with IN6 time-of-flight spectrometer. The resolution was about 50 μeV at zero energy. The experiments were performed at different temperatures between 1.5 and 300K.

The experimental scattering law $S(Q,\omega)$ is treated by standard procedures [14,15], to extract the generalized frequency distribution, or vibrational density of states $G(\mathcal{E})$.   The data were corrected from the multiphonon scattering contribution with the help of an algorithm developped by A.J. Dianoux

RESULTS

a) The generalized frequency distributions, in the highest attainable energy range, are displayed in fig 1. The comparison of data from the four species indicates that a strong and very wide maximum exists at about 92-95 meV ($\sim$ 750-770 cm-1), in sample /1/ and /3/ only.

This fact, and the absence of the same maximum in the two other samples, shows that the associated vibration involves the proton of the hydrogen-bond.

This maximum appears as the main difference between /1/ and /2/ (and between /3/ and /4/) INS spectra. It is therefore possible to assign this maximum to the N-H out-of-plane bending mode whose frequency is expected in this energy region, but never before identified . Fig 1b shows the temperature evolution of this wide maximum for sample /3/.

b) Fig. 2 gives the experimental values of the phonon density

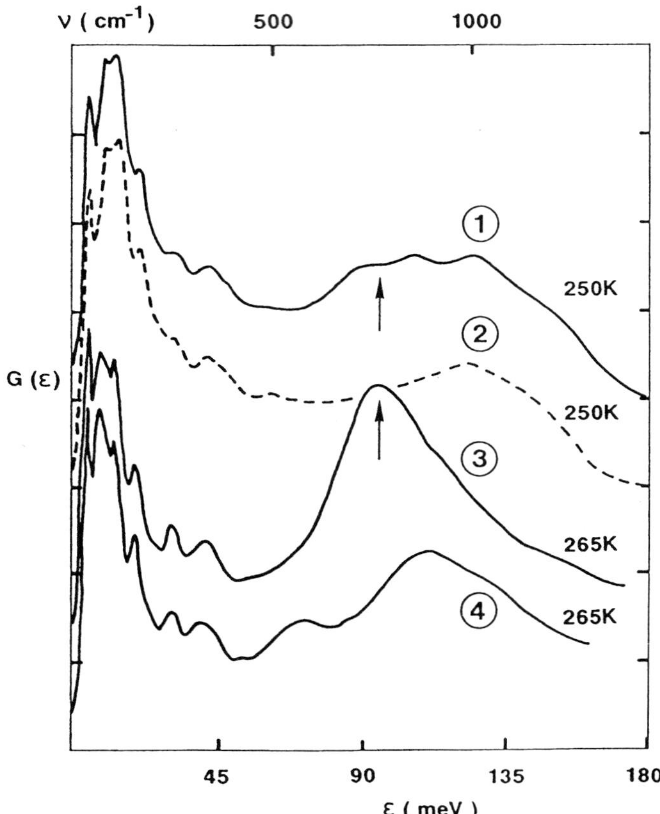

Fig. 1   Generalized Frequency Distribution for the four ACN derivatives. The arrows indicate the maximum at 92 - 95 meV in samples /1/ and /3/.

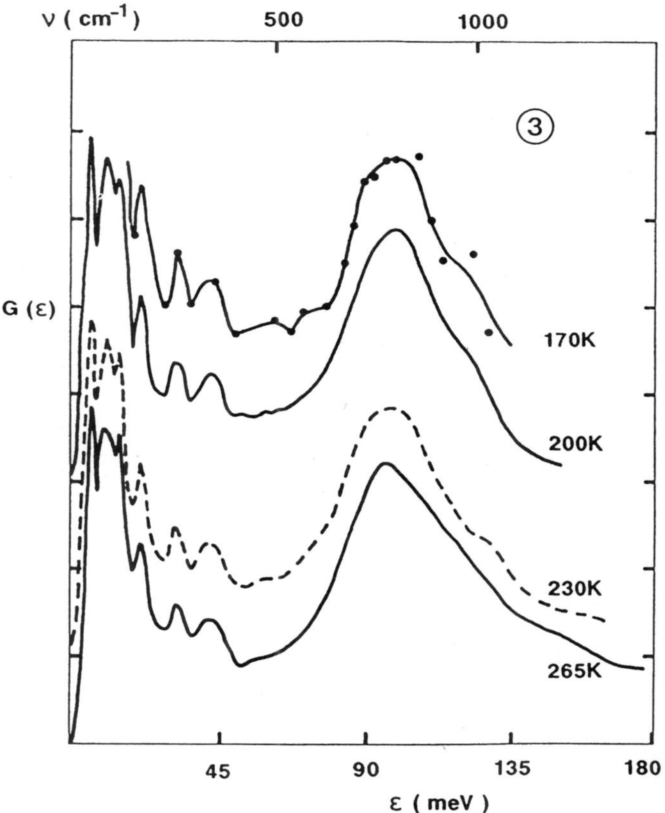

Fig. 1b   Temperature dependence of the wide maximum in G(ε) at 92 – 95 meV.

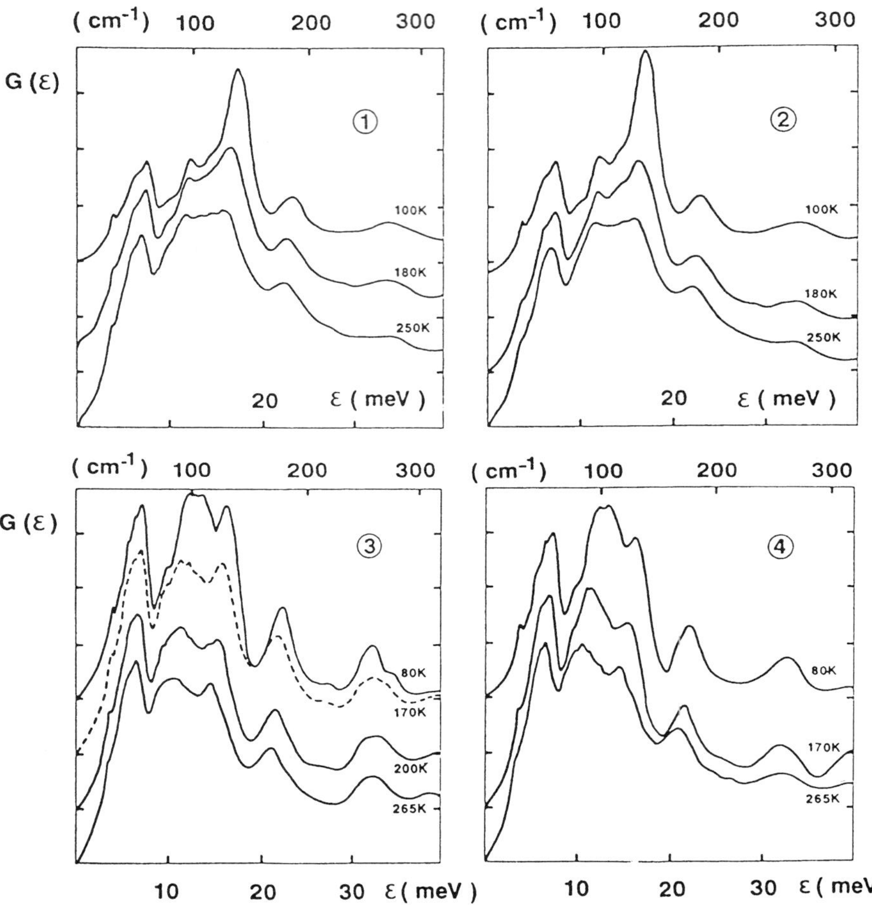

Fig. 2   Vibrational Density of States of the four ACN specimens, at different temperatures ,between 0 and 40 meV.

of states at different temperatures, for the four selectively deuterated samples, in a smaller range of energy (0-40 meV or 320 cm-1). Mainly, normal external modes, librations and acoustic phonons contribute to this part of the frequency distribution. However a detailed assignment of all the maxima is beyond the scope of the present paper. Only the anomalous features will be analysed, and their possible implication in a non -linear coupling mechanism with higher frequency modes will be considered.

It is clear from fig .2 that, among the strong maxima, the peak situated at $\sim$ 16-17 meV in samples /1/ and /2/ - and the maximum at $\sim$ 10-12 meV in samples /3/ and /4/ -   exhibit a different behaviour from others :

- larger energy shift at low temperature
- faster broadening with temperature than the other peaks in the phonon density of states
- significant intensity decrease by deuteration
- very large isotopic shift :

$$\Delta = \frac{\mathcal{E}_{/1/}}{\mathcal{E}_{/3/}} = \frac{\mathcal{E}_{/2/}}{\mathcal{E}_{/4/}} \simeq 1.3$$

These properties are precisely the same as the special characteristics of the methyl torsional modes [16,17]. In particular, the observed isotopic ratio, Δ , may be identified as follows :

$$\Delta \simeq \left(\frac{\theta_{CD_3}}{\theta_{CH_3}} \right)^{1/2}$$

where θ_i is the momentum of inertia of the methyl (and CD3) group. So, the energy of this motion scales with the mass of the hydrogen atom (and not with the mass of the whole molecule, or parts of it).

Moreover, the librationnal excitations have usually little dispersion, and therefore lead to a pronounced peak in the inelastic neutron scattering spectra. Their strong intensity is also a consequence of the large amplitude motion of the protons. The modes under consideration in ACN display also a significant intensity. It is thus reasonable to assign them to the torsional transitions of the methyl, and CD3 groups, in consistency with their specific behaviour. It is obvious from the close similarities of their properties in samples /1/ and /2/ in one hand, in /3/ and /4/ in the other hand ,that they could not result from motions involving the amide-group .

Librations involve very small dipole moment change or polarizability ,and therefore are infrared and Raman inactive or weak. In the case of ACN,no unambiguous evidence of their activity in Raman scattering has been found[12], though the existence of a band at about the same frequency , and exhibiting a strong energy shift versus temperature , has been observed in a fully deuterated single crystal.

From the low energy part of the frequency distributions, it is also possible to deduce another physical information : no obvious anomaly is observable in the range of acoustic phonons, for none of the four species, thus confirming the previously acquiered data of our coherent inelastic scattering study [13].

C) The quasi-elastic scattering (QES) and its temperature dependence ,have also been studied. Samples /1/ and /2/ exhibit the same broadening and the same temperature evolution (fig.3). A weaker quasi-elastic broadening is also observed in samples/3/ and /4/, with again, the same temperature dependence for both. In the four species, the quasi-elastic broadening is thermally activated. It disappears below 100K in samples /3/ and /4/. Fig.4 shows a plot of the temperature dependence of the full width at half-maximum, Γ, estimated at each temperature after correction for the elastic peak [15], and fitting the QES with a lorentzian.

As in the preceding case , the similarities between the QES of /1/ and /2/ in one hand, the similarities between the QES of /3/ and /4/ in the other, demonstrate that the broadening may be attributed to some diffusive motion involving either the methyl group or the phenyl ring, but in no case the amide-group (because the activation energy is different for /1/ and /3/).

Though the accuracy on the Γ estimation is not very large, fig.4 however shows that the QES temperature dependence is consistent with a thermally-activated process, with an activation energy E_0 of about 500K (43 meV, or $\sim$ 350 cm-1) for samples /1/ and /2/, and about 630K (54 meV, or $\sim$ 440 cm-1) for /3/ and /4/ .

As observed in other methylated compounds [15,16,17] above 50K, the quasi-elastic broadening reflects the thermally activated hopping of protons of the methyl group between adjacent wells of the rotational potential. The activation energy gives a measurement of the hindering barrier of the methyl rotation. The most commonly used rotational model for this motion involves random reorientations over the barrier by instantaneous jumps of 120° around the threefold axis. The time τ between jumps is related to the temperature by an Arrhenius law.

If we assume that the observed QES in ACN has the same origin , we may estimate the time between two random jumps of the proton ($\tau \simeq$ 10 picoseconds). The high activation energies indicate a high hindering barrier and as a consequence, the tunnel splitting for the methyl protons is presumably very small, below the resolution of back scattering neutron spectrometers.

It is important to note that the activation energy E_0 is of the same order of magnitude as those deduced from the temperature

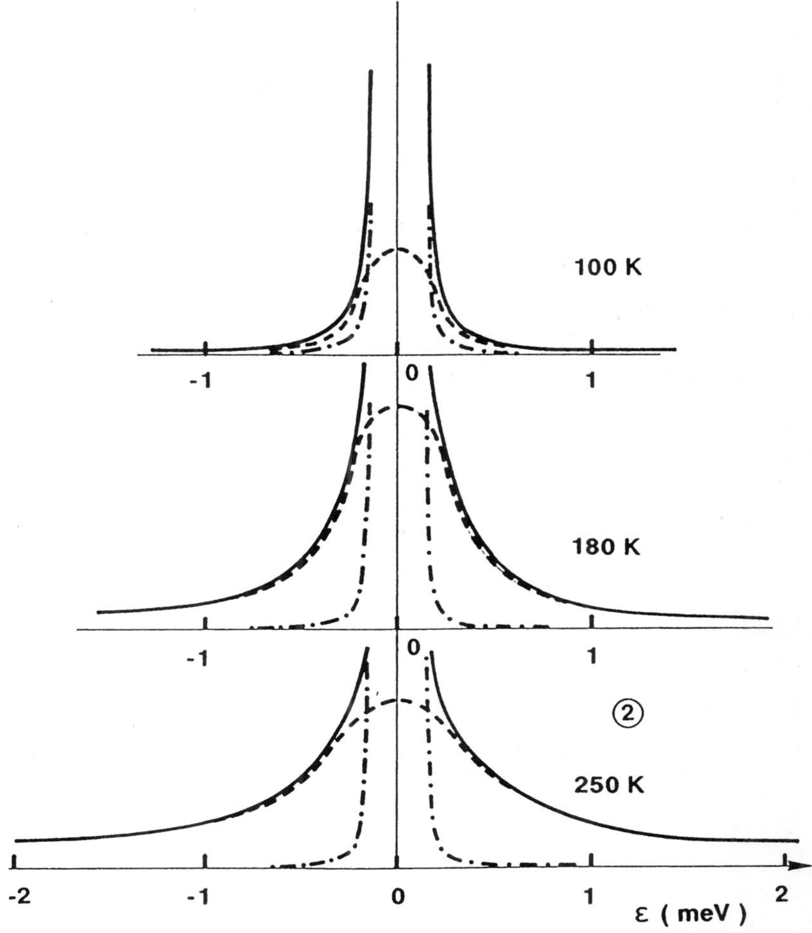

Fig. 3   Quasi-elastic broadening measured in sample /2/ (dashed line). The Q.E.S. of sample /1/ is the same, and may be exactly superimposed (not shown).The solid line is $S(Q,\omega)$; the dash-dotted line represents the elastic contribution,measured by the scattering of vanadium .

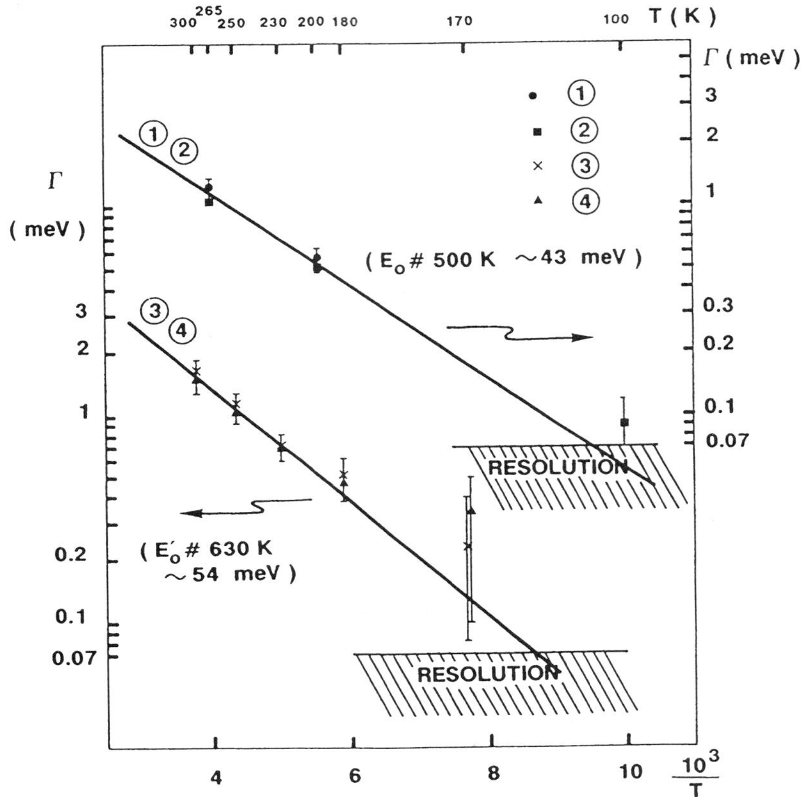

Fig. 4   Temperature dependence of the full width at half maximum of the quasi elastic scattering, for the four samples.

dependence of the relaxation time T_1 of protons in NMR measurements.[18]

As a summary, INS measurements have given the new following results :

- a thermally activated quasi-elastic broadening above 100 K, identified as diffusive motions of the protons of the methyl group,is observed. The measured activation energy is related to the height of the hindering barrier of the methyl rotation .The large values of E_0 (40 to 50 meV) in the present case,indicate that at low temperature,the quantum tunneling of the methyl protons has a very small energy .

- the methyl torsional transitions,at 16-17 meV in /1/ and /2/ , at 10-12 meV in /3/ and /4/, are identified owing to their specific temperature dependence and isotopic shift .

- no obvious anomaly exists , in the energy range of the acoustic modes,for none of the four species, confirming the data of our previous inelastic scattering study of acoustic branches in a fully deuterated single crystal.

- a new,very broad band,at 92-95 meV, is only found in sample /1/ and /3/, and is assigned to a movement of the proton of the hydrogen bond,namely the N-H out-of-plane bending mode.This band is unusually wide (> 40 meV,or 300 cm-1) . Such an effect has been already observed in the case of KH_2PO_4 [19] . The existence of very broad hydrogen modes ,that appear prominently only in neutron scattering,supports the picture that there exists an asymetric double minimum potential well where the proton of the hydrogen bond tunnels .

However,in the case of ACN there is ,as yet, no direct experi mental indication of quantum tunneling of this proton, though this effect has been invoked to explain the fine structure of the amide-I mode[4] . High resolution neutron scattering investigations (below 50 μeV) will be done to check this important point.

The relationship of this mode at 92-95 meV with the optical anomalies in ACN will be discussed in more details in the second part. ·However,considering its anomalous width, and its intensity and energy dependence versus temperature,the question of the coupling of this mode with the vibrations of the nearby groups and particularly to the methyl torsional transitions is open. In the same way , an effect of the methyl deuteration on the components of the γ-NH bending has recently been observed in in INS spectra of N-methylacetamide [20] .

2   INFRA-RED MEASUREMENTS

The observation of a wide band at 92-95meV in INS spectra of ACN /1/ and /3/ has naturally led us to re-investigate the same energy range with infra-red technics and with the same powdered specimens.

New measurements of infra-red absorption,using samples /1/ , /3/ ,and /4/ ,in the energy range 600-900 cm-1 are first presented and then the results in the range 1400-1700 cm-1 for sample /3/ .

EXPERIMENTS AND RESULTS

The infra-red spectra were taken with a 113 V - Brucker Fourier transform infra-red spectrometer.Data were collected with a 1 cm-1 resolution.The sample was thermostated in an helium circulation cryostat (from Air Liquide,Grenoble),at fourteen different temperatures comprised between 18K and 300K.

The data in the 600-800 cm-1 frequency range,for the extreme temperatures,and the three species,are shown in fig. 5 . These measure ments reveal the existence of a wide absorption band, centered at about 750 cm-1,in sample /1/ and /3/ only,consistently with incoherent neu tron scattering results . (Note on fig.1 that the scale of the absorbance is multiplied by five for the fully deuterated sample,/4/ and that no anomaly appears in its spectra) . It is the proof that this band is associated to some excitation involving the hydrogen bond,and not,as previously assigned,to a C-H out-of-plane bending mode[22].(otherwise it would exist in sample /1/ and not in /3/;the corresponding maximum in INS spectra would also have been seen in /1/ and /2/ but not in /3/,in contradiction with the observations.)

It is thus possible to assign this mode to the N-H out-of-plane bending mode,the frequency of which is expected in this range . This attri bution is consistent with all experimental features on the four samples . On the other hand, in N-methylacetamide also, the N-H out-of-plane bending mode (γ -NH) is observed in this energy region[23].

Detailed measurements of the temperature dependence of this IR absorption band indicate an important energy shift to higher

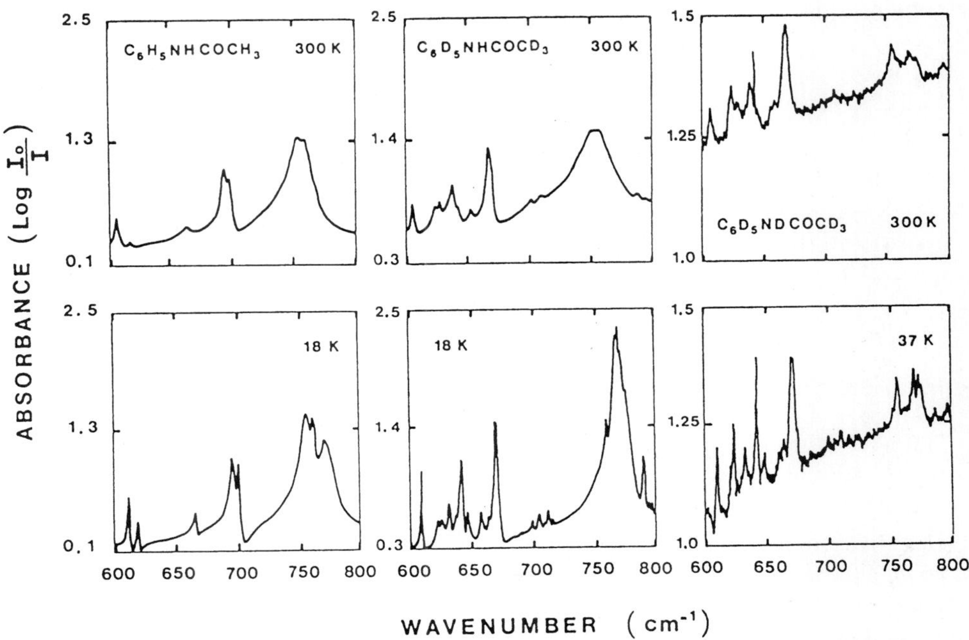

Fig. 5   Infra-red spectra of ACN/1/, /3/, and /4/ in the range 600 – 800 cm-1.The anomalous peak is absent in /4/ (C6D5NDCOCD3). Note the increase of the scale of absorbance (X 5) for sample/4/

frequencies with decreasing temperature , (fig. 6), in the sense expected for the γ -NH mode[23] . However, this shift to higher energy at low temperature is anomalously large in ACN . Indeed, the γ-NH mode frequency of an amide group involved in a hydrogen bonding is classically expected to shift in the opposite sense of the N-H stretching mode, as observed, but the amount of the γ-NH shift is expected to be very small, only a fraction of those of the N-H stretching[24]. It is not the case in ACN : the observed shift of the γ-NH mode is of the same order of magnitude as those of the N-H stretching[10], or perhaps higher . It is also worth to note that the shift of the γ- NH has the same sense and a comparable value to those of the methyl torsional modes observed in INS spectra.

The integrated intensity of the absorption band corresponding to the $\bar{\gamma}$-NH vibration increases also significantly (of about 25%) with decreasing temperature (fig. 7). This anomalous temperature dependence was already observed by G.CARERI et al. in the fully hydrogenated sample (see fig. 4 of ref. 2 , p.4692) . No significant change with temperature was observed in the corresponding absorption bands in sample /4/, down to 500 cm-1.

Fig. 7 also indicates that,within the experimental accuracy, the integrated intensity may be described by a law

$$\frac{I(T)}{I(0)} \propto \exp(-T^2/\theta^2)$$

as proposed in the first model of a "polaronic" mode,in which the anomalous band is treated as an analogous to the zero-phonon line of a localized mode . (θ, here, is a constant).

A qualitative comparison of this intensity increase,with the temperature variation of the intensity of the methyl libration (fig. 8) indicates parallel behaviours , and suggests a possible coupling of both.

To go further in comparing experimental results with available theories, we have plotted, for each temperature, the frequency of the anomalous γ-NH mode as a function of the measured linear birefringence[25] (Fig. 9) . It is well known[26] that the energy shift versus temperature of the zero-phonon line of a localized mode is linearly dependent on the internal energy of the crystal . The plot of fig. 9 is consistent with this expected relationship, the birefringence being,in general, proportionnal to the internal energy.

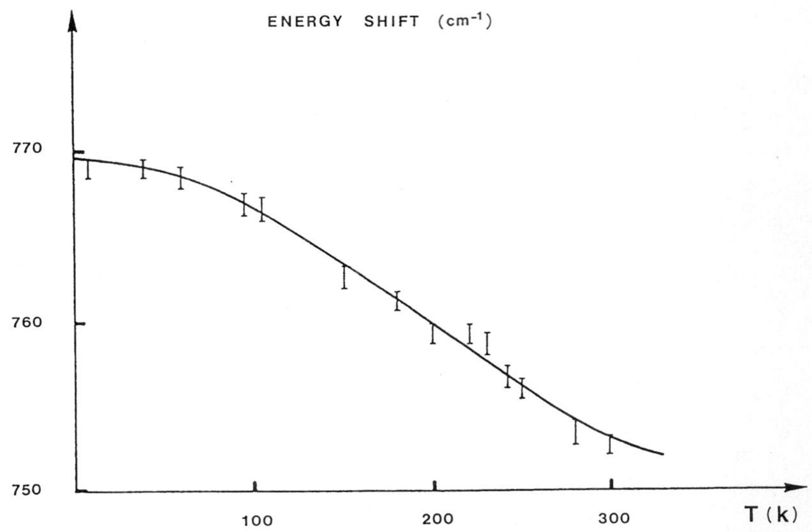

Fig. 6   Energy of the γ-NH bending mode, as a function of the temperature .

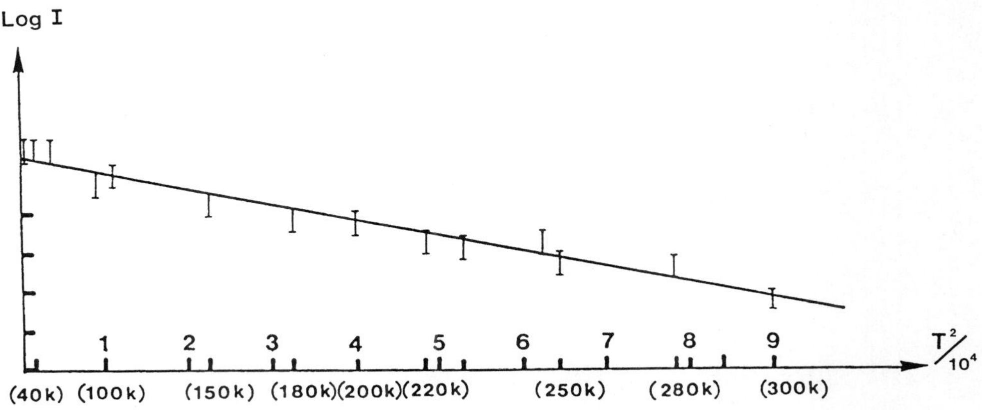

Fig. 7   Temperature dependence of the integrated intensity of the -NH bending mode .

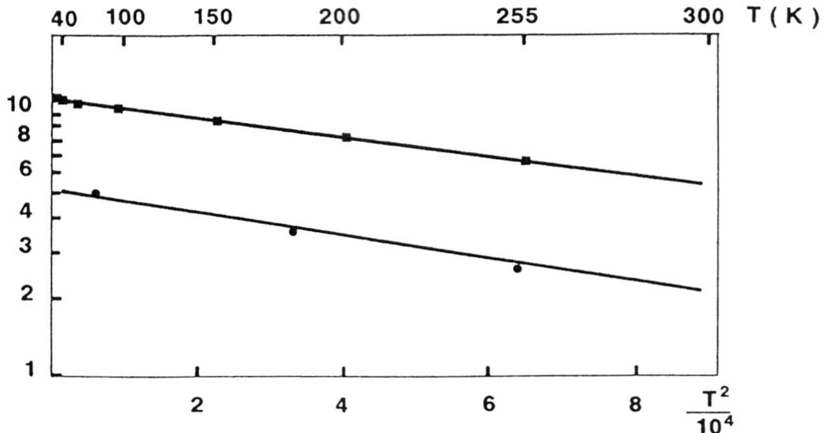

Fig. 8   Comparison of the temperature dependence of the intensity of the γ-NH mode,(IR data) with the intensity of the methyl torsional mode (INS data).

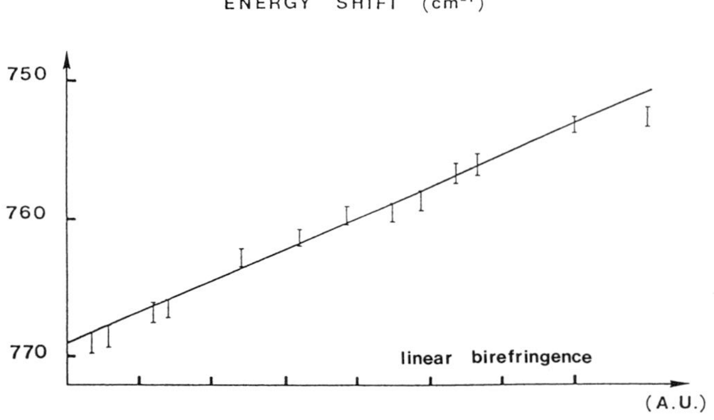

Fig. 9   Relation between the energy of the mode at 750 – 770 cm–1 and the linear birefringence of ACN.

The low temperature absorption band associated to the γ-NH bending mode has a different fine structure for samples /1/ and /3/, (fig.5) : Three components appear at 18K in /1/,whereas six multiplets may be observed in the corresponding band,with the same resolution,in /3/ .This effect clearly shows that the γ-NH mode is sensitive to the deuteration of the other parts of the molecule. A similar influence of the deuteration of the methyl groups on the splitting of the γ-NH mode has been also observed in N-methylacetamide [20].

Our data in the 1400 - 1700 cm-1 range confirm the presence of an anomalous absorption band at 1650 cm-1,in ACN /1/ [2] .Moreover,they show that no anomalous band exists in the same region of the spectrum of sample /4/,in consistency with previous measurements on sample /2/[1,11]. It seems that the deuteration of the proton of the hydrogen bond prevents all anomalous effects to occur.

In addition, the unconventional band at 1650 cm-1 is not observed, at any temperature , in sample /3/ . Surprisingly , a new , previously unobserved , anomalous absorption band is present in ACN /3/, at about 1480 cm-1 (fig.10) , in the region of amide-II modes (NH - in-plane bending mode and coupling between NH bending and CN stretching vibrations).The intensity of this band increases,and its frequency shifts to higher energy when the temperature is lowered.A detailed study of this new mode will be published elsewhere.

3   DISCUSSION

It is necessary to examine how these new experimental features could or not, be taken into account by the existing theoretical mechanisms of the nonlinear excitations in ACN :

- The lack of any observable quasi-elastic broadening of INS spectra of /3/ and /4/ at low temperature,within a resolution of about 50μeV,confirming previous data[13], allows to rule out the assumption of topological solitons,which would have created some structural disorder.

- The lack of any anomaly on acoustic phonons confirms that a Davydov-like model involving a nonlinear coupling of the C=O stretching mode to some acoustic phonons is unlikely.

On the contrary,no experimental fact ,as yet, contradicts the possibility of a nonlinear coupling between high energy internal modes and low lying optical phonons or librations . So a Davydov soliton created by coupling to low energy optical phonons,or a polaronic defect as described in [9] are not in contradiction with the experimental findings.

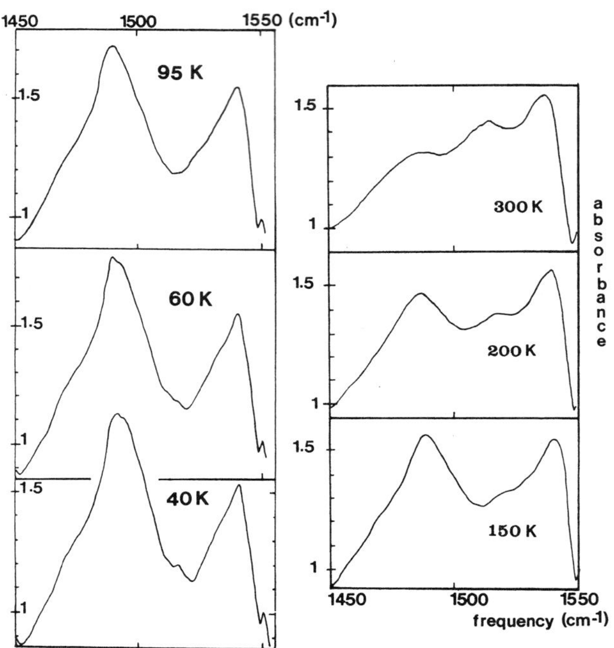

Fig. 10   Infra-red spectra of ACN/3/ (C6D5NHCOCD3),at different temperatures,in the 1400 - 1600 cm-1 energy range.

- The existence of a very wide structure in the INS spectra at 92-95 meV , related to an infra-red mode at the same frequency displaying an anomalous behaviour , brings a new argument against the explanation of anomalies by means of a Fermi resonance, (in consistency with the observations of overtones [3] and of Raman experiments under pressure[21]). Indeed, an accidental temperature tuning of a Fermi resonance, would not give rise to a strong maximum in the vibrational frequency distribution, especially at room temperature. Moreover, it would probably appear at different frequencies in sample /1/ and /3/.

The presence of anomalous modes at 1650 cm-1, 3250 cm-1,... are well explained in the frame of a model of "polaronic" localised mode. The classical localised modes in crystals containing impurities or F-centers are also well known to exhibit a large number of overtones. The temperature dependence of the integrated intensity of the mode at 750-770 cm-1 ,and the linear relationship between its energy and the internal energy of the crystal are also consistent with this class of theoretical models .

Now,to give a complete description of ACN,it will be necessary to take into account the sequence 750 cm-1, 1650 cm-1, 3250 cm-1, ...,of anomalous modes for ACN/1/ ,   and the sequence 750 cm-1, 1480 cm-1, 3250 cm-1,.... , in ACN /3/. It is an alternative possibility that the localized mode may be created by coupling of the CONH out-of-plane vibrations to vibrations of nearby groups, and that the upper anomalous frequencies   result from overtones .

Moreover,some features exhibited by the γ-NH mode (very broad maximum in the INS spectra, splitting sensitive to the deuteration of the nearby groups, temperature dependence of its intensity and energy) suggest a coupling to the methyl torsional transitions, and the possible existence of a asymetric double-well potential in which the proton of the N-H bond tunnels[4] ,in a similar way to the case of N-methylacetamide[23]. This assumption will be checked by means of high resolution backscattering neutron spectra at low temperature.As yet,however,it is clear from our INS measurements that no tunnel splitting exists at energy above 50 μ eV,in samples /3/ and /4/.

The molecular structure of ACN is such that the proton of the hydrogen bond may well be situated at comparable distances not only of the N and O atoms of the N-H···O bond , but also of one proton of the rotating methyl. So,an interaction between the N-H bending and the CH_3 hindered rotation is not unreasonable . A crystal structure done by neutron diffraction technics will give the mean positions of the protons and help to clarify this point .

Regarding the assumption of a coupling of the N-H bending to the methyl librations,it is also worth to note that,in the first model [2] (Davydov-like soliton) of the anomalies in ACN,the best fit with the temperature dependence of the anomalous band intensity was obtained assuming that the low-energy phonon has the frequency ω = 131 cm-1, which is precisely the energy of the methyl libration.Moreover ,the authors indicated that the fit could be improved if ω was allowed to be a decreasing function of the temperature,and that is ,again, the case for the frequency of the torsional transition.

The hypothesis of the coupling of the amide group internal vibrations to the methyl modes, and the existence of the double -well structure for the amide proton , could be an interesting starting point to understand the remaining unexplained properties of ACN , such that the Raman measurements under hydrostatic pressure [21] . However,new experimental informations are necessary before going further in these speculations.

The present data do not allow to decide whether the self-trapped state is a mobile,"solitonic like" excitation ,or a "polaronic like", pinned on the molecular chain, defect.Owing to the high elastic constants of ACN[13],the propagation of a large amplitude distorsion asso - ciated to a Davydov soliton is unlikely.
A polaronic-like defect,pinned on the lattice, is the formal analogous to a localized mode created by the interaction of an impurity with a host crystal lattice[26].So, it is important to underline that the anomalous IR and Raman modes in ACN are the first experimental mani festation of the existence of such localized modes in _pure crystals_ . This so observed energy localisation arises from quantized "extrinsic nonlinearities"[27] in the dynamics of the molecular chain , a new way to consider the anharmonicity in crystals and self-trapping of vibrational energy.

REFERENCES

1 - G.CARERI,U.BUONTEMPO,F.CARTA,E.GRATTON and A.C.SCOTT - Phys.Rev. Lett. 51,304,1983

2 - G.CARERI,U.BUONTEMPO,F.GALLUZZI,A.C.SCOTT,E.GRATTON and E.SHYAMSUNDER - Phys.Rev.B 30,4689,1984.

3 - A.C.SCOTT,E.GRATTON,E.SHYAMSUNDER and G.CARERI - Phys.Rev.B 32, 5551,1985.

4 - G.GARERI, E.GRATTON AND E.SHYAMSUNDER   - Phys.Rev.A 37, 4048, 1988.

5 - J.C.EILBECK, P.S.LOMDAHL AND A.C.SCOTT Phys.Rev. B 30, 4703, 1984.

6 - S. TAKENO - Prog.Theor.Phys. 75,1,1986.

7 - D.M. ALEXANDER -Phys.Rev.Lett.54,138,1985

8 - D.M.ALEXANDER and J.A.KRUMHANSL - Phys.Rev.B 33,7172,1986.

9 - A.C. SCOTT,I. J.BIGIO and C.T. JOHNSTON - Phys.Rev.B 39,15 june 1989

10 - G. BLANCHET and C. FINCHER - Phys.Rev.Lett. 54, 1310,01985.

11 - C.T.JOHNSTON, B.J.SWANSON - Chem.Phys.Lett. 114, 547,1985.

12 - J.L.SAUVAJOL, R.ALMAIRAC, M.BARTHES ,J.MORET and J.L.RIBET - J.of
    Raman spectroscopy , (in press)

13 - M.BARTHES ,R.ALMAIRAC, J.L.SAUVAJOL,R.CURRAT,J.MORET and
    J.L.RIBET - Europhysics Letters 7,55, 1988

14 - P.EGELSTAFF and P.SCHOFIELD - Nucl. Sci. Eng. ,12,260,1962.

15 - A.J. DIANOUX - "The Time Domain in Surface and Structural Dynamics"
    Ed. G.LONG and F.GRANDJEAN - Kluwer Acad.Publ. 1988

16 - M.PRAGER, J.STANISLAWSKI and W.HAUSLER - J.Chem. Phys. 89, 2563,
    1987

17 - D.CAVAGNAT J.Chimie Phys.82, 239,1985

18 - G.GUSMAN , F.MASIN and P. BROEKAERT - 6th Interdisciplinary
    Workshop on Nonlinear Coherent Structures in Physics,Mechanics and
    Biological Systems - June 1989 - Montpellier.

19 - "Molecular Spectroscopy with Neutrons" - J.BOUTIN and M.YIP ,
    MIT Press, 1968.

20 - F.FILLAUX and J.TOMKINSON - Annual ISIS Report - 3,A140,1987-1988

21 - C.JOHNSTON ,unpubl. results - J.L. SAUVAJOL ,unpubl. results.

22 - N. ABBOTT and A. ELLIOT - Proc. R. Soc. London A.234,247,1956.

23 - F. FILLAUX and M. BARON - Chem. Phys. 62,275,1981.

24 - J. COLTHUP - "Introduction to I-R and Raman Spectroscopy" - Acad.
    Press - 1975 .

25 - M. BARTHES,R. ALMAIRAC,J.L.SAUVAJOL,R.CURRAT,J.MORET,J.L.RIBET -
    J. Phys. 50 ,coll.C3,209,1989

26 - D.B. FITCHEN in "Physics of Color Centers" Ed. by W.B.FOWLER - Acad.
    Press - 1968 .

27 - A.SCOTT - 6th Int. Workshop N.L.C.S. - June 1989 - Montpellier .

SPECTROSCOPY OF THE AMIDE-I MODES OF ACETANILIDE

Irving J. Bigio,* Alwyn C. Scott,[†] and Clifford T. Johnston[‡]

*Los Alamos National Laboratory
Los Alamos, NM 87545

[†]Department of Mathematics
The University of Arizona
Tucson, AZ 85721

[‡]University of Florida
Gainsville, FL 32611

INTRODUCTION

In 1973, Careri published evidence of an "unconventional" amide-I (CO stretching) band at 1650 cm^{-1} in crystalline acetanilide (CH$_3$CONHC$_6$H$_5$), abbreviated ACN.[1] Following the "soliton" theory of Davydov,[2] this band has been assigned to a "self-trapped state" in which CO vibrational energy is localized near a single ACN molecule through interactions with lattice phonons.[3,4] A complementary "polaron" picture[5] of this self-trapped state has been presented recently by Alexander and Krumhansl.[6,7]

In this contribution best available data are presented of the integrated intensity of the 1650 cm^{-1} band as a function of temperature. The experimental procedures and data reduction were highly rigorous, and the data are believed to be the most reliable available. A concise theory of polaron states is presented and used to interpret the data.

SPECTROSCOPIC MEASUREMENTS

In our early experimental work we performed both infrared absorption spectroscopy and Raman spectroscopy. We soon focused on Raman spectroscopy, however, for several reasons. For one, it was easier to control sample temperature since ACN is essentially transparent at the wavelength of the illumination source. Also, the method itself afforded us an additional technique for determining the sample temperature in the illuminated volume, by the Stokes/anti-Stokes ratio technique described below. Moreover, when Raman spectra are taken with oriented single crystals and with polarized illumination and detection, more detailed information about the

Davydov's Soliton Revisited, Edited by P.L. Christiansen and A.C. Scott
Plenum Press, New York, 1990

crystal structure and symmetry of vibrational centers is provided.

The Raman measurements reported here were made on single crystals of acetanilide prepared as follows. Primary standard grade ACN (N-phenylacetamide), 99.995% pure by assay, was purified using a multiple-pass zone refiner. The zone refined material was placed in a glass tube and the vessel was sealed under vacuum. Single crystals were grown from the melt as the vessel was lowered through a modified Bridgeman-Stockbarger furnace at a rate of 2 cm/day; the temperature gradient at the tip of the furnace was approximately 110 K/cm. The crystal of ACN is [100] tabular; thus cleavage was nearly perfect parallel to the [100] face and fairly good cleavage was obtained along the [001] face. The crystals used were, on average, 10 mm along the y, 4 mm on the z, and 1 mm along the x axis. Single crystals were cleaved and their orientations were determined using x-ray precession methods and a cross-polarizing microscope. The crystals were mounted in an oxygen-free copper cold cell, which enclosed the crystal in an inert atmosphere (He) to protect it from water vapor and pump oil. The temperature of the Raman cell was measured using an Au-cromel thermocouple.

In addition to the thermocouple measurements, the crystal temperature was determined directly by measuring the Stokes to anti-Stokes ratio of the intensities for a particular low-frequency mode by scanning over the -200 cm^{-1} to +200 cm^{-1} region in a single scan with an incident power of the laser below 50 mW. This allowed assessment of the amount of heating within the focal volume of the laser. Under these conditions local heating of the crystal was observed to be less than 10 K.

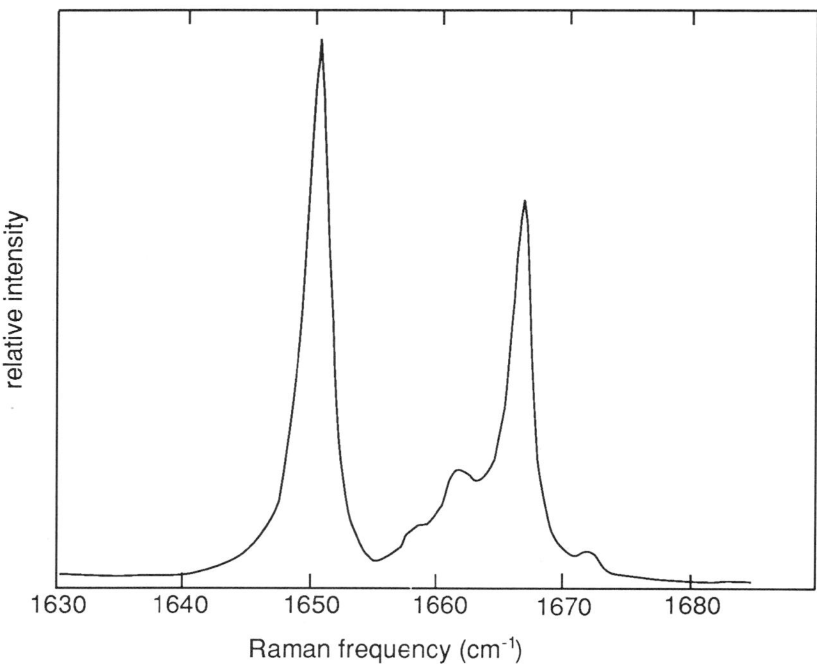

Fig. 1 Raman spectrum of the amide-I region of ACN at 21 K. Both the illumination polarization and the selected observation polarization were parallel to the crystal x axis.

Raman spectra were obtained on a 3/4-meter double monochromator (Spex model 1403) interfaced to a Nicolet 1180E computer. The 514.5 nm line of an argon-ion laser (Spectra Physics Model 171) was used as the exciting line. An interference filter was used to suppress plasma frequencies and a quartz crystal wedge was used at the entrance slit of the monochromator as a polarization scrambler. The accuracy of the frequency measurements was checked periodically by scanning over the attenuated laser line using a reduced slit width on the monochromator. Although all data was recorded digitally, for illustrative purposes a typical low-temperature spectrum is shown in Fig. 1.

In Table 1 are listed the actual numerical values for measurements of the intensity (i.e. integrated area) of the Raman line at 1650 cm^1 for a range of temperatures. These results are in general agreement with infrared absorption measurements over the same temperature range.[8] The reader should notice, however, that these data differ somewhat from those in Table III of Ref. 3, which were discussed in detail in Ref. 6. The difference is that *peak* intensities were recorded in Ref. 3 while *integrated* intensities are listed here. This distinction is important since integrated intensities are what the theories calculate. The integrated intensities result from fitting Lorentzian lineshapes to the spectra. Any apparent asymmetry of the lineshape (see Fig. 1) is actually dealt with properly by accounting for the real contributions of weaker amide-I features at 1659 cm^{-1} and 1662 cm^{-1}, as well as the tail of the 1665-cm^{-1} "normal" amide-I feature itself.

Table 1. Normalized integrated Raman intensity values of the 1650-cm^{-1} band. Temperature uncertainties are due primarily the the method for estimating the degree of heating caused by absorption in the focal volume of the laser beam.

Temp	σ	Relative Intensity	σ
21	4	1.00	.016
21	4	.996	.016
53	4	.892	.019
53	4	.895	.019
100	10	.710	.022
100	10	.726	.022
149	6	.496	.025
149	6	.507	.025
227	12	.356	.028
227	12	.346	.028
305	8	.170	.032
305	8	.169	.032

THEORY

The theory presented in this section is taken almost literally from Ref. 9.

As a theoretical model we use a linear coupling (or Fröhlich[10]) Hamiltonian operator of the form

$$\widehat{H} = \widehat{H}_0 + J\,\widehat{V} \quad , \tag{1}$$

where

$$\widehat{H}_0 = \sum_{q=1}^{R} \left\{ \Omega_0\,\widehat{B}_q^\dagger\,\widehat{B}_q + \sum_{j=1}^{M} \left[\omega_j\,\widehat{b}_{qj}^\dagger\,\widehat{b}_{qj} + \chi_j\left(\widehat{b}_{qj} + \widehat{b}_{qj}^\dagger\right)\widehat{B}_q^\dagger\,\widehat{B}_q \right] \right\} \tag{2}$$

$$\widehat{V} = \sum_{q=1}^{R} \left(\widehat{B}_q^\dagger\,\widehat{B}_{q+1} + \widehat{B}_q^\dagger\,\widehat{B}_{q-1} \right) . \tag{3}$$

The parameters in Eqs. (1) - (3) are defined as follows:

J   is the nearest neighbor, electromagnetic coupling energy which was calculated as 4 cm^{-1} in Ref. 4,

Ω_0 is the undressed amide-I vibrational energy,

R   is the number of ACN molecules in a linear chain,

M   is the number of phonon modes coupled to each ACN molecule,

$\{\omega_j\}$ with j = 1, 2, ...M is the set of phonon energies, and

$\{\chi_j\}$ is the corresponding set of coupling energies.

To begin our analysis, we assume CO vibrational energy to be localized at a single ACN molecule. We do this for two reasons: i) this assumption is approximately true, and ii) it provides the basis for a perturbation expansion (in small J) which includes the effects of $\widehat{V}$ to which we shall return later. Thus we are led to consider the operator

$$\widehat{h} = \Omega_0\widehat{B}^\dagger\,\widehat{B} + \sum_{j=1}^{M} \left[\omega_j\,\widehat{b}_j^\dagger\,\widehat{b}_j + \chi_j\left(\widehat{b}_j + \widehat{b}_j^\dagger\right)\widehat{B}^\dagger\,\widehat{B} \right] \tag{4}$$

where the subscripts, q, have been temporarily dropped for typographical convenience.

The eigenvalue equation

$$\widehat{h}|\,\psi\,\rangle = E|\,\psi\,\rangle \tag{5}$$

is satisfied exactly by the eigenvalue-eigenstate pair

$$E = N\,\Omega_0 + \sum_{j=1}^{M}\left[n_j\,\omega_j - N^2\,\frac{\chi_j^2}{\omega_j}\right] \tag{6}$$

$$|\psi\rangle = |N\rangle \prod_{j=1}^{M}|\phi_j\rangle \tag{7}$$

where

$$|\phi_j\rangle = \sqrt{n_j!}\,\exp\left(-\frac{1}{2}N^2\frac{\chi_j^2}{\omega_j^2}\right)\sum_{m_j=0}^{\infty}\left[\frac{\left(-N\frac{\chi_j}{\omega_j}\right)^{m_j-n_j}}{\sqrt{m_j!}}\;L_{n_j}^{m_j-n_j}\left(N^2\frac{\chi_m^2}{\omega_j^2}\right)|m_j\rangle\right] \tag{8}$$

and where $\hat{B}^\dagger\hat{B}\,|N\rangle = N\,|N\rangle$ and $\hat{b}_j^\dagger\hat{b}_j\,|m_j\rangle = m_j\,|m_j\rangle$.

At this point, three comments are appropriate:

i) $L_n^m(\cdot)$ is an associated Laguerre polynomial using the normalization in Gradshteyn and Ryzhik[11] and not (for example) that in Morse and Feshbach.[12]

ii) The expression for $|\phi_j\rangle$ in Eq. (8) is not new knowledge. It is the number representation of the displaced Hermite polynomial eigenstate (more commonly seen in the position representation) that has been employed in the theory of color centers.[13-18]

iii) Equation (6) predicts an overtone spectrum that has been observed in ACN.[19] Thus, it is an experimental fact that

$$\sum_{j=1}^{M}\frac{\chi_j^2}{\omega_j} = 24.7\text{ cm}^{-1}. \tag{9}$$

As has been pointed out by Alexander and Krumhansl,[6,7] a correct calculation of the temperature dependent intensity of the 1650 band in ACN follows that for a "zero phonon" band of a color center. This point is made with particular clarity by Krumhansl in reference.[7] Thus one seeks the sum of all transitions from the ground states $|m_j\rangle$ to first excited states with $m_j = n_j$. Since the ground states are thermally populated with probabilities

$$P_j = \left[1 - \exp\left(-\frac{\hbar\omega_j}{kT}\right)\right]\exp\left(-m_j\frac{\hbar\omega_j}{kT}\right) \tag{10}$$

the desired temperature dependence is

$$W(T) = \prod_{j=1}^{M}W_j(T), \tag{11}$$

where

$$W_j(T) = \sum_{m_j = 0}^{\infty} P_j \, |\langle m_j | \phi_j \rangle|^2 \tag{12}$$

and $|\phi_j\rangle$ is calculated from Eq. (8) with $n_j = m_j$. Using the identity number 8.976 from Gradshteyn and Ryzhik,[11] the sum in Eq. (12) is readily computed to be

$$W_j(T) = \exp\left[-\frac{\chi_j^2}{\omega_j^2}\coth\left(\frac{\hbar\omega_j}{2kT}\right)\right] I_0\left[\frac{\chi_j^2}{\omega_j^2}\operatorname{csch}\left(\frac{\hbar\omega_j}{2kT}\right)\right] \tag{13}$$

where $I_0[\bullet]$ is the modified Bessel function of the first kind, of order zero.

Returning to our original Hamiltonian, we perform a first order perturbation calculation as follows. A zero order estimate of an eigenstate that exhibits the translational symmetry of the model is

$$|\psi^{(0)}\rangle = \frac{1}{\sqrt{R}}\sum_{q=1}^{R} \exp(ikq)|\widetilde{\psi_q}\rangle \tag{14}$$

where

$$k = \frac{2\pi}{R}\nu \tag{15}$$

and $\nu = 0, \pm 1, ..., \frac{R}{2}\left(\pm\frac{R-1}{2}\right)$ for R even(odd). In Eq. (14), $|\widetilde{\psi_q}\rangle$ is calculated from Eq. (7) and Eq. (8) with N = 1. Then to first order in J the energy is

$$E(k) \doteq \Omega_0 - \sum_{j=1}^{M}\chi_j^2/\omega_j + 2J\prod_{j=1}^{M}\exp(-\chi_j^2/\omega_j^2)\cos k . \tag{16}$$

A localized solution can be constructed as a wave packet of the eigenstates approximated in Eq. (14). Thus the linewidth, ΔE, of such a wave packet must satisfy the inequality

$$\Delta E < 4J\prod_{j=1}^{M}\exp\left(\frac{-\chi_j^2}{\omega_j^2}\right) . \tag{17}$$

Finally we note from Eq. (11) and Eq. (13) that

$$W(0) = \prod_{j=1}^{M}\exp\left(-\chi_j^2/\omega_j^2\right) . \tag{18}$$

Thus the same (Franck-Condon) factor that reduces the low-temperature intensity of the 1650-cm^{-1} band also reduces its linewidth, thereby increasing the lifetime of a localized state.

COMPARISON OF THEORY WITH DATA

In Fig. 2 Raman spectra in the range 20 − 200 cm^{-1} are presented. We have observed a total of 30 distinct features in the (xx), (yy), and (xy) Raman spectra between 20 and 200 cm^{1} at 21 K. This is in qualitative agreement with the data of Gerasimov[20] where 27 bands were observed at 113 K. A more complete analysis of these data will be presented by one of us (CTJ) in a forthcoming publication. In the present discussion we consider whether coupling to these phonon modes can account for the temperature dependence of the 1650-cm^{-1} band. Ideally we would like to assign an ω_j and χ_j in Eq. (2) to each phonon mode that contributes to self-trapping of the 1650-cm^{-1} band. Since this is impractical, we assume that each ACN molecule couples to M phonon modes, all having the identical frequency ω_1 and coupling strength χ_1. In other words we view ω_1 and χ_1 as the average frequency and coupling strength of the M phonon modes that contribute to the self-trapping.

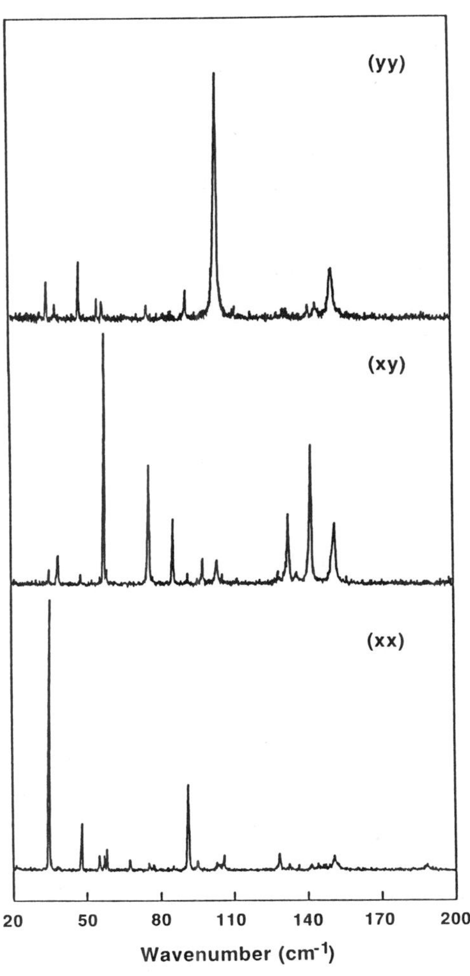

Fig. 2. Raman spectra of crystalline acetanilide in the optical phonon region (20 − 200 cm^{-1}) at 21K in the (a) xx, (b) yy, and (c) xy orientations.

Then, from Eq. (9),

$$\frac{\chi_1^2}{\omega_1} = \frac{24.7}{M} \text{ cm}^{-1} \qquad (19)$$

and from Eqs. (11) and (13), and expressing ω_1 in wave numbers,

$$\frac{W(T)}{W(0)} = \left\{ \frac{\exp\left[-\frac{24.7}{M\omega_1}\coth\left(0.7174\frac{\omega_1}{T}\right)\right] I_0\left[\frac{24.7}{M\omega_1}\operatorname{csch}\left(0.7174\frac{\omega_1}{T}\right)\right]}{\exp\left[-\frac{24.7}{M\omega_1}\right]} \right\}^M . \qquad (20)$$

The solid curve in Fig. 3 is plotted from Eq. (20) with M and ω_1 adjusted to give the least squares fit to the data. These optimum values are found to be

$$M = \text{any value} \geq 12$$

$$\omega_1 = 71 \text{ cm}^{-1}.$$

The effect of varying the parameter M, the number of coupled modes, is illustrated by Fig. 4 in which Eq. (20) is plotted for a range of M values while keeping ω_1 constant at 71 cm^{-1}. The curves quickly converge after $M \approx 8$.

CONCLUDING COMMENTS

1. We have presented a concise theory of polaron states in crystalline acetanilide (ACN). Being in the number representation, this theory complements that developed for color centers,[13-18] and it is in agreement with the point of view expressed recently by Krumhansl and Alexander.[6,7]

2. We display here (Fig. 1) our best Raman-spectroscopic measurements of the relative intensity of the 1650-cm^{-1} polaron state in ACN as a function of absolute temperature.

3. We show (Fig. 2) a spectrum of the Raman-active optical phonon modes in the range between 20 and 200 cm^{-1}. These measurements were made on carefully prepared single crystals of ACN.

4. We conclude that the data in Fig. 1 are consistent with the assumption of polaron formation through interactions with a band of 10-20 optical phonon modes clustered near 72 cm^{-1}. We note that this conclusion is at variance with Alexander and Krumhansl who suggest[6] the data require acoustic phonon modes in addition to the optical mode near 70 cm^{-1} in order to explain both the temperature dependence and the energy shift of the 1650-cm^{-1} band. We emphasize that we cannot exclude the possibility that acoustic phonons participate in the self-trapping, but we assert that they are not necessary to explain the data.

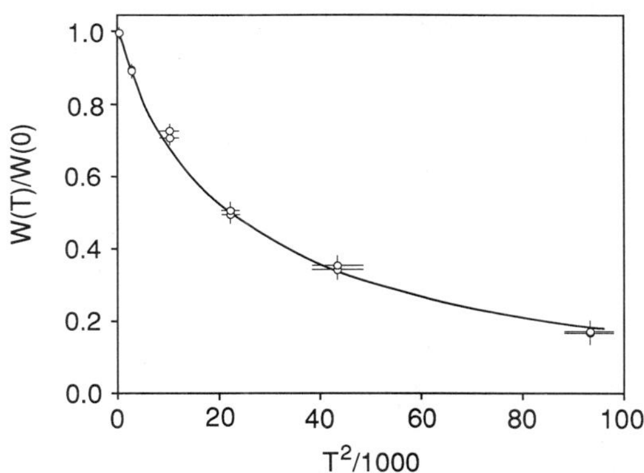

Fig. 3. Data points "o" indicate measurements of relative integrated intensity of the 1650 cm^{-1} line in crystalline acetanilide. The solid curve is calculated from Eq. (20) with M = 20 and ω_1 = 71 cm^{-1}.

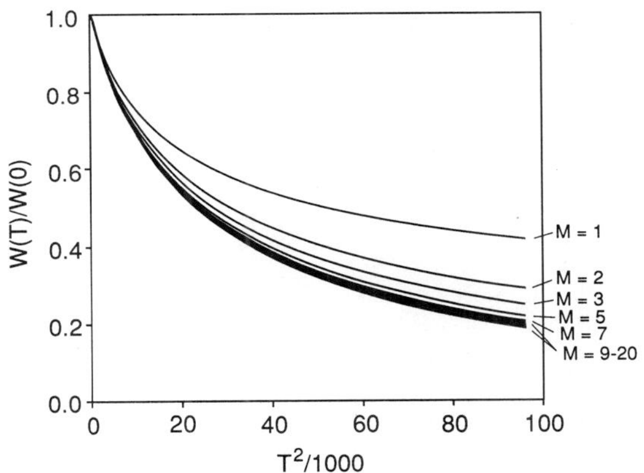

Fig. 4. Plots of Eq. (20) with different values of M, and for ω_1 = 71 cm^{-1}.

5. From Eq. (17) our theory predicts a polaron linewidth that is less than 11.5 cm^{-1}. This is consistent with our experimental observations as well as those of Ref. 3.

6. We note from Eq. (17) that the polaron linewidth is proportional to the Franck-Condon factor (FCF) as defined in Eq. (18). Thus, as has previously been suggested,[21] one might consider the following scenario: i) An optical photon excites a polaron, which is self-trapped by optical-mode phonons. Since the FCF is relatively large, this localized polaron is unstable. ii) The unstable polaron relaxes into a wave packet that is self-trapped by acoustic-mode phonons. From Eq. (18) this would have a much smaller FCF and from Eq. (17) a much smaller linewidth and, therefore, greater stability. The final state, in this scenario, would correspond to the "soliton" proposed by Davydov.[22]

REFERENCES

1. G. Careri, in *Cooperative Phenomena*, edited by H. Haken and M. Wagner (Springer, Berlin, 1973) p. 391.
2. A.S. Davydov, Sov. Phys. Usp. **25**, p.898 (1982; Usp. Fiz. Nauk **138**, p.603 (1982).
3. G. Careri, U. Buontempo, F. Galluzzi, A. C. Scott, E. Gratton, and E. Shyamsunder, Phys. Rev. B **30**, 4689 (1984).
4. J. C. Eilbeck, P. S. Lomdahl, and A. C. Scott, Phys. Rev. B **30**, 4703 (1984).
5. T. Holstein, Ann. Phys. (N.Y.) **8**, p.325 (1959).
6. D. M. Alexander and J. A. Krumhansl, Phys. Rev. B **33**, 7172 (1986).
7. J. A. Krumhansl, in *Energy Transfer Dynamics*, edited by T. W. Barrett and H. A. Pohl (Springer, Berlin, 1987) p. 174.
8. G. Careri, E. Gratton and E. Shyamsunder, Phys. Rev. B **37**, p.4048 (1988).
9. Alwyn C. Scott, Irving J. Bigio and Clifford T. Johnston, Phys. Rev. B 39, p. 12883 (1989).
10. H. Fröhlich, Adv. Phys. **3**, 325 (1954).
11. I. S. Gradshteyn and I. M. Ryzhik, *Table of Integrals Series and Products* (Academic, New York 1965).
12. P. M. Morse and H. Feshbach, *Methods of Theoretical Physics* (McGraw-Hill, New York, 1953) p. 784.
13. K. Huang and A. Rhys, Proc. Roy. Soc. (London) **204A**, 406 (1950).
14. M. Lax, J.. Chem. Phys. **20**, 1752 (1952).
15. R. C. O'Rourke, Phys. Rev. **91**, 265 (1953).
16. J. J. Markham, Rev. Mod. Phys. **31**, 956 (1959).
17. T. H. Keil, Phys. Rev. **A601,** 140 (1965).
18. D. B. Fitchen, in *Physics of Color Centers*, edited by W. B. Fowler (Academic, New York, 1968) p. 293.
19. A. C. Scott, E. Gratton, E. Shyamsunder, and G. Careri, Phys. Rev. B **32**, 5551 (1985).
20. V. P. Gerasimov, Opt. Spektrosk. **43**, 705 (1977) [Opt. Spectros. (USSR) 43, 417 (1978)].
21. A. C. Scott, in *Synergetics of the Brain*, edited by E. Bazar, F. Flohr, H. Haken, and A. J. Mandell (Springer, Berlin, 1983) p. 345.
22. A. S. Davydov, *Biology and Quantum Mechanics* (Pergamon, New York, 1982).

28

BIOMOLECULAR DYNAMICS STUDIED BY VIBRATIONAL SPECTROSCOPY

Ole Faurskov Nielsen

Department of Chemical Physics
University of Copenhagen
The H.C. Ørsted Institute
Universitetsparken 5
DK-2100 Copenhagen Ø., Denmark

INTRODUCTION

Since A.S. Davydov[1,2] proposed the soliton model for the alpha-helix of a protein many theoretical approaches have appeared. These are covered by other reviews in this book and they shall not be referenced here. The Davydov work was an inspiration for a more general view on the importance of non-linear dynamics in many different fields. However, a direct evidence for the existence of Davydov solitons is very diffi-cult to obtain. A theoretical explanation of protein dynamics in terms of Davydov solitons was published by Scott[3]. Studies of the carbonyl-stretching vibrations of acetanilide (ACN) and N-methylacetamide (NMA) reveal evidence of effects due to non-linear dynamics[4-6]. ACN and NMA are examples of model compounds for real proteins. A direct correlation between observed spectra and theories are much more difficult to obtain for macromolecules with a biological significance. Experimental studies in the amide-I region will mainly be covered by other contributors to this book[7]. In the present paper is shown how the low-frequency part of the vibrational spectrum (10-400 cm^{-1}) can be of importance in breaking of hydrogen bonds through a coupling to energy trapped in vibrational modes. Furthermore the carbonyl-stretching vibration of liquid formamide in dif-ferent Raman scattering configurations will be discussed.

SOME EXPERIMENTAL PROBLEMS IN STUDIES OF BIOMOLECULES

Biological macromolecules contain a large number of atoms, and thus a description of the dynamics in terms of atomic dis-placements is complicated. This can be reduced if symmetry occurs in the molecule. The symmetry operations may be either of translational or point origin (or combinations of these). In fact for an idealised biological macromolecule symmetry can occur. However, real proteins most often contain only smaller regions with symmetry-related configurations (α-helix, β-sheet, etc). Hydrogen bonding plays an important role in keeping biomolecules in fixed configurations. The hydrogen

Davydov's Soliton Revisited, Edited by P.L. Christiansen and A.C. Scott
Plenum Press, New York, 1990

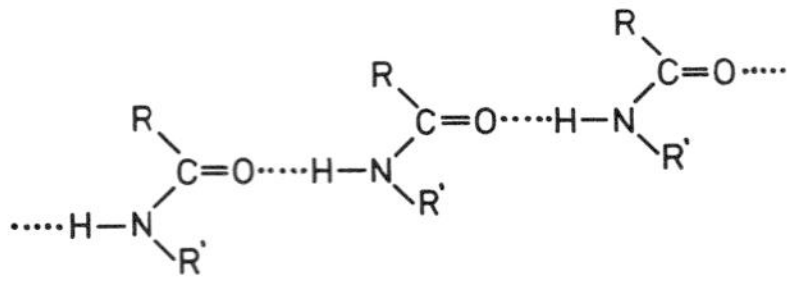

$$R=H,CH_3 \quad ; \quad R'=H,CH_3$$

Figure 1   A survey of the simple amides used as model systems for proteins. As shown in the figure all molecules form "linear" hydrogen bonded associates in the liquid state. The structure in the figure is idealised.

Table I   Names of the investigated hydrogen bonded amides. R and R' refers to figure 1.

R	R'	Name
H	H	Formamide
H	CH_3	N-methylformamide
CH_3	H	Acetamide
CH_3	CH_3	N-methylacetamide

bond can be intra- or inter-molecular. In biological macromolecules many different types of hydrogen bonds may occur. In order to perform a systematic investigation of hydrogen bonding some model systems must be selected. For proteins and nucleic acids the simpler amides can be chosen, because these are the simplest hydrogen bonded systems with hydrogen bonding similar to findings in real biological macromolecules.

A survey of different possible molecules are given in figure 1. By combination of R and R' the molecules in Table I are obtained.

Formamide, N-methylformamide, and N-methylacetamide are liquids of room temperature. The melting point for acetamide is 82°C. Another amide which has been used as a model compound for proteins is acetanilide (ACN)[4,5]. In the crystalline state all these molecules form hydrogen bonded linear chains (fig. 1). The real crystalline structures are of course different, but the main feature of a long hydrogen bonded

chain exists in all the molecules mentioned above. Linear
chains are also present in the liquid state and in solutions
of hydrogen bonded amides[8,9].

One serious problem in dealing with model compounds is to
decide which properties are expected to be common for the
model substance and the real biological polymer. Some conclusions may be drawn based on structural similarities. However,
hydrogen bonding causes serious problems, because the most
common experimental technique for structure determination is
X-ray diffraction, which usually does not give the position of
hydrogen atoms. In this context neutron diffraction is more
suited. Vibrational amplitudes are complications which make
precise structure determination by diffraction methods difficult to perform. Especially anharmonic vibrations give rise to
problems. Thus errors might be introduced in structure determination of hydrogen bonded systems. The conclusion is, that even
if it seems straightforward to compare structures it may in
practice be difficult, because uncertainties in the experimental results must be considered.

Another possibility in finding correlations within different model systems and between a model system and a macromolecule is to look for similar dynamical features. The carbonyl-
stretching vibration of the amides (amide-I) has been used in
this context. These investigations will be covered by other
contributions to this book[7]. Very interesting experimental
results for NMA has recently been obtained by F. Fillaux[10].

A study of low-frequency vibrations from 10-400 cm^{-1}
reveals a direct information about interactions between molecules or between different parts of a macromolecule. The
present paper shall try to use the region from 10-400 cm^{-1} to
compare results for different amides, a polypeptide and a
protein. These investigations provide a direct evidence for
similarities between the model compounds and proteins, and
support in general transferability of results involving
hydrogen bonding, obtained experimentally or theoretically for
the simpler amides to more complicated systems, like proteins.

LOW-FREQUENCY VIBRATIONS (10-400 cm^{-1})

The low-frequency part of the vibrational spectrum from
10-400 cm^{-1} is rather difficult to investigate. Fourier
transform infrared spectroscopy (FTIR) is today a very powerful
technique, but even with the best equipment the far-infrared
(FIR) region causes problems for liquids with a high absorption
coefficient. Unfortunately, water and many other hydrogen
bonded systems show high absorption coefficients in the FIR-
region. Aqueous solutions of biological molecules are thus
difficult to investigate due to the absorption from the
solvent. In the future synchroton radiation might be useful in
the FIR-region, because this light source is much more intense
than conventionally used sources.

Inelastic neutron scattering is also a powerful technique
in low-frequency studies. However, the neutron source is not
easily accessible and in general this method is not frequently
used.

Raman spectroscopy is another scattering technique. In crystals conventional monochromators can be used down to a few wavenumbers. In principle this technique is applicable to even lower frequencies by use of interferometric techniques (depolarized Rayleigh and Brillouin spectroscopy). In liquids the high intensity of the very intense Rayleigh line is a serious problem. One way to solve this is to use the so-called $R(\bar{\nu})$-representation[11-14]. This representation has several advantages as compared to the ordinary Raman spectrum. The next section will be devoted to a short description of the $R(\bar{\nu})$-representation.

The R($\bar{\nu}$)representation   The $R(\bar{\nu})$-spectrum is defined in the following way[11-15]:

$$R(\bar{\nu}) \propto (\bar{\nu}_L - \bar{\nu})^{-4} \bar{\nu} [1 - \exp(-h\bar{\nu}c/kT)] I(\bar{\nu}) \qquad (1)$$

where $\bar{\nu}_L$ is the frequency of the laser, $\bar{\nu}$ is the Raman shift, $I(\bar{\nu})$ is the intensity at a Raman shift on $\bar{\nu}$ wavenumbers. h is Planck's, and k is Boltzmann's constant, c is the velocity of light and T is the absolute temperature. In our original work the fourth power term was omitted, because the effect of this is negligible within a small frequency interval[11-13]. The $R(\bar{\nu})$-spectrum is an example of a reduction procedure for a Raman spectrum[14,15].

AMIDES

The region from 10 to 400 cm^{-1}.   In a series of publications we have investigated low-frequency Raman spectra of liquid amides[8,9, 16-18]. In general the molecules given in figure 1 show a band with a maximum around 100 cm^{-1}. This is illustrated for N-methylacetamid (NMA) in figure 2. The band at 100 cm^{-1} is easily recognized in the $R(\bar{\nu})$-spectrum, where the influence of the Rayleigh-line is diminished. Table II gives a survey over frequency maxima observed in the liquid state for different amides. Note that intermolecular hydrogen bonding is impossible for N,N-dimethylformamide and N,N-dimethylacetamide.

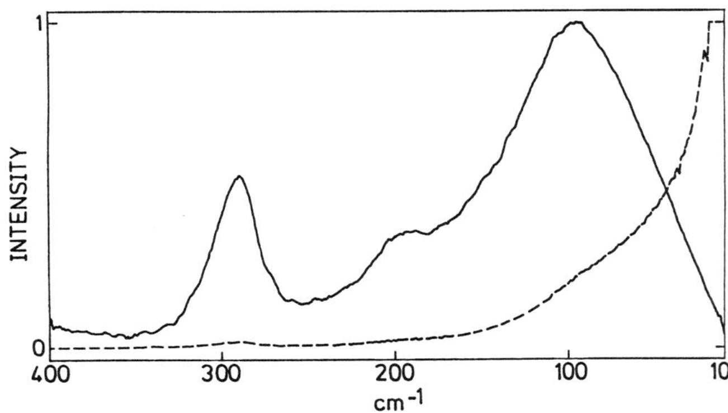

Figure 2   I($\bar{\nu}$)-spectrum (broken curve) and R($\bar{\nu}$)-spectrum (full curve) of liquid N-methylacetamide (NMA) at room temperature.

Table II   Frequency maxima observed in the $R(\bar{\nu})$-
           spectra of liquid amides*

Molecule	Freq. maximum (cm^{-1})
Formamide	110
N-methylformamide	112
N-ethylformamide	105
Acetamide	103
N-methylacetamide	100
N,N-dimethylformamide	65
N,N-dimethylacetamide	60

*The spectrum of acetamide was obtained at 90°C,
 the spectrum of N-methylacetamide at 30°C, all
other spectra at room temperature.

The amides without a possibility for hydrogen bonding,
N,N-dimethylformamide and N,N-dimethylacetamide, show no band
at 100 cm^{-1}. Based on many different isotopically molecules
we have assigned the band at 100 cm^{-1} to the mode illustrated
in figure 3[18].

The main features of this mode is supported by ab initio
calculations on linear dimeric formamide molecules[19].

We have performed experiments as a function of temperature for some of the amides[8,9]. The intensity variations
confirm an assignment to linear hydrogen bonded associates,
although it should be emphasized that it is difficult to
separate the influence of intermolecular interactions from
nonlinear effects.

All the hydrogen bonded amides show similar features due
to hydrogen bonding in the low-frequency part of the spectrum.
Hydrogen bonding is most often investigated in the N-H stretching region above 3000 cm^{-1} or in the C=O stretching region
between 1600 and 1700 cm^{-1}. However, the effect due to hydrogen
bonding is only small perturbations of the stiff bonds.
Furthermore over- and combination tones make it difficult to
draw clear conclusions from these frequency regions. The low
frequency region reflects directly vibrations due to hydrogen

Figure 3   Approximate description of the mode
           giving rise to the 100 cm^{-1} band in
           hydrogen bonded amides.

bonding, and should be more representative of the hydrogen
bonds in a molecule.

   Amide-I and amide-II regions. The carbonyl-stretching
vibration in amides corresponds to the amide-I band in a
protein, and the N-H bending vibration to the amide-II band.
The latter is usually weak in Raman spectra of proteins, but
rather intense in the IR-spectrum. The amide-I band in the
solid state has been intensively studied for ACN[4,5] and some
investigations have been performed for NMA[6,10]. The linear
structure in the liquid hydrogen bonded amides may be used
as a model for real proteins, and some of the vibrational features of the carbonyl stretching and N-H bending vibrations
will be discussed.

   Figure 4A shows the ordinary Raman spectra of liquid formamide in two different scattering configurations, I_{VV} and I_{VH}.
The intensities are not comparable, because the maximum intensity in both spectra was normalized to the same value. The
band around 1650 cm^{-1} corrresponds to the carbonyl-stretching
vibration (amide-I) and the band around 1600 cm^{-1} to the bending

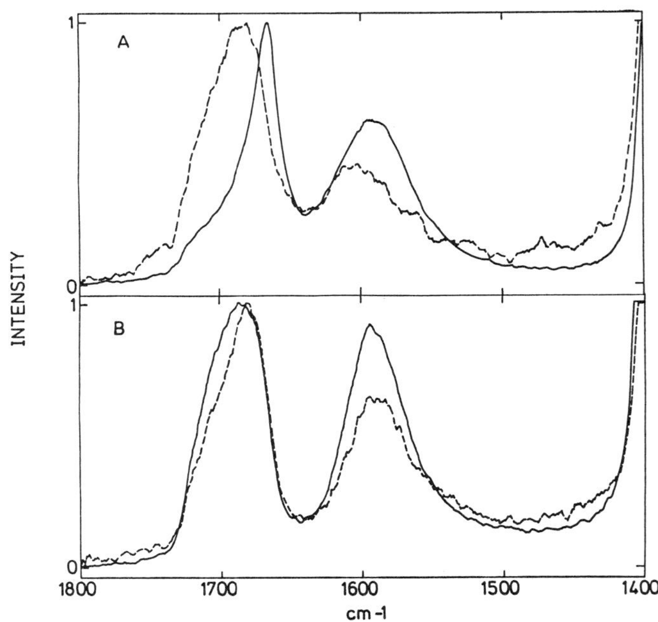

Figure 4   In figure A is shown Raman spectra of
            liquid formamide. Full curve I_{VV}, broken
            I_{VH}. Horizontal scattering plane.
            Figure B shows similar results for the
            lithium complex with formamide. The maxi-
            mum intensity in a given curve is nor-
            malized and the relative intensities
            should not be compared.

vibration (amide-II). Evidently the maxima for a given vibration are found at different frequencies in the two scattering configurations. This fact might support a cyclic structure of formamide in the liquid state, because a coupling of the vibrations in the dimer is expected to give a symmetric and an antisymmetric vibration. However, N,N-dimethylformamide shows similar features[20], and in this molecule no hydrogen bonding is possible. In fact this splitting between bands observed in different scattering configurations has been known for more than 15 years, and a mechanism including a transition dipole-dipole coupling has been proposed[20]. This mechanism is similar to the theories proposed by A.S. Davydov for crystals[21].

Recently we have used a lithium salt in order to break the hydrogen bonds in liquid formamide. Figure 4B shows the spectrum of a solution of LiCl in formamide at room temperature. A complex is formed between the lithium ions and the formamide molecules. Different models have been discussed[22,23]. Figure 4B shows that the frequencies of a given band in the two scattering configurations are very similar for the complex. In the low frequency part of the spectrum the band with a maximum at 110 cm^{-1} in neat formamide disappears in the lithium salt[24].

A computational study of N-methylacetamide (NMA). Computational results using a nonlinear quantum mechanical model have been performed for NMA[25]. In the 100 cm^{-1} region, there is a qualitative agreement between experimental and computational results. In the amide-I region a new low frequency band appears at low temperature[6] and it is in the computational study suggested, that this result can be explained in terms of self-trapped states, which occur in the theoretical model[25].

Irradiation experiments. The mode shown in figure 3 has been proposed to have a significant importance in biological systems[26]. A coupling between low-frequency modes at 100 cm^{-1} was propqsed to be important for the action of the anti-cancer drug N-methylformamide[26]. The mode shown in figure 3 may also be essential in a breaking of a hydrogen bond. In this sense a coupling (linear or nonlinear) to energy trapped in other vibrations is important. We have tried to see whether a coupling to the amide-I mode in NMA could be proven experimentally. A carbon monoxide laser in resonance with the carbonyl stretching vibration was used as exciting source while the R($\bar{\nu}$)-spectrum was simultaneously recorded[9,17]. The preliminary results showed changes in the low-frequency spectrum by irradiation[17]. However, later experiments showed that these results most probably was caused by local temperature changes in the sample[9], but we believe that experiments along these lines might be fruitful in experimental justifications of the importance of nonlinear effects, including trapping of energy.

POLYPEPTIDES AND PROTEINS

In general polypeptides and proteins are difficult to investigate. They are solids at room temperature and in order to mimic the situation in living systems low concentrations in aqueous solutions should be investigated. Fluorescense from unavoidable impurities is another serious problem, because the fluorescense obscures the Raman spectrum. This problem can be solved by use of Fourier-transform (FT)-Raman spectroscopy,

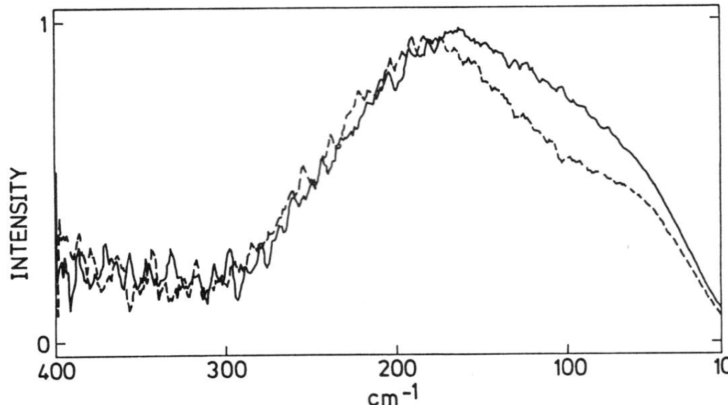

Figure 5 $R(\bar{\nu})$-spectrum of poly-L-lysine (27% w/w)
            in buffer solution (full curve). The
            broken curve shows the $R(\bar{\nu})$-spectrum of
            the buffer (Tris/HCl, pH 7.5, ionic
            strength 10 mM).

where a laser with a frequency in the near infrared region is
used as the exciting source. We hope in the future by this
method to be able to investigate proteins along the outlines
given above for amides. Here is a presentation of some preliminary results.

Figure 5 shows the $R(\bar{\nu})$-spectrum of an aqueous poly-L-
lysine solution (27% w/w) together with the spectrum of neat
liquid water. A band from poly-L-lysine is observed with a
maximum around 100 cm⁻¹. This band is believed to arise from
hydrogen bonds in accordance with results from the amides. The
band might be assigned either to an intramolecular mode or to
a mode involving both atoms in the polypeptide and water,
hydrogen bonded to the peptide.

We have recently reported similar results for lysozyme[24,27].
A band around 100 cm⁻¹ is observed. We are now trying to
investigate the polypeptides and proteins in aqueous solution
in some more detail by temperature variations, and denatura-
tion in different ways. We wish to study the amide-I bands of
polypeptides and proteins in order to see whether features
like those in the amide-I band of ACN and NMA are observed.
We also hope to obtain IR-spectra of these molecules.

CONCLUSION

Proteins are difficult to investigate in detail on a
molecular or atomic scale. It is important to choose model
compounds which can mimic some of the essential features in
the proteins. The simple amides are used as model systems. The
low-frequency $R(\bar{\nu})$-representation of the Raman spectrum shows
a band at 100 cm⁻¹ in liquid amides and aqueous solutions of

proteins, which indicate similarities in these molecules. This
band is assigned to a mode involving atoms in a hydrogen bond.
Energy trapping may be of importance in a breaking of the
hydrogen bond.

The amides show differences between Raman spectra obtained
in different scattering configurations for the amide-I and
amide-II bands. We believe that the hydrogen bonds are broken
in a complex with lithium ions. This fact is supported by
observations in the low-frequency spectrum.

A more general investigation of amides and proteins is
needed in both the low-frequency part of the spectrum and in
the amide-I and amide-II region. We plan to do this in the
near future.

Acknowledgements   The author is grateful to Prof. F. Fillaux,
France and Dr. B. Pierce, USA for making their results available prior to publication. I wish to thank Prof. G. Dodin,
France for permission to use the results on poly-L-lysine,
which is part of a joint project.

REFERENCES

1. A. S. Davydov, The Theory of Contraction of Proteins under
    their Excitation, J. Theor. Biol. 38:559 (1973).
2. A. S. Davydov and N.I. Kislukha, Solitary Excitons in One-
    Dimensional Molecular Chains, Physica Status Solidi,
    59:465 (1973).
3. A. L. Scott, Dynamics of Davydov Solitons, Phys. Rev. A
    26:578 (1982).
4. G. Careri, V. Buontempo, F. Carta, E. Gratton and A. C.
    Scott, Infrared Absorption in Acetanilide by Solitons,
    Phys. Rev. Lett. 51:304 (1983).
5. G. Careri, V. Buontempo, F. Galluzzi, A. C. Scott, E.
    Gratton and E. Schyamsunder, Spectroscopic Evidence for
    Davydov-Like Solitons in Acetanilide, Phys. Rev. B30:
    4689 (1984).
6. I. J. Bigio, O. Faurskov Nielsen, and C. T. Johnston,
    Laser Techniques to Study Nonlinear Dynamics in Bio-
    polymers, in: "Computer Analysis for Life Science",
    C. Kawabata and A. R. Bishop, eds., Omsha, Ltd. (1986).
7. See contributions to this book by I. J. Bigio and G. Careri.
8. O. Faurskov Nielsen, P.-A. Lund and E. Praestgaard,
    Hydrogen Bonding in Liquid Formamide, J. Chem. Phys.,
    77:3878 (1982).
9. O. Faurskov Nielsen, I. J. Bigio, I. Olsen and J.-M.
    Berquier, Low Frequency (20-400 cm^{-1}) Vibrational
    Spectra of N-Methylacetamide in the Liquid State,
    Chem. Phys. Lett., 132:502 (1986).
10. F. Fillaux, Private communication.
11. O. Faurskov Nielsen, D. H. Christensen, P.-A. Lund and
    E. Praestgaard, Short Time Molecular Dynamics in
    Molecular Liquids Studied by Depolarized Raman and
    Rayleigh-Wing, and Infrared Absorption, Proc. 6th
    Intern. Conf. Raman Spectrosc., E. D. Schmid, R. S.
    Krishnan, W. Kiefer and H. W. Schrötter, eds., Heyden,
    London - Philadelfia-Rheine, Vol. 2 1978 p. 208.

12. P.-A- Lund, O. Faurskov Nielsen and E. Praestgaard, Comparison of Depolarized Rayleigh-Wing Scattering and Far Infrared Absorption in Molecular Liquids, Chem. Phys., 28:167 (1978).

13. O. Faurskov Nielsen, P.-A. Lund and E. Praestgaard, Comments on the $R(\bar{\nu})$-Spectral Representation of the Low-Frequency Raman Spectrum, J. Chem. Phys., 75:1586 (1981).

14. M. H. Brooker, O. Faurskov Nielsen and E. Praestgaard, Assesment of Correction Procedures for Reduction of Raman Spectra, J. Ram. Spectrosc., 19:71 (1988).

15. W. F. Murphy, M. H. Brooker, O. Faurskov Nielsen, E. Praestgaard and J. E. Bertie, Further Assessment of Reduction Procedures for Raman Spectra, J. Ram. Spectrosc., in press.

16. O. Faurskov Nielsen and P.-A. Lund, Low-Frequency Raman Spectra of Liquid Formamide and Aqueous Solutions of Formamide, Chem. Phys. Lett., 78:626 (1981).

17. O. Faurskov Nielsen, I. J. Bigio, C. Johnston and S. P. Layne, An Experimental Search for Nonlinear Dynamics in Amides and Proteins, A. J. P. Alix, L. Bernard and M. Manfait, eds., J. Wiley and Sons, Chichester (1985).

18. O. Faurskov Nielsen, Hydrogen Bonding in Liquid Amides Studied by Low-Frequency Raman Spectroscopy, J. Mol. Struct., 175:251 (1988).

19. B. Pierce, Private contribution. See also the contribution by B. Pierce to this book.

20. V. M. Shelley, A. Talintyre, J. Yarwood and R. Buchner, Spectroscopic Studies of Solute-Solute and Solute-Solvent Interactions in Solutions Containing N,N-Dimethylform-amide, Faraday Discuss. Chem. Soc., 85:1 (1988) and references cited therin.

21. A. S. Davydov, Theories of Molecular excitations, M. Kasha and M. Oppenheimer, jr., eds., McGraw-Hill, Dover, London (1962).

22. J. Bukowska, Raman and Infrared Studies of Interactions between Amides and Ions, J. Mol. Struct., 98:1 (1983).

23. J. Bukowska, Intermolecular Coupling of the CO Stretching Vibrations in Electrolyte Solutions of Carbonyl Compounds, J. Mol. Struct., 143:309 (1986).

24. O. Faurskov Nielsen, D. H. Christensen and O. Have Rasmussen, Interactions in Amides and Proteins Studied by Low-Frequency Vibrational Spectroscopy, Proc. 3rd Eur. Conf. Raman Spectrosc. Biol. Mol., Rimini (1989), in press.

25. J. Halding Jensen, P. L. Christiansen, O. Skovgaard, O. Faurskov Nielsen and I. J. Bigio, Experimental and Computational Study of N-Methylacetamide, Phys. Lett., A117:123 (1986).

26. O. Faurskov Nielsen, P.-A. Lund, L.S. Nielsen and E. Praestgaard, Aspects of Low-Frequency Vibrations (20-350 cm^{-1}) from Watson-Crick Base Pairing in an Aqueous Solution of tRNA, Biochem. Biophys. Res. Com. III, 120 (1983).

27. O. Faurskov Nielsen, D. H. Christensen and E. Praestgaard, Low-Frequency Vibrations (20-400 cm^{-1}) in Liquids and Aqueous Solutions, Biochem. (Life Sci. Adv.), 7:57 (1988).

29

Molecular Crystals and Localized Vibrational States

A.Migliori, A.M.Clogston, P.M.Maxton[*],
J.R.Hill, D.S.Moore, H.K.McDowell

Los Alamos National Laboratory, Los Alamos, New Mexico
[*]University of California at Los Angeles, Los Angeles Ca.

The normal modes arising from a simple linear model of a crystalline solid
predict many important properties such as the specific heat, the sound
velocity, and the Raman and infrared response. Such a model fails,
however, to produce thermal expansion and finite thermal conductivity. The
introduction of weak non-linear couplings between the normal modes
immediately corrects these simpler problems, while preserving the usual
phonon representation of the purely linear system as a useful first
approximation for small phonon amplitudes. What we wish to consider here
is the next step in the process, taken by keeping the weak non-linearities
but introducing large phonon amplitudes. Putting aside, for the moment,
the question of how to achieve the necessary amplitudes, we can consider
the consequences if certain not unusual conditions obtain which put the
phonon equation of motion in correspondence with the non-linear
Schrodinger equation. We can then be guided by classical solutions of
this equation which predict spontaneous symmetry breaking and vibrational
localization. To focus our thoughts, we outline in physical terms the
processes leading to the localization of vibrational energy. We use the
insight gained to guide our choice of materials, and we describe the
measurements we have made to begin the establishment of a rigourous
connection between real solids and the theoretical constructs related to
this problem.

Introduction

To see in physical terms how the process of localization can work, we
begin with a solid having an optical phonon branch with a minimum at the
Brillouin zone center. Although not usually the case in simple solids, we
take this to be the lowest optical phonon branch, a situation occasionally
found in more complex structures. The dispersion curves then look
something like Fig.1. Next consider a small non-linear term in the
Hamiltonian that couples the optical phonon to the acoustic phonon such
that the energy of the solution decreases. Such a term is easy to imagine
if we consider a rubber band. The optical branch is a twist in the band,
the acoustic branch is a stretch. Clearly a large twist in either
direction contracts the rubber band, but a small twist, to first order in
the twist angle, has no effect. In a region of the solid large compared
to a lattice spacing but small compared to a wavelength of light we
imagine the optical mode is sufficiently well populated to have shifted
the eigenfrequency downward by some small amount as shown in Fig.1. Thus
the new frequency now resides below the bottom of the low-amplitude phonon
band, and therefore, below the lowest propagating frequency in the balance
of the crystal. The vibrations are, therefore, self-trapped. Again

Davydov's Soliton Revisited, Edited by P.L. Christiansen and A.C. Scott
Plenum Press, New York, 1990

referring to Fig.1, we see that there is a spread of k-vectors in the
high-amplitude dispersion curve, all of which lie below the bottom of the
low-amplitude band, and this spread must determine the width of the
localized state. The more strongly excited the state, the bigger the
spread in k-space, and the smaller the width in real space.

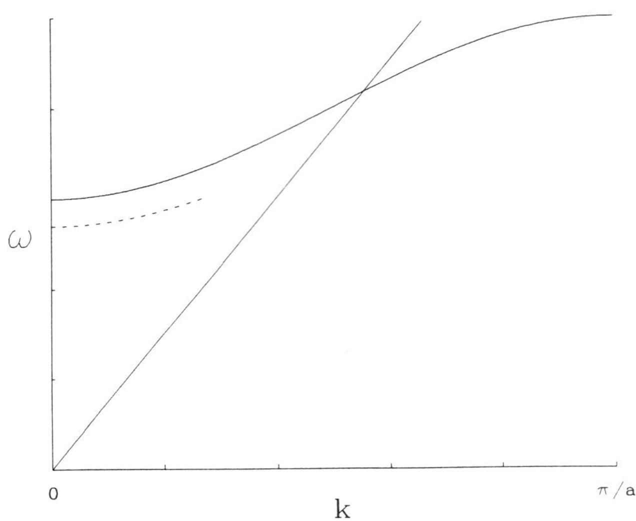

FIG. 1. Schematic dispersion curve of a nonlinear lattice mode. The
solid line represents the "eigenmodes" at low vibrational amplitudes. The
dashed curve shows the shifted frequencies of these modes at large
amplitudes, and extends out to the largest k value for which the shifted
frequency is lower than the lowest low-amplitude frequency. A typical
acoustic branch that does not couple linearly to the optical mode is the
lower solid line.

The possibility of such localized states was first brought into
prominence by Davydov and Kislukha[1] in a 1973 paper which proposed that
such states might exist on the α-helix because of the interaction of the
intramolecular carbon-oxygen amide I stretching modes with low frequency
acoustic phonons. This original model has spawned much related work on
refined Hamiltonians[2-5] and computer simulations[6] which have called into
question the likelihood of Davydov solitons in polypeptide chains living
long enough to be experimentally significant at temperatures much above
10K (although simulations[7] based on another model by Yomosa[8] predict
stable long lived lattice solitons in these systems). Nevertheless, work
on the α-helix system has inspired experimental and theoretical studies of
other systems that may stand a better chance of supporting long-lived
localized states.
One class of promising systems consists of the pseudo one-dimensional
molecular crystals such as l-alanine. Through the non-linear interaction
between longitudinal acoustic phonons and low-lying vibrational states, we
derive equations of motion appropriate to optical lattice solitons, but
with parameters significantly different from the α-helix system. This is
because the vibrational spectrum in many crystals composed of long chains
of hydrogen-bonded molecular units arises from hindered rotation of the
molecules. These librations have much lower frequencies and, in
principle, can have a much stronger nonlinear coupling to acoustic
phonons. In the molecular crystal l-alanine, two such modes lying near 49

cm^{-1} are particularly interesting because they have an unusually long
lifetime even for low level excitations.

A Simplified Model

To study molecular crystals in general as potential candidates for
localization, and to help guide and interpret our experimental
measurements on 1-alanine, we have set up a simple model of a one
dimensional chain of molecules based on linear spring constants that
exhibit both phonon and vibron degrees of freedom and non-linear coupling
between them. The model is designed to preserve only certain relevant
features of the 1-alanine crystal and can be thought of as a simple
molecular dynamics model with all except the essential degrees of freedom
suppressed. A form of this model was originally proposed and studied by
John C. Wheatley. It turns out to lead to a particular case of the Takeno
hamiltonian and therefore includes the possibility of localized states.
One advantage of the model is that it enables an estimation of the
characteristic parameters of the phonon and vibron systems and the
non-linear coupling constant. We can, therefore, use it to make contact
with the experimental work.

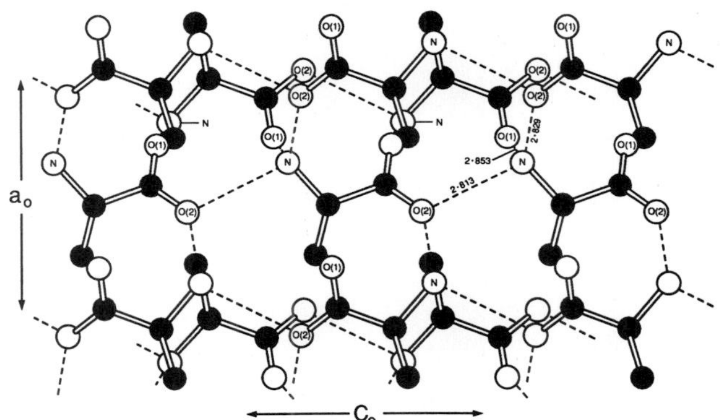

FIG. 2. The crystal structure of 1-alanine (following Simpson and Marsh).
The dark circles represent carbon atoms. The dashed lines are hydrogen
bonds (hydrogen atoms not shown).

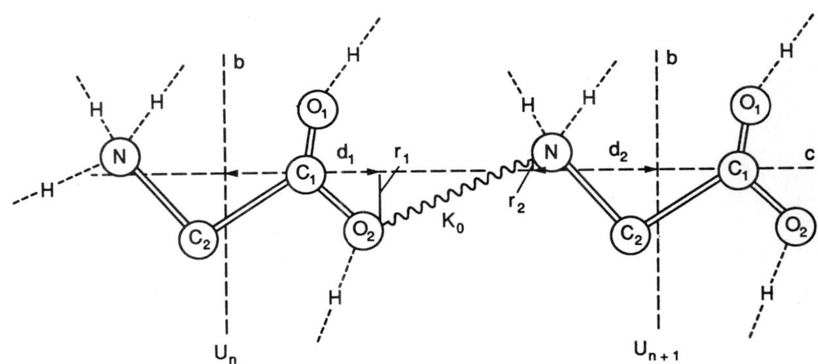

FIG. 3. A section of the ribbon-like chain of molecules in crystaline
1-alanine showing the geometric features that enable a nonlinear coupling
between a longitudinal acoustic mode and a libration about the c-axis. A
hydrogen atom and a -CH$_3$ group have been deleted for clarity.

The l-alanine crystal consists of long chains of hydrogen bonded l-alanine molecules with the formula $C_3H_7O_2N$ (Fig.2)[9]. The oxygen atoms O_1 and O_2, the carbon atoms C_1 and C_2 and the nitrogen atom N are approximately coplanar. In addition C_2 is bonded to a hydrogen atom and CH_3 groups lying respectively above and below the plane formed by N-C_2-C_1-O_1-O_2. Each nitrogen atom is hydrogen bonded to 3 oxygen atoms, O_1 is bonded to one nitrogen atom and O_2 to two nitrogen atoms.

To model the l-alanine crystal, we retain only the essential features. All internal coordinates of the molecule are assumed to be fixed. The only degrees of freedom allowed for the nth molecule are longitudinal motion along the c-axis with coordinates u_n, and rotation about the principle axis C with coordinate Θ_n. Referring to figure 3, the kinetic energy associated with longitudinal motion of the nth molecule is $\frac{1}{2}M\dot{u}_n^2$. In addition to this energy and the potential energy associated with the spring K_o, we assume that hindering potentials act on the molecules that give rise to energies $(\frac{1}{2}I_c\dot{\Theta}^2 + \frac{1}{2}I_c\omega_\Theta^2\Theta_n^2)$ and $(\frac{1}{2}I_a\dot{\phi}_n^2 + \frac{1}{2}I_a\omega_\phi^2\phi_n^2)$ for rotation about the c-axis and a-axis respectively.

To sketch how we have proceeded, consider rotation about the c-axis. The potential energy associated with the spring K_o from longitudinal motion and rotation of the molecules is straightforward. For the rotations we introduce a staggered order parameter $Q_n = (-1)^n\Theta_n\frac{1}{2}(r_1+ r_2)$. This places out-of-phase rotation of adjacent molecules, for which the strongest non-linear effects are expected, at the center of the one-dimensional Brillouin zone. The axial motion of the molecules is measured in terms of displacement from their equilibrium position by the variable $q_n = u_n$-na. The potential energy function is then found to be to cubic order in the displacements,

$$U = \frac{K_0}{2}\sum_n\left\{\left(1-\frac{\ell_0}{\ell_1} + \frac{\ell_0 D^2}{\ell_1^3}\right)(q_{n+1}-q_n)^2 - \left(1 - \frac{\ell_0}{\ell_1}\right)\frac{4r_1 r_2}{(r_1+2)^2}(Q_{n+1}-Q_n)^2\right.$$

$$\left. - \frac{\ell_0 D}{\ell_1^3}\frac{4r_1 r_2}{(r_1+r_2)^2}(q_{n+1}-q_n)(Q_{n+1}+Q_n)^2\right\} \tag{1}$$

If the kinetic energies and potential energy of hindered rotation are added, this leads to a Hamiltonian of the form

$$H = \sum_n\left[\frac{1}{2M}P_n^2 + \frac{K}{2}(q_{n+1}-q_n)^2\right] + \frac{m\omega_0}{4\hbar}\times\sum_n(Q_n + Q_{n+1})^2(q_{n+1}-q_n)$$

$$+ \sum_n\left[\frac{1}{2m}P_n^2+ \frac{m\omega_0^2}{2}Q_n^2 - \frac{m\omega_0}{\hbar}J Q_n(Q_{n+1}-Q_{n-1})\right] \tag{2}$$

where

$$K = K_0\left(1 - \frac{\ell_0}{\ell_1} + \frac{\ell_0 D^2}{\ell_1^3}\right) \simeq K_0\left(\frac{D}{\ell_1}\right)^2 \tag{3}$$

$$\chi = -\frac{\hbar}{m\omega_0}K_0\frac{\ell_0 D}{\ell_1^3}\frac{4r_1 r_2}{(r_1+r_2)^2} \tag{4}$$

$$\omega_0^2 = \omega_\Theta^2 - \frac{2K_0}{m}\left(1-\frac{\ell_0}{\ell_1}\right)\frac{4r_1 r_2}{(r_1+r_2)^2} \tag{5}$$

$$J = \frac{\hbar}{m\omega_0} \frac{K_0}{2}(1-\frac{\ell_0}{\ell_1}) \frac{4r_1 r_2}{(r_1+r_2)^2} \tag{6}$$

and

$$m = I_c \frac{4}{(r_1+r_2)^2} \tag{7}$$

This hamiltonian is a form of the Takeno hamiltonian and establishes that l-alanine is a useful material for studying the existence and lifetime of localized states in solids. It should be noted that the vibron spectrum determined by ω_0 and J is strongly dependent on whether the spring K_0 is compressed or extended in its equilibrium position. That is, if ℓ_1 is less than or greater than ℓ_0. If the sping is extended ($\ell_1 > \ell_0$) J is positive and the vibron spectrum then has a minimum at $k = 0$, the center of the Brillouin zone (as we observed experimentally, described below). This is also a necessary condition for the formation of localized modes, given the usual softening non-linearities.

All the constants in the hamiltonian (2) can be derived from a knowledge of the geometric factors ℓ_0, ℓ_1, r_1, r_2, and D, the mass M and moment of inertia I_c of the molecule, and the longitudinal velocity of sound $v = \sqrt{\frac{K}{M}}$. The non-linear coupling constant does not depend on any intrinsically non-linear forces.

<u>L-Alanine</u>

The above model can now be used to justify our choice of l-alanine (CH_3-$CHNH_3$-CO_2), one of the simplest amino acid crystals, as an appropriate material. From the above and other considerations, we can list the important properties and their relevance to l-alanine:

1. Low symmetry. L-alanine is an orthorhombic crystal which, although it certainly exhibits a three dimensional structure, can be approximately modeled as a chain of molecules (oriented along the c-axis of the crystal) coupled by hydrogen bonds at oblique angles. Thus we have at least some justification for treating this as a 1-D system.

2. Simple to model. Although this crystal has 13 atoms per molecule with 4 molecules per unit cell (153 optical phonon modes) the higher energy intra-molecular modes are separated from the 21 lattice modes so that the unit cell can be modeled as four rigid molecules. These low-frequency lattice modes, characteristic of molecular crystals, are quite amenable to simpler models such as the one described above, where all high-frequency vibrations are considered to be unpopulated.

3. Anharmonicity in phonon fields. This is typically demonstrated by large coefficients of thermal expansion. Figure 4 shows the unit cell dimensions of l-alanine as a function of temperature.[10] The steep slopes for the 'a' and 'b' dimensions exhibit the large coefficients of linear expansion expected of a hydrogen bonded molecular crystal. The small, negative slope of the c-axis curve suggests an even more interesting nonlinearity. Remember that the crystal is characterized in all directions by similar hydrogen bonds, and the bonds themselves are expected to show large, positive expansion with increasing temperature. Therefore, the only mechanism for a negative thermal expansion coefficient is a geometric structure in which vibrational or librational motion perpendicular to the c-axis forces a contraction of the lattice along that dimension. It is this type of mechanism that provides the non-linearties of our models.

4. Pure single crystals should be readily available to minimize the presence of impurity- or defect-induced localization. L-alanine is relatively simple to grow in large, defect free, single crystals. The crystals are also transparent and colorless, a valuable property as our most useful probes will be optical.

5. Long lived optical phonon modes. In order to allow enough time

for true, dynamically stable, localized states to distinguish themselves
from very-short-mean-free-path phonons as well as to allow for the
possibility of observing the dynamics of phonon to local-mode transitions,
a long lifetime is important. The lifetimes of the lowest lying optical
phonons (usually the longest lived because of a reduced number of decay
channels) have been measured by Kosic et al[11] to be of order nanoseconds
at low temperatures. This corresponds to a quality factor (Q) in excess of
10^3.

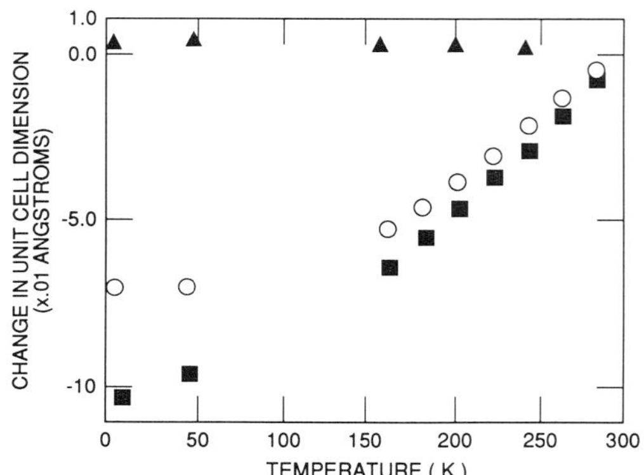

FIG. 4. Thermal expansion of 1-alanine. Boxes=Δa, circles=Δb, and
triangles=Δc. Data above 150 K are from Forss. Low temperature data are
from Vergimini et al[10].

<u>Phonon Dispersion</u>

     Crucial to the study of localization described above is knowledge of
the dispersion curves of the phonon fields expected to participate in the
localization process. As indicated in the discussion of figure 1,
knowledge of the curvature at the Brillouin zone center is particularly
important. The only way of measuring this is via inelastic neutron
scattering, a time-consuming process only to be undertaken after all other
criteria have been met. We measured the relevant dispersion curves at the
Brookhaven High Flux Beam Reactor[12]. The sample was a fully deuterated,
single crystal of 1-alanine, grown by slow evaporation. The resulting
data for the a and c directions at low temperature are shown in figure 5.
Although the c direction curves are somewhat tangled, some important
observations can be made. First, for the lowest lying optical mode, there
is a minimum at the zone center for the dispersion curves in both the a
and c directions. This is a crucial point as it is a condition for
zone-center localization for systems in which the non-linear term softens
the optical mode at high amplitude. Second, the total dispersion in the a
direction corresponds to an inter-cell coupling constant, J, of
approximately .71 meV (5.7 cm^{-1}). Finally, the initial slope of the
c-axis longitudinal acoustic branch of 3.9 x 10^{-12}eV m $\pm$.2 (5.9 km/sec)
agrees well with our pulse-echo ultrasound measurements of the speed of
sound in undeuterated 1-alanine of 6.2 km/sec in the c direction.

<u>Raman Scattering</u>

     To further characterize 1-alanine, and to test for phase transitions
that might complicate our interpretation of the measurements, we conducted
a temperature dependent study of the low-frequency Raman modes.
Deuterated and undeuterated samples were grown by slow evaporation or slow
cooling of a saturated solution, yielding optically clear crystals with

easily identifiable growth habits. The symmetry axes of the crystals were
confirmed by Laue diffraction--the well defined diffraction patterns
obtained, as well as the optical clarity of the crystals imply a well
ordered structure with minimal defects and no twinning.

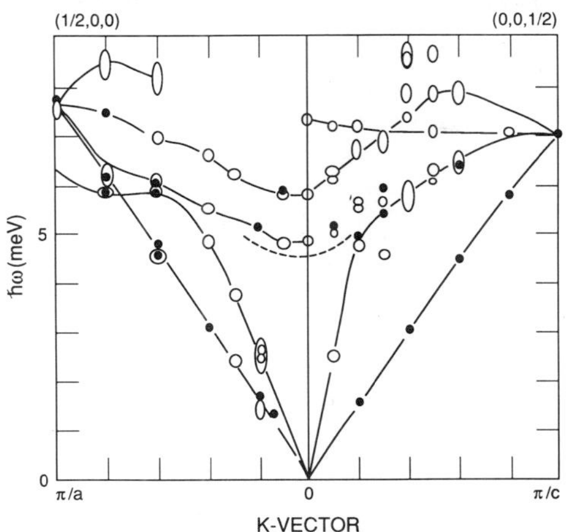

FIG. 5. Inelastic neutron scattering data on fully deuterated l-alanine,
showing the low-lying optical modes and the acoustic branches.

The low-frequency Raman spectrum consists of 21 optical lattice
modes, most of which are concentrated in overlapping bands between 90 and
150 cm^{-1}, but the two lowest lying modes, the ones of interest here, are
well isolated (figure 6). It is in the intensity of these modes versus
temperature (figure 7a) that we observe anomalies that suggest that the
two modes are in some way related, and that the appearance of the
lower-frequency mode is associated only with high vibrational amplitude.

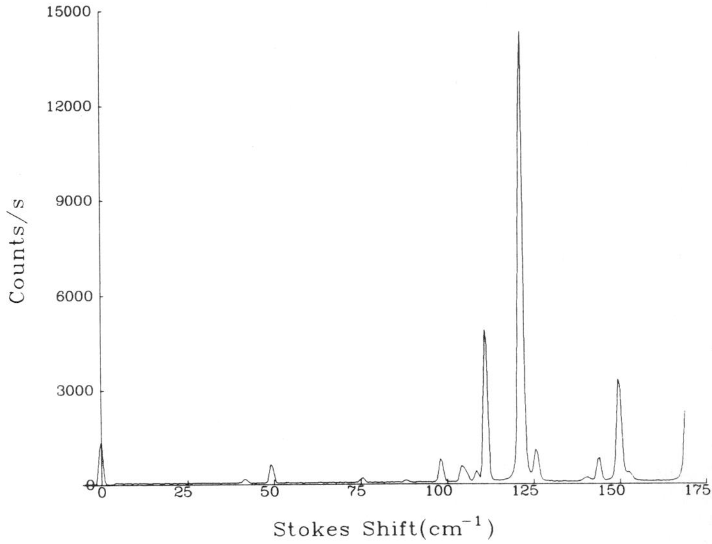

FIG. 6. Low temperature Raman spectra of l-alanine in c(bb)a orientation.

Note that in the usual picture of Raman scattering off a typical phonon,
the phonon is detectable at low temperatures. For a mode to disappear
when cold, something unusual must happen.

To analyze the raw data, intensities were normalized to that of an
851 cm^{-1} mode, well fit by a Lorentzian, which is expected to have an
intensity independent of temperature for the temperatures of interest.
The two lowest-energy modes were reduced using least-squares fits to
Gaussians, with estimated relative intensity errors of ±3%. What is
surprising is that we could fit the obviously anomalous intensities of
both lines to a single very simple model having only one adjustable
parameter. To see how this works, note that the sum of the intensities of
the 42 cm^{-1} and 49 cm^{-1} bands, shown in figure 7b, obey Maxwell-Boltzmann
statistics to well within our expected errors. Thus it must be that the
two modes are sharing a single thermal occupation factor. The
energy-level scheme for this Maxwell-Boltzmann fit was constructed with a
spacing of 49 cm^{-1} between levels from the ground state to the m th
excited state of one harmonic potential and a spacing of 49 cm^{-1} for a
second harmonic potential from level n up, suggesting that some change or
an instability occurs at excitation level m. Such an energy level scheme
produces a fit essentially identical to the usual Maxwell-Boltzman
function for a harmonic oscillator with level spacing of 45.5 cm^{-1}, the
average value of the two bands. We conclude from this fit that the 49
cm^{-1} and 42 cm^{-1} bands can be considered as representing the same degree
of freedom. However, by using two level spacings, two Raman frequencies
are present, and the same model predicts the temperature-dependent
intensities of each, shown as the solid curves in figure 7a, with n the
only adjustable parameter.

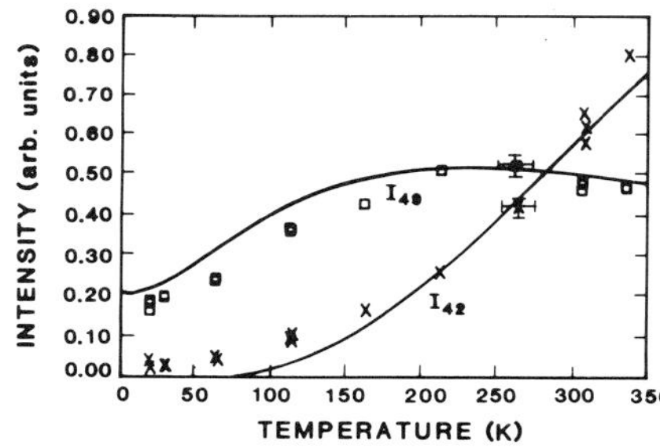

FIG. 7a The normalized, integrated
intensities of the 49cm^{-1} (boxes)
and the 42cm^{-1} (X). The solid
curves are a fit to a model describing the bands as one degree of
freedom with an instability that
shifts the oscillator frequency from
49 cm^{-1} at low amplitude to 42 cm^{-1}
at high amplitude.

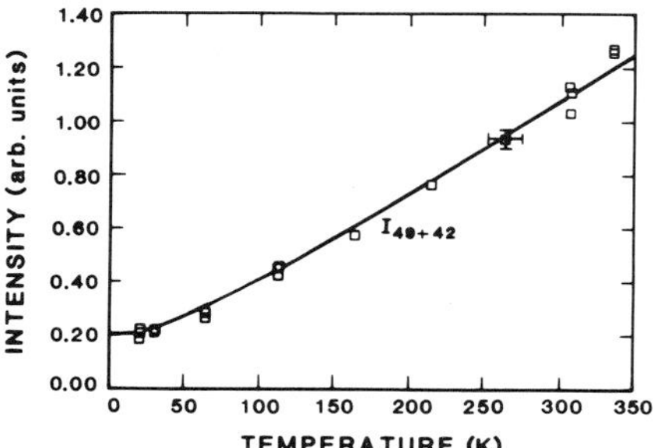

FIG. 7b The sum of the normalized
intensities of the 42cm^{-1} and 49cm^{-1}
Raman bands. The result of the
model (solid curve) is not
detectably different from a
Bose distribution for 45.5cm^{-1}.

The physical basis for this fit is not entirely clear to us, but
having made it, we also see that it has one very interesting imperfection.
It is that the predicted intensity of the 42cm^{-1} line at 20K is well below
our minimum detectable intensity, but we do indeed observe it to be
present, although at a level impossible to explain by a simple
Maxwell-Boltzman distribution. Because the model constrains the 42cm^{-1}
intensity to be present only at high temperatures we can conclude that it
is large-amplitude motion that makes it appear, and hence the intensity of

this mode is a measure of vibrational amplitude. High vibrational
amplitude is, of course, generated by high temperatures in normal modes.
But high amplitudes can also be generated if a normal mode at low
temperatures localizes. The localization concentrates an energy
insufficient by itself to push the normal mode over the 42cm^{-1} threshold.
But by concentrating this energy in a few unit cells, large motions can be
achieved cold. This evidence for localization is, of course, somewhat
tenuous. However, when combined with all the other positive indications
that l-alanine is a good candidate for localization, we find it compelling
to intensify our studies, eventually leading to the rather complex
picosecond coherent Raman measurement we plan, and which is described
below.

     To prepare for the coherent Raman work, we have attempted to make the
best possible study of the non-linear effects observable with conventional
means. Figure 8 shows the temperature dependence of the frequency of the
49 cm^{-1} mode (the 42 cm^{-1} mode shows similar behavior). These data were
obtained by scanning through the laser line and then continuously through
the band of interest to obtain a repeatable, absolute measure of the
frequency. Two important points must be made about this plot:

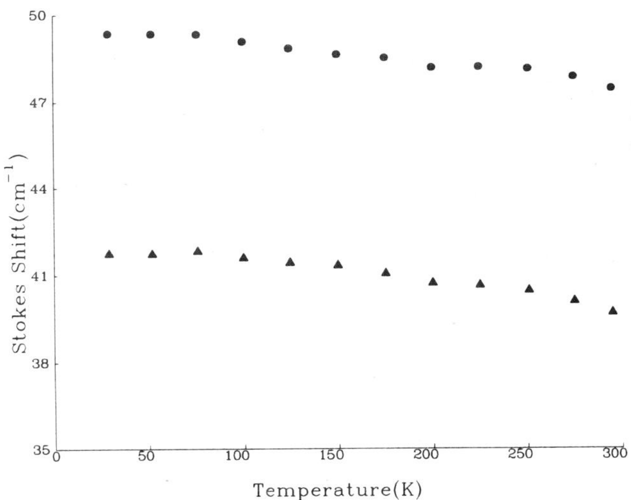

FIG. 8 The temperature dependence of the frequencies of the 49 cm^{-1}
(circles) and 42 cm^{-1} (triangles) Raman bands in l-alanine.

     First, the temperature dependence of the frequency shift is
approximately Bose-like, with a characteristic energy of approximately 180
cm^{-1} implying that the dominant nonlinearities for this mode involve
couplings to other lattice modes rather than intra-mode non-linearity or
coupling to acoustic phonons. This suggests that a strict Davydov model
is inappropriate for this system, that coupling to optical modes is
important, and that the models we propose are valid only below 100 K.

     Second, the total frequency shift may be used as an estimate for the
eigenfrequency softening and thus provide a very rough limit on the
maximum degree of localization of vibrational energy possible in this
system. The total frequency shift from 0 K to 295 K is approximately 2
cm^{-1}. Because the lattice breaks up near 450 K, we may take 2 cm^{-1} as an
order-of-magnitude estimate of the maximum available softening of these
low-energy lattice-mode eigenfrequencies. In the introduction to this
work, we demonstrated in figure 1 how a softening of the dispersion curve
of an excitation allows for the possibility of self trapping of that
excitation. The dotted line under the lowest-energy modes near zone
center in figure 5 represents a dispersion curve softened by 2 cm^{-1} (.25

meV). This softened curve is presumed to have the same shape as the low
temperature curve represented by the data points. This figure is the
experimental analog of the schematic figure 1. From figure 5 we can now
see that the spread in k-states used to form a localized state must extend
in k-space no further than approximately one tenth of the Brillouin zone.
Thus the minimum possible width of a dynamically stable localized state
is of the order of 10 lattice constants in this system. It should be
reiterated that since the 2 cm^{-1} softening is a result of all nonlinear
couplings and that not all of these couplings may be useful in self
trapping the excitation, this estimate is very much a lower bound on the
real space size of the localized state and the actual state, if it exists,
will almost certainly be larger.

<u>Picosecond Coherent Raman Scattering</u>

     In order to obtain greater insight into the detailed behavior of the
49 cm^{-1} mode of l-alanine at both high and low excitation levels,
time-resolved measurements using the technique of ps CARS can be
performed. Ps CARS (picosecond coherent anti-stokes raman spectroscopy) is
a four-wave mixing technique whereby three laser beams are overlapped in a
solid in order to generate a fourth[14] (figure 9). This technique can be
utilized in either the time-domain or the frequency domain. In the time
domain, two laser pulses (from two sync-pumped, cavity-dumped tunable dye
lasers driven by a mode-locked, Q-switched, frequency-doubled Nd-YAG
laser) are overlapped in the sample to pump coherently the phonon (Vibron)
of interest. At some later time another dye-laser pulse is used to
interrogate the coherently excited ensemble. This probe pulse interacts
with the coherence remaining in the ensemble to produce a fourth pulse,
spectrally and spatially resolved. In the time domain, the intensity of
the fourth pulse is a measure of the decay rate of the phonon. Because
decay times as long as several nanoseconds are easily resolved, the
corresponding linewidth resolution is many orders of magnitude smaller
than the resolution limit of the best spectrometer[15].

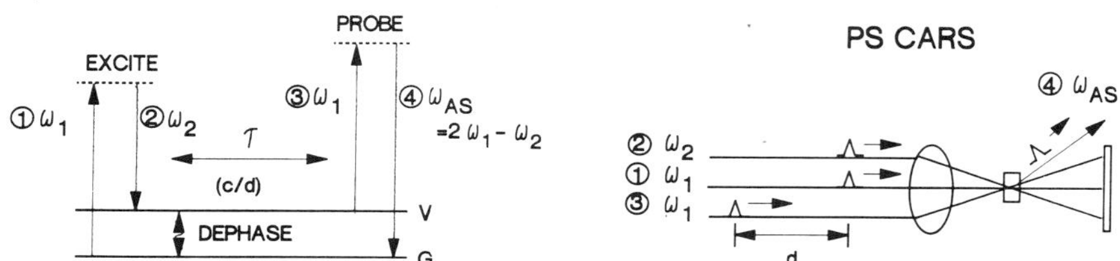

Fig. 9 Spatial arrangement and transitions excited in a time delayed CARS
experiment. Simultaneous pulses ω_1 and ω_2 excite a Raman active vibration
V. At time τ later a probe pulse at ω_1 stimulates anti-Stokes emission
$\omega_{AS}=2\omega_1 - \omega_2$. The conservation of both energy as well as momentum (for a
system having translational invariance) causes the emitted ω_{AS} pulse to be
spatially as well as spectrally resolved.

     By using Ps CARS, we hope that the onset of dynamic localization can
be followed in time, perhaps enabling us to observe the growth of
localization from an extended (phonon) initial state, and then the
subsequent decay back to that state. Because the generated optical signal
is spatially resolved only if k-vector (crystal momentum) is conserved, a
local mode that breaks the translational symmetry of the solid, thereby
relaxing the condition for k-vector conservation, may cause the fourth

398

beam to spread from a narrow beam to a cone, the cone angle inversely
proportional to the spatial extent of the mode. The use of two sets of
detection optics, one on axis and one off axis, can be used to measure the
cone angle and hence the degree of localization. Within limits mostly
determined by lifetimes and local-mode density, this measurement is
completely unambiguous. For 1-alanine, we know that the lifetimes of the
excited states at $42cm^{-1}$ and at $49cm^{-1}$ exceed nanoseconds at $10K^{16}$, a long
enough time for anything we might hope to observe. We can also estimate
the density by noting that the inverse k-vector for light in 1-alanine is
about 30 unit cells (including the effects of the index of refraction),
and as discussed above, the neutron and Raman data imply a minimum soliton
size of approximately 10 unit cells. Thus the local mode, if it is there
and can be optically excited, is somewhat bigger than 10 unit cells and
smaller than 30. If (for example only) it were formed from regions 30 unit
cells in size, then the density of these objects would be of order $(1/30)^3$
or about 0.003%, a value easily low enough to prevent the formation of
coherent arrays. This is important because coherent arrays would produce
beam-like scattering instead of cone-like scattering. This experiment,
though not yet running, is in the final stages of development.

This work was supported by a grant from the Defense Advanced Research
Projects Administration/DSO and performed under the auspices of the United
States Department of Energy. P.M.Maxton was also supported by the
Associated Western Universities.

References

1)   A.S. Davydov and N.K. Kislukha, Phys. Stat. Sol. (b) 59, 465 (1973);
     A.S. Davydov, J. Theor. Biol. 38, 559 (1973); *Biology and Quantum
     Mechanics* (Pergamon, New York, 1982).

2)   S. Takeno, Prog Theor. Phys. 73 No. 4, 853 (1985).

3)   D.W. Brown, B.J. West, and K. Lindenberg, Phys. Rev. A, 33 No. 6,
     4110, (1986); X. Wang, D.W. Brown, K. Lindenberg and B.J. West, Phys.
     Rev. A 37, 3557 (1988).

4)   W. Rhodes, in press.

5)   B. Mechtly and P.B. Shaw, Phys. Rev. B 38 No. 5, 3075 (1988).

6)   P.S. Lomdahl and W.C. Kerr, Phys. Rev. Lett. 55, 1235 (1985).

7)   D. Hochstrasser, F.G. Mertens, and H. Buttner, Phys. Rev. A 40 No.
     5, 2602 (1989).

8)   S. Yomosa, J. Phys. Soc. Jpn. 53, 3692 (1984); Phys. Rev. A 32, 1752
     (1985).

9)   H. Simpson, Jr. and R. E. Marsh, Acta. Cryst. 20, 550 (1966).

10) The low temperature unit cell dimensions are from P. Vergimini, P.
     Maxton, and A. Migliori (unpublished), Data above 150K are from S.
     Forss, J. Raman Spec. 12 No. 3, 266 (1982).

11) T.J. Kosic, R. E. Cline Jr. and D.D. Dlott, Chem. Phys. Lett. 103 No.
     2, 109 (1983).

12) P. Maxton, J. Eckert, G. Kwei, A. Migliori, M. Field, in preparation.

13) A. Migliori, P. Maxton, A. M. Clogston, E. Zirngiebl, and M. Lowe,
     Phys. Rev. B 38 No. 18, 13464 (1988).

14) Y. R. Shen, 1984, *The Principles of Non-Linear Optics*, New
     York,Wiley.

15) D. D. Dlott, Ann. Rev. Phys. Chem. 37, 157, (1986).

16) T. J. Kosic, R. E. Cline Jr and D. D. Dlott, J. Chem. Phys. 81, 4932
     (1984).

30

SEARCH FOR REMOTE TRANSFER OF VIBRATIONAL ENERGY IN PROTEINS

R. S. Knox,* S. Maiti,* and P. Wu**,§

University of Rochester

Rochester, NY 14627-0011

ABSTRACT

We describe the preliminary stages of an experiment designed to search for long-range transfer of vibrational energy. The experiment, motivated by theoretical interest in solitons and other modes of coherent transport, is of the pump-probe variety and is intended as a means of measuring the time of flight of a vibrational packet, whether or not it has dispersion. While no results can yet be reported, this note highlights some of the practical and fundamental problems associated with the measurement and the application of the principle of coherent transport to real biological systems.

INTRODUCTION

In the context of elementary reaction rate theory, the enhancement of biochemical rates by enzymes is so enormous that highly specific structures and mechanisms must be assumed to be involved. Among the necessary and possible considerations are static and dynamic protein conformations, charge states and specific molecular properties of reactants, and concerted action of two or more enzymes in series or parallel. Nonetheless, after all reasonable things are taken into consideration, the disparity between theory and reality is usually too large. This disparity in typical rate predictions has been estimated [1] as a factor of 10^9 to 10^{12}. The modeling required to attack this general problem is extensive (see, *e.g.*, [2]).

---

* Department of Physics and Astronomy
** Department of Biophysics
§ Present address: Department of Biology, Johns Hopkins University, Baltimore, MD 21218

Davydov's Soliton Revisited, Edited by P.L. Christiansen and A.C. Scott
Plenum Press, New York, 1990

In addition to those based on purely configurational and bonding considerations, mechanisms involving the disposition of excess vibrational energy have been proposed by theorists to help account for high rates of biological activity. A recurring theme is that vibrational energy can somehow avoid the usual diffusive and dispersive processes and bridge a gap in space or time, or, as in the case of allosteric enzymes, cause conformational changes. The accrued advantage in either of these bridging or conformation processes, simply put, would be an increase in the probability of reactants coming together.

In connection with a theory of muscle action Davydov [3] proposed a mechanism for transport of vibrational energy in coherent packets. His theory, which involves nonlinear coupling of the amide-I vibrational modes of protein peptide groups with acoustic vibrations of spines of hydrogen-bonded peptide groups along an alpha helix, has been subjected to increasingly intense development, beginning with the early work of Scott and colleagues[4], and scrutiny, as the contents of the present volume show clearly.

In the Davydov-Scott theory, a nondispersive packet exists only for specific values of the above-mentioned non-linear coupling, and, as long as acoustic vibrations constitute the carrier wave, the speed of propagation was predicted [4] to range from zero to 1.7 nm/ps, which is about 3/8 the speed of sound in the protein. Scott made direct estimates of the size of the nonlinear coupling constants and they seemed attainable in practice. However, interactions of these excitations with the environment of the alpha helix were not included in most of the Davydov-Scott work. Numerous calculations have now been made, and the most recent have rather pessimistic results: "...the lifetime of the Davydov soliton at 300K appears to be at least 2 orders of magnitude too small to allow this soliton to be biologically useful" [5]. Other calculations have reached similar but less drastic conclusions [6-8]. Davydov himself had earlier included thermal effects, using a different approximation [9], reaching a less pessimistic conclusion.

An alternative to the Davydov approach was introduced by Yomosa in 1984 [10]. His lattice-soliton model has been developed by Hochstrasser *et al.* [11], who find "The lifetime of the solitons is finite due to the emission of optical phonons. However, using α-helix parameters, this effect is negligible." We note that packet lifetimes recently predicted are in the range of 1-20 ps, adequate for travel over a mean length of about 2-40 nm if the packet moves at its highest speed.

We are not prepared to claim that any specific enzymatic rate will be explained by invoking the soliton mechanisms. Theoretical work to date has shown only that under ideal conditions they are *possible* mechanisms, physically. If they can be discovered or demonstrated experimentally, they should become the object of wider biophysical attention. Consider, for example, a case in which it is important to maintain a proper sequence of events in a concerted enzymatic process without benefit of direct physical contact between reactants. It would then be es-

sential for one region of a protein to 'know' that a partial reaction had
been completed in another region. The energy being delivered along with
the 'knowledge' might or might not be important in itself.

RATIONALE AND DESIGN OF A TIME-OF-FLIGHT MEASUREMENT

In a variety of amino acid and peptide crystals [12] Raman-active
modes in the region 30-1600 cm^{-1} are seen to decay within 10 ps. Scott
[13] has suggested that some libron modes in protein IR spectra may have
a soliton interpretation. But purely spectroscopic experiments do not
deal with the question of whether energy is taken *from point A to point
B* in a coherent package. Simply finding a line in a spectrum does not locate either point A or point B. If a reaction catalyzed at point B is enhanced by infrared excitation at a spectral feature tentatively identified as a soliton line, point A is still not located, because as an ab-
sorber of light the soliton is *a fortiori* delocalized.

The central theme here is that "points A and B" must be defined at
the outset and a search for an A-to-B propagation effect then made. For
example, A and B might represent chromophores bonded tightly to the alpha
helix at the two ends of a transmembrane protein segment. A membrane is
not essential to the experiment, but it might be needed to stabilize and
disperse the protein. The transmembrane configuration represents one of
the interesting possibilities available: directed energy or information
transfer without transmembrane mass or charge transport.

PHYSICAL CONSIDERATIONS

It is not known how nature excites coherent vibrational packets, if
at all, so the input mode and site are ours to design. Chemical disposition of energy, although originally considered by Davydov as the natural source [3], is not readily "tunable." Optical deposition of the required amount of vibrational energy (0.2 to 0.5eV or 1600-4000 cm^{-1}) at
a single kind of site rules out an infrared short pulse, because IR will
be absorbed at a large variety of sites and modes in a biologically interesting system. We therefore suggest working with optical excitation
in the visible region, where spectral windows exist in proteins. In the
following, we develop one version of the experimental concept.

As shown in Fig. 1(a), a pump pulse which places a donor chromophore,
A, in a highly vibrational excited region of an excited electronic state
is proposed as the source. A packet of vibration is injected into the
transport medium (*e.g.*, alpha helix) with probability g_a and travels to
chromophore B, where the inverse process occurs and the ground state of B
is placed in a high vibrational stated with probability g_b.

Considering the A-B pair as a single unit, one can interpret the
scheme of Fig. 1 as a search for a particular transient hot ground state
population. Such a population might be detected by transient absorption
measurement after excitation, by collecting the fluorescence of B induced by the absorption, or by examining the ground-state excitations

directly by Raman spectroscopy ($e.g.$, [14]). The output would have the
general form of a time-dependent background with the coherent effect superimposed on it, as shown in Fig. 2, where the fluorescence version of
the experiment is assumed. The peak of the signal occurs at the packet
time-of-flight (TOF), which, for a typical transmembrane distance, would
be of the order of at least a few picoseconds (7.0 nm/(1.7 nm/ps)).

In the absorption version of the experiment, the measured quantity
is the varying absorptivity of the sample at the probe pulse wavelength.
We do not develop this method here because the transient absorption of B
is likely to be extremely small in comparison with the background absorption of A in the region of λ_b. Rather we concentrate on the fluorescence
induced in B which can be separated from that of A by use of a cutoff filter for wavelengths longer than λ_b.

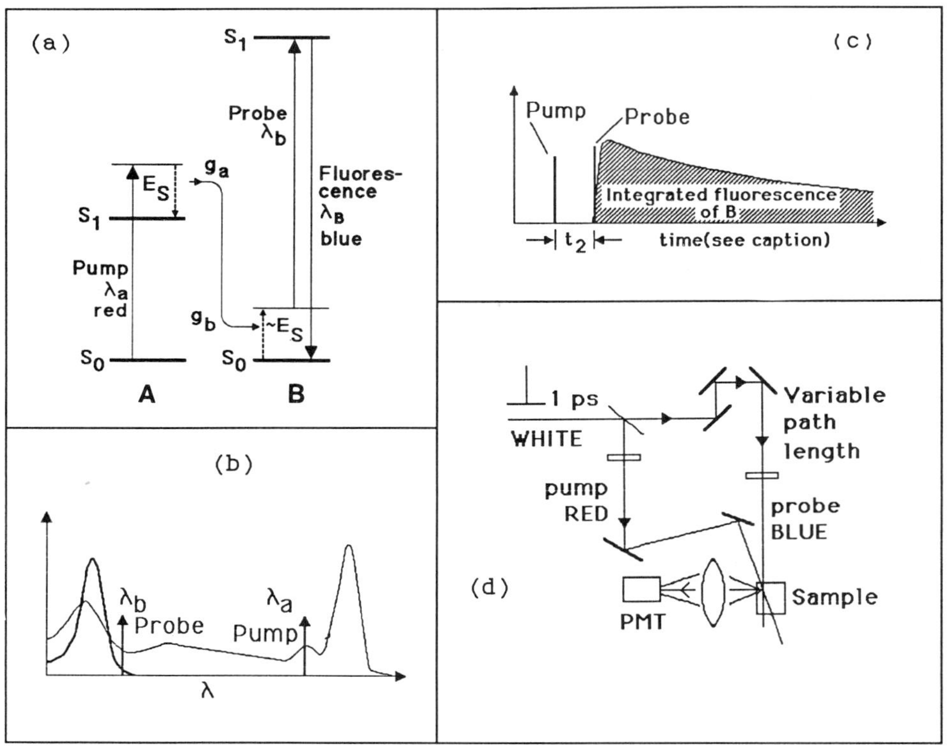

Figure 1. (a) Energy levels of two chromophores A and B (see text). After
the 'red' pump pulse, a 'blue' probe pulse is applied at the time of arrival of
energy E_S which has temporarily excited the ground state of B vibrationally. The
pump may also excite B and the probe certainly reexcites A, both producing background effects. (b) Absorption of A (light curve) and B (heavy curve) in an idealized situation. Blue region at left, red region at right. (c) Real-time sequence of the experiment. The time scale is distorted for convenience. The lifetime of B's fluorescence is very long compared with the pump-probe separation.
(d) The essentials of the optical layout needed to accomplish the task outlined
in 1a through 1c. Pump and probe wavelengths are selected by filters or a dichroic
mirror.

404

The total fluorescence of B will clearly be proportional to B's absorption cross-section σ_B at the probe wavelength λ_b. As vibrational excitation of the ground electronic state of B occurs, the absorption in the tail increases [15] by some factor $h_B \gg 1$. Thus at a time t after the pump pulse

$$\sigma_B(t) = \sigma_B^0[1 + f_A g_a g_b h_B S(t)] \tag{1}$$

where σ_B^0 is the normal absorption cross-section of B at the probe wavelength, $S(t)$ is a shape function of order unity at the peak, f_A is the fractional concentration of excited donors ($\sim 1/2$ at optimum), g_a is the probability of coupling the excess donor energy into the transport mode, and g_b is the probability of coupling it from the transport mode into the acceptor. The delay-time dependence of σ_B can be shown to carry over directly to the total fluorescence induced from B provided that $\sigma_A > \sigma_B$ (A acts as a "buffer"). Therefore, if t_2 is the time of the probe (see Fig. 1c), we will have

$$I_f(t_2) = I_f(\infty)[1 + f_A g_a g_b h_B S(t_2)], \tag{2}$$

where $I_f(\infty)$ is the background of fluorescence from B (*i.e.*, when excited by the probe with no pump prepulse).

Since a reasonable expectation is to detect a signal 0.2% above background (see below), we require

$$f_A g_a g_b h_B > 0.002. \tag{3}$$

In their work on anthracene, Gottfried *et al.* [15] found $h_B \sim 10$. Under optimum conditions $f_A \sim 1/2$, so Eq. (3) implies $g_a g_b > 0.0004$, or, in words, the geometric mean of the packet-to-chromophore coupling factors must be at least 2% for observability. One will certainly not

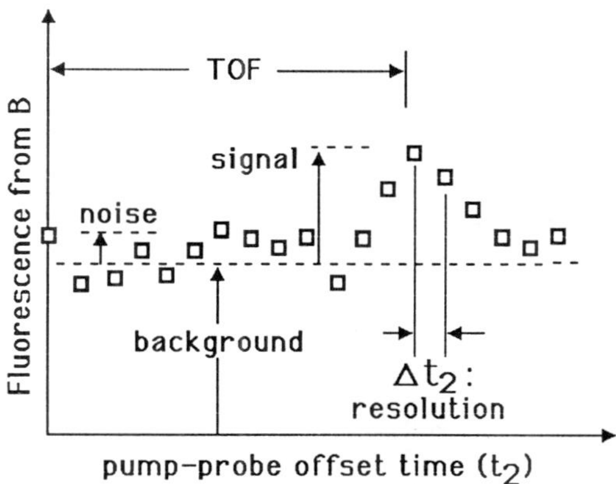

Figure 2. Idealized response for the coherent transport experiment. TOF - time of flight of the vibrational packet. The shape of the signal ideally will be related to the shape of the packet itself, regarded as a probability of coherent excitation at the energy corresponding to the λ_B - λ_b difference. The terms signal, noise, background, and resolution are explained in the text.

immediately find such excellent coupling. The sensitivity can be increased considerably, however, in principle, by lowering the temperature. The factor h_B is the ratio of an artificially induced population to a thermal population. In a homogeneously broadened system it will vary as $\exp(\Delta E/kT)$, where ΔE is the vibrational energy of about 1700 cm^{-1} [4], and therefore, since $h_B \sim 10$ is a room-temperature value, at 100K it would be about 10^8. More realistically, the anthracene in [15] may have been inhomogeneously broadened and in practice we should look for an improvement of an order of magnitude of two over the room-temperature sensitivity. (Work on crystalline anthracene [16] shows that there is an appreciable thermal component in the absorption tail.)

The pulse-probe sequence proposed here is immune from simple resonance energy transfer from A to B in two separate ways. first, the excited state donor-acceptor energies are in reverse order of magnitude, making Förster overlap negligible. Förster transfer will enter in an unspecific way, simple back-transfer from B to A, leading to quenching of B's fluorescence, a problem largely at small distances. Second, the buildup of fluorescence from B we are looking for is a unique temporal phenomenon. Any direct resonant transfer between A and B, small or large, will have a completely different effect as s function of probe offset time.

Note on sensitivity. Each point in the simulated data of Fig. 2 is taken at a particular probe delay time t_2. If there is no signal present, *i.e.*, for a fluorescer otherwise unprepared by a prepulse, the range of fluorescence intensity from B will be determined by the stability of the laser pulse train. In a 10-second run at a 1 kHz repetition rate, there will be 10,000 fluorescent pulses averaged, hence, a statistical improvement of 100 over the intrinsic laser stability. A typical average stability in the blue is 10%, rms, so one can therefore expect a normal variation of (10% /100)=0.1%. For a signal to be seen, let us assume that two standard deviations are required; hence the 0.2% quoted above.

Another consideration is that the time between data points Δt_2 (see Fig. 2) should be less than the expected width of the signal pulse; that width is at least as great as the time it takes for the excitation packet to pass by one amino acid site, or 0.5 nm/(1.7 nm/ps))=0.3 ps. This requires an optical path-length increment of $\Delta L(2) = c\Delta t(2) \sim 100\ \mu m$, easily attainable with good accuracy.

It is not at all clear that the best acceptor molecules will have adequate fluorescence. However, the general principles of this measurement are valid for alternatives based on ultrafast Raman spectroscopy [14] or transient grating methods [17].

The Helix

Emphasis on the protein alpha helix in the theoretical literature
focussed our attention on that structure as a primary experimental tar-
get. Although the natural protein alpha helix is a very difficult transport medium to prepare, it has an extremely important advantage for our
method. We want to be very specific about the beginning and end points
of the energy transport. By using alpha-helix segments whose length and
sequence are known and by performing specific attachments, one might obtain something very close to a sample having a uniform A-B distance.

We search particularly for a stable helical segment of length ∼3 to
10 nm. We also prefer that no proline residues be present in the helical
region since they would tend to bend or break the helix. (Crystal structure, CD, or NMR information can assess the segment helicity.) The ideal
segment will have only one tryptophan or tyrosine residue to act as an
intrinsic acceptor B. We also specify an independently reactive attachment site for chromophore A. We now discuss some possibilities.

Natural Proteins

Glycophorin A, derived from human erythrocyte membranes [18], is
a protein that spans a cell membrane. Its amino acid sequence has been
determined by Tomita and Marchesi [19]. Although to our knowledge the
three-dimensional structure has not yet been determined, the intramem-
branous region is suspected to exist in an alpha-helical conformation.
The helix has been proposed to span anywhere from twenty-three [20] to
thirty-two [19] residues and is bordered at either end by a clustering
of charged amino acids. The isolation of the helix is accomplished by
trypsin cleavage, which clips off much of the carbohydrate-rich N-terminus
section of the protein and some of the COOH terminus section. The section that remains, called the T6A segment, has been shown by CD spectra
to maintain a helical conformation even when isolated from the protein
[18a]. This despite the fact that one of the residues (no. 71) is now
known [21] to be a proline, so that one continuous helix is probably not
present.

There is only one tyrosine residue in T6A, position 93, and no tryptophan present to mask the tyrosine's absorption. The other tyrosine
residues are removed by the trypsin cleavage. The two most probable sites
for the covalent linkage of chromophore A are Met-81 and His-66.

The M13 coat protein spans the cytoplasmic membrane of the M13 bac-
teriophage virus [22-25]. This protein may have a helical conformation
in its hydrophobic region. Under certain conditions, chymotryptic cleavage of the protein produces four fragments when the protein is reconsti-
tuted into micelles (see [23], p. 249). We are interested in the micelle-
bound region consisting of residues Asp-12 through Phe-42 [23]. This
fragment contains only one tryptophan residue [22] to act as chromophore

B. A single fragment phenylalanine residue is conveniently located at
the end of the segment as the probable attachment site for chromophore
A. We have not yet isolated or worked with the M13 protein, but its properties seem quite appropriate [26]. The main difference between the use
of this protein segment and glycophorin is that the original membrane or
lipid vesicle will be necessary to maintain the conformation of the helix
and circumvent its insolubility. The use of vesicles should be avoided
because of their short lifetime and the light scattering they will probably induce.

Synthetic Helix Systems

We have had some preliminary success in isolating the glycophorin
T6A segment and have confirmed its helicity. However, the intricate problem of chromophore attachment and a desire to begin work with a homogeneous amino acid composition have led us to adopt the following strat-
egy. Techniques for attachment are to be perfected using a synthetic homogeneous polypeptide of random lengths; then the techniques are to be
transferred to the case of a similar sample with a specific length. After
assessing the optical results, work is to be continued with the natural
samples.

Our ongoing attempts to produce suitable samples are based on employing a porphyrin as the donor A and either tyrosine or tryptophan as
the acceptor B. At first a "zero-length" sample was constructed from
A and B alone. Methylpyrroporphyrin XXI ethyl ester (designated XXI-
COOEt) was coupled to tyrosine by the DCCI method [27], which may be used
to couple a carboxyl group with an amino group. XXI-COOEt was hydrolyed
to XXI-COOH and combined with o-t-butyl-tyrosine methyl ester (Serva
Biochemicals) to produce the assumed compound shown in Fig. 3. We observe porphyrin fluorescence between 600 and 700 nm upon 280-nm excitation, some impurity fluorescence (400-500 nm) from trifluoroacetic
acid involved in the preparation, and some fluorescence (300-400 nm)
that could be attributed to the spectroscopic-grade chloroform solvent.
No tyrosine fluorescence is detected, probably indicating efficient
Förster transfer to the porphyrin, as expected for short distances. A
second preparation designed to eliminate stacking between the phenol
ring and the porphyrin produced similar results.

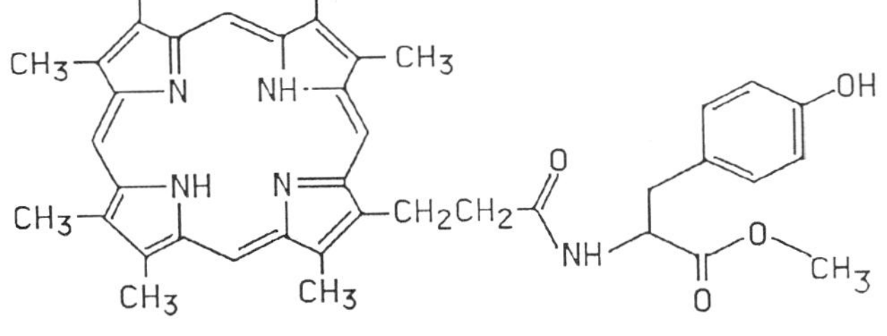

Figure 3. Synthesized compound, representing a combined donor (the porphyrin),
acceptor (the tryosyl), and minimal helical segment (one peptide bond).

The DCCI method has also been used to couple tyrosine to poly-L-
alanine (research grade, average molecular weight 3000, Serva Biochemicals) and then the combination to XXI-COOH. These syntheses were apparently successful but again no fluorescence from the tyrosine was observed. We therefore attempted the DCCI attachment of tryptophan instead
of tyrosine. Unfortunately tryptophan decomposition occurred because of
the strong acidic conditions under which polyalanine is soluble. As this
work continues, we are planning use of a different polypeptide such as
poly(ϵ-protected-lysine).

PROGRESS REPORT AND REMARKS

All experimental efforts, particularly concentrated in the last
year, have gone into sample preparation as described above. As of the
writing of this note, we are not yet sufficiently confident of any sample to initiate any pump-probe experiment. Indeed, we have found at this
stage that obtaining a realistic sample to test the theory is probably at
least as daunting as finding a universally accepted theory.

In our view, the transfer probabilities g_a and g_b persist as a major
theoretical challenge. Figure 4 is meant to help bring out their formal
connection to existing theory of the Davydov soliton and related excitations. It shows two models, collisional transfer (top) and static transfer (bottom). Existing theory concentrates on sites 1 through N, their
primary vibrational excitation (such as the amide-I mode) and associated
phonons which relate to the trapping and localization of that excitation. Coupling to the outside world is generally included indirectly,
for estimates of damping. But it is clear that coupling to the outside
world must eventually be introduced explicitly to produce a model for a
useful soliton in the sense of one which moves energy or information.

Each of the methods of coupling the energy of "A" into the chain
has its advantages and disadvantages. For dispersonless transport, the
state prepared by injection should have a localized distortion that follows the excitation coherently, the whole having non-zero group veloc-

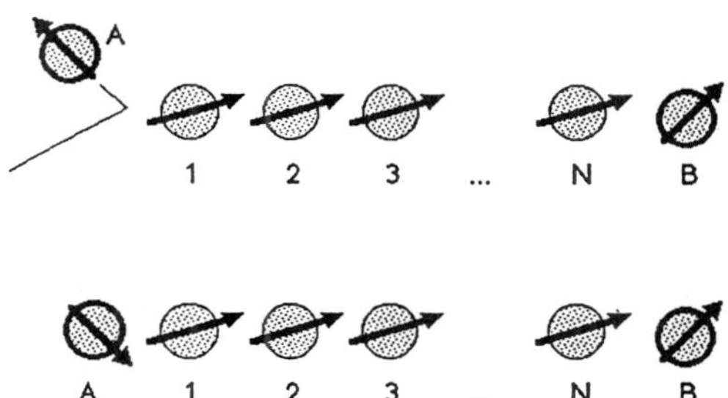

Figure 4. Illustrating the two dynamical methods by which a donor molecule
(A) might impart energy to a chain of molecular sites having coupled localized
excitations indicated by arrows. Upper case, collisional injection. Lower case,
simple resonant injection with no overall motion of "A".

ity. For collisional excitation, we can more easily imagine the launch
of a packet with non-zero velocity because there is a specific initial
momentum to transfer. Moreover, this is the mode of excitation first
envisioned by Davydov. One disadvantage to the collisional excitation
method is that the excitation and the local distortion must enter coherently, a very low-probability event. Another is that if chemical energy
is to be converted, the amount of energy transferred will not be highly
specific in such an inelastic process. Static injection has the advantage of providing, in principle, a specific energy quantum and is "more
easily" tested. However, the disadvantage here is that a local distortion in the vicinity of molecules 1 and 2 must develop spontaneously and
with a finite group velocity.

The extraction of the coherent packet at site N must necessarily involve the inverse of static injection; even the greatest optimist cannot
expect any fundamental process to depend on the accidental arrival of a
colliding "B" at just the right time. It seems essential, therefore, to
evaluate the static transfer process first in any complete theory.

While assessing different mechanisms of coupling, we computed the
Förster resonance transfer rate between C=O bonds. It is frequently remarked that this quantity is small, but we have failed to locate such a
calculation. In order of magnitude, this rate w can be estimated from
$w = (1/\tau)(R_0/R)^6$, where R_0^6 is the mean value of the product $(\lambda/2\pi)^4\sigma$ over
the emission spectrum of the donor and σ is the absorption cross sec-
tion of the acceptor (for this particular formulation see [28]). Inter-
estingly, in the infrared the wavelength is 10 times as great as in the
visible and σ is typically 10^{-4} as large as for electronic transitions;
therefore R_0 tends to have the same order of magnitude as for electronic
excitation transfer ($\sim$ 1 to 5 nm). However, $1/\tau$, the natural radiative
rate, is very small, as estimated from the Ladenburg relation [29]. We
have done a full Förster calculation using existing absorption [30] and
emission [31] data from formic acid and find that $w = 10$ ns^{-1} at $R = 0.54$
nm. This mechanism is therefore not viable for transport, injection, or
extraction, if it must compete with loss processes whose rates are in the
ps^{-1} region.

We believe that the injection and extraction questions raised here
deserve as much theoretical attention as those concerning localized excitation lifetimes. At first glance it appears that the outcome will
do little to raise confidence in the Davydov and related mechanisms for
useful transport. However, this should not deter us from experimental
search for their effects in reasonably conceived systems.

ACKNOWLEDGEMENTS

We are indebted to numerous colleagues for their advice and encour-
agement. We also acknowledge initial support by the Naval Air Systems
Command, contract MDA903-86-C-0057, and further support by the National
Science Foundation, grant DMB-87-03695. Finally, we thank Prof. A. C.

Scott for encouraging the preparation of this article for inclusion in
the Workshop proceedings.

REFERENCES

1. D. E. Koshland, Jr., and E. Neet, Ann. Rev. Biochem. **37**, 359 (1968);
   A. L. Lehninger, *Biochemistry: The Molecular Basis of Cell
   Structure and Function* (Worth Publishers, New York, NY, 2nd ed.,
   1975, pp. 222-230
2. B. S. Green, Y. Ashani, and D. Chipman, eds. *Chemical Approaches
   to Understanding Enzyme Catalysis: Biomimetic Chemistry
   and Transition-State Analogs* (Elsevier, Amsterdam, 1982)
3. A. S. Davydov and N. I. Kislukha, Phys. Stat. Sol. **(b)59**, 465 (1973);
   A. S. Davydov, J. Theoret. Biol. **38**, 559 (1973)
4. J. M. Hyman, D. W. McLaughlin, and A. C. Scott, Physica **30**, 23 (1981);
   A. C. Scott, Phys. Rev. **A26**, 578 (1982)
5. J. P. Cottingham and J. W. Schweitzer, Phys. Rev. Lett. **62**, 1792
   (1989)
6. P. S. Lomdahl and W. C. Kerr, Phys. Rev. Lett. **55**, 1235 (1985)
7. A. F. Lawrence, J. C. McDaniel, D. B. Chang, and R. R. Birge, Phys.
   Rev. Lett. **55**, 1235 (1985)
8. L. Cruzeiro *et al.*, Phys. Rev. **A33**, 4110 (1986)
9. A. S. Davydov, Sov. Phys. JETP **51**, 397 (1980)
10. S. Yomosa, J. Phys. Soc. Japan **53**, 3692 (1984); Phys. Rev. **A32**,
    1752 (1985)
11. D. Hochstrasser, F. G. Mertens, and H. Buettner, Phys. Rev. **A40**,
    2602 (1989)
12. See the summary by T. J. Kosic, E. L. Chronister, R. E. Cline, Jr.,
    J. R. Hill, and D. D. Dlott, in K. B. Eisenthal et al., eds., *Pi-
    cosecond Phenomena III* (Springer-Verlag, N. Y., 1982), pp. 452-
    455
13. A. C. Scott, Physica Scripta **25**, 651 (1983)
14. J. W. Petrich and J.-L. Martin, Chem. Phys. **131**, 31 (1989)
15. N. H. Gottfried, A. Selmeier, and W. Kaiser, Chem. Phys. Lett. **111**,
    326 (1984)
16. I. Nakada, J. Phys. Soc. Japan **20**, 346 (1965)
17. L. Genberg, Q. Bao, S. Gracewski, and R. J. D. Miller, Chem. Phys.
    131, 81 (1989)
18. (a) T. H. Schulte and V. T. Marchesi, Biochemistry **18**, 275 (1979);
    (b) J. P. Segrest et al., Arch. Biochem. Biophys. **155**, 167 (1973)
19. M. Tomita and V. T. Marchesi, Proc. Natl. Acad. Sci. (US) **72**, 2964
    (1975)
20. J. P. Segrest, R. L. Jackson, and V. T. Marchesi, Biochim. Biophys.
    Res. Comm. **49**, 964 (1972)
21. V. T. Marchesi, quoted in L. Stryer, *Biochemistry* (W. H. Freeman,
    San Francisco, third edition, 1988), p. 303
22. See, *e.g.*, D. Kimmelman et al., Biochemistry **18**, 5874 (1979)
23. H. D. Bettman, J. H. Weiner, and B. D. Sykes, Biophys. J. **37**, 243
    (1982)
24. V. F. Asbeck et al., Hoppe-Seyler's Z. Physiol. Chem. **350**, 1047
    (1969)

25. Y. Nakashima and W. Konigsberg, J. Mol. Biol. **88**, 598 (1974)

26. R. Knippers and H. Hoffman-Berling, J. Mol. Biol. **21**, 281 (1966)

27. D. H. Rich and J. Singh, in *The Peptides. Analysis, Synthesis, and Biology* (E. Ross and J. Meienhofer, eds., Academic Press, N.Y., 1979), vol. 1, p. 242

28. R. S. Knox, in *Encyclopedia of Plant Physiology* (L. A. Staehelin and C. J. Arntzen, eds., Springer Verlag, Berlin and Heidelberg, 1986), new series Volume 19, *Photosynthesis III*, p. 286, esp. section 2.2

29. See, *e.g.*, S. J. Strickler and R. A. Berg, J. Chem. Phys. **37**, 814 (1962)

30. R. H. Pierson, A. N. Fletcher, and E. St. Clair Gantz, Analyt. Chem. **28**, 1218 (1956)

31. R. M. Hailey, H. M. Barnes, C. Woodward, and J. W. Robinson, Anal. Chim. Acta **56**, 161 (1971)

Section V

Related Topics

I seen my opportunities and I took 'em.
George Washington Plunkitt

These proceedings are almost entirely devoted to studies of Davydov's soliton which was first proposed in 1973 as a component in the mechanism of muscular contraction [1]. Five years earlier, Fröhlich had suggested [2,3]:

> If the energy fed into the branch of longitudinal electric modes exceeds a critical rate then, under stationary conditions, the excitation energy is channelled into the mode with lowest frequency in a manner typical for Bose condensation.

In both cases, the authors suggest nonlinear and nonequilibrium effects that might help to explain the mysteries of biological dynamics. It is interesting to read these early works and to sense the concepts that were circulating two decades ago. The subsequent influence of Fröhlich's ideas on biodynamics can be appreciated by looking through the **Festschrift** that recently appeared in honor of his eightieth birthday [4].

In the first chapter of this section, Tuszyński considers the relationships that exist between the works of these two scientists. It is a slippery question because Fröhlich's theory is in some sense a "meta-theory." Thus in Reference [2] he asks the reader to

> Assume the system to consist of Z components suspended in a heat bath, and consider each component capable of dipolar electric oscillations. Such a component

may be a repeated dipolar part of a giant molecule (e.g. H-bond) or it may be a
larger unit.

Davydov, on the other hand, considered the specific structure of contractile protein [1]. Thus
it may be that the Davydov model is one example of the class of theories proposed by Fröhlich.

The three following chapters are on a variety of subjects. Brizhik considers the possibility
of detecting Davydov's solitons (both the biological and the superconducting varieties) by slow
neutron scattering. Eremko discusses a mechanism through which Davydov solitons might
be dissociated by electromagnetic radiation. This effect is biologically significant because it
may explain the sharp resonances observed in biological growth rates caused by nonthermal
microwave radiation [5]. It is interesting to note that these biological experiments confirm
work going back to 1968 [6] which was inspired by Fröhlich. The chapter by Giansanti et
al. describes ongoing studies of crystalline acetanilide (ACN) by molecular dynamics. This
work should be considered in the context of experimental results reported in the chapters by
Careri, Barthes, and by Bigio et al.

The last three chapters deal with the dynamics of two resonant dipoles sharing a single
quantum of excitation. The resulting excited dimer, or "excimer", can convert dipole field
energy to mechanical energy and vice versa. The first two papers, by Wu et al. and Kenkre
et al. discuss the effect in the context of chemical experiments on crystals of alpha-perylene.
The last paper, by Scott, considers the exciton as a means for converting the rest energy of a
Davydov soliton into mechanical work.

References

[1] A.S.Davydov, "The theory of contraction of proteins under their excitation," J.theor. Biol.
38, 559-569 (1973) .

[2] H. Fröhlich, "Bose condensation of strongly excited longitudinal electric modes," Phys.
Lett. A, 402-403 (1968).

[3] H. Fröhlich, "Long-range coherence and energy storage in biological systems," Int. J.
Quant. Chem. **2**, 641-649 (1968).

[4] T.W. Barrett and H.A. Pohl (Editors), **Energy Transfer Dynamics**, Springer-Verlag
(1987).

[5] W. Grundler and F. Keilmann, "Sharp resonances in yeast growth prove nonthermal sensitivity to microwaves," Phys. Rev. Lett. **51**, 1214-1216 (1983).

[6] S.J. Webb and D.D. Dodds, Nature (London) **218** 374 (1968); S.J. Webb, ibid. **222**, 1199
(1969).

31

DAVYDOV'S SOLITON AND FRÖHLICH'S CONDENSATION:

IS THERE A CONNECTION?

J.A. Tuszyński

Department of Physics
The University of Alberta
Edmonton, Alberta, Canada, T6G 2J1

ABSTRACT

In this paper we present the physicist's view of a biological cell as envisaged by Fröhlich, from the point of view of membrane organization, and by Davydov, from the energy and particle transport viewpoint. An investigation into the possible interrelationships between these two approaches is given which involves formal similarities and functional dependences. In the former case effective Hamiltonians for the two models are derived and demonstrated to lead to the creation of coherent structures. In the latter case, nonequilibrium considerations result in a scenario where the two phenomena may be mutually functionally dependent. It is also outlined how Fröhlich's dipolar mode coherence could be seen as a Davydov soliton in momentum space.

1. INTRODUCTION

The molecular membrane bounds the cytoplasm of the cell and facilitates communication with the environment. The membrane is essentially a lipid bilayer with a variety of other molecules such as globular proteins and cholesterol, embedded in it.[1] In particular, such macromolecules as α-helices and β-sheets are active channels of the vital trans-membrane transport.[2] The main membrane constituents are phospholipids which consist of a hydrophylic polar head group based on a phosphate and attached through a glycerol moiety to two hydrophobic hydrocarbon chains containing 14-20 carbon atoms. The thickness of the bilayered lamina is 40 to 50 Å and it is a stable, closed structure. It is believed that most of the metabolic activity of the cell is affected by the state of the membrane and that includes such processes as growth, division and energy intake. The actual role played by the membrane and its constituents is still subject to intensive investigations and some of the questions posed are very difficult to answer even in a most qualitative fashion.[3] The first set of questions is related to the cell's organization, function and cell-cell communication and has been addressed by Fröhlich[4] and his followers who employed modern physical concepts of coherence, condensation and long-range interactions to elucidate the biological problem at hand. The second set of questions is related to bioenergetics and the mystery of efficient transport of energy and mass along peptide chains, DNA and other structures which are so important in the cell's activities.

Davydov's Soliton Revisited, Edited by P.L. Christiansen and A.C. Scott
Plenum Press, New York, 1990

This aspect has been investigated by Davydov[5] and many other researchers who picked up his ideas of soliton formation.

The present paper is devoted to the analysis of the question whether the two seemingly independent and uncorrelated models of co-operative phenomena in biological cells are indeed connected. We shall describe both the formal similarities of the underlying mathematical structure as well as some functional relations which pertain to the nonequilibrium conditions for the onset of these effects.

II. THE FRÖHLICH MODEL

Fröhlich[3,4] suggested in his theory of membrane coherence that if the components of a biological system (e.g. membrane) undergo coherent elastic vibrations, then long-range interactions can be strongly excited. This theory applies to systems which satisfy the following three prerequisites:

(1) The existence of a high degree of organization which requires collective properties at the microscopic level.

(2) The existence of a sufficiently strong trans-membrane potential difference which will electrically polarize molecules within the membrane (which itself is an excellent insulator). A potential difference of about (10-100) mV is known to exist across biological membranes with thickness of (4-10) nm. This results in an electric field of $(1-25)10^6$ V/m very strongly polarizing the membrane and possibly inducing an electret state.

(3) The rate of supply of metabolic energy S, must exceed a critical value S_0. It should also be mentioned that ion pumps are also needed to maintain the internal composition of the membrane distinct from that of the environment.[2]

As a starting point for the development of microscopic models of long-range order in biomembranes consider the longitudinal dipole oscillations u_i of their segments (such as the dipolar head groups) described using the Hamiltonian[6]

$$H_o = \frac{1}{2} \sum_{i=1}^{N} m_i (\dot{u}_i^2 + \omega_i^2 u_i^2) - \frac{1}{2} \sum_{i \neq j = i}^{N} u_i V_{ij} u_j \tag{1}$$

where the interaction term was given by Fröhlich[7] as

$$V_{ij} = \gamma^2 e^2 \frac{\sqrt{z_i/m_i} \ \sqrt{z_j/m_j}}{R_{ij}^3 \varepsilon'(\omega)} \tag{2}$$

where z_i is the number of bound charged particles on each head group, R_{ij} is the instantaneous spacing $R_{ij} = r_{ij}^o + u_i - u_j$ and $\varepsilon'(\omega)$ is the real part of the frequency-dependent dielectric constant of the whole medium. Precisely because $\varepsilon' = \varepsilon'(\omega)$, Fröhlich[7] showed for two interacting dipoles, and it was also demonstrated[6] for a regular chain of dipoles, that when dipole oscillations are at resonance ($\omega_i = \omega_o$ for all $1 \leq i \leq N$) the effective interaction energy is of long-range type $(1/R^3)$ while in the other situations it is only of a Van der Waals type $(1/R^6)$. Assuming equal masses of dipoles and expanding R_{ij}^{-3} around equilibrium r_{ij}^o according to

$$R_{ij}^{-3} \simeq (r_{ij}^o)^{-3} [1 - 3(\Delta u_{ij}/r_{ij}^o) + 6(\Delta u_{ij}/r_{ij}^o)^2 + \ldots],$$

where $\Delta u_{ij} = u_i - u_j$, yields for the nearest neighbor approximation the following nonlinear differential-difference equation

$$m_i \ddot{u}_i = f_1 (u_{i+1} + u_{i-1} - 2u_i) + f_2 [(u_{i+1} - u_i)^2 - (u_i - u_{i-1})^2]$$

$$+ f_3 [(u_{+1} - u_i)^3 - (u_i - u_{i-1})^3] + \ldots \tag{3}$$

When $f_3 = 0$ this represents the equation of the famous Fermi-Pasta-Ulam problem which displays a remarkable lack of thermalization of input energy into various modes but, rather, a storage of it in the lowest mode. A quantum mechanical treatment of this problem has been carried out[8] and resulted in a KdV equation derived from the evolution equation for the expectation value of the annihilation operator within a coherent state. Hence, the one-soliton solutions interpreted as a Poisson-distributed superposition of phonons. Another recent study[9] showed that depending on the values of coupling constants f_1, f_2, f_3 one obtains a full range of behavior between the normal modes and local modes which includes sinusoidal, periodic, quasi-periodic and chaotic solutions. Furthermore, Pnevmatikos[10] demonstrated that in the long-wavelength limit eq. (7) with $f_3 \neq 0$ reduces to the generalized Boussinesq equation for u_x. Again, depending on the coefficients of the equation one finds among its solutions both topological and non-topological solitons. In our case, however, not only do we have to evaluate the parameters f_1, f_2, f_3 from ab initio calculations but also investigate the role of the dielectric constant since all of these parameters depend on it. This, therefore, suggests a self-consistent calculation where the solutions of (3) depend on $\varepsilon'(\omega)$ and, in turn, $\varepsilon'(\omega)$ depends on the regime of the solutions since the dipoles are a part of the whole system. A calculation of $\varepsilon'(\omega)$ for biological membranes was undertaken before, but it envisaged the dipoles as harmonic oscillators which are coupled to a heat bath and an energy pump and lacked direct coupling amongst themselves. A more realistic approach must include nonlinear coupling between dipoles using the Hamiltonian (1)-(2).

We already see that within this, rather oversimplified model of the membrane seen as a collection of strongly interacting head group dipoles, there is a scope for condensation phenomena. In fact, both Fröhlich and the score of researchers that followed him (see for example Ref. 11) demonstrated the possibility for a nonequilibrium Bose condensation in the space of dipolar modes.

In terms of a quantum-mechanical description, Wu and Austin[12] introduced the following Hamiltonian to adequately describe the Fröhlich model:

$$H = \sum_i \omega_i a_i^\dagger a_i + \sum_i \Omega_i b_i^\dagger b_i + \sum_i \theta_i p_i^\dagger p_i + \frac{1}{2} \sum_{i,j,k} (\chi a_i^\dagger a_j^\dagger b_k + \chi^* a_j a_i^\dagger b_k)$$

$$+ \sum_{i,j} (\lambda b_i a_j^\dagger + \lambda^* b_i^\dagger a_j) + \sum_{i,j} (\xi p_i a_j^\dagger + \xi^* p_i^\dagger a_j) \tag{4}$$

where $(a_i^\dagger, a_i)$, $(b_i^\dagger, b_i)$ and $(p_i^\dagger, p_i)$ are the cell, heat-bath and energy-pump creation and annihilation operators, respectively. All of these operators are assumed to be of Bose-Einstein type.

It has been demonstrated through a sequence of canonical transformations[13] that the Wu-Austin Hamiltonian can be brought to an effective form

$$H_{eff} = \sum_k W_k a_k^\dagger a_k + \sum_q \Omega_q b_q^\dagger b_q + \sum_i \theta_i p_i^\dagger p_i + \sum_{kk'q} \Delta_{kk'q} a_{k+q}^\dagger a_{k-q}^\dagger a_k a_{k'} \tag{5}$$

where the coefficients W_k, Ω_q, θ_i and $\Delta_{kk'q}$ are "dressed" to include the effects of the various interactions. The effective Hamiltonian H_{eff} basically decoupled the interacting degrees of freedom and included the role of interactions in the dressed coefficients. The effect on the dipole modes is to produce the two-body term that describes mode-mode coupling.

III. THE DAVYDOV MODEL

The other aspect of the cell's activity is the energy and mass transport across its membrane. Davydov[5] attempted to provide an explanation of the remarkably efficient transfer along peptide chains $(H-N-C=0)_n$ of energy produced in an ATP-ADP reaction. In order to explain the apparent lack of thermalization of the $C=0$ excitation due to the dipole-dipole forces between the molecules, Davydov[5] postulated a model involving a strong exciton-lattice coupling in the peptide chain. The Hamiltonian postulated in this context is:

$$H = \frac{1}{2} \sum_n [M\dot{x}_n^2 + K\Delta x_n^2] - \sum_n [DB_n^\dagger B_n + J(B_{n+1}^\dagger B_n + B_{n+1} B_n^\dagger)] \qquad (6)$$

where M is the mass of the peptide group, K is the elastic modulus, $\Delta x_n = r - (x_n - x_{n-1})$ is the deviation from the equilibrium separation r between two adjacent peptide groups. The electronic excitation operators of the $C=0$ bond, $B_n^\dagger$ and B_n, are assumed to obey the Bose-Einstein commutation relations $(B_n, B_m^\dagger) = \delta(n,m)$. The summation over n runs over all sites of the peptide chain. Both exciton constants D and J depend strongly on the instantaneous separation between adjacent sites $(x_n - x_{n-1})$. It has been demonstrated[14] that the normal mode representation for phonons:

$$x_q \equiv (2\pi N)^{-1/2} \sum_n x_n \exp(iqn); \quad p_q \equiv (2\pi N)^{-1/2} \sum_n p_n \exp(iqn) \qquad (7)$$

followed by the introduction of phonon creation and annihilation operators $b_q^\dagger$ and b_q, as

$$b_q \equiv (2M\Omega_q)^{-1/2} (M\Omega_q x_q^\dagger - ip_q); \quad b_q \equiv (2M\Omega_q)^{-1/2} (M\Omega_q x_q + ip_q^\dagger); \qquad (8)$$

and similarly for excitons:

$$a_p^\dagger \equiv (2\Pi N)^{-1/2} \sum_n B_n^\dagger \exp(-ipn); \quad a_p \equiv (2\Pi N)^{-1/2} \sum_n B_n \exp(ipn) \qquad (9)$$

gives the Davydov Hamiltonian as

$$H = V + \sum_q \Omega_q b_q^\dagger b_q + \sum_p \omega_p a_p^\dagger a_p + \frac{1}{2} \sum_{pq} [\chi_{pq}^* a_p^\dagger a_q b_{q-p}^\dagger + \chi_{pq} a_q^\dagger a_p b_{q-p}] \qquad (10)$$

Here, Ω_q is the dispersion relation for the "bare" lattice phonons; ω_p is approximated by:

$$\omega_p \cong D_o - 2J_o(1 + p^2/2) \qquad (11)$$

in the long-wavelength limit and the coupling constant is

418

$$\chi_{pq} = 2(\Omega_{q-p})^{-1/2} \left[1 - e^{i(q-p)}\right] \cdot \left[D_1 + J_1(e^{iq} + e^{-ip})\right] \tag{12}$$

It is important to note that the form of the Hamiltonian H of eq. (10) is identical to that of the Frohlich model[13] when the pump mechanism has been decoupled through a unitary transformation. Moreover, applying a unitary transformation to the Davydov Hamiltonian in normal form, i.e. eq. (10), yields:[14]

$$H_{eff} = V + \sum_q \Omega_q b_q^\dagger b_q + \sum_p W_p a_p^\dagger a_p + \sum_{pqr} \Delta_{pqr} a_{q+r}^\dagger a_{p-r}^\dagger a_p a_q \tag{13}$$

where the exciton eigenfrequency has been shifted to

$$W_p = \omega_p - \sum_q \frac{|\chi_{pq}|^2}{\Omega_o^2}$$

and the effective coupling constant is now

$$\Delta_{pqr} = |\chi_{pq}|^2 \Omega_r^2 / \left(\chi(\omega_p - \omega_q)^2 - \Omega_r^2\right)$$

This form of the Hamiltonian is exactly the same as the one obtained for the Fröhlich problem suggesting important similarities between these two models.

IV. THE FORMATION OF COHERENT STRUCTURES

We have seen in the previous two sections that both the Fröhlich and Davydov Hamiltonians can be effectively described using a generic form of the normal-mode Hamiltonian as:

$$H_{eff} = \sum_{\underline{k},\underline{\ell}} \omega_{\underline{k},\underline{\ell}} q_{\underline{k}}^\dagger q_{\underline{\ell}} + \sum_{\underline{k},\underline{\ell},\underline{m}} \Delta_{\underline{k},\underline{\ell},\underline{m}} q_{\underline{k}}^\dagger q_{\underline{\ell}}^\dagger q_{\underline{m}} q_{\underline{k}+\underline{\ell}-\underline{m}} \tag{14}$$

where the second-quantized operators $q_{\underline{k}}^\dagger$ and $q_{\underline{k}}$ obey Bose-Einstein commutation rules and the interaction coefficients are model dependent. In the Fröhlich model the operators q refer to dipolar modes while in the Davydov model to the exciton modes.

Recently[15] a novel approach has been presented to analyze many-body Hamiltonians of this type, especially close to criticality. In the context of present applications this would mean biological coherence as manifested by a nonequilibrium long-range order in k-space in the membrane or, by the formation of a Davydov soliton in a peptide chain, respectively.

The first step in this method is to derive Heisenberg's equation of motion for $q_{\underline{k}}^\dagger$ and $q_{\underline{k}}$. A standard technique is then used where a quantum field operator is defined by

$$\psi(\underline{r}) = \Omega^{-1/2} \sum_{\underline{k}} e^{-i\underline{k}\cdot\underline{r}} q_{\underline{k}} \tag{15}$$

Each of the coefficients in equation (14) is then expanded in a Taylor series about the critical point to second order and the result written in terms of the field ψ. There are four main cases which have been found:

(i) <u>Non-interacting particles</u>

Here Δ in (14) is zero and ω is expanded to second order. Following a co-ordinate transformation and subsequent rescaling we find the equation of motion in the form of a linear Schrödinger equation. No coherence is resulting in this case.

(ii) <u>Zeroth order</u>

In this case Δ is expanded to zero order only (taken as a constant) and ω again to second order. Non-linearity first appears in this situation as a result of the interaction. Following a similar procedure as in (i) the equation of motion becomes the cubic non-linear Schrödinger equation in either Euclidean or Minkowski space depending on the case

$$i\hbar\partial_t\psi = \nu_0\psi + i\underline{\nu}_1\cdot\nabla_\varepsilon\psi - \frac{1}{2}\nabla_\varepsilon^2\psi + \Omega f(\underline{\eta}_0,\underline{k}_0,\underline{m}_0)\,\psi^\dagger\psi\psi \tag{16}$$

where $(\underline{\eta}_0,\underline{k}_0,\underline{m}_0)$ is the position of the critical point in reciprocal space and for convenience we have defined

$$f(\underline{\eta}_0,\underline{k}_0,\underline{m}_0) = 2\Delta_{\underline{\eta}+\underline{m}-\underline{k},\underline{k},\underline{m}}$$

In (16) we have used the Laplace-Beltrami operator ∇_ε^2, defined in three dimensions, by

$$\nabla_\varepsilon^2 = \varepsilon_1\partial_{xx}^2 + \varepsilon_2\partial_{yy}^2 + \varepsilon_3\partial_{zz}^2$$

and $\varepsilon_i = \mp 1$ with a similar definition for ∇_ε. In (16) ν_0 is a constant and $\underline{\nu}_1$ a constant vector determined by the expansion.

(iii) <u>First order</u>

Following (i) and (ii) Δ is here expanded to first order and we find

$$i\hbar\partial_t\psi = \nu_0\psi + i\underline{\nu}_1\cdot(\nabla_\varepsilon\psi) - \frac{1}{2}\nabla_\varepsilon^2\psi + \nu_2\psi^\dagger\psi\psi + i\underline{\nu}_3\cdot\psi^\dagger\big(\psi(\nabla\psi) + (\nabla\psi)\psi\big) \tag{17}$$

(iv) <u>Second order</u>

We obtain in this case

$$i\hbar\partial_t\psi = \mu_0\psi + i\underline{\mu}_1\cdot(\nabla_\varepsilon\psi) + \mu_2\nabla_\varepsilon^2\psi + \sum_{i,j} R_{ij}\partial_{x_ix_j}^2\psi - 2(\nabla_\varepsilon\psi^\dagger)\psi(\nabla_\varepsilon\psi)$$

$$+ \mu_3\psi^\dagger\psi\psi + i[\psi^\dagger\psi(\underline{\mu}_4\cdot\nabla_\varepsilon)\psi + \psi^\dagger\big((\underline{\mu}_4\cdot\nabla_\varepsilon)\psi\big)\psi] + \big((\nabla_\varepsilon^2\psi^\dagger)\psi\psi + \psi^\dagger\psi\nabla_\varepsilon^2\psi\big) \tag{18}$$

where R_{ij} is a small residual term which vanishes identically in spherical or cubic symmetry, μ_0, μ_2 and μ_3 are constants and $\underline{\mu}_1$ and $\underline{\mu}_4$ are constant vectors resulting from the expansion we have performed.

It should be noted that the demonstrated direct connection between two-body interactions and non-linearity, and especially the resultant generalized nonlinear Schrödinger equation has been hinted at before.[13,14] Moreover, the effective Hamiltonian, when recast in terms of the field ψ produces the celebrated Landau-Ginzburg-Wilson Hamiltonian in the zeroth order.

$$H_{LGW} = a\psi^{\dagger}\psi + b\psi^{\dagger}\psi^{\dagger}\psi\psi + c\nabla\psi^{\dagger}\cdot\nabla\psi \tag{19}$$

which has played a very prominent role in the development of field theoretic approaches to critical phenomena. First and second order Hamiltonians would, in addition, have terms of the type

$$(\nabla\psi^{\dagger})\psi, \quad (\nabla\psi^{\dagger})\psi^{\dagger}\psi\psi, \quad (\nabla^2\psi^{\dagger})\psi, \quad (\nabla^2\psi^{\dagger})\psi^{\dagger}\psi\psi, \quad (\nabla\psi^{\dagger})\cdot(\nabla\psi^{\dagger})\psi\psi$$

and their Hermitian conjugates.

The approach to solving the above equations of motion is a standard one in quantum field theory[16] and consists in assuming that $\psi = \phi + \Lambda$ where ϕ is the classical component of the field operator ψ and Λ is a small quantum correction. Thus, initially we retain only the classical part of ψ and solve the resulting equations for ϕ.

It has been demonstrated[15] that localised solutions (solitons) may exist in all the orders starting from the zeroth under special conditions. Moreover, quantum corrections can readily be found by linearizing the equation of motion for ψ about the classical field ϕ. The result is a Schrodinger equation for Λ which is analytically solvable in a great number of situations.

Finally, it is transparent that sufficiently strong interactions among the particles lead to strong non-linearity which in turn produces effects of localization and coherence.

V. FRÖHLICH'S CONDENSATION AS A DAVYDOV SOLITON

In a very recent paper a model Hamiltonian which is a generalization of Fröhlich's concepts, has been postulated for a biological membrane.[17] This accounts for the presence of both the hydrocarbon tails and the lipid head groups with their vastly different orders of relaxation times. The tail Hamiltonian is given as:

$$H_T \cong \sum_n \sum_\alpha \left[\frac{P^2_{n,\alpha}}{2m_n} + \frac{k_\alpha}{2}(x_{n,\alpha} - x_{n\pm1,\alpha})^2 + (\frac{E_{o\alpha}}{a^4_\alpha})(x^2_{n,\alpha} - a^2_\alpha)^2 \right] \tag{20}$$

where n enumerates tail, α runs over their degrees of freedom, m_n is the tail's mass, $p_{n,\alpha}$ its momentum and $x_{n,\alpha}$ its position coordinates. It is subsequently shown that above the gel-liquid crystalline phase transition (i.e. within physiological temperatures) one can approximate:

$$H_T \cong \sum_{q,\alpha} \Omega_{q,\alpha}\left(A^{\dagger}_{q,\alpha}A_{q,\alpha} + \frac{1}{2} \right) \tag{21}$$

where $A_{q,\alpha}$ and $A^{\dagger}_{q,\alpha}$ are chain phonon annihilation and creation operators, respectively, and $\Omega_{q,\alpha}$ are the eigenfrequencies. Within the head groups, the O-N dipoles vibrate at microwave frequencies and interact through Eq. (1) where V is given by (2). It is then demonstrated that the head group Hamiltonian becomes:

$$H_H = \sum_k \tilde{\omega}_k a^{\dagger}_k a_k + \sum_{k'k''k'''} \chi_{k'k''k'''}(A_{k'}a_{k''}a^{\dagger}_{k'''} + A^{\dagger}_{k'}a^{\dagger}_{k'''}a_{k''}) \tag{22}$$

where the first term is the energy of normal "bare" dipole modes in the head groups while the second describes the coupling between tails and heads for the expansion of Hamiltonian (1) in the normal modes.

It is subsequently shown that with the inclusion of dispersion of energy in the head groups through hopping of strength J the equation of motion for the field translation

$$\alpha_k \equiv \langle\psi|a_k|\psi\rangle$$

where $|\psi\rangle$ is the ansatz wavefunction for the membrane is

$$i\hbar\,\dot{\alpha}_k = \tilde{\omega}_k\alpha_k - J(\alpha_{k-1} + \alpha_{k+1}) - \frac{2\chi_k^2}{\Omega_k}\alpha_k^*\alpha_k\alpha_k \tag{23}$$

which is virtually the discrete version of the nonlinear Schrödinger equation with a continuum limit solution of

$$\alpha_k(0) = \alpha_o \, \text{sech}[\lambda(k - k_o)] \tag{24}$$

where

$$\alpha_o = \pm[-(E_s - \tilde{\omega}_o)\,\Omega_o/\chi_o^2]^{1/2}$$

E_s is the soliton energy and

$$\lambda = [-(E_s - \tilde{\omega}_o)J]^{1/2}$$

This can be interpreted as a Davydov soliton for the membrane in the momentum   space. Its width is proportional to $\sqrt{J}$ and it is centred at k_o. This can be seen as a manifestation of Bose Condensation amongst the dipole modes of the membrane. This is a nonequilibrium effect very similar to the laser action and it requires continuum pumping at or above a critical rate.

VI. FUNCTIONAL INTERDEPENDENCE

In an independent series of papers Del Giudice et al[18,19] presented both the Fröhlich and Davydov models using modern concepts of Quantum Field Theory. They have shown that the two models may correspond to two regimes of a quantum field and each is a manifestation of a broken symmetry in the system. This closely corresponds to the presentation given in Sec. IV of the present paper.

A very interesting idea that also agrees with some of our results is that both the Davydov and Fröhlich phenomena may be interdependent. Del Giudice et al[18,19] pointed out that the activation energy of 0.2 eV for Davydov solitons is in fact of the same order of magnitude as the energy of an electret state in biopolymers.[20] They concluded that the coherent electric dipole wave in a membrane, as predicted by Fröhlich, may in fact be brought about by the presence of Davydov solitons on the peptide chains located within the membrane, through the induction of an electret state in the water of hydration.

Conversely, one may imagine that condensation of excitonic modes into a Davydov soliton may be greatly facilitated by the presence of an electret state of the water and a polarized state of membrane dipoles, provided the nonequilibrium transition has occurred into a Fröhlich state of the membrane. We therefore see that the two phenomena may be interdependent mutually facilitating (or controlling) the existence of its counterpart.

In a particular calculation[21] of coagulation rates of red blood
cells, it was found that a possible cause of anomalous rouleau formation
could be an occupation inversion of the dipole energy levels as advocated
by Fröhlich. Using the numbers presented in that paper[21] it is easy to
calculate the energy required for the occupation inversion to be on the
order of 10^{-15} J. This is not an unreasonably high amount of energy for
a biological cell to accumulate. It could, for example, be obtained
through the supply of ATP resulting via the process of hydrolysis in the
creation of approximately 10^5 elementary units of energy (0.205 eV each).
The energy released in the ATP-ADP transformation can be transmitted
losslessly through the membrane via Davydov solitons which may polarize
the membrane and create an electret state as pointed out by Del Giudice
et al.[18,19] This would indicate that the Fröhlich theory and the Davydov
theory play complementary roles in inducing a metabolic state in the
biological cell. Within this scenerio a drop of the coagulation rate
would come about as a result of insufficient supply of energy in the form
of Davydov solitons leading to depolarization of the membrane, a subsequent reversal of the occupation levels to their thermodynamic equilibrium values, and eventual elimination of the long-range force.

Conversely, the role of thermal and dissipative effects in the
existence and stability of Davydov solitons, which is in the focus of
this conference, may be significantby influenced by such conditions as
the state of surrounding biomolecules. If, for example the peptide
chains considered are within a polarized membrane and surrounded by water
of hydration in an electret state, the net effect may be that the "effective" temperature "felt" by the chain is much lower than the actual temperature. In other words, the question of soliton stability should be
addressed within the molecular environment of the chains along which they
propagate.

V. CONCLUSIONS

In this paper we have briefly examined the question of a possible
connection between the Fröhlich model of membrane coherence and the
Davydov model of soliton formation in molecular chains. We have shown
that there exists a unifying mathematical formalism in the form of the
effective Hamiltonian. It has also been pointed out that the Fröhlich
condensed state in the space of membrane dipoles may be viewed as a
Davydov soliton in momentum space. Finally, a functional interdependence
has been postulated which may be very important in nonequilibrium phenomena of metabolic processes in a living cell.

ACKNOWLEDGEMENTS

This research has been supported by a grant from NSERC.

REFERENCES

1.   B. Chance, P. Mueller, D. De Vault and L. Power, Phys. Today 80:32
     (1980).
2.   D.A. Eisner and S.C. Wray, Contemp. Phys. 26:3 (1985).
3.   H. Fröhlich, Adv. Electron. Electron. Phys. 53:85 (1980).
4.   F. Fröhlich, IEEE Trans., MIT 26:613 (1978).
5.   A.S. Davydov, "Solitons in Molecular Systems," D. Reidel, Dordrecht
     (1985).
6.   J.A. Tuszyński, Phys. Lett. A107:225 (1985).
7.   H. Fröhlich, Phys. Lett. A39:153 (1972).

8.   Y.H. Ichikawa, N. Yajima and K. Takano, <u>Prog. Theor. Phys.</u> 55:1723
     (1976).
9.   A.C. Scott, P.S. Lomdahl and J.C. Eilbeck, <u>Chem. Phys. Lett.</u> 113:29
     (1985).
10. S.N. Pnevmatikos, Solitons in nonlinear atomic chains, in: "Singu-
     larities and Dynamical Systems," S.N. Pnevmatikos, ed., Elsevier
     Science, Amsterdam (1985).
11. R. Paul, <u>Phys. Lett.</u> A96:263 (1983).
12. T.M. Wu and S. Austin, <u>Phys. Lett.</u> 64A:151 (1977).
13. J.A. Tuszyński, R. Paul, R. Chatterjee and S.R. Sreenivasan, 30:2666
     (1984).
14. J.A. Tuszyński, <u>Int. J. Quant. Chem.</u> 29:379 (1986).
15. J.A. Tuszyński and J.M. Dixon (1989), submitted to <u>Phys. Lett. A.</u>
16. R. Jackiw, <u>Rev. Mod. Phys.</u> 49:681 (1977).
17. H. Bolterauer and J.A. Tuszyński (1989), submitted to <u>J. Biol. Phys.</u>
18. E. Del Giudice, S. Doglia, M. Milani and G. Vitiello, <u>Nucl. Phys.</u>
     B251:375 (1985).
19. E. Del Giudice, S. Doglia, M. Milani and G. Vitiello, <u>Nucl. Phys.</u>
     B275:185 (1986).
20. J.B. Hasted, H.M. Millany and D. Rose, <u>J. Chem. Soc. Farraday Trans.</u>
     77:2289 (1981).
21. J.A. Tuszyński, <u>J. Biol. Phys.</u> (1989), in press.

Question by Kenkre: You presented as one of your conclusions, that Fröhlich condensation corresponded to having a large number of Davydov solitons. However, as support you mentioned the need, for Fröhlich condensation, of high pumping. Does is not mean at it is not a large number of Davydov solitons but only a large value of Q (in the notation of Kerr in his talk at this conference, or of Lindenberg), i.e. a large number of excitations, that Fröhlich condensation requires? How can this be a real convection between Fröhlich and Davydov (except that similar or identical Hamiltonians are involved)?

Reply: Indeed, one of the requirements for Fröhlich's condensation is a sufficient supply of energy and it does not necessarily mean it to be in the form of Davydov solitons. However, as this workshop's proceedings demonstrated, it is rather difficult to come up with an alternative mechanism of energy transfer in molecular chains which would not imply very short relaxation times. On the question of a relationship between the Fröhlich and Davydov models we see first of all mathematical similarities through the use of very similar Hamiltonians. Secondly, Fröhlich's Bose condensation of dipolar modes may be envisaged as a Davydov soliton in momentum space. Finally, the occurrence of one phenomena may be a prerequisite for the other.

32

THE SOLITON AND BISOLITON INPUT INTO THE ELASTIC SCATTERING OF SLOW NEUTRONS

Larisa Brizhik

Institute for Theoretical Physics

Academy of Sciences of the Ukrainian SSR, Kiev

The wide use of the Davydov soliton model [1] in bioenergetics [2] and superconductivity [3,4] raises a question of direct experimental evidence for the existence of solitons and bisolitons in quasi-one-dimensional systems and investigation of their properties. As far as both solitons and bisolitons are localized electrons or intermolecular excitations [1] or their bound state [5], respectively, moving along the chain with its local deformation, it is natural to expect their influence on the scattering spectra of extra radiations, including slow neutrons. Here elastic coherent scattering of ultracold neutrons by the Davydov soliton or bisoliton propagating with velocity V along the chain of similar subunits (peptide groups, e.g.) is studied in the assumption of small role of thermal vibrations. If the neutron energy is of order 10^{-5}–10^{-7} eV, then its de Broglie wavelength (10^{-7}–10^{-8}m) exceeds the (bi)soliton size, which is usually several chain units. That is why nuclei of the same isotope from different subunits scatter neutrons coherently provided the (bi)soliton velocity is small. The second important circumstance arises because the (bi)soliton kinetic energy is large compared with that of an ultracold neutron. So the (bi)soliton velocity change in the result of a collision with a neutron may be neglected, and one can study elastic coherent scattering only.

The Hamiltonian of the system being studied has the form of a sum

$$H = H_0 + V, \tag{1}$$

the first of which is the sum of Hamiltonians of a free neutron, virtual phonons interacting with an electron or exciton and resulting in a (bi)soliton, and nuclei thermal vibrations; and the second one describes the neutron interaction with the chain:

$$V = -\frac{2\pi\hbar^2}{m_n} \sum_{\ell m} A_\ell \sigma(\vec{r} - \vec{R}_{\ell m}). \tag{2}$$

Here $\sum_\ell$ means the summation over atoms of type ℓ, belonging to the same unit cell, $\sum_m$ means the summation over chain units, m_n is the neutron mass, A_ℓ is the amplitude of neutron coherent scattering by atoms of type ℓ, and $\vec{R}_{\ell m}$ is the site of the ℓth atom in the region of the mth unit. Thus

$$\vec{R}_{\ell m} = m\vec{a} + \vec{a}_\ell + \frac{\beta\vec{a}}{a}[1 - \tanh(\xi_{\ell m} - \mu V t)], \qquad \xi_{\ell m} = [(m - m_0)a + a_\ell], \tag{3}$$

in which $\vec{a} = (a, 0, 0)$ (a is the equilibrium distance between nearest unit cells of the chain oriented along the x-axis), and $\vec{a}_\ell = (a_\ell, 0, 0)$ (a_ℓ is the ℓth atom site in equilibrium). (Bi)soliton

Davydov's Soliton Revisited, Edited by P.L. Christiansen and A.C. Scott
Plenum Press, New York, 1990

parameters β and μ are determined through the chain parameters, namely,

$$
\begin{aligned}
\beta &= \beta_0, \quad \mu = g/2 \quad \text{for the soliton} \\
\beta &= 2\beta_0, \quad \mu = g \quad\;\; \text{for the bisoliton}
\end{aligned}
\tag{4}
$$

where

$$
\beta_0 = \frac{\chi}{w(1-s^2)}, \qquad g = \frac{\chi^2}{2Jw(1-s^2)}, \qquad s = \frac{V}{V_{ac}}
\tag{5}
$$

(χ is the electron-phonon interaction constant, w is the chain elasticity, J is the resonance or exchange integral, V_{ac} is the sound velocity in the chain). The part of the displacement caused by nuclei thermal vibrations is omitted in (3) for, as is well known [6], it leads to the Debye-Waller factor $\exp(-2W)$ in the expression for scattering cross-section.

Thus, according to (1)–(3), the system Hamiltonian depends on time and we shall use in what follows the time-dependent theory of scattering [7]. Its application to the case of neutron scattering by a soliton was described in detail in [8], so we shall write here only its scheme.

The differential cross-section of elastic neutron scattering by the chain is given by the formula

$$
\Delta\sigma = \sum_b \frac{P_{ba}(t)}{F_n}
\tag{6}
$$

in which $P_{ba}(t)$ is the probability of the system to change its state from initial $\psi_a(t)$ into finite $\varphi_b(t)$ at time t: $P_{ba}(t) = |\langle\varphi_b(t)|\psi_a(t)\rangle|^2$, $\sum_b$ means summation over the finite states of scattered neutrons, and F_n is the density of incident neutron flux. The wave function $\psi_a(t)$ is determined by the system evolution operator

$$
\psi_a(t) = \exp(-iH_0t/\hbar)U(t,t_0)\psi_a(0),
\tag{7}
$$

which satisfies the integral equation

$$
U(t,t_0) = 1 - \frac{i}{\hbar}\int_{t_0}^{t} dt'\tilde{V}U(t',t_0)
\tag{8}
$$

where $\tilde{V}(t)$ is the operator (2) in the interaction representation.

Calculating the amplitude of transition probability to first order in the perturbation theory under the assumption that for the time of neutron passing through the chain (bi)soliton will change its position slightly, and substituting this amplitude into (6)–(8) gives

$$
\Delta\sigma = \Delta\Omega_b \exp\left[-2W\right]\left|\sum_{\ell m} A_\ell\left\{\exp\left[i\varphi q(ma + a_\ell + \beta)\right] + \right.\right.
$$
$$
\left.\left. + \frac{\hbar\beta\psi}{2a\varphi}\mu V\frac{dq}{dE}\operatorname{sech}^2\xi_{\ell m}(1 - iq\beta\varphi\tanh\xi_{\ell m})\right\}\exp\left[-iq\beta\varphi\tanh\xi_{\ell m}\right]\right|^2.
\tag{9}
$$

Here q is the modulus of neutron impulse, $\Delta\Omega_b$ is the angular spread of scattered neutrons, φ and ψ are values defined by incident (α) and scattering (ϑ) angles by the relations

$$
\varphi = \cos\alpha - \cos(\vartheta + \alpha),
\tag{10}
$$

$$
\psi = \cos\alpha + \cos(\vartheta + \alpha).
\tag{11}
$$

The first term in (9) is responsible for the scattering by a (bi)soliton at rest. One can show that it consists of two parts, $\Delta\sigma_0 = \Delta\sigma_B + \Delta\sigma_{s,0}$, one of which, $\Delta\sigma_B$, describes neutron scattering at Bragg angles by a chain without the (bi)soliton, and another, $\Delta\sigma_{s,0}$, is the input from scattering by a (bi)soliton at rest. It is possible to sum it in a way similar to [9], thus

$$
\Delta\sigma_{s,0} = \left|\sum_\ell A_\ell\right|^2 \tfrac{1}{6}\beta^2 q^2\varphi^2\left[5 + \cosh\frac{\hbar u}{\mu}\right]\left\{1 + \left[\frac{2}{u} - \frac{\hbar}{\mu\sinh\frac{\hbar u}{2\mu}}\right]^2\right\}\exp[-2W]\Delta\Omega_b
\tag{12}
$$

where $u = qa\psi - 2\pi m$ (m is the number of the Bragg peak).

The second term in (9) is proportional to the (bi)soliton velocity. It can be estimated in the continuum approximation as

$$\Delta\sigma_{s,V} = \sum_{\ell} |A_\ell|^2 \frac{\hbar m_n \psi}{\hbar q^2 \mu \varphi^2 a} V \exp[-2W] \coth[\cosh^{-1} \sqrt{\beta\mu}]\Delta\sigma_B. \tag{13}$$

It follows from (12)–(13), that the (bi)soliton input into the differential cross-section does not depend on the (bi)soliton initial position and increases with increasing localization. The part of the cross-section, $\Delta\sigma_{s,0}$, determined by a (bi)soliton at rest, has extra maxima situated between the Bragg ones, and it increases exponentially with the divergence of scattering direction from the Bragg angle. Another part of cross-section, $\Delta\sigma_{s,V}$, as can be seen from (13), has a sharp maximum in the direction of neutron mirror reflection from the chain.

Thus, one can expect that quasi-one-dimensional systems sustaining Davydov solitons or bisolitons have the special features of slow neutron scattering spectra indicated above. In this connection it is interesting to consider Reference [10] in which an extra peak in the spectrum of neutron scattering on small angles by the superconductor $ErRh_4B_4$ at low temperatures (< 1 K) was registered.

References

[1] A.S. Davydov and N.I. Kislukha, Solitons in one-dimensional molecular chains, **phys. stat. sol. (b)** 75: 735 (1976).

[2] A.S. Davydov, Solitons in molecular systems, **Physica Scripta** 20: 387 (1979).

[3] L.S. Brizhik and A.S. Davydov, Nonlinear theory of conductivity in quasi-one-dimensional molecular crystals, **Sov. Fiz. Nizk. Temp.** 10: 358 (1984).

[4] L.S. Brizhik and A.S. Davydov, Soliton mechanism of superconductivity in organic quasi-one-dimensional crystals, **phys. stat. sol. (b)** 143: 689 (1987).

[5] L.S. Brizhik and A.S. Davydov, Binding of electrosolitons in soft molecular chains, **Sov. Fiz. Nizk. Temp.** 10: 748 (1984).

[6] A. Akhiezer and I.Pomeranchuk, "Some aspects of the nuclei theory", Cos. Izd. Techn.-Teoret. Lit., Moscow-Leningrad (1950).

[7] M. Goldberger and K. Watson, "The theory of collisions", Mir, Moscow (1967).

[8] L.S. Brizhik, The theory of scattering of ultracold neutrons by Davydov's solitons, **Ukr. Fiz. Zh.** 29: 492 (1984).

[9] V.M. Adamyan and A.L. Mitler, The calculation of soliton input in cross-section of neutron scattering by molecular chains, **Ukr. Fiz. Zh.** 26: 205 (1981).

[10] D.E. Moncton, D.B. Whan, and P.H. Schmidt, Oscillatory magnetic fluctuations near the superconductor-to-ferromagnet transition in $ErRh_4B_4$, **Phys. Rev. Lett.** 45: 2060 (1980).

33

DISSOCIATION OF DAVYDOV SOLITONS BY ELECTROMAGNETIC WAVES

A.A. Eremko

Institute for Theoretical Physics
Academy of Sciences of the Ukrainian SSR, Kiev

INTRODUCTION

In the 1970-s Davydov proposed a soliton mechanism of energy transport in biological macromolecules [1]. At present the concept of the "Davydov soliton" has been used in many studies. In particular, the Davydov soliton has been put into operation as a mechanism useful in the description of a variety of biological phenomena [2,3]. Numerical studies by Scott and co-workers [2-4] supported the appearance of Davydov solitons in Alpha-helix protein molecules. Therefore an experimental detection of solitons appear to be natural.

In the theoretical research reported in this article the absorption spectra of electromagnetic waves by a molecular chain, in which a soliton excitation is present, has been calculated.

THEORY

According to a general perturbation theory the probability (per time unit) of electromagnetic radiation absorption summarized over all finite states of a chain is as follows

$$P=\frac{2}{\hbar}\mathrm{Re}\int \langle \Psi_1(t) | H_{int}^{+}(t)e^{-i\frac{H}{\hbar}(t-t')}H_{int}(t') | \Psi_i(t')\rangle dt' \qquad (1)$$

where $|\Psi_i(t)\rangle$ is the initial state of a molecular chain described by Hamiltonian H, and H_{int} is the operator of interaction between the

Davydov's Soliton Revisited, Edited by P.L. Christiansen and A.C. Scott
Plenum Press, New York, 1990

molecular chain and the electromagnetic field.

If $\vec{E}(t)=\vec{E}_0 \exp(-i\,wt)+compel.conj.$ is an alternating electric field of electromagnetic radiation and $\overline{M}$ is the dipole moment of the system than in the dipole approximation, the interaction operator H_{int} is

$$H_{int}(t)=-\vec{E}(t)\vec{M}. \qquad (2)$$

Let us consider only one isolated nondegenerated band of Amid vibrations in Alpha-helix. In this case the system may be described by Fröhlich-type Hamiltonian operator of the form

$$H = \sum_k \mathcal{E}(k) B_k^+ B_k + \frac{1}{\sqrt{N}} \sum_{kq} \chi(q) B_k^+ B_{k-q} (b_q + b_{-q}^+) + \sum_q \hbar\Omega_q b_q^+ b_q. \qquad (3)$$

In this expression

$$\mathcal{E}(k) = \mathcal{E}_0 + 4J\sin^2(ka/2) \approx \mathcal{E}_0 + \hbar^2 k^2/2m, \quad m = \hbar^2/2Ja^2 \qquad (4)$$

is the energy of excitons corresponding to intrapeptide vibration (for example Amid-I) and

$$\Omega(q) = 2\sqrt{w/M}\ \left|\sin\frac{qa}{2}\right| \approx v_0|q|\ , \quad v_0 = a\sqrt{w/M} \qquad (5)$$

is the frequency of longitudinal acoustic phonons (k is a wave number).

The interaction between intrapeptide excitations and acoustic phonons is characterized by a coupling function

$$\chi(q) = 2i\chi \left[\frac{\hbar}{2M\Omega_q}\right]^{1/2} \sin qa \approx 2i\chi \left[\frac{\hbar}{2Mv_0}\right]^{1/2} \frac{q}{\sqrt{|q|}}\ . \qquad (6)$$

The exciton creation operator B_k^+ is connected with the creation operator of the intrapeptide excitation on site n by the unitary transformation

$$B_k^+ = \frac{1}{\sqrt{N}} \sum_n e^{ikan} B_n^+\ .$$

We assume that at the initial time the polypeptide chain is in the soliton state described by the wave function [5]

$$|\Psi_i(t)\rangle \equiv |\Psi_s(t)\rangle = e^{-i\mathscr{E}_s t/\hbar} \frac{1}{\sqrt{N}} \sum_k e^{-ikvt} \Psi_s(k) B_k^+ U|0\rangle \qquad (7)$$

where

$$U \equiv e^{\sigma(t)}; \quad \sigma(t) = \frac{1}{\sqrt{N}} \sum_q (f_q e^{-iqvt} b_q^+ - f_q^* e^{iqvt} b_q), \qquad (8a)$$

$$\Psi_s(k) = \frac{1}{\sqrt{2\mu}} \operatorname{sech}\frac{\pi a(\hbar k - mv)}{2\mu\hbar},$$

$$\qquad (8b)$$

$$f_q = i\sqrt{M/2\hbar\Omega_q}\;\frac{\pi J(\Omega_q + vq)}{\chi\,\operatorname{sh}\frac{\pi qa}{2\mu}}, \quad \mu = \frac{\chi^2}{Jw(1-v^2/v_o^2)}.$$

This wave function describes the soliton state of a chain in which the intrapeptide excitation is localized in some region and moves along the chain with velocity v. The soliton energy

$$\mathscr{E}_s = \mathscr{E}_o - \Delta + m_s v^2/2 \quad \text{at} \quad v^2 \ll v^2 \qquad (9)$$

is separated from the exciton band (4) by the energy gap

$$\Delta = \chi^4/3Jw^2 \qquad (10)$$

where w is the elasticity of hydrogen bonds and χ is the coupling parameter in (6). Since the motion of the soliton is accompanied by the motion of the local chain deformation, its effective mass

$$m_s = m + \delta, \quad \delta = 4\chi^4/3Jw^2v_o^2 \qquad (11)$$

is larger then that of the free exciton (4) by the value δ.

The dipole moment $\vec{M}$ of Alpha-helix depends on coordinates of all atoms and may by expanded in powers of their displacement from the equilibrium. If we take into account only the Amid vibration coordinate $Q_k \sim (B_k + B_{-k}^+)$ in explicit form than the interaction operator (2) will be represented in the form

$$H_{int} = -\vec{E}(t)\left[\vec{M}'(0)Q_{k=0} + \frac{1}{2}\sum_k \vec{M}''(-k,k)Q_{-k}Q_k\right] =$$

$$= -\vec{E}(t)\left[\vec{m}'(0)\left(B^+_{k=0}+B_{-k=0}\right) + \frac{1}{2}\sum_k \vec{m}''(k)\left(B^+_k B^+_{-k} + B_k B_{-k} + 2B^+_k B_k\right)\right] \qquad (12)$$

The first term in (12) describes an absorption (emission) of the light with a creation (annihilation) of one quantum of the Amid vibration. A process of spontaneous emission of light from the soliton states of a chain was considered in [5]. The second term determines an optical anharmonicity of molecules and stipulates the appearance of Amid overtones (line at double frequencies) in the infrared spectra of biological macromolecules[6]. These overtones are experimentally observed and therefore $\vec{m}''(k)=\vec{m}''^*(-k)$ is not equal zero. We are interested in the case when the frequency ω of the external field is considerably less than that of the Amid vibration $\mathscr{E}_o/\hbar$. In this case it is possible to take into account in (12) only the term

$$H_{int} = -\sum_k \vec{E}_o\vec{m}''(k)B^+_k B_k e^{-i\omega t} \qquad (12a)$$

Substituting (12a) and the explicit form of $|\Psi_i(t)\rangle$, (7) into (1) we find

$$P(\omega) = \frac{2}{\hbar N}\,\mathrm{Re}\sum_{k,k'}\left(\vec{E}_o\vec{m}''(k)\Psi_s(k)\right)^*\left(\vec{E}_o\vec{m}''(k')\Psi_s(k')\right)\int_{-\infty}^t dt' \times \qquad (13)$$

$$\times\, e^{i(\omega+\mathscr{E}_s/\hbar)(t-t')}\,\langle 0|B_k e^{ikvt}e^{-\sigma(t)}e^{-iH(t-t')/\hbar}e^{\sigma(t')}e^{-ik'vt'}B^+_{k'}|0\rangle$$

To calculate the correlation function in (13) we take into consideration the following equalities

$$e^{ikvt}B_k = e^{-i\mathscr{P}vt/\hbar}B_k e^{i\mathscr{P}vt/\hbar}\,, \quad e^{iqvt}b_q = e^{-i\mathscr{P}vt/\hbar}b_q e^{i\mathscr{P}vt/\hbar}\,, \qquad (14)$$

where $\mathscr{P} = \sum_k \hbar k(B^+_k B_k + b^+_k b_k)$. Then, taking into account that $[H,\mathscr{P}]=0$ and $\mathscr{P}|0\rangle=0$, it is possible to rewrite the expression (13) in the form

$$P(\omega) = \frac{2}{\hbar N}\,\mathrm{Re}\sum_{k,k'}\left(\vec{E}_o\vec{m}''(k)\Psi_s(k)\right)^*\left(\vec{E}_o\vec{m}''(k')\Psi_s(k')\right)\int_0^\infty dt\, e^{i(\omega+\mathscr{E}_s/\hbar)t}G_{k,k'}(t)$$

where

$$G_{k,k'}(t) = \langle 0|B_k e^{-\sigma} e^{-i(H-\mathcal{P}v)t/\hbar} e^{\sigma} B_{k'}^+ |0\rangle = \langle 0|B_k e^{-i\tilde{H}t/\hbar} B_{k'}^+ |0\rangle \tag{15}$$

$$\tilde{H} = e^{-\sigma}(H-\mathcal{P}v)e^{\sigma} \quad , \quad \sigma = \frac{1}{\sqrt{N}} \sum_q (f_q b_q - f_q^* b_q^+) \quad . \tag{15a}$$

The transformation (14) and the appearance of the energy operator $H-\mathcal{P}v$ instead of the Hamiltonian operator H may be treated as a transition to a frame moving with the soliton. The unitary transformation (15a) describes a particular deformation of the chain in the soliton state (7).

It is straightforward to calculate that

$$\tilde{H} = \mathcal{E}_d + \sum_k \left[\mathcal{E}(k)-\hbar v k\right]B_k^+ B_k - \frac{1}{N}\sum_{k,k'} W(k-k')B_k^+ B_{k'} +$$

$$+ \sum_q \hbar(\Omega_q - qv)b_q^+ b_q - \frac{1}{\sqrt{N}}\sum_q \varphi_q(b_q + b_{-q}^+) + \frac{1}{\sqrt{N}}\sum_{k,q} \chi(q)B_k^+ B_{k-q}(b_q + b_{-q}^+) \tag{16}$$

where

$$W(k) = \frac{2\pi J k \alpha}{\mathrm{sh}(\pi k \alpha/2\mu)} \quad ; \quad \varphi_q = i\left(\hbar M/2\Omega_q\right)^{1/2} \frac{\pi J \Omega_q^2(1-v^2/v_o^2)}{\chi\,\mathrm{sh}(\pi q \alpha/2\mu)} \tag{17}$$

and $\mathcal{E}_d = \sum_q \hbar(\Omega_q - vq)|f_q|^2 = \dfrac{2\chi^4}{3Jw^2(1-v^2/v_o^2)}$ is the energy of the lattice

deformation.

The operator $\tilde{H}$ is partially diagonalized by the unitary transformation

$$B_k = \frac{1}{\sqrt{N}} \sum_\nu \Psi_\nu(k)A_\nu \quad ; \quad \frac{1}{N}\sum_k \Psi_\nu^*(k)\Psi_{\nu'}(k) = \delta_{\nu,\nu'} \tag{18}$$

where transformation coefficients $\Psi_\nu(k)$ are found from equations

$$[\mathcal{E}(k) - \hbar v k]\Psi_\nu(k) - \frac{1}{N}\sum_{k'} W(k-k')\Psi_\nu(k') = \mathcal{E}_\nu \Psi_\nu(k) \tag{19}$$

In order to solve this equation we make the usual longwave approximation in (4)-(6) and introduce the function

$$\Psi_\nu(\xi) = \frac{1}{\sqrt{N\alpha}} \sum_k e^{ik\xi}\,\Psi_\nu(k) \tag{20a}$$

For $\Psi_\nu(k)$ we have

$$\Psi_\nu(k) = \frac{1}{\sqrt{N\alpha}} \int_{-N\alpha/2}^{N\alpha/2} e^{-ik\xi}\Psi_\nu(\xi)\,d\xi \tag{20b}$$

Take this into account one obtains the equation for $\Psi_\nu(\xi)$:

$$\left\{ \frac{\hbar^2}{2m} \frac{d^2}{d\xi^2} + i\hbar v \frac{d}{d\xi} + V(\xi) + \mathcal{E}_0 \right\} \Psi_\nu(\xi) = \mathcal{E}_\nu \Psi_\nu(\xi)$$

here $\quad V(\xi) = -\frac{1}{N}\sum_k e^{ik\xi} W(k) = -2J\mu^2 \operatorname{sech}^2(\mu\xi/a)$.

For our case of $V(\xi)$ there is only one bond state

$$\psi_s(\xi) = \sqrt{\mu/2} \; e^{imv\xi/\hbar} \operatorname{sech}(\mu\xi/a) \tag{21}$$

with the energy

$$\mathcal{E}_s = \mathcal{E}_0 - J\mu^2 - mv^2/2 \tag{22}$$

and unbounded (delocalized) states

$$\Psi_\varkappa(\xi) = \frac{1}{\sqrt{Na}} \frac{\varkappa a + i\mu \operatorname{th}(\mu\xi/a)}{\varkappa a + i\mu} \exp\left[i(\varkappa + mv/\hbar)\xi \right] \tag{23}$$

with energies

$$\mathcal{E}_\varkappa = \mathcal{E}_0 + \frac{\hbar^2 \varkappa^2}{2m} - mv^2/2 \tag{24}$$

Hence, in the new representation (18)

$$\tilde{H} = \mathcal{E}_s A_s^+ A_s + \sum_\varkappa \mathcal{E}_\varkappa A_\varkappa^+ A_\varkappa + \sum_q \hbar(\Omega_q - qv) b_q^+ b_q - \frac{1}{\sqrt{N}} \sum_q \varphi_q (b_q + b_{-q}^+) +$$

$$+ \frac{1}{\sqrt{N}} \sum_{\varkappa,q} F(\varkappa,q) A_\varkappa^+ A_{\varkappa-q} (b_q + b_{-q}^+) + \frac{1}{N} \sum_{\varkappa,q} \Phi(\varkappa,q)(A_s^+ A_{-\varkappa} + A_\varkappa^+ A_s)(b_q + b_{-q}^+) \tag{25}$$

where

$$F(\varkappa,q) = \frac{\mu^2 + \varkappa(\varkappa-q)}{\mu^2 + \varkappa(\varkappa-q) + i\mu q} \chi(q)$$

$$\Phi(\varkappa,q) = \sqrt{2/\mu} \; \frac{iq\chi(q)}{\mu + i\varkappa} \operatorname{sech}\left[\pi(\varkappa-q)/2\mu \right] \tag{26}$$

Note, that the soliton state (7) is not an exact eigenstate of the chain, owing to the presence of the term in $\tilde{H}$ with $A_k^+ A_s$. Only when this term is sufficiently small, the state (7) is a good approximation to the real autolocalized state. It is possible to show that the term with $A_k^+ A_s$

may be neglected on condition that

$$\varkappa^2 \gg \pi \hbar \nu_o w, \quad \nu_o = \sqrt{w/M} \tag{27}$$

In the case when inequality is not fulfilled it is necessary to use another anzats for an eigenstate instead of (7).

Substituting (18) and (25) into (15) and (13a) one obtains

$$P(\omega) = \frac{2}{\hbar} \sum_\varkappa |\vec{E}_o \vec{d}_\varkappa|^2 \, G_\varkappa(\omega), \tag{28}$$

where

$$G_\varkappa(\omega) = \mathrm{Re} \int_0^\infty e^{i(\omega + \mathscr{E}_s/\hbar)t} G_\varkappa(t)dt \; ; \quad G_\varkappa(t) = \langle 0|A_\varkappa e^{-i\tilde{H}t/\hbar} A_\varkappa^+|0\rangle \tag{29}$$

$$\vec{d}_\varkappa = \frac{1}{N} \sum_k \Psi_\varkappa^*(k)\vec{m}''(k)\Psi_s(k) \tag{30}$$

The expression (28) describes the absorption of the electromagnetic radiation which is accompanied by the transition of the polypeptide chain from the soliton state into exciton states with all wavenumbers $\varkappa$. In the longwave approximation $\vec{m}''(k) \approx i\vec{d}_o ka$. Than from (30) and (21), (23), (20) we obtain the following expression for the dipole momentum of the transition into the exciton state with wavenumber $\varkappa$:

$$\vec{d}_\varkappa = -\pi \vec{d}_o \frac{\mu - i\varkappa a}{\sqrt{2\mu N}\ \mathrm{ch}(\pi \varkappa a/2\mu)} \tag{31}$$

In order to calculate the correlation function in zero approximation we may omit last two terms in operator (25). In this case one obtains

$$G_\varkappa(\omega) = \mathrm{Re} \int_0^\infty e^{i(\omega - \Delta_\varkappa)t + g(t)} dt, \tag{32}$$

where

$$\hbar\Delta_\varkappa = \mathscr{E}_\varkappa - \mathscr{E}_s = \Delta + \hbar^2\varkappa^2/2m + \delta v^2/2, \tag{33}$$

$$g(t) = \frac{1}{N}\sum_q |f_q|^2 \left[e^{-i(\Omega_q - vq)t} - 1 \right] \equiv -\Gamma(t) - iR(t) =$$

$$= -\frac{2}{N}\sum_q |f_q|^2 \sin^2\frac{\Omega_q - vq}{2} t - \frac{i}{N}\sum_q |f_q|^2 \sin(\Omega_q - vq)t \tag{34}$$

For small t, when $\pi v_o t/a < 1$, we have

$$\Gamma(t) \approx B^2 t^2/2, \quad B^2 = \frac{1}{N} \sum_q |f_q|^2 (\Omega_q - vq)^2 = 14.4 \chi^2 v_o \mu^2 / \hbar a w \tag{35}$$

and

$$R(t) = \mathscr{E}_d t/\hbar.$$

For $\pi v_o t/a < 1$, $\Gamma(t)$ is well approximated by

$$\Gamma(t) = [\gamma + \ln(\pi v_o t/a)] 2\chi^2 / \pi \hbar \nu_o w ,$$

where $\gamma = 0.577$ is Euler constant.

Since, when the unequality (27) is fulfilled, the major contribution to the integral (32) is made by small values of t. In this case, taking into account (35) one will obtain

$$G_\varkappa(\omega) = \sqrt{\pi/2B^2} \, e^{-(\omega - \omega_\varkappa)^2/2B^2} \tag{36}$$

where

$$\hbar\omega_\varkappa = \hbar\Delta_\varkappa + \mathscr{E}_d = \mathscr{E}_\varkappa - \mathscr{E}_s + \mathscr{E}_d . \tag{37}$$

In other case, when unequality

$$\chi^2 / w \hbar \nu_o \ll 1$$

is fulfilled, it is easy to show that

$$G_\varkappa(\omega) = \frac{\gamma_\varkappa(\omega)}{(\omega - \omega_\varkappa + \delta_\varkappa)^2 + \gamma_\varkappa^2(\omega)} ,$$

where $\delta_\varkappa(\omega)$ and $\gamma_\varkappa(\omega)$ are correspondingly the real and the imagenary parts of the mass-operator

$$M_\varkappa(\omega) = \frac{\hbar}{N} \sum_q \frac{|f_q|^2 (\Omega_q - vq)^2}{\hbar\omega - \mathscr{E}_\varkappa + \mathscr{E}_s - \hbar(\Omega_q - vq) + i\varepsilon} , \quad \varepsilon \to +0$$

In particular for $\gamma_\varkappa(\omega)$ we have

$$\gamma_\varkappa(\omega) = \begin{cases} \dfrac{\pi^2 J^2 w[(\hbar\omega - \mathscr{E}_\varkappa + \mathscr{E}_s)/\hbar\nu_o]^2}{2\hbar\chi^2 \, sh^2[\pi(\hbar\omega - \mathscr{E}_\varkappa + \mathscr{E}_s)/2\mu\hbar\nu_o]} & \text{at } \hbar\omega - \mathscr{E}_\varkappa + \mathscr{E}_s \geq 0 , \\[12pt] 0 & \text{at } \hbar\omega - \mathscr{E}_\varkappa + \mathscr{E}_s < 0. \end{cases}$$

Note that appearance of the energy $\mathscr{E}_d$ in the expression (37) for the resonance frequency $\omega_{\varkappa}$ is the result of the Frank-Condon principle.

Therefore, the absorption of electromagnetic radiation (EMR) at the frequency (37) can be evidence of the presence of a soliton in the molecular chain.

In the literature there is wide agreement on the following values [2-4]

$$J = 1.55 \cdot 10^{-22} \text{ J} , \quad w = 19.5 \text{ N/m} , \quad \chi = (3.4 - 4) \cdot 10^{-11} \text{ N} .$$

Using values in these ranges yields values for $\omega_{\varkappa=0}/2\pi$ between 34 GHz and 65 GHz, or for the wavelength between 4.6 mm and 8.8 mm. These values allow to connect the biological effect of low intensity millimeter EMR with Davydov theory [7].

A number of authors [8-10] have discussed some problems concerning the effect of EMR on living organisms and some general regularities of these effects were formulated. For example, the resonant dependence of the effect of EMR frequency has been observed. However, this effect has not yet been explained convincingly.

According to Davydov[1], the formation of stable solitons guaranties the high efficiency of the energy transfer along α-helix. In a living organism a number of organs and systems operate simultaneously and in concord in energy metabolism. The changes in energy metabolism can greatly affect the functioning of the entire organism. In the alternating field of EMR with the frequency (37) soliton dissociation takes place. The excitons due to to their smearing and more intense interaction with disordered thermal vibrations transfer energy with less efficiency. Consequently, if Davydov solitons indeed play an important role in biological processes, than EMR affect the functioning of the living organism through the affect the energy transport processes. Thus the theory of Davydov solitons seems to elucidate the mechanism of the biological effect of EMR of the millimeter range at the molecular level.

REFERENCES

1. A. S. Davydov, J. Theor. Biol., 38:559 (1973); Int. J. Quant. Chem., 16:5 (1979).

2. A. C. Scott, Phys. Rev. A, 26:578 (1982).

3. P. S. Lomdahl, L. MacNeil, A. C. Scott, M. E. Stoneham, S. J. Webb, Phys. Lett. A, 92:207 (1982).

4. M. Hyman, D. W. McLaughlin, A. C. Scott, Physica (Amsterdam) D, 3:23 (1981).

5. A. S. Davydov, A. A. Eremko, <u>Ukr. Phys. J.</u>, 22:881 (1977).

6. S. N. Timasheff and J. D. Fasman, eds., "Structure and Stability of Biological Macromolecules", Marcel Dekker, New York, 1969.

7. A. A. Eremko, <u>DAN</u> <u>Ukr. SSR,</u> A, N3:52 (1984).

8. The Science Session of Branch of General Physics and Astronomy of Academy of Sciences of the USSR (17 - 18 January, 1973), <u>UFN,</u> 110:452 (1973).

9. H. Fröhlich, <u>Phys. Lett.</u> <u>A</u>, 51:21 (1975).

10. N. D. Devyatkov and M. B. Golant, <u>Letters in ZhTF</u>, 8:39 (1982).

34

VIBRATIONAL PROPERTIES AND ENERGY TRANSPORT IN

ACETANILIDE BY MOLECULAR DYNAMICS

Andrea Giansanti[@], Alessandro Campa[@],
Decio Levi[*], Orlando Ragnisco[*],
Alexander Tenenbaum[@]

Dipartimento di Fisica.
Universita' degli Studi di Roma "La Sapienza"
P.le A. Moro 2 - 00185 ROMA (Italy)

* INFN - Sezione di Roma - @ GNSM/CISM

INTRODUCTION

Soon after Careri, Scott and co-workers assigned the anomalous low-temperature amide I band of ACN to a self trapped state similar to a Davydov soliton[1,2], we started computer experiments on a model of ACN with at least three goals:

a) to set up molecular models of the interactions present in the real crystal, simple but sufficient to elucidate the degrees of freedom involved in some relevant nonlinear effects;
b) to explore the occurence of a stochastic transition, also to be related to peculiarities of the ACN dynamics;
c) to study the transport of energy along chains of ACN molecules.

It was our aim to complement the well established picture based on the Davidov soliton with information obtained via explorative molecular dynamics (MD) i.e., from nonlinear classical mechanics. Point a) above has been achieved. Some preliminary results have been presented at the 1986 MIDIT workshop on Dynamical Systems[3], a more complete report has been published afterwards[4]. First results on point b) are presented here while on point c) we only outline what we are doing at present.

Davydov's Soliton Revisited, Edited by P.L. Christiansen and A.C. Scott
Plenum Press, New York, 1990

REVIEW OF PREVIOUS WORK

Let us briefly review some of our published results. In ref.[4] we considered a chain segment made of five molecules with fixed ends. In each molecule, the center of mass was kept also fixed at the equilibrium position, and we considered the three degrees of freedom involved in the hydrogen bond: the C=O stretching, and the stretching and bending of the N-H bond. We considered in this first simple model harmonic forces and anharmonic forces, derived from intermolecular and intramolecular potentials, expanded up to 2nd order. The intermolecular hydrogen bond is modeled by an electrostatic interaction between excess charges on hydrogen and oxygen atoms. The intramolecular force in the peptide group is provided by a dipole-charge interaction, by attributing a polarizability to the (rigid) C-N bond. At different temperatures (i. e. total energies) we simulated the dynamics of this model with short (8 ps) and long (400 ps) molecular dynamics computer runs, and studied energy exchanges between the degrees of freedom. From these experiments we concluded that: i) Energy is widely exchanged between the stretching and the bending of a N-H bond, with a characteristic period of about 0.2 ps. This exchange is maximum when both degrees of freedom have initially nearly the same energy, and is strongly reduced when the two degrees of freedom have initially very different energies; ii) an energy transfer through the hydrogen bond takes place between an H atom and the related O atom; this exchange is periodic (period of about 0.4 ps), is of small amplitude, and is most effective when the two neighbor molecules have similar energies, iii) no energy transfer across the peptide group, via the dipole-charge interaction, was observed; a lower bound to characteristic times for this process to take place was at least 400 ps, the longest simulated time. Points ii) and iii) above indicated that our first model shows a tendency toward self trapping of energy; but, although the model had evident nonlinear effects, these were definitely too weak to ensure any energy transfer out of the region where this energy was initially confined. This weakness of the nonlinearity in the model was confirmed (unpublished result) via a perturbative asymptotic analysis, outlined below.

OCCURRENCE OF A STOCHASTIC TRANSITION

We then set up a second model, in which the molecules as a whole were allowed to vibrate in three dimensions, but not to rotate. The forces in the model were modified to include also a dipole-dipole interaction between adjacent molecules. During the work on this model we obtained a temperature effect resembling the appearance of the anomalous band in ACN; in this effect a coupling between the hydrogen bonded protons and the low-frequency vibrations of the centers of mass was clearly involved. The interested reader can find a discussion and details of the model in ref. [5].

In this second model we began studying the occurrence of a stochastic transition (i.e. order/chaos) with temperature (this was objective (b)). As a diagnostic tool for the onset of chaos we computed the autocorrelation functions of the energy in the degrees of freedom involved in the hydrogen bond and the related power spectra (in the first model, where the centers of mass were fixed and no dipole-dipole interaction was present, we observed the signature of regular motions even at high temperature i. e., narrow peaks at the proper vibration frequencies, and peaks corresponding to the observed frequency of energy exchange between the degrees of freedom). In figures 1,2 and 3 power spectra and autocorrelation functions of the energy for the three degrees of freedom involved in the hydrogen bond at three different temperatures are shown. At low temperature the motion is ordered and energy is periodically exchanged between the different degrees of freedom: in particular, autocorrelations for the N-H stretching and bending are almost coincident, both at T=100 K and T=350 K, indicating a localized exchange of energy between the two degrees of freedom; at a temperature of 700 K these power spectra are populated by Fourier components of similar magnitude over the whole spectrum, and the correlation functions go almost immediately to zero, indicating a chaotic situation. In figure 4 power spectra for the N-H stretching are reported over a frequency interval extending up the 3300 cm^{-1} region; at low temperature there are essentially two channels of energy exchange: the dominant one centred at the natural frequency of the vibration , the other one at the frequency of the C=O stretching. At high temperature the spectrum is invaded. As a conclusion we can say that in our model there is a stochastic transition located at a temperature above 350 K. This is a quite high threshold, but it must be considered that we take into account a small number of degrees of freedom: increasing the number of vibrations should lower this threshold. The observation of an order/chaos transition is interesting for we generally expect to have coherent dynamical behavior at temperatures in which the dynamics of the system separate in regular motions of the various degrees of freedom. One of the aims of our project is to determine the transition temperature on the basis of a sufficiently realistic model.

THE CURRENT MODEL

From the point of view of the energy transfer the second model did not show a marked difference from the previous one; if we had initially two regions of the chain differently 'heated', their energies tended to equilibrate with a very long relaxation time. The natural choice to improve the energy transfer along the chain is to allow for the C-N bond to vibrate and so we developed a new Hamiltonian for the model we are now working with.
Let us discuss here this Hamiltonian of the model. In figure 5 the coordinates are shown; a configuration of a molecule is given when the four vectors $\vec{r}_2$, $\vec{r}_3$, $\vec{r}_4$, $\vec{r}_5$, are given or when the center of mass coordinates r_G are given together with those of the three vectors $\vec{u}$, $\vec{v}$, $\vec{p}$., the two sets of coordinates being related by a linear substitution.
In principle we have 12 degrees of freedom per molecule in the chain, but we introduce constraints, to maintain the right molecular structure.

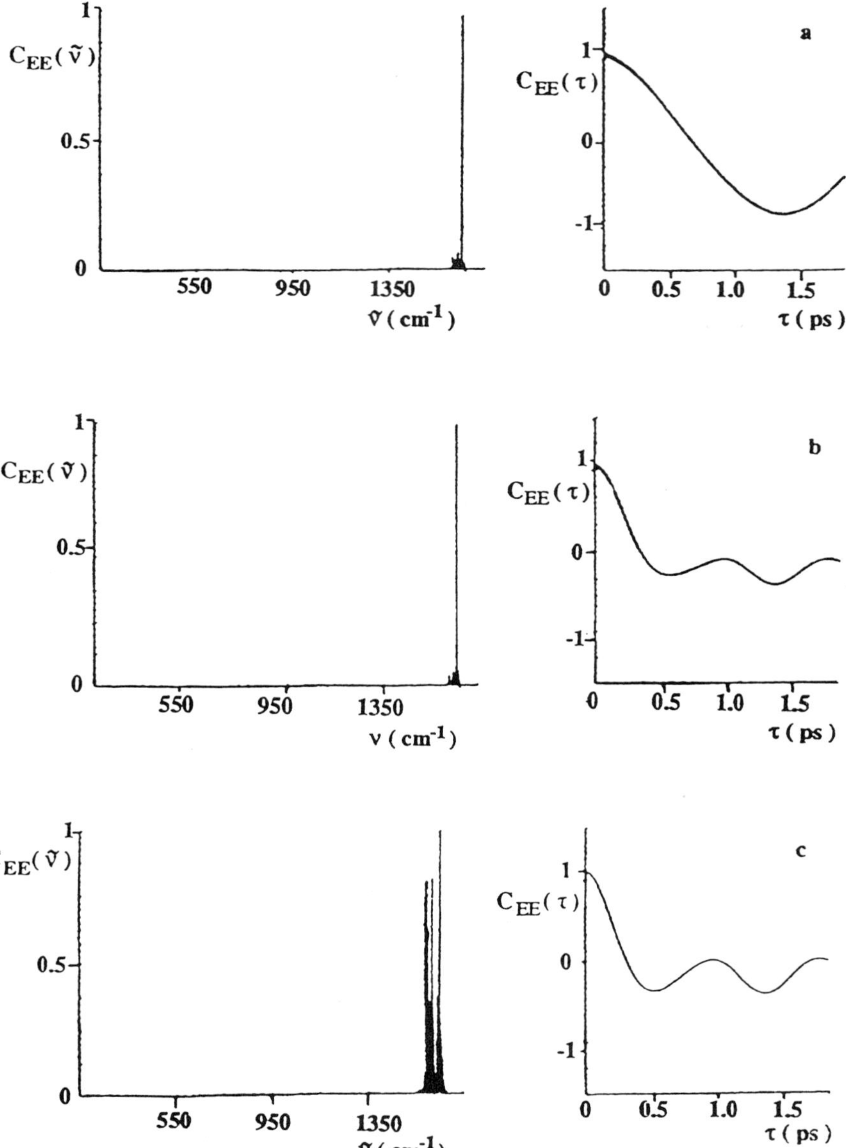

Fig.1. Autocorrelation functions (right) and power spectra (left) of the energy in the C=O stretching (a), N-H stretching (b) and N-H bending (c). T=100 K.

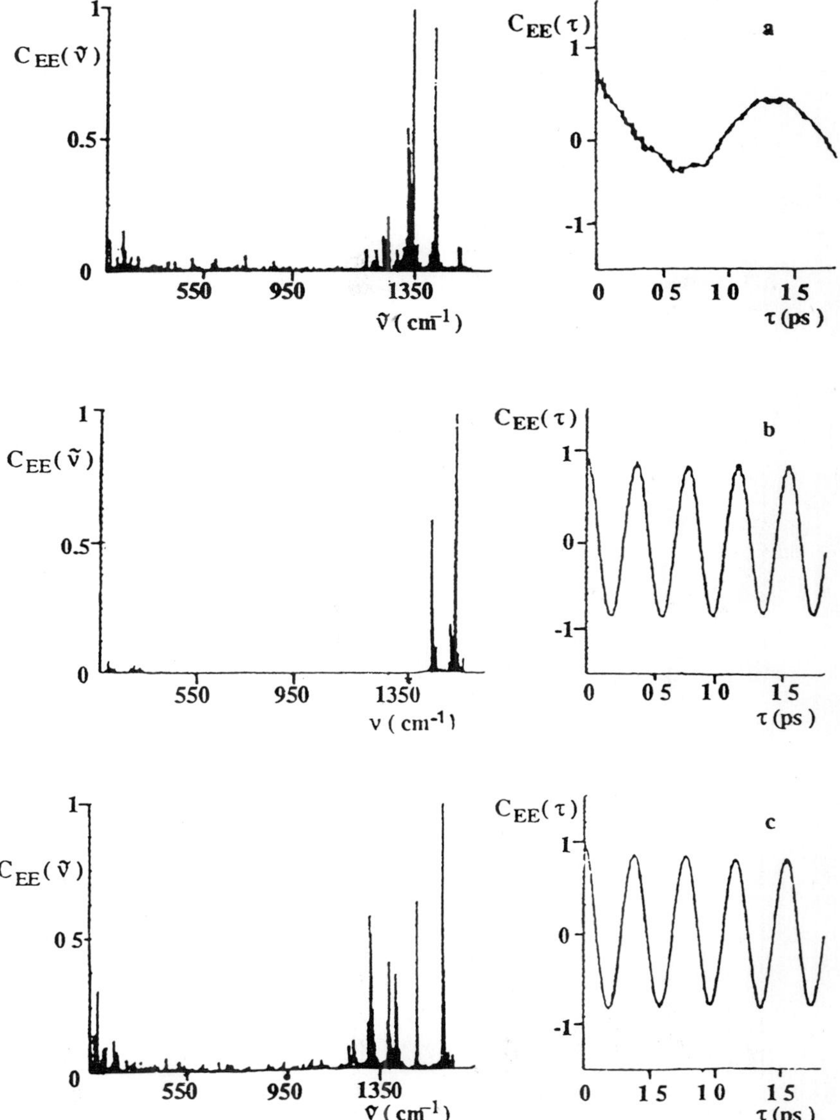

Fig.2.   Same as fig. 1, but at T=350 K.

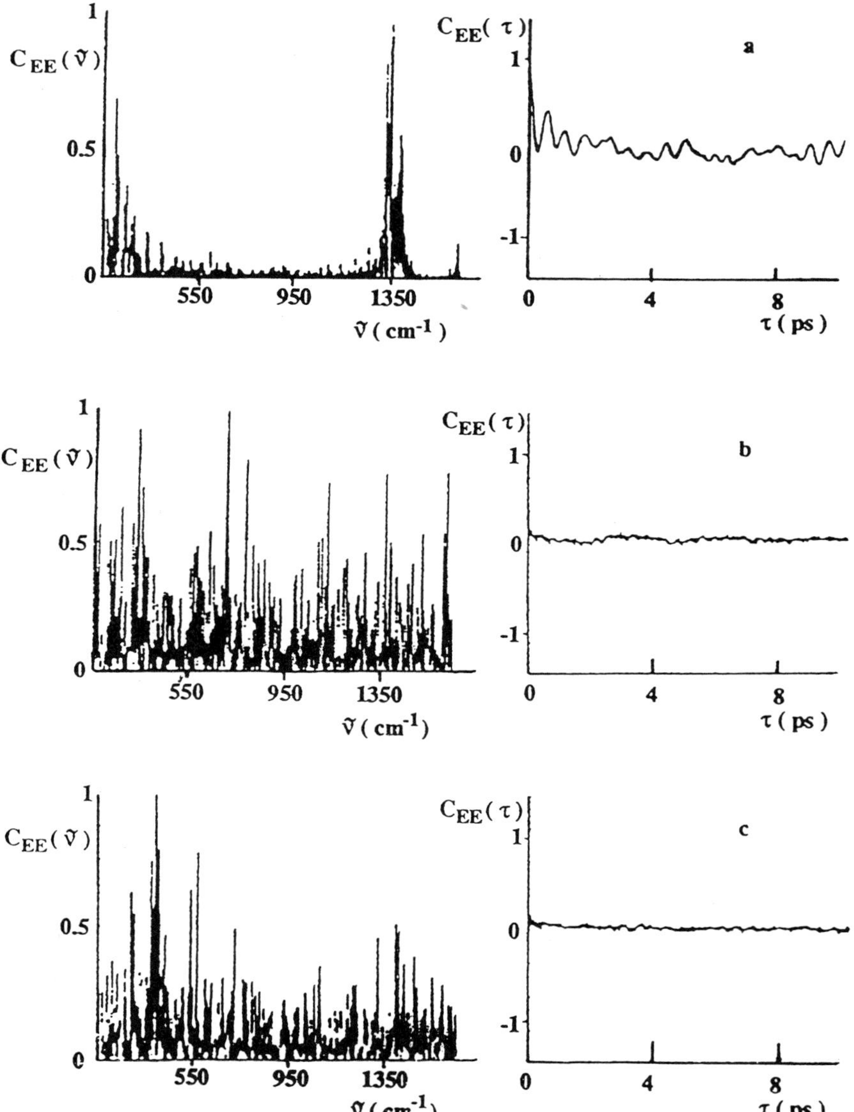

Fig.3.   Same as fig. 1, but at T=700 K.

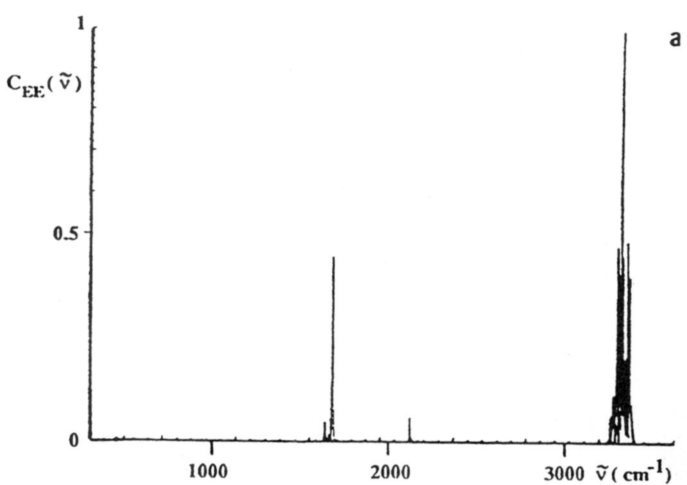

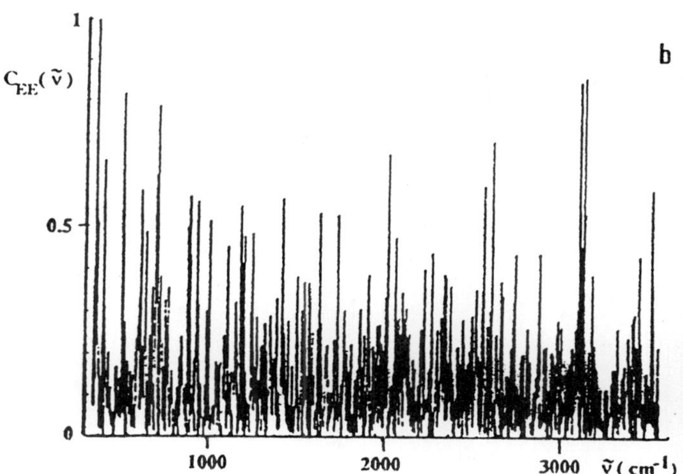

Fig.4.   Extended power spectra of the energy in the N–H stretching: (a) T=100 K, (b) T=700 K.

445

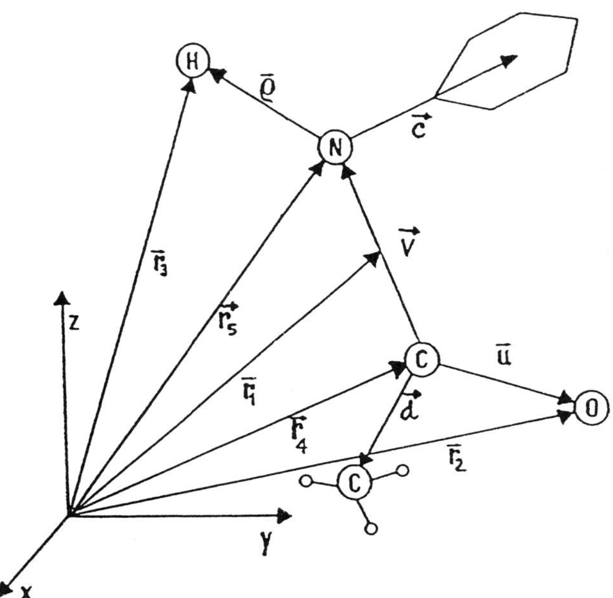

Fig.5.Set of coordinates in the ACN model we are using to study energy transfer: see text for details of the Hamiltonian.

We consider a harmonic potential of the form:

$$H_1 = \left\{ \sum_{i=1}^{N} \frac{1}{2} K_1 \left[\left(U_{x,i} - U_{x',i}^0\right)^2 + \left(U_{y,i} - U_{y',i}^0\right)^2 + \left(U_{z,i} - U_{z',i}^0\right)^2 \right] \right.$$

$$+ \frac{1}{2} K_2 \left\{ \left(\rho_{x,i} - \rho_{x',i}^0\right)^2 + \left[\hat{\rho}_0 \cdot \hat{y}\left(\rho_{y,i} - \rho_{y',i}^0\right) + \hat{\rho}_0 \cdot \hat{z}\left(\rho_{z,i} - \rho_{z',i}^0\right)\right]^2 \right\} +$$

$$+ \frac{1}{2} K_3 \left[-\hat{\rho}_0 \cdot \hat{z}\left(\rho_{y,i} - \rho_{y',i}^0\right) + \hat{\rho}_0 \cdot \hat{y}\left(\rho_{z,i} - \rho_{z',i}^0\right)\right]^2 +$$

$$+ \frac{1}{2} K_4 \left[\left(V_{z,i} - V_{z',i}^0\right)^2 + \left(V_{y,i} - V_{y',i}^0\right)^2 + \left(V_{z,i} - V_{z',i}^0\right)^2 \right] +$$

$$+ \frac{1}{2} K_5 \left[\left(X_{G,i} - X_{G',i}^0\right)^2 + \left(Z_{G,i} - Z_{G',i}^0\right)^2 \right] \right\} +$$

$$\sum_{i=1}^{N-1} \frac{1}{2} K_5 \left(y_{G,i+1} - y_{G',i+1}^0 - y_{G,i} + y_{G',i}^0\right)^2$$

from which we derive forces, linear in the displacements from the equilibrium configuration. The vibration of the hydrogen atom is represented attributing the bending consant K_3 to the component which is normal to the direction of the vector $\vec{\rho}$ at equilibrium. The masses, constants and equilibrium coordinates entering the model were obtained from published structural data of ACN.

The dipole-charge intramolecular interaction can be derived from a potential of the form: $p^2/(2a)$; where $\vec{p}$ is the induced dipole:

$$\vec{p}_i = a|q| \frac{\vec{r}_{2,i} - \vec{r}_{1,i}}{\left|\vec{r}_{2,i} - \vec{r}_{1,i}\right|} - a|q'| \frac{\vec{r}_{3,i} - \vec{r}_{1,i}}{\left|\vec{r}_{3,i} - \vec{r}_{1,i}\right|}$$

a, q, q' are the polarizability of the C-N bond and the excess charges on O and H, respectively. The hydrogen bond is represented by an electrostatic potential plus a repulsive LJ potential:

$$U_{Hb}^{(i,i+1)} = \frac{-|qq'|}{\left|\vec{r}_{3,i+1} - \vec{r}_{2,i}\right|} + \frac{b^{12}}{\left|\vec{r}_{3,i+1} - \vec{r}_{2,i}\right|^{12}}$$

with $b = 10^{-10}$ cgs units.

Finally a dipole-dipole potential is introduced between the induced dipoles in adjacent molecules:

$$U_{dd}^{(i,i+1)} = \frac{1}{\left|\vec{r}_{1,i+1} - \vec{r}_{1,i}\right|^3} \left[\vec{p}_i \cdot \vec{p}_{i+1} - 3 \left(\frac{\vec{r}_{1,i+1} - \vec{r}_{1,i}}{\left|\vec{r}_{1,i+1} - \vec{r}_{1,i}\right|} \cdot \vec{p}_i \right) \left(\frac{\vec{r}_{1,i+1} - \vec{r}_{1,i}}{\left|\vec{r}_{1,i+1} - \vec{r}_{1,i}\right|} \cdot \vec{p}_{i+1} \right) \right]$$

From the nonharmonic parts of the potential, which are similar to those in the second model, non linear forces have been derived and included as perturbative terms in the equations of motion. With this model we will explore the existence of peculiar energy transport properties along a chain of several molecules, to relate the transport properties to the occurrence of a stochastic transition.

We have almost completed the computer code to numerically integrate the equations of motion with constraints; instead of fixed ends we shall use in the simulation periodic boundary conditions. To give our simulations a solid theoretical ground we are making a perturbative asymptotic analysis also of this more complex system. On a preliminary basis we have used the MACSYMA symbolic manipulation system to derive the perturbative structure of the system of discrete nonlinear equations we shall integrate in the Molecular Dynamics runs. The asymptotic analysis we are thinking about is inspired by a technique recently revived by Calogero and Eckhaus for nonlinear PDEs'[6]. For short times and small amplitudes the system has a natural oscillatory character due to the linear part of the forces. The nonlinear part can be parametrized, via rescalings, by a set of small parameters. These small parameters are used to introduce slow variables, for which a nonlinear asymptotic evolution can be consistently derived. Working out the symmetries, the continuum limit and the qualitative properties of the asymptotic evolution, one can get informations on the original model, on the various time scales involved, and to evidentiate possible peculiar behaviors.

This is our present project, and we hope to be able to report new results in a next future.

REFERENCES

1.   G. Careri, U. Buontempo, F. Galluzzi, A. C. Scott, E. Gratton, E. Shyamsunder, _Phys. Rev._, B30: 4689 (1984).

2.   J. C. Eilbeck, P. S. Lomdahl, A. C. Scott., _Phys. Rev._, B30: 4703 (1984).

3.   A. Tenenbaum, _in_: "Structure, Coherence and Chaos in Dynamical Systems," P.L. Christiansen and R. Parmentier, eds., Manchester University Press, Manchester and New York (1989).

4.   A. Campa, A. Giansanti, A. Tenenbaum, _Phys. Rev._, B36: 4394 (1987).

5.   A. Tenenbaum, A. Campa, A. Giansanti, _Physics Letters_, A121: 126 (1987).

6.   F. Calogero and W. Eckahaus, _Inverse Problems_, 3: 229 (1987).

35

ON THE POSSIBLE ROLE OF PHONON-MODULATED TUNNELING IN EXCIMER FORMATION

Ten-Ming Wu[a], David W. Brown[b], and Katja Lindenberg[b,c]

[a] Department of Physics, B-019
[b] Institute for Nonlinear Science, R-002, and
[c] Department of Chemistry, B-040
University of California at San Diego, La Jolla, CA 92093 U.S.A.

ABSTRACT

We present a model of excimer formation based only on the most generic features of dimerized lattices such as those found in the excimer-forming molecular crystal α-perylene. The modulation of resonance integrals by lattice vibrations is found to cause significant effects consistent with the known characteristics of excimers and the related Y states. A specific calculation of the thermally-activated transition rate from the Y state to the excimer state is in good agreement with experimental observations.

One of the mechanisms of interest in bioenergetic processes (such as may be involved in muscle contraction, for example) is that of excimer formation as proposed by McClare [1]. McClare envisioned an attractive force between two molecular units arising from a resonance between "monomer" energy levels on the interacting units. There is nothing mysterious about the basic idea, which is simply that when two monomers having degenerate energy levels are brought close enough together, wave function overlap generally results in quantum tunneling between the monomer levels. The tunneling between monomers is one sign of the hybridization of the two monomer orbitals into two dimer orbitals -- one anti-bonding orbital having an energy higher than the separated monomers, and one bonding orbital having an energy lower than the separated monomers. The energy lowering which is made possible in this way provides a mechanical force which may affect the interaction of molecular units. What we are really addressing, of course, is the problem of chemical bonding. The only twist is that the bonding we consider takes place only when one of our monomer units is in an excited state -- for our purposes, the unexcited monomers are essentially noninteracting. In the parlance of chemical physics, the excited-state complex we refer to is an excited dimer or *excimer*.

In the following, we are not concerned with bioenergetic processes, nor with the interaction of macromolecules as was McClare. Instead, we focus on the process of excimer formation in the solid state, between monomers sufficiently rigid as to be regarded for our purposes as essentially structureless. The systems we select for study are sufficiently simple as to be almost ideal for the study of excimer formation. To be specific, we focus on the molecular crystal α-perylene. The unit cell of α-perylene contains four molecules so arranged as to form two essentially-independent sandwich-like dimers. Comparison of α-perylene fluorescence spectra and those of the related crystal modification β-perylene suggests that it is the sandwich-dimer structure which gives rise to excimer formation, so as a simplifying model assumption we ignore the presence of the second dimer in the unit cell. The sandwich-dimer configuration allows the two perylene molecules involved to have substantial wave function overlap, particularly in excited states where wave functions typically occupy a larger effective volume than in the ground state. The quantum tunneling allowed by this overlap is a sensitive function of distance between the two monomers comprising the sandwich dimer, so we should expect the quantum tunneling to be strongly modulated by lattice vibrations. We focus on these two material properties: 1) the dimerized crystal structure which predisposes α-perylene to excimer formation, and 2) the strong interaction between quantum tunneling and lattice vibrations which this dimerized structure suggests.

Davydov's Soliton Revisited, Edited by P.L. Christiansen and A.C. Scott
Plenum Press, New York, 1990

A number of experimental facts are available to guide us in developing theoretical models. Room temperature fluorescence spectra of α-perylene clearly display the broad, structureless and highly Stokes-shifted emission which is the signature of excimer states. As temperatures are below 77 K, however, the excimer or "E" emission is gradually replaced by a weakly-structured fluorescence, which has been called the "Y" emission. The experimentally determined lifetime, Stokes shift and the halfwidth of Y-state spectra are found to be intermediate between those of the E-state fluorescence and the monomeric fluorescence of β-perylene (cf. Figure 1).

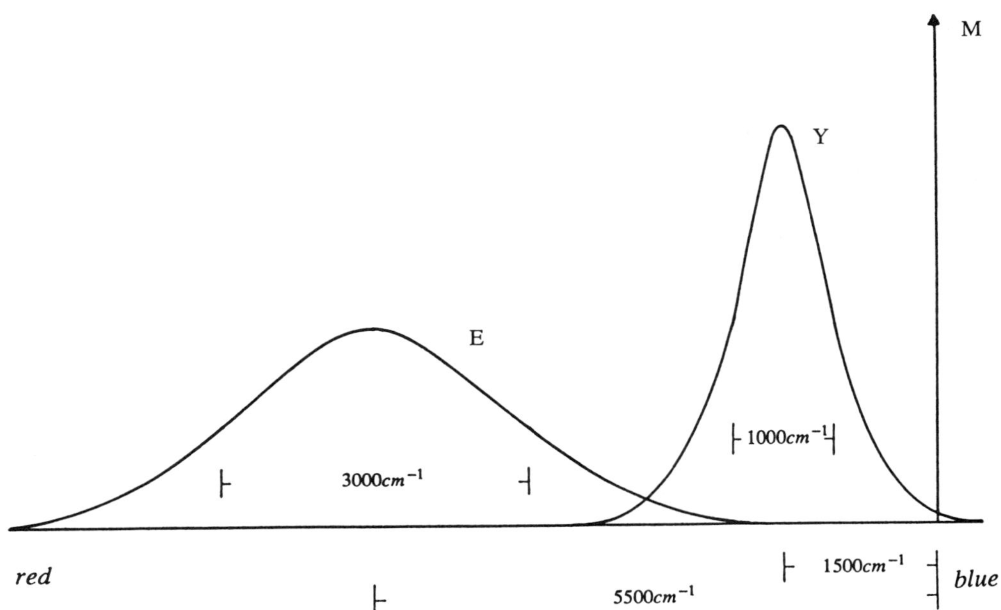

Fig. 1   Excimer spectra, schematic. "E": typical excimer fluorescence, Stokes shift $\approx 5500cm^{-1}$, full width $\approx 3000cm^{-1}$. "Y": typical Y fluorescence, Stokes shift $\approx 1500cm^{-1}$, full width $\approx 1000cm^{-1}$. "M": typical monomer fluorescence, width negligible, Stokes shift irrelevant. Numerical values have been chosen to approximate the relative magnitudes of the important quantities; all values depend on one's choice of material and the temperature.

For such reasons, the Y state is often said to be an incompletely formed excimer. The physical origin of the "incomplete excimer" has remained somewhat mysterious since the Y state was first observed by Tanaka [2], though a number of experimental results have ruled out several possible explanations, including impurity effects, structural defects and surface effects [3].

Recently, Walker *et al.* [4] found that the decay time of the Y state and the rise time of the E state have the same activated temperature dependence. Moreover, they found that the decay time of the E fluorescence is increased by deuteration, suggesting, perhaps, sensitivity to vibronic coupling characteristic of a largely-forbidden optical transition; however, the decay time of the Y fluorescence was found to be essentially unaffected by isotopic substitution. Walker *et al.* proposed a kinetic model to describe the experimental results. In this model, the Y state is a precursor of the E state and is separated from the E state by a (phenomenological) potential barrier. That is, the Y state is populated directly from the optically excited monomeric state, but the E state is populated only via thermal activation from the Y state. Unfortunately, while this simple kinetic picture consistently organizes the known experimental facts, this *description* does not provide an explanation of the mechanism leading to the formation of the Y state, nor of the associated activation barrier, nor of the isotope effect found in the decay times.

We assume that the Frenkel exciton created by optical absorption in a dimeric crystal can transfer resonantly across the "short bond" between monomers of one and the same unit cell, but not across the "long bond" between dimers in distinct unit cells. We assume that vibrations of the lattice around its ground state equilibrium are sufficiently small that we can expand both the monomer energies and the resonance integral in Taylor series and truncate at linear order in the lattice coordinates. We refer to the modulation of monomer energy levels by lattice vibrations as "local" coupling, and to modulations of the resonance integral as "nonlocal" coupling. We find both local and nonlocal exciton-phonon interactions essential to a unified description of Y and E states.

In the interest of brevity, we pose our model Hamiltonian directly in the "Bloch" basis, where states are labeled by symmetry indices: $(+)$ indicating the symmetric linear combination of monomer levels (gerade), $(-)$ indicating the antisymmetric linear combination (ungerade). The rigid-lattice Hamiltonian H_d describing the excited dimer is diagonal in this basis.

$$H = H_d + H_{ph} + H_{mod} + H_{tran} , \tag{1}$$

$$H_d = E_+ a_+^\dagger a_+ + E_- a_-^\dagger a_- , \tag{2}$$

$$H_{vib} = \sum_{q\alpha} \hbar\omega_{q\alpha} b_{q\alpha}^\dagger b_{q\alpha} , \tag{3}$$

$$H_{mod} = \sum_{q\alpha} e^{iqnl} \hbar\omega_{q\alpha} (b_{q\alpha}^\dagger + b_{-q\alpha})(\chi_+^{q\alpha} a_+^\dagger a_+ + \chi_-^{q\alpha} a_-^\dagger a_-) , \tag{4}$$

$$H_{tran} = \sum_{q\alpha} e^{iqnl} \hbar\omega_{q\alpha} \chi_{+-}^{q\alpha} (b_{q\alpha}^\dagger + b_{-q\alpha})(a_+^\dagger a_- + a_-^\dagger a_+) , \tag{5}$$

$a_\pm$ ($a_\pm^\dagger$) annihilates (creates) an exciton in the $\pm$ Bloch state. The rigid-lattice eigenvalues for these states are $E_\pm = E \pm J$, where E is the energy of the monomer level of interest and J is the resonance integral, both evaluated in the ground-state equilibrium configuration. $b_{q\alpha}^\dagger$ ($b_{q\alpha}$) creates (annihilates) a phonon of mode q of branch α and frequency $\omega_{q\alpha}$. The coupling interaction in the Bloch representation can be decomposed into two parts; H_{mod} describes the modulation of the bare symmetric and antisymmetric excitations by phonons and H_{tran} describes phonon-induced transitions between these excitations.

This general model is applicable to dimerized lattices of any spatial dimension; however, for simplicity, we take as our model lattice a dimerized harmonic linear chain having two alternating sublattice constants d and $l-d$ and corresponding stiffness coefficients κ_1 and κ_2. We assume that direct interaction of the excimer pair with the rest of the lattice extends only to the unit cells immediately adjacent to the excited cell. In all, there are 10 local and 2 nonlocal interactions among the molecules of these three unit cells. These can all be given (approximately) in terms of the three local force constants F_0, F_1, δF and the one nonlocal force constant G (cf. Figure 2). In terms of these elementary force constants, the coupling coefficients in (4) and (5) take the form

$$\chi_\pm^{q\alpha} = \frac{1}{2} \left[\frac{1}{\hbar MN \omega_{q\alpha}^3} \right]^{\frac{1}{2}} \left[\left\{ F_0 \pm G - \delta F [1 + \cos(ql)] \right\} (\eta_1^{q\alpha} - \eta_2^{q\alpha}) \right. \tag{6}$$

$$\left. + i\, 2F_1 \sin(ql)\, (\eta_1^{q\alpha} + \eta_2^{q\alpha}) \right] ,$$

$$\chi_{+-}^{q\alpha} = \frac{\delta F}{2} \left[\frac{1}{\hbar MN \omega_{q\alpha}^3} \right]^{\frac{1}{2}} [1 - \cos(ql)] (\eta_1^{q\alpha} + \eta_2^{q\alpha}) , \tag{7}$$

where M is the mass of a single molecule and N is the total number of unit cells. The dimerized lattice structure requires $\eta_1^{q\alpha} = (\kappa_1 + \kappa_2 e^{iql})/|\kappa_1 + \kappa_2 e^{iql}|$, $\eta_2^{q\alpha} = +1$ for the acoustic branch and $\eta_2^{q\alpha} = -1$ for the optical branch. Two features of these coupling coefficients are of particular importance: one is that $(\chi_+^{q\alpha} - \chi_-^{q\alpha})$ depends only the nonlocal force G, and is mostly derived from optical phonons; the

a

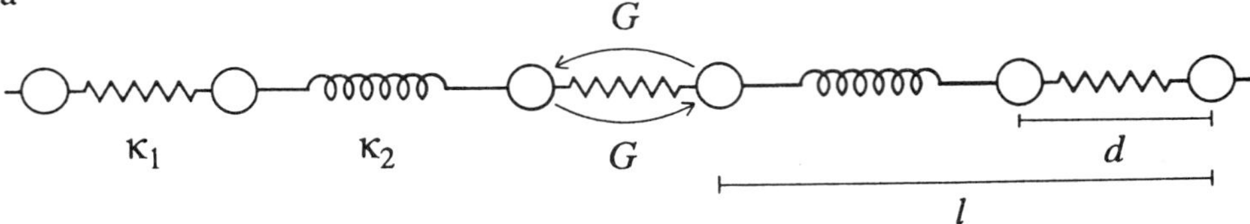

b

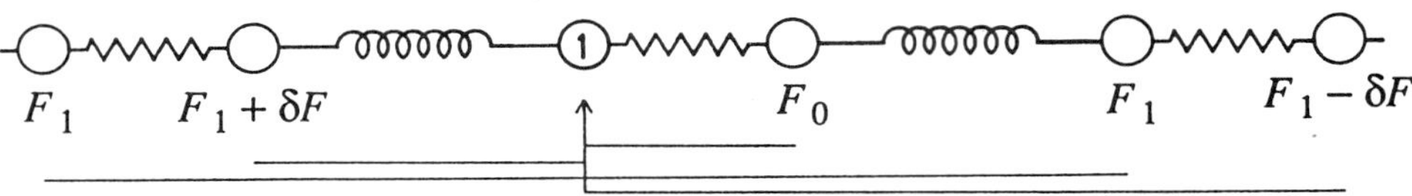

c

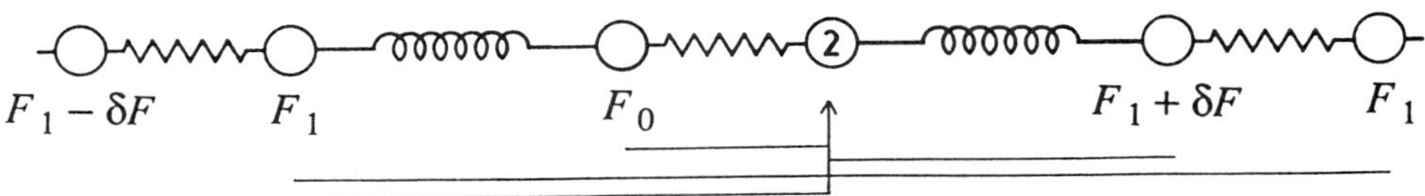

Fig. 2   Geometry of exciton-phonon coupling. a) κ_1, κ_2: Stiffness coefficients, $\kappa_1 > \kappa_2$; G: nonlocal coupling constant; d, l: dimer spacing in the ground state, lattice constant, $l > d$. b) Forces affecting monomer level 1; $|F_0| > |F_1 + \delta F| > |F_1| > |F_1 - \delta F|$. b) Forces affecting monomer level 2; while the magnitudes of corresponding forces are the same as in (b), the difference in symmetry affects the complex phases of coupling constants.

other is that $\chi_{+-}^{q\alpha}$ depends only on the local force differential δF, and is mostly derived from acoustic phonons.

The full Hamiltonian can be divided into two parts, H_0 and H_{tran}, where $H_0 = H_d + H_{ph} + H_{mod}$ can be diagonalized exactly by a canonical transformation [5] and H_{tran} acts as a perturbation. We call the eigenstates of H_0 "Bloch polarons" (cf. Figure 3). The energies of the Bloch polarons are

$$\bar{E}_{\pm} = E_{\pm} - E_{\pm}^{bind} \, , \tag{8}$$

$$E_{\pm}^{bind} = \sum_{q\alpha} \hbar\omega_{q\alpha} |\chi_{\pm}^{q\alpha}|^2 \, . \tag{9}$$

Due to the nonlocal force G, the polaron binding energy in the (+) state typically is larger than that in the (−) state. Characteristic lattice distortions in the (+) and (−) states are given approximately by

$$\delta D_{\pm} \approx - \frac{(F_0 \pm G) + 2F_1}{\kappa_1} \, . \tag{10}$$

We see that in the symmetric state, local and nonlocal forces act constructively to compress the excimer bond. On the other hand, in the antisymmetric state, local and nonlocal forces compete, with the local forces generally acting to compress the excimer bond while nonlocal forces act in the opposite sense; as a result the excimer bond is compressed less in the antisymmetric state than in the symmetric state. It is possible, if the nonlocal force is greater than the net local force, for the excimer bond to be extended rather than compressed in the antisymmetric state.

a

b

c

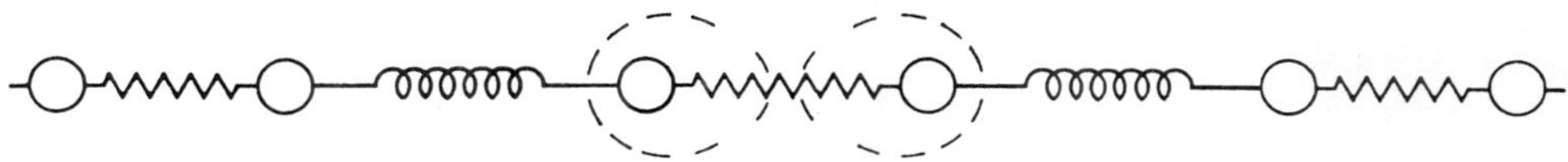

Fig. 3 Bloch polaron schematic. a) Unexcited chain. b) Symmetric Bloch polaron, excimer. c) Antisymmetric Bloch polaron, Y state.

The bare (undistorted) symmetric state is already lower in energy than the bare antisymmetric state ($E_+ < E_-$); moreover, in the presence of strong nonlocal coupling, the polaron binding energy in the symmetric state may be much larger than the polaron binding energy in the antisymmetric state ($E_+^{bind} > E_-^{bind}$) reflecting the greater distortion in the symmetric state ($\delta D_+ > \delta D_-$). Since these latter factors should lead to the appearance of greater spectral widths and Stokes shifts in the symmetric state, we identify the (+) state with the excimer and the (−) state with the Y state.

Optical selection rules forbid direct one-photon absorption into the symmetric pair state from the ground state; thus, one-photon absorption can only populate the antisymmetric excited state.[†] Vibrational relaxation times are of the order of ω_α^{-1}, which in organic molecular crystals is of the order of *ps*. Because the observed lifetime of the Y state in α-perylene is of the order of *ns*, several orders of magnitude larger than vibrational relaxation times, we can safely assume that thermal equilibrium is quickly established in the Y state vibrational manifold.

There are several channels through which an excitation in the Y state may decay. The excitation may return to the ground state through radiative processes (producing the Y-emission observed in the experiment) or through nonradiative processes. The combined rate k_0 for these processes is found experimentally to be essentially temperature-independent. Instead of decaying into the ground state, however, an excitation in the Y state may undergo a nonradiative transition into the E state vibrational manifold at a rate we denote by $k_{+-}(T)$. Thus, the experimentally observed lifetime of the Y state is given by

$$\tau(T) = \frac{1}{k_0 + k_{+-}(T)} \; . \tag{11}$$

We calculate $k_{+-}(T)$ from our model and compare the resulting lifetime $\tau(T)$ with experimental observations.

[†] This provides a possible explanation of the difference in the isotope effects observed in the lifetimes of the Y and E states.

We use the Fermi Golden rule to calculate the transition rate from the Y state to the E state taking H_{trans} as the perturbation interaction on the eigenstates of H_0, the latter being direct products of the Bloch polaron states and many-mode products of the dressed phonon number states. Thus

$$k_{+-}(T) = \frac{2\pi}{\hbar^2} \sum_{\{m_{q\alpha}\},\{n_{q\alpha}\}} P(\{n_{q\alpha}\},T)$$

$$\times \; |<+,\{m_{q\alpha}\}|H_{tran}|-,\{n_{q\alpha}\}>|^2 \; \delta\left(\bar{E}_{-}\{n_{q\alpha}\} - \bar{E}_{+}\{m_{q\alpha}\}\right) , \tag{12}$$

where $|-,\{n_{q\alpha}\}>$ is the initial antisymmetric Bloch-polaron state augmented by a set of dressed phonon states having occupation numbers $\{n_{q\alpha}\}$ and $|+,\{m_{q\alpha}\}>$ is the final symmetric Bloch-polaron state similarly augmented. The H_0 eigenvalues $\bar{E}_{\pm}\{n_{q\alpha}\}$ are given by

$$\bar{E}_{\pm}\{n_{q\alpha}\} = \bar{E}_{\pm} + \sum_{q\alpha} \hbar\omega_{q\alpha} \, n_{q\alpha} , \tag{13}$$

and $P(\{n_{q\alpha}\},T)$ denotes the thermal equilibrium probability associated with the initial-state dressed phonon distribution set $\{n_{q\alpha}\}$.

Proceeding in standard fashion, the delta function enforcing conservation of energy can be changed to a time integration which is subsequently evaluated approximately. Because the nonlocal force causes the lattice distortions for the Y and E states to be different, the transition matrix elements contain a complex Debye-Waller factor describing the reduced overlap of phonon wave functions for the initial and final states. To proceed with the evaluation of the time integral, we use a saddle-point approximation in a manner which parallels the usual polaron problem [6]. In our model, this saddle-point approximation is best in the regime where nonlocal coupling is strong. After the time integration, the transition rate formula has a structure familiar from the usual polaron problem and is valid over a wide temperature range; however, due to its complexity, this expression will not be given here (cf. [7]). At the *highest* temperatures, however, the formula for the transition rate approaches the simple asymptotic form

$$k_{+-}(T) \propto T^{1/2} e^{-E_{act}/k_B T} \qquad k_B T \; \gg \; \hbar\omega_a , \hbar\omega_E \tag{14}$$

where ω_a and ω_E are the acoustic phonon bandwidth and the Einstein frequency for the optical branch, respectively, and

$$E_{act} = (\Delta\bar{E} - \Gamma)^2/4\Gamma , \tag{15}$$

$$\Delta\bar{E} = \bar{E}_{-} - \bar{E}_{+} ,$$

$$\Gamma = \sum_{q\alpha} \hbar\omega_{q\alpha} \, |\chi_{+}^{q\alpha} - \chi_{-}^{q\alpha}|^2 . \tag{16}$$

Thus, in the high temperature limit, the transition process is thermally activated with the activation energy E_{act}. In this essentially classical limit we may think of E_{act} as the *average* activation energy for the transition. The potential energy surfaces implicit in this semiclassical picture of the transition process are indicated in Figure 4. In addition to the thermal-activation exponential, a temperature-dependent prefactor persists as evidence that thermal activation is not the end of the story. Unlike the usual polaron hopping rate, for which the lowest order process is a "diagonal" or "zero-phonon" transfer, the lowest order process contributing to our transition rate is a "non-diagonal" one-phonon process. The practical consequence of this distinction is that while the high-temperature prefactor in the usual polaron hopping rate is proportional to $T^{-1/2}$, we find $T^{1/2}$ instead. Our transition rate is not only phonon assisted, but is also phonon *controlled*.

In order to fit the data of Walker *et al.* [4] for α-perylene, we must recognize that the high-temperature limit (14) is not strictly applicable. We therefore retreat to the more general saddle-point formula (not shown, cf. [7]) which follows more directly from (12). The value for k_0 has been

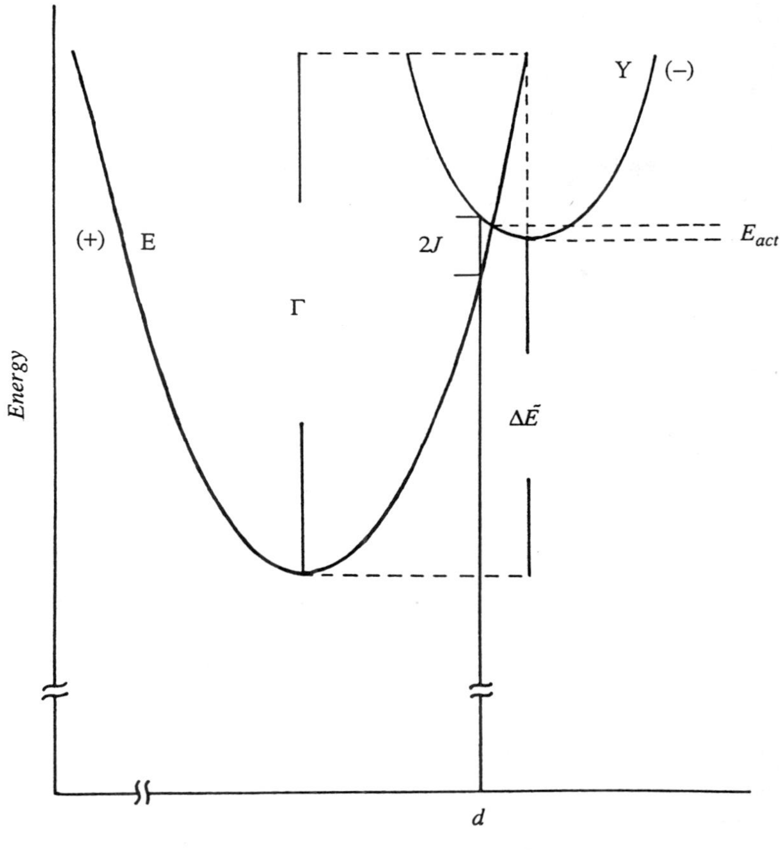

Fig. 4   Semiclassical potential energy surfaces implicit in the thermally activated Y-to-E transition (cf. (14) - (16)).

experimentally determined to be $2.25 \times 10^7 s^{-1}$ [4]. The optical phonons expected to participate in excimer formation are the B_g and A_g modes having frequencies of $95 cm^{-1}$ and $104 cm^{-1}$ respectively [8]. Unfortunately, reliable estimates of the corresponding acoustic frequencies appear not to be available; scaling from the isomorphic naphthalene crystal, a value of $18 cm^{-1}$ appears appropriate. To test the appropriateness of this value, we allowed the acoustic phonon bandwidth to vary as a fit parameter, and found best fit values in the range $15.7–16.7 cm^{-1}$, in reasonable agreement with the scaling argument. The energy separation of the bare pair states ($2J$) was taken to be $550 cm^{-1}$, the value reported by Tanaka [2] for the Davydov splitting. The result of fitting our model to the experimental data is shown in Figure 5. Best-fit values for the elementary force constants are $F_0 = 1.29–1.42 \times 10^{-5} dyne$, $G = 1.18–1.30 \times 10^{-5} dyne$, $-F_1 = 2.27–2.52 \times 10^{-6} dyne$ and $\delta F = 4.62–2.74 \times 10^{-8} dyne$. The negative value of F_1 implies that the excimer repels its nearest neighbor dimers. For these parameters, the excimer bond is compressed by approximately $0.98–1.08 Å$ in the E state, but is extended by about $0.27–0.30 Å$ in the Y state.

Considering that we have approximated the four Davydov components of α-perylene by only two levels, and have replaced the three-dimensional α-perylene crystal lattice by a one-dimensional harmonic chain, what is most meaningful about these numbers is not their absolute values but their relative magnitudes; perforce, it should be noted that throughout the fitting procedure, and under many trial assumptions, the relative values of the best-fit elementary force constants were roughly

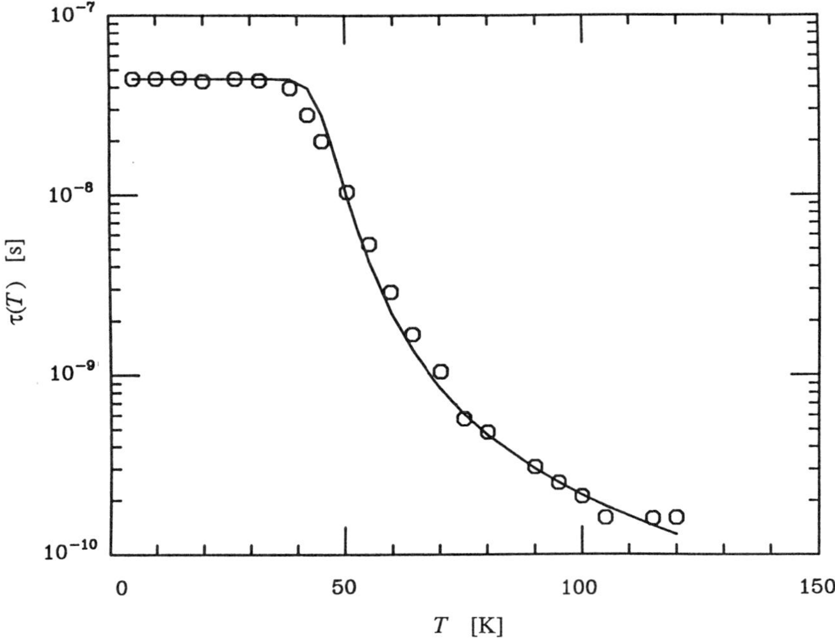

Fig. 5 Fit of the complete theoretical calculation of the transition rate to the experimental data of Walker *et al.* [4].

maintained. The fact that a satisfactory fit of the α-perylene data should require such a large relative strength of the nonlocal exciton-phonon interaction is perhaps the most significant single observation for future modelling considerations.

ACKNOWLEDGEMENT

This work was supported in part by DARPA Contract No. DAAG 29-85-K-0246 and by NSF Grant No. DMR 86-19650-A1

REFERENCES

1.   C. W. F. McClare, J. Theor. Biol. **35**, 569 (1972).

2.   J. Tanaka, Bull. Chem. Soc. Japan **36**, 1237 (1963).

3.   E. Von Freydorf, J. Kinder, M. E. Michel-Beyerle, Chem. Phys. **27**, 199 (1978).

4.   B. Walker, H. Port and H. C. Wolf, Chem. Phys. **92**, 177 (1985).

5.   I. G. Lang and Yu. A. Firsov Zh. Eksp. Teor. Fiz. **43**, 1834 (1962).

6.   K. D. Schotte, Z. Phys. **196**, 393 (1966).

7.   Ten-Ming Wu, David W. Brown and Katja Lindenberg, unpublished.

8.   T. J. Kosic, C. L. Schosser, and D. D. Dlott, Chem. Phys. Lett. **96**, 57 (1983).

36

EXCIMERS IN MOLECULAR CRYSTALS:

THE RELAXATION OF A NONLINEAR OSCILLATOR

V. M. KENKRE, D. H. DUNLAP, and P. GRIGOLINI

Department of Physics and Astronomy
University of New Mexico
Albuquerque, NM 87131, USA

1. INTRODUCTION

Excimer physics is intimately related to nonlinear physics in general. We describe below a simple classical model which two of the present authors developed several years ago. When extended, as we have done recently, the model has the capability of treating both the normal excimer state, and the so-called Y-state of α-perylene.

An excimer is an excited complex consisting of a pair of identical monomers with an equilibrium separation which corresponds to a strongly repulsive interpair interaction upon deexcitation. Excimers were first recognized in a solution of pyrene in n-heptane[1,2]. At low concentrations of pyrene, the absorption and emission spectra show vibrational structure, and are observed to be essentially mirror images of each other. As the concentration of pyrene is increased, the structured fluorescence is replaced by a broad structureless emission, shifted to the red[3] by about 3000 cm^{-1}. Excimer formation has been observed in many other solutions of aromatic hydrocarbons as well, including solutions of naphthalene, perylene, anthracene, and benzene[4]. For reasons which we will not discuss for want of space here, excimer formation in solids is much more surprising[5,6] and is the focus of our own studies. The excimer in pyrene is a prototypical example[5] in aromatic hydrocarbon crystals. The proportionality of the spectral width and temperature at high temperatures[7], activated

Davydov's Soliton Revisited, Edited by P.L. Christiansen and A.C. Scott
Plenum Press, New York, 1990

transport[8-10], a large Stokes shift of roughly 6000 cm^{-1}, a long lifetime, and a shift of the excimer peak to the blue with increasing temperature[11] are quite general[3]. Additional properties are observed in α-perylene. The most interesting is an emission band which, at low temperatures, is displaced to the blue[12-14] and is attributed to the "Y-state", considered a precursor to the excimer state[15-20].

The pyrene excimer spectrum in solution is understood on the basis of a potential diagram for a single configuration coordinate which represents the pair separation[1,2]. In the ground state, no bound pair exists. During an excitation, the excimer pair is completely separated, and only the vibrations of an individual monomer characterize the spectrum. However, if the molecular migration results in the excited pair achieving a separation where the potential is steeply attractive, the pair forms a bound (excimer) state. Emission to the steeply repulsive part of the (unbounded) ground state pair-potential gives a broad (structureless) contribution to the spectrum. In the case of the crystal, the ground state excimer potential is bounded, and there is little change in the frequencies of the excimer pair in the ground and excited states[3]. However, the absorption and emission spectra of pyrene crystals are similar to those in solution[5], so it is also likely that the interaction potential in the solid and in solution are also similar. The transition from the ground state to the excited state must be to the flat tail of the excited state pair potential so that the absorption spectrum will be nearly monomeric. It follows that the forces initially drawing the excited molecular pair together must be weak, in spite of the fact that the difference between their initial separation and their excimer state separation is rather small (0.2 A for pyrene). Therefore, the pair potential in the crystal must fall off extremely quickly, thus demanding that the contracting force be highly nonlinear.

On the basis of the above considerations, and motivated by the work of Collins and Craig[21] who introduced the study of the classical vibrational relaxation in a chain in the excimer problem (although in the context of an intermediate or precursor state), we proposed[22,23] that excimer formation could be understood simply as the relaxation of an oscillator which becomes suddenly highly nonlinear on absorption. This is one connection between excimer phenomena and nonlinear physics. We present our model and some essential results[22] in section

2, provide, on the basis of the model, a possible explanation of the formation of the peculiar Y-state[23] in section 3, and conclude in section 4.

2. OUR METHOD OF ATTACK: RELAXATION OF A NONLINEAR OSCILLATOR

The model of an excimer forming crystal we consider in this section is, as stated above, an infinite linear chain of masses (molecules) interacting via nearest-neighbor linear springs (Hooke's law forces). We assume that, on electronic excitation, one of the intermolecular springs is replaced by a nonlinear spring. If the equilibrium length of the new nonlinear spring is smaller than that of the old (linear) spring, the two neighboring molecules will contract. In our model, the excimer formation process is the contraction of this defect pair, and the time for excimer formation is the time for the oscillations caused by this contraction to damp out.

The displacement from equilibrium of the mth molecule in the chain, $x_m(t)$, obeys the Euler equation

$$M\ddot{x}_m = -k_o\left(2x_m - x_{m+1} - x_{m-1}\right) - \left(\delta_{m,r} - \delta_{m,r+1}\right)$$
$$\times\left[k_o\left(x_r - x_{r+1}\right) - U'\left(x_{r+1} - x_r\right)\right] \tag{2.1}$$

where M is the mass of each molecule, k_o is the intermolecular force constant, and $U'(x)$ is the derivative of the anharmonic excimer potential $U(x)$ with respect to its argument. The term with the kronecker deltas in (2.1) represents the replacement of the linear interaction by the nonlinear one for the excimer pair residing at sites r and $r+1$. The difference, $x_{m+1} - x_m$, which will be termed ζ_m, obeys

$$\ddot{\zeta}_m = \omega_o^2\left(\zeta_{m-1} + \zeta_{m+1} - 2\zeta_m\right) + \left(2\delta_{m,r} - \delta_{m,r-1} - \delta_{m,r+1}\right)\left[\omega_o^2\zeta_r - \frac{U'\left(\zeta_r\right)}{M}\right] \tag{2.2}$$

where $\omega_o^2 = \dfrac{k_o}{M}$. By using a discrete Fourier transform over the sites m, and using Laplace transform techniques, we find

$$\tilde{\zeta}_r(\varepsilon) = \tilde{\eta}_r(\varepsilon) + \tilde{\phi}(\varepsilon)\int_0^\infty dt\, e^{-\varepsilon t}\left[\omega_o^2\zeta_r(t) - \frac{U'\left(\zeta_r(t)\right)}{M}\right] \tag{2.3}$$

where tildes denote Laplace transforms, ε is the Laplace variable, $\tilde{\eta}_r(t)$ is the homogeneous solution of (2.1) in the absence of the defect spring, $\tilde{\phi}(\varepsilon)$ is the Laplace transform of $\phi(t) = \dfrac{2}{\omega_o} J_1(2\omega_o t)$, and J_1 is the ordinary Bessel function of the first kind of order 1. Inverting (2.3) yields

$$\zeta_r(t) = \eta_r(t) + \int_0^t dt'\, \tilde{\phi}(t-t')\left[\omega_o^2\, \zeta_r(t') - \frac{U'\big(\zeta_r(t')\big)}{M}\right]. \qquad (2.4)$$

Taking the initial velocity and separation of the excimer pair to be zero, (2.4) can be solved numerically for $\zeta_r(t)$ no matter how nonlinear $U\big(\zeta_r\big)$ is. In this manner, we have studied several cases of the nonlinearity including the Morse, Rydberg, and Beckel-Findley[24] potentials, and determined the relaxation in the time domain and spectra in the frequency domain. In order to model excimer formation in crystals such as dichloroanthracene[10] which possess a stack structure, we have studied linear chains, and to study crystals such as pyrene[25] which consist of preformed dimers, we have studied chains in which there are two sites per unit cell. We refer the reader to ref. 22 for details.

3. Y-STATES: NONLINEAR RELAXATION IN CONSTRAINED SITUATIONS

Our model provides a natural quantification of arguments that the peculiar Y-state observed in α-perylene is caused by the incomplete relaxation of the excimer pair coordinate due to the hindering of the lattice[13-19,26]. Whereas, a 1-dimensional lattice hinders excimer formation only in the sense that it provides an effective viscosity in which the excimer must contract, in higher dimensions, more restrictive forces occur, there being horizontal, vertical, and diagonal interactions. We take the spring constants in the horizontal and vertical directions to be k_0, and the spring constant in the diagonal directions to be k_1. Equation (2.4) results again, with the modification that the kernel $\varphi(t)$ is complicated by the interplay between the horizontal and vertical interactions and the diagonal interactions. Rich behavior emerges from a numerical solution of (2.4) for several values of k_1/k_0. The equilibrium excimer-pair separation increases monotonically with k_1/k_0 because the neighboring molecules are interconnected via an infinite number of indirect diagonal paths. These interactions tend to oppose the lattice distortion which accompanies the contraction for the excimer pair, and their effect increases with the strength of the diagonal spring. The striking result is that the equilibrium separation of the pair undergoes a sudden transition as k_1/k_0 is varied. This phenomenon is indicative of the Y-state.

To understand the source of the transition, we return to (2.3) and rewrite the equation in the following form

$$\widetilde{\kappa}(\varepsilon)\,\widetilde{\zeta}(\varepsilon) = -\int_0^\infty dt\, e^{-\varepsilon t}\, u'\big(\zeta(t)\big) \tag{3.1}$$

where $\widetilde{\kappa}(\varepsilon) = \dfrac{1}{\widetilde{\varphi}(\varepsilon)} - \omega_o^2$ and $u'\big(\zeta\big) = \dfrac{U'(\zeta)}{M}$. From (3.1), exact conclusions concerning $\zeta(\infty)$, the equilibrium value of the excimer coordinate $\zeta(\infty)$ may be obtained. On multiplying (3.1) by ε and taking the limit that $\varepsilon \to 0$ and invoking an Abelian theorem concerning asymptotic values, we get

$$\widetilde{\kappa}(0)\,\zeta(\infty) = \left[\frac{1}{\widetilde{\varphi}(0)} - \omega_o^2\right]\zeta(\infty) = -u'\big(\zeta(\infty)\big) \tag{3.2}$$

Equation (3.2) can be solved directly to obtain the equilibrium value of the excimer coordinate once the time integral of the propagator $\widetilde{\varphi}(0)$, and the nonlinear force $U'(\zeta)$ are known. It also provides us with immediate insight into the formation of the Y-state: the Y-state arises only when (3.2) has more than one solution for $\zeta(\infty)$. The physical meaning of (3.2) is as follows. At equilibrium, the excimer pair is acted upon by two forces. The nonlinear excimer pair force which tends to pull the pair together and is represented by the right-hand side of (3.2), and an effective restoring force exerted by the lattice which tends to oppose the contraction of the excimer pair. The latter is represented by the left-hand side of (3.2), and corresponds to an effective spring constant $\left[\dfrac{1}{\widetilde{\varphi}(0)} - \omega_o^2\right]$. When, as in the 1-dimensional chain, hindering interactions arising from the (diagonal) k_1 springs are absent, $\widetilde{\varphi}(0)$ equals $\dfrac{1}{\omega_o^2}$, and the effective spring constant $\widetilde{\kappa}(0)$ vanishes. The equilibrium state of the excimer pair is then at the minimum of the nonlinear potential $U(\zeta)$, and there is only a single excimer state, the E-state. This situation also occurs for excimers in solution for which the interactions with the rest of the "lattice" are negligible. When, however, those

interactions are present, two effects can occur. One is a simple
shift of the E-state equilibrium coordinate caused by the addition of
the left-hand side term in (3.2). The other, which only occurs when
the effective spring constant $\widetilde{\kappa}(0)$ exceeds a critical value, is that
multiple minima appear in the total potential $u(\zeta) + \frac{1}{2}\widetilde{\kappa}(0)\zeta^2$. The
additional minimum in the effective potential represents the Y-state.

4. SIMPLIFIED MODEL AND THERMAL FLUCTUATIONS

Recently, two of the present authors[27] have extended the above
ideas to take into account thermal fluctuations and the passage from
the Y-state to the E-state on the basis of a simplification of the
evolution equation. Equation (3.2) above is an exact asymptotic
consequence of the dynamics of our excimer model. However, its form is
compatible with the consequence of a simplified model in which the
coordinate ζ follows the time evolution

$$\frac{d^2\zeta}{dt^2} + \gamma\frac{d\zeta}{dt} + \left[\frac{1}{\widetilde{\phi}(\varepsilon)} - \omega_0^2\right]\zeta = -u'(\zeta) \tag{4.1}$$

Equation (4.1) has been constructed to give the same equilibrium
solution as (3.2), a damping agent γ having been introduced to ensure
the approach to that equilibrium. The damping arises in the actual
system from the interaction of the ζ coordinate with the rest of the
lattice. Thermal fluctuations can be introduced[27] into the model by
appending to (4.1) a noise term which converts (4.1) into a Langevin
equation and allows the calculation such quantities as the escape rate
from the Y-state to the E-state induced by the stochastic interactions
of the excitation with the lattice. This work and the application to
observations may be found in ref. 27. Here we will show how an
estimate of γ can be made. It is clear that such an estimate is
necessary both for (4.1) to be useful and for the random force to be
introduced (since it must obey the fluctuation dissipation relation
with γ.

One simple way of estimating γ is to fit (4.1) to the results of
(2.2) suitably generalized to higher dimensions, and to obtain γ from
these fits. This is a straightforward procedure. It is successful
because the qualitative features of the evolution as described by
(4.1) is found to be not very different from that of the generalized
(2.2). However, it is also possible to generalize (4.1) as

$$\frac{d^2\zeta}{dt^2} + \int_0^t ds\ \Gamma(t-s)\frac{d\zeta}{ds} + \left[\frac{1}{\tilde{\phi}(\varepsilon)} - \omega_0^2\right]\zeta = -u'(\zeta) \tag{4.2}$$

where $\Gamma(t)$ is a "memory-function" or non-Markoffian damping agent related to γ through

$$\gamma = \int_0^\infty dt\ \Gamma(t) \tag{4.3}$$

Equating the Laplace transform of ζ as obtained from (4.1) and (3.1) respectively yields an explicit expression for the non-Markoffian damping memory Γ in the Laplace domain. A Markoffian approximation, in cases where applicable, then gives the damping rate γ.

4. CONCLUSION

In this article we have summarized our model of the formation of an excimer in a crystal as the relaxation of an oscillator turned nonlinear by the absorption of light. A high degree of nonlinearity is required by the features of the observed spectra. We explicitly include in our model interactions with the rest of the lattice which tend to oppose complete relaxation of the excimer. Our analysis has shown that the Y-state is separated discontinuously from the fully relaxed state E-state, that it does not arise in the absence of the hindering interactions, and that, even when the latter are present, it arises only when the strength of those interactions exceeds a threshold value. The underlying picture of the Y-state, made quantitative here, is thus precisely that prevalent in the literature[13-19,26].

REFERENCES

1. Th. Foerster and K. Kasper, Z. Physik. Chem. N.F. 1, 275 (1954).
2. Th. Foerster, Angew. Chem. Internat. Edit. 8, 333 (1969).
3. J. B. Birks, Photophysics of Aromatic Molecules (Wiley-Interscience, London, 1970), Chapter 7.
4. J. B. Birks and L. G. Christophorou, Proc. Roy. Soc. A 277, 571 (1964).
5. J. Ferguson, J. Chem. Phys. 28, 765 (1958).
6. B. Stevens and E. Hutton, Nature 186, 1045 (1960).
7. J. B. Birks and A. A. Kazzaz, Proc. Roy. Soc. A 304, 291 (1968).

8. W. Kloepffer, H. Bauser, F. Dolezalek, and G. Naundorf, Mol.Cryst. Liq. Cryst. 16, 229 (1972).

9. Z. Lebovitz, S. Mansour, and A. Weinreb, J. Chem. Phys. 69, 647 (1978).

10. U. Mayer, H. Auweter, A. Braun, H. C. Wolf, and D. Schmid, Chem. Phys. 59, 449 (1981).

11. V. Yakhot, M. D. Cohen, and Z. Ludmer, Advances in Theor. Chem. 13, 489 (1979).

12. J. Ferguson, J. Chem. Phys. 44, 2677 (1966).

13. J. Tanaka, Bull. Chem. Soc. Jpn. 36, 1237 (1963).

14. R. M. Hochstrasser, Can. J. chem. 39, 451 (1961); J. Chem. Phys.40 , 2559 (1964).

15. E. Von Freydorf, J. Kinder, and M. E. Michel-Beyerle, Chem.Phys. 27, 199 (1978).

16. H. Auweter, D. Ramer, B. Kunze, and H. C. Wolf, Chem. Phys. Lett. 85, 325 (1982).

17. T. Kobayashi, J. Chem. Phys. 69, 3570 (1978).

18. A. Matsui and H. Nishimura, J. Phys. Soc. Japan 49, 657 (1980).

19. B. Walker, H. Port, and H.C. Wolf, Chem. Phys. 92, 177 (1985).

21. M. A. Collins, and D. P. Craig, Chem. Phys. 54, 305 (1981).

22. D. H. Dunlap and V. M. Kenkre, Chem. Phys. 105, 51 (1986)

23. D. H. Dunlap and V. M. Kenkre, Phys. Rev. B 37, 4390 (1988).

24. P. M. Morse, Phys. Rev. 34, 57 (1929); R. Rydberg, Z. Physik 73, 376 (1931); C. L. Beckel and P. R. Findley, J. Chem. Phys. 73, 3517 (1980).

25. H. Port, R. Seyfang, and H. C. Wolf, J. Phys. Colloq. C7, 391 (1985); . Maatsui and H. Nishimura, J. Phys. Soc. Japan 51, 1711 (1982).

26. M.D.Cohen, R. Haberkorn, E. Huler, Z. Ludmer, M. E. Michel-Beyerle, D.Rabinovich,R.Sharon, A.Warshel,V.Yakhot,Chem.Phys. 27, 211 (1978).

27. V. M. Kenkre and P. Grigolini, University of New Mexico preprint.

37

THE NONRESONANT DST EQUATION AS A MODEL FOR McCLARE'S EXCIMER

Alwyn Scott

Department of Mathematics, University of Arizona
Tucson, Arizona 85721 U.S.A.
    and
Laboratory of Applied Mathematical Physics
The Technical University of Denmark
DK-2800 Lyngby, Denmark

In the contribution by Knox et al. to these proceedings, it was emphasized that the mechanism for injecting energy into—and extracting it from—the alpha helix region has not yet been adequately considered. They suggest that this question deserves as much theoretical effort as the problem of calculating soliton lifetime. A similar concern was noted by the present author several years ago [1]. Introducing an arbitrary perturbation of the sound system in Davydov's equations, it was shown that only the (kinetic) energy of motion would be converted to mechanical energy, not the rest energy. Other mechanisms, it was concluded, must be supposed to make use of the main portion of soliton energy. The aim here is to consider one such mechanism: McClare's "excimer" [2].

The basic idea can be explained by considering the classical *nonresonant* discrete self-trapping (DST) system [3]

$$\left\{ i\frac{d}{dt} + \begin{bmatrix} (\gamma|A_1|^2 - \omega_1) & \varepsilon \\ \varepsilon & (\gamma|A_2|^2 - \omega_2) \end{bmatrix} \right\} \bar{A} = \bar{0} \tag{1}$$

where $\bar{A} = \mathrm{col}(A_1, A_2)$ and γ is an anharmonicity parameter. The dispersion parameter, ε, is the interaction energy of the two dipoles shown in Figure 1. It is assumed to have the functional dependence

$$\varepsilon = \varepsilon(x) \tag{2}$$

where x is the extension of a linear spring, W. Briefly, an attraction between the two dipoles converts electric field energy into potential energy of the spring.

This might be a mechanism for energy transfer out of or into a Davydov soliton. In the first case, one dipole would represent a soliton and the other might be a closely tuned local mode in a protein structure. The spring would then be some elastic region of the protein. Since we cannot expect the soliton and the local mode to be exactly in resonance, it is important to consider how energy transfer depends on the "detuning" parameter

$$\Delta \equiv \omega_2 - \omega_1. \tag{3}$$

Davydov's Soliton Revisited, Edited by P.L. Christiansen and A.C. Scott
Plenum Press, New York, 1990

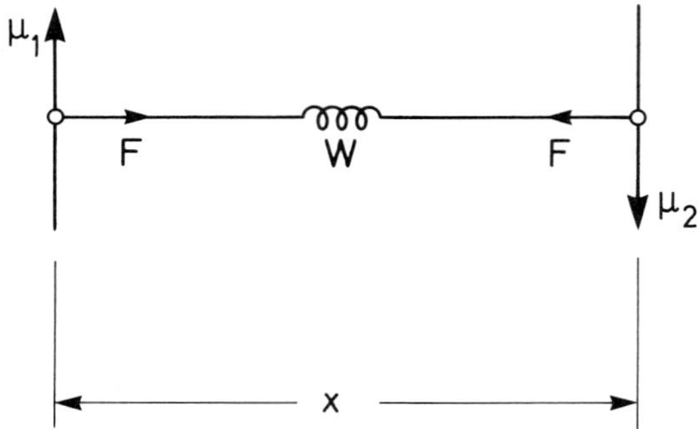

Figure 1: A sketch of McClare's excited dimer or *excimer*. An attractive force, F, acts on a linear spring (W) to convert vibrational energy into mechanical work.

This system is described by the quantum mechanical energy operator

$$\hat{H} = \hat{H}_1 + \hat{H}_2 + \hat{H}_{int} \tag{4}$$

where

$$\hat{H}_1 = (\omega_1 - \tfrac{1}{2}\gamma)(\hat{B}_1^\dagger \hat{B}_1 + \tfrac{1}{2}) + (\omega_2 - \tfrac{1}{2}\gamma)(\hat{B}_2^\dagger \hat{B}_2 + \tfrac{1}{2}) -$$
$$-\tfrac{1}{2}\gamma(\hat{B}_1^\dagger \hat{B}_1 \hat{B}_1^\dagger \hat{B}_1 + \hat{B}_2^\dagger \hat{B}_2 \hat{B}_2^\dagger \hat{B}_2) \tag{5}$$

$$\hat{H}_2 = \tfrac{1}{2}W(\hat{q} - x_0)^2 \tag{6}$$

$$\hat{H}_{int} = -\varepsilon(\hat{q})(\hat{B}_1^\dagger \hat{B}_2 + \hat{B}_2^\dagger \hat{B}_1). \tag{7}$$

As has been discussed in earlier chapters of this section, $\hat{B}_i^\dagger$ ($\hat{B}_i$) are creation (annihilation) operators for the two degrees of freedom, $\hat{q}$ is a quantum mechanical displacement operator corresponding to x, and $\hat{H}_2$ is the energy operator of the spring. Clearly, $\hat{H}$ commutes with the number operator $\hat{B}_1^\dagger \hat{B}_1 + \hat{B}_2^\dagger \hat{B}_2$.

To find an approximate solution to this problem, the technique introduced by Davydov is appropriate [4,5]. Assume that the motion of the spring is slow enough so that it is well described by the coherent state $|\phi\rangle$ where

$$\hat{q}\,|\phi\rangle = x\,|\phi\rangle. \tag{8}$$

If the vibrational system is at the first excited level, then an approximate wave function is the product

$$|\psi_1\rangle = [\tilde{a}_1(t)\,|1\rangle\,|0\rangle + \tilde{a}_2(t)\,|0\rangle\,|1\rangle]\,|\phi\rangle \tag{9}$$

for which

$$\langle\psi_1|\hat{H}|\psi_1\rangle = (\tfrac{3}{2}\omega_1 + \tfrac{1}{2}\omega_2 - \tfrac{3}{2}\gamma)|\tilde{a}_1|^2 + (\tfrac{3}{2}\omega_2 + \tfrac{1}{2}\omega_1 - \tfrac{3}{2}\gamma)|\tilde{a}_2|^2 -$$
$$-\varepsilon(x)(\tilde{a}_1^*\tilde{a}_2 + \tilde{a}_1\tilde{a}_2^*) + \tfrac{1}{2}W(x - x_0)^2. \tag{10}$$

Variation of this average Hamiltonian with $\tilde{a}_1$, $\tilde{a}_2$, and x gives the dynamical system

$$i\dot{\tilde{a}}_1 = (\tfrac{3}{2}\omega_1 + \tfrac{1}{2}\omega_2 - \tfrac{3}{2}\gamma)\tilde{a}_1 - \varepsilon(x)\tilde{a}_2 \tag{11}$$

$$i\dot{\tilde{a}}_2 = (\tfrac{3}{2}\omega_2 + \tfrac{1}{2}\omega_1 - \tfrac{3}{2}\gamma)\tilde{a}_2 - \varepsilon(x)\tilde{a}_1 \tag{12}$$

$$W(x - x_0) = \varepsilon'(x)(\tilde{a}_1^*\tilde{a}_2 + \tilde{a}_1\tilde{a}_2^*). \tag{13}$$

With the gauge transformation

$$\tilde{a}_j = A_j \exp\left[-\frac{i}{2}t(\omega_1 + \omega_2 - 3\gamma)\right], \qquad j = 1,2 \tag{14}$$

Equations (11,12,13) become

$$i\dot{A}_1 = \omega_1 A_1 - \varepsilon(x)A_2 \tag{15}$$

$$i\dot{A}_2 = \omega_2 A_2 - \varepsilon(x)A_1 \tag{16}$$

$$W(x - x_0) = \varepsilon'(x)[A_1^* A_2 + A_1 A_2^*]. \tag{17}$$

Since the system is assumed to be in the first excited state, the motion of Equations (15,16,17) is constrained by the condition

$$|A_1|^2 + |A_2|^2 = 1. \tag{18}$$

Note also that the derivative $\varepsilon'(x) = d\varepsilon/dx$ in Equations (15,16,17) is the "resonance" exciton-phonon coupling parameter discussed by Kuprievich, Pierce and Østergård in Section II of these proceedings.

To solve Equations (15,16,17), it is convenient to let

$$A_1(t) = a_1 \exp(i\theta_1) \tag{19}$$

$$A_2(t) = a_2 \exp(i\theta_2) \tag{20}$$

where $a_1(t)$, $a_2(t)$, $\theta_1(t)$ and $\theta_2(t)$ are all real and, from Equation (18), $a_1^2 + a_2^2 = 1$. Defining $N_1 \equiv a_1^2$ and $\theta \equiv \theta_1 - \theta_2$, Equations (15,16,17) become

$$\dot{N}_1 = 2\varepsilon(x)\sqrt{N_1 - N_1^2}\,\sin\theta \tag{21}$$

$$\dot{\theta} = \omega_2 - \omega_1 + \varepsilon(x)\left[\frac{1 - 2N_1}{\sqrt{N_1 - N_1^2}}\right]\cos\theta \tag{22}$$

$$W(x - x_0) = 2\varepsilon'(x)\sqrt{N_1 - N_1^2}\,\cos\theta \tag{23}$$

with the Hamiltonian

$$H = -(\omega_2 - \omega_1)N_1 - 2\varepsilon(x)\sqrt{N_1 - N_1^2}\,\cos\theta + \tfrac{1}{2}W(x - x_0)^2. \tag{24}$$

The mechanical force acting on the spring is

$$F = 2\varepsilon'(x)\sqrt{N_1 - N_1^2}\,\cos\theta. \tag{25}$$

The motion of Equations (21,22,23) can be rather complicated, but to simplify matters it is convenient to assume a stationary state ($\dot{N}_1 = 0$ and $\dot{\theta} = 0$). This implies that $\theta = 0$ and therefore

$$N_1 = \tfrac{1}{2}\left[1 \pm \sqrt{1 - \left(1 + \frac{a^2}{4}\right)^{-1}}\,\right] \tag{26}$$

where $a \equiv (\omega_2 - \omega_1)/\varepsilon \equiv \Delta/\varepsilon$. Thus $(N_1 - N_1^2) = 1/(4 + a^2)$ and

$$F = \frac{\varepsilon'(x)}{\sqrt{1 + \dfrac{a^2}{4}}}\,. \tag{27}$$

As indicated in Figure 1, this force is an attraction between the two dipoles. We have, therefore, a mechanism that can transfer energy in two directions: i) As the two dipoles come closer together, field energy is converted into mechanical energy of the spring, and ii) As the dipoles are pulled apart, the mechanical energy that is doing the pulling is converted into dipole field energy. If the mechanical coordinate decreases from x_1 to x_2, the resulting mechanical energy will be

$$U = \int_{x_1}^{x_2} F\, dx$$

$$= \sqrt{\varepsilon^2(x_2) + \frac{\Delta^2}{4}} - \sqrt{\varepsilon^2(x_1) + \frac{\Delta^2}{4}}. \tag{28}$$

Under the most favorable circumstances, $x_1 = \infty$ and $\Delta = 0$. Then

$$U = \frac{\mu_1\mu_2}{4\pi\varepsilon_0 x_2^3} \tag{29}$$

where ε_0 $(= 10^{-9}/36\pi$ farads/meter$)$ is the dielectric permittivity of free space. If both dipoles have the transition dipole moment of an amide-I oscillation [6] (i.e. $\mu_1 = \mu_2 = 0.3$Debye $= 1.3 \times 10^{-30}$coulomb meters) and $x_2 = 2$Å, then Equation (29) indicates that about 6% of the amide-I vibrational energy can be converted to mechanical work. This is an order of magnitude less than measured values of muscular efficiency [7].

To increase this efficiency, larger dipole moments must be employed. On the one hand, this might be achieved by assuming that a vibrational soliton interacts with a low lying electronic state (so $\mu_2 \gg \mu_1$). On the other hand, as Davydov has suggested in the first chapter of these proceedings, the soliton itself might be a self-trapped electron.

This work has been supported by the (US) National Science Foundation and the Air Force Office of Scientific Research.

References

[1] A.C. Scott, Phys. Rev. A **26**, 578 (1982).

[2] C.W.F. McClare, J. Theor. Biol. **35**, 569 (1972).

[3] A.C. Scott, "A nonresonant discrete self-trapping system" Physica Scripta (in press).

[4] A.S. Davydov, Physica Scripta **20**, 387 (1979).

[5] A.S. Davydov, *Biology and quantum mechanics*, Pergamon, New York (1982).

[6] Yu.N. Chirgadze and N.A. Nevskaya, Dok. Akad. Nauk USSR **208**, 447 (1973).

[7] B. Alberts, D. Bray, J. Lewis, M. Raff, K. Roberts, and J.D. Watson, **Molecular Biology of the Cell**, Garland, New York (1983) pp. 550- 609.

Section VI

The Discrete Self-Trapping Equation

The better part of valour is discretion.
William Shakespeare

In the final section of these proceedings, several chapters on the discrete self-trapping (DST) equation are included. It can be written in the form

$$i\dot{A}_j = \frac{\partial H}{\partial A_j^*} \qquad j = 1, 2, \ldots, f \tag{1}$$

where

$$H = \sum_{j=1}^{f} \left[\omega_j |A_j|^2 - \frac{1}{2}\gamma |A_j|^4 \right] - \varepsilon \sum_{j \neq k} m_{jk} A_j^* A_k \tag{2}$$

and $m_{jk} = m_{kj} = m_{jk}^*$. Considering A_j to be a component of the vector $\bar{A} = \mathrm{col}(A_1, A_2, \ldots, A_f)$ with the L^2 norm

$$N = \sum_{j=1}^{f} |A_j|^2, \tag{3}$$

it is easily seen that $\dot{N} = 0$.

Why should one be interested in the DST equation? Because it is a rich mathematical object with many physical applications. Through the parameters γ and ε, the DST equation exhibits the interplay of anharmonicity and dispersion that characterizes much modern research in applied mathematics. Its physical applications include the following.

Classical description of molecular vibration. The DST equation provides a useful model for the vibrations of small molecules [1]. In this case, $A_j = (p_j - i\omega_j q_j)/\sqrt{2}$ where p_j and q_j are the standard momentum and position of the jth atom. The anharmonicity, scaled by γ, is called *intrinsic* because it arises from the intrinsic nonlinear character of an atomic bond. As the chapter by Bernstein shows, it is important in this application to consider quantum effects.

Quantum description of classical anharmonicity. The quantum DST equation is unusually simple because the conserved norm, N, is identical to the energy of f uncoupled, harmonic oscillators. Thus wave functions are readily constructed on a basis of boson number states [2]. The details of such quantization are described below in the chapters by Eilbeck, Bernstein and Salerno. The chapter by Bernstein discusses the quantum description of energy localization, while those by Feddersen et al. and by Salerno consider the quantum description of chaos.

Quantum description of extrinsic anharmonicity. This is the main subject of these proceedings, and it was discussed in detail in the chapters of Section I. The anharmonicity is called *extrinsic* since it arises through interaction with a lower frequency phonon field. The DST system arises as an adiabatic description of the parameters in Davydov's product trial wave function for the Fröhlich Hamiltonian, with anharmonicity parameter

$$\gamma = \frac{\chi^2}{w} \tag{4}$$

where χ and w are defined in the introductory remarks to Section I.

Nonlinear Anderson localization. With $\gamma = 0$, Equations (1) and (2) are just those used to describe linear localization of eigenfunctions in a random medium [3]. By including $\gamma \neq 0$, the DST equation provides a means for studying the effects of (both intrinsic and extrinsic) anharmonicity on Anderson localization. This question is particularly important when one considers the nonlinear eigenstates of a globular protein. Such a study, which is currently being pursued by Henrick Feddersen, can be viewed as a generalization of Davydov's original alpha-helix theory to more realistic protein geometries.

Nonlinear quantum theory. In this application of the DST equation, A_j in Equation (1) is considered to be a true quantum mechanical probability amplitude. The idea is to test whether quantum mechanics itself is nonlinear [4]. This application requires a different Hamiltonian from that defined in Equation(2).

With all these applications for the DST equation, it is not surprising that there are some problems of nomenclature. If in Equations (1) and (2) one assumes: i) only nearest neighbor interactions along a linear chain, and ii) all the ω_j are equal and scaled out of the problem, the system reduces to

$$i\dot{A}_j + \varepsilon(A_{j+1} + A_{j-1}) + \gamma|A_j|^2 A_j = 0. \tag{5}$$

Physicists often refer to Equation (5) as the discrete nonlinear Schrödinger (DNLS) equation. However there is a seemingly similar equation

$$i\dot{A}_j + \varepsilon(A_{j+1} + A_{j-1}) + \frac{1}{2}\gamma|A_j|^2(A_{j+1} + A_{j-1}) = 0 \tag{6}$$

which is also called the discrete nonlinear Schrödinger (DNLS) equation by applied mathematicians.

Although Equations (5) and (6) reduce to the exactly integrable nonlinear Schrödinger (NLS) equation [5] in the continuum limit, Equation (6) with periodic boundary conditions is also exactly integrable [6] while Equation (5) is not. Furthermore the L^2 norm is conserved for Equation (5) but not for Equation (6). Thus Equation (5) is more readily quantized. The

470

two equations have recently been compared as a basis for numerical integration of the NLS equation [7].

Our preference is to call Equation (5) the discrete nonlinear Schrödinger (DNLS) equation recognizing that it is a special case of the discrete self-trapping (DST) equation defined in Equations (1) and (2). Equation (6), on the other hand, can be called the Ablowitz-Ladik (AL) equation.

It is interesting to note that both (the DNLS) Equation (5) and (the AL) Equation (6) can be expected to arise in practice. To see this, consider the simplest description of an exciton in a linear chain [8]:

$$i\dot{A}_j - E_0 A_j + J(A_{j+1} + A_{j-1}) = 0 \qquad j = 1, 2, \ldots, f. \tag{7}$$

where A_j is the probability amplitude for finding the exciton at site j, E_0 is the site energy and J is the nearest neighbor coupling. Following the ideas outlined by Kuprievich in Section II, extrinsic anharmonicity can be expected to manifest itself in two ways: i) Modulation of E_0 by the phonon field (through χ) leading to (the DNLS) Equation (5), and ii) Modulation of J by the phonon field (through χ^{res} leading to (the AL) Equation (6). In this development, it is important to remember that the A_j appearing in Equations (5) and (6) is the time dependent parameter in Davydov's product trial wave function and no longer a true probability amplitude.

It is also interesting to consider generalizations of Equations (1) and (2) that preserve time independence of the norm defined in Equation (3). The reason for this requirement, as was noted above, is that it allows quantum wave functions to be constructed on a basis of boson number states. Since the norm is the conserved quantity related (via Noether's theorem) to gauge symmetry, it is sufficient to require the Hamiltonian

$$H = H(\bar{A}, \bar{A}^*) \tag{8}$$

to be invariant under the transformation $\bar{A} \to \bar{A} \, exp(i\alpha)$ [9].

Motivated by the study of chaos in a periodically perturbed, polarized optical beam, David et al. [10] have considered norm conserving DST system with two degrees of freedom and

$$H = A_j^* \chi_{jk}^{(1)} A_k + \frac{3}{2} A_j^* A_k \chi_{jk\ell m}^{(3)} A_\ell A_m^* \tag{9}$$

where the χ's are susceptibility tensors satisfying the conditions: $\chi_{jk}^{(1)} = \chi_{kj}^{(1)*}$, $\chi_{jk\ell m}^{(3)} = \chi_{jk\ell m}^{(3)*}$, and $\chi_{jk\ell m}^{(3)} = \chi_{mk\ell j}^{(3)} = \chi_{j\ell km}^{(3)}$, $\quad (j, k, \ell, m = 1, 2)$.

For his test of nonlinearity in quantum mechanics, Weinberg assumed two degrees of freedom and chose [4]

$$H = N F \left(\frac{|A_2|^2}{N} \right) \tag{10}$$

where $F(\cdot)$ is an arbitrary differentiable function. This Hamiltonian conserves the L^2 norm and also satisfies the homogeneity condition

$$H(\lambda \bar{A}, \bar{A}^*) = H(\bar{A}, \lambda \bar{A}^*) = \lambda H(\bar{A}, \bar{A}^*) \tag{11}$$

which insures that if $\bar{A}(t)$ is a solution of Equation (1) then $\lambda \bar{A}(t)$ is also.

Several teams are currently searching for experimental evidence of nonlinearity in quantum mechanics [11]. To date, none has been found [12].

References

[1] A.C. Scott, P.S. Lomdahl and J.C. Eilbeck, Chem. Phys. Lett. **113**, 29 (1985).

[2] A. C. Scott and J.C. Eilbeck, Phys. Lett. A **119**, 60 (1986).

[3] P.W. Anderson, Phys. Rev. **109**, 1492 (1958); D.J. Thouless, Physics Reports **13** (no. 3), 94 (1974); S. Yoshino and M. Okazaki, J. Phys. Soc. Japan, **43**, 415 (1977); P.W. Anderson, Rev. Mod. Phys., **50** (no. 2), 191 (1978).

[4] S. Weinberg, Phys. Rev. Lett. **62**, 485 (1989).

[5] V.E. Zakharov and A.B. Shabat, Zh. Eks. Teor. Fiz. **61**, 118 (1971) [Soviet Phys. JETP **34**, 62 (1972)].

[6] M.J. Ablowitz and J.F. Ladik, J. Math. Phys. **17**, 1011 (1976); Stud. Appl. Math. **55**, 213 (1976).

[7] B.M. Herbst and M.J. Ablowitz, Phys. Rev. Lett. **62**, 2065 (1989).

[8] We are indebted to Nitant Kenkre for pointing this out.

[9] We are indebted to Graham Morrison and to Darryl Holm for pointing this out.

[10] D. David, D.D. Holm and M.V. Tratnik, Phys. Lett. A **138**, 29 (1989).

[11] B.G. Levi, "Does quantum mechanics have nonlinear terms?" Physics Today, October 1989, pp. 20-21.

[12] J.J. Bollinger, D.J. Heinzen, W.M. Itano, S.L. Gilbert and D.J. Wineland, Phys. Rev. Lett. **63**, 1031 (1989).

38

INTRODUCTION TO THE
DISCRETE SELF-TRAPPING EQUATION

J.C. Eilbeck

Department of Mathematics, Heriot-Watt University

Edinburgh EH14 4AS Scotland

Abstract

The discrete self-trapping (DST) equation models a coupled system of classical or quantum anharmonic oscillators. In this paper we review the physical motivations for this model, and describe some of the known solutions of the equation. The aim of this paper is to provide a basic introduction to other contributions to this volume covering recent results on the DST equation and its applications.

1   Introduction

The discrete self-trapping (DST) equation [11] can be written as

$$(i\frac{d}{dt} - \omega_0)\mathbf{A} + \gamma \mathrm{diag}(|A_1|^2, ..., |A_f|^2)\mathbf{A} + \epsilon M \mathbf{A} = 0. \tag{1}$$

Here the column vector

$$\mathbf{A} \equiv (A_1, A_2, ..., A_f)' \tag{2}$$

represents the complex mode amplitudes of f anharmonic, interacting degrees of freedom. Linear dispersive interactions between the degrees of freedom are introduced through the real, symmetric, $f \times f$ matrix $M = [m_{ij}]$ and anharmonicity through the diagonal matrix in the second term. If the real parameter $\gamma = 0$ but the real parameter $\epsilon \neq 0$, Eq. (1) represents a system of coupled linear oscillators. If, on the other hand, $\epsilon = 0$ but $\gamma \neq 0$, Eq. (1) represents a system of noninteracting but anharmonic oscillators. In molecular chemistry these limits correspond to the normal mode and local mode limits respectively [32]. In the general case, both ϵ and γ are allowed to be nonzero.

The DST system was motivated by two different models of nonlinear modes in molecules. One was a Davydov-like theory for the interaction of amide-I excitons with optical phonons, and leads after some approximations to Eq. (1), where in this case the A are the *quantum* c- numbers of the Davydov modes. The other approach considered a *classical* model for certain aspects of the dynamical interactions of anharmonic modes in small molecules. In this application, $A_j = (p_j - i\omega_0 q_j)/\sqrt{2}$ is the complex amplitude of a particular molecular bond, γ represents the (anharmonic) softening of the bonds under extension, and ϵM represents the electromagnetic and mechanical coupling between the bonds. Both these approaches are

Davydov's Soliton Revisited, Edited by P.L. Christiansen and A.C. Scott
Plenum Press, New York, 1990

described in more detail in §2. In §3 we describe some of the solutions of Eq. (1) for various values of f.

In the second approach described above, since the DST equation describes classical modes, it can be quantised. Such quantization is remarkably straightforward, and the results are briefly discussed in §4.

2 Physical Origins

2.1 Quantum c-number equations

In the Fall of 1983, Peter Lomdahl and I began working on a Davydov-style model of self-trapping in crystalline acetanilide (ACN) under the direction of Alwyn Scott. An approximate theory led to the following set of equations for the amplitudes of the amide-I vibrational quanta A_k and the optical phonons q_k of the form (see [12] for notation and details)

$$i\hbar \frac{d}{dt} A_k = (E - \chi q_k A_k) + \sum_{m=1}^{m=f} \epsilon m_{km} A_m, \quad k = 1, \cdots, f \tag{3}$$

$$\frac{1}{\omega^2} \ddot{q}_k + q_k = -\frac{\chi}{W} |A_k|^2 \tag{4}$$

here A_k and q_k are quantum c-numbers. Making an adiabatic approximation for the optical phonons, i.e. neglecting the $\ddot{q}_k$ term in (4) allows one to eliminate q_k, and after some rescaling we get the DST equation (1). In this derivation the approximations are due to a number of effects: (a) in this version of the theory only nearest neighbour unit cell interactions are considered along the ACN chain, and other interactions are neglected, including interactions with other parallel chains, (b) the usual approximations due to the Davydov ansatz (the D_2 ansatz in the notation of Brown *et al.* [3,4]), discussed elsewhere in this volume, and (c) the adiabatic approximation discussed above.

Brown *et al.* have argued that very localized solutions to these semi-classical equations are better described by approximations based on the "small" polaron solution to the full quantum equations, where "small" means localized on one site. They have constructed a dressed polaron model to extrapolate between the Davydov and polaron limits. Although it is likely that such models will give more accurate results in some cases, we argue that the DST model is useful as the simplest possible model covering all ranges of nonlinearity from the linear limit to the strongly self-trapped nonlinear region. Phenomena observed in solutions to the DST equation can be used to stimulate research into the same behaviour in more complicated theories. Of course once a simple model such as the DST equations is accepted as a starting point, it is instructive to see how far it can be pushed to explain experimental results before breaking down.

2.2 Classical model for coupled anharmonic operators

In studying the DST model for ACN, in which $f \approx 200$, many different solutions were observed. In order to understand the large f case better, a search for solutions of the DST equation (1) for small f values was begun. At the same time, Scott observed that the DST equation for small f could be used as a simple model system for the interaction of classical modes of excitation in small molecules, such as the C-H stretch mode in benzene. In this theory, the complex mode A_k represents the complex mode described by classical (p, q) coordinates in the form $A_k = (p_k - i\omega_0 q_k)/\sqrt{2}$. Expanding the Hamiltonian in powers of A_k, and neglecting all nonlinear terms except self-interaction quartic terms of the form $|A_k|^4$ gives the DST equation. In particular we neglect terms of the type A_k^4 and A_k^{*4}. These terms are not neglected in, for example, the Takeno Hamiltonian [26,34]. This particular approximation is important because it leads in the DST case to a number-conserving Hamiltonian.

3 General Properties and Known Solutions

3.1 General Properties

We note that Eq. (1) is invariant under the gauge transformation $A \to Ae^{i\sigma}$ for any real σ. The constant ω_0 can be scaled out of the DST equation (1) by the transformation $A \to Ae^{i\omega_0 t}$, and we shall assume in the rest of this paper that this transformation has been carried out to give

$$i\frac{d}{dt}\mathbf{A} + \gamma\text{diag}(|A_1|^2, ..., |A_f|^2)\mathbf{A} + \epsilon M \mathbf{A} = 0. \tag{5}$$

Since any constant diagonal terms in M can be incorporated into the ω_0, this means that if the modes have identical environments, we can assume the diagonal elements of M are effectively zero. The case where the diagonal elements of M are different, or equivalently the ω_0 in (1) are replaced by $\omega_1, \omega_2, \ldots, \omega_f$, can be called the *non-resonant* DST equation and is considered in [24,30]. We do not consider this case further here. Since M is an interaction matrix, it is usual to assume that it is real and symmetric. However we note in passing that if we look for solutions to (5) with some partial symmetry, i.e. $A_2 = A_3$ in the $f = 3$ case, we get a reduced DST equation which has an unsymmetrical M with nonzero diagonal elements.

We can rescale time in (5) to make ϵ unity if required, and we can rescale the A's to make γ unity also. Such scalings are useful in some calculations (c.f. [14]), but we will retain these parameters to clarify the taking of the linear and strong nonlinear limits respectively. In general we will restrict ourselves to the case that γ and ϵ are both positive: only their relative signs are important. We note however that for even f, changing ϵ to $-\epsilon$ is equivalent to the transformation $A_i = -A_i, i = 1, 3, \ldots, f - 1$. Note also that Kenkre and co-workers, in their study of the dimer equations ($f = 2$), use a positive $V = -\epsilon$, so our in-phase solutions become out-of- phase in their description and *vice versa*.

Solutions of the Eq. (5) have two constants of the motion: the *number*

$$N = \sum_{i=1}^{f} |A_i|^2 \tag{6}$$

and the *energy*

$$H = \omega_0 N - \frac{1}{2}\gamma\sum_{i=1}^{f} |A_i|^4 - \epsilon\sum_{i\neq j} m_{ij} A_i^* A_j. \tag{7}$$

We will normally fix N to be unity, since it can easily transformed to any other positive value by rescaling A and/or γ.

An important sub-class of solutions to (5) are the so-called *stationary solutions*, which are defined to be those with time dependence

$$A(t) = \Phi e^{i\omega t} \tag{8}$$

where Φ is the constant vector $(\phi_1, \phi_2, \cdots, \phi_f)'$. For such solutions, $|A_i(t)|^2$ is of course time-independent. These solutions can be expected to be important when the interaction of the system with electromagnetic waves is considered. Also Kenkre and Wu [23] have shown that when some of the assumptions used in §2.1 are removed, a system of equations are generated which always relax into these stationary solutions. From a mathematical point of view, the importance of stationary solutions is that they are easy to find, either analytically or numerically. In the space of solutions to this nonlinear equation, they act as islands of knowledge from which other solutions can be constructed by perturbation theory or otherwise. In this respect they play the same role as exact soliton solutions in the theory of nonlinear wave equations.

By substitution in (5) it can be seen that the stationary solutions satisfy the nonlinear eigenvalue equation for (ω, Φ)

$$-\omega\Phi + \gamma\mathrm{diag}(|\phi_1|^2, ..., |\phi_f|^2)\Phi + \epsilon M\Phi = 0. \tag{9}$$

For large γ there are a large number of solutions of this equation, since in this limit we have $\omega\phi_i = \gamma|\phi_1|^2\phi_i$, $i = 1, \ldots, f$, and hence if γ is real we have $\phi_i = 0$ or $\pm\sqrt{(\omega/\gamma)}$. Thus we have at least three possibilities for each ϕ_i and 3^f real solutions. This limiting case for large γ is used to label branches for *all* γ values. If $\phi_i \to \pm\sqrt{(\omega/\gamma)}$ as $\gamma \to \infty$, we use "↑" or "↓" respectively to label the solution curve. If $\phi_i \to 0$ as $\gamma \to \infty$, we use a "·", whereas if $\phi_i = 0$ for *all* γ we use a "0". If ϕ_i is complex we use a "*". (A solution is said to be complex if there is no gauge transformation that will make it real). Curves of stationary solutions that cannot be determined analytically can be found by numerical path-following techniques [11].

For non-stationary solutions, only the $f = 2$ case has been studied in detail (see §3.2 below). For higher values of f it is sometimes possible to find solutions expressible in terms of Jacobian elliptic functions by imposing some extra symmetry on the problem. In general solutions for $f \geq 3$ are chaotic and must be studied numerically: some analytic results on the chaotic region have been found by Morrison [27].

3.2 The $f = 2$ (dimer) case

Although mathematically straightforward, the $f = 2$ case reveals a wealth of new features due to nonlinearity. In stationary solutions case the solutions are easily found:

$$\uparrow\uparrow : \phi_i = 1/\sqrt{2}, \quad i = 1, 2 \tag{10}$$

$$\uparrow\downarrow : \phi_i = \pm 1/\sqrt{2} \tag{11}$$

$$\uparrow\cdot : \phi_i = \{\tfrac{1}{2}[1 \pm (1 - 4\epsilon/\gamma^2)]\}^{\frac{1}{2}}, \quad (\gamma/\epsilon \geq 2) \tag{12}$$

The stability of such solutions can easily be calculated [5]. The $\uparrow\uparrow$ solution is stable for $\gamma/\epsilon \leq 2$ and unstable otherwise. The $\uparrow\downarrow$ solution is always stable, as is the $\uparrow\cdot$ solution. The $\uparrow\cdot$ solution branch bifurcates from the $\uparrow\uparrow$ solution at $\gamma/\epsilon = 2$ (by symmetry there is also a $\cdot\uparrow$ branch). The energy of each solution is easily calculated [11] and is graphed as a function of γ in Fig 1.

Note that the $\uparrow\cdot$ solution has the lowest energy for $\gamma/\epsilon > 2$: as $\gamma \to \infty$, more and more of the energy is concentrated on one of the bonds or sites, i.e. the distribution of energy becomes more and more asymmetrical. This concentration of energy is the dimer equivalent of a sharp soliton or "small" polaron solution. For those more used to linear theory, it is worth pointing out that the number of solutions *changes* as the nonlinearity parameter γ is increased through 2ϵ.

The non-stationary solutions in the dimer case can also be found in a straightforward manner. It was pointed out in [11] that this case was integrable, and in [19] that the $f = 2$

476

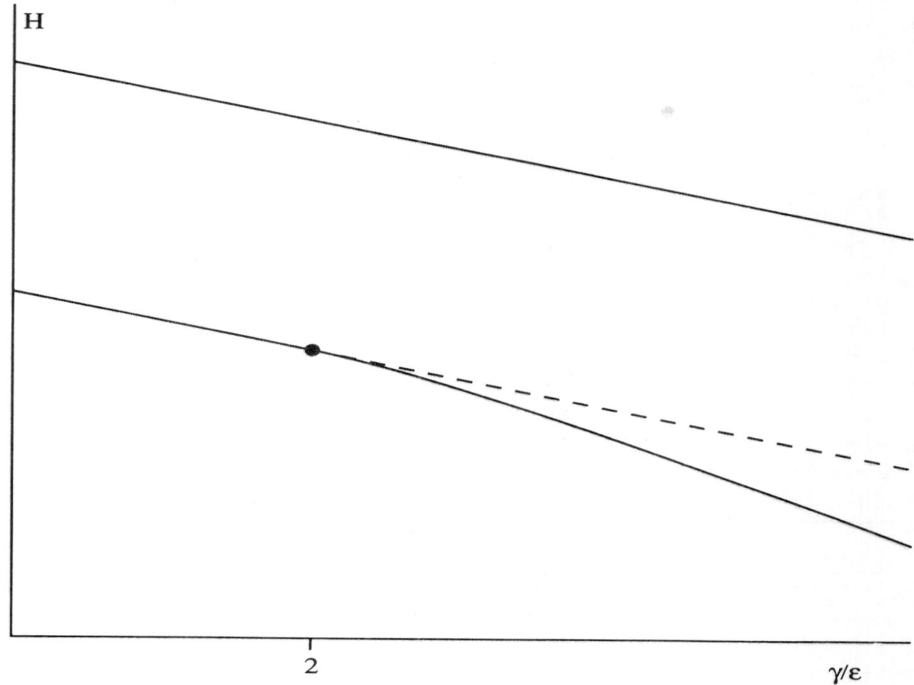

Figure 1. H v. γ for the dimer

DST equation could be transformed into the pendulum equation. Kenkre and Campbell [22] showed that this case could also be transformed into a Duffing's equation

$$d^2 p/dt^2 = Ap - Bp^3. \tag{13}$$

The connection between all three equations was discussed in [7]. The reader should be warned that the transformations connecting these three equations are nonlinear and the resulting equations have parameters which depend on the initial conditions in the DST equation. This can sometimes confuse rather then clarify: for example the single $p = 0$ solution to (13) corresponds to *both* the $\uparrow\uparrow$ and the $\uparrow\downarrow$ solutions of the $f = 2$ DST equation.

Kenkre and Campbell [22] gave the complete solution to the $f = 2$ DST equation in terms of Jacobian elliptic functions, and the reader is referred to that and later papers by Kenkre and co-workers for details [20,35]. However, much insight can be gained by studying the phase plots for this case. The phase plane figures given in [11] (Figs. 6–8 in that paper) are useful except for initial conditions near $(A_1, A_2) = (1,0)$ or $(0,1)$. The appropriate phase space is the sphere S^2 [27], and the projections given in [11] distort the trajectories near these "poles".

A better way of plotting this phase space is as follows. If we put $A_i = a_i e^{\theta_i}$ for $i = 1, 2$ then we can eliminate a_2 using $a_1{}^2 + a_2{}^2 = N = 1$ and consider only the phase difference $\theta = \theta_1 - \theta_2$. These reduced coordinates allow us to write the Hamiltonian (7) in the form

$$H = -\frac{1}{2}\gamma(1 + 2a_1{}^4 - 2a_1{}^2) - 2a_1(1 - a_1{}^2)^{\frac{1}{2}} \cos\theta \tag{14}$$

A natural choice of spherical coordinates is

$$\phi' = \theta, \qquad \cos(\theta'/2) = a_1 \tag{15}$$

which gives H in the simple form (note there is no connection between θ and θ' or between the ϕ's)

$$H(\theta',\phi') = -\frac{1}{2}\gamma\left(1 - \frac{1}{2}\sin^2\theta'\right) - \sin\theta'\cos\phi' \tag{16}$$

In Fig 2. we plot a contours of H on a different projection of this phase space for various values of γ in units where $\epsilon = 1$. The left-hand of each figure shows $-\pi/2 \leq \phi' \leq \pi/2$ and the right-hand $\pi/2 \leq \phi' \leq 3\pi/2$. For small γ (Fig 2a), the two stationary solutions $\uparrow\uparrow$ and $\uparrow\downarrow$ are the two fixed points in the phase space in these reduced co-ordinates, and all the other orbits oscillate about these points. At $\gamma = 2\epsilon$ the fixed point corresponding to $\uparrow\uparrow$ bifurcates into a saddle point and spawns two new fixed points corresponding to the $\uparrow\cdot$ and $\cdot\uparrow$ solutions respectively, as shown in Fig 2b. The separatrix through the saddle point does not encircle the North and South poles which correspond to the solutions $(A_1, A_2) = (1,0)$ or $(0,1)$. Hence solutions starting at one pole will undergo an oscillation which takes in the other pole, as observed by [35]. The "dynamic phase transition" observed by those authors occurs at $\gamma = 4$, when the separatrix crosses the poles. Now (cf. Fig 2c) any orbit starting from one of the poles will oscillate about the fixed point in the upper or lower half hemisphere.

The discussion in [21] concerning out-of-phase initial conditions with $a_1 \neq 1$ correspond to in-phase ($\pi = 0$) conditions in our scheme, i.e. a point on the line $\phi' = 0$ through the poles and the two fixed points in the left-hand plots of Figs 2b and 2c. The "repulsion effect" of Kenkre can now be easily understood: for smaller values of γ the initial conditions lie at the *top* of an orbit encircling one of the $\uparrow\cdot$ fixed points. As γ increases, this fixed point passes through the point of initial conditions, so the initial conditions now lie at the *bottom* of an orbit encircling the fixed point. Orbits corresponding to other choices of initial conditions can be read off the phase diagrams in a similar way.

For large values of γ, the $\uparrow\cdot$ fixed points get closer and closer to the poles, and the separatrix gets squeezed closer to the equator. In this limit $\theta' = \frac{1}{2}\pi + \Delta$, where $\Delta \approx \pm\frac{2}{\gamma}(1 - \cos\phi')$ for $0 \leq \phi' < 2\pi$.

The dimer equations and their solutions have also been discussed in a theory of particle transfer in molecular complexes [1]. Lax representations can be found both for the resonant dimer equations [18] and the nonresonant case (V Enol'skii, private communication).

3.3   Other small n results

In [11] the stationary solutions for the cases $f = 3,4$ were extensively discussed. For the case $f = 3$, with an equal interaction matrix of the form $m_{ij} = 1$ if $i \neq j$, 0 if $i = j$, all the stationary solutions are known in analytic form. This case corresponds to a planar triangular molecule. The new effect with $f \geq 3$ is that some solutions are chaotic. This chaos has been studied by a number of authors, both in the classical case [11,16,17,19,27] and the quantum case [8,15].

For the case $f = 4$ with equal interactions, corresponding to a tetrahedron molecule, all known stationary solutions can be expressed in analytic form. The case $f = 4$ with nearest neighbour interactions, corresponding to a square molecule has also been studied. In this case some of the stationary solution branches need to be calculated numerically. The case $f = 6$, corresponding to a hexagonal molecule, has been studied in [9,13], and the results applied to the spectra of benzene in [31]. In this case also, some stationary solutions must be calculated numerically.

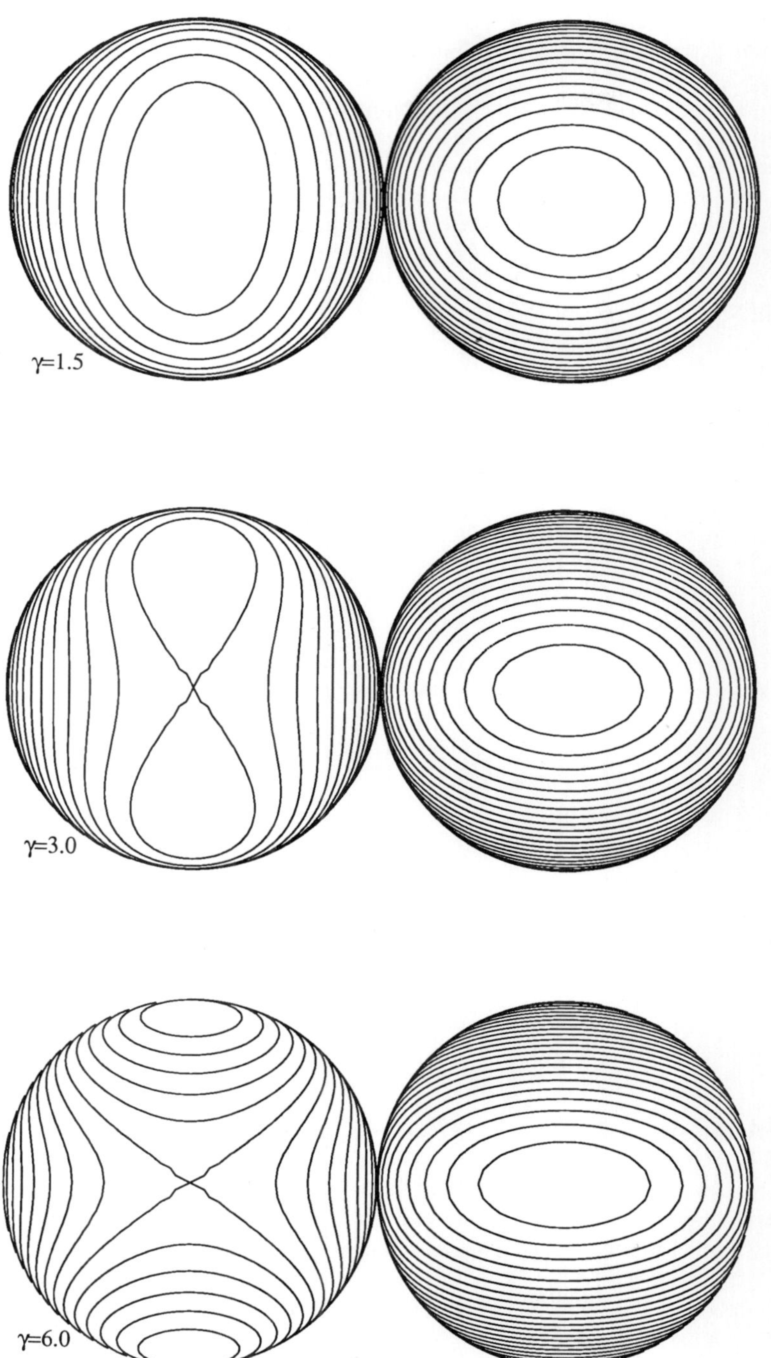

Figure 2. Phase plots for the $f = 2$ case

3.4 Results for large n

The case of an infinite chain with nearest neighbor interactions was first studied by Scott and MacNeil [33]. They found the solution corresponding to a single soliton by numerical techniques. This type of solution was also studied by Kuprievich [25] using variational and numerical methods, and in the continuum approximation. Some results about the general f case are given in [11,5], and the special case discussed in the original ACN study [12] corresponds to $f \approx 200$. Nussbaum and Fischer [29] discussed an analytic perturbation theory to treat the single soliton solution and this was generalised by Nussbuam [28] to deal with doubly periodic solutions of the DST equation. Some preliminary numerical studies of travelling solitary waves on a large ($f = 50$) lattice were reported in [10].

In addition to the applications to ACN and other molecular lattice systems, Christodoulides and Joseph [6] have used the DST equation to model discrete self-focussing effects in arrays of coupled waveguides.

4 Quantum DST

Since, as mentioned in §2, in one incarnation the DST equation can be thought of as a classical equation, we can quantize it and investigate the quantum solutions. If we introduce the usual boson operators $\hat{B}_i$ and $\hat{B}_i^\dagger$, we can write the number and energy operators as

$$\hat{N} = \sum_{i=1}^{f} (\hat{B}_i^\dagger \hat{B}_i + \frac{1}{2}) \tag{17}$$

$$\hat{H} = \hbar(\omega_0 + \frac{1}{2}\gamma)N - \frac{1}{2}\hbar\gamma \sum_{i=1}^{f} \hat{B}_i^\dagger \hat{B}_i \hat{B}_i^\dagger \hat{B}_i - \hbar\epsilon \sum_{i\neq j} m_{ij} \hat{B}_i^\dagger \hat{B}_j \tag{18}$$

Note that $\hat{N}$ and $\hat{H}$ commute - we can find states which are eigenstates of both operators simultaneously. This makes calculations of eigenstates simple, accurate and efficient. As a simple introductory example, consider the dimer case again, with the two sites labeled a and b. The first excited state can be written as

$$|\Psi_1\rangle = c_1|1\rangle_a|0\rangle_b + c_2|0\rangle_a|1\rangle_b \tag{19}$$

where $|i\rangle_a|j\rangle_b$ means i quanta on site a and j quanta on site b. The Hamiltonian is (putting $\hbar = 1$)

$$< \Psi_1|\hat{H}|\Psi_1\rangle = \begin{pmatrix} 2\omega_0 - \frac{3}{2}\gamma & -\epsilon \\ -\epsilon & 2\omega_0 - \frac{3}{2}\gamma \end{pmatrix} \tag{20}$$

with eigenvalues and eigenstates

$$E_1^{(1)} = 2\omega_0 - \frac{3}{2}\gamma - \epsilon, \quad |\Psi_1\rangle_s = \frac{1}{\sqrt{2}}(|1\rangle_a|0\rangle_b + |0\rangle_a|1\rangle_b) \tag{21}$$

$$E_1^{(2)} = 2\omega_0 - \frac{3}{2}\gamma + \epsilon, \quad |\Psi_1\rangle_{as} = \frac{1}{\sqrt{2}}(|1\rangle_a|0\rangle_b - |0\rangle_a|1\rangle_b) \tag{22}$$

for the symmetric and antisymmetric states respectively. For the second excited state we have

$$|\Psi_2\rangle = c_1|2\rangle_a|0\rangle_b + c_2|1\rangle_a|1\rangle_b + c_3|0\rangle_a|2\rangle_b \tag{23}$$

with Hamiltonian

$$< \Psi_2|\hat{H}|\Psi_2\rangle = \begin{pmatrix} 3\omega_0 - \frac{7}{2}\gamma & -\sqrt{2}\epsilon & 0 \\ -\sqrt{2}\epsilon & 3\omega_0 - \frac{5}{2}\gamma & -\sqrt{2}\epsilon \\ 0 & -\sqrt{\epsilon} & 3\omega_0 - \frac{7}{2}\gamma \end{pmatrix} \tag{24}$$

with eigenvalues and eigenstates

$$E_2^{(1)} \;=\; 3\omega_0 - \frac{7}{2}\gamma, \quad |\Psi_2\rangle_{as} = \frac{1}{\sqrt{2}}[1,0,-1] \tag{25}$$

$$E_2^{(2)} \;=\; 3\omega_0 - 3\gamma - \frac{1}{2}\sqrt{\gamma^2 + 16\epsilon^2}, \quad |\Psi_2\rangle_{s-} \propto [\epsilon, \frac{\gamma - \sqrt{\gamma^2 + 16\epsilon^2}}{2\sqrt{2}}, \epsilon] \tag{26}$$

$$E_2^{(3)} \;=\; 3\omega_0 - 3\gamma + \frac{1}{2}\sqrt{\gamma^2 + 16\epsilon^2}, \quad |\Psi_2\rangle_{s+} \propto [\epsilon, \frac{\gamma + \sqrt{\gamma^2 + 16\epsilon^2}}{2\sqrt{2}}, \epsilon] \tag{27}$$

As $\gamma \to \infty$,

$$E_2^{(2)} = 3\omega_0 - \frac{7}{2}\gamma, \quad |\Psi_2\rangle_{s-} = \frac{1}{\sqrt{2}}[1,0,1] \tag{28}$$

In this limit the $E_2^{(2)}$ symmetric solution is degenerate with the antisymmetric solution $E_2^{(1)}$, so we can form a *localized* state

$$|2\rangle_a |0\rangle_b = \frac{1}{\sqrt{2}}(|\Psi_2\rangle_{s+} + |\Psi_2\rangle_{as}) \tag{29}$$

This is *not* an exact eigenstate for finite γ, but very long lived since the energy splitting $E_2^{(1)} - E_2^{(2)}$ is $O(\epsilon^2/\gamma)$. For higher excited states, the splitting is even smaller, as discussed in the paper by Bernstein [2]. This near-degeneracy is the quantum analogue of the classical "soliton" or "polaron" solution.

Moving away from the stationary solutions, the chaotic solutions in the $f = 3$ case have been studied in the quantum case as well as the classical case, as discussed in §3.

Acknowledgements

I am grateful to the NATO Special Programme Panel on Chaos, Order and Patterns for support for a collaborative programme, and to the SERC for research funding under the Nonlinear System Initiative.

References

[1] P.Baňacký and A Zajac. Theory of particle transfer dynamics in solvated molecular complexes: analytic solutions of the discrete time-dependent nonlinear Schrödinger equation. I. conservative system. *Chem Phys.*, 123:267–276, 1988.

[2] L Bernstein. Local modes and degenerate perturbation theory. 1989. (these proceedings).

[3] D W Brown. When is a soliton? 1989. (these proceedings).

[4] D W Brown and Z Ivic. Unification of polaron and soliton theories of exciton transport. 1989. (San Diego preprint).

[5] J Carr and J C Eilbeck. Stability of stationary solutions of the discrete self-trapping equation. *Phys. Letts. A*, 109:201–204, 1985.

[6] D N Christodoulides and R I Joseph. Discrete self- focusing in nonlinear arrays of coupled waveguides. *Optics Letters*, 13:794-796, 1988.

[7] L Cruzeiro-Hansson, P L Christiansen, and J N Elgin. Comment on "Self-trapping on a dimer: time-dependent solutions of a discrete nonlinear Schrödinger equation". *Phys. Rev. B.*, 37:7896-7897, 1988.

[8] L Cruzeiro-Hansson, H Feddersen, R Flesch, P L Christiansen, M Salerno, and A C Scott. Classical and quantum analysis of chaos in the discrete self-trapping equation. 1989. (submitted for publication).

[9] J C Eilbeck. Nonlinear vibrational modes in a hexagonal molecule, pages 41-51. Volume 12, Ukrainian Academy of Sciences, Kiev, 1987.

[10] J C Eilbeck. Numerical simulations of the dynamics of polypeptide chains and proteins. In Chikao Kawabata and A R Bishop, editors, *Computer Analysis for Life Science— Progress and Challenges in Biological and Synthetic Polymer Research*, pages 12- 21, Ohmsha, Tokyo, 1986.

[11] J C Eilbeck, P S Lomdahl, and A C Scott. The discrete self-trapping equation. *Physica D: Nonlinear Phenomena*, 16:318-338, 1985.

[12] J C Eilbeck, P S Lomdahl, and A C Scott. Soliton structure in crystalline acetanilide. *Phys. Rev. B*, 30:4703- 4712, 1984.

[13] J C Eilbeck and A C Scott. Theory and applications of the discrete self-trapping equation. In P L Christiansen and R D Parmentier, editors, *Structure, Coherence and Chaos in Dynamical Systems*, pages 139–159, Manchester University Press, 1989.

[14] H Feddersen. Quantum and classical descriptions of chaos in the DST equation. 1989. (these proceedings).

[15] S De Filippo, M Fusco Girard, and M Salerno. Avoided crossing and nearest-neighbour level spacings for the quantum DST equation. *Nonlinearity*, 2:477–487, 1989.

[16] S De Filippo, M Fusco Girard, and M Salerno. Lyapunov exponents for the $n = 3$ discrete self-trapping equation. *Physica D*, 26:411–414, 1987.

[17] S De Filippo, M Fusco Girard, and M Salerno. Numerical evidence of a sharp order window in a Hamiltonian system. *Physica D*, 29:421–426, 1988.

[18] V I Inozemtsev and N A Kostov. New integrable systems of interacting nonlinear waves. 1988. Dubna preprint E5- 88-622.

[19] J H Jensen, P L Christiansen, J N Elgin, J D Gibbon, and O Skovgaard. Correlation exponents for trajectories in the low- dimensional discrete self-trapping equation. *Phys. Lett. A*, 110:429–431, 1985.

[20] V M Kenkre. The discrete nonlinear Schrödinger equation: nonadiabatic effects, finite temperature consequences, and experimental manifestations. 1989. (these proceedings).

[21] V M Kenkre. The quantum nonlinear dimer and extensions. In S Pnevmatikos, T Bountis, and St. Pnevmatikos, editors, *Singular behaviour and nonlinear dynamics*, World Publishers, 1989.

[22] V M Kenkre and D K Campbell. Self-trapping on a dimer: time-dependent solutions of a discrete nonlinear Schrödinger equation. *Phys. Rev. B*, 34:4959–4961, 1986.

[23] V M Kenkre and H-L Wu. Interplay of quantum phases and nonlinearity in the nonadiabatic dimer. *Phys. Lett. A*, 135:120–124, 1989.

[24] E W Knapp and S F Fischer. A unified theory of electron transfer and internal conversion based on solitary electronic states. *J. Chem. Phys.*, 90:354–365, 1988.

[25] V A Kuprievich. On autolocalization of the stationary states in a finite molecular chain. *Physica D*, 14:395–402, 1985.

[26] K Lindenberg. Vibron solitons. 1989. (these proceedings).

[27] G J Morrison. Homoclinic chaos in the DST equation. 1989. (in preparation).

[28] I Nussbaum. Non-steady solutions of the discrete self-trapping equation. *Phys. Lett. A*, 118: 127–130, 1986.

[29] I Nussbaum and S F Fischer. Analytic treatment of localized, stationary states of the discrete self-trapping equation. *Phys. Lett. A*, 115:268–270, 1986.

[30] A C Scott. A nonresonant discrete self-trapping system. *Physica Scripta.* (in press).

[31] A C Scott and J C Eilbeck. On the CH stretch overtones of benzene. *Chem. Phys. Lett.*, 132:23–28. 1986.

[32] A C Scott, P S Lomdahl, and J C Eilbeck. Between the local mode and normal mode limits. *Chem. Phys. Letts.*, 113:21–36, 1985.

[33] A C Scott and L MacNeil. Binding energy versus nonlinearity for a "small" stationary soliton. *Phys. Lett. A.*, 98:87–88, 1983.

[34] S Takeno. A classical and quantum-mechanical theory of vibron solitons and kinks in open systems and their implication of biological energy transfer. 1989. (these proceedings).

[35] G P Tsironis and V M Kenkre. Initial condition effects in the evolution of a nonlinear dimer. *Phys. Lett. A.*, 127:209–212, 1988.

Note added in proof: Since this paper was written, F. Kh. Abdullaev has drawn my attention to the work of A.A. Maĭer, Sov. J. Quantum Electron., **14**, 101-104 (1984), where the $n = 2$ DST equation appears as a special case of a model in nonlinear optics, and its cnoidal wave solutions are discussed.

39

ENERGY LOCALIZATION IN SMALL BIOMOLECULES

Peter L. Christiansen

Laboratory of Applied Mathematical Physics
The Technical University of Denmark
DK-2800 Lyngby, Denmark

ABSTRACT

A generalized discrete self-trapping model of biomolecules which allows adjustment of the degree of nonlinearity is investigated both classically and quantum mechanically. With two degrees of freedom, the system is closely related to the Feynman top. A dramatic effect of the nonlinearity on the localization of energy is observed.

1. INTRODUCTION

In Davydov's theoretical model of protein an extrinsic nonlinear coupling to the phonon system is introduced [1-2]. In the discrete self-trapping (DST) equation this coupling is reduced to an intrinsic anharmonicity arising from nonlinear force constants in the molecular bonds [3]. Yet the equation is able to interpolate between normal modes and local modes. DST has been used to model anharmonic vibrations of small molecules like water, ammonia, methane, and benzene [4-5] and in this connection a quantum theory for DST has been developed [6-9]. For two degrees of freedom the classical DST equation is integrable [3,10-11] while for three and higher degrees of freedom chaos occurs [6,12-14]. The signature of chaos in the corresponding quantum problem ("quantum chaology") has also been studied in the case of DST [15-16].

The Hamiltonian for the classical DST in generalized form can be written

$$H_\sigma = \sum_{f=1}^{N} \left(\omega_f |A_f|^2 - \frac{\gamma_f}{\sigma+1} |A_f|^{2\sigma+2} \right) - \varepsilon \sum_{f,g} m_{f,g} A_f^* A_g \ . \tag{1}$$

The Hamiltonian operator for the corresponding quantum DST becomes

$$\hat{H}_\sigma = \sum_{f=1}^{N} \left[\omega_f \, P_o(\hat{B}_f^\dagger \hat{B}_f) - \frac{\gamma_f}{\sigma+1} P_\sigma (\hat{B}_f^\dagger \hat{B}_f) \right] - \varepsilon \sum_{\substack{f,g \\ f \neq g}} m_{f,g} \, \hat{B}_f^\dagger \hat{B}_g \ . \tag{2}$$

Davydov's Soliton Revisited, Edited by P.L. Christiansen and A.C. Scott
Plenum Press, New York, 1990

Here $A_1(t), A_2(t), \cdots, A_N(t)$ are complex mode amplitudes for the N degrees of freedom. γ_f and ε (real) are the anharmonicity parameter of the f'th oscillator and the dispersion parameter in the system while σ (non-negative integer) determines the degree of nonlinearity of the system. $m_{f,g}$ is a symmetric dispersion matrix describing the interaction between the N sites. ω_f is the eigenfrequency of the f'th oscillator in the absence of anharmonicity and dispersion ($\gamma_f = \varepsilon = 0$ for $f = 1,2,\cdots,N$). In the quantum model A_f^* and A_f are replaced by creation and annihilation operators $\hat{B}_f^+$ and $\hat{B}_f$, satisfying boson commutation rules

$$\left[\hat{B}_f, \hat{B}_g^\dagger\right] = \delta_{fg} \ , \qquad \left[\hat{B}_f, \hat{B}_g\right] = \left[\hat{B}_f^\dagger, \hat{B}_g^\dagger\right] = 0 \ . \tag{3}$$

Averaging of all possible orderings of the operators yields the polynomials P_σ given in Table 1 [17].

Table 1. Polynomials $P_\sigma(\hat{x})$ of operator $\hat{x}$.

σ	$P_\sigma(\hat{x})$
0	$\hat{x} + \dfrac{1}{2}$
1	$\hat{x}^2 + \hat{x} + \dfrac{1}{2}$
2	$\hat{x}^3 + \dfrac{3}{2}\hat{x}^2 + 2\hat{x} + \dfrac{3}{4}$
3	$\hat{x}^4 + 2\hat{x}^3 + 5\hat{x}^2 + 4\hat{x} + \dfrac{3}{2}$
4	$\hat{x}^5 + \dfrac{5}{2}\hat{x}^4 + 10\hat{x}^3 + \dfrac{25}{2}\hat{x}^2 + \dfrac{23}{2}\hat{x} + \dfrac{15}{4}$

In the classical DST system not only the energy, H_σ, but also the number

$$N = \sum_{f=1}^{N} |A_f|^2 \tag{4}$$

is a conserved quantity. The corresponding number operator may be defined as

$$\hat{N} = \sum_{f=1}^{N} P_\sigma(\hat{B}_f^\dagger \hat{B}_f) \tag{5}$$

In [3-16] the cases in which $\sigma = 1$, $\omega_f = 0$, and $\gamma_f = \gamma$ ($f = 1,2,\cdots,N$) are considered. Recently, higher degrees of nonlinearity ($\sigma = 2,3,\cdots$) and detuning ($\omega_1 \neq \omega_2$) were investigated in the case of two degrees of freedom (N = 2), [17,18] and [19] respectively. The present contribution summarizes results for higher degrees of non-linearity in the case N = 2, $\omega_1 = \omega_2 = 0$, $\gamma_1 = \gamma_2 = \gamma$.

2. CLASSICAL THEORY

We consider the generalized DST system

$$i\dot{A}_1 + \gamma|A_1|^{2\sigma} A_1 + \varepsilon A_2 = 0$$

$$i\dot{A}_2 + \gamma|A_2|^{2\sigma} A_2 + \varepsilon A_1 = 0 \tag{6a,b}$$

with the conserved quantities

$$H_\sigma = -\frac{\gamma}{\sigma+1} \left[|A_1|^{2\sigma+2} + |A_2|^{2\sigma+2}\right] - \varepsilon(A_1^* A_2 + A_1 A_2^*) \tag{7}$$

and

$$N = |A_1|^2 + |A_2|^2 . \tag{8}$$

Evidently, Eq. (6) is integrable for any non-negative integer value of σ. Also Eq. (6) is a Hamiltonian for which various Hamiltonian formulations can be given. We can introduce the canonical variables

$$q_f = \frac{1}{\sqrt{2}} (A_f^*+A_f) \quad \text{and} \quad p_f = \frac{1}{i\sqrt{2}} (A_f^*-A_f) , \qquad f = 1,2 \tag{9a,b}$$

However, also

$$\phi = \phi_2 - \phi_1 \quad \text{and} \quad N_1 = A_1^2 , \tag{10a,b}$$

yielding the Hamiltonian

$$\tilde{H}_\sigma = -\frac{\gamma}{\sigma+1} \left[N_1^{\sigma+1}+(N-N_1)^{\sigma+1}\right] - 2\varepsilon(NN_1-N_1^2)^{1/2} \cos\phi , \tag{11}$$

are canonical variables, since the equations

$$\dot{N}_1 = \frac{\partial\tilde{H}_\sigma}{\partial\phi}$$

$$\dot{\phi} = -\frac{\partial\tilde{H}_\sigma}{\partial N_1} \tag{12a.b}$$

are identical with the dynamical equations for N_1 and ϕ, namely

$$\dot{N}_1 = 2\varepsilon(NN_1-N_1^2)^{1/2} \sin\phi$$

$$\dot{\phi} = \gamma[N_1^\sigma - (N-N_1)^\sigma] + \varepsilon \frac{N-2N_1}{(NN_1-N_1^2)^{1/2}} \cos\phi . \tag{12'a,b}$$

Eliminating ϕ between Eqs. (11) and (12) yields

$$(\dot{N}_1)^2 = 4\varepsilon^2(NN_1-N_1^2) - \left[H + \frac{\gamma}{\sigma+1} (N_1^{\sigma+1} + (N-N_1)^{\sigma+1})\right]^2 \tag{13}$$

where N and H are to be interpreted as constants determined by initial conditions. We shall discuss the solutions of Eq. (13) below.

The generalized DST system with two degrees of freedom, Eq. (6), can also be written in the form of the Feynman top

$$\dot{\bar{r}} = \bar{\Omega}_\sigma \times \bar{r}$$ (14)

Here [20]

$$\begin{aligned}
\bar{r} &= (r_1, r_2, r_3) \\
r_1 &= A_1 A_2^* + A_2 A_1^* \\
r_2 &= i(A_1 A_2^* - A_2 A_1^*) \\
r_3 &= |A_1|^2 - |A_2|^2
\end{aligned}$$ (15a,b,c,d)

and in the generalized DST system

$$\bar{\Omega}_\sigma = (-2\varepsilon, 0, \gamma(|A_2|^{2\sigma} - |A_1|^{2\sigma})) \ .$$ (16)

Expressed in terms of the Feynman variables r_1, r_2, r_3 the conserved norm becomes

$$N = (r_1^2 + r_2^2 + r_3^2)^{1/2}$$ (17)

and the Hamiltonian

$$\tilde{\tilde{H}}_\sigma = - \frac{\gamma}{\sigma+1} \left[\left(\frac{N+r_3}{2}\right)^{\sigma+1} + \left(\frac{N-r_3}{2}\right)^{\sigma+1} \right] - \varepsilon r_1 \ .$$ (18)

Introducing r_3 and its conjugate momentum as canonical variables and assuming that the kinetic energy, E_{kin}, is a function of p_3 alone, we get

$$E_{kin} = - \varepsilon r_1(p_3)$$ (19a)

and the potential energy, a function of r_3 alone,

$$E_{pot} = - \frac{\gamma}{\sigma+1} \left[\left(\frac{N+r_3}{2}\right)^{\sigma+1} + \left(\frac{N-r_3}{2}\right)^{\sigma+1} \right] \ .$$ (19b)

Also in this formulation Hamilton's equations

$$\begin{aligned}
\dot{p}_3 &= - \frac{\partial \tilde{\tilde{H}}_\sigma}{\partial r_3} \\
\dot{r}_3 &= \frac{\partial \tilde{\tilde{H}}_\sigma}{\partial p_3}
\end{aligned}$$ (20a,b)

reproduce the dynamical equations of the system (14).

Eliminating r_1 between Eqs. (14), (17), and (18) yields

$$(\dot{r}_3)^2 = 4\varepsilon^2(N^2 - r_3^2) - \left[2H + \frac{\gamma}{(\sigma+1)2^\sigma}\left((N+r_3)^{\sigma+1} + (N-r_3)^{\sigma+1}\right) \right]^2 \ ,$$ (21)

which, like Eq. (13), is a first-order differential equation for the
Hamiltonian variable. The right-hand side polynomials of Eqs. (13) and (21)
are of order ν (in N_1 and r_3), which is related to σ as follows: For $\sigma = 0$,
$\nu = 2$, N_1 and r_3 are a sinusoid of frequency 2ε. For $\sigma = 1$ and 2, $\nu = 4$
and N_1 and r_3 are a Jacobi elliptic function. For $\sigma = 3$ and 4, $\nu = 8$ and N_1
and r_3 are the inverse of a hyperelliptic integral of the first kind of
class 3. For $\sigma = \mu$ and $\mu + 1$, where μ is an odd integer, $\nu = 2(\mu + 1)$ and N_1
and r_3 are the inverse of a hyperelliptic integral of the first kind of
class μ [21]. These results are summarized in column denoted "Feynman variable" in Table 2. Note also that differentiation of Eq. (21) with respect
to time yields a second-order differential equation for r_3 of the form $\ddot{r}_3 =$
polynomial of r_3 of degree $\nu - 1$. For $\sigma = 1$ and 2 the result is $\nu - 1 = 3$
leading to the space-independent ϕ^4-equation as found for $\sigma = 1$ in [10].

Table 2. Feynman variable and its conjugate momentum versus degree of
nonlinearity. (Terminology as defined in [21]).

Degree of nonlinearity	Feynman variable	Equation for conjugate momentum
σ	$r_3(t)$	$p_3(t)$
0	sinusoid	$p_3 = $ const.
1,2	Jacobi elliptic function	Pendulum equation with sinusoid
3,4	Inverse of hyperelliptic integral of 1st kind, 3rd class	Generalized pendulum equation with inverse of elliptic integrals of 3rd kind
5,6	Inverse of hyperelliptic integral of 1st kind, 5th class	Generalized pendulum equation with inverse of hyperelliptic integrals of 3rd kind, 2nd class
	etc.	etc.

From the first canonical equation (20a) follows

$$\dot{p}_3 = \frac{\gamma}{2^{\sigma+1}} \left[(N+r_3)^\sigma - (N-r_3)^\sigma \right] , \tag{22}$$

leading to

$$\overline{\Omega}_\sigma = (-2\varepsilon, 0, -2\dot{p}_3) \tag{23}$$

in the Feynman picture. The second canonical equation (20b) leads to the
set of equations

$$\left. \begin{array}{l} r_1 = f(q) \\[2ex] r_2 = -f'(q) \\[2ex] r_3 = \dfrac{\dot{p}_3}{\varepsilon} (f(q) + f''(q)) \end{array} \right\} \tag{24a,b,c}$$

where $\dot{p}_3$ is given by (22). Here the unknown function $f(q)$, where $q = 2p_3$
has been introduced for convenience, has to be determined. (Prime denotes
differentiation with respect to q). The equation for the second derivative
with respect to time, $\ddot{q}$, is also of interest.

For $\sigma = 0$ Eq. (22) gives

$$\dot{p}_3 = 0 \quad \Rightarrow \quad q \text{ const.} \tag{25}$$

For $\sigma = 1$ we get

$$\left.\begin{aligned}
\dot{p}_3 &= \frac{\gamma}{2} \, r_3 \\[2mm]
f(q) &= \frac{2\varepsilon}{\gamma} + \alpha \cos(q-q_o) \\[2mm]
\ddot{q} &= 2\gamma\varepsilon\alpha \, \sin(q-q_o)
\end{aligned}\right\} \quad , \tag{26a,b,c}$$

where α and q_o are real integration constants. Thus we get a pendulum equation [11] for $q = 2p_3$.

For $\sigma = 2$ we get

$$\left.\begin{aligned}
\dot{p}_3 &= \frac{\gamma N}{2} \, r_3 \\[2mm]
f(q) &= \frac{2\varepsilon}{\gamma N} + \alpha \cos(q-q_o) \\[2mm]
\ddot{q} &= 2\gamma N\varepsilon\alpha \, \sin(q-q_o)
\end{aligned}\right\} \quad , \tag{27a,b,c}$$

where N is given by (17). Here (27c) is again a pendulum equation.

For $\sigma = 3$ Eq. (22) gives

$$\left.\begin{aligned}
\dot{p}_3 &= \frac{\gamma}{2} \left(\frac{3}{4} N^2 + \frac{1}{4} r_3^2 \right) r_3 \\[2mm]
f(q) &= -\frac{\gamma}{32\varepsilon} F^2 + \frac{\gamma N^2}{4\varepsilon} F - \alpha \\[2mm]
\ddot{q} &= -\frac{3}{2} \gamma\varepsilon(2N^2-F) \sqrt{-\left(\frac{\gamma}{32\varepsilon} F^2 - \frac{\gamma}{4\varepsilon} N^2 F + \alpha\right)^2 + F}
\end{aligned}\right\} \tag{28a,b,c}$$

where $F = F(q)$ satisfies

$$F'(q) = \frac{4\varepsilon}{\gamma} \frac{1}{N^2-\frac{1}{4}F} \sqrt{-\left(\frac{\gamma}{32\varepsilon} F^2 - \frac{\gamma}{4\varepsilon} N^2 F + \alpha\right)^2 + F} \tag{28d}$$

Thus the inverse of F is an elliptic integral of third kind [21]. Eq. (28c) is a generalized pendulum equation for $q = 2p_3$.

In [17-18] higher values of σ are considered. For $\sigma = \mu$ and $\mu + 1$, where μ is an odd integer, we find hyperelliptic integrals of the third kind of class $(\mu-1)/2$ [21]. The results are summarized in the column denoted "Equation for conjugate momentum" in Table 2. For a given degree of non-linearity the Feynman variable, r_3, is seen to be a more complicated periodic function than the function occurring in the corresponding generalized pendulum equation for the conjugate momentum, p_3.

3. QUANTUM THEORY

Any number state, $|n_1\rangle|n_2\rangle$, is an eigenfunction of $\hat{N}$, redefined for convenience as

$$\hat{N} = \hat{B}_1^\dagger \hat{B}_1 + \hat{B}_2^\dagger \hat{B}_2 \tag{29}$$

with eigenvalue $n = n_1 + n_2$. For a fixed value of n, the most general wavefunction is

$$\psi_n = \sum_{j=0}^{n} c(n,j)\,|(n-j)\rangle|j\rangle \tag{30}$$

where $c(n,j)$ are possibly complex coefficients. The $c(n,j)$ are determined by requiring also that ψ_n be an eigenfunction of

$$\hat{H}_\sigma = - \frac{\gamma}{\sigma+1}\left[P_\sigma(\hat{B}_1^\dagger \hat{B}_1) + P_\sigma(\hat{B}_2^\dagger \hat{B}_2) \right] - \varepsilon(\hat{B}_1^\dagger \hat{B}_2 + \hat{B}_1 \hat{B}_2^\dagger) \tag{31}$$

where P_σ is given in Table 1. This requirement

$$\hat{H}_\sigma \psi_n = E \psi_n \tag{32}$$

leads to the following matrix equation for the $c(n,f)$:

$$H_n(\sigma)\overline{c} = E\,\overline{c} \tag{33}$$

where

$$\overline{c} \equiv \mathrm{col}\ (c(n,0),c(n,1),\cdots,c(n,n)) \tag{34}$$

and $H_n(\sigma)$ is a real, $(n+1) \times (n+1)$, symmetric matrix. For $\sigma = 1$, it was explicitly constructed in [6].

Considering an orthonormal set of eigenfunctions with number eigenvalues n and energy eigenvalues $E(n,p)$, where $p = 0,1,\cdots,n$, it is found that the two levels of lowest energy are separated by a small energy difference ΔE. From a perturbation theory in small ε/γ it was calculated (in [8,17])

$$\frac{\Delta E}{\varepsilon} = \frac{2n}{\Pi_\sigma(n)(n-1)!}\left(\frac{\varepsilon}{\gamma}\right)^{n-1} + 0\!\left(\frac{\varepsilon^n}{\gamma^n}\right) \tag{35}$$

where expressions for $\Pi_\sigma(n)$ are listed in Table 3. From this table it is seen that a conservative estimate of the denominator polynomial is

$$\Pi_\sigma(n) \sim n^{(\sigma-1)(n-1)} . \tag{36}$$

Thus we see that ΔE becomes very small with increasing σ. This leads to a dramatic increase in the time that the quantum wave packet primarily composed of the two states of lowest energy (corresponding to an unsymmetric, classical stationary solution) remains localized.

Table 3. Denominator polynomials appearing in Eq. (35).

σ	$\Pi_\sigma(n)$
1	1
2	$(n+1)^{(n-1)}$
3	$\displaystyle\prod_{k=1}^{n-1}\left[n^2 + \left(\frac{3}{2}-\frac{1}{2}k\right)n + \left(\frac{5}{2}+\frac{1}{2}k^2\right)\right]$
4	$\displaystyle\prod_{k=1}^{n-1}\left[n^3 + (2-k)n^2 + (6-k+k^2)n + (5+k^2)\right]$

Table 4 shows detailed calculations of the coefficients of the number states for $n = 3$, $\gamma/\varepsilon = 1$, and $\sigma = 0,1$. In the two states of lowest energy, which are $(n,p) = (3,3)$ and $(3,2)$, the coefficients of the states $|3\rangle|0\rangle$ and $|0\rangle|3\rangle$ increase with increasing σ, indicating increased localization of the three quanta in one of the two nonlinear oscillators.

Table 4. Coefficients for number states $|3\rangle|0\rangle$, $|2\rangle|1\rangle$, $|1\rangle|2\rangle$, $|0\rangle|3\rangle$ for $n = 3$, $\gamma/\varepsilon = 1$, and $\sigma = 0$ (upper value), $\sigma = 1$ (middle value), and $\sigma = 2$ (lower value).

n,p	$|3\rangle|0\rangle$	$|2\rangle|1\rangle$	$|1\rangle|2\rangle$	$|0\rangle|3\rangle$	E(n,p)
	0.353	-0.612	0.612	-0.353	-1
3,0	0.247	-0.662	0.662	-0.247	8.23
	0.117	-0.697	0.697	-0.117	16.46
	0.612	-0.353	-0.353	0.612	-3
3,1	0.5	-0.5	-0.5	0.5	7.15
	0.183	-0.683	-0.683	0.183	16.29
	0.612	0.353	-0.353	-0.612	-5
3,2	0.662	0.247	-0.247	-0.662	4.77
	0.697	0.117	-0.117	-0.697	9.54
	0.353	0.612	0.612	0.353	-7
3,3	0.5	0.5	0.5	0.5	1.85
	0.683	0.183	0.183	0.683	5.71

4. CONCLUSION

A new nonlinear ordinary differential equation system, the generalized discrete self-trapping equation, has been investigated. Various Hamiltonian formalisms are discussed. One of them leads to a generalization of the classical equations for the "Feynman" top. In the context of this picture, we have catalogued the functional representations of periodic solutions with increasing degree of nonlinearity. The localization of an unsymmetric stationary has been shown to increase dramatically with the degree of nonlinearity.

ACKNOWLEDGEMENTS

The support from the National Science Foundation, Julie Damms Studie-
fond, and the Danish Technical Research Council is acknowledged. The author
expresses thanks for the warm hospitality of the Institute of Mathematics
and its Applications, University of Minnesota (funds provided by the
National Science Foundation), the Program in Applied Mathematics, University of Arizona, and the Center for Nonlinear Studies, Los Alamos National
Laboratory.

REFERENCES

1. A.S. Davydov, Biology and Quantum Mechanics, Pergamon, Oxford (1981).
2. A.S. Davydov, Sov. Phys. Usp. $\underline{25}$ (1982) 899; Usp. Fiz. Nauk $\underline{138}$
    (1982) 603.
3. J.C. Eilbeck, P.S. Lomdahl, and A.C. Scott, Physica D $\underline{16}$ (1985) 318.
4. A.C. Scott, P.S. Lomdahl, and J.C. Eilbeck, Chem. Phys. Lett. $\underline{113}$
    (1985) 29.
5. A.C. Scott and J.C. Eilbeck, Chem. Phys. Lett. $\underline{132}$ (1986) 23.
6. A.C. Scott and J.C. Eilbeck, Phys. Lett. A $\underline{119}$ (1986) 60.
7. A.C. Scott, L. Bernstein, and J.C. Eilbeck, J. Biol. Phys. $\underline{17}$ (1989) 1.
8. L.J. Bernstein, J.C. Eilbeck, and A.C. Scott, "The Quantum Theory of
    Local Modes in a Coupled System of Nonlinear Oscillators", Nonlinear-
    ity (accepted).
9. A.C. Scott, L. Bernstein, and J.C. Eilbeck, "Local Modes in Molecules",
    J. Molecular Liquids $\underline{41}$ (1989) 105.
10. V.M. Kenkre and D.K. Campbell, Phys. Rev. B $\underline{34}$ (1986) 4959.
11. L. Cruzeiro-Hansson, P.L. Christiansen, and J.N. Elgin, Phys. Rev.
    B $\underline{37}$ (1988) 7896
12. J.H. Jensen, P.L. Christiansen, J.N. Elgin, J.D. Gibbon, and O.
    Skovgaard, Phys. Lett. A $\underline{110}$ (1985) 429.
13. S. De Filippo, M. Fusco Girard, and M. Salerno, Physica D $\underline{26}$ (1987)
    411; ibid. D $\underline{29}$ (1988) 421.
14. H. Feddersen, R. Flesch, M. Salerno, and P.L. Christiansen, "Comment
    on 'Theory of the Liapunov Exponents of Hamiltonian Systems and a
    Numerical Study of the Transition from Regular to Irregular Classical
    Motion'", J. Chem. Phys. $\underline{92}$ (1990) (to appear).
15. S. De Filippo, M. Fusco Girard, and M. Salerno, "Avoided Crossing and
    Next Neighbour Spacings for the Quantum DST Equation" (to appear).
16. L. Cruzeiro-Hansson, H. Feddersen, R. Flesch, P.L. Christiansen, M.
    Salerno, and A.C. Scott, "Classical and Quantum Analysis of Chaos in
    the discrete Self-trapping Equation" (to appear).
17. A.C. Scott and P.L. Christiansen, "A generalized discrete self-trap-
    ping equation", Physica Scripta (to appear).
18. P.L. Christiansen, "On a generalized discrete self-trapping equation",
    J. Molecular Liquids $\underline{41}$ (1989) 113-121.
19. A.C. Scott, "A non-resonant discrete self-trapping system", Physica
    Scripta (to appear).
20. R.P. Feynman, F.L. Vernon, and R.W. Hellwarth, J. Appl. Phys. $\underline{28}$
    (1957) 49.
21. P.F. Byrd and M.D. Friedman, "Handbook of Elliptic Integrals for En-
    gineers and Physicists", Springer, Berlin (1971).

40

LOCAL MODES AND DEGENERATE PERTURBATION THEORY

Lisa Bernstein

Program in Applied Mathematics, University of Arizona

Tucson, Arizona 85721 U.S.A.

1 Introduction

Nonlinearity can lead to the localization of energy. Beyond this simple statement lies a myriad of models, mechanisms, theories, and approximations of theories directed at understanding energy localization in specific physical systems. Successful application of these mathematical models requires understanding the distinguishing features of the behavior each describes. As discussed in [11] and [9], the Discrete Self-Trapping Equation (DST)

$$(i\frac{d}{dt} - \omega_0)\bar{A} + \gamma\text{diag}(|A_1|^2, ..., |A_f|^2)\bar{A} + \epsilon M\bar{A} = 0 \tag{1}$$

is one such mathematical model, used to describe the anharmonic bond vibrations of polyatomic molecules. In this paper, the nonlinear dynamical system of Eq. (1) is considered to represent the time evolution of the complex amplitudes of *classical* oscillators and thus it can be quantized. The aim here is to present, in the context of other well-known energy-localization models, the particular characteristics of the mechanism for energy localization in the quantum-mechanical version of Eq. (1), the *Quantized* DST system (QDST).

As for any quantum-mechanical system, the governing equations for the QDST are linear equations. The discussion here is part of a more general theme in modern mathematical physics – the effort to understand the relationship of inherently nonlinear classical effects to systems with significant quantum-mechanical character. The Discrete Self-Trapping model is particularly suited to studies of this nature due to the simple connection between its classical and quantum versions [28]. In the case of energy localization, the connection between the energy-localizing solutions of the classical DST and the local modes of the QDST is captured in a simple formula, to be presented here [3].

This paper is organized as follows. The physical chemistry phenomenon of local modes in small polyatomic molecules is discussed in Section 2. The QDST model is introduced in Section 3. Section 4 contains a description of the mechanism for energy localization in the QDST system. In Section 5, the QDST localization mechanism is compared with the mechanisms of other models, and comments are made concerning energy localization in biological proteins.

2 Local Modes in Molecules

It is well-known in the literature of physical chemistry that the vibrational energy in some

Davydov's Soliton Revisited, Edited by P.L. Christiansen and A.C. Scott
Plenum Press, New York, 1990

small, polyatomic molecules can become localized [1,13,14,15,20,21,32,33]. The evidence for this localization comes from the striking resemblance of the infrared absorption spectra of systems with several bond oscillators of a particular type to the spectra of systems with a single such oscillator.

Consider, for example, the experimentally-measured spectra of benzene (C_6H_6) and pentadeuterobenzene (C_6HD_5) [22]. It is possible to identify the peaks in the spectra of these molecules that correspond to CH stretch vibrations. It is found that the positions and intensities corresponding to CH bond vibrations in benzene, a system of six identical CH bond oscillators, are nearly identical to the positions and intensities of the CH vibration peaks observed for the single CH oscillator in C_6HD_5.

The positions (energies) of the CH peaks in these spectra are well-described by the Birge-Sponer relation

$$E(n) = An - Bn^2 \qquad (2)$$

where A and B are constants and n is the quantum level. For C_6HD_5, this indicates that the single CH oscillator is anharmonic. The existence of lines at these same positions in the spectra of benzene suggests that the CH bonds of benzene also vibrate anharmonically and are coupled to each other very weakly. The absence of any additional CH lines in the spectra of benzene indicates that benzene does not have excited levels in which energy is distributed over the six CH oscillators. The only observed overtone levels are those in which all the CH vibrational energy is on one of benzene's six CH bonds.

There are two quantum-mechanical descriptions of the overtone states of benzene consistent with the observed positions of spectral peaks. Quantum states could be either those that localize energy on a single, specific bond, or those for which there is some probability of energy being localized on any of the six bonds. At first consideration, the latter possibility appears to be a better quantum-mechanical description because *stationary* quantum wavefunctions for a system of identical, interacting oscillators must be properly symmetrized with respect to oscillator exchanges. However, attributing the observed spectral lines of benzene not to stationary states but to asymmetric *nonstationary wavepackets* that localize energy on a specific site gives a much better explanation of the intensities of those lines. The high intensities of spectral lines observed for benzene and C_6HD_5 indicate that the corresponding quantum states have large transition dipole moments. In many cases, properly-symmetrized stationary states of benzene do not have this property, but asymmetric superpositions of stationary states do.

Like the local mode theories of physical chemistry, the QDST model describes local modes in molecules as such nonstationary wavepackets. The stationary-state components of such a wavepacket must be properly symmetrized, but the wavepacket itself may initially localize energy on a single, specific bond. These non-symmetric quantum wavepackets are an example of the non-symmetric "quasi-modes" that are described in semiclassical theories by Arnol'd [2,4]. The fully quantum-mechanical QDST description of local mode dynamics is uniquely simple, exact and quantitative, and leads to analytic results that link the quantum theory to the corresponding classical dynamics.

3   The Quantized DST

The transformation from the classical DST system to the exact quantum-mechanical problem is very straightforward. The two conserved quantities of the classical model, the *energy* and the *number*, are transformed to the corresponding quantum-mechanical operators by the substitutions

$$A_i \to \hat{B}_i \qquad (3)$$

and

$$A_i^* \to \hat{B}_i^\dagger. \tag{4}$$

The linear operators $\hat{B}_i(\hat{B}_i^\dagger)$ are annihilation(creation) operators for bosons of the ith bond. Products of classical amplitudes such as $|A_i|^4$ are transformed to the averages over all possible orderings of the non-commuting quantum operators [28]. One obtains from the above transformations the energy operator

$$\hat{H} = (\omega_0 - \frac{1}{2}\gamma)\sum_{i=1}^{f} \hat{B}_i^\dagger \hat{B}_i - \frac{1}{2}\gamma \sum_{i=1}^{f} (\hat{B}_i^\dagger \hat{B}_i)^2 - \epsilon \sum_{i \neq j} m_{ij} \hat{B}_i^\dagger \hat{B}_j \tag{5}$$

and the number operator

$$\hat{N} = \sum_{i=1}^{f} \hat{B}_i^\dagger \hat{B}_i + \frac{f}{2}. \tag{6}$$

Because the $\hat{H}$ commutes with $\hat{N}$, it is possible to find simultaneous eigenstates of both operators.

Each term in Eq. (6) has the form of the hamiltonian for a single independent harmonic oscillator and has the number states $|n_i\rangle$ as eigenfunctions [18]. Thus an eigenfunction of $\hat{N}$ with eigenvalue n is the product

$$|\phi\rangle = |n_1\rangle|n_2\rangle|n_3\rangle \ldots |n_{f-1}\rangle|n_f\rangle \tag{7}$$

where

$$\sum_{i=1}^{f} n_i = n. \tag{8}$$

The product states $|\phi\rangle$ have the same *spatial* character as the stationary states of a system of noninteracting harmonic oscillators. For given values of n and f, there are

$$p = \frac{(n+f-1)!}{(f-1)!\,n!} \tag{9}$$

distinct such product states, where p is the number of ways that n indistinguishable quanta can be placed onto f degrees of freedom. The most general eigenfunction of $\hat{N}$ with eigenvalue n is the linear combination of these product states

$$|\Psi_n\rangle = \sum_{j=1}^{p} c_j|\phi_j\rangle. \tag{10}$$

It is possible to choose the coefficients $\{c_j\}$ in Eq. (10) so that $|\Psi_n\rangle$ is also an eigenfunction of $\hat{H}$,

$$\hat{H}|\Psi_n\rangle = E|\Psi_n\rangle. \tag{11}$$

Satisfying Eq. (11) requires that the column vector $\bar{c} = \mathrm{col}(c_1, c_2, c_3, \ldots, c_p)$ satisfy the matrix equation

$$H_n \bar{c} = E\bar{c} \tag{12}$$

The full, infinite-dimensional hamiltonian operator $\hat{H}$ is block-diagonal in the basis of number eigenstates $|\phi\rangle$, with each block corresponding to a particular quantum level n. The matrix H_n in Eq. (12) is the nth such block. It is a symmetric $p \times p$ matrix with real components.

The energy levels corresponding to a given value of n may be computed numerically to arbitrary precision from the $p \times p$ matrix H_n. Stationary states and stationary state energies computed in this manner are *exact*. No approximations have been used in the deriving these quantum equations from the Eq. (1).

By fitting the parameters ϵ, γ, and ω_0 and the dispersion matrix elements m_{ij}, the QDST system has been used to model the spectra of benzene [27,26] and the three dihalomethanes [3]. In each case, simple, small-scale computations of the eigenvalues of the matrices H_n yielded energies in good agreement with experiment.

4 Local Modes for the QDST

In many cases, solutions of the classical DST equation (1) have been found that have harmonic time dependence and concentrate energy on a single bond oscillator for all time [11,24]. These local modes of the classical model are a direct result of the nonlinearity in the system. Here it is shown how the nonlinearity in the classical model is made manifest in the structure of the corresponding linear quantum-mechanical problem.

4.1 Two Degrees of Freedom

Consider first the quantum theory of local modes for a system of two coupled bond oscillators. The system consists of two sites, a and b. Energy-localizing solutions of the classical system are known in terms of elliptic functions [9,11,17,19,25]. We seek here the quantum-mechanical state that corresponds to those classical local solutions.

For a two-freedom system, such as the two CH stretch vibrations of dichloromethane, stationary quantum states must be either symmetric or antisymmetric with respect to exchanges of the two oscillators. At the nth excited level, the two stationary states of lowest energy are the symmetric state

$$|\Psi_n\rangle_{s+} = \frac{1}{\sqrt{2}}(|n\rangle_a|0\rangle_b + |0\rangle_a|n\rangle_b) + O(\epsilon) \tag{13}$$

and the antisymmetric state

$$|\Psi_n\rangle_{s-} = \frac{1}{\sqrt{2}}(|n\rangle_a|0\rangle_b - |0\rangle_a|n\rangle_b) + O(\epsilon). \tag{14}$$

The wavefunctions above are written according to the notation in [9], where $|i\rangle_a|j\rangle_b$ means i quanta on site a and j quanta on site b. The stationary states $|\Psi\rangle_{s+}$ and $|\Psi\rangle_{s-}$ evolve harmonically in time according to the corresponding energy eigenvalues, E_{s+} and E_{s-}. An experimental measurement of a system in one of these two states would find all n quanta at a single site with an equal probability of finding the quanta at either site. It is possible to localize energy on a *specific site* at some initial time by the nonstationary wavepacket

$$|\Psi_{local}(t)\rangle \frac{1}{\sqrt{2}}(|\Psi_n\rangle_{s+} + |\Psi_n\rangle_{s-}). \tag{15}$$

For example, at $t = 0$,

$$|\Psi_{local}(0)\rangle = |n\rangle_a|0\rangle_b + O(\epsilon), \tag{16}$$

and there is a high probability of finding all the quanta on site a. The energy localized by this nonstationary wavepacket will disperse from site a in a time

$$\Delta t \propto \frac{1}{\Delta E} \tag{17}$$

where

$$\Delta E = |E_{s+} - E_{s-}|. \tag{18}$$

For a two-freedom system at the first ($n = 1$) excited level, the matrix equation for the energy eigenvalues is particularly simple. Here

$$H_1 = \begin{pmatrix} \omega_0 - \gamma & \epsilon \\ \epsilon & \omega_0 - \gamma \end{pmatrix}$$

and

$$\Delta E = 2\epsilon. \tag{19}$$

The lifetime of a local mode at the first excited level is independent of the magnitude of the nonlinearity.

The effect of the nonlinearity becomes pronounced in the QDST system for quantum levels $n \geq 2$. It is not generally possible to solve the Hamiltonian eigenvalue problem analytically for these overtone levels. However, in the physically relevant case $\epsilon < \gamma$, useful results have been obtained using perturbation theory [3]. For a two-freedom QDST system at the nth excited level, the energy splitting between the stationary state wavefunctions in Eqs. (13) and (14) satisfies

$$\Delta E = \frac{2n\epsilon^n}{(n-1)! \, \gamma^{n-1}} + O(\frac{\epsilon^{n+1}}{\gamma^n}). \tag{20}$$

Eq. (20) suggests that as the classical limit of large n is approached, the two stationary states that comprise a local wavepacket become quasi-degenerate. It follows that the amount of time during which energy can remain localized becomes arbitrarily long.

The $(n-1)!$ in the denominator is an important feature of Eq. (20); it indicates that ΔE decreases quite rapidly as a function of n. It follows that the lifetimes of quantum local modes become extremely long even for small values of n. In this manner, the localized quantum wavepackets resemble the local modes of the classical system, even very close to the quantum-mechanical ground state.

Numerical calculations were used to test the significance of the higher order terms in the Eq. (20). The first term of Eq. (20) was compared with exact values of ΔE for a range of physically realistic values of the ratio ϵ/γ and n. For $n \geq 2$, the first term of Eq. (20) was found to be a good estimate of the exact ΔE. Further, ΔE appears to converge to this first term for large n. Similar results were obtained for the six-freedom model of benzene.

4.2   Many Freedoms

The energy-localizing mechanism described in the previous section can be shown to exist in systems with an arbitrary number of oscillators. It is essential that all the oscillators in the system are identical to each other in fundamental frequency and in their environments due to coupling. Specific restrictions on the class of dispersion matrices that satisfy this requirement are given in [3].

For a QDST system of f identical oscillators, there exist f properly-symmetrized stationary states of the form

$$|\Psi_n\rangle = c_1|n\rangle|0\rangle|0\rangle \ldots |0\rangle|0\rangle + c_2|0\rangle|n\rangle|0\rangle \ldots |0\rangle|0\rangle +$$

$$c_3|0\rangle|0\rangle|n\rangle \ldots |0\rangle|0\rangle + \cdots + c_{f-1}|0\rangle|0\rangle|0\rangle \ldots |n\rangle|0\rangle +$$

$$c_f|0\rangle|0\rangle|0\rangle \ldots |0\rangle|n\rangle + O(\epsilon).$$

Of the p stationary state wavefunctions $|\Psi_n\rangle$ that satisfy Eq. (10), the f states of this form have the lowest energies. At some initial time, the energy in the system can be localized, up to $O(\epsilon^2)$, on a specific one of the f oscillators in the system by a superposition of these f stationary states. The time evolution of each stationary state component in this nonstationary wavepacket is governed by the corresponding energy eigenvalue $E^{(j)}$. The magnitude ΔE of the difference between the largest and smallest of these energy levels determines the lifetime of the energy localization. In a time interval

$$\Delta t \propto \frac{1}{\Delta E}, \tag{21}$$

the contributions to the amplitude on the original site from different stationary states become out of phase and energy is no longer localized there.

As in the case of two degrees of freedom, the value of ΔE is independent of the magnitude of nonlinearity at the first excited level. Energy that is initially localized disperses quickly, as it would for a linear system.

For higher quantum levels, n^{th} order degenerate perturbation theory can be used calculate a general bound on ΔE in terms of system parameters. For an f-freedom QDST system at the nth excited level,

$$\Delta E \leq \frac{f n \epsilon^n}{(n-1)! \, \gamma^{n-1}} + O(\frac{\epsilon^{n+1}}{\gamma^n}). \tag{22}$$

As the excited level of the system is increased, the value of ΔE decreases "factorially" and the energy that is localized by quantum-mechanical wavepackets can remain localized for increasingly long times. Thus, the quantum mechanical description of the system resembles the classical dynamics in the classical limit, and even for energy levels close to the ground state.

For specific oscillator systems, a precise estimate of ΔE can be calculated in terms of the elements of the corresponding dispersion matrix M [3]. An interesting special case, motivated by the Davydov soliton theory [5,6,7], is a long oscillator chain with periodic boundary conditions and coupling of magnitude ϵ between nearest neighbors. For any system in this class,

$$\Delta E = \frac{4 n \epsilon^n}{(n-1)! \, \gamma^{n-1}} + O(\frac{\epsilon^{n+1}}{\gamma^n}). \tag{23}$$

The time for energy localization in this nearest-neighbor chain system is *independent of the number of sites in the chain.*

5   Conclusions

The specific mechanism for the localization of energy in the QDST model of indistinguishable coupled anharmonic oscillators has been described. The particular characteristics of this mechanism are the following:

1. Quantum-mechanical local modes are *nonstationary states* that localize energy for some finite time interval.

2. At the first excited level, the nonlinearity in the system has no effect on the duration of energy localization.

3. The lifetimes of quantum local modes lengthen rapidly with increasing quantum level n, and are described by a simple formula.

The first item listed above is a direct consequence of the symmetry in QDST model. Symmetry requirements on stationary quantum states do not exist for similar but asymmetric systems such as the *nonresonant* quantum DST model

$$\hat{H} = \sum_{i=1}^{f} (\omega_i - \frac{1}{2}\gamma)\hat{B}_i^\dagger \hat{B}_i - \frac{1}{2}\gamma \sum_{i=1}^{f} (\hat{B}_i^\dagger \hat{B}_i)^2 - \epsilon \sum_{i \neq j} m_{ij} \hat{B}_i^\dagger \hat{B}_j, \tag{24}$$

where each bond oscillator has a distinct fundamental frequency ω_i. As shown in [30], there are significant differences in the quantum theory for this model. In particular, energy can be localized to some extent by *stationary* quantum states.

The "linear" appearance of the QDST local modes at the first excited level is a result of the QDST having *intrinsic* nonlinearity, that is, nonlinearity that comes about because the oscillators in the system vibrate in nonlinear potentials. A different type of nonlinearity is the *extrinsic* type of nonlinearity found in the Fröhlich Hamiltonian [8]

$$\hat{H} = \sum_{i=1}^{f} \left[E_0 \hat{B}_i^\dagger \hat{B}_i - J(\hat{B}_i^\dagger \hat{B}_{i+1} + \hat{B}_i^\dagger \hat{B}_{i-1}) \right] + \sum_{q=1}^{f} \Omega_q \hat{b}_q^\dagger \hat{b}_q + \sum_{i,q=1}^{f} \chi_q (\hat{b}_q^\dagger + \hat{b}_q)(\hat{B}_i^\dagger \hat{B}_i). \tag{25}$$

In Eq. (25), nonlinear effects are produced by interactions between the exciton and phonon systems. Although both *extrinsic* and *intrinsic* nonlinearity can lead to the localization of energy, there is a significant difference in the corresponding quantum dynamics. Perturbation theory in the weak-coupling limit for the Fröhlich model [29] shows that the energy level splitting ΔE at the first excited level is diminished by the Frank-Condon factor $\exp(-\chi^2/\omega_0^2)$. Due to this factor, the magnitude of the nonlinearity parameter χ in a Fröhlich system has a significant effect at the first excited level, while the corresponding parameter γ in the QDST model does not.

A consideration of these different energy-localization mechanisms suggests two specific comments concerning energy localization in biological proteins. First, the QDST model shows clearly that the presence or absence of symmetry is important in characterizing localization phenomena. A biological protein molecule is clearly an asymmetric structure, yet many current theories are based on highly symmetric model systems. This point should be considered in evaluating present and future work. Second, a significant feature of the QDST mechanism for energy localization is that the nonlinear effects develop starting at the second excited level. Although the nonlinearity does appear at the first excited level in the Fröhlich system, it is possible that nonlinear effects in that model also become significantly stronger at the second excited level. This idea is reinforced by the fact that the proposed source for the energy in localization theories based on the Fröhlich model is the hydrolysis of ATP, a reaction producing *two* quanta of amide-I energy [23,16]. Investigating the case $n = 2$ may be important in understanding the dynamics of real proteins.

Acknowledgements

Support for this work from the National Science Foundation, NATO, and the nonlinear systems research program of the SERC is gratefully acknowledged.

References

[1] Ahmed M K and Henry B R 1987 Intensity distribution in the overtone spectra of methyl halides: A local mode analysis of the spectra of methyl halides and methyl cyanide *J. Chem. Phys.* **87** 3724-30

[2] Arnol'd V I 1972 Modes and Quasimodes *Funct. Anal. Applic.* **6** 94-101

[3] Bernstein L J, Eilbeck J C and Scott A C 1990 The quantum theory of local modes in a coupled system of nonlinear oscillators *Nonlinearity* (to appear)

[4] Berry M V 1977 Remarks on degeneracies of semiclassical energy levels *J. Phys. A* **10** L193-L194

[5] Davydov A S 1981 *Biology and Quantum Mechanics* (Oxford: Pergamon)

[6] Davydov A S 1979 *Phys. Scr.* **20** 387

[7] Davydov A S 1982 *Sov. Phys. Usp.* **25** *899; 1982 Usp. Fiz. Nauk* **138** 603

[8] Fröhlich 1954 *Adv. Phys.* **3** 325

[9] Eilbeck J C Introduction to the Discrete Self-Trapping Equation 1989 (these proceedings)

[10] Eilbeck J C, Lomdahl P S and Scott A C 1984 Soliton structure in crystalline acetanilide *Phys. Rev. B* **30** 4703-12

[11] Eilbeck J C, Lomdahl P S and Scott A C 1985 The discrete self trapping equation *Physica D* **16** 318-38

[12] Feddersen H 1989 Quantum and classical descriptions of chaos in the DST equation (these proceedings)

[13] Henry B R and Siebrand W 1968 Anharmonicity in Polyatomic Molecules: The CH-Stretching Overtone Spectrum of Benzene *J. Chem. Phys.* **49** 5369-76

[14] Henry B R 1976 Local Modes and Their Application to the Analysis of Polyatomic Overtone Spectra *J. Phys. Chem.* **80** 2160-4

[15] Henry B R 1981 The Local Mode Model *Vibrational Spectra and Structure* ed J R Durig (Amsterdam: Elsevier) 269-319

[16] Hock F L 1971 *Energy Transformations in Mammals: Regulatory Mechanisms* (Philadelphia: Saunders) 1-5

[17] Kenkre V M and Campbell D K 1986 Self-trapping on a dimer: time-dependent solutions of a discrete nonlinear Schrödinger equation *Phys. Rev. B* **34** 4959-61

[18] Louisell W H 1977 *Radiation and Noise in Quantum Electronics* (New York: Krieger)

[19] Mayer A A 1984 Self-switching of light in a directional coupler *Sov. J Quantum Electron.* **14** 101-4

[20] Mortensen O S, Henry B R and Mohammadi M A 1981 The effects of symmetry within the local mode picture: a reanalysis of the overtone spectra of the dihalomethanes *J. Chem. Phys.* **75** 4800-8

[21] Mortensen O S, Ahmed M K, Henry B R and Tarr A W 1985 Intensities in local mode overtone spectra: Dichloromethane and deuterated dichloromethane *J. Chem. Phys.* **82** 3903-11

[22] Reddy K V, Heller D F and Berry M J 1982 Highly vibrationally excited benzene: Overtone spectroscopy and intramolecular dynamics of C_6H_6, C_6D_6, and partially deuterated or substituted benzenes *J. Chem. Phys.* **76** 2814-37

[23] Scott A C 1982 Dynamics of Davydov Solitons *Phys. Rev. A* **26** 578-95

[24] Scott A C and MacNeil L 1983 Binding energy versus nonlinearity for a "small" stationary soliton *Phys. Lett.* **98A** 87-88

[25] Scott A C, Lomdahl P S and Eilbeck J C 1985 Between the local-mode and normal-mode limits *Chem. Phys. Lett.* **113** 29-36

[26] Scott A C and Eilbeck J C 1986 On the CH stretch overtones of benzene *Chem. Phys. Lett.* **132** 23-8

[27] Scott A C, Bernstein L and Eilbeck J C 1989 Energy levels of the quantized discrete self-trapping equation *J. Biol. Phys.* **17** 1-17

[28] Scott A C and Eilbeck J C 1986 The quantized discrete self-trapping equation *Phys. Lett. A* **119** 60-4

[29] Scott A C, Bigio I J, and Johnston C T 1989 Polarons in acetanilide *Phys. Rev. B* **39** 12883-7

[30] Scott A C 1989 A non-resonant discrete self-trapping system *Physica Scripta* (to appear)

[31] Scott A C, Bernstein L and Eilbeck J C 1989 Local Modes in Molecules *J. Molec. Liquids* **41** 105-111

[32] Wallace R 1975 *J. Chem. Phys.* **11** 189

[33] Watson I A, Henry B R and Ross I G 1981 Local Mode Behavior: the Morse oscillator model *Spectrochemica Acta* **37a** 857-65

41

QUANTUM AND CLASSICAL DESCRIPTIONS OF CHAOS IN THE DST EQUATION

H.K. Feddersen, L. Cruzeiro-Hansson[1], R. Flesch[2],
P.L. Christiansen, M. Salerno[3], and A.C. Scott

Laboratory of Applied Mathematical Physics
The Technical University of Denmark
DK-2800 Lyngby

1. INTRODUCTION

In this paper we will consider the discrete self-traping (DST) equation [1] with only three freedoms, the simplest case in which the DST equation may show chaotic dynamics. The dispersion matrix $\underline{\underline{M}}$ is throughout taken to be

$$M = \begin{bmatrix} 0 & 1 & 1 \\ 1 & 0 & 1 \\ 1 & 1 & 0 \end{bmatrix} . \tag{1}$$

With this dispersion matrix the Hamiltonian of the DST equation reads

$$H = -\frac{\gamma}{2} \left(|A_1|^4 + |A_2|^4 + |A_3|^4 \right)$$
$$- \varepsilon \left(A_1^* A_2 + A_2^* A_1 + A_1^* A_3 + A_3^* A_1 + A_2^* A_3 + A_3^* A_2 \right) \tag{2}$$

where A_j are the complex mode amplitudes. This Hamiltonian leads to the equations of motion $i\dot{A}_j = \partial H / \partial A_j^*$ or

$$i\dot{A}_1 = -\gamma |A_1|^2 A_1 - \varepsilon (A_2 + A_3)$$
$$i\dot{A}_3 = -\gamma |A_2|^2 A_2 - \varepsilon (A_1 + A_3) \qquad . \tag{3}$$
$$i\dot{A}_3 = -\gamma |A_3|^2 A_3 - \varepsilon (A_1 + A_2)$$

Besides the Hamiltonian (energy) the DST system has the conserved quantity, the norm (or the number)

$$N = |A_1|^2 + |A_2|^2 + |A_3|^2 \qquad . \tag{4}$$

Note that the equations of motion (3) and the quantities γN and γH are invariant under the scale transform [17]

$$A_j \rightarrow \alpha A_j , \quad N \rightarrow \alpha^2 N , \quad \gamma \rightarrow \gamma / \alpha^2 , \quad H \rightarrow \alpha^2 H \qquad . \tag{5}$$

Hence the solution to (3) does not depend upon N, γ, and H separately; rather on the products γN and γH.

The DST equation is well suited for a comparison between a classically chaotic system and its quantum mechanical counterpart ("quantum chaos") because it is so easily quantized [2]. Under quantization Eq. (2) becomes the Hamiltonian operator:

Davydov's Soliton Revisited, Edited by P.L. Christiansen and A.C. Scott
Plenum Press, New York, 1990

$$\hat{H} = -\frac{\gamma}{2} \hat{N} - \frac{\gamma}{2} (\hat{B}_1^\dagger \hat{B}_1 \hat{B}_1^\dagger \hat{B}_1 + \hat{B}_2^\dagger \hat{B}_2 \hat{B}_2^\dagger \hat{B}_2 + \hat{B}_3^\dagger \hat{B}_3 \hat{B}_3^\dagger \hat{B}_3)$$
$$- \varepsilon (\hat{B}_1^\dagger \hat{B}_2 + \hat{B}_2^\dagger \hat{B}_1 + \hat{B}_1^\dagger \hat{B}_3 + \hat{B}_3^\dagger \hat{B}_1 + \hat{B}_2^\dagger \hat{B}_3 + \hat{B}_3^\dagger \hat{B}_2) \qquad (6)$$

where $\hat{B}_j^\dagger$ and $\hat{B}_j$ are boson creation and annihilation operators, respectively.

To see how chaotic a classical dynamical system is, one may study the
distribution of the spacings between nearest neighbour energy levels.
This distribution (after a scaling of the energies which makes the density
of states unity) varies from Poisson when the classical system is
completely integrable to Wigner when it is completely chaotic [3].

The aim of this paper is to study what happens in the case where the
classical DST equation (3) has coexisting regular and chaotic trajectories. The degree of chaos of the classical DST equation is given as an
estimate of the fraction, ρ_c, of the phase space which is occupied by
chaotic trajectories. This estimate is compared with the energy level
spacing distribution of the quantized DST (QDST) equation from which a
similar parameter, ρ_q, can be calculated.

2. QUANTUM CHAOS IN THE DST EQUATION

It has been proved [7] that if a classical system is completely
integrable then the distribution of the scaled [5] nearest neighbour
energy level spacings of the corresponding quantized system is Poisson,
$P(s) = e^{-s}$, in the semiclassical limit, i.e. when $\hbar \to 0$. In the DST
system this is equivalent to the excitation level (eigenvalue of the
number operator $\hat{N}$) $n \to \infty$. If the classical system is completely chaotic
there is strong support (theoretical as well as numerical) to the
conjecture that the distribution is Wigner, $P(s) = \frac{\pi}{2} s e^{-\pi s^2/4}$ [8-11].
When the classical system has coexisting regular and chaotic trajectories
Berry and Robnik [3] has suggested that the (scaled) energy level
spacings will follow the distribution

$$P(s) = (1-\rho)^2 e^{-(1-\rho)s} \, \mathrm{erfc} \left(\frac{\sqrt{\pi}}{2} \rho s\right)$$
$$+ (2\rho(1-\rho) + \pi \rho^3 s/2) \; e^{-(1-\rho)s - \pi \rho^2 s^2/4} \qquad (7)$$

where it for simplicity has been assumed that the phase space can be
divided into one connected chaotic region and one regular region (this is
not necessarily true for the three freedom DST equation). In this case the
parameter ρ can be interpreted as the fraction of the phase space which
is chaotic. Thus (7) interpolates between the Poisson distribution
($\rho = 0$) and the Wigner distribution ($\rho = 1$), see Fig. 1. It should be
noted that other interpolation formulas than (7) have been proposed, e.g.
the Brody distribution [12].

It is clear from the Wigner distribution that degenerate energy levels
should be avoided in order to satisfy $P(0) = 0$. Therefore we consider
only the symmetric states [6,13]. The number of these states is for the
n'th excited level $1 + \mathrm{int} \left(\frac{n(n+6)}{12}\right)$.

3. THE CHAOTIC FRACTION OF THE PHASE SPACE

The parameter ρ in Eq. (7) can be interpreted as the chaotic fraction
of the classical phase space, i.e. $\rho = \rho_c$ is defined as [14]

$$\rho_c = \frac{\int X(\underline{r}) \, \delta[H(\underline{r}) - E] d\underline{r}}{\int \delta[H(\underline{r}) - E] d\underline{r}} \qquad (8)$$

where the integrals are evaluated over the entire phase space, $d\underline{r}$ being
the volume element, and $X(\underline{r})$ is 0 or 1 for quasiperiodic and chaotic
trajectories, respectively.

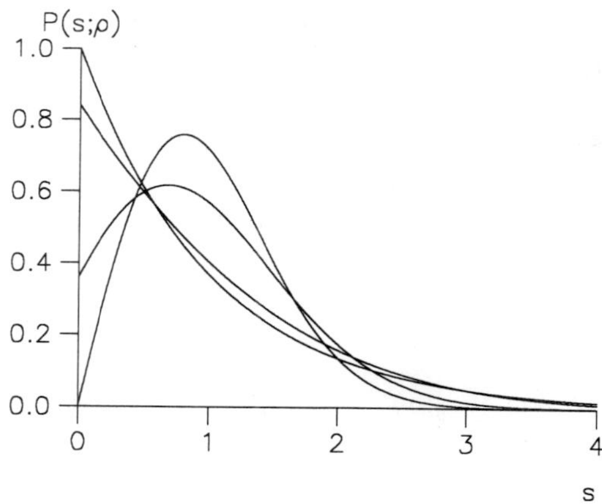

Fig. 1   Distributions obtained from the interpolation
          formula (7) for $\rho = 0$ (Poisson), $\rho = 0.4$,
          $\rho = 0.8$, and $\rho = 1$ (Wigner).

Since Eq. (8) is inconvenient for numerical calculations we choose to estimate ρ_c as [6,15]

$$\rho_c \approx \frac{\text{number of chaotic trajectories}}{\text{number of initial conditions}} \tag{9}$$

where the initial conditions are picked at random from a uniform distribution on the energy surface, $H = E$. The details of this procedure can be found in [6].

In expression (9) it is sufficient to consider a reduced version of the DST equation (3) which is obtained from the canonical transformation

$$A_1 = \sqrt{P_1}\, e^{-i(\phi_1 + \phi_3)}, \qquad 0 < P_1 < N$$

$$A_2 = \sqrt{P_2}\, e^{-i(\phi_2 + \phi_3)}, \qquad 0 < P_2 < N \tag{10}$$

$$A_3 = \sqrt{N - P_1 - P_2}\, e^{-i\phi_3}, \qquad 0 < P_1 + P_2 < N$$

leading to the Hamiltonian

$$H(\underline{\phi},\underline{P}) = -\tfrac{1}{2}\,\gamma[P_1{}^2 + P_2{}^2 + (N - P_1 - P_2)^2]$$

$$- 2\varepsilon[\sqrt{P_1 P_2}\,\cos(\phi_1 - \phi_2) + \sqrt{P_1(N - P_1 - P_2)}\,\cos\phi_1$$

$$+ \sqrt{P_2(N - P_1 - P_2)}\,\cos\phi_2] \tag{11}$$

where P_i is the moment conjugate to ϕ_i. In this description ϕ_3 is an ignorable coordinate [16] conjugate to the conserved quantity N.

The determination of whether or not a trajectory is chaotic is based on an examination of a Poincaré section which is defined as the plane $(\phi_2,\ P_2)$ with $P_1 = $ constant and ϕ_1 given by $H(\underline{\phi},\ \underline{P}) = E$. We now use the fact that a regular trajectory intersects the Poincaré section in points lying on a (one-dimensional) curve, whereas the intersections of a chaotic trajectory is area filling (two-dimensional) [4], see the examples in Fig. 2.

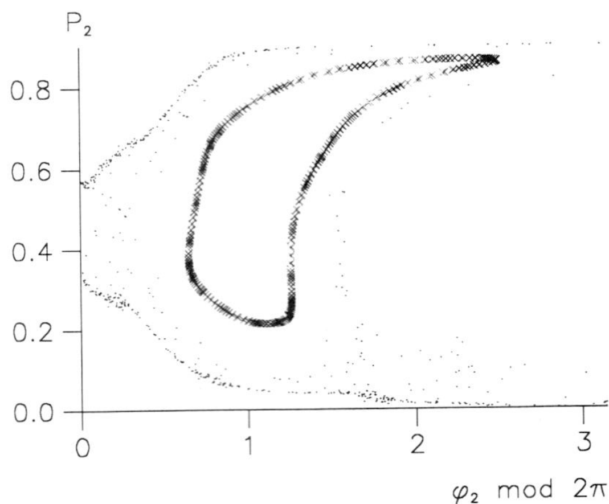

Fig. 2   Poincare section of a regular (x) and a chaotic
(•) trajectory ($P_1 = 0.1$, $\gamma N = 3$, $\gamma E = -2.4$).

4.   NUMERICAL RESULTS

In this section we compare values of ρ determined by i) fitting the
distribution (7) to the histograms of the energy level spacings obtained
numerically: $\rho = \rho_q$, and ii) estimating the chaotic fraction of the
classical phase space: $\rho = \rho_c$, using (9).

The fitting is done by calculating the second moment from the data and
by comparing this with the second moment of the distribution (7) [3] a ρ_q
is determined. Fig. 3 shows two examples of histograms and the fitted
distribution (7). In Fig. 3a the distribution is Poisson-like ($\rho_q = 0.28$)
and in Fig. 3b it is more Wigner-like ($\rho_q = 0.67$).

In Fig. 4 the main results are shown. For n = 310 and $\gamma n = 3$, a number
of ρ_q's have been calculated. These are indicated as horizontal bars.
Also shown in Fig. 4 as square dots and error bars are classical
calculations of ρ with $\gamma n = 3$. The determination of these ρ_c's is based
on 50-100 trajectories.

To check how many levels are needed to give reliable statistics we
have increased the number of levels per histogram and n until we got
convergence. This way we find that 2000 energy levels seem to be about
the least we can use. With n = 310 it takes about 20 CPU minutes on an
AMDAHL VP1100 super computer to compute the 8164 energy eigenvalues.

5.   CONCLUSIONS

The main results are presented in Fig. 4. In the intermediate range of
energy, where $\rho_c \approx \rho_q \approx 1/2$, the agreement seems to be within the error
bars. At high energies, however, where ρ_c is small the agreement is not
good. Similar discrepancies between ρ_c and ρ_q for other integrable or
near-integrable Hamiltonian systems have been observed in numerical
calculations [18,19]. This may indicate that for small ρ_c's to reach
the semiclassical limit requires more energy levels (larger n) than for
bigger ρ_c's. The difference between ρ_c and ρ_q for high energies might
seem rather big, but here one should bear in mind that ρ_q is extremely
sensitive to changes in the second moment for small ρ_q's (cf. Fig. 1;
there is only little difference between the distributions for $\rho = 0$ and
$\rho = 0.4$).

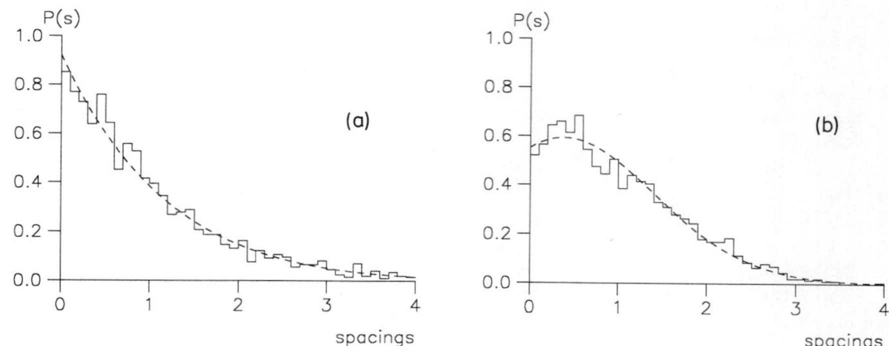

Fig. 3   Histograms obtained numerically (2000 energy level
spacings) and the corresponding distribution from
the interpolation formula (7). The parameters in
the DST equation and Eq. (7) are $\gamma n = 3$, $\varepsilon = 1$, and
a) $\gamma E = -0.18$, $\rho_q = 0.28$; b) $\gamma E = -5.40$, $\rho_q = 0.67$.

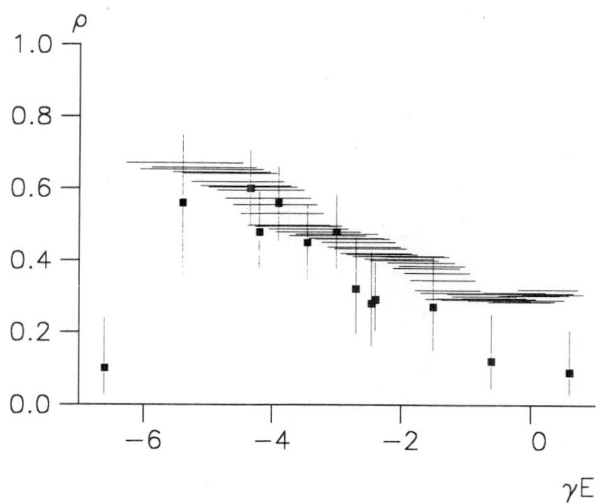

Fig. 4 ρ_q (-) and ρ_c (□) as functions of energy γE for
fixed $\gamma n = 3$. The ρ_q's have been calculated from
the QDST equation with n = 310 giving a total of
8164 symmetric energy levels. They are indicated
as horizontal bars where the length of one bar
equals the standard deviation of the energy levels
from which the statistics is calculated. Here we
use 2000 levels as in Fig. 3. The vertical bars
indicate 95% confidence intervals for the estimate
(9) of the classical ρ_c's.

The EEC Science Programme under grant no (89 100079/JU1) and the NATO Special Programme on Chaos, Order and Patterns for support for a collaborative programme are acknowledged for financial support during this work.

6. REFERENCES

[1] J.C. Eilbeck, P.S. Lomdahl, and A.C. Scott, Physica 16D (1985) 318 and the contributions to these proceedings by J.C. Eilbeck and M. Salerno.

[2] A.C. Scott and J.C. Eilbeck, Phys. Lett. A119 (1986) 60.

[3] M.V. Berry and M. Robnik, J. Phys. A17 (1984) 2413.

[4] O. Bohigas and M.-J. Giannoni, "Chaotic motion and random matrix theories" in Mathematical and computational methods in nuclear physics (eds. J.S. Dehesa, J.M.G. Gomez and A. Polls), Lect. Notes in Physics 209, Springer, Berlin (1984) 1.

[5] E. Haller, H. Köppel and L.S. Cederbaum, Chem. Phys. Lett. 101 (1983) 215.

[6] L. Cruzeiro-Hansson, H. Feddersen, R. Flesch, P.L. Christiansen, M. Salerno and A.C. Scott, "Classical and quantum analysis of chaos in the discrete self-trapping equation" (submitted)

[7] M.V. Berry and M. Taber, Proc. R. Soc. London A356 (1977) 375.

[8] P. Pechukas, Phys. Rev. Lett. 51 (1983) 943.

[9] O. Bohigas, M.J. Giannoni and C. Schmit, Phys. Rev. Lett. 54 (1984) 1.

[10] T.H. Seligman, J.J.M. Verbaarschot and M.R. Zirnbauer, Phys. Lett. 53 (1984) 215.

[11] D. Wintgen and H. Friedrich, Phys. Rev. A35 (1987) 1464.

[12] T.A. Brody, J. Flores, J.B. French, P.A. Mello, A. Pandey and S.S.M. Wong, Rev. Mod. Phys. 53 (1981) 385.

[13] S. De Filippo, M. Fusco Girard and M. Salerno, "Avoided crossing and next neighbour level spacings for the quantum DST equation", Nonlinearity (to appear)

[14] H.-D. Meyer, J. Chem. Phys. 84 (1986) 3147.

[15] T.H. Seligman and J.J.M. Verbaarschot, J. Phys. A18 (1985) 2227.

[16] H. Goldstein, "Classical Mechanics", Addison-Wesley, Reading, Massachusetts (1980).

[17] S. De Filippo, M. Fusco Girard and M. Salerno, Physica 29D (1988) 421.

[18] G. Casati, B.V. Chirikov and I. Guarneri, Phys. Rev. Lett. 54 (1985) 1350.

[19] T. Zimmerman, H.-D. Mayer, H. Köppel and L.S. Cederbaum, Phys. Rev. A 33 (1986) 4334.

[1] Present address: Department of Crystallography, Birkbeck College, Univ. London, Malet Street, London WC1E 7HX, U.K.

[2] Present address: Department of Mathematics, Heriot-Watt Univ., Riccarton, Edinburgh EH14 4AS, U.K.

[3] Present and permanent address: Dipartimento di Fisica Teorica, Università di Salerno, I-84100 Salerno, Italy.

42

EIGENVALUE STATISTICS AND EIGENSTATE WIGNER FUNCTIONS

FOR THE DISCRETE SELF TRAPPING EQUATION

Mario Salerno
Dipartimento di Fisica Teorica
Universita' di Salerno
I-84100 Salerno
Italy

In some recent papers [1,2] the discrete self-trapping (DST) equation [3] was shown to be an effective working model to analyse the semiclassical limit of the quantum version of classically non-integrable dynamics [4]. Canonical quantisation of the DST eq. with m freedoms leads in fact to the Hamiltonian

$$\hat{H} = \Gamma \hbar^2 \sum_{j=1}^{m} (\hat{A}_j^+ \hat{A}_j + 1/2)^2 + \hbar \,\varepsilon \sum_{j,k=1}^{m} M_{jk} (\hat{A}_j^+ \hat{A}_k + \hat{A}_k \hat{A}_j^+)/2 \tag{1}$$

which commutes with the norm operator

$$\hat{N} = \hbar \,(m/2 + \sum_{j=1}^{m} \hat{N}_j) \equiv \hbar \,(m/2 + \sum_{j=1}^{m} \hat{A}_j^+ \hat{A}_j), \tag{2}$$

whose eigenspaces are finite-dimensional. Here M is an arbitrary hermitean matrix and $\hat{A}$'s and $\hat{A}^+$'s are annihilation and creation operators satisfying the usual commutation relations

$$[\hat{A}_j , \hat{A}_k^+] = \delta_{jk} , \quad [\hat{A}_j^+ , \hat{A}_k^+] = [\hat{A}_j , \hat{A}_k] = 0 . \tag{3}$$

The considered system has two (both classically and quantum) integrable limits, respectively corresponding to $\Gamma = 0$ and $\varepsilon = 0$. Unfortunately both of them are non-generic [4], which makes the system poorly relevant in the nearly integrable regime. This naturally led in ref. [5] to consider a generalised DST Hamiltonian

$$H = H_0(\hbar(\hat{N}_1 + 1/2) , . \; . \; . , \hbar(\hat{N}_m + 1/2)) + \hbar \,\varepsilon \sum_{j,k=1}^{m} M_{jk} (\hat{A}_j^+ \hat{A}_k + \hat{A}_k \hat{A}_j^+), \tag{4}$$

H_0 being an arbitrary function, which is the canonical quantum version of the classical Hamiltonian

$$H = H_0(J_1, J_2, ..., J_m) + \frac{\varepsilon}{2} \sum_{j,k=1}^{m} M_{j,k}(q_j q_k + p_j p_k) =$$

$$H_0(\mathbf{J}) + \frac{\varepsilon}{2} \sum_{j,k=1}^{m} \sqrt{2J_j J_k} \, \cos(\vartheta_j - \vartheta_k) \tag{5}$$

where J's and ϑ's are action angle variables of the harmonic oscillator, with unit mass

---

The material in this lecture summarizes results of S. De Filippo, M. Fusco Girard, and myself which are described in more details in ref.s [1], [5], [12].

and frequency, i.e. $J_j = (q_j + p_j^2)/2$ and $\vartheta_j = \tan^{-1}(q_j/p_j)$. Here the $\varepsilon = 0$ limit classically describes an arbitrary integrable system, since H_0 in the classical limit gives the expression of the hamiltonian function in action variables (1/2 terms should in general be replaced by the proper Maslov indices to obtain the corresponding semiclassical hamiltonian operator). As in the mentioned paper, the minimal modification of the DST Hamiltonian is here chosen in order to get genericity, i.e.

$$H_0 = \Gamma \hbar \sum_{j=1}^{m} \alpha_j \, (\hat{N}_j + 1/2)^2 \tag{6}$$

with the α's being real numbers with irrational ratios. The possibility of separate diagonalisation of $\hat{H}$ in the finite-dimensional eigenspaces of $\hat{N}$ (thus avoiding truncation of infinite wave function expansions), together with the possibility of independently varying the anharmonic parameter Γ in eq. (4-6) as one approaches the correspondence limit, makes this system particularly suitable to analyse the semiclassical limit of the quantum version of classically non-integrable dynamics. In this paper we use statistical properties of quantum spectra to study quantum ergodicity of the DST Hamiltonian far from the uncoupled limit and in the near integrable regime for the case of three freedoms, while in the integrable case (two freedoms) we use Wigner functions of eigenstates to reproduce the semiclassical behaviour of non-degenerate integrable systems.

To this end the Hamiltonian (4) ,(6) is diagonalised by exploiting the decomposition of the Hilbert space of quantum states as a direct sum

$$K = K_0 \oplus K_1 \oplus \dots \oplus K_n \oplus \dots \tag{7}$$

where K_n denotes the eigenspace of $\hat{N}$ with $\hbar(n+m/2)$ as an eigenvalue. Denote by F_k the Fock space corresponding to the creation operator $\hat{A}_k^+$, and by $|0\rangle_k, |1\rangle_k, \dots, |j\rangle_k, \dots$ its usual basis defined by

$$\hat{A}_k |0\rangle_k = 0$$

$$\hat{A}_k^+ |j\rangle_k = (j+1)^{1/2} |j+1\rangle_k \cdot \tag{8}$$

The Hilbert space K can be identified with the tensor product

$$K = F_1 \otimes F_2 \otimes \dots \otimes F_m \tag{9}$$

and

$$|j_1, j_2, \dots, j_m\rangle \equiv |j_1\rangle_1 \otimes |j_2\rangle_2 \otimes \dots \otimes |j_m\rangle_m \tag{10}$$

denotes the generic element of the corresponding product basis. As explained in more details in ref.s [1,5] the representation of eigenstates of the considered Hamiltonian belonging to each K_n in basis (10) can be easily obtained by diagonalisation of the corresponding (finite) matrix representation of the restriction of the hamiltonian to K_n. Eigenvalue spectra for the considered generalised DST Hamiltonian operator were then evaluated for m=3, N = 28 and $\alpha_1 = 1$, $\alpha_2 = (\sqrt{5} - 1)/2$, $\alpha_3 = \sqrt{\alpha_2}$ in order to be as far as possible from rational ratios, while the matrix M is chosen as in refs. [1,3] . A further bonus of modification (4,6) of the DST Hamiltonian comes from the breaking of the invariance under the symmetric group S_3, which forced to separately consider subspaces corresponding to different irreducible representations [1] in order to desymmetrise the spectrum [4]; now on the contrary the whole spectrum of (28+1)(28+2)/2=435 eigenvalues can be considered at once in order to study its statistical properties. In particular the nearest-neighbour level spacing distribution was considered once the eigenvalues were scaled according to the unfolding procedure [6]

$E_k = N_{AVG}(E_k)$ where N_{AVG} denotes the average integrated mean density evaluated from the numerical stair case function by fifth order polynomial interpolation. Numerical data were fitted just as in ref. [1] by the density distribution

$$P(S,\mu)=\mu^2\exp(-\mu S)\,\mathrm{erfc}(\pi^{1/2}\mu S/2)+(2\mu\bar\mu+\pi\bar\mu^3 S/2)\exp(-\mu S-\pi\bar\mu^2 S^2/4)\,, \qquad (11)$$

conjectured by M.V. Berry and Robnik in ref. [7], in the semiclassical limit and for an infinitesimal energy interval, for a classical phase space consisting of a regular and just one connected irregular region. Here S is the normalised spacing, i.e. the ratio between spacing and mean spacing, erfc as usual denotes the error function complement and μ is the fraction (in measure) of the energy hypersurface for which the motion is regular, $\bar\mu \equiv 1 - \mu$; $\mu=0$ and $\mu=1$ respectively reproduce Wigner's (fully developed chaos) and Poisson's (generic integrable system) distributions. It should be remarked that distribution (11) follows in the hypothesis that the irregular region is connected, which is not the case here in principle since the presence of a further constant of motion prevents, for $m=3$, Arnold diffusion. However this simplifying assumption was already seen to produce a reasonable fitting in ref. [1], so it will be retained here.

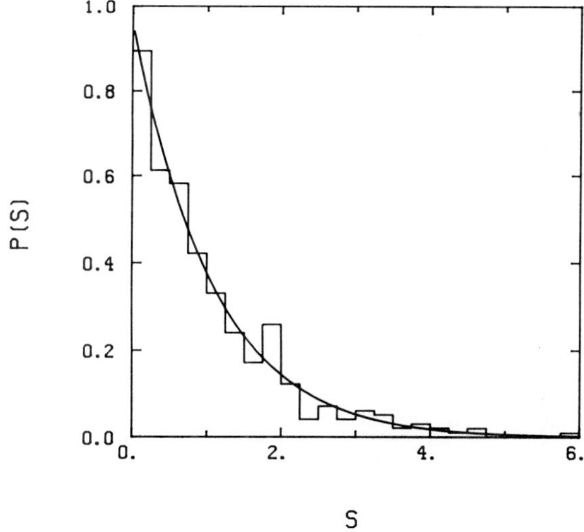

Fig. 1 The stepped line gives the population of nearest–neighbour level spacings for spacing intervals of 0.25 corresponding to a sample of 400 eigenvalues for the Hamiltonian operator given by eq. (4) with H_O as in eq. (6), obtained for $N=28$, $\Gamma=4.0$ and $\varepsilon=0.05$ The full line fits data according to formula (11), with a μ value of 0.79.

To fit data μ was evaluated by solving the equation

$$\langle S^2\rangle = \int P(S,\mu)S^2 dS = (2/\mu)[1-\exp(\mu^2/\pi\bar\mu^2)\,\mathrm{erfc}(\mu/\pi^{1/2}\bar\mu)] \qquad (12)$$

with the l.h.s. replaced by the numerical second moment. The reliability of the fitting was tested by comparing the third and fourth moments $\langle S^3\rangle$ and $\langle S^4\rangle$ of the

numerical samples with the ones obtained by distribution (11) with the fitting values of
μ. In what follows all eigenvalue spectra refer to the value $\Gamma = 4$.

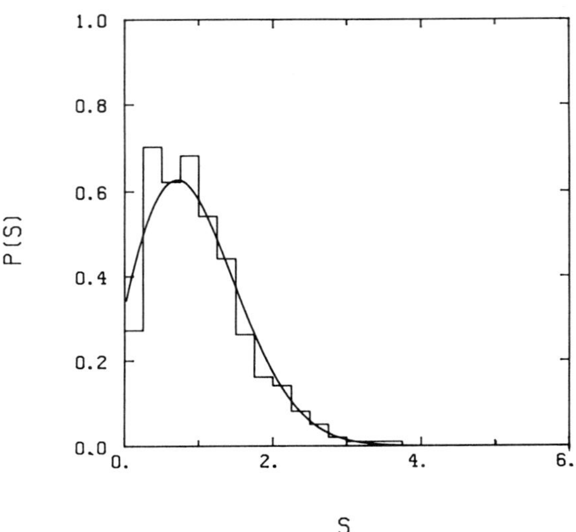

Fig. 2   Same as in fig. (1) but for $\varepsilon = 5.0$ and $\mu = 0.175$.

Fig.s (1),(2) refer to the spectra of the generalised DST Hamiltonian with $N = 28$ and
$\varepsilon = 0.05$, $\varepsilon = 5.0$ respectively. The continuous curves, represent distribution (11) with μ
evaluated by solving eq. (12) with respect to μ ($\langle S^2 \rangle$ refers to numerical samples)
[1]. These figures show a good agreement between experimental data and theoretical
predictions both in the quantum ergodic regime far from $\varepsilon = 0$ and in the near
integrable limit $\varepsilon = 0$. On the other hand, while statistical properties of quantum spectra
are better understood, and therefore widely used, due to extensive numerical and
analytical investigations, results for eigestates behaviour in phase space are quite poor.
In this respect the considered system will be shown to be particularly suitable in order
to elucidate the semiclassical features of Wigner functions of stationary states in 2m
dimensional phase space. To this end we recall that Wigner's function [8,9] W_ψ of a
pure state $|\psi\rangle$, which is obtained by the Weyl formula for the phase space function
corresponding to the quantum operator $|\psi\rangle\langle\psi|/\hbar^m$, is given in general by the real
normalised function

$$W_\psi(\mathbf{q}, \mathbf{p}) = \frac{1}{(\pi\hbar)^m} \int d\mathbf{X} \, \exp\left(-\frac{2i}{\hbar} \mathbf{p} \cdot \mathbf{X}\right) \langle \mathbf{q} + \mathbf{X}|\psi\rangle\langle\psi|\mathbf{q} - \mathbf{X}\rangle, \tag{13}$$

where bold characters denote m dimensional vectors. It is well known that, although it
is not positive definite, it gives after smearing by averaging on quantum cells, a
probability density in classical phase space. On the other hand its evolution is given in
the classical limit by the Liouville equation. Since ergodic properties of classical
systems are generally analysed in phase space, Wigner's function is a natural
candidate for a detailed study of their correspondent properties in the semiclassical
limit [9]. For generic classically integrable motions with m freedoms the classical limit

514

of the Wigner's function is a delta function concentrated on invariant tori; in the semiclassical limit there are peaks of the order $\hbar^{-2m/3}$ corresponding to classical tori [9]. Berry and Voros conjectured that also for irregular motions each semiclassical eigenstate has a Wigner function concentrated near the region explored by a typical orbit over infinite times [10]. A revised version of this conjecture was proved in ref. [11]. While Wigner's functions for simple systems (maps, one freedom systems) have been extensively analysed, the situation is quite different for Hamiltonian systems with several freedoms, due to both numerical and analytical difficulties.

Here the general expression of the Wigner's function for a state vector in the considered representation is derived; some preliminary analysis of the so built Wigner's functions is performed for the simplest case of two freedoms, for which the system is integrable. To this end consider the representation of state vectors $|\psi\rangle \in K_n$ in the form

$$|\psi\rangle = \sum_{j_1+j_2+\ldots+j_m=n} |j_1,j_2,\ldots,j_m\rangle\langle j_1,j_2,\ldots,j_m|\psi\rangle. \tag{14}$$

By inserting this expression in eq.(13) one obtains

$$W_\psi(\mathbf{q},\mathbf{p}) = \sum_{\substack{\mu_1+\mu_2+\ldots+\mu_m=n \\ \nu_1+\nu_2+\ldots+\nu_m=n}} \langle\mu_1,\mu_2,\ldots,\mu_m|\psi\rangle\langle\psi|\nu_1,\nu_2,\ldots,\nu_m\rangle \, W_{(\mu_1,\mu_2,\ldots,\mu_m;\nu_1,\nu_2,\ldots,\nu_m)}(\mathbf{q},\mathbf{p}) \tag{15}$$

where

$$W_{(\mu_1,\mu_2,\ldots,\mu_m;\nu_1,\nu_2,\ldots,\nu_m)}(q_1,q_2,\ldots,q_m; p_1,p_2,\ldots,p_m) =$$

$$\frac{1}{(\pi\hbar)^m} \int d\mathbf{X} \exp\left(-\frac{2i}{\hbar}\,\mathbf{p}\cdot\mathbf{X}\right) \langle\mathbf{q}+\mathbf{X}|\mu_1,\mu_2,\ldots,\mu_m\rangle\langle\nu_1,\nu_2,\ldots,\nu_m|\mathbf{q}-\mathbf{X}\rangle =$$

$$\prod_{j=1}^{m} \frac{1}{(\pi\hbar)} \int dX \exp\left(-\frac{2i}{\hbar}\,p_j X\right) \langle q_j+X|\mu_j\rangle\langle\nu_j|q_j-X\rangle \equiv \prod_{j=1}^{m} W_{\mu_j,\nu_j}(q_j,p_j), \tag{16}$$

$\langle q|\mu\rangle$ denoting the μth normalised eigenfunction of the harmonic oscillator with unit mass and frequency

$$\langle q|\mu\rangle = (-1)^\mu (\hbar)^{-1/4} (2^\mu \pi^{1/2} \mu!)^{-1/2} e^{q^2/(2\hbar)} (\sqrt{\hbar}\, \partial/\partial q)^\mu e^{-q^2/\hbar}. \tag{17}$$

On the other hand it can be shown that

$$W_{\mu,\nu}(q,p) = f_{\mu,\nu}(J,\vartheta) = \frac{(-1)^\mu}{\pi\hbar} \sqrt{\frac{\mu!}{\nu!}} \left(\frac{4J}{\hbar}\right)^{\frac{\nu-\mu}{2}} e^{-(2J/\hbar)+i(\mu-\nu)\vartheta} L_\mu^{\nu-\mu}(4J/\hbar), \tag{18}$$

with L_k^α denoting as usual generalised Laguerre polynomials, as Bartlett and Moyal derived in their phase space theory of the harmonic oscillator [13].

In the specific case of two freedoms eq.s (15), (18) then give

$$W_\psi(J_1,J_2,\vartheta_1,\vartheta_2) = \frac{(-1)^n}{(\pi\hbar)^2} \sum_{\mu,\nu=0}^{n} c_\mu c_\nu \sqrt{\frac{\mu!(n-\mu)!}{\nu!(n-\nu)!}} \left(\frac{J_1}{J_2}\right)^{\frac{\nu-\mu}{2}}$$

$$e^{-2(J_1+J_2)/\hbar+i(\mu-\nu)(\vartheta_1-\vartheta_2)} L_\mu^{\nu-\mu}(4J_1/\hbar) L_{n-\mu}^{\mu-\nu}(4J_2/\hbar) \tag{19}$$

where $c_\mu = \langle \mu, n-\mu | \psi \rangle$. Equation (19) is applied by replacing the c's by the values obtained after numerical diagonalisation of the Hamiltonian operator (4),(6) with Γ and ε fixed so that $\hbar^2 \Gamma = 1$ and $\hbar\varepsilon = 32$ for n=32, on the eigenspace K_n of $\hat{N}$. The α's in eq.(6) were chosen as $\alpha_1 = 1$, $\alpha_2 = (\sqrt{5} - 1)/2$ in order to be as far as possible from rational ratios, while Γ, ε were fixed by requiring that the two terms in the Hamiltonian have comparable weights. In order to study the semiclassical behaviour, the eigenvalue of $\hat{N}$, i.e. $(n+m/2)\hbar$, should be kept constant. This implies that $\hbar$ and n have to be varied in inverse proportion. In the presentation of numerical results $\hbar$ and $1/n$ will be then identified. The obtained Wigner function is studied by fixing $\vartheta_1 - \vartheta_2 = \pi/2$ and $N = J_1 + J_2 = 1$ so that W_ψ becomes a function of the single variable $J_1 \equiv J$.

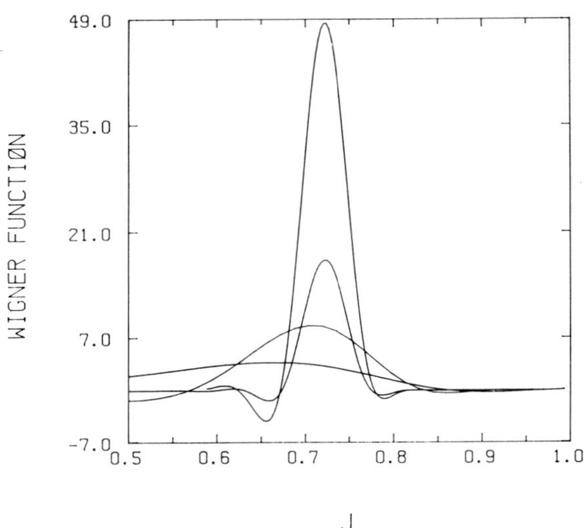

Fig. 3 Wigner functions vs J for eigestates of the m=2 DST Hamiltonian whose energy eigenvalues best approximate the fixed value E=591. The different curves refer to doubling values of n starting from n=32 up to n=256, the upper one corresponding to n=256.

In order to compare homogeneous results, eigenstates are considered whose energy eigenvalues best approximate the fixed value E=591. Results are reported in figs. 3 where Wigner functions are plotted vs J for doubling values of n starting from n=32 (lower curve) up to n=256 (upper curve). For the fixed values of $\vartheta_1 - \vartheta_2$ and N, equation H=E, with the classical Hamiltonian given by eq.s (5), (6) with the before chosen α's, gives two values of J both in the allowed range [0,1]

$$J = \left(\alpha_2 n \pm \sqrt{(\alpha_1 + \alpha_2)E - \alpha_1 \alpha_2 n^2} \right) / (\alpha_1 + \alpha_2). \tag{20}$$

Due to the numerical difficulties in achieving the semiclassical limit, we reported in fig. 3 only the peaks of Wigner functions corresponding to the highest value of J. From this figure we clearly see that Wigner functions peak around the classically expected value of J given by eq. (20) as $\hbar$ goes to zero. In ref. [9] it was also shown that for a generic integrable system with m freedoms the peak height of the Wigner function in correspondence of classically invariant tori in the semiclassical limit is of the order $\hbar^{-2m/3}$. In order to approximate this semiclassical limit one should properly

516

avoid numerical difficulties. In fact, with growing n values the evaluation of the Wigner function becomes more critical due to the exploding character both of Laguerre polynomials of higher order and of the factorial prefactor in eq. (19) . However, while the former numerical obstruction can be overcome by directly evaluating the product of the real exponential factor and the Laguerre polynomials in a proper way, the latter can simply be avoided by considering the case $\varepsilon=0$, which with the present modification of the DST Hamiltonian , classically still represents, in spite of its simplicity, an integrable system with generally non-resonant invariant tori. In this case ($\varepsilon=0$) indeed , state vectors $|\mu,n-\mu\rangle$ are energy eigenvectors and equation (19) for $|\psi\rangle=|\mu,n-\mu\rangle$ simplifies to

$$W_\mu(J_1,J_2) = \frac{(-1)^n}{(\pi\hbar)^2}\, e^{-2(J_1+J_2)/\hbar}\, L_\mu^0(4J_1/\hbar)\, L_{n-\mu}^0(4J_2/\hbar). \tag{21}$$

from which it is easy to obtain explicit asymptotic formulas for $\hbar \to 0$ near the corresponding invariant torus. In fact by using the asymptotic formula for Laguerre polynomials given in ref. [14]

$$L_n^{(\alpha)}(x)=\frac{(-1)^n e^{x/2}}{2^{\alpha+1/3}n^{1/3}}\left\{\mathrm{Ai}\left(\frac{x-\nu}{(4\nu)^{1/3}}\right)+\widetilde{\mathrm{Ai}}\left(\frac{x-\nu}{(4\nu)^{1/3}}\right)\left[O\!\left(\frac{1}{\nu}\right)+O\!\left(\frac{x-\nu}{\nu}\right)+O\!\left(\frac{(x-\nu)^{5/2}}{\nu^{3/2}}\right)\right]\right\}, \tag{22}$$

which holds for x near ν, where $\nu=4n+2\alpha+2$ is the peak location, Ai denoting the usual Airy function and $\widetilde{\mathrm{Ai}}$ being given in terms of Airy functions as in ref. [14], the leading asymptotic term of eq. (21) is given by

$$W_\mu(J_1,J_2)\approx\frac{1}{(\pi\hbar)^2}\,\frac{2^{-2/3}}{\mu^{1/3}(n-\mu)^{1/3}}\left\{\mathrm{Ai}\left(\frac{4J_1/\hbar-4\mu-2}{[4(4\mu+2)]^{1/3}}\right)\mathrm{Ai}\left(\frac{4J_2/\hbar-4(n-\mu)-2}{[4(4(n-\mu)+2)]^{1/3}}\right)\right\}. \tag{23}$$

Remembering that when considering semiclassical limit, $\mu\hbar$ and $n\hbar$ are to be taken constant, eq. (23) exhibits the $\hbar^{-4/3}$ behaviour, it being a specific realisation of formula (8.13) in ref. [9]. Similarly one can derive an asymptotic formula holding far from invariant tori, in agreement with the analogous one obtained in ref. [9]. It is worth to remark however, that while general results were derived in ref. [9] by means of the semiclassical wave function, the exact eigenstates were here used. In conclusion, the presented analysis shows that the expected semiclassical features of Wigner's function of stationary states of integrable systems are easily reproduced for the two freedoms DST system. An analogous study for three or more freedoms, although more complex would still be more approachable than for other integrable systems for which it is not so easy to find explicit analytical and numerically computable expressions for stationary state Wigner functions.

Acknowledgement

The author would like to thank prof. J.H. Hannay for stimulating discussions and for pointing out to him ref [9]. This work was partially supported by the G.N.S.M., C.I.S.M., C.N.R. Italy.

References

[1] S. De Filippo, M. Fusco Girard, M. Salerno, *Avoided Crossing and Nearest Neighbour Level Spacings for The Quantum DST Equation.* Nonlinearity (1989);
[2] L. Cruzeiro-Hansson, H. Feddersen, R. Flesch, P.L. Christiansen, M. Salerno, A.C. Scott, *Classical and Quantum Analysis of Chaos in The DST Equation* Preprint, Lyngby (Denmark) (1989);

[3] J.C. Eilbeck, P.S. Lomdahl and A.C. Scott, Physica $\underline{16D}$ (1985) 318,
A.C. Scott, P.S. Lomdahl and J.C. Eilbeck, Chem. Phys. Letters $\underline{113}$ (1985) 29,
A.C. Scott and J.C. Eilbeck, Phys. Lett. $\underline{119A}$ (1986) 60,
S. De Filippo, M. Fusco Girard and M. Salerno, Physica $\underline{26D}$ (1987) 411,
S. De Filippo, M. Fusco Girard and M. Salerno, Physica $\underline{29D}$ (1988) 421;

[4] M.V. Berry, in: Chaotic Behaviour of Deterministic Systems, eds. G. Iooss,
R.H.G. Helleman, R. Stora (North-Holland, Amsterdam, 1983)
pp. 171-271 and references quoted therein;

[5] S. De Filippo, M. Salerno, *A Generalised Discrete Self-Trapping Equation as a Model for Quantum Chaology*, Preprint Salerno (1989);

[6] O. Bohigas, M.J. Giannoni and C. Schmit, in *Quantum Chaos and Statistical Nuclear Physics*, T.H. Seligman and H. Nishioka eds.
Lecture Notes in Phys. 263, Springer Verlag 1986

[7] M.V. Berry and M. Robnik, J. Phys. $\underline{17A}$ (1984) 2413;

[8] E.P.Wigner, Phys. Rev. $\underline{40}$ (1932) 749;

[9] M.V.Berry, Phil. Trans. Roy. Soc. of London $\underline{287}$ (1977) 30;

[10] M.V. Berry, J. Phys. $\underline{A10}$ (1977) 2083,
A.Voros in *Stochastic Behaviour in Classical and Quantum Hamiltonian Systems* eds.
G.Casati and J.Ford, Lectures Notes on Physics 93. (Springer,Berlin 1979) 326;

[11] J.W.Helton and M.Tabor, Physica $\underline{14D}$ (1985) 409;

[12] S. De Filippo, M.Fusco Girard and M. Salerno, preprint (1989)

[13] M.S. Bartlett, J.E. Moyal, Proc. Camb. Phyl. Soc. Math. Phys. Sci. $\underline{45}$ (1949) 545.

[14] W.Magnus, F.Oberhettinger, R.P.Soni, *Formulas and Theorems for the Special Functions of Mathematical Physics*, Springer, Berlin (1966).

43

THE DISCRETE NONLINEAR SCHROEDINGER EQUATION:
NONADIABATIC EFFECTS, FINITE TEMPERATURE CONSEQUENCES,
AND EXPERIMENTAL MANIFESTATIONS

V. M. KENKRE

Department of Physics and Astronomy

University of New Mexico

Albuquerque, NM 87131, U S A

Exciting and powerful new ways of describing quasiparticle transport in solids have recently appeared in the literature[1,2]. Thus, strong interactions of moving quasiparticles with vibrations can give rise to the discrete nonlinear Schroedinger equation which, in other physical contexts has gone under the name of the discrete self-trapping equation. Our recent theoretical work[3,4] has shown that the nonlinearity in such an equation can have profound effects, some expected and others novel, in the evolution of experimental observables. That work, which has been based on exact solutions for the time evolution for small systems and on numerical solutions for large systems, is described with reference to observables such as fluorescence depolarization in molecular dimers and chains, muon spin relaxation and the neutron scattering function. The rotational polaron and the random nonlinear dimer are also described. Going beyond the usual adiabatic limit, striking effects of *slow* vibrational relaxation[5,6] are pointed out, and a simple theory of the Brownian motion in the nonadiabatic limit[7,8,9] is presented in the context of the discrete nonlinear Schroedinger equation. The temperature behaviour of the Kramers escape rate and the stability of nonlinear excitations are discussed. Comments are also made concerning the interesting problem of the microscopic derivation of these equations. In particular, it is pointed out, in reference to the dimer, what is

Davydov's Soliton Revisited, Edited by P.L. Christiansen and A.C. Scott
Plenum Press, New York, 1990

the precise place where the semiclassical assumption (on the phonon
field), which converts the exact dynamics into the dynamics of the
nonlinear Schroedinger equation, enters into the analysis.

REFERENCES

1.   A. S. Davydov, Usp. Fiz. Nauk. <u>138</u>, 603 (1982).

2.   A. Scott, Phil. Trans. Royal Soc. A <u>315</u>, 423 (1985).

3.   V. M. Kenkre in <u>Disorder and Nonlinearity</u>, ed. A. Bishop, D. K.
     Campbell and St. Pnvematikos, (SPRINGER, 1989) and refs therein.

4.   V. M. Kenkre in <u>Singular Behaviour and Nonlinear Dynamics</u>, vol II,
     ed. St. Pnevmatikos, T. Bountis and S. Pnevmatikos, (WORLD
     publishers, 1989), p. 698, and refs therein.

5.   V. M. Kenkre and H. -L. Wu, Phys. Rev. B <u>39</u>, 6907, (1989).

6   V. M. Kenkre and H. -L. Wu, Phys. Lett. A <u>135</u>, 120 (1989).

7.   P. Grigolini, H. -L. Wu, and V. M. Kenkre, Phys. Rev. B, <u>40</u>, 7045
     (1989).

8.   V. M. Kenkre and P. Grigolini, Phys. Rev. B, submitted.

9.   V. M. Kenkre and P. Lomdahl, unpublished.

Participants

Mariette Barthes Université des Sciences et Techniques du Languedoc, G.D.P.C. 34060 Montpellier Cedex, France

Lisa Bernstein Program in Applied Mathematics, University of Arizona, Tucson, Arizona 85721, U.S.A.

Irving Bigio CLS-5, MS-E543, Los Alamos National Laboratory, Los Alamos, New Mexico 87545, U.S.A.

Hannes Bolterauer Inst. für Theoretische Physik, Justus-Liebig-Universität Giessen, D-6300 Giessen, F.R.G.

Kiron C. Bordoloi Department of Electrical Engineering, University of Louisville, Louisville, Kentucky 40292 U.S.A.

Larisa Brizhik Institute for Theoretical Physics, Academy of Sciences of the Ukrainian SSR, Kiev 252130, U.S.S.R.

David W. Brown Institute for Nonlinear Science, R-002, University of California at San Diego, La Jolla, California 92093-0402 U.S.A.

Giorgio Careri Dipartimento di Fisica, Università di Roma I, I-00185 Roma, Italy

Peter Leth Christiansen Laboratory of Applied Mathematical Physics, The Technical University of Denmark, Building 303, DK-2800 Lyngby, Denmark

Leonor Cruzeiro-Hansson Department of Crystallography, Birkbeck College, Malet Street, London WC1E 7HX, England

Alexandr S. Davydov Institute for Theoretical Physics, Academy of Sciences of the Ukrainian SSR, 252130 Kiev, U.S.S.R.

J. C. Eilbeck Department of Mathematics, Heriot-Watt University, Riccarton, Edinburgh EH14 4AS, Scotland

Victor Enol'skii Institute of Metal Physics, Kiev-142, 252680 U.S.S.R

Alexander A. Eremko Institute for Theoretical Physics, Academy of Sciences of the Ukrainian SSR, Kiev 252130, U.S.S.R.

Henrik Feddersen Laboratory of Applied Mathematical Physics, The Technical University of Denmark, Building 303, DK-2800 Lyngby, Denmark

Sigart F. Fischer Department Theoretische Physik T38, Technische Univ. München, Boltzmannstrasse, D-8046 Garching, F.G.R.

Randy Flesch Department of Mathematics, Heriot-Watt University, Riccarton, Edinburgh EH14 4AS, Scotland

Wolfgang Förner Chair for Theoretical Chemistry and Laboratory of the National Foundation for Cancer Research of the Friedrich-Alexander-University, D-8520 Erlangen, F.R.G.

Andrea Giansanti Dipartimento di Fisica, Università di Roma I, I-00185 Roma, Italy

Demet Gülen Middle East Technical University, Physics Department, Ankara 06531, Turkey

Darko Kapor Institute of Physics, Faculty of Sciences, University of Novi Sad, Dr I. Djuričića 4, YU-21000 Novi Sad, Yugoslavia

V. M. Kenkre Department of Physics and Astronomy, University of New Mexico, Albuquerque, New Mexico 87131, U.S.A.

William Kerr Olin Physical Laboratory, Wake Forest University, Winston-Salem, North Carolina 27109-7507, U.S.A.

Victor A. Kuprievich Institute for Theoretical Physics, Academy of Sciences of the Ukrainian SSR, 252130 Kiev, U.S.S.R.

Katja Lindenberg Department of Chemistry B-040, University of California at San Diego, La Jolla, California 92093, U.S.A.

Peter Lomdahl Theoretical Division, T-11, MS B262, Los Alamos National Laboratory, Los Alamos, New Mexico, 87545 U.S.A.

Ljiljana Mašković Institute of Physics, Faculty of Sciences, University of Novi Sad, Dr. I. Djuričića 4, YU-21000 Novi Sad, Yugoslavia

Graham Morrison Department of Mathematics, Heriot-Watt University, Riccarton, Edinburgh EH14 4AS, Scotland

Ole Faurskov Nielsen Department of Chemical Physics, University of Copenhagen, The H.C. Ørsted Institute, Universitetsparken 5, DK-2100 Copenhagen Ø, Denmark

Lewis Rothberg AT& T Bell Laboratories, 600 Mountain Avenue, Murray Hill, New Jersey, 07974 U.S.A.

Eigil Præstgaard Roskilde Universitetscenter, Naturvidenskabelig Overbygning 181, Postbox 260, DK-4000 Roskilde, Denmark

Brian M. Pierce Hughes Aircraft Company, Building A-1, MS 3C924, P.O. Box 9399, Long Beach, California 90810-0399 U.S.A.

Stephanos Pnevmatikos Research Center of Crete, P.O. Box 1527, 71110 Heraklio, Crete, Greece

Mario Salerno Department of Theoretical Physics, University of Salerno, I-84100 Salerno, Italy

Miljko Satarić Faculty of Technical Sciences, University of Novi Sad, YU-21000 Novi Sad, Yugoslavia

John W. Schweitzer Department of Physics and Astronomy, The University of Iowa, Iowa City, Iowa 52242-1479 U.S.A.

Alwyn Scott Department of Mathematics, University of Arizona, Tucson, Arizona 85721 U.S.A.   and
Laboratory of Applied Mathematical Physics, The Technical University of Denmark, Building 303, DK-2800 Lyngby, Denmark

Mario Škrinjar Institute of Physics, Faculty of Sciences, University of Novi Sad, Dr. I. Djuričića 4, YU-21000 Novi Sad, Yugoslavia

Shozo Takeno Physics Laboratory, Faculty of Engineering and Design, Kyoto Institute of Technology, Kyoto 606, Japan

J.A. Tuszyński Department of Physics, The University of Alberta, Edmonton, Alberta T6G 2J1, Canada

Xidi Wang Department of Physics B-019, University of California at San Diego, La Jolla, California 92093 U.S.A.

Alexander Zolotaryuk Institute for Theoretical Physics, Academy of Sciences of the Ukrainian SSR, 252130 Kiev, U.S.S.R.

Niels Østergård Laboratory for Applied Mathematical Physics, The Technical University of Denmark, Building 303, DK-2800 Lyngby, Denmark

1. Tuszyński
2. Cruzeiro-Hansson
3. Davydov
4. Takeno
5. Gudmandsen
6. Gülen
7. Eilbeck
8. Barthes
9. Kuprievich
10. Kenkre
11. Lindenberg
12. Careri
13. Faurskov Nielsen
14. Christiansen
15. Bolterauer
16. Bigio
17. Scott
18. Feddersen
19. Bernstein
20. Lomdahl
21. Wang
22. Fischer
23. Pierce
24. Brizhik
25. Bordoloi
26. Brown
27. Schweitzer
28. Salerno
29. Rothberg
30. Giansanti
31. Østergård
32. Kerr
33. Enol'skii
34. Eremko
35. Pnevmatikos
36. Förner
37. Škrinjar
38. Zolotaryuk
39. Mašković
40. Satarić
41. Kapor
42. Flesch
43. Morrison

"Self-trapping of Vibrational Energy in Protein"

NATO-MIDIT Advanced Research Workshop
July 30 to August 5, 1989 at Hanstholm, Denmark

Optical phonons, 22, 169, 395

Pairing energy, 18
Parachloro-acetanilide, 339
Parameters for alpha-helix protein, 3, 227,
    245, 260, 293
Partition function, 85, 313
Pauli spin operators, 40–43
Pekar's polaron, 169, 174–177
Pentadeuterobenzene, 496
Permanent, 25, 29, 30
Perturbation theory, 288–291, 372, 429,
    491, 499, 501
Perylene, 449, 454–460
Phenylalanine residue, 408
Phonon cloud, 64, 67
Phonon density of states, 349, 353
Picosecond coherent anti-stokes Raman,
    398
Poincaré section, 507, 508
Poisson distribution, 506, 513
Polaron, 22, 63, 64, 76, 177, 253, 452, 453
    binding energy, 95
    effective mass, 75, 178
    large, 76
    linewidth, 378
    small, 71, 76, 95, 474
Poly-ℓ-alanine, 409
Poly-ℓ-lysine, 386
Porphyrin, 408
Prebiotic life, 346
Proline residues, 407
Protoenzyme, 346
Proton, 12
Pump-probe experiment, 398, 401–410
Pyrene, 457, 458, 460

Quantum chaos, 435, 505, 506–508
Quantum DST system, 480–481
Quantum effects, 99–100
Quantum fluctuations, 84, 91
Quantum lifetime, 6–7, 100
Quantum Monte Carlo, 6, 83–97, 265, 293
Quantum trajectories, 67
Quasi-elastic scattering, 354, 356
Quasi-modes, 7, 8, 247, 496

R(ν)-representation, 382–386
Raman scattering, 379, 394
Raman spectra of optical phonons, 375,
    395
Raman spectroscopy, 369–371, 382, 406
Rapid pulse heating, 341
Reaction rate theory, 401
Red shift, 339, 340, 344

Reflectionless potential, 135
Remote transfer of vibrational energy,
    401–410
Resonant dipoles, 3, 4, 199, 414, 467
Resonant injection, 409, 465–468
Rotating-wave approximation, 40, 54
Rouleau formation, 423

Scattering data, 134
Schrödinger equation, 7, 66, 110, 269
Schrödinger equation, nonlinear 6, 7, 32,
    36–38, 40, 71, 133–134, 164, 173,
    420
Semiclassical analysis, 7
Semiclassical limit, 508, 511, 514
Shifting operator, 100, 171, 318
Single channel parameters, 2, 3, 246, 260
Slater-type orbital, 214
Softening of lattice-mode, 397
Soliton, 6, 12, 134, 150, 172, 247, 253
    binding energy, 6, 35, 57, 58, 95, 255,
        263
    detector, 261–263
    effective mass, 11, 12, 14, 57, 58, 95,
        173, 174, 178, 256
    excitation efficiency, 138
    excitation thresholds, 7, 133, 247
    generation, 133–141
    initial conditions, 191
    size, 3, 5
    symmetric and antisymmetric, 157
Solitonic window, 281, 304
Spring constant of the hydrogen bonds, 4,
    227, 260, 268
Stability of solitons, 12, 18, 59–62, 91–96,
    194, 251–256
Static injection, 410
Static protein conformations, 401
Stationary solutions, 476
Steepest descent technique, 181, 188
Stochastic transition, 441
Stokes shift, 458
Stokes/anti-Stokes ratio, 369, 370
Substrate potential, 181–194
Synthetic helix systems, 408–409

Takeno Hamiltonian, 38–43, 47–62, 235,
    250, 391, 393, 475
Temperature stability, 245–335
Thermal equilibrium, 83
Thermal fluctuations, 8, 194, 247
Thermal lifetime, 249, 259–265, 273–281,
    293, 322, 327–330
Three channel model, 3, 245, 248

BIN TRAVELER FORM

#9

Cut By: Daphne Williams   **Qty** 30   **Date** 8-3-26

Scanned By: _______________ **Qty** _______ **Date** _______

Scanned Batch ID's

_______________ _______________ _______________

Notes / Exceptions

____